Properties of Exponents and Logarithms

For $y \neq 0$, $b > 0$, $b \neq 1$, $M > 0$, and $N > 0$:

	Exponential Form	Logarithmic Form
	$b^l = M$	$\log_b M = l$
Product Rule	$x^m x^n = x^{m+n}$	$\log_b MN = \log_b M + \log_b N$
Quotient Rule	$\dfrac{x^m}{x^n} = x^{m-n}$	$\log_b \dfrac{M}{N} = \log_b M - \log_b N$
Power Rule	$(x^m)^n = x^{mn}$	$\log_b N^p = p \log_b N$
Product to a Power	$(xy)^m = x^m y^m$	$\log_b (MN)^p = p(\log_b M + \log_b N)$
Quotient to a Power	$\left(\dfrac{x}{y}\right)^m = \dfrac{x^m}{y^m}$	$\log_b \left(\dfrac{M}{N}\right)^p = p(\log_b M - \log_b N)$
Change of Base Formula	$M^x = b^{x \log_b M}$	$\log_a M = \dfrac{\log_b M}{\log_b a}$
Special Identities	$b^0 = 1$	$\log_b 1 = 0$
	$b^1 = b$	$\log_b b = 1$
	$b^{-1} = \dfrac{1}{b}$	$\log_b \dfrac{1}{b} = -1$
	$b^{\log_b M} = M$	$\log_b b^x = x$

Properties of Radicals

For $x, y > 0$:

$$\sqrt[n]{xy} = \sqrt[n]{x}\sqrt[n]{y}$$

$$\sqrt[n]{\dfrac{x}{y}} = \dfrac{\sqrt[n]{x}}{\sqrt[n]{y}}$$

$$\sqrt[n]{x^m} = (\sqrt[n]{x})^m = x^{m/n}$$

$$\sqrt[n]{x^n} = x$$

$$\sqrt[m]{\sqrt[n]{x}} = \sqrt[mn]{x}$$

For $x < 0$:

$$\sqrt[n]{x^n} = \begin{cases} |x| & \text{for } n \text{ even} \\ x & \text{for } n \text{ odd} \end{cases}$$

Conic Sections

	Equation	Features
Parabola	$y = ax^2 + bx + c$	x-coordinate of the vertex is $\dfrac{-b}{2a}$; graph opens up for $a > 0$, down for $a < 0$
Circle	$(x-h)^2 + (y-k)^2 = r^2$	Center is (h, k), radius is r
Ellipse	$\dfrac{(x-h)^2}{a^2} + \dfrac{(y-k)^2}{b^2} = 1$	Center is (h, k), length of major axis is $2a$, length of minor axis is $2b$ for $a > b$
Hyperbola	$\dfrac{(x-h)^2}{a^2} - \dfrac{(y-k)^2}{b^2} = 1$	Center is (h, k); graph opens left and right

THE PRINDLE, WEBER & SCHMIDT SERIES IN MATHEMATICS

Althoen and Bumcrot, *Introduction to Discrete Mathematics*
Boye, Kavanaugh, and Williams, *Elementary Algebra*
Boye, Kavanaugh, and Williams, *Intermediate Algebra*
Burden and Faires, *Numerical Analysis*, Fourth Edition
Cass and O'Connor, *Fundamentals with Elements of Algebra*
Cullen, *Linear Algebra and Differential Equations*, Second Edition
Dick and Patton, *Calculus, Volume I*
Dick and Patton, *Calculus, Volume II*
Dick and Patton, *Technology in Calculus: A Sourcebook of Activities*
Eves, *In Mathematical Circles*
Eves, *Mathematical Circles Adieu*
Eves, *Mathematical Circles Squared*
Eves, *Return to Mathematical Circles*
Fletcher, Hoyle, and Patty, *Foundations of Discrete Mathematics*
Fletcher and Patty, *Foundations of Higher Mathematics*, Second Edition
Gantner and Gantner, *Trigonometry*
Geltner and Peterson, *Geometry for College Students*, Second Edition
Gilbert and Gilbert, *Elements of Modern Algebra*, Third Edition
Gobran, *Beginning Algebra*, Fifth Edition
Gobran, *Intermediate Algebra*, Fourth Edition
Gordon, *Calculus and the Computer*
Hall, *Algebra for College Students*
Hall, *Beginning Algebra*
Hall, *College Algebra with Applications*, Third Edition
Hall, *Intermediate Algebra*
Hartfiel and Hobbs, *Elementary Linear Algebra*
Humi and Miller, *Boundary-Value Problems and Partial Differential Equations*
Kaufmann, *Algebra for College Students*, Fourth Edition
Kaufmann, *Algebra with Trigonometry for College Students*, Third Edition
Kaufman, *College Algebra*, Second Edition
Kaufmann, *College Algebra and Trigonometry*, Second Edition
Kaufmann, *Elementary Algebra for College Students*, Fourth Edition
Kaufmann, *Intermediate Algebra for College Students*, Fourth Edition
Kaufmann, *Precalculus*, Second Edition
Kaufman, *Trigonometry*
Kennedy and Green, *Prealgebra for College Students*
Laufer, *Discrete Mathematics and Applied Modern Algebra*
Nicholson, *Elementary Linear Algebra with Applications*, Second Edition
Pence, *Calculus Activities for Graphic Calculators*
Pence, *Calculus Activities for the TI-81 Graphic Calculator*
Plybon, *An Introduction to Applied Numerical Analysis*
Powers, *Elementary Differential Equations*
Powers, *Elementary Differential Equations with Boundary-Value Problems*
Proga, *Arithmetic and Algebra*, Third Edition
Proga, *Basic Mathematics*, Third Edition
Rice and Strange, *Plane Trigonometry*, Sixth Edition

Schelin and Bange, *Mathematical Analysis for Business and Economics*, Second Edition
Strnad, *Introductory Algebra*
Swokowski, *Algebra and Trigonometry with Analytic Geometry*, Seventh Edition
Swokowski, *Calculus*, Fifth Edition
Swokowski, *Calculus*, Fifth Edition (Late Trigonometry Version)
Swokowski, *Calculus of a Single Variable*
Swokowski, *Fundamentals of College Algebra*, Seventh Edition
Swokowski, *Fundamentals of Algebra and Trigonometry*, Seventh Edition
Swokowski, *Fundamentals of Trigonometry*, Seventh Edition
Swokowski, *Precalculus: Functions and Graphs*, Sixth Edition
Tan, *Applied Calculus*, Second Edition
Tan, *Applied Finite Mathematics*, Third Edition
Tan, *Calculus for the Managerial, Life, and Social Sciences*, Second Edition
Tan, *College Mathematics*, Second Edition
Trim, *Applied Partial Differential Equations*
Venit and Bishop, *Elementary Linear Algebra*, Third Edition
Venit and Bishop, *Elementary Linear Algebra*, Alternate Second Edition
Wiggins, *Problem Solver for Finite Mathematics and Calculus*
Willard, *Calculus and Its Applications*, Second Edition
Wood and Capell, *Arithmetic*
Wood and Capell, *Intermediate Algebra*
Wood, Capell, and Hall, *Developmental Mathematics*, Fourth Edition
Zill, *Calculus*, Third Edition
Zill, *Differential Equations with Boundary-Value Problems*, Second Edition
Zill, *A First Course in Differential Equations with Applications*, Fourth Edition
Zill and Cullen, *Advanced Engineering Mathematics*

THE PRINDLE, WEBER & SCHMIDT SERIES IN ADVANCED MATHEMATICS

Brabenec, *Introduction to Real Analysis*
Ehrlich, *Fundamental Concepts of Abstract Algebra*
Eves, *Foundations and Fundamental Concepts of Mathematics*, Third Edition
Keisler, *Elementary Calculus: An Infinitesimal Approach*, Second Edition
Kirkwood, *An Introduction to Real Analysis*
Ruckle, *Modern Analysis: Measure Theory and Functional Analysis with Applications*
Sieradski, *An Introduction to Topology and Homotopy*

College Algebra with Applications

Third Edition

James W. Hall
Parkland College

PWS-KENT Publishing Company □ Boston

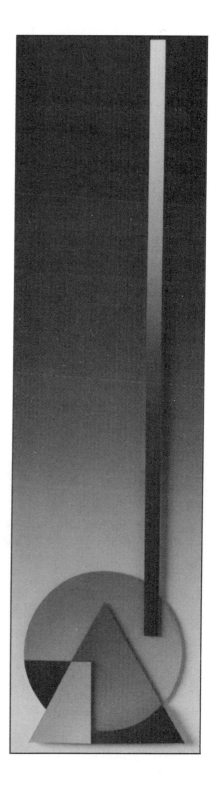

PWS-KENT
Publishing Company

20 Park Plaza
Boston, Massachusetts 02116

This book is dedicated to my friend and colleague Richard D. Bennett, who made many contributions to the first two editions of this book.

Copyright © 1992, 1989 by PWS-KENT Publishing Company. Copyright © 1985 PWS Publishers.

All rights reserved. No part of this book may be reproduced, stored in a retrieval system, or transcribed, in any form or by any means—electronic, mechanical, photocopying, recording, or otherwise—without the prior written permission of the publisher, PWS-KENT Publishing Company.

PWS-KENT Publishing Company is a division of Wadsworth, Inc.

Permission has been granted for images of mathematicians for Chapters 1, 2, 3, and 5 by The Historical Picture Service, for Chapters 4 and 8 by The Smithsonian Institution, Photo Nos. 60277D and 46834M, and Chapter 7 courtesy of The Bettmann Archive.

Library of Congress Cataloging-in-Publication Data
Hall, James W.
  College algebra with applications/James W. Hall.—3rd ed.
    p. cm.
  Includes index.
  ISBN 0-534-92788-2
  1. Algebra. I. Title.
  QA154.2.H35 1992   91-23168 CIP
  512.9—dc20

Printed in the United States of America
92 93 94 95 96—10 9 8 7 6 5 4 3 2 1

Sponsoring Editor: Timothy Anderson
Developmental Editor: Barbara Lovenvirth
Production Editor: S. London
Manufacturing Coordinator: Peter Leatherwood
Interior Designer: S. London
Interior Illustrator: Scientific Illustrators
Chapter Opening Artist: Joyce Culkin
Cover Designer: S. London
Cover Artist: Joyce Culkin
Cover Photographer: Greg Bowl Studio
Typesetter: Techset Composition Ltd.
Cover Printer: John P. Pow Company, Inc.
Printer and Binder: R. R. Donnelley & Sons

Contents

CHAPTER ONE
Review of the Basic Concepts of Algebra 1

- **1-1** The Real Number Line 2
- **1-2** Properties of the Real Number System 10
- **1-3** Integral Exponents and Algebraic Expressions 18
- **1-4** Operations with Polynomials 27
- **1-5** Factoring Polynomials 36
- **1-6** Rational Expressions 43
- **1-7** Rational Exponents and Radicals 52
- **1-8** Complex Numbers 63
- Key Concepts 71
- Mastery Test 74

CHAPTER TWO
Equations and Inequalities 76

- **2-1** Linear Equations 77
- **2-2** Using Formulas and Forming Equations 88
- **2-3** Quadratic Equations 99
- **2-4** Equations That Result in Quadratic Equations 109
- **2-5** Applications of Equations 114
- **2-6** Linear and Absolute Value Inequalities 126
- **2-7** Nonlinear Inequalities 134
- Key Concepts 140
- Mastery Test 145

CHAPTER THREE
Functions, Relations, and Graphs 147

- **3-1** Introduction to Functions and Relations 148
- **3-2** Lines 163
- **3-3** Translations of Graphs and Parabolas 176
- **3-4** Symmetry and the Distance and Midpoint Formulas 191
- **3-5** Combining Functions 204
- **3-6** Inverse Functions 217
- **3-7** Circles and Ellipses 229
- ***3-8** Hyperbolas and a Summary of Conic Sections 238
- Key Concepts 250
- Mastery Test 259

* This is an optional section.

CHAPTER FOUR

Polynomial and Rational Relations 262

- **4-1** Graphs of Polynomial Functions 264
- **4-2** Synthetic Division and the Factor Theorem 276
- **4-3** Complex Zeros of Polynomials 287
- **4-4** Rational Zeros of Polynomials 294
- *__4-5__ Shortcuts for Finding Zeros 303
- *__4-6__ Approximation of Zeros 313
- **4-7** Rational Functions 318
- Key Concepts 335
- Mastery Test 340

CHAPTER FIVE

Exponential and Logarithmic Functions 341

- **5-1** Exponential Functions 342
- **5-2** Logarithmic Functions 352
- **5-3** Common and Natural Logarithms 359
- **5-4** Exponential and Logarithmic Equations 369
- Key Concepts 377
- Mastery Test 382

CHAPTER SIX

Systems of Equations and Inequalities 384

- **6-1** The Substitution and Elimination Methods 385
- **6-2** Solving Linear Systems Using Augmented Matrices 396
- **6-3** Applications of Linear Systems 407
- **6-4** Nonlinear Systems of Equations 415
- **6-5** Systems of Inequalities 425
- **6-6** Linear Programming 431
- Key Concepts 438
- Mastery Test 441

CHAPTER SEVEN

Matrices and Determinants 443

- **7-1** Matrix Operations 444
- **7-2** The Inverse of a Square Matrix 456
- **7-3** Determinants and Cramer's Rule 465
- **7-4** Properties of Determinants 474
- Key Concepts 480
- Mastery Test 485

* This is an optional section.

CHAPTER EIGHT

Sequences and Counting Problems 486

- 8-1 Sequences and Series 487
- 8-2 Arithmetic Sequences and Series 495
- 8-3 Geometric Sequences and Series 503
- 8-4 Mathematical Induction 513
- 8-5 Binomial Theorem 520
- 8-6 Counting: Permutations and Combinations 525
- 8-7 Probability 535
- Key Concepts 542
- Mastery Test 545

APPENDIX A

Calculators 547

Answers to Odd-Numbered Section Exercises, Review Exercises, and All Mastery Tests A-1

Index A-54

About the Author

James W. Hall is an experienced author, having written several mathematics textbooks for PWS-KENT Publishing Company. In addition to *College Algebra with Applications*, Third Edition, his texts include *Beginning Algebra*; *Algebra for College Students*; and *Intermediate Algebra*. He is also the co-author of *Developmental Mathematics*, now in its fourth edition.

Dr. Hall received his BS and MA from Eastern Illinois University, and Ed.D. from Oklahoma State University. He has more than 20 years experience teaching college mathematics and has been a member of the Department of Mathematics at Parkland College for the past 15 years. Dr. Hall spent the 1989–1990 academic year teaching at Dandenong College in Australia as part of a professional exchange program. He is a member of MAA, IMACC, and a past vice-president of AMATYC.

Preface

College Algebra with Applications, Third Edition, was written for college students who have completed a course in intermediate algebra. The topics included in this text prepare students for more advanced courses, such as trigonometry, statistics, computer science, linear algebra, finite mathematics, business calculus, and traditional calculus.

Three main guidelines were followed in the writing of this text. First, the material had to be thorough and accurate. Second, the development had to exhibit thoughtful and consistent pedagogy. Third, the structure of the text had to incorporate abundant student aids to clearly emphasize key points and to focus students' efforts.

Changes in the Third Edition

- Graphing has been emphasized throughout the text.
- Examples and exercises throughout the text have been reworked and updated to accommodate the changing needs of the user.
- Text material has been reorganized to facilitate the incorporation of optional material on graphing calculators.
- Optional exercises have been added for calculus-bound students.
- Exercises for business students have been designated with the symbol $ for easy access.
- Topics have been reordered, particularly in Chapters 3 and 4, to focus on families of functions to a greater extent.

Specific Text Priorities

Graphing
The introduction to graphing emphasizes recognizing features common to families of functions and includes a discussion on the use of symmetry and translations. Graphing techniques are developed gradually throughout the book, with many review exercises incorporating a wide variety of types of questions. The coverage of the graphing of polynomial and rational functions is designed to provide a solid preparation for calculus.

Since there is an advantage to seeing both the traditional approach and the graphical approach within the same text, we have included optional material on graphics calculators. The graphics calculator illustrated in this edition is the TI-81.

Problem Solving
Throughout the text, developing a consistent strategy for solving word problems is emphasized. This strategy involves the use of tables and principles to organize problems and the use of *word equations* to form algebraic equations. In keeping with national recommendations, students are encouraged to think critically, then write the answers to word problems in complete sentences. Exercises designed to develop the ability to use the language of mathematics and to make mental approximations are included.

Careful Development of the Concept of a Function
The development of the concept of a function starts intuitively with the mapping notation. Next this notation is related to the other notations used to represent functions. The text then gradually develops the skills needed by calculus-bound students, including the writing of functions from statements given in word problems.

Concepts and Skill Building in Exercises
There are more than 3,500 carefully selected exercises in the Third Edition. These exercises are graded in levels of difficulty from the **A** exercises that are drill and practice, to the slightly higher level of difficulty in the **B** exercises, and to the **C** exercises that offer some challenge to the student. Many exercises are designed to monitor the student's progress in areas where errors are common. The optional exercises for the calculus-bound student, new to this edition, are designed to give additional practice on the algebraic manipulations that are required in calculus. All are designed to build problem-solving skills.

Features

- Objectives for each section are listed at the beginning of each chapter and at the beginning of each section.

PREFACE

- Self-check exercises are interspersed throughout each section so that students can monitor their own progress.
- Key points and procedures are boxed for easy reference.
- The examples are presented with side-bar explanations. This format provides both a model for the student to emulate and the explanation needed to understand the steps.
- Key concepts are summarized at the end of each chapter.
- Review exercises are presented for each chapter.
- Mastery tests are presented for each chapter.
- The answers to all odd-numbered exercises are in the back of the book, as well as all answers to mastery tests, review exercises, and optional exercises.

Supplements

For Instructors

Instructors' Manual with Chapter Tests. Contains answers to even-numbered exercise, and three sample chapter tests for each chapter.

EXPTest. A testing program for IBM-PCs and compatibles that allows users to view and edit all tests—that is, add to, delete from, and modify existing questions. Any number and variety of tests can be created. A graphics importation feature permits display and printing of graphics, diagrams, and maps provided with the test banks. A demo disk is available.

Exam Builder. This Macintosh testing program allows users to view, edit, and create tests. A demo disk is available.

For Students

Student Solutions Manual. (Vicki Beitler, Parkland College.) Contains complete and worked-out solutions for all odd-numbered exercises in the text and contains complete solutions to all review exercises.

College Algebra Activities for the TI-81 Graphing Calculator. (Lawrence R. Huff and David R. Peterson, University of Central Arkansas.) This manual is designed to supplement any textbook for college algebra. It offers concise instructions on how to use the TI-81 graphing calculator, provides examples on how to use it, and demonstrates techniques on how to use it as a problem-solving tool.

Acknowledgments

I wish to express my appreciation to the following reviewers for their careful and thoughtful criticisms and suggestions:

Fred Alderman
Valencia Community College

C. Leary Bell
Columbus College

Timothy D. Cavanagh
University of Northern Colorado

Ben Cornelius
Oregon Institute of Technology

Ruth Edidin
Morgan State University

Alice W. Essary
University of Southern Mississippi

Eunice F. Everett
Seminole Community College

Margaret J. Greene
Florida Community College at Jacksonville

Mary Lou Hart
Brevard Community College

Lawrence R. Huff
University of Central Arkansas

Ann Lael
Westminster College

Pamela E. Matthews
Mt. Hood Community College

Lynda S. Morton
University of Missouri

J. Douglas Mountain
Lakewood Community College

Ken Peters
University of Louisville

Ron Steffani
South Oregon State College

Beverly Taylor
Valencia Community College

Cliff Tremblay
Pembroke State University

John L. Whitcomb
University of North Dakota

John Venables
Lakewood Community College

Special thanks go to Linda Hermann, Caroline Goodman, Ann Webbink, and Vicki Beitler for their work in rechecking the answers and solutions to the exercises. Peggy Hall, as usual, assisted in the typing, proofing, and numerous other tasks that made it possible to complete this manuscript on schedule.

Finally, it is a pleasure to acknowledge the professional attitude and cooperation of Tim Anderson and Barbara Lovenvirth and my ever-efficient production editor, Susan London, of PWS-KENT Publishing Company.

James W. Hall

CHAPTER ONE

Review of the Basic Concepts of Algebra

Chapter One Objectives

1. Identify the relationships among important subsets of the real numbers.
2. Use interval notation for bounded and unbounded intervals.
3. Use the basic properties of the real number system.
4. Use integral exponents.
5. Simplify expressions using the properties of exponents.
6. Evaluate expressions with grouping symbols using the correct order of operations.
7. Use scientific notation.
8. Determine the degree of a polynomial and classify it according to the number of terms.
9. Add, subtract, multiply, and divide polynomials.
10. Factor polynomials.
11. Reduce a rational expression to lowest terms.
12. Add, subtract, multiply, and divide rational expressions.
13. Use fractional exponents.
14. Simplify radical expressions.
15. Add, subtract, multiply, and divide radical expressions.
16. Add, subtract, multiply, and divide complex numbers.
17. Represent complex numbers graphically.

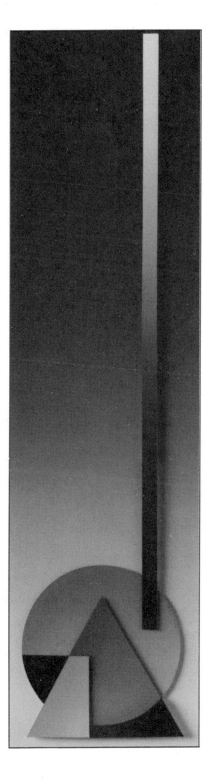

Leonard Euler, 1707–1783

Euler, born in Switzerland, was hired by Catherine the Great of Russia to write the elementary mathematical textbooks for the Russian schools. He wrote prolifically on various mathematical topics. From Euler's textbook *Introductio* came many symbols, such as *i* for $\sqrt{-1}$, π for the ratio of a circumference of a circle to its diameter, and *e* for the base of natural logarithms.

The word "algebra" comes from the book *Hisab al-jabr w'almugabalah*, which dealt with the solving of equations. It was written by an Arab mathematician in A.D. 830, and translations of the text became widely known in Europe as *al-jabar*. Algebra, however, is concerned with more than solving equations; it is a generalization of arithmetic.

Algebra as we study it today is the culmination of hundreds of years of effort by some of the best minds in the history of civilization. What we see is an efficient system, a polished product that has been carefully organized, unified, and condensed into an orderly body of knowledge.

This chapter will review many of the topics usually covered in courses in intermediate algebra. The text then proceeds to extend many of these topics and to develop new concepts and applications for them.

SECTION 1-1

The Real Number Line

Section Objectives

1 Identify the relationships among important subsets of the real numbers.

2 Use interval notation for bounded and unbounded intervals.

Algebra uses both variables and constants to represent numbers; the numbers we use most frequently are the real numbers. The number line is a geometric model of the real numbers that we use to clarify many algebraic concepts. A one-to-one correspondence can be established between points on the line and the real numbers.

The **number line** is an infinite straight line with a fixed point of reference called the **origin**. Once a convenient unit of measure has been chosen, equally spaced points can be labelled as illustrated in Figure 1-1. The number associated with a point on the number line is called the **coordinate** of the

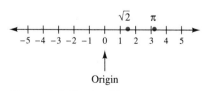

Figure 1-1 Number Line

SECTION 1-1 THE REAL NUMBER LINE

point. Points to the right of the origin have **positive** coordinates, and those to the left of the origin have **negative** coordinates. Zero is neither negative nor positive.

Word problems such as those involving length may require positive answers. Similarly, the problem of calculating the number of bags of fertilizer to buy for a lawn may require a natural number as an answer, since most stores will not sell a fractional portion of a bag. Thus, we should be familiar with many special subsets of the real numbers. Some of the sets used most frequently are listed below.

$\mathbb{N}$ = Natural numbers $\mathbb{N} = \{1, 2, 3, 4, 5, 6, 7, \ldots\}$

$\mathbb{W}$ = Whole numbers $\mathbb{W} = \{0, 1, 2, 3, 4, 5, 6, \ldots\}$

$\mathbb{I}$ = Integers $\mathbb{I} = \{\ldots, -3, -2, -1, 0, 1, 2, 3, \ldots\}$

$\mathbb{Q}$ = Rational numbers $\mathbb{Q} = \left\{\dfrac{a}{b} : a, b \text{ are integers and } b \neq 0\right\}$

$\tilde{\mathbb{Q}}$ = Irrational numbers $\tilde{\mathbb{Q}}$ = The set of real numbers that are not rational

$\mathbb{R}$ = Real numbers $\mathbb{R}$ = The set of coordinates of points on the number line

In the set notation above, note that:

1 Braces are used to enclose the numbers (elements or members) of a set.
2 Capital letters are usually used to name sets.
3 Commas are used to separate elements in a set.
4 Three dots (the ellipsis notation) indicate that elements have been omitted in the listing or that the elements continue indefinitely.
5 The order in which the elements are listed is not significant.
6 Each of the sets listed above is infinite, since the counting of the elements in each set would never terminate.

Figure 1-2 illustrates many of the relationships among the set of real numbers and its important subsets. In particular, note that every natural number is a whole number, every whole number is an integer, every integer is a rational number, and all the rational numbers are real numbers.

The first five letters of "rational" spell "ratio"; this is exactly what a **rational number** is—the ratio of two integers. (The symbol $\mathbb{Q}$ is for quotient.) Every real number is either rational or irrational. The symbols $\mathbb{R} = \mathbb{Q} \cup \tilde{\mathbb{Q}}$ mean that $\mathbb{R}$ equals the **union** of $\mathbb{Q}$ and $\tilde{\mathbb{Q}}$. The symbols $\mathbb{Q} \cap \tilde{\mathbb{Q}} = \emptyset$ mean that the **intersection** of the rationals and the irrationals is the null set.

$\cup$ for union
$\cap$ for intersection

Real numbers are often written in decimal form. This decimal form can be used to characterize the number as either rational or irrational.

4 CHAPTER 1 REVIEW OF THE BASIC CONCEPTS OF ALGEBRA

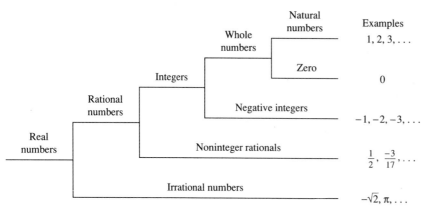

Figure 1-2 The relationships among the set of real numbers and its subsets

Rational and Irrational Numbers ▼

Rational:	The decimal form of a rational number is either
	a. a terminating decimal or
	b. an infinite repeating decimal.
Irrational:	The decimal form of an irrational number is an infinite nonrepeating decimal.

EXAMPLE 1 Classify each of these real numbers as either
(a) a rational number or
(b) an irrational number.

$$0.5, \quad 1.027, \quad 1.01001000100001\ldots, \quad 2\tfrac{1}{3}, \quad 3.14, \quad 3.\overline{14}, \quad \pi$$

SOLUTIONS

(a) Rational: 0.5, 1.027, $2\tfrac{1}{3}$, 3.14, and $3.\overline{14}$.

The terminating decimals 0.5, 1.027, and 3.14 can be written, respectively, as $\tfrac{1}{2}$, $\tfrac{1027}{1000}$, and $\tfrac{157}{50}$.

Note that $\tfrac{314}{100}$ reduces to $\tfrac{157}{50}$.

The infinite repeating decimals $2.\overline{3}$ and $3.\overline{14}$ can be written, respectively, as $\tfrac{7}{3}$ and $\tfrac{311}{99}$.

$2\tfrac{1}{3}$ in decimal form is the infinite repeating decimal $2.333\ldots$, or $2.\overline{3}$. Repeating decimals such as $3.\overline{14}$ (also denoted as $3.141414\ldots$) are converted to fractional form in Chapter 8.

(b) Irrational: $1.01001000100001\ldots$ and π.

Both have an infinite nonrepeating decimal form. They cannot be written as the ratio of two integers.

π does not equal 3.14, $3.\overline{14}$, $\tfrac{22}{7}$, or any other rational number. It can be approximated by these values, but they are not exactly π.
$1.01001000100001\ldots$ does not repeat, since the number of 0 digits between successive 1's changes.

SECTION 1-1 THE REAL NUMBER LINE

Two subsets of the natural numbers play such an important role in arithmetic and algebra that they deserve special attention. A **prime number** is a natural number greater than 1 that has exactly two factors, 1 and itself. A **composite number** is a natural number greater than 1 that has factors other than itself and 1.

EXAMPLE 2 Given $S = \{18, 19, 20, \ldots, 29\}$,

SOLUTIONS

(a) List the prime numbers in S.   19, 23, 29
(b) List the odd composite numbers in S.   21, 25, 27

▶ **Self-Check** ▼

State whether each of the following is true (T) or false (F).

1 1.732 equals $\sqrt{3}$.
2 1.732 is an approximation of $\sqrt{3}$.
3 1.732 is a rational number.
4 $\sqrt{3}$ is a rational number.
5 1.732 is a rational approximation of the irrational number $\sqrt{3}$.

The number line offers an excellent visual representation of the inequalities < (less than) and > (greater than). Because they describe the order of numbers on the number line, inequalities are also known as the **order relations**. The statements $a < b$ and $b > a$ are equivalent, since they have exactly the same meaning.

EXAMPLE 3 Determine the order relationship between the given numbers.

SOLUTIONS

(a) $\dfrac{47}{59}$ and $\dfrac{135}{147}$

$\dfrac{47}{59} \approx 0.80$ and $\dfrac{135}{147} \approx 0.92$

Thus $\dfrac{47}{59} < \dfrac{135}{147}$.

A calculator can be used to form the decimal approximation of a fraction. Both $\approx$ and $\doteq$ are used to denote approximately equal to.

(b) $\dfrac{5}{12}$ and $\dfrac{7}{18}$

$\dfrac{5}{12} = \dfrac{5}{12} \cdot \dfrac{3}{3} = \dfrac{15}{36}$ and $\dfrac{7}{18} = \dfrac{7}{18} \cdot \dfrac{2}{2} = \dfrac{14}{36}$

Thus $\dfrac{5}{12} > \dfrac{7}{18}$.

(c) $\dfrac{22}{33}$ and $\dfrac{34}{51}$

$\dfrac{22}{33} = \dfrac{2 \cdot 11}{3 \cdot 11} = \dfrac{2}{3}$ and $\dfrac{34}{51} = \dfrac{2 \cdot 17}{3 \cdot 17} = \dfrac{2}{3}$

Thus $\dfrac{22}{33} = \dfrac{34}{51}$.

Compound inequalities are frequently written in a condensed form. For example, $2 < x \leq 3$ means that x is greater than 2 *and* less than or equal to 3. In set notation this is written $\{x \mid 2 < x \leq 3\}$. This interval can be graphed on the number line as either

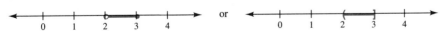

Figure 1-3 $2 < x \leq 3$ as graphed on number lines

▶ **Self-Check Answers** ▼

1 F   2 T   3 T   4 F   5 T

We will use the latter notation, which is consistent with interval notation. **Interval notation** for $2 < x \leq 3$ is $(2, 3]$. Parentheses denote that an "endpoint" *is not* included in the interval, and brackets denote that it *is* included in the interval.

> ▶ **Self-Check** ▼
>
> Fill in each blank with $=$, $<$, or $>$ to make a true statement.
>
> 1 $\frac{4}{5}$ ___ $\frac{12}{15}$ 2 $\frac{2}{7}$ ___ $\frac{2}{5}$
>
> 3 $\frac{2}{3}$ ___ $\frac{7}{11}$

Interval Notation ▼

	Meaning	Graph
Bounded Intervals		
(a, b)	$a < x < b$	open at a, open at b
$(a, b]$	$a < x \leq b$	open at a, closed at b
$[a, b)$	$a \leq x < b$	closed at a, open at b
$[a, b]$	$a \leq x \leq b$	closed at a, closed at b
Unbounded Intervals		
$(a, +\infty)$	$x > a$	open at a, extends right
$[a, +\infty)$	$x \geq a$	closed at a, extends right
$(-\infty, a)$	$x < a$	extends left, open at a
$(-\infty, a]$	$x \leq a$	extends left, closed at a

Note the infinite intervals in the last four cases. The **infinity symbol**, ∞, is *not* a real number; rather, it signifies that the values continue without any end or bound. Since ∞ is not a real number, operations such as $2 + \infty$ and $6 - \infty$ have no meaning in the real number system.

EXAMPLE 4   Represent each interval using interval notation.

SOLUTIONS

(a) $-2 < x \leq 2$ $(-2, 2]$

(b)   the nonnegative real numbers $[0, +\infty)$

(c)   all real numbers less than -1 or greater than 1 (Use $\cup$ to denote the union of these two intervals.) $(-\infty, -1) \cup (1, +\infty)$

> The union of two sets is the set whose members belong to at least one of these two sets.

▶ **Self-Check Answers** ▼

1 $=$ 2 $<$ 3 $>$

SECTION 1-1 THE REAL NUMBER LINE

EXAMPLE 5 Graph the solution for $(-\infty, 3) \cap [2, 5]$.

SOLUTION The intersection of these intervals, denoted by $\cap$, is the set of points in both intervals. The overlap is the interval shown in color.

Answer $[2, 3)$

The absolute value of x, denoted by $|x|$, is defined by

$$|x| = \begin{cases} x & \text{if } x \geq 0 \\ -x & \text{if } x < 0 \end{cases}$$

Thus $|4| = 4$ and $|-4| = -(-4) = 4$. The number line provides an intuitive means for examining the concept of absolute value, since the absolute value of a number is its distance from the origin. For example, $|x| < 3$ represents the interval of points less than 3 units from zero.

▶ **Self-Check** ▼

Represent these intervals using interval notation.
1. $-3 \leq x < -1$
2. the set of all real numbers
3. all reals less than 3 or greater than or equal to 5
4. $(-3, 5] \cap [2, +\infty)$

Write each expression in interval notation.
5. $|x| < 3$ 6. $|x| \geq 5$
7. $|x| \leq 2$

Absolute Value Expressions ▼

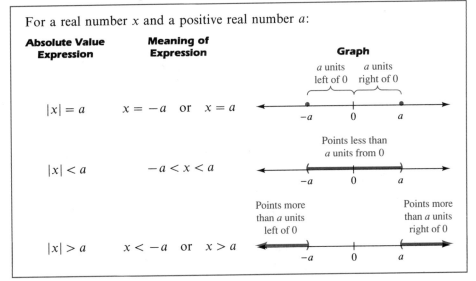

For a real number x and a positive real number a:

Absolute Value Expression	Meaning of Expression	Graph
$\|x\| = a$	$x = -a$ or $x = a$	(points at $-a$ and a; a units left of 0, a units right of 0)
$\|x\| < a$	$-a < x < a$	Points less than a units from 0
$\|x\| > a$	$x < -a$ or $x > a$	Points more than a units left of 0; Points more than a units right of 0

We will use this concept of absolute value when we solve equations and inequalities in Section 2-6.

▶ **Self-Check Answers** ▼

1. $[-3, -1)$ 2. $(-\infty, +\infty)$ 3. $(-\infty, 3) \cup [5, +\infty)$ 4. $[2, 5]$ 5. $(-3, 3)$ 6. $(-\infty, -5] \cup [5, +\infty)$ 7. $[-2, 2]$

EXERCISES 1-1

A

In Exercises 1 and 2, simplify each expression.

1. **a.** $|-5.23|$ **b.** $|5.23|$ **c.** $|-6+6|$ **d.** $|-6|+|6|$
2. **a.** $|3-13|$ **b.** $|3|-|13|$ **c.** $-|-15|$ **d.** $|-15|-|15|$

In Exercises 3–6, correctly fill in each blank with =, <, or >. Use a calculator, if necessary, to approximate each number.

3. **a.** $5.4 \_\_ 5.4$ **b.** $-5.4 \_\_ -4.5$
4. **a.** $\frac{4}{5} \_\_ \frac{8}{9}$ **b.** $-\frac{4}{5} \_\_ -\frac{8}{9}$
5. **a.** $1.41 \_\_ \sqrt{2}$ **b.** $1.42 \_\_ \sqrt{2}$
6. **a.** $\frac{22}{7} \_\_ \pi$ **b.** $-\frac{1}{3} \_\_ -0.\overline{3}$

7. Which of the following is *not* a natural number?

$$-7, \quad 7, \quad 77, \quad 777$$

8. Which of the following is *not* a whole number?

$$51, \quad 15, \quad 5, \quad 11, \quad 0, \quad -5$$

9. What whole number is *not* a natural number?
10. Which of the following integers is a natural number?

$$-9, \quad -6, \quad -3, \quad 0, \quad 3$$

11. Which of the following numbers is rational?

$$\sqrt{3}, \quad \sqrt{5}, \quad \sqrt{16}, \quad \sqrt{17}$$

12. Which of the following numbers is irrational?

$$1.2, \quad 1.2222\ldots, \quad 1.21121112111112111112\ldots, \quad 1.\overline{2}$$

Use the set $S = \{10, 11, 12, 13, 14, 15, 16, 17, 18, 19, 20, 21, 22, 23\}$ to answer Exercises 13 and 14.

13. List all prime numbers in S.
14. List all composite numbers in S.
15. What real number is neither negative nor positive?
16. What prime number is even?
17. What natural number is neither prime nor composite?

In Exercises 18–28, fill in the chart by checking the sets to which each of the numbers belongs.

		$\mathbb{N}$	$\mathbb{W}$	$\mathbb{I}$	$\mathbb{Q}$	$\tilde{\mathbb{Q}}$	$\mathbb{R}$
18	-25						
19	$3\frac{1}{7}$						
20	π						
21	37						
22	2.236						

SECTION 1-1 THE REAL NUMBER LINE

23   0   ___ ___ ___ ___ ___ ___
24   -4.7 ___ ___ ___ ___ ___ ___
25          $\sqrt{5}$ ___ ___ ___ ___ ___ ___
26          $\sqrt{36}$ ___ ___ ___ ___ ___ ___
27          $3.\overline{147}$ ___ ___ ___ ___ ___ ___
28          $3.14114111411114\ldots$ ___ ___ ___ ___ ___ ___

In Exercises 29 and 30, rewrite each rational number as the ratio of two integers.

29   **a.** $3\frac{2}{5}$ **b.** 3.2 **c.** $0.\overline{6}$ **d.** $\sqrt{25}$
30   **a.** 0 **b.** -0.25 **c.** $-0.\overline{3}$ **d.** $\sqrt{81}$

In Exercises 31–34, graph the given set of points.

31   **a.** $[2, 4)$ **b.** $(2, 4]$ **c.** $(-\infty, 2)$ **d.** $[2, +\infty)$
32   **a.** $(-1, 3)$ **b.** $[-3, 1]$ **c.** $(-\infty, 1]$ **d.** $(-3, +\infty)$
33   **a.** $|x| > 3$ **b.** $[-4, 2) \cap (1, 3)$ **c.** $(-2, 0] \cup (1, 4]$ **d.** $(-\infty, 7) \cap [3, +\infty)$
34   **a.** $|x| \leq 2$ **b.** $(-4, 3] \cap [-1, 6)$ **c.** $(-4, 3] \cup (-1, 6]$ **d.** $(-\infty, 2) \cap [1, +\infty)$

In Exercises 35 and 36, write each expression in interval notation.

35   **a.** $-5 \leq x < 3$ **b.** $x < 5$ **c.** $|x| \geq 4$
36   **a.** $2 < x \leq 5$ **b.** $x \geq -5$ **c.** $|x| < 6$

In Exercises 37 and 38, write the interval notation for each graph.

37   **a.**    **b.** 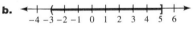   **c.**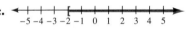

38   **a.** 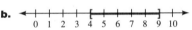   **c.**

B

39   Is there a smallest integer greater than one? If so, what is this integer? If not, explain why not.

40   Is there a smallest rational number greater than one? If so, what is this number? If not, explain why not.

41   Is there a smallest real number greater than one? If so, what is this number? If not, explain why not.

42   Is there a smallest real number in the interval $[1, +\infty)$? If so, what is this number? If not, explain why not.

43   Is there a smallest real number in the interval $(1, +\infty)$? If so, what is this number? If not, explain why not.

44   Is there a largest real number in the interval $[1, +\infty)$? If so, what is this number? If not, explain why not.

45   Is there a largest real number in the interval $[1, 3)$? If so, what is this number? If not, explain why not.

46   Is there a largest real number in the interval $[1, 3]$? If so, what is this number? If not, explain why not.

For Exercises 47 and 48, state whether each statement is true (T) or false (F).

47 You can give the exact length of any line segment with an integer.

48 You can give the exact length of any line segment with a rational number.

C

Answer each part of Exercises 49 and 50. Some answers based on computations with calculators may vary from model to model. See the material on calculators in the Appendix if you have questions on proper calculator usage.

49 **a.** Can all rational numbers be represented exactly on a calculator? Explain why or why not.

b. Using the reciprocal key `1/x` or `x⁻¹`, evaluate $\frac{1}{3} - 0.33333333$ on your calculator. (For a calculator with more than an eight-digit display, enter as many threes as possible.) Explain the result. Does the value shown for $\frac{1}{3}$ really equal $\frac{1}{3}$?

50 **a.** Are numbers shown on calculators rational or irrational? Why?

b. Is π rational or irrational?

c. Is the number you obtain by pressing the `π` key on a calculator rational or irrational?

d. Write and then compare the values you obtain from your calculator for the following expressions:

  i. entering 3.14   **ii.** finding $\frac{22}{7}$

  iii. finding $\frac{355}{113}$ **iv.** `π` key on your calculator

e. Are any of the values in part (d) equal to π? If so, which one? If not, which value is closest?

f. What conclusion can you form about using calculators for calculations with irrational numbers?

SECTION 1-2

Properties of the Real Number System

Section Objective

3 Use the basic properties of the real number system.

Algebra is sometimes referred to as a universal language. It is a system of written communication for transmitting information about our mathematical world. Part of the power of this algebraic language comes from the conciseness of the notation. Each symbol and its location convey a precise meaning. For example, 32, 23, 2^3, 3^2, $\frac{3}{2}$, and $\frac{2}{3}$ all have distinct meanings.

The reason that we can all obtain the same answer to a problem is that we follow the same rules, or **properties**. Everyone, even those who do not know the names of the properties, must obey them. The basic properties, or **axioms**, are not proven but are assumptions that are made to characterize or describe how the real number system and equality behave. Using these basic properties, mathematicians have discovered other facts that are a logical consequence of previous information. Important new statements are called **theorems**; a well-known example is the Pythagorean Theorem, which states that the sum of the squares of the lengths of the legs of a right triangle is equal to the square of the length of the hypotenuse ($a^2 + b^2 = c^2$).

We will now list those properties which serve as the cornerstone of our real number system. As an aid to remembering the names of these properties, compare them to the common words "reflect," "symmetrical," "commute," and so on. Note how each name describes the property.

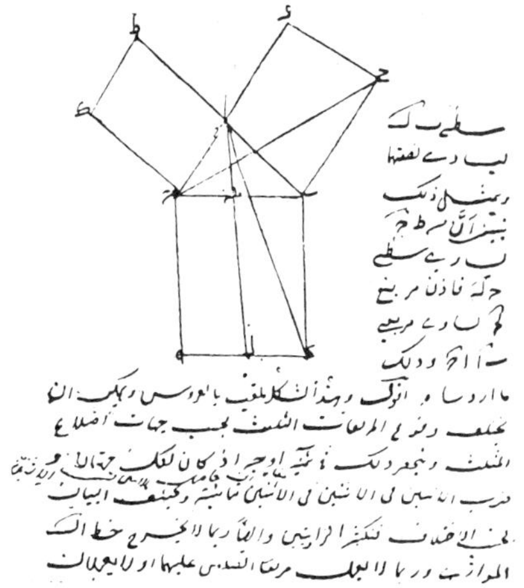

Figure 1-4 Theorem of Pythagoras in an Arabic mathematical manuscript of the 14th century. This illustration appears in The VNR Concise Encyclopedia of Mathematics, edited by Küstner and Kästner, © 1977. Reprinted with permission of Van Nostrand Reinhold Company, NYC.

Properties of Equality ▼

For all real numbers a, b, and c:

Reflexive property:	$a = a$
Symmetric property:	If $a = b$, then $b = a$.
Transitive property:	If $a = b$ and $b = c$, then $a = c$.
Substitution property:	If $a = b$, then a may be substituted for b in any expression without changing the value of that expression.

EXAMPLE 1 Identify the property that justifies each equality.

                                          SOLUTIONS

(a) $4y^2 - 7y = 4y^2 - 7y$.   Reflexive property

(b) If $\dfrac{1}{3} = 0.\bar{3}$, then $0.\bar{3} = \dfrac{1}{3}$.   Symmetric property

(c) If $7w + 3z = x$ and $x = 11$, then $7w + 3z = 11$.   Transitive property

(d) If $5 = z$ and $w + z = 11$, then $w + 5 = 11$.   Substitution property   ☐

Properties of Inequality ▼

For all real numbers a, b, and c:

Trichotomy property:	Exactly one of the statements is true: $a = b$, $a > b$, or $a < b$.
Transitive property:	If $a < b$ and $b < c$, then $a < c$.

EXAMPLE 2 Identify the property that justifies each statement.

SOLUTIONS

(a) $0.\bar{9} = 0$, $0.\bar{9} > 0$, or $0.\bar{9} < 0$.   Trichotomy property

(b) If $y + z > w$ and $w > 19$, then   Transitive property
$y + z > 19$.

Note that the transitive property holds for all the order relationships: $<$, $\leq$, $>$, and $\geq$.   ☐

Properties of the Real Numbers ▼

For all real numbers a, b, and c:

Closure: $a + b$ is also a real number.   The real numbers are closed with respect to addition.

ab is also a real number.   The real numbers are closed with respect to multiplication.

Commutative: $a + b = b + a$ Commutative property of addition

$ab = ba$ Commutative property of multiplication

Associative: $a + (b + c) = (a + b) + c$ Associative property of addition

$a(bc) = (ab)c$ Associative property of multiplication

Distributive: $a(b + c) = ab + ac$ Multiplication distributes over addition.

Note that the distributive property is the only one of these properties that involves two operations; it relates multiplication and addition.

EXAMPLE 3 Identify the property that justifies each statement.

SOLUTIONS

(a) $(x + 2y) + z = x + (2y + z)$ Associative property of addition

(b) $x(y + z) = (y + z)x$ Commutative property of multiplication

(c) $x(y + z) = xy + xz$ Distributive property

(d) $3x^2 + 6xy = 3x(x + 2y)$ Distributive property   ☐

We say that a set is **closed** with respect to an operation if performing the operation on elements in the set always yields a unique answer in that set. For example,

- The even numbers are closed under addition, since the sum of any two even numbers is even.

SECTION 1-2 PROPERTIES OF THE REAL NUMBER SYSTEM

- Subtraction of the natural numbers is not closed, since the difference of the natural numbers $5 - 7$ is not a natural number.
- The real numbers are not closed under division, since $8 \div 0$ is not a real number. The expression $8 \div 0$ is undefined.

The expression $8 \div 0$ is undefined since $\frac{8}{0} \neq x$ for any real number x. There is no number x that can be multiplied by 0 to give a product of 8. Since $0 \cdot x$ cannot equal 8, the expression $\frac{8}{0}$ cannot equal x for any real number x. Thus we say that division of a nonzero number by zero is undefined. The expression $\frac{0}{0}$ is called indeterminate since any real number x will satisfy $0 = 0 \cdot x$. (There is no reason to select or determine one value of x in preference to any of the other values of x.) The indeterminate expression $\frac{0}{0}$ is also undefined since there is no unique value to assign to $\frac{0}{0}$. For a set to be closed, there must not only be an answer for an operation on two elements, but also this answer must be unique.

Division by Zero is Undefined ▼

$\dfrac{x}{0}$ is undefined for all real numbers x.

Identities and Inverses ▼

Additive identity:	0	0 is the only real number for which $a + 0 = a$ for every real number a.
Multiplicative identity:	1	1 is the only real number for which $1 \cdot a = a$ for every real number a.
Additive inverse:	$-a$	For each real number a, there is an opposite, $-a$, such that $a + (-a) = 0$.
Multiplicative inverse:	$\dfrac{1}{a}$	For each real number $a \neq 0$, there is a real number $\dfrac{1}{a}$ such that $a\left(\dfrac{1}{a}\right) = 1$.

▶ **Self-Check ▼**

Complete these statements using the property specified.

1. $a(x + y) + b(x + y) = (?)(x + y)$
   Distributive property
2. $a(x + y) = a(?)$
   Commutative property of addition
3. $(2a)y = ?$
   Associative property of multiplication

Inverses of some selected numbers are illustrated below.
- The additive inverse of $-\frac{3}{4}$ is $+\frac{3}{4}$, since $-\frac{3}{4} + (+\frac{3}{4}) = 0$.
- The multiplicative inverse of $-\frac{3}{4}$ is $-\frac{4}{3}$, since $-\frac{3}{4}(-\frac{4}{3}) = 1$.
- The multiplicative inverse of 1 is 1, since $1 \cdot 1 = 1$.
- The additive inverse of 1 is -1, since $1 + (-1) = 0$.

▶ **Self-Check Answers ▼**

1 $a + b$ **2** $y + x$ **3** $2(ay)$

Subtraction is usually defined in terms of addition. That is, $a - b$ means $a + (-b)$. Likewise, division can be defined in terms of multiplication by defining $a \div b$ to mean $a \cdot \dfrac{1}{b}$ for $b \neq 0$.

The information in the following box is a direct result of the meaning of a multiplicative inverse.

The Multiplicative Identity ▼

If A is an algebraic expression, then $\dfrac{A}{A} = 1$ for $A \neq 0$.

Caution: $\dfrac{0}{0}$ is undefined.

Thus each of the following numbers is equal to 1:

$$\dfrac{2}{2}, \quad \dfrac{-31}{-31}, \quad \dfrac{\pi}{\pi}, \quad \dfrac{3.73}{3.73}, \quad \dfrac{\sqrt{3}}{\sqrt{3}}, \quad \dfrac{1+\sqrt{2}}{1+\sqrt{2}}, \quad \text{and} \quad \dfrac{x-y}{x-y} \text{ for } x \neq y$$

The identity elements and the inverses play key roles in the simplification and manipulation of algebraic expressions. For example, we can use the multiplicative identity $\dfrac{11}{11}$ to justify that $\dfrac{22}{33} = \dfrac{2 \cdot \cancel{11}}{3 \cdot \cancel{11}} = \dfrac{2}{3}$.

We have listed only a few axioms; however, we are now free to combine these assumptions into many new and powerful observations. Some of these observations are so important that they are called **theorems** and are listed for future usage. Although this text does not emphasize proofs of theorems, it is still important that you understand the role of theorems and proofs in mathematics. Theorems generally consist of a **hypothesis** (the "if" part) and a **conclusion** (the "then" part). The **proof** is a step-by-step justification that the hypothesis always forces the conclusion to follow. The next theorem illustrates how the axioms we have presented can be used in proofs.

Zero-Factor Theorem

If a is any real number, then $0 = a \cdot 0$.
            hypothesis   conclusion

SECTION 1-2 PROPERTIES OF THE REAL NUMBER SYSTEM

Proof

Step 1	$0 = a + (-a)$	Additive inverse
Step 2	$= a \cdot 1 + (-a)$	Multiplicative identity 1
Step 3	$= a \cdot (0 + 1) + (-a)$	Additive identity 0
Step 4	$= (a \cdot 0 + a \cdot 1) + (-a)$	Distributive property
Step 5	$= (a \cdot 0 + a) + (-a)$	Multiplicative identity
Step 6	$= a \cdot 0 + [a + (-a)]$	Associative property of addition
Step 7	$= a \cdot 0 + [0]$	Additive inverse
Step 8	$= a \cdot 0$	Additive identity
Step 9	$0 = a \cdot 0$	Transitive property of equality

Double-Negative Theorem If a is any real number, then $-(-a) = a$. The proof of this theorem is similar to the previous proof and is left as an exercise at the end of this section.

The results below follow from the Zero-Factor Theorem and the Double-Negative Theorem:

$$7 \cdot 0 = 0, \quad 0 \cdot 9 = 0, \quad -3 \cdot 0 = 0, \quad 0 \cdot (\tfrac{1}{2}) = 0, \quad 0 \cdot 0 = 0$$

$$-(-7) = 7, \quad -(-\tfrac{1}{2}) = \tfrac{1}{2}, \quad -(-\pi) = \pi, \quad -[-(-2)] = -2$$

Addition Theorem of Equality If a, b, and c are real numbers and $a = b$, then $a + c = b + c$.

Multiplication Theorem of Equality If a, b, and c are real numbers and $a = b$, then $ac = bc$.

Most proofs in algebra are not written in the format given in the Zero-Factor Theorem; rather, they are given with the steps and reasons combined into paragraph form. This is illustrated by the proof of the Zero-Factor Principle, which is given below.

Zero-Factor Principle If a and b are real numbers and if $ab = 0$, then $a = 0$ or $b = 0$.

Proof
If both a and b are 0, then the statement is trivially satisfied. Thus we shall now examine the case where one of them, say b, is not 0. Since $b \neq 0$, multiply both sides of $ab = 0$ by $\dfrac{1}{b}$ to obtain $ab\left(\dfrac{1}{b}\right) = 0\left(\dfrac{1}{b}\right)$. Hence $a\left(b \cdot \dfrac{1}{b}\right) = 0$, which leads to $a(1) = 0$. Thus $a = 0$, which is what we wanted to prove.

EXERCISES 1-2

A

In Exercises 1–16, complete the statement of the given property by replacing the question mark with the correct expression. Assume that all the variables used represent real numbers.

1. $r + (s + t) = ?$ — Associative property of addition
2. $[r + (s + t)] + 0 = ?$ — Additive identity
3. $r + (s + t) = (s + t) + ?$ — Commutative property of addition
4. $(vw)x = (?)x$ — Commutative property of multiplication
5. $(vw)x = x(?)$ — Commutative property of multiplication
6. $(vw)x = v(?)$ — Associative property of multiplication
7. $-x + x = ?$ — Additive inverse
8. For $y \neq 0$, $y\left(\dfrac{1}{y}\right) = ?$ — Multiplicative inverse
9. If $a > x$ and $x > z$, then ? — Transitive property of inequality
10. $7(a + 11) = 7a + ?$ — Distributive property
11. $a \not< b$, $a \not> b$, thus ? — Trichotomy property
12. $(a + b)(x + y) = (a + b)(?)$ — Commutative property of addition
13. $(a + b)(x + y) = (a + b)x + ?$ — Distributive property
14. $(a + b)(x + y) = (x + y)?$ — Commutative property of multiplication
15. $18xy + 9xw = 9x(2y + ?)$ — Distributive property
16. $7ab + 7ac = (?)(b + c)$ — Distributive property

In Exercises 17–28, identify the property that justifies the statement.

17. $e\left(\dfrac{1}{e}\right) = 1$
18. $1 \cdot e = e$
19. $\pi(a + b) = \pi(b + a)$
20. $\pi(a + b) = \pi a + \pi b$
21. $(x + 3y) + z = x + (3y + z)$
22. $x + 3 > 7$, or $x + 3 < 7$, or $x + 3 = 7$
23. $t(ap) = (ta)p$
24. $t(ap) = (ap)t$
25. $(x + y)(z + w) = (z + w)(x + y)$
26. $(x + y)(z + w) = (y + x)(z + w)$
27. $(x + y)(z + w) = (x + y)z + (x + y)w$
28. $(x + y)(z + w) = x(z + w) + y(z + w)$

In Exercises 29–34, simplify each expression. Be ready to justify your steps orally.

29. $8(5w)$
30. $6 + (7 + c)$
31. $(7 + c) + 8$
32. $5(x + 2)$
33. $-[-(-y)]$
34. $(x + y) \cdot 0$
35. What is the additive inverse of -1?
36. What is the multiplicative inverse of -1?
37. Does every real number x have an additive inverse? Explain your answer.
38. Does every real number x have a multiplicative inverse? Explain your answer.

SECTION 1-2 PROPERTIES OF THE REAL NUMBER SYSTEM

B

39 Give an example of two real numbers a and b such that $a + b$ is positive and $a - b$ is negative.

40 Give an example of two real numbers a and b such that $a + b$ is negative and $a - b$ is positive.

41 Give an example of two real numbers a and b such that $a + b$ is neither positive nor negative.

In Exercises 42–48, state whether each statement is true (T) or false (F).

42 **a.** $\sqrt{3}$ is irrational.
  b. $1 + \sqrt{3}$ is irrational
  c. $1 - \sqrt{3}$ is irrational.
  d. $(1 + \sqrt{3}) + (1 - \sqrt{3})$ is irrational.
  e. The irrationals are closed under addition.

43 **a.** $\sqrt{2}$ is irrational.
  b. $\sqrt{18}$ is irrational.
  c. $(\sqrt{2})(\sqrt{18})$ is irrational.
  d. The irrationals are closed under multiplication.

44 Subtraction is commutative.

45 Division is associative.

46 The negative real numbers are closed under addition.

47 The negative real numbers are closed under multiplication.

48 The negative real numbers are closed under subtraction.

C

49 **a.** What is the hypothesis of the Double-Negative Theorem?
  b. What is the conclusion of the Double-Negative Theorem?

50 **a.** What is the hypothesis of the Addition Theorem of Equality?
  b. What is the conclusion of the Addition Theorem of Equality?

51 Give the justification for each step of the following proof of the Multiplication Theorem of Equality.

   Theorem
   If a, b, and c are real numbers and $a = b$, then $ac = bc$.

   Proof
   Step 1 $ac = ac$
   Step 2 $ac = bc$

52 Prove the Double-Negative Theorem.

53 Prove the Addition Theorem of Equality.

CHAPTER 1   REVIEW OF THE BASIC CONCEPTS OF ALGEBRA

SECTION 1-3

Integral Exponents and Algebraic Expressions

Section Objectives

4 Use integral exponents.

5 Simplify expressions using the properties of exponents.

6 Evaluate expressions with grouping symbols using the correct order of operations.

7 Use scientific notation.

Exponential notation is one of the most concise and convenient notations ever devised by mathematicians. Originally natural numbers were used as exponents as a means of denoting repetitive multiplication.

$$a^n = \underbrace{a \cdots a}_{n \text{ factors of } a}$$

Exponent n is a natural number. Base a.

Using this notation one can easily multiply or divide expressions with the same base. Because of this power, the concept of exponents was eventually extended beyond repetitive multiplication so that any real number could be used as an exponent. In this section we will review integer exponents.

Zero and Negative Exponents ▼

For any nonzero* real number x and natural number n:

1 $x^0 = 1$

2 $x^{-n} = \dfrac{1}{x^n}$

* Note that 0^0 is undefined.

EXAMPLE 1   Simplify each expression.

SOLUTIONS

(a) $(-5)^2$ $(-5)^2 = (-5)(-5) = 25$ The base is -5.

(b) -5^2 $-5^2 = -(5 \cdot 5) = -25$ The base is 5, not -5.

(c) $(-5)^0$ $(-5)^0 = 1$

SECTION 1-3 INTEGRAL EXPONENTS AND ALGEBRAIC EXPRESSIONS

(d) -5^0 　　　　　　　　　　$-5^0 = -(1) = -1$

(e) $(3x + 3y)^0 + (2x)^0$ 　　$(3x + 3y)^0 + (2x)^0 + (3y)^0 + 2x^0 + 3y^0$
　　$+ (3y)^0 + 2x^0 + 3y^0$ 　$= 1 + 1 + 1 + 2 + 3 = 8$

(f) $(2 + 3)^{-1} + 2^{-1} + 3^{-1}$ 　$(2 + 3)^{-1} + 2^{-1} + 3^{-1}$
$$= \frac{1}{5} + \frac{1}{2} + \frac{1}{3} = \frac{31}{30}$$
□

The correct order for performing operations is important since algebraic expressions often involve several operations. The conventional order of operations has also been adopted by most computer languages and by the manufacturers of most calculators. You should verify this hierarchy on your calculator before undertaking an important calculation. Grouping symbols that are used are parentheses ( ), brackets [ ], braces { }, and the vinculum, or bar —.

Order of Operations ▼

Step 1 Start with the expression within the innermost pair of grouping symbols.

Step 2 Perform all exponentiations.

Step 3 Perform all multiplications and divisions, working from left to right.

Step 4 Perform all additions and subtractions, working from left to right.

▶ Self-Check ▼

Simplify each expression.

1 $2^3 + 3^2$ 　　**2** $5^0 + 0^5$

3 $\left(\frac{3}{5}\right)^{-4}$ 　　**4** $-1^{50} - 1^{47}$

EXAMPLE 2 Simplify $3 + 5[2 - 3(6 - 7) - 5]$.

SOLUTION $3 + 5[2 - 3(6 - 7) - 5] = 3 + 5[2 - 3(-1) - 5]$ 　Start with the innermost grouping symbols.
$$= 3 + 5[2 + 3 - 5]$$
$$= 3 + 5[0]$$
$$= 3 + 0$$ 　　　Multiplication by 5 has priority over addition.
$$= 3$$
□

▶ Self-Check Answers ▼

1 17 　　**2** 1 　　**3** $\frac{625}{81}$ 　　**4** -2

EXAMPLE 3 Simplify $\dfrac{3^2 + 4^2 + (3+4)^2}{29 - 2[3 + 5 \cdot 6]}$.

SOLUTION
$$\dfrac{3^2 + 4^2 + (3+4)^2}{29 - 2[3 + 5 \cdot 6]} = \dfrac{9 + 16 + 7^2}{29 - 2[3 + 30]}$$

Notice the distinct meaning of $3^2 + 4^2$ and $(3+4)^2$. Also notice the fraction bar serves as a grouping symbol to separate the numerator of a fraction from the denominator.

$$= \dfrac{9 + 16 + 49}{29 - 2(33)}$$

$$= \dfrac{74}{29 - 66}$$

$$= \dfrac{74}{-37}$$

$$= -2 \qquad \square$$

The properties of exponents that are used most frequently are summarized in the following box.

Properties of Exponents ▼

For any nonzero real numbers x and y and integer exponents m and n:

Product rule: $\quad x^m x^n = x^{m+n}$

Quotient rule: $\quad \dfrac{x^m}{x^n} = x^{m-n}$

Power rule: $\quad (x^m)^n = x^{mn}$

Rule for a product to a power: $\quad (xy)^m = x^m y^m$

Rule for a quotient to a power: $\quad \left(\dfrac{x}{y}\right)^m = \dfrac{x^m}{y^m}$

$$\left(\dfrac{x}{y}\right)^{-n} = \left(\dfrac{y}{x}\right)^n$$

EXAMPLE 4 Simplify the following expressions. Write the answer using only positive exponents. Assume that each base is nonzero.

SOLUTIONS

(a) $\dfrac{(2x + 3y)^5}{2x + 3y}$ $\qquad \dfrac{(2x + 3y)^5}{2x + 3y} = (2x + 3y)^{5-1}$ $\qquad$ In the denominator, $2x + 3y = (2x + 3y)^1$.

$\qquad\qquad\qquad\qquad\qquad = (2x + 3y)^4$

(b) $x^7 \cdot x^{-12}$ $\qquad x^7 \cdot x^{-12} = x^{7+(-12)} = x^{-5} = \dfrac{1}{x^5}$

(c) $[(x - y)^{-3}]^{-5}$ $\qquad [(x - y)^{-3}]^{-5} = (x - y)^{(-3)(-5)} = (x - y)^{15}$

(d) $\left(\dfrac{x^2}{y^5}\right)^{-3}$ $\qquad \left(\dfrac{x^2}{y^5}\right)^{-3} = \left(\dfrac{y^5}{x^2}\right)^3 = \dfrac{y^{15}}{x^6}$ $\qquad \square$

SECTION 1-3 INTEGRAL EXPONENTS AND ALGEBRAIC EXPRESSIONS

EXAMPLE 5 Simplify these expressions.

SOLUTIONS

(a) $\left(\dfrac{12x^5 y^2 z^7}{18x^4 y^2 z^9}\right)^3$ 　 $\left(\dfrac{12x^5 y^2 z^7}{18x^4 y^2 z^9}\right)^3 = \left(\dfrac{2x}{3z^2}\right)^3$ Reduce the fraction.

$= \dfrac{8x^3}{27z^6}$ Cube.

(b) $\left(\dfrac{(3x^{-1} y^5)^2}{(2x^{-2} y^4)^3}\right)^{-1}$ 　 $\left(\dfrac{(3x^{-1} y^5)^2}{(2x^{-2} y^4)^3}\right)^{-1} = \left(\dfrac{9x^{-2} y^{10}}{8x^{-6} y^{12}}\right)^{-1}$ Square numerator and cube denominator.

$= \dfrac{8x^{-6} y^{12}}{9x^{-2} y^{10}}$ Exponent of -1

$= \dfrac{8y^{12-10}}{9x^{-2-(-6)}}$ Quotient rule: Subtracting the smaller exponent from the larger exponent will put the result in a form that has positive exponents.

$= \dfrac{8y^2}{9x^4}$

EXAMPLE 6 Evaluate these algebraic expressions for $a = -1$, $b = -2$, and $c = -3$.

SOLUTIONS

(a) $a - bc$ 　 $a - bc = -1 - (-2)(-3) = -1 - 6 = -7$

(b) $a - b^c$ 　 $a - b^c = -1 - (-2)^{-3}$

$= -1 - \left(-\dfrac{1}{8}\right) = -\dfrac{7}{8}$

(c) $(a - c)^b$ 　 $(a - c)^b = [-1 - (-3)]^{-2} = [2]^{-2} = \dfrac{1}{4}$

(d) $b^a + c^a - (b + c)^a$ 　 $b^a + c^a - (b + c)^a$

$= (-2)^{-1} + (-3)^{-1} - [-2 + (-3)]^{-1}$

$= -\dfrac{1}{2} + \left(-\dfrac{1}{3}\right) - \left(-\dfrac{1}{5}\right)$

$= \dfrac{-15 - 10 + 6}{30}$

$= -\dfrac{19}{30}$

▶ **Self-Check** ▼

Simplify each expression.

1 $(5^2) + (-3^2) + (5 - 3)^2$
2 $2^3 + 2^{-1} + 2^0$
3 $-48 + 24 \div 2^2 \cdot 3 + 9$
4 $\dfrac{6 + 2(4)}{17 - 5[8 - 25 \div 5]}$

Simplify the following expressions.

5 $(2x)^3 - 2x^3$ 　 6 $\dfrac{2x^{-1} y^2}{6x^2 y^{-3}}$

7 $(3x^{-4} y^3)^{-2}$

▶ **Self-Check Answers** ▼

1 20 　 2 9.5 　 3 -21 　 4 7 　 5 $6x^3$ 　 6 $\dfrac{y^5}{3x^3}$ 　 7 $\dfrac{x^8}{9y^6}$

Scientific Notation

Calculations with very large numbers, such as 9,454,177,440,000, or very small numbers, such as 0.000000005, appear more difficult than they really are because there are so many zeros that serve only to locate the decimal point. Scientists have used scientific notation for years to simplify their calculations, and now this notation is also used in scientific calculators and many computer languages. A number is in **scientific notation** if it is written as a product of a number between 1 and 10 (or -1 and -10 if negative) and an appropriate power of 10. The following numbers are written in scientific notation (see Table 1-1).

Table 1-1 Converting to Scientific Notation

Number		Scientific Notation
312	3.12×100	3.12×10^2
31.2	3.12×10	3.12×10^1
0.312	$3.12 \times (0.1)$	3.12×10^{-1}
0.000312	$3.12 \times (0.0001)$	3.12×10^{-4}
-3120	-3.12×1000	-3.12×10^3

One Light-year

A light-year is the distance that light, traveling at 299,790 kilometers per second, travels in one year. The conversion of a light-year to kilometers is shown below.

299,790	km per second
$\times 60$	
17987400	km per minute
$\times 60$	
1079244000	km per hour
$\times 24$	
25901856000	km per day
$\times 365$	
9454177440000	km per year

The number of positions the decimal point must be moved determines the correct power of 10 to use. Magnitudes larger than 1 indicate a positive exponent on 10 while numbers smaller than 1 indicate a negative exponent on 10.

EXAMPLE 7 Write these numbers in scientific notation.

SOLUTIONS

(a) 71,489,000 $71,489,000. = 7.1489 \times 10^7$
   7 positions to the left

(b) 0.000035 $0.000035 = 3.5 \times 10^{-5}$
   5 positions to the right

▶ **Self-Check** ▼

Convert each of these numbers from scientific notation to standard decimal notation.

1 5.091×10^{-3}
2 5.091×10^6

The display on a calculator can show only a few digits; hence, most scientific models are designed to accept input and to display answers in scientific notation. The scientific notation key is usually labeled **EE** or **EXP**. The actual label on this key varies from model to model.

▶ **Self-Check Answers** ▼

1 0.005091   **2** 5,091,000

SECTION 1-3 INTEGRAL EXPONENTS AND ALGEBRAIC EXPRESSIONS

EXAMPLE 8 The display of a calculator shows `1.23  11`. What value does this represent?

SOLUTION This represents 1.23×10^{11} or, in standard decimal form, the 12-digit number 123,000,000,000. □

Scientific Calculator

EXAMPLE 9 Enter the value 0.0000000072 into a calculator.

SOLUTION In scientific notation, the number is 7.2×10^{-9}. Thus a typical key sequence is:

Keystrokes   **Display**

S: `7` `.` `2` `EE` `9` `+/-`   `7.2   -09`

or G: `7` `.` `2` `EE` `(-)` `9` `ENTER`   `7.2E   -9`

S is used to denote a typical key sequence for a scientific calculator and G is used to denote the key sequence for a TI-81. □

Graphics Calculator

The value shown on a calculator display may not be exact. Calculators make errors because of rounding and truncating, and they also display digits that may not be significant.

Accuracy and Precision

The **accuracy** of a number refers to the number of **significant digits** (those digits considered reasonably trustworthy in a measurement). For a written numeral, all nonzero digits are significant and zeros are significant except when their only purpose is to locate the decimal point (as in 0.00073 or 730,000). It is easier to identify the accuracy of a number written in scientific notation. In scientific notation, all digits written are considered significant.

The **precision** of a measurement refers to the smallest unit used in the measuring device and thus to the position of the last significant digit. The accuracy and precision of some sample measurements are given in Table 1-2.

Table 1-2 Accuracy and Precision

Number	Scientific Notation	Accuracy (number of significant digits)	Precision
123.04 m	1.2304×10^2 m	5	Hundredths of a meter
0.04 m	4×10^{-2} m	1	Hundredths of a meter
73,000 m	7.3×10^4 m	2	Thousands of meters
0.073 m	7.3×10^{-2} m	2	Thousandths of a meter
0.0730 m	7.30×10^{-2} m	3	Ten-thousandths of a meter

The answer to a problem cannot be assumed to be more accurate than the measurements that produce that answer. Thus we have standardized some rules to use for recording answers based on measurements.

Rounding ▼

> For calculations involving multiplication, division, or exponentiation, round the answer to the same number of significant digits as the measurement having the least number of significant digits.
>
> For calculations involving addition and subtraction, round the answer to the same precision as the least precise measurement.

Many real-life problems contain approximations based on measurements. We will also consider theoretical problems for which all values are assumed to be exact. Numbers found by counting rather than by measurement are also exact. Some problems will involve both approximate and exact values. In these problems, we can use as many digits from the exact values as is necessary.

EXAMPLE 11 The radius of a cylindrical tank is 12.2 meters and the height is 7.5 meters.

(a) Find the surface area of the liquid at the top of this tank.

(b) To paint this tank, the painter will want to know the lateral area. Calculate this area.

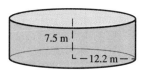

SOLUTIONS

(a) For surface area,

$$A = \pi r^2$$ Theoretical model; π is exact.

$$A \approx \pi (12.2 \text{ m})^2$$ Use the ▮ π ▮ key on a calculator with as many significant digits as possible.

$$A \approx 468 \text{ m}^2$$ Round the answer to three significant digits (as in 12.2 m).

(b) For lateral area,

$$A = 2\pi rh$$ Theoretical model; both 2 and π are exact.

$$A \approx 2\pi(12.2)(7.5)$$ Use 2 (exact) and the ▮ π ▮ key on a calculator.

$$A \approx 570 \text{ m}^2$$ Round the answer to two significant digits (as in 7.5 m).

SECTION 1-3 INTEGRAL EXPONENTS AND ALGEBRAIC EXPRESSIONS

EXERCISES 1-3

A

In Exercises 1–24, simplify each expression

1. -2^4
2. 4^{-2}
3. $(-2)^0 + 2^0$
4. $7^0 + (-7)^0$
5. $9^2 - 7^2 - (9-7)^2$
6. $8^2 - 5^2 - (8-5)^2$
7. $(3x+5y)^0 + (3x)^0 + (-5y)^0 + 3x^0 - 5y^0$
8. $(7a-4b)^0 + (7a)^0 - (4b)^0 + 7a^0 - 4b^0$
9. $\left(\dfrac{1}{2} + \dfrac{1}{5}\right)^{-1}$
10. $\left(\dfrac{1}{3} + \dfrac{1}{4}\right)^{-1}$
11. $6^9 \cdot 6^{-7} + \dfrac{9^{-17}}{9^{-19}}$
12. $8^{11} \cdot 8^{-8} + \dfrac{7^0}{7^{-1}}$
13. $(3^{-2})^2 + (2^{-3})^{-2} - (2^{-3} \cdot 4^0)^{-2}$
14. $\dfrac{5^{-8}}{5^{-5}} + \dfrac{2^{-7}}{2^{-11}} - (-3^{-2} \cdot 5^0)^{-1}$
15. $-3 + 4(8-2) + 16 \div 4 \times 3 + 1$
16. $12 - 9(11+3) + 10 \div 5[4 + 2(4-6)]$
17. $12 - 3[1 + 2(3-4)]$
18. $17 - 4[1 - 2(8-11) + 6]$
19. $6 - 2(3^2 - 5)^3 + 7 \cdot 2$
20. $8 + 5(7^2 - 4)^3 - 8 \cdot 6$
21. $-3^2 - 4[8 - 7(1+3)^2 - 9]^0$
22. $-5^2 + 3[11 + 75(6-83) - 101]^0$
23. $\dfrac{6(83-84)^7 - 4}{(2+3)^2 - 5 \cdot 8^0}$
24. $\dfrac{1 + 5 \cdot 3 + 24 \div 8 - 7^0}{5[2(36) - 71]^8 + 3(5-4)}$

In Exercises 25–28, evaluate each algebraic expression for $u = 5$, $v = 6$, $x = -3$, $y = 2$, and $z = 7$.

25. $(v+y)^{-1} - (v^{-1} + y^{-1})$
26. $(x+y)^2 - x^2 - y^2$
27. $\left(\dfrac{u+x}{v+y}\right)^{-3}$
28. $\left(\dfrac{x+z}{z+u}\right)^{-2}$

In Exercises 29–42, simplify each expression. Write the answers using only positive exponents. Assume that all variables are nonzero real numbers.

29. $y^{-5} y^3 y^{-7}$
30. $z^8 z^{-13} z^{-5}$
31. $(x-y)^{-5}(x-y)^4$
32. $(x+y)^{-3}(x+y)^5$
33. $(v^{-3} w^{-2} z^0)^{-4}$
34. $(2a^3 b^{-4} c^0)^2$
35. $\left(\dfrac{24a^{-3} b^4 c^{-7}}{30a^{-4} b^{-5} c^6}\right)^{-2}$
36. $\left(\dfrac{42a^5 b^{-3} c^{-2}}{35a^{-2} b^{-4} c^3}\right)^{-3}$
37. $\left(\dfrac{(4x^2 y^{-3})^2}{32(2x^{-1} y^4)^{-3}}\right)^{-2}$
38. $\dfrac{(p^2 q^3 r^4)^2 (p^4 q^3 r^{-2})^{-3}}{(pq^2 r^{-1})^4}$
39. $2v^{-1} + 3w^{-1}$
40. $5(v+w)^{-1}$
41. $[(2x^2 y^3)^4 z^5]^0$
42. $[(-3ab^3)^4 c^7]^0$

In Exercises 43–50, simplify each expression. Assume that each base is nonzero and that m and n are natural numbers.

43. $x^{3m+1} x^{2m-1}$
44. $y^{4n+2} y^{2n-3}$
45. $\dfrac{x^{3m+1}}{x^{2m-1}}$
46. $\dfrac{y^{4n+3}}{y^{2n-3}}$
47. $\left(\dfrac{12a^{3m-1}}{18a^{4m+3}}\right)^{-2}$
48. $\left(\dfrac{14b^{2n-1}}{21b^{5n-3}}\right)^{-3}$
49. $\dfrac{(x-2y)^m (x-2y)^{-3}}{(x-2y)^{m-4}}$
50. $\dfrac{(3a-2b)^m (3a-2b)^4}{(3a-2b)^{m-5}}$

26 CHAPTER 1 REVIEW OF THE BASIC CONCEPTS OF ALGEBRA

In Exercises 51–54, convert each number given in scientific notation to standard decimal form.

51 The CRAY-2 computer built in 1984 could perform as many as 3.0×10^9 floating-point operations per second.

52 The ENIAC computer built in 1946 could perform 5.0×10^3 operations per second.

53 The number of grams of mass in a hydrogen atom is 1.673×10^{-24}.

54 In 1981, for the first time, the national debt exceeded one trillion (10^{12}) dollars.

In Exercises 55–58, express each number in scientific notation.

55 The speed of light is approximately 29,980,000,000 centimeters per second.

56 The Earth is approximately 149,000,000 kilometers from the sun.

57 Avogadro's number is 602,000,000,000,000,000,000,000.

58 The design for the CRAY-3 computer specifies switching speeds of 300 picoseconds. A picosecond is one-trillionth of a second; 300 picoseconds = 0.000 000 000 3 seconds.

59 The display of a calculator shows `4.917  -13`. Write this number in standard decimal notation.

60 The display of a calculator shows `-5.001  +12`. Write this number in standard decimal notation.

In Exercises 61 and 62, give the standard decimal form of the number represented by each computer printout. The number following the letter E is the exponent on 10 in scientific notation.

61 $-1.20987\text{E} + 09$

62 $7.91053\text{E} - 07$

B

63 Write the keystrokes needed to enter 8,765,000,000,000 into your calculator.

64 Write the keystrokes needed to enter 0.000 000 000 000 432 into your calculator.

65 A scientific calculator key sequence and a graphics calculator key sequence are shown below. Which of the following expressions would be calculated by these key sequences?

S: `8` `-` `3` `×` `5` `÷` `2` `=`

G: `8` `-` `3` `×` `5` `÷` `2` `ENTER`

a. $(8-3)\left(\dfrac{5}{2}\right)$ **b.** $8 - \dfrac{3 \cdot 5}{2}$ **c.** $8 - \dfrac{3 \cdot 2}{5}$ **d.** $\dfrac{8 - 3 \cdot 5}{2}$

66 A scientific calculator key sequence and a graphics calculator key sequence are shown below. Which of the following expressions would be calculated by these key sequences?

S: `7` `÷` `(` `4` `+` `6` `÷` `π` `)` `=`

G: `7` `÷` `(` `4` `+` `6` `÷` `2nd` `π` `)` `ENTER`

a. $\dfrac{7}{4 + \dfrac{6}{\pi}}$ **b.** $\dfrac{7}{4 + 6}{\pi}$ **c.** $\dfrac{7}{4} + \dfrac{6}{\pi}$ **d.** $7 \cdot \left(4 + \dfrac{6}{\pi}\right)$

SECTION 1-4 OPERATIONS WITH POLYNOMIALS

In Exercises 67–70, complete the table.

		Scientific Notation	Accuracy	Precision
67	0.0032 g			
68	1200 g			
69	1.234 g			
70	42.001 g			

C

In Exercises 71–74, use a calculator to simplify each expression. Use the appropriate rule for rounding each answer.

71 $(5.234 \times 10^{11} \text{ m})(8.1 \times 10^{15})$ **72** $(7.29 \times 10^{-14})^8$ m

73 The circumference of a circle with radius 2.932×10^5 meters.

74 The area of a circle with radius 2.932×10^5 meters.

In Exercises 75–78, use a calculator to evaluate each expression accurately to four significant digits.

75 $(\pi - 1)^4$ **76** $(97{,}840{,}000{,}000)^{-2}$ **77** $\left(\dfrac{7\pi}{12}\right)^3$ **78** $76.23^2 - 32.45^2 - (76.23 - 32.45)^2$

79 An IRA investment of $2000 at 8% compounded annually for 20 years amounts to $A = 2000(1 + 0.08)^{20}$. Record the answer, accurate to the nearest penny.

80 The formula used to compute the monthly payment P needed to amortize a load of amount A at an interest rate r in n years is

$$P = \frac{A\left(\dfrac{r}{12}\right)}{1 - \left(1 + \dfrac{r}{12}\right)^{-12n}}$$

Compute the monthly payment needed to amortize a $50,000 loan at 11.25% in 30 years.

SECTION 1-4

Operations with Polynomials

Section Objectives

8 Determine the degree of a polynomial and classify it according to the number of terms.

9 Add, subtract, multiply, and divide polynomials.

A **monomial** is an algebraic expression formed by taking the product of constants and variables. Division by variables is not permitted. Thus the variables in a monomial can contain only positive integral exponents. **Polynomials** are algebraic expressions that contain only monomials as terms (or addends). Polynomials containing one, two, and three terms are called, respectively, monomials, **binomials**, and **trinomials**.

The following examples illustrate these classifications:

- $6, \pi, m, 3v^2, \frac{1}{2}x^2y, \pi r^2$, and $-2.73v^2w^3x^4$ are all monomials.
- $\frac{5}{w}, 3z^{-4}$, and $x^{1/2}$ are not monomials, since the exponents are not positive integers.
- $2x^2 - 7x$ is a binomial.
- $6v^3 + 5v^2 - 4v + 7$ is a polynomial in v with four terms.
- $5x^4y^2 - 6x^3y^5 + 11$ is a trinomial.

The constant factor in a monomial is called the **numerical coefficient**, or just the coefficient. The **degree of a monomial** is the sum of the exponents of the variables. A nonzero constant has degree zero, but no degree is assigned to 0, the **zero polynomial**. The **degree of a polynomial** is the degree of the term of the highest degree.

Polynomials are frequently written in standard form to facilitate comparisons and computations. A polynomial is in **standard form** if (1) the variables are written in alphabetical order in each term and (2) the terms are arranged in decreasing powers of the first variable. If there is a constant term, it is written last.

These polynomials have been rewritten in standard form, and the degree of each is given.

$2y^3w^4x = 2w^4xy^3$ Monomial of degree eight

$3x^2 - 5y^2 + 2yx = 3x^2 + 2xy - 5y^2$ Trinomial of degree two

$cab - a^2b + 17 - c^2b = -a^2b + abc - bc^2 + 17$ Polynomial of degree three

Polynomials are added and subtracted by combining like terms. **Like terms** may contain different numerical coefficients, but the variable factors —including their exponents—must be identical. For simple problems the work can be done using a horizontal format; for more complicated problems a vertical format can be used to clarify the correct steps.

EXAMPLE 1   Perform the indicated operations.

(a) $(2a^2 - 2ab + b^2) - (a^2 - ab + 3b^2)$

(b) $(3x^3 - 5x^2y + y^3) + (2x^3 + 6xy^2 - 3y^3) - (x^3 - 3x^2y + xy^2)$

SOLUTIONS

(a) In horizontal format,
$$(2a^2 - 2ab + b^2) - (a^2 - ab + 3b^2)$$
$$= 2a^2 - a^2 - 2ab + ab + b^2 - 3b^2$$
$$= a^2 - ab - 2b^2$$

(b) In vertical format,

$$\begin{array}{l} 3x^3 - 5x^2y + y^3 \\ 2x^3 + 6xy^2 - 3y^3 \\ -x^3 + 3x^2y - xy^2 \\ \hline 4x^3 - 2x^2y + 5xy^2 - 2y^3 \end{array}$$

Only similar terms are aligned.

Note change of signs due to subtraction. □

The vertical format is also a convenient format for multiplying polynomials with three or more terms.

EXAMPLE 2 Find the product $(a^2 - a + 1)(a^2 + a - 1)$.

SOLUTION
$$\begin{array}{l} a^2 - a + 1 \\ a^2 + a - 1 \\ \hline a^4 - a^3 + a^2 \\ a^3 - a^2 + a \\ - a^2 + a - 1 \\ \hline a^4 - a^2 + 2a - 1 \end{array}$$

Answer $(a^2 - a + 1)(a^2 + a - 1) = a^4 - a^2 + 2a - 1$ □

Since the product of two binomials occurs frequently, it is advantageous to be able to form these products quickly without using the vertical format. You should practice until you can write the last step without writing the intermediate work. The steps are shown here only to convey the thought process.

EXAMPLE 3 Find the product $(x - 7)(x - 8)$.

SOLUTION $(x - 7)(x - 8) = x^2 - 15x + 56$

Think:

- **F** First terms $(x)(x)$
- **O** Outer terms $+(x)(-8)$
- **I** Inner terms $+(-7)(x)$
- **L** Last terms $+(-7)(-8)$

▶ **Self-Check** ▼

Perform the indicated operations.

1 $(3r^2 - 4rs - 5s^2) + (8rs - 7s^2)$
   $-(r^2 - 11s^2)$

2 $(x^2 - 5x + 8)(x^2 + 4x - 9)$

▶ **Self-Check Answers** ▼

1 $2r^2 + 4rs - s^2$ **2** $x^4 - x^3 - 21x^2 + 77x - 72$

Certain types of products occur so often that it is worthwhile to memorize them so that the answers can be written by inspection.

Special Products ▼

The square of a sum:	$(a + b)^2 = a^2 + 2ab + b^2$
The square of a difference:	$(a - b)^2 = a^2 - 2ab + b^2$
A sum times a difference:	$(a + b)(a - b) = a^2 - b^2$

EXAMPLE 4 Form these special products by inspection.

SOLUTIONS

(a) $(w + 4)^2$

$(w + 4)^2 = w^2 + 8w + 16$

Think:
Square of the first term, w
Twice the product of 4 and w
Square of the last term, 4

(b) $(2s + 11)(2s - 11)$

$(2s + 11)(2s - 11) = 4s^2 - 121$

Think:
Square of the first term, $2s$
Minus the square of the second term, 11

(c) $(7x^3 - 9y)(7x^3 + 9y)$

$(7x^3 - 9y)(7x^3 + 9y) = (7x^3)^2 - (9y)^2$
$= 49x^6 - 81y^2$ □

The expansion of $(a + b)^2$ is one specific binomial expansion of an expression of the form $(a + b)^n$. Optional material on binomial expansions is given at the end of this section. Further information on binomial expansions is given in Chapter 8. For now we will examine the expansions of $(a + b)^3$ and $(a - b)^3$.

$$\begin{array}{r} (a+b)^3 = (a+b)(a+b)^2 \\ a^2 + 2ab + b^2 \\ a + b \\ \hline a^3 + 2a^2b + ab^2 \\ a^2b + 2ab^2 + b^3 \\ \hline a^3 + 3a^2b + 3ab^2 + b^3 \end{array}$$

Expansions of $(a + b)^3$ and $(a - b)^3$ ▼

The cube of a sum:	$(a + b)^3 = a^3 + 3a^2b + 3ab^2 + b^3$
The cube of a difference:	$(a - b)^3 = a^3 - 3a^2b + 3ab^2 - b^3$

$$\begin{array}{r} (a-b)^3 = (a-b)(a-b)^2 \\ a^2 - 2ab + b^2 \\ a - b \\ \hline a^3 - 2a^2b + ab^2 \\ - a^2b + 2ab^2 - b^3 \\ \hline a^3 - 3a^2b + 3ab^2 - b^3 \end{array}$$

EXAMPLE 5 Expand $(x + 5)^3$.

SOLUTION $(x + 5)^3 = (\ )^3 + 3(\ )^2(\ ) + 3(\ )(\ )^2 + (\ )^3$
$= (x)^3 + 3(x)^2(5) + 3(x)(5)^2 + (5)^3$
$= x^3 + 15x^2 + 75x + 125$

First set up the blank form for $(a + b)^3$, and then substitute x for a and 5 for b. Now simplify each term.

□

SECTION 1-4 OPERATIONS WITH POLYNOMIALS

The division of one polynomial by another is often denoted by fractional notation. To divide polynomials, remember to first write them in standard form.

EXAMPLE 6 Simplify $\dfrac{9x^2 + 6x^3 - 12x}{3x}$.

SOLUTION
$$\dfrac{9x^2 + 6x^3 - 12x}{3x} = \dfrac{6x^3 + 9x^2 - 12x}{3x}$$ First write the numerator in standard form.
$$= \dfrac{6x^3}{3x} + \dfrac{9x^2}{3x} - \dfrac{12x}{3x}$$
$$= 2x^2 + 3x - 4 \qquad \square$$

The familiar long-division algorithm can be used to divide by polynomials of more than one term. Similar terms should be aligned and space provided for missing terms in the dividend.

EXAMPLE 7 Perform the division $\dfrac{x^5 + 7x^2 - 4x + 20}{x^2 + 2}$.

SOLUTION

$$\begin{array}{r} x^3 + 0x^2 - 2x + 7 \\ x^2 + 2 \overline{\smash{\big)}\, x^5 + 0x^4 + 0x^3 + 7x^2 - 4x + 20} \\ \underline{x^5 + 2x^3} \\ -2x^3 + 7x^2 - 4x \\ \underline{-2x^3 - 4x} \\ 7x^2 + 0x + 20 \\ \underline{7x^2 + 14} \\ 6 \end{array}$$

Writing zero coefficients for the missing terms facilitates the alignment of similar terms.

Answer $x^3 - 2x + 7 + \dfrac{6}{x^2 + 2}$

You can check the answer to the division in Example 7 by multiplying. We leave it for you to verify that

$$(x^2 + 2)\left(x^3 - 2x + 7 + \dfrac{6}{x^2 + 2}\right) = x^5 + 7x^2 - 4x + 20$$

▶ **Self-Check** ▼

Form these products without writing any intermediate steps.
1 $(2x - y)(x + 3y)$
2 $(2x - y)^2$
3 $(x + 3y)^2$
4 $(2x - y)(2x + y)$

Find the quotients indicated.
5 $\dfrac{10m^2n^2 - 25mn^3}{5mn}$
6 $\dfrac{10a^2 - 11a + 3}{2a - 1}$

▶ **Self-Check Answers** ▼

1 $2x^2 + 5xy - 3y^2$ 2 $4x^2 - 4xy + y^2$ 3 $x^2 + 6xy + 9y^2$ 4 $4x^2 - y^2$ 5 $2mn - 5n^2$ 6 $5a - 3$

Table 1-3 Binomial Expansions

$(a + b)^n$	Expansion of $(a + b)^n$	The Coefficients (Pascal's triangle)*
$(a + b)^0 =$	1	1
$(a + b)^1 =$	$a + b$	$1 \quad 1$
$(a + b)^2 =$	$a^2 + 2ab + b^2$	$1 \quad 2 \quad 1$
$(a + b)^3 =$	$a^3 + 3a^2b + 3ab^2 + b^3$	$1 \quad 3 \quad 3 \quad 1$
$(a + b)^4 =$	$a^4 + 4a^3b + 6a^2b^2 + 4ab^3 + b^4$	$1 \quad 4 \quad 6 \quad 4 \quad 1$
$(a + b)^5 =$	$a^5 + 5a^4b + 10a^3b^2 + 10a^2b^3 + 5ab^4 + b^5$	$1 \quad 5 \quad 10 \quad 10 \quad 5 \quad 1$

* This triangle of coefficients is named for the French mathematician Blaise Pascal (1623–1662), whose use of this triangle made it well known.

Optional Material on Binomial Expansions

There are many patterns that one can observe in binomial expansions. Before you read further, examine the expansions in Table 1-3 and try to discover some of these patterns for the terms of $(a + b)^n$. Specifically, try to answer the list of questions in Table 1-4. Then we will use these patterns to perform some binomial expansions.

Table 1-4

Questions About the Expansion of $(a + b)^n$	Answers
1. How many terms are in this expansion?	1. $n + 1$
2. What is the degree of each term in the expansion?	2. n
3. Is there any pattern for the exponents on a and b in the terms?	3. As the exponents on a decrease by 1, from a^n to a^0, the exponents on b increase by 1, from b^0 to b^n
4. What is the coefficient of the first term and the last term?	4. 1
5. What is the coefficient of the second term and the last term?	5. n. The coefficients of $(a + b)^n$ form a symmetric pattern that starts with the first term and ends with the last term.
6. What is the relationship between the coefficients and the exponents?	6. After the first term, obtain the coefficient of the next term by multiplying the current coefficient by the exponent on a and dividing by one more than the exponent on b.

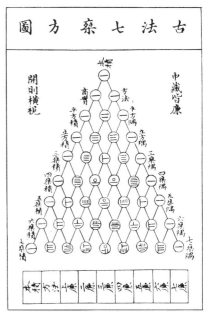

Figure 1-5 From an Ancient Chinese manuscript dated 1303, the triangle of numbers later named after Pascal. This illustration appears in <u>The VNR Concise Encyclopedia of Mathematics</u>, edited by Küstner and Kästner, © 1977. Reprinted with permission of Van Nostrand Reinhold Company, NYC.

By taking advantage of these observations, we can directly write the expansion of $(a + b)^n$.

SECTION 1-4 OPERATIONS WITH POLYNOMIALS

Expanding $(a + b)^n = a^n + na^{n-1}b + \cdots + nab^{n-1} + b^n$ ▼

Step 1 Write the exponents on all $(n + 1)$ terms: Start with a^n, decreasing the exponents on a by 1 and increasing the exponents on b by 1 until the last term is b^n.

Step 2 Write the coefficients of each term:
  a. Make the first coefficient 1 and the second n.
  b. Calculate each new coefficient by multiplying the current coefficient by the exponent on a and dividing by one more than the exponent on b.
  c. Use the symmetric pattern of coefficients, which starts with the first term and ends with the last term.

EXAMPLE 8 Write the expansion of $(a + b)^6$.

SOLUTION

1. $(a + b)^6 = a^6 + \_\_a^5b + \_\_a^4b^2 + \_\_a^3b^3 + \_\_a^2b^4 + \_\_ab^5 + b^6$ — Write the exponents on all seven terms.

2. $(a + b)^6 = a^6 + 6a^5b + 15a^4b^2 + \cdots + 6ab^5 + b^6$ — Write the first two coefficients, 1 (understood) and 6. Then calculate the third coefficient, 15.

Coefficient of the second term → $\dfrac{6 \cdot 5}{2}$ ← Exponent on a in the second term

One more than the exponent on b in the second term

$(a + b)^6 = a^6 + 6a^5b + 15a^4b^2 + 20a^3b^3 + 15a^2b^4 + 6ab^5 + b^6$ — Calculate the fourth coefficient, 20, and then fill in the rest of the coefficients using the symmetry of coefficients.

Coefficient of the third term → $\dfrac{15 \cdot 4}{3}$ ← Exponent on a in the third term

One more than the exponent on b in the third term ☐

EXAMPLE 9 Expand $(2x^2 - 3y)^4$.

SOLUTION

$(a + b)^4 = (a)^4 + 4(a)^3(b) + 6(a)^2(b)^2 + 4(a)(b)^3 + (b)^4$ — First set up this form for $(a + b)^4$.

$(2x^2 - 3y)^4 = (2x^2)^6 + 4(2x^2)^3(-3y) + 6(2x^2)^2(-3y)^2$
$\quad + 4(2x^2)(-3y)^3 + (-3y)^4$ — Then substitute in $2x^2$ for a and $-3y$ for b to expand $[2x^2 + (-3y)]^4$.

$\quad = 16x^8 - 96x^6y + 216x^4y^2 - 216x^2y^3 + 81y^4$ — Now simplify each term.

☐

EXERCISES 1-4

A

In Exercises 1–4, classify each polynomial as a monomial, binomial, or trinomial.

1. $7x^2 - 9x + 11$
2. $-6x^2y^3z^4$
3. $-6x^2 + 4y^3$
4. $7v - 9w^2$

In Exercises 5–8, write each polynomial in standard form.

5. $13 + 5v^2 - 8v$
6. $9a - 3 + 7a^2$
7. $2y^2x + 9yx^2$
8. $12b^2a^3c$

Exercises 9–14 refer to the polynomial $3w^4 - w^3 + \frac{3}{7}w^2 - w - 27$.

9. What is the constant term of this polynomial?
10. What is the coefficient of the second term?
11. What is the coefficient of the third term?
12. What is the degree of the second term?
13. What is the degree of the polynomial?
14. What is the degree of the fifth term?

In Exercises 15–58, perform the indicated operations.

15. $(x^2 - 5xy + 9y^2 - 17) - (2x^2 - 7xy - 3y^2 - 11)$
16. $(3a^2 - 4ab - b^2) - (a^2 - 5ab + 7b^2)$
17. $[(2x - 3y - 7z) + (-9x - 10y + z)] - (-4x + 7y + 9z)$
18. $(-12a + 7b - 2c) - [(a - b + 3c) + (4a - b - c)]$
19. $5(w^4 - 11w^3 + 13) - 3(w^3 + 5w^2 + 7w - 9) + 2(w^4 - 2w^2 + 7w - 8)$
20. $-2(7x^2 - 9xy - 13y^2) + 3(-8x^2 - 2xy + y^2) + 7(11x^2 + 4xy + 12y^2)$
21. $[(m^2 - n^2) - (m^2 - 2mn + n^2)] - [(m^2 + n^2) - (m^2 + mn - n^2)]$
22. $3a - b - [a - (2a + 3b) - (a - 2b)]$
23. $(2x^3y^4)(-5x^6y^8) + (4x^3y^4)^3$
24. $(-3ab^3c)(7a^3b^5c^3) + (-5a^2b^4c^2)^2$
25. $-5n^2m(3m^3n - 2m^2n + 3mn)$
26. $-2r^2s^3(6r^3 - 6r^2s - 3rs^2 + s^3)$
27. $(3v + 4w)(5v - 7w)$
28. $(8a - 7b)(3a + 2b)$
29. $(3a^2 + 5a + 7)(2a - 3)$
30. $(-2b^2 - 5b + 7)(3b - 4)$
31. $4y(y^2 - 7y + 5) - y(9 - y^2 + 3y)$
32. $(a - 1)(a + 3) - (a + 2)(a + 1)$
33. $(x + y)^2 - (x^2 + y^2)$
34. $(r - s)^2 - (r^2 - s^2)$
35. $(2a + 3b)^2 - (2a - 3b)^2$
36. $(5v - 3z)(5v + 3z) - [(5v)^2 + (3z)^2]$
37. $(x - 2)^3$
38. $(y + 4)^3$
39. $(7x^{-1} + 5y^{-1})(4x^{-1} - 3y^{-1})$
40. $(9v^{-2} + 8)(5v^{-1} + 2)$
41. $\dfrac{y^3 - 2y^2 + 3y}{y}$
42. $\dfrac{z^3 + 5z^2 + 7z}{z}$
43. $\dfrac{39x^3yz^2 - 65xy^2z^3 - 78x^2y^2z^2}{13xyz^2}$
44. $\dfrac{-18m^5 - 6m^4 + 15m^3}{-3m^2}$
45. $\dfrac{b^2 + 4b - 12}{b + 6}$
46. $\dfrac{y^2 - 7y + 10}{y - 5}$
47. $\dfrac{10n^2 - 11n + 8}{2n - 1}$
48. $\dfrac{12x^3 + 10x^2 - 36x + 21}{3x - 2}$
49. $\dfrac{21x^4 - 7x^3 + 62x^2 - 9x + 45}{3x^2 - x + 5}$
50. $\dfrac{2z^5 + 4z^4 - z^3 - 5z + 1}{z^2 + 2z - 1}$
51. $\dfrac{a^5 - 1}{a - 1}$
52. $\dfrac{b^4 - 16}{b - 2}$
53. $\dfrac{18x^{5m-3} - 27x^{6m+4}}{9x^{4m-5}}$
54. $\dfrac{-36a^{2n}b^{m+2} - 60a^{n+7}b^{m+3}}{12a^nb^m}$
55. $(x^{2n} - 1)(x^{2n} + 1)$
56. $(x^n - 1)(x^{2n} + x^n + 1)$
57. $\dfrac{27x^{3n} - 1}{3x^n - 1}$
58. $\dfrac{x^{4n} - 625}{x^n + 5}$

OPTIONAL EXERCISES

In Exercises 59-62, use the diagram to the right to write a polynomial to describe the area or perimeter specified.

59 The perimeter of the entire figure.
60 The perimeter of the triangular portion of the figure.
61 The area of the square region.   **62** The area of the entire region.

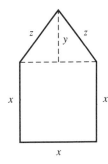

Figure for Exercises 59-62

B

63 Give an example of two second-degree polynomials whose sum is a first-degree polynomial.
64 Give an example of two third-degree polynomials whose sum is a constant.
65 If a second-degree binomial is multiplied by a third-degree trinomial, of what degree will the product be?
66 If a fourth-degree trinomial is squared, of what degree will the result be?
67 If a trinomial in x of degree five is divided by a trinomial in x of degree three, what is the degree of the quotient?
68 If the dividend is a fourth-degree polynomial in x and the quotient is a first-degree polynomial in x, what is the degree of the divisor?

C

69 The profit in dollars made by selling t units is given by $P = -t^2 + 27t - 50$. Evaluate and interpret each expression.
  a. P for $t = 0$ **b.** P for $t = 1$ **c.** P for $t = 4$ **d.** P for $t = 25$

70 The production cost P and the marketing cost M for t units of a product are given. Find the total cost C of t units of this product.
  a. $P = 5t^3 - 2t^2 + 7t + 30$ **b.** $P = 4t^3 - t^2 - 7t + 35$
  $M = 3t^2 + 8t + 13$ $M = 2t^2 - 6t + 11$

71 The revenue produced by selling t units of a product is given by R. The cost of these units is given by C. Calculate the profit made by selling t units.
  a. $R = 10t^2 - 15t$ **b** $R = t^3 - t^2 + t$
  $C = 2t^2 + 3t + 8$ $C = 2t^2 + 25$

OPTIONAL EXERCISES

1 How many terms are in the expansion of $(x - 3y)^8$?
2 What is the degree of each term of $(x - 3y)^8$?

In Exercises 3-10, write the binomial expansion of each expression.

3 $(x + 2y)^4$ **4** $(3x - y)^4$ **5** $(x - 2)^5$ **6** $(x + 3)^6$
7 $(5v + 2)^6$ **8** $(3v - 5)^5$ **9** $(x^2 - 1)^4$ **10** $(x^2 + y^3)^5$
11 Write the first four terms of $(v + w)^{15}$.   **12** Write the first four terms of $(v + 3)^{11}$.

SECTION 1-5

Factoring Polynomials

Section Objective

10 Factor polynomials.

Factoring is the process of expressing a polynomial as a product; the polynomial is **factored completely** when each factor is prime. A polynomial is **prime over the integers** if its only factorization is a trivial factorization involving 1 or -1.

The following examples illustrate both factorable and prime polynomials.

- $x^2 - 1$ is factorable, since it can be written as the product of polynomials of lower degree: $x^2 - 1 = (x - 1)(x + 1)$. Note that each factor is a prime polynomial.
- $3y - 6$ is factorable, since the GCF (greatest common factor) is 3; $3y - 6 = 3(y - 2)$.
- $z^2 - 7$ is prime over the integers. You could factor $z^2 - 7$ as $(z + \sqrt{7})(z - \sqrt{7})$, but this would not be a factorization over the integers because $\sqrt{7}$ and $-\sqrt{7}$ are not integers.
- $w^2 + 1$ is prime over the integers because the only factorizations over the integers are the trivial ones: $1(w^2 + 1)$ and $-1(-w^2 - 1)$. In Section 4-2 we will note that $w^2 + 1$ factors over the complex numbers.

The following table outlines the methods of factoring presented in this section. It is usually best to factor out the greatest common factor first, even though there may be other ways to obtain the same prime factors.

Factoring a Polynomial Over the Integers ▼

Step 1 Factor out the GCF (Greatest Common Factor).

Step 2 **Binomials:**

Factor special forms:

a. difference of two squares

$$a^2 - b^2 = (a + b)(a - b)$$

b. sum of two cubes

$$a^3 + b^3 = (a + b)(a^2 - ab + b^2)$$

c. difference of two cubes

$$a^3 - b^3 = (a - b)(a^2 + ab + b^2)$$

(continued)

SECTION 1-5 FACTORING POLYNOMIALS

Factor by completing the square on some polynomials of degree four or more.

Step 3 **Trinomials:**

Factor forms that are perfect squares:

a. square of a sum
$$a^2 + 2ab + b^2 = (a + b)^2$$

b. square of a difference
$$a^2 - 2ab + b^2 = (a - b)^2$$

Factor by trial and error.
Factor by completing the square on some polynomials of degree four or more.

Step 4 **Polynomials of four or more terms:**

Factor forms that are perfect cubes:

a. cube of a sum
$$a^3 + 3a^2b + 3ab^2 + b^3 = (a + b)^3$$

b. cube of a difference
$$a^3 - 3a^2b + 3ab^2 - b^3 = (a - b)^3$$

Factor by grouping.

EXAMPLE 1 Factor $5x^5y - 5xy^5$ completely.

SOLUTION $5x^5y - 5xy^5 = 5xy(x^4 - y^4)$ Use the distributive property to factor out $5xy$.
$ = 5xy[(x^2)^2 - (y^2)^2]$ A difference of two perfect squares
$ = 5xy(x^2 + y^2)(x^2 - y^2)$ Another difference of two perfect squares
$ = 5xy(x^2 + y^2)(x + y)(x - y)$

Answer $5x^5y - 5xy^5 = 5xy(x^2 + y^2)(x + y)(x - y)$

EXAMPLE 2 Factor $7a^3 - 42a^2b + 63ab^2$.

SOLUTION $7a^3 - 42a^2b + 63ab^2 = 7a(a^2 - 6ab + 9b^2)$ The GCF is $7a$.
$ = 7a[a^2 - 2(3ab) + (3b)^2]$ Factor the perfect square trinomial.
$ = 7a(a - 3b)^2$

Answer $7a^3 - 42a^2b + 63ab^2 = 7a(a - 3b)^2$

EXAMPLE 3   Factor $8w^3 - 27z^3$.

SOLUTION $8w^3 - 27z^3 = (2w)^3 - (3z)^3$ Express the polynomial as the difference of two cubes.

$\qquad\qquad\qquad = (2w - 3z)(?)$

$\qquad\qquad\qquad = (2w - 3z)(4w^2 + 6wz + 9z^2)$ Write the binomial factor, the difference of the two cube roots. Use the binomial factor to obtain each term of the trinomial factor.

- Square of the last term, $-3z$
- Opposite of the product of the two terms
- Square of the first term, $2w$

Answer $8w^3 - 27z^3 = (2w - 3z)(4w^2 + 6wz + 9z^2)$ □

EXAMPLE 4   Factor $8a^3b^6 + 125c^3$.

SOLUTION $8a^3b^6 + 125c^3 = (2ab^2)^3 + (5c)^3$ Express as the sum of two cubes.

$\qquad\qquad\qquad = (2ab^2 + 5c)[(2ab^2)^2 - (2ab^2)(5c) + (5c)^2]$ Write the binomial factor, and use it to obtain the trinomial factor.

$\qquad\qquad\qquad = (2ab^2 + 5c)(4a^2b^4 - 10ab^2c + 25c^2)$

Answer $8a^3b^6 + 125c^3 = (2ab^2 + 5c)(4a^2b^4 - 10ab^2c + 25c^2)$ □

Trinomials of the form $ax^2 + bx + c$ either are prime or can be factored into binomials of the form $(a_1x + c_1)(a_2x + c_2)$, with $a_1a_2 = a$ and $c_1c_2 = c$. The trinomial is factorable if the factors a_1, a_2, c_1, and c_2 can be selected so that a_1c_2 plus a_2c_1 equals b. Since there may be several possible factors to examine, this process is referred to as trial and error. Nonetheless, the possibilities should be examined in a systematic manner rather than by random guessing. An important first step is using the signs of b and c to determine the sign pattern for the binomial factors.

EXAMPLE 5   Factor $6x^2 - 7x - 5$.

SOLUTION   The blank form is $6x^2 - 7x - 5 = (\_\_x - \_\_)(\_\_x + \_\_)$ Make a blank form using opposite signs, since -5 is negative.

Possible Factors	Resulting Linear Terms
$(6x + 1)(x - 5)$	$-29x$
$(6x - 1)(x + 5)$	$29x$
$(3x + 1)(2x - 5)$	$-13x$
$(3x - 1)(2x + 5)$	$13x$
$(2x + 1)(3x - 5)$	$-7x$
$(2x - 1)(3x + 5)$	$7x$
$(x + 1)(6x - 5)$	x
$(x - 1)(6x + 5)$	$-x$

All possible combinations of the factors of 6 and -5 (given below) are used to fill in the blank sign pattern.

Factors of 6	Factors of -5
6, 1	1, -5
3, 2	-1, 5
2, 3	
1, 6	

Obviously, in actual problems we stop the list when the correct factors are found.

$6x^2 - 7x - 5 = (2x + 1)(3x - 5)$ □

SECTION 1-5 FACTORING POLYNOMIALS

EXAMPLE 6 Factor $26m^4 + 19m^2n - 7n^2$.

SOLUTION $(\_\_m^2 + \_\_n)(\_\_m^2 - \_\_n)$ is the blank form because the terms must be of opposite signs. Mentally list the possibilities using the factors of 26 and 7. The correct choice is

$$26m^4 + 19m^2n - 7n^2 = (26m^2 - 7n)(m^2 + n) \qquad \square$$

▶ **Self-Check** ▼

Factor these polynomials.
1 $10x^2y - 15xy^2$
2 $9v^2 - 16w^2$
3 $x^2 - 6z + 9$
4 $27x^3 + 1$

EXAMPLE 7 Factor $21x^2 - 30x + 10$.

SOLUTION The blank form is $(\_\_x - \_\_)(\_\_x - \_\_)$, since the product c_1c_2 must equal $+10$ and $a_1c_2 + a_2c_1$ must equal -30. We will systematically fill in this form with all possible factors of 21 and 10.

Possible Factors	Resulting Linear Terms	Factors of 21	Factors of 10
$(21x - 10)(x - 1)$	$-31x$	1, 21	1, 10
$(21x - 1)(x - 10)$	$-211x$	3, 7	2, 5
$(21x - 5)(x - 2)$	$-47x$		
$(21x - 2)(x - 5)$	$-107x$		
$(7x - 10)(3x - 1)$	$-37x$		
$(7x - 1)(3x - 10)$	$-73x$		
$(7x - 5)(3x - 2)$	$-29x$		
$(7x - 2)(3x - 5)$	$-41x$		

Answer The polynomial is prime over the integers, since none of these possibilities yields the correct linear term. $\square$

Experienced students often discover shortcuts for expediting the trial-and-error process. You may wish to discuss your own shortcuts with your teacher. The following discussion may trigger some ideas for successful shortcuts. One method is to use the size of the middle term of the trial factorization as a guide for selecting the next trial factors. A second method is to use the properties of the even and odd integers to eliminate factors $(a_1x + c_1)(a_2x + c_2)$ that cannot produce the proper even-odd pattern for a, b, and c in $ax^2 + bx + c$. (For an excellent article on this topic, see John R. Geyer's "Even-Odds in Factoring" in the February 1982 *Mathematics Teacher*.) For example, $21x^2 - 30x + 10$ must factor so that a_1 and a_2 are odd, since $a = 21$. Also, c_1 and c_2 must both be even because $b = -30$ and $c = 10$ are even. (Think about this claim.) However, 10 factors only as $1 \cdot 10$ and $2 \cdot 5$, both of which contain an odd factor. Thus, as noted in Example 7, $21x^2 - 30x + 10$ is prime. Another helpful observation is that the trinomial $ax^2 + bx + c$ must be irreducible if a, b, and c are all odd. (See the C Exercises at the end of this section.)

▶ **Self-Check Answers** ▼

1 $5xy(2x - 3y)$ 2 $(3v + 4w)(3v - 4w)$ 3 $(z - 3)^2$ 4 $(3x + 1)(9x^2 - 3x + 1)$

Many polynomials with four or more terms can be factored if some of the terms are grouped together so that the group of terms is factorable. This technique of factoring is called **factoring by grouping**. Start by trying to identify groups that share a common factor or groups that have special forms. As some groupings may be useful and others fruitless, be sure to examine all possibilities for grouping the terms.

> **▶ Self-Check ▼**
>
> Factor each trinomial.
>
> **1** $3m^2 + 5m + 2$
> **2** $5n^2 - 17n + 6$
> **3** $4a^2 + 29ab + 25b^2$

EXAMPLE 8 Factor $2ax^2 + 2ax + 2ay - 2ay^2$.

SOLUTION
$$\begin{aligned} 2ax^2 + 2ax + 2ay - 2ay^2 &= 2a(x^2 + x + y - y^2) && \text{The GCF is } 2a. \\ &= 2a(x^2 - y^2 + x + y)] && \\ &= 2a[(x^2 - y^2) + (x + y)] && \text{Group the first two and the last two terms.} \\ &= 2a[(x+y)(x-y) + (x+y)(1)] && \text{Factor the difference of two squares.} \\ &= 2a(x + y)(x - y + 1) && \text{Factor out the common factor } (x+y). \end{aligned}$$

EXAMPLE 9 Factor $b^3 + x + x^3 + b$.

SOLUTION
$$\begin{aligned} b^3 + x + x^3 + b &= (b^3 + x^3) + (x + b) && \text{Reorder the terms.} \\ &= (x^3 + b^3) + (x + b) && \text{Group the first two and the last two terms.} \\ &= (x+b)(x^2 - xb + b^2) + (x+b)(1) && \text{Factor the sum of two cubes.} \\ &= (x + b)(x^2 - xb + b^2 + 1) && (x+b) \text{ is a common factor.} \end{aligned}$$

EXAMPLE 10 Factor $z^2 - a^2 + 4ab - 4b^2$.

SOLUTION
$$\begin{aligned} z^2 - a^2 + 4ab - 4b^2 &= z^2 - (a^2 - 4ab + 4b^2) && \text{Group the last three terms.} \\ &= z^2 - (a - 2b)^2 && \text{Factor the perfect square trinomial.} \\ &= [z + (a - 2b)][z - (a - 2b)] && \text{Factor the difference of two squares.} \\ &= (z + a - 2b)(z - a + 2b) \end{aligned}$$

Even though $a^3 + 3a^2b + 3ab^2 + b^3$ and $a^3 - 3a^2b + 3ab^2 - b^3$ can be factored by grouping, it may be more efficient to note that these forms are the expansions of $(a + b)^3$ and $(a - b)^3$. The polynomial in the next example involves the expanded form of $(a - b)^3$, the cube of a difference.

EXAMPLE 11 Factor $7x^4 - 21x^3y + 21x^2y^2 - 7xy^3$.

SOLUTION
$$\begin{aligned} 7x^4 - 21x^3y + 21x^2y^2 - 7xy^3 &= 7x(x^3 - 3x^2y + 3xy^2 - y^3) && \text{Factor out the GCF } 7x. \\ &= 7x(x - y)^3 && \text{Then note that the second factor is cube of a difference.} \end{aligned}$$

We can determine if second-degree trinomials are prime by listing all possible binomial factors. Although this method works for second-degree trinomials, it does not ensure that higher-degree trinomials are prime. Some of these polynomials can be rewritten so that the expression can then be

---

▶ Self-Check Answers ▼

1 $(3m + 2)(m + 1)$ **2** $(5n - 2)(n - 3)$ **3** $(4a + 25b)(a + b)$

SECTION 1-5 FACTORING POLYNOMIALS

factored as a difference of two squares. For this to be possible:

1. The first and last terms of the original polynomial must be perfect squares.
2. The degree of the first term and the dgree of the last term must both be either 0, 4, 8, or some other multiple of 4.
3. A perfect square must be added to the polynomial in order to form a perfect square trinomial. This perfect square is then subtracted so that the value of the polynomial remains unchanged.

Polynomials of this form are factored by **completing the square**. The key step is to determine what middle term is necessary to have a perfect square trinomial. Recall that the middle term of both $a^2 + 2ab + b^2$ and $a^2 - 2ab + b^2$ is twice the product of the square roots of the first and last terms.

EXAMPLE 12 Factor $w^4 - w^2 + 16$.

SOLUTION
$$w^4 - w^2 + 16 = (w^2)^2 - w^2 + 4^2$$
$$= (w^2)^2 \underbrace{- w^2 + 4^2}_{}$$
$$= \frac{\overbrace{+9w^2}\quad\overbrace{-9w^2}}{[(w^2)^2 + 8w^2 + 4^2] - 9w^2}$$
$$= (w^2 + 4)^2 - (3w)^2$$
$$= [(w^2 + 4) + 3w][(w^2 + 4) - 3w]$$
$$= (w^2 + 3w + 4)(w^2 - 3w + 4)$$

The first and last terms are perfect squares.

Also note that the degree of the first term, w^4, is 4 and the degree of the last term, 16, is 0. The middle term needed is $\pm 2(w^2)(4) = \pm 8w^2$. In this case, we add and then subtract the perfect square $9w^2$ to produce $+8w^2$.

A difference of two squares

EXAMPLE 13 A student's answer to a problem in calculus is $x^{-1} - 4x^{-3}$, but the book's answer is $\dfrac{(x+2)(x-2)}{x^3}$. Factor the variable with the smallest exponent from the student's answer and simplify to obtain the book's answer.

SOLUTION
$$x^{-1} - 4x^{-3} = x^{-3}(x^2 - 4)$$
$$= \frac{x^2 - 4}{x^3}$$
$$= \frac{(x+2)(x-2)}{x^3}$$

Factor x^{-3} from each term since -3 is the smallest exponent.

▶ **Self-Check** ▼

Factor each polynomial.
1. $rt - st + rv - sv$
2. $6xz - 21yz - 22x + 77y$
3. $y^2 - z^2 - 6z - 9$
4. $5a^3 + 15a^2b + 15ab^2 + 5b^3$

Given $w^4 - 12w^2 + 16$:

5. What middle term is needed for a perfect square trinomial?
6. What term must be added to and subtracted from this polynomial to produce each of the middle terms in part 1?
7. Which of the two answers in part 6 is a perfect square?
8. Use this information to factor $w^2 - 12w^2 + 16$.

▶ **Self-Check Answers** ▼

1. $(r-s)(t+v)$
2. $(2x-7y)(3z-11)$
3. $(y+z+3)(y-z-3)$
4. $5(a+b)^3$
5. $\pm 8w^2$
6. $20w^2$ or $4w^2$
7. $4w^2$
8. $(w^2 + 2w - 4)(w^2 - 2w - 4)$

EXERCISES 1-5

A

In Exercises 1–58, factor each polynomial completely over the integers. If the polynomial cannot be factored, indicate that it is prime.

1. $25a^2 - 16b^2$
2. $(v + 2)w - (v + 2)$
3. $25a^2 - 10a + 1$
4. $4a^2 - 28a + 49$
5. $x^2 + 6x + 5$
6. $x^2 + 5x + 6$
7. $12m^2 - 27m + 15$
8. $10m^2 - 6m - 21$
9. $4ay^2 - 4ay$
10. $25m^2 - 144n^2$
11. $(5c - 7)z + 5c - 7$
12. $49 - 36v^2$
13. $4x^4 - 500x$
14. $27a^3 - 125b^3$
15. $23ax^4 - 23a$
16. $3ax^2 + 3ay^2$
17. $8x^3 - y^3$
18. $12x^3y - 12xy^3$
19. $25y^2 - 30yz + 9z^2$
20. $49b^2 + 126bc + 81c^2$
21. $3ax^2 + 33ax + 72a$
22. $5a^2bc - 6b^3c$
23. $4m^2 + 3m + 1$
24. $12ax^2 - 10axy - 12ay^2$
25. $4bx^3 - 32b$
26. $x^7 + 4x^4y^2 + 4xy^4$
27. $4x^{10} + 12x^5y^3 + 9y^6$
28. $a(x - y) - b(x - y)$
29. $x(a - b) + y(a - b)$
30. $63x^2 + 30x - 72$
31. $7s^5t - 7st^5$
32. $ax^2 + ax + bxy + by$
33. $cx + cy + dx + dy$
34. $-6ax^4 + 48ax$
35. $27x^3y + 72x^2y^2 + 48xy^3$
36. $18x^3 - 21x^2y - 60xy^2$
37. $4a^7 + 32ab^3$
38. $7a^2d - 28b^2c^2d$
39. $c(c + d)^2 + 2(c + d)^3$
40. $(3v - 1)^2 + 5v(3v - 1)$
41. $4t(2t + 3)^4 + 4(2t + 3)^3$
42. $3(2r - 5)^5 + r(2r - 5)^6$
43. $18x^2 - 3xy - 10y^2$
44. $100s^4 + 120s^3t + 36s^2t^2$
45. $63a^3b - 175ab$
46. $x^6 + 4x^3y + 4y^2$
47. $8a - 8ax^2$
48. $5x^2 - 55$
49. $6kx - 6k + 6jx - 6j$
50. $7abx + 35ax + 7bx + 35x$
51. $x^3 - y^3 + x - y$
52. $x^2 + 2xy + y^2 - 16z^2$
53. $9x^2 - 6x + 1 - 25y^2$
54. $2x^2 + 2x + 2y - 2y^2$
55. $3ax^2 - 3ay^2 + 6ay - 3a$
56. $3s^2 + 3s + 3t - 3t^2$
57. $v^3 + 3v^2 + 3v + 1$
58. $c^3 - 3c^2d + 3cd^2 - d^3$

B

In Exercises 59–62, factor each polynomial completely.

59. $3a^3b + 9a^2b^2 + 9ab^3 + 3b^4$
60. $5a^3x - 15a^2bx + 15ab^2x - 5b^3x$
61. $x^3 + 3x^2(6) + 3x(6)^2 + 6^3$
62. $m^3 - 3m^2(11) + 3m(11)^2 - (11)^3$

In Exercises 63 and 64, fill in the blank with the term that must be added to and then subtracted from this polynomial in order to factor it by completing the square.

63. $x^4 - 11x^2 + 1 = (x^4 - 11x^2 + \underline{} + 1) - \underline{}$
64. $x^4 - 7x^2 + 1 = (x^4 - 7x^2 + \underline{} + 1) - \underline{}$

In Exercises 65–70, factor each polynomial by completing the square.

65. $v^4 + 5v^2 + 9$
66. $y^4 - 14y^2 + 25$
67. $x^4 + 4$
68. $w^4 + 64$
69. $y^4 + y^2 + 25$
70. $v^4 - 4v^2 + 36$

In Exercises 71–74 a student's answer to a calculus problem is given along with the book's answer. Factor the variable with the smallest exponent from the student's answer and simplify to obtain the book's answer.

Student's Answer	Book's Answer
71. $x^{-2} - 9x^{-4}$	$\dfrac{(x + 3)(x - 3)}{x^4}$
72. $2a^{-3} - 8a^{-5}$	$\dfrac{2(a + 2)(a - 2)}{a^5}$

SECTION 1-6 RATIONAL EXPRESSIONS

Student's Answer	Book's Answer
73 $2u^{-2} - 7u^{-3} + 3u^{-4}$	$\dfrac{(2u-1)(u-3)}{u^4}$
74 $1 - s^{-1} - 6s^{-2}$	$\dfrac{(s-3)(s+2)}{s^2}$

In Exercises 75 and 76, use division to determine the factorization of each polynomial, given that $a - b$ is one factor.

75 $a^5 - b^5$ 76 $a^7 - b^7$

C

In Exercises 77–80, factor each polynomial completely.

77 $x^{2m} - y^{2n}$ 78 $81y^{2n} - 16$ 79 $4x^{2m} + 20x^m y^n + 25y^{2n}$ 80 $z^{4n} - z^{2n} - 12$

81 Use either "even" or "odd" to complete these statements about $(ax^2 + bx + c) = (a_1 x + c_1)(a_2 x + c_2)$.

  a. If a is odd, then a_1 and a_2 must both be _____.
  b. If c is odd, then c_1 and c_2 must both be _____.
  c. If a is even, then a_1 and a_2 cannot both be _____.
  d. If a_1 is odd and c_2 is odd, then $a_1 c_2$ is _____.
  e. If a_2 is odd and c_1 is even, then $a_2 c_1$ is _____.
  f. If a_1 is even and c_2 is even, then $a_1 c_2$ is _____.
  g. If $a_1 c_2$ is odd and $a_2 c_1$ is odd, then $a_1 c_2 + a_2 c_1 = b$ is _____.
  h. If $a_1 c_2$ is odd and $a_2 c_1$ is even, then $a_1 c_2 + a_2 c_1 = b$ is _____.

82 Explain why $ax^2 + bx + c$ must be irreducible if a, b, and c are all odd. (*Hint*: See Exercise 81.)

SECTION 1-6

Rational Expressions

Section Objective

11 Reduce a rational expression to lowest terms.

12 Add, subtract, multiply, and divide rational expressions.

A fraction that is the ratio of two polynomials is called a **rational expression**. For example,

$$\frac{v}{2w - 4}, \quad \frac{3x^2 - 7}{8x^3 + 9x}, \quad \text{and} \quad \frac{s^2 + 5st - t^2}{s^4 + 3s^2 t^2 - t^4}$$

are rational expressions. A fraction that contains at least one variable in the denominator is called an **algebraic fraction**. Since the fraction $\frac{43}{97}$ contains no variable in the denominator, it is an arithmetic fraction; in fact, it is a rational number. The algebraic fraction $\dfrac{3x-5}{\sqrt{x}-\sqrt{y}}$ is not a rational expression because the denominator is not a polynomial.

Since division by zero is undefined, values of the variable that make the denominator zero are not allowed. These values are called **excluded values**. The only excluded value for x in $\dfrac{x-2}{x+5}$ is $x = -5$. There are two excluded values for y in $\dfrac{2y-3}{(y+4)(y-7)}$; they are $y = -4$ and $y = 7$. If no values are excluded, then we will assume that the replacement set for each variable is the set of real numbers.

A rational expression is in **lowest terms** when the numerator and the denominator have no common factor other than 1 or -1. To reduce a rational expression to lowest terms, we use the multiplicative identity 1 in the form $\dfrac{A}{A}$ $(A \ne 0)$.

Reducing a Rational Expression to Lowest Terms ▼

Step 1 Factor both the numerator and the denominator of the rational expression.

Step 2 Divide the numerator and the denominator by any common nonzero factors.

EXAMPLE 1 Reduce $\dfrac{5x^3 + 5y^3}{10x^2 - 10xy + 10y^2}$.

SOLUTION
$$\dfrac{5x^3 + 5y^3}{10x^2 - 10xy + 10y^2} = \dfrac{5(x^3 + y^3)}{10(x^2 - xy + y^2)}$$

Factor the numerator and the denominator and then divide them by their common factors.

$$= \dfrac{\overset{1}{\cancel{5}}(x+y)\cancel{(x^2 - xy + y^2)}}{\underset{2}{\cancel{10}}\cancel{(x^2 - xy + y^2)}}$$

$$= \dfrac{x+y}{2} \qquad \square$$

EXAMPLE 2 Reduce $\dfrac{x^2 - 5x + 6}{4 - x^2}$.

SECTION 1-6 RATIONAL EXPRESSIONS

SOLUTION

$$\frac{x^2 - 5x + 6}{4 - x^2} = \frac{\overset{-1}{\cancel{(x-2)}}(x-3)}{\cancel{(2-x)}(2+x)}$$ Note that $(x - 2) = -(2 - x)$.

$$= \frac{-(x-3)}{x+2} \quad \text{or} \quad \frac{3-x}{x+2}$$

▶ **Self-Check** ▼

Reduce each rational expression to lowest terms.

1 $\dfrac{x^2 - y^2}{3x + 3y}$ **2** $\dfrac{5r - 10s}{2s - r}$

Since the variables in a rational expression represent real numbers, the rules and procedures for performing operations with rational expressions are the same as those for performing operations with arithmetic fractions. For multiplication and division, remember to examine the numerators and denominators for common factors so that the answer can be reduced before the product is actually formed.

EXAMPLE 3 Multiply $\dfrac{3a - 3b}{x^2 + 2xy + y^2} \cdot \dfrac{x^2 - y^2}{a^2 - b^2}$.

SOLUTION

$$\frac{3a - 3b}{x^2 + 2xy + y^2} \cdot \frac{x^2 - y^2}{a^2 - b^2} = \frac{3\cancel{(a-b)}\cancel{(x+y)}(x-y)}{\cancel{(x+y)}(x+y)\cancel{(a-b)}(a+b)}$$

Indicate the product of the factors of the numerators and denominators, respectively.

$$= \frac{3(x-y)}{(x+y)(a+b)}$$

Reduce this fraction, and indicate the product of the remaining factors.

Note The answer is intentionally left in factored form because this form is often the most useful in applications.

EXAMPLE 4 Simplify $\dfrac{8ax^3 - 8ay^3}{5ax^2 - 5ay^2} \div \dfrac{16x^2 + 16xy + 16y^2}{10bx + 10by}$.

SOLUTION

$$\frac{8ax^3 - 8ay^3}{5ax^2 - 5ay^2} \div \frac{16x^2 + 16xy + 16y^2}{10bx + 10by} = \frac{8a(x^3 - y^3)}{5a(x^2 - y^2)} \div \frac{16(x^2 + xy + y^2)}{10b(x + y)}$$

$$= \frac{\overset{2}{\cancel{8a}}\cancel{(x-y)}\cancel{(x^2+xy+y^2)}}{\cancel{5a}\cancel{(x-y)}\cancel{(x+y)}} \cdot \frac{\overset{}{\cancel{10b}\cancel{(x+y)}}}{\underset{2}{\cancel{16}\cancel{(x^2+xy+y^2)}}}$$

Indicate the product of the dividend and the reciprocal of the divisor.

$$= b$$

▶ **Self-Check Answers** ▼

1 $\dfrac{x-y}{3}$ **2** -5

EXAMPLE 5 Simplify $\dfrac{12x^3y^2}{x+y} \div \dfrac{5x-5}{x^2-y^2} \cdot \dfrac{xy-y}{3x^3y^3}$.

SOLUTION
$$\dfrac{12x^3y^2}{x+y} \div \dfrac{5x-5}{x^2-y^2} \cdot \dfrac{xy-y}{3x^3y^3} = \dfrac{12x^3y^2}{x+y} \cdot \dfrac{x^2-y^2}{5x-5} \cdot \dfrac{xy-y}{3x^3y^3}$$ Reciprocate the divisor.

$$= \dfrac{(12x^3y^2)(x-y)(x+y)(y)(x-1)}{(x+y)(5)(x-1)(3x^3y^3)}$$ Factor.

$$= \dfrac{4(x-y)}{5}$$ Reduce.

We add and subtract rational expressions in exactly the same way that we add and subtract arithmetic fractions—that is, the fractions must first be expressed in terms of a common denominator. Remember to reduce the result to lowest terms.

EXAMPLE 6 Simplify $\dfrac{5x}{24} + \dfrac{7x}{24}$.

SOLUTION
$$\dfrac{5x}{24} + \dfrac{7x}{24} = \dfrac{5x+7x}{24}$$

$$= \dfrac{12x}{24}$$

$$= \dfrac{x}{2}$$

EXAMPLE 7 Simplify $\dfrac{x^2}{x^2-1} - \dfrac{2x-1}{x^2-1}$.

SOLUTION
$$\dfrac{x^2}{x^2-1} - \dfrac{2x-1}{x^2-1} = \dfrac{x^2-(2x-1)}{x^2-1}$$ Use parentheses on this step to avoid an error in sign.

$$= \dfrac{x^2-2x+1}{x^2-1}$$

$$= \dfrac{(x-1)(x-1)}{(x-1)(x+1)}$$

$$= \dfrac{x-1}{x+1}$$

A common denominator of a set of rational expressions is a polynomial that is exactly divisible by each of the given denominators. The **least common denominator** (LCD) is the common denominator that has the lowest degree and the smallest coefficient on the high-order term. It is calculated by the same procedure used for arithmetic fractions.

SECTION 1-6 RATIONAL EXPRESSIONS

Finding the LCD ▼

Step 1 Factor each denominator completely, including constant factors. Express repeated factors in exponential form.

Step 2 Determine the greatest number of times each factor occurs in any single factorization. Then form the LCD by indicating the product of these factors.

▶ **Self-Check** ▼

Simplify each expression.

1 $\dfrac{2a - 2b}{6a} \cdot \dfrac{a^2 - b^2}{a^2 - 2ab + b^2}$

2 $\dfrac{14x^2 - 21x}{42x - 63} \div \dfrac{24x - 16}{12x - 8}$

EXAMPLE 8 Simplify $\dfrac{5v}{v^2 + vw - 6w^2} + \dfrac{2v - w}{v^2 - 3vw + 2w^2} + \dfrac{v + 3w}{v^2 + 2vw - 3w^2}$.

SOLUTION

$v^2 + vw - 6w^2 = (v - 2w)(v + 3w)$
$v^2 - 3vw + 2w^2 = (v - 2w)(v - w)$
$v^2 + 2vw - 3w^2 = (v + 3w)(v - w)$
$\text{LCD} = (v - 2w)(v + 3w)(v - w)$

Calculate the LCD; do this by inspection whenever possible.

$\dfrac{5v}{v^2 + vw - 6w^2} + \dfrac{2v - w}{v^2 - 3vw + 2w^2} + \dfrac{v + 3w}{v^2 + 2vw - 3w^2}$

$= \dfrac{5v(v - w)}{(v - 2w)(v + 3w)(v - w)} + \dfrac{(2v - w)(v + 3w)}{(v - 2w)(v + 3w)(v - w)}$

$+ \dfrac{(v + 3w)(v - 2w)}{(v - 2w)(v + 3w)(v - w)}$

Convert to a common denominator.

$= \dfrac{5v(v - w) + (2v - w)(v + 3w) + (v + 3w)(v - 2w)}{(v - 2w)(v + 3w)(v - w)}$

$= \dfrac{(5v^2 - 5vw) + (2v^2 + 5vw - 3w^2) + (v^2 + vw - 6w^2)}{(v - 2w)(v + 3w)(v - w)}$

$= \dfrac{8v^2 + vw - 9w^2}{(v - 2w)(v + 3w)(v - w)}$

$= \dfrac{(8v + 9w)(v - w)}{(v - 2w)(v + 3w)(v - w)}$

Factor the numerator and reduce this rational expression.

$= \dfrac{8v + 9w}{(v - 2w)(v + 3w)}$ ☐

▶ **Self-Check Answers** ▼

1 $\dfrac{a + b}{3a}$ 2 $\dfrac{x}{6}$

CHAPTER 1 REVIEW OF THE BASIC CONCEPTS OF ALGEBRA

EXAMPLE 9 Simplify $\dfrac{2x+3}{x^2-y^2} + \dfrac{x^2+xy+y^2}{6xy} \cdot \dfrac{18y}{x^3-y^3}$.

SOLUTION
$$\dfrac{2x+3}{x^2-y^2} + \dfrac{x^2+xy+y^2}{6xy} \cdot \dfrac{18y}{x^3-y^3}$$

$$= \dfrac{2x+3}{(x-y)(x+y)} + \dfrac{\overset{3}{\cancel{18y}}\cancel{(x^2+xy+y^2)}}{\cancel{6xy}(x-y)\cancel{(x^2+xy+y^2)}}$$

Note the correct order of operations—multiplication before addition.

$$= \dfrac{2x+3}{(x-y)(x+y)} + \dfrac{3}{x(x-y)}$$

$$= \dfrac{x(2x+3)}{x(x-y)(x+y)} + \dfrac{3(x+y)}{x(x-y)(x+y)}$$

LCD is $x(x-y)(x+y)$.

$$= \dfrac{(2x^2+3x)+(3x+3y)}{x(x-y)(x+y)} = \dfrac{2x^2+6x+3y}{x(x-y)(x+y)} \quad \square$$

A **complex rational expression** is a rational expression whose numerator or denominator (or both) is also a rational expression. Each of the following fractions contains more than one fraction bar and is therefore a complex fraction:

$$\dfrac{\dfrac{1}{x}}{y}, \quad \dfrac{\dfrac{b}{a}}{a-b}, \quad \dfrac{\dfrac{2}{x}}{\dfrac{y}{3}}$$

▶ **Self-Check** ▼

Simplify each expression.

1. $\dfrac{x}{yz} + \dfrac{y}{xz} + \dfrac{z}{xy}$

2. $\dfrac{3x}{3x-y} - \dfrac{3x}{3x+y} - \dfrac{2y^2}{9x^2-y^2}$

Some problems are worked more easily by the first method shown in the following box, whereas others are worked more easily by the second method. Example 10 is worked by both methods so that you can compare them.

Simplifying Complex Fractions ▼

Method I

Step 1 Calculate the LCD of all the fractions in the numerator and the denominator of the complex fraction.

Step 2 Multiply both the numerator and the denominator of the complex fraction by this LCD, and simplify the result.

Method II

Step 1 Express both the numerator and the denominator as single fractions.

Step 2 Divide the simplified numerator by the simplified denominator.

▶ **Self-Check Answers** ▼

1. $\dfrac{x^2+y^2+z^2}{xyz}$

2. $\dfrac{2y}{3x+y}$

SECTION 1-6 RATIONAL EXPRESSIONS

EXAMPLE 10 Simplify $\dfrac{x - \dfrac{1}{y}}{y - \dfrac{1}{x}}$ by Method I and then by Method II.

SOLUTION By Method I,

$$\frac{x - \dfrac{1}{y}}{y - \dfrac{1}{x}} = \frac{x - \dfrac{1}{y}}{y - \dfrac{1}{x}} \cdot \frac{xy}{xy}$$

Multiply both the numerator and the denominator by xy, the LCD of all the fractions in this expression.

$$= \frac{x^2 y - x}{xy^2 - y}$$

$$= \frac{x(xy - 1)}{y(xy - 1)} = \frac{x}{y}$$

By Method II,

$$\frac{x - \dfrac{1}{y}}{y - \dfrac{1}{x}} = \frac{\dfrac{xy - 1}{y}}{\dfrac{xy - 1}{x}}$$

Express both the numerator and the denominator as single fractions.

$$= \frac{xy - 1}{y} \div \frac{xy - 1}{x}$$

Then divide the numerator by the denominator.

$$= \frac{xy - 1}{y} \cdot \frac{x}{xy - 1}$$

$$= \frac{x}{y} \qquad \square$$

Since negative exponents are an alternate notation used to represent reciprocals, we can also use these methods to simplify some expressions with negative exponents.

EXAMPLE 11 Simplify $\dfrac{1 + x^{-1}}{x - x^{-1}}$.

SOLUTION $\dfrac{1 + x^{-1}}{x - x^{-1}} = \dfrac{1 + x^{-1}}{x - x^{-1}} \cdot \dfrac{x}{x}$

Multiply both the numerator and the denominator by x, since x^{-1} is the lowest power of x in either one. What we are observing is that x will be the LCD if this fraction is rewritten in the form

$$\frac{1 + \dfrac{1}{x}}{x - \dfrac{1}{x}}.$$

$$= \frac{x + 1}{x^2 - 1}$$

$$= \frac{(x + 1)}{(x + 1)(x - 1)}$$

$$= \frac{1}{x - 1} \qquad \square$$

EXERCISES 1-6

A

In Exercises 1–4, determine the excluded value(s) of the variable in each expression.

1. $\dfrac{4x + 24}{3x - 24}$

2. $\dfrac{6x + 2}{6x - 3}$

3. $\dfrac{12x - 11}{(x - 2)(x + 5)}$

4. $\dfrac{5x + 13}{(x + 3)(x - 7)}$

In Exercises 5–18, reduce each of the rational expressions to lowest terms.

5. $\dfrac{33a^3b^4}{44a^2b^5}$

6. $\dfrac{15a^7b^2}{35a^4b^4}$

7. $\dfrac{30x^2y^3 - 95xy^4}{5xy^2}$

8. $\dfrac{46x^4y^2z^2 - 69x^3y^2z^3}{23x^2yz^2}$

9. $\dfrac{25x^2 - 4}{14 - 35x}$

10. $\dfrac{33ay - 88ax}{64x^2 - 9y^2}$

11. $\dfrac{x^3 + y^3}{13x^2 - 13xy + 13y^2}$

12. $\dfrac{5m^2 + 25m + 30}{7m^2 - 7m - 42}$

13. $\dfrac{10a^2 - 70a + 60}{2a^2 - 72}$

14. $\dfrac{x(a + b) - y(a + b)}{5x - 5y}$

15. $\dfrac{s(x + y) - t(x + y)}{s^2 - t^2}$

16. $\dfrac{7x - 14y + ax - 2ay}{3x - 6y}$

17. $\dfrac{a^2 + 2a + 1 + ab + b}{9a + 9b + 9}$

18. $\dfrac{x^2 + x + y - y^2}{2x^3 + 2y^3}$

In Exercises 19–34, find the product or quotient. Express all answers in reduced form.

19. $\dfrac{10y - 14x}{x - 1} \cdot \dfrac{x^2 - 2x + 1}{21x - 15y}$

20. $\dfrac{a^2 - 9b^2}{4a^2 - b^2} \cdot \dfrac{4a^2 - 4ab + b^2}{2a^2 - 7ab + 3b^2}$

21. $\dfrac{-69x^5y}{-35x^2y^3} \div \dfrac{46x^4y^4}{-84x^6y}$

22. $\dfrac{20x^2yz^2}{-27wz} \div \dfrac{60x^3z}{9wy^2}$

23. $\dfrac{4x^2 + 12x + 9}{5x^3 - 2x^2} \div \dfrac{14x^2 + 21x}{10x^2y^2}$

24. $\dfrac{x^3 - y^3}{6x^2 - 6y^2} \div \dfrac{2x^2 + 2xy + 2y^2}{9xy}$

25. $\dfrac{(-2xy)^3}{3x^3z} \cdot \dfrac{-42x}{112x^2z^2}$

26. $\dfrac{(-3a^2b^3)^4}{a^{12}b^9} \div \dfrac{(-3ab^2)^3}{(2a^2b)^2}$

27. $\dfrac{8x^3 + 27}{x^3 - 64} \div \dfrac{10x + 15}{3x^2 + 12x + 48}$

28. $\dfrac{121ab}{33a^2 + 44a} \div \dfrac{3a^2 + a - 4}{36a^2b + 96ab + 64b}$

29. $\dfrac{x^4 - y^4}{-2x - 3y} \div \dfrac{7x^2 + 7y^2}{2x + 3y} \cdot \dfrac{12xy}{8x^2 + 8xy}$

30. $\dfrac{x^3 - y^3}{18x^2y^3} \div \dfrac{x^2 + xy + y^2}{36y^4} \cdot \dfrac{9x^2y + 9xy^2}{x^2 - y^2}$

31. $\dfrac{3x - 6y}{7x^2y} \div \dfrac{17x^3 - 17y^3}{34xy^2} \div \dfrac{3x}{2x^2 + 2xy + 2y^2}$

32. $\dfrac{5a - b}{a^2 - 5ab + 4b^2} \div \left(\dfrac{6ab}{3a - 12b} \cdot \dfrac{b^2 - 5ab}{4a - 4b} \right)$

33. $\dfrac{x^2 + x - y^2 - y}{3x^2 - 3y^2} \div \dfrac{5x + 5y + 5}{7x^2y + 7xy^2}$

34. $\dfrac{x(a - b + 2c) - 2y(a - b + 2c)}{x^2 - 4xy + 4y^2} \div \dfrac{2a^2 - 2ab + 4ac}{7xy^2 - 14y^3}$

In Exercises 35–38, fill in the missing numerator, so that the rational expressions will be equal.

35. $\dfrac{x - y}{x + y} = \dfrac{?}{3x + 3y}$

36. $\dfrac{2s + t}{s + 2t} = \dfrac{?}{12s + 24t}$

37. $\dfrac{3}{s + t} = \dfrac{?}{s^3 + t^3}$

38. $\dfrac{2x - y}{x + 3y} = \dfrac{?}{x^2 + 5xy + 6y^2}$

In Exercises 39–48, perform the indicated operations and reduce the results to lowest terms.

39. $\dfrac{14}{x - 7} - \dfrac{2x}{x - 7}$

40. $\dfrac{3x^2 + 1}{8x^3} - \dfrac{1 - 3x^2}{8x^3}$

41. $\dfrac{x + 1}{x - 1} + \dfrac{x - 1}{x + 1}$

42. $\dfrac{x}{x - 1} + \dfrac{x - 3}{x + 2}$

43. $\dfrac{2x + 1}{x - 3} - \dfrac{x - 2}{x + 4}$

44. $\dfrac{2z + 11}{z^2 + z - 6} + \dfrac{2}{z + 3} - \dfrac{3}{z - 2}$

SECTION 1-6 RATIONAL EXPRESSIONS

45. $\dfrac{p+6}{p^2-4} - \dfrac{4}{p+2} - \dfrac{2}{p-2}$

46. $\dfrac{2w-7}{w^2-5w+6} - \dfrac{2-4w}{w^2-w-6} + \dfrac{5w+2}{4-w^2}$

47. $\dfrac{1}{a+b} - \dfrac{3b^2}{a^3+b^3} + \dfrac{b-a}{a^2-ab+b^2}$

48. $\dfrac{9w+2}{3w^2-2w-8} - \dfrac{7}{4-w-3w^2}$

In Exercises 49–55, simplify each expression.

49. $\dfrac{\dfrac{3x^2-27x+42}{34x^4 y^5}}{\dfrac{5x^2-20}{51x^5 y^3}}$

50. $\dfrac{\dfrac{5s^2+5st+5t^2}{8s^3-8t^3}}{\dfrac{7s^2+14st+7t^2}{14s^2-14t^2}}$

51. $\dfrac{2-\dfrac{1}{x}}{4-\dfrac{1}{x^2}}$

52. $\dfrac{\dfrac{1}{x^2}-49}{7-\dfrac{1}{x}}$

53. $\dfrac{1}{a+\dfrac{1}{a+\tfrac{1}{2}}}$

54. $\dfrac{\dfrac{1}{x}-\dfrac{8}{x^2}+\dfrac{15}{x^3}}{1-\dfrac{5}{x}}$

55. $\dfrac{1-\dfrac{1}{x^3}}{1+\dfrac{1}{x}+\dfrac{1}{x^2}}$

B

In Exercises 56–64, simplify each expression.

56. $\dfrac{32}{2+\dfrac{1}{2+\dfrac{1}{2+\tfrac{1}{2}}}}$

57. $\dfrac{\dfrac{x^2+b^2}{x^2-b^2}}{\dfrac{x-b}{x+b}+\dfrac{x+b}{x-b}}$

58. $\dfrac{\dfrac{w-a}{w+a}-\dfrac{w+a}{w-a}}{\dfrac{w^2+a^2}{w^2-a^2}}$

59. $(m-n)^{-1}(m^{-1}-n^{-1})$

60. $\dfrac{6x^{-2}-5x^{-1}+1}{1-4x^{-2}}$

61. $\dfrac{x-y}{xy} + \dfrac{x^3+y^3}{x^3 y^3} \cdot \dfrac{x^2 y^2}{4x^2-4xy+4y^2}$

62. $\left(3-\dfrac{b^2}{3a^2}\right)\left(\dfrac{9a^2-ab}{3a-b}-a\right)$

63. $\left(\dfrac{4}{5x+25} + \dfrac{1}{5x-25}\right) \div \dfrac{x^2-x-6}{x+5}$

64. $\dfrac{x+2}{x+3} - \dfrac{x^3-8}{x^2-9} \div \dfrac{x^2+2x+4}{x-3}$

In Exercises 65–68, determine whether each statement is T (true) or F (false).

65. $\dfrac{a-b}{7} = \dfrac{b-a}{-7}$

66. $\dfrac{a-b}{7} = -\dfrac{b-a}{7}$

67. $\dfrac{1}{x}+\dfrac{1}{y} = \dfrac{1+1}{x+y} = \dfrac{2}{x+y}$

68. $\dfrac{x+4}{y+4} = \dfrac{x}{y}$

C

In Exercises 69–74, simplify each expression.

69. $\dfrac{a^3+3a^2 b+3ab^2+b^3}{a^3+b^3}$

70. $\dfrac{x^3-1}{x^3-3x^2+3x-1}$

71. $\dfrac{v^4-v^2+16}{7v^2+21v+28}$

72. $\dfrac{w^4+3w^2+4}{5w^2-5w+10}$

73. $\dfrac{v^2-2v-2}{2v^2-13v+6} - \dfrac{7v-2}{2v^2+v-1} + \dfrac{v-20}{v^2-5v-6}$

74. $\dfrac{3+\dfrac{9}{x}}{\dfrac{1}{x}+\dfrac{8+\tfrac{15}{x}}{x^2}}$

SECTION 1-7

Rational Exponents and Radicals

Section Objectives

13 Use fractional exponents.

14 Simplify radical expressions.

15 Add, subtract, multiply, and divide radical expressions.

Radicals occur frequently in the solution of many types of equations. Two formulas containing radicals are the quadratic formula and Cardan's formula for solving certain cubic equations. The ability to simplify radicals and to perform calculations with radical expressions is therefore an important tool for those who need to solve equations.

Radical Notation ▼

Index → $\sqrt[n]{R}$ denotes the principal nth root of R.
          ↑ ↑
          | └ Radicand
          └── Radical sign

Each positive real number has two square roots. The square roots of 9 are $+3$ and -3, since $3^2 = 9$ and $(-3)^2 = 9$. The positive square root is called the **principal square root**. The symbol $\sqrt{R}$ represents the principal square root of R. ($\sqrt{9} = 3$, but $\sqrt{9} \neq -3$.) To represent the negative root, we can write $-\sqrt{9}$; $-\sqrt{9} = -3$. Likewise, $\sqrt[4]{16} = 2$ but not -2. Each real number has only one real cube root. This real root is the principal root.

Principal Root ▼

$\sqrt[n]{R} = x$ indicates that x is the principal nth root of R if $x^n = R$.

For $R \geq 0$,

$\sqrt[n]{R}$ is a nonnegative real number for all natural numbers n.

For $R < 0$,

$\sqrt[n]{R}$ is a negative real number if n is odd.

$\sqrt[n]{R}$ is an imaginary number (see Section 1–8) if n is even.

EXAMPLE 1 Find the principal root of each expression.

SOLUTIONS

(a) $\sqrt{36}$ $\quad \sqrt{36} = 6$ because $6^2 = 36$ and $6 \geq 0$. $\quad$ Even though $(-6)^2 = 36$, -6 is not the principal square root of 36.

(b) $\sqrt[n]{0}$ $\quad \sqrt[n]{0} = 0$ because $0^n = 0$.

SECTION 1-7 RATIONAL EXPONENTS AND RADICALS

(c) $\sqrt{(-8)^2}$ $\quad \sqrt{(-8)^2} = \sqrt{64} = 8.$ Note that $\sqrt{(-8)^2}$ is not -8; it is $|-8|$.

(d) $\sqrt[3]{-27}$ $\quad \sqrt[3]{-27} = -3$ because $(-3)^3 = -27.$ Note that $\sqrt[3]{(-3)^3} = -3.$

Example 1(c) points out an important subtlety of algebra concerning $\sqrt[n]{x^n}$. Since $\sqrt{(-8)^2} = |-8|$ and $\sqrt{8^2} = 8$, we can state for any radicand (positive or negative) that $\sqrt{x^2} = |x|$. In fact, for any *even* number n, $\sqrt[n]{x^n} = |x|$.

For any odd number n, $\sqrt[n]{(x^n)} = x$, as in Example 1(d), where $\sqrt[3]{(-3)^3} = -3$.

$\sqrt[n]{x^n}$ ▼

For any real number x,

$$\sqrt[n]{x^n} = |x| \quad \text{if } n \text{ is an even natural number}$$

$$\sqrt[n]{x^n} = x \quad \text{if } n \text{ is an odd natural number}$$

In particular, $\sqrt[2]{x^2} = |x|.$

We will now examine the relationship between radicals and rational exponents. This relationship is important because many of the properties of radicals are restatements of the laws of exponents. As noted in Section 1–3, the definition of exponents was extended from natural numbers and repetitive multiplication so that exponentiation would be meaningful for all real exponents. The motivation behind this extension is to have the same properties of exponents hold for all real numbers.

For example, we want $\dfrac{x^a}{x^a}$ to equal 1 for $x \neq 0$. Thus we define x^0 so that the quotient rule will yield this result. That is,

$$\frac{x^a}{x^a} = x^{a-a} = x^0 = 1 \qquad \text{Quotient rule: } \frac{x^m}{x^n} = x^{m-n}$$

The same reasoning leads us to define x^{-a} as $\dfrac{1}{x^a}$. We will now examine fractional exponents. If the product rule holds, then

$$x^{1/2} \cdot x^{1/2} = x^{1/2 + 1/2} = x^1 \qquad \text{Product rule: } x^m x^n = x^{m+n}$$

Thus we define $x^{1/2} = \sqrt{x}$. Similarly, we define $x^{1/3} = \sqrt[3]{x}$, so that $(x^{1/3})^3 = x$.

nth Roots in Exponential Form ▼

For each natural number n,
$$x^{1/n} = \sqrt[n]{x}$$
For n an even number, x is restricted to nonnegative real numbers.
For n an odd number, x can be any real number.

EXAMPLE 2 Simplify each expression.

SOLUTIONS

(a) $64^{1/2}$ $\quad 64^{1/2} = \sqrt{64} = 8$

(b) $-64^{1/2}$ $\quad -64^{1/2} = -\sqrt{64} = -8$

(c) $(-64)^{1/3}$ $\quad (-64)^{1/3} = \sqrt[3]{-64} = -4$

(d) $32^{1/5}$ $\quad 32^{1/5} = \sqrt[5]{32} = 2$

By the power rule, $(x^m)^{1/n} = x^{m/n}$ and $(x^{1/n})^m = x^{m/n}$. Thus we have the following definition.

Rational Exponents ▼

$x^{m/n} = (\sqrt[n]{x})^m$, or, equivalently, $x^{m/n} = \sqrt[n]{x^m}$.

We will usually use the first form for arithmetical computations and the second form to express algebraic results.

EXAMPLE 3 Simplify each expression.

SOLUTIONS

(a) $8^{2/3}$ $\quad 8^{2/3} = (\sqrt[3]{8})^2 = 2^2 = 4$ $\qquad x^{m/n} = (\sqrt[n]{x})^m$

(b) $(-32)^{3/5}$ $\quad (-32)^{3/5} = (\sqrt[5]{-32})^3 = (-2)^3 = -8$

(c) $81^{3/4}$ $\quad 81^{3/4} = (\sqrt[4]{81})^3 = 3^3 = 27$

EXAMPLE 4 Convert each expression to exponential notation.

SOLUTIONS

(a) $(\sqrt[4]{x})^3$ $\quad (\sqrt[4]{x})^3 = x^{3/4}$ $\qquad (\sqrt[n]{x})^m = x^{m/n}$

(b) $\sqrt[3]{x^5}$ $\quad \sqrt[3]{x^5} = x^{5/3}$ $\qquad \sqrt[n]{x^m} = x^{m/n}$

(c) $\dfrac{1}{\sqrt[3]{x^2}}$ $\quad \dfrac{1}{\sqrt[3]{x^2}} = \dfrac{1}{x^{2/3}} = x^{-2/3}$

SECTION 1-7 RATIONAL EXPONENTS AND RADICALS

Since all the properties of exponents also hold for fractional exponents, we can use these properties to simplify the expressions in Example 5.

▶ **Self-Check** ▼

Simplify these expressions.
1. $64^{3/2}$
2. $64^{2/3}$
3. $(-8)^{5/3}$
4. $(-1)^{4/3}$

EXAMPLE 5 Simplify each expression.

SOLUTIONS

(a) $x^{1/2}x^{1/3}$ $\quad x^{1/2}x^{1/3} = x^{1/2 + 1/3} = x^{5/6}$ Product rule: $x^m x^n = x^{m+n}$

(b) $(z^{-2/3})^{-3}$ $\quad (z^{-2/3})^{-3} = z^{(-2/3)(-3)} = z^2$ Power rule: $(x^m)^n = x^{mn}$

(c) $(a^{-9}b^{-6})^{-1/3}$ $\quad (a^{-9}b^{-6})^{-1/3} = a^{9/3}b^{6/3} = a^3 b^2$ Power rule

(d) $\dfrac{c^{1/2}}{c^{1/3}}$ $\quad \dfrac{c^{1/2}}{c^{1/3}} = c^{1/2 - 1/3} = c^{1/6}$ Quotient rule: $\dfrac{x^m}{x^n} = x^{m-n}$

□

The special products can be used to simplify expressions such as those in Example 6.

EXAMPLE 6 Simplify each expression.

SOLUTIONS

(a) $(x^{1/2} + y^{1/2})(x^{1/2} - y^{1/2})$ $\quad (x^{1/2} + y^{1/2})(x^{1/2} - y^{1/2})$
$\quad = x - y$

This product is of the form $(a + b)(a - b) = a^2 - b^2$.

(b) $(x^{1/3} + y^{1/3})(x^{2/3} - x^{1/3}y^{1/3} + y^{2/3})$ $\quad (x^{1/3} + y^{1/3})$
$\quad \times (x^{2/3} - x^{1/3}y^{1/3} + y^{2/3})$
$\quad = x + y$

This product is of the form $(a + b)(a^2 - ab + b^2) = a^3 + b^3$.

□

Because radical expressions can be simplified in different ways, we will summarize the conditions that a radical expression must meet to be in simplified form.

Simplified Form for Radical Expressions ▼

A radical expression is in simplified form if all of the following conditions are satisfied.

1. The radicand is as small as possible: $\sqrt[n]{x^m}$ has $m < n$.
2. The index is as small as possible: $\sqrt[n]{x^m}$ has the fractional power $\dfrac{m}{n}$ irreducible.
3. There are no fractions in the radicand.
4. There are no radicals in the denominator.

▶ **Self-Check Answers** ▼

1. 512   2. 16   3. −32   4. 1

To simplify a radical, we will use the properties given in the box. Note that these properties are just restatements of the properties of exponents.

Properties of Radicals ▼

For positive values of a and b:

Radical Form	Exponential Form
$\sqrt[n]{ab} = \sqrt[n]{a}\sqrt[n]{b}$	$(ab)^{1/n} = a^{1/n}b^{1/n}$
$\sqrt[n]{\dfrac{a}{b}} = \dfrac{\sqrt[n]{a}}{\sqrt[n]{b}}$	$\left(\dfrac{a}{b}\right)^{1/n} = \dfrac{a^{1/n}}{b^{1/n}}$
$\sqrt[n]{a^n} = a$	$(a^n)^{1/n} = a$

▶ **Self-Check** ▼

Simplify each expression.

1 $x^{3/4}x^{1/5}$ **2** $(x^{-3/4})^{-2/3}$

To simplify square roots, try to recognize perfect square factors, such as 1, 4, 9, 16, and 25. To simplify cube roots, think of the perfect cubes 1, 8, 27, 64, 125, and so on.

EXAMPLE 7 Simplify each radical, assuming that x and y represent only positive real numbers.

SOLUTIONS

(a) $\sqrt{24x^3}$

$\sqrt{24x^3} = \sqrt{4x^2}\sqrt{6x}$
$= 2x\sqrt{6x}$ $4x^2$ is a perfect square. $\sqrt{ab} = \sqrt{a}\sqrt{b}$

Note that if x were allowed to be negative, we would write $\sqrt{24x^3} = 2|x|\sqrt{6x}$.

(b) $\sqrt[3]{54x^5y^7}$

$\sqrt[3]{54x^5y^7} = \sqrt[3]{(27x^3y^6)(2x^2y)}$
$= \sqrt[3]{27x^3y^6}\sqrt[3]{2x^2y}$ $27x^3y^6$ is a perfect cube.
$= 3xy^2\sqrt[3]{2x^2y}$ $\sqrt[3]{ab} = \sqrt[3]{a}\sqrt[3]{b}$ □

If a radical expression contains a fraction, then we can use the property that $\sqrt[n]{\dfrac{a}{b}} = \dfrac{\sqrt[n]{a}}{\sqrt[n]{b}}$ to remove fractions from the radicand.

We can add like radicals in exactly the same way we add like terms of a polynomial. **Like radicals** have the same index and the same radicand. For example, $-\sqrt{xy^2}$ and $4\sqrt{xy^2}$ are like radicals, but $\sqrt{x}$ and $\sqrt[3]{x}$ are unlike because they are of different order. The radicals $\sqrt[4]{xy^2}$ and $\sqrt[4]{x^2y}$ are also unlike, since the radicands xy^2 and x^2y are unequal. To add like radicals, we use the distributive property. We can rewrite $5\sqrt{3} + 7\sqrt{3}$ as $(5 + 7)\sqrt{3} = 12\sqrt{3}$. Thus we add or subtract like radicals by adding their coefficients.

▶ **Self-Check Answers ▼**

1 $x^{19/20}$ **2** $x^{1/2}$

SECTION 1-7 RATIONAL EXPONENTS AND RADICALS

EXAMPLE 8 Perform the following additions and subtractions, assuming that x and y represent only positive real numbers.

> **▶ Self-Check ▼**
>
> Simplify each radical.
> 1. $\sqrt{12x^7y^2}$, for $x > 0$
> 2. $\sqrt[3]{81x^6y^8}$

SOLUTIONS

(a) $5\sqrt{11} - 2\sqrt[3]{11} + 7\sqrt{11}$

$$5\sqrt{11} - 2\sqrt[3]{11} + 7\sqrt{11}$$
$$= (5 + 7)\sqrt{11} - 2\sqrt[3]{11}$$
$$= 12\sqrt{11} - 2\sqrt[3]{11}$$

(b) $7\sqrt{20} - 2\sqrt{45}$

$$7\sqrt{20} - 2\sqrt{45} = 7\sqrt{4}\sqrt{5} - 2\sqrt{9}\sqrt{5}$$
$$= 7(2\sqrt{5}) - 2(3\sqrt{5})$$
$$= 14\sqrt{5} - 6\sqrt{5}$$
$$= 8\sqrt{5}$$

(c) $\sqrt{8x} + 2\sqrt{18x}$

$$\sqrt{8x} + 2\sqrt{18x}$$
$$= \sqrt{4}\sqrt{2x} + 2\sqrt{9}\sqrt{2x}$$
$$= 2\sqrt{2x} + 6\sqrt{2x}$$
$$= 8\sqrt{2x}$$ □

To multiply radicals, we use the property

$$\sqrt[n]{a}\,\sqrt[n]{b} = \sqrt[n]{ab}$$

Thus radicals must be of the same order to be multiplied under the radical sign. Since multiplication of radical expressions is similar to multiplication of polynomials, try to recognize forms that you can multiply by inspection.

EXAMPLE 9 Find each product. Assume x and y are nonnegative.

SOLUTIONS

(a) $\sqrt{6xy}\,\sqrt{2x}$

$$\sqrt{6xy}\,\sqrt{2x} = \sqrt{12x^2y} \qquad \sqrt[n]{a}\,\sqrt[n]{b} = \sqrt[n]{ab}$$
$$= \sqrt{4x^2}\,\sqrt{3y}$$
$$= 2x\sqrt{3y}$$

(b) $2\sqrt{3}(3\sqrt{3} - 5)$

$$2\sqrt{3}(3\sqrt{3} - 5)$$
$$= (2\sqrt{3})(3\sqrt{3}) - 5(2\sqrt{3}) \qquad \text{Distribute the factor } 2\sqrt{3} \text{ to both terms.}$$
$$= 6(\sqrt{3})^2 - 10\sqrt{3}$$
$$= 18 - 10\sqrt{3}$$

(c) $(\sqrt{7} + \sqrt{5})(\sqrt{7} - \sqrt{5})$

$$(\sqrt{7} + \sqrt{5})(\sqrt{7} - \sqrt{5})$$
$$= (\sqrt{7})^2 - (\sqrt{5})^2 \qquad \text{This product is of the form } (a+b)(a-b) = a^2 - b^2.$$
$$= 7 - 5$$
$$= 2 \qquad \square$$

▶ Self-Check Answers ▼

1. $2x^3|y|\sqrt{3x}$ 2. $3x^2y^2\sqrt[3]{3y^2}$

Multiplications of the form

$$(\sqrt{a} + \sqrt{b})(\sqrt{a} - \sqrt{b}) = a - b$$

play an important role in simplifying radical expressions since this product is free of radicals. The expressions $\sqrt{a} + \sqrt{b}$ and $\sqrt{a} - \sqrt{b}$ are called **conjugates** of each other.

Some sample conjugates illustrate the concept:

- The conjugate of $\sqrt{7} - 2$ is $\sqrt{7} + 2$.
- The conjugate of $3 + 5\sqrt{11}$ is $3 - 5\sqrt{11}$.
- The conjugate of $2\sqrt{x} - \sqrt{y}$ is $2\sqrt{x} + \sqrt{y}$.

If a radical expression contains a fraction, then we can use the property that $\sqrt[n]{\dfrac{a}{b}} = \dfrac{\sqrt[n]{a}}{\sqrt[n]{b}}$ to remove fractions from the radicand.

EXAMPLE 10 Simplify each radical, assuming that all variables represent only positive real numbers.

SOLUTIONS

(a) $\sqrt{\dfrac{4}{9}}$ $\sqrt{\dfrac{4}{9}} = \dfrac{\sqrt{4}}{\sqrt{9}} = \dfrac{2}{3}$ $\sqrt{\dfrac{a}{b}} = \dfrac{\sqrt{a}}{\sqrt{b}}$

(b) $\sqrt{\dfrac{6x}{25y^2}}$ $\sqrt{\dfrac{6x}{25y^2}} = \dfrac{\sqrt{6x}}{\sqrt{25y^2}} = \dfrac{\sqrt{6x}}{5y}$

(c) $\sqrt[3]{\dfrac{5}{8}}$ $\sqrt[3]{\dfrac{5}{8}} = \dfrac{\sqrt[3]{5}}{\sqrt[3]{8}} = \dfrac{\sqrt[3]{5}}{2}$ □

The process of simplifying an expression by removing the radical from the denominator is called **rationalizing the denominator**. Historically, radicals have often been rationalized since forms like $\dfrac{1}{\sqrt{3}}$ were easier to approximate when first expressed in the form $\dfrac{\sqrt{3}}{3}$. (It is easier to compute $\dfrac{1.732}{3}$ by pencil and paer than it is $\dfrac{1}{1.732}$.) Although the availability of calculators has diminished the need to approximate constants using this technique, rationalizing denominators is still a useful algebraic simplification that is used with expressions with variables. If the denominator contains a single-term square root, try to produce a perfect square; likewise for a cube root, try to produce a perfect cube. If the denominator contains a binomial term with a square root, conjugates can be used to rationalize the denominator.

SECTION 1-7 RATIONAL EXPONENTS AND RADICALS

EXAMPLE 11 Rationalize the denominator of each expression.

SOLUTIONS

(a) $\dfrac{1}{\sqrt{3}}$

$\dfrac{1}{\sqrt{3}} = \dfrac{1}{\sqrt{3}} \cdot \dfrac{\sqrt{3}}{\sqrt{3}}$ Multiply both the numerator and the denominator by $\sqrt{3}$.

$= \dfrac{\sqrt{3}}{3}$

(b) $\sqrt{\dfrac{9}{11}}$

$\sqrt{\dfrac{9}{11}} = \dfrac{\sqrt{9}}{\sqrt{11}}$ $\sqrt{\dfrac{a}{b}} = \dfrac{\sqrt{a}}{\sqrt{b}}$

$= \dfrac{3}{\sqrt{11}}$

$= \dfrac{3}{\sqrt{11}} \cdot \dfrac{\sqrt{11}}{\sqrt{11}}$ Multiply both the numerator and the denominator by $\sqrt{11}$.

$= \dfrac{3\sqrt{11}}{11}$

(c) $\dfrac{1}{\sqrt[3]{2x^2}}$

$\dfrac{1}{\sqrt[3]{2x^2}} = \dfrac{1}{\sqrt[3]{2x^2}} \cdot \dfrac{\sqrt[3]{4x}}{\sqrt[3]{4x}}$ Note that the denominator is multiplied by a value that will produce a radicand that is a perfect cube.

$= \dfrac{\sqrt[3]{4x}}{\sqrt[3]{8x^3}}$

$= \dfrac{\sqrt[3]{4x}}{2x}$

(d) $\dfrac{6}{\sqrt{7}-\sqrt{5}}$

$\dfrac{6}{\sqrt{7}-\sqrt{5}} = \dfrac{6}{\sqrt{7}-\sqrt{5}} \cdot \dfrac{\sqrt{7}+\sqrt{5}}{\sqrt{7}+\sqrt{5}}$ Multiply the numerator and denominator by the conjugate of $\sqrt{7}-\sqrt{5}$.

$= \dfrac{6(\sqrt{7}+\sqrt{5})}{7-5}$

$= \dfrac{6(\sqrt{7}+\sqrt{5})}{2}$

$= 3\sqrt{7}+3\sqrt{5}$

(e) $\dfrac{\sqrt{x}}{\sqrt{x}+\sqrt{y}}$

$\dfrac{\sqrt{x}}{\sqrt{x}+\sqrt{y}} = \dfrac{\sqrt{x}}{\sqrt{x}+\sqrt{y}} \cdot \dfrac{\sqrt{x}-\sqrt{y}}{\sqrt{x}-\sqrt{y}}$ The conjugate of $\sqrt{x}+\sqrt{y}$ is $\sqrt{x}-\sqrt{y}$.

$= \dfrac{x-\sqrt{xy}}{x-y}$

▶ **Self-Check** ▼

Perform the indicated operations.
1. $\sqrt{45} - 2\sqrt{20} + 3\sqrt{125}$
2. $(2\sqrt{x} - \sqrt{y})(2\sqrt{x} + \sqrt{y})$

▶ **Self-Check Answers** ▼

1 $14\sqrt{5}$ **2** $4x - y$

Reducing the order of a radical simplifies the radical by making the index as small as possible. Perform the simplification mentally if you wish; the steps are shown in the next example to illustrate that the reasoning follows from the properties of exponents.

EXAMPLE 12 Simplify $\sqrt[4]{9x^2y^2}$ by reducing the order, assuming that x and y represent only positive real numbers.

SOLUTION
$$\sqrt[4]{9x^2y^2} = [(3xy)^2]^{1/4}$$
$$= (3xy)^{2/4}$$
$$= (3xy)^{1/2}$$
$$= \sqrt{3xy}$$

▶ **Self-Check** ▼

Simplify each radical.

1. $\sqrt{\dfrac{36}{169}}$
2. $\dfrac{2}{\sqrt{11}}$
3. $\dfrac{18}{\sqrt[3]{2}}$
4. $\dfrac{12}{\sqrt{9} - \sqrt{5}}$

EXERCISES 1-7

A

In Exercises 1 and 2, write each radical expression in exponential form.

1. a. $\sqrt{7}$ b. $\sqrt[3]{5}$ c. $\sqrt{x}$ d. $-\sqrt[5]{y^3}$
2. a. $\sqrt{11}$ b. $\sqrt[4]{6}$ c. $\sqrt[3]{x}$ d. $\sqrt[5]{x^4}$

In Exercises 3 and 4, write each exponential expression in radical form.

3. a. $11^{1/2}$ b. $13^{1/4}$ c. $x^{2/3}$ d. $y^{3/4}$
4. a. $5^{1/2}$ b. $17^{1/3}$ c. $x^{3/5}$ d. $y^{2/7}$

In Exercises 5–7, simplify each expression.

5. a. $\sqrt{9} + \sqrt{16}$ b. $\sqrt{9+16}$ c. $\sqrt[3]{8}$ d. $\sqrt[3]{-8}$
6. a. $\sqrt{25} + \sqrt{144}$ b. $\sqrt{25+144}$ c. $\sqrt[5]{32}$ d. $\sqrt[5]{-32}$
7. a. $36^{1/2}$ b. $81^{1/4}$ c. $\left(\dfrac{25}{144}\right)^{1/2}$ d. $\left(\dfrac{8}{27}\right)^{-2/3}$
8. a. $81^{1/2}$ b. $81^{1/4}$ c. $\left(\dfrac{8}{125}\right)^{1/3}$ d. $\left(\dfrac{1}{32}\right)^{-2/5}$

In Exercises 9–18, simplify each expression using the properties of exponents. Assume that all variables represent positive real numbers. Write your answers without using negative exponents.

9. $z^{1/2}z^{3/10}$
10. $w^{1/3}w^{2/5}$
11. $(z^{1/2})^{3/10}$
12. $(w^{1/3})^{2/5}$
13. $\dfrac{z^{1/2}}{z^{3/10}}$
14. $\dfrac{w^{2/5}}{w^{1/3}}$
15. $(9v^{-2}w^{-6})^{-1/2}$
16. $(25s^{-4/9}t^{2/15})^{-3/2}$
17. $\dfrac{y^{3/8}z^{-2/5}}{y^{-5/8}z^{3/5}}$
18. $\dfrac{a^{-1/4}b^{5/3}}{a^{3/4}b^{-1/3}}$

▶ **Self-Check Answers** ▼

1. $\dfrac{6}{13}$ 2. $\dfrac{2\sqrt{11}}{11}$ 3. $9\sqrt[3]{4}$ 4. $3\sqrt{9} + 3\sqrt{5}$

SECTION 1-7 RATIONAL EXPONENTS AND RADICALS

In Exercises 19–22, simplify each expression, first assuming x and y are positive and then assuming that x and y can be any real number.

19 $\sqrt{16x^2}$ **20** $\sqrt{121y^6}$ **21** $\sqrt{12x^6y^4}$ **22** $\sqrt{96x^4y^2}$

In Exercises 23–30, find the indicated sum or difference. Assume that all variables represent positive real numbers.

23 $8\sqrt{2} - 5\sqrt{2}$ **24** $9\sqrt{5} + 13\sqrt{5}$ **25** $4\sqrt[3]{x} + 7\sqrt[3]{x}$ **26** $8\sqrt{y} + 4\sqrt{y}$
27 $2\sqrt{27} - \sqrt{75}$ **28** $5\sqrt{8} - 3\sqrt{50}$ **29** $3\sqrt{12x} - 5\sqrt{27x}$ **30** $2\sqrt{20y} + 3\sqrt{45y}$

In Exercises 31–36, find the indicated products. Use the special products covered in Sections 1-4 and 1-5 wherever appropriate. Assume that all variables represent positive real numbers.

31 $(\sqrt{11} + \sqrt{7})(\sqrt{11} - \sqrt{7})$ **32** $(\sqrt{17} - \sqrt{2})(\sqrt{17} + \sqrt{2})$ **33** $(2\sqrt{x} - 3)^2$
34 $(3\sqrt{y} + 5)^2$ **35** $(a^{1/3} - 1)(a^{2/3} + a^{1/3} + 1)$ **36** $(\sqrt[3]{x} + 2)(\sqrt[3]{x^2} - 2\sqrt[3]{x} + 4)$

In Exercises 37–44, perform each indicated division by rationalizing the denominator and then simplifying. Assume that all variables represent positive real numbers.

37 $\dfrac{15}{\sqrt{5}}$ **38** $\dfrac{\sqrt{3}}{\sqrt{x}}$ **39** $\dfrac{2}{2 + \sqrt{3}}$ **40** $\dfrac{4}{2 - \sqrt{x}}$

41 $\dfrac{-20}{\sqrt{7} - \sqrt{2}}$ **42** $\dfrac{-6}{\sqrt{5} - \sqrt{3}}$ **43** $\dfrac{\sqrt{b}}{\sqrt{a} - \sqrt{b}}$ **44** $\dfrac{\sqrt{c}}{\sqrt{c} + \sqrt{d}}$

In Exercises 45–52, simplify each radical expression. Assume that all variables represent positive real numbers.

45 $\sqrt{50x^3}$ **46** $\sqrt{12x^5y^3}$ **47** $\sqrt[3]{54x^8y^{13}}$ **48** $\sqrt[3]{24w^7x^{11}}$
49 $\sqrt{\dfrac{3x}{4y^2}}$ **50** $\sqrt{\dfrac{14y}{50x^2}}$ **51** $\sqrt[3]{\dfrac{2v}{5w^2}}$ **52** $\sqrt[3]{\dfrac{3y}{7z^2}}$

In Exercises 53–56, simplify each radical expression by reducing the order of the radical. Assume that all variables represent positive real numbers.

53 $\sqrt[4]{25v^2w^2}$ **54** $\sqrt[6]{9a^2b^4}$ **55** $\sqrt[12]{a^3b^9}$ **56** $\sqrt[6]{8x^3y^3}$

B

57 Give a real number x for which $\sqrt{x^2} \neq x$.

58 The best mental estimate of $\dfrac{2 + \sqrt{15.99}}{3}$ is
  a. 1 **b.** 2 **c.** 3 **d.** 4 **e.** 5

59 The best mental estimate of $\dfrac{11 - \sqrt{25.03}}{2}$ is
  a. 1 **b.** 2 **c.** 3 **d.** 4 **e.** 5

60 The best mental estimate of $\dfrac{-7 + \sqrt{36.15}}{2}$ is
  a. -2 **b.** -1 **c.** -0.5 **d.** 0.5 **e.** 1

61 The best mental estimate of $\dfrac{-8 - \sqrt{63.87}}{4}$ is

a. -4 **b.** -2 **c.** 0 **d.** 2 **e.** 4

62 A scientific calculator key sequence and a graphics calculator key sequence are shown below. Which of the following expressions would be calculated by these key sequences?

S: [7] [y^x] [4] [1/x] [=]

G: [(] [7] [^] [4] [)] [x^{-1}] [ENTER]

a. $\dfrac{1}{7^4}$ **b.** $\dfrac{1}{4^7}$ **c.** 7^4 **d.** $\sqrt[4]{7}$ **e.** $\sqrt[7]{4}$

63 A scientific calculator key sequence and a graphics calculator key sequence are shown below. Which of the following expressions would be calculated by these key sequences?

S: [5] [x^y] [3] [+/−] [=]

G: [5] [^] [(−)] [3] [ENTER]

a. -5^3 **b.** $\dfrac{1}{5^3}$ **c.** $-\dfrac{1}{5^3}$ **d.** -5^{-3} **e.** $\dfrac{1}{5^{-3}}$

64 The time t for the period of a simple pendulum is given by $t = 2\pi\sqrt{l/g}$, where l is the length of the pendulum and g represents the acceleration of gravity. By observing changes in t, geologists in search of oil can detect changes in g caused by minerals beneath the surface. Find t if the length of the pendulum is 1.000 meters and g is 9.780 m/sec².

65 A company study revealed that the demand for its brand of energy control systems was related to the price of these systems by the following formula. For t units and a price of p dollars per unit, $t = (32{,}000 - p)^{2/3}$ for $0 < p < 32{,}000$. Find the demand for systems with a price of $5{,}000 each.

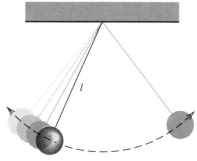

Figure for Exercise 64

C

In Exercises 66–73, simplify each expression. Assume that all variables represent positive real numbers.

66 $\sqrt{a^2 + 2ab + b^2} - \sqrt{a^2} - \sqrt{b^2}$

67 $\dfrac{4}{\sqrt{3}} - \sqrt{\dfrac{4}{3}}$

68 $\left(\dfrac{16a^{-8/9}}{81b^{-4/3}}\right)^{-3/4}$

69 $6\sqrt{45} - 7\sqrt{320} + 5\sqrt[3]{16} - 4\sqrt[3]{54}$

70 $\sqrt[3]{135w^7 x^9 y^{12}}$

71 $2\sqrt[3]{10xy}\,(7\sqrt[3]{25x^2} - 5\sqrt[3]{4y^2})$

72 $\dfrac{\sqrt{x-1} + \sqrt{x+1}}{\sqrt{x-1} - \sqrt{x+1}}$

73 $\dfrac{1}{\sqrt[3]{3} + 5}$ (*Hint:* Remember the special form $(a+b)(a^2 - ab + b^2) = a^3 + b^3$.)

SECTION 1-8 COMPLEX NUMBERS

In Exercises 74 and 75, use a calculator to evaluate each expression, accurate to four significant digits.

74 $\dfrac{\sqrt{7}-\sqrt{3}}{\sqrt{5}+\sqrt{2}}$

75 $(19.573)^{2/3}$

76 Heron's formula for the area A of a triangle with sides a, b, and c is

$$A = \sqrt{s(s-a)(s-b)(s-c)}$$

where $s = \dfrac{a+b+c}{2}$. Use this formula to compute the area of the triangle shown to the right.

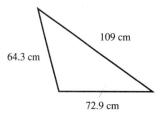

Figure for Exercise 76

SECTION 1-8

Complex Numbers

Section Objectives

16 Add, subtract, multiply, and divide complex numbers.

17 Represent complex numbers graphically.

The desire to solve problems is a primary motivation for new ideas. The history of our number system provides many examples of new numbers that were initially regarded with great suspicion because they had never been used before. Without the invention of rational numbers such as $\tfrac{7}{5}$ there would be no solution for the equation $5x = 7$. Similarly, we need irrationals such as $\sqrt{2}$ to solve $x^2 = 2$.

Even the set of all real numbers does not provide solutions for all equations. The equation $x^2 = -1$ has no real solution, since the square of a real number cannot be negative. The desire to solve these equations led to the definition of a number i so that $i^2 = -1$. Because they mistrusted these new numbers, mathematicians called them **imaginary numbers**—a hedge against committing themselves to this unproven idea. Although these numbers have been very useful, we unfortunately are still stuck with the misleading name "imaginary."

In 1893, Charles P. Steinmetz used imaginary numbers in his theory of alternating currents. Today imaginary numbers are used routinely in analyzing electrical circuits. We also use imaginary numbers to perform mathematical computations, even though the final answer may be a real number. This is analogous to using fractions in the computations for a problem whose answer is a natural number.

The Imaginary Number i ▼

$$i^2 = -1, \quad i = \sqrt{-1}$$

For $a > 0$, we define $\sqrt{-a}$ to be $i\sqrt{a}$. Then, using i, the real numbers, and the operations of addition, subtraction, multiplication, and division, we obtain numbers that can always be written in the form $a + bi$, where a and b are real numbers. Any number that has this form is called a **complex number** (a complex number is a combination of a real number and an imaginary number). If $b = 0$, then $a + bi$ is just the real number a. If $b \neq 0$, then $a + bi$ is called **imaginary**. If $a = 0$ and $b \neq 0$, bi is called **pure imaginary**. (Some texts use the term imaginary in a slightly different manner and do not use the phrase "pure imaginary.")

Complex Number ▼

If a and b are real numbers and $i = \sqrt{-1}$, then

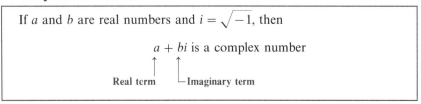

The examples in Table 1-7 illustrate how to classify complex numbers:

Table 1-7 Classifications of complex numbers

	Real Term	Coefficient of Imaginary Term	Classification
$3 - 4i$	3	-4	Imaginary
$6 = 6 + 0i$	6	0	Real
$-7i = 0 - 7i$	0	-7	Pure imaginary
$0 = 0 + 0i$	0	0	Real
$-\sqrt{25} = -5 + 0i$	-5	0	Real
$\sqrt{-25} = 0 + 5i$	0	5	Pure imaginary

The relationship between the real numbers and the complex numbers is illustrated in Figure 1-6.

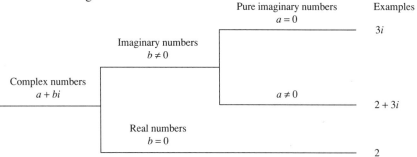

Figure 1-6 Subsets of the complex numbers

SECTION 1-8 COMPLEX NUMBERS

The operations that are defined for complex numbers $a + bi$ are similar to those used in working with binomials of the form $a + bx$. Although the rules are given for your reference, you should continue to perform the operations as if you were working with binomials rather than try to memorize these rules.

If $a + bi$ and $c + di$ are complex numbers, then

Equality: $a + bi = c + di$ if and only if $a = c$ and $b = d$; that is, both the real terms and the imaginary terms must be equal.

Addition: $(a + bi) + (c + di) = (a + c) + (b + d)i$

Subtraction: $(a + bi) - (c + di) = (a - c) + (b - d)i$

Multiplication: $(a + bi)(c + di) = (ac - bd) + (bc + ad)i$

Division: $\dfrac{a + bi}{c + di} = \dfrac{ac + bd}{c^2 + d^2} + \dfrac{bc - ad}{c^2 + d^2} i$

EXAMPLE 1 Simplify $(3 - 4i) + 6(8 - 9i) - 11(3 + 2i)$.

SOLUTION
$$
\begin{array}{rl}
3 - 4i \rightarrow & 3 - 4i \\
+ 6(8 - 9i) \rightarrow & 48 - 54i \\
- 11(3 + 2i) \rightarrow & -33 - 22i \\ \hline
& 18 - 80i
\end{array}
$$

Answer $(3 - 4i) + 6(8 - 9i) - 11(3 + 2i) = 18 - 80i$

EXAMPLE 2 Determine real values of a and b so that
$$a + bi = \sqrt{9} - \sqrt{-16} + \sqrt{-9}$$

SOLUTION $\sqrt{9} - \sqrt{-16} + \sqrt{-9} = 3 - \sqrt{16}\sqrt{-1} + \sqrt{9}\sqrt{-1}$
$$= 3 - 4i + 3i$$
$$= 3 - i$$

Answer $a = 3$ and $b = -1$.

▶ **Self-Check** ▼

Determine a and b so that these complex numbers will be equal.

1 $a + bi = \sqrt{-36} - \sqrt{64}$

2 $a + bi = 3(4 - 7i) - 8(5 - 2i)$

▶ **Self-Check Answers** ▼

1 $-8 + 6i$; $a = -8$, $b = 6$ **2** $-28 - 5i$; $a = -28$, $b = -5$

66 CHAPTER 1 REVIEW OF THE BASIC CONCEPTS OF ALGEBRA

Since $i^2 = -1$, higher powers of i can always be simplified to i, -1, $-i$, or 1. The first four powers of i are keys to simplifying higher powers to standard form and should therefore be memorized.

First Four Powers of i ▼

$$i^1 = i$$
$$i^2 = i \cdot i = -1$$
$$i^3 = i^2 \cdot i = (-1)i = -i$$
$$i^4 = i^2 \cdot i^2 = (-1)(-1) = 1$$

The powers of i repeat in cycles of four: i^4; i^8; i^{12}, and so on, all equal 1. We will use this fact to simplify i^n, where n is any integral exponent.

EXAMPLE 3 Simplify the following powers of i.

SOLUTIONS

(a) i^5 $i^5 = i^4 \cdot i = 1 \cdot i = i$

(b) i^6 $i^6 = i^4 \cdot i^2 = (1)(-1) = -1$

First extract the largest multiple of 4 from each exponent. Then simplify, replacing i^4, i^8, i^{12}, and so on, by 1.

(c) i^{45} $i^{45} = i^{44} \cdot i = (i^4)^{11}(i) = 1^{11} \cdot i = i$

(d) i^{-14} $i^{-14} = \dfrac{1}{i^{14}} = \dfrac{1}{i^{12}i^2} = \dfrac{1}{(1)(-1)} = -1$

Note that $i^{12} = (i^4)^3 = 1^3 = 1$.

(e) i^{1026} $i^{1026} = i^{1024} \cdot i^2$
$= (i^4)^{256} i^2$
$= (1)(-1)$
$= -1$

Note:
```
    256
4)1026
    8
    ‾‾
    22
    20
    ‾‾
    26
    24
    ‾‾
     2
```
$1026 = 4(256) + 2$

$i^{45} = \dfrac{i^{45}}{1} = \dfrac{i^{45}}{(i^4)^{11}} = i$

Because multiplication of complex numbers is so similar to multiplication of binomials, these products should be formed using the horizontal format.

EXAMPLE 4 Find these products.

SOLUTIONS

(a) $(5 + i)(6 + 2i)$ $(5 + i)(6 + 2i) = 30 + 16i + 2i^2$
$= 30 + 16i - 2$
$= 28 + 16i$

The middle term $16i$ is the sum of $5(2i) = 10i$ and $6i$.

SECTION 1-8 COMPLEX NUMBERS

(b) $(2 + i)^2$

$(2 + i)^2 = (2 + i)(2 + i)$
$= 4 + 4i + i^2$
$= 4 + 4i - 1$
$= 3 + 4i$

This product is the square of a sum. Replace i^2 with -1.

(c) $(11 - 8i)(11 + 8i)$

$(11 - 8i)(11 + 8i) = 121 - 64i^2$
$= 121 - 64(-1)$
$= 185$

A sum times a difference equals the difference of the squares. Replace i^2 with -1.

The **conjugate** of $a + bi$ is $a - bi$. Note that the product $(a + bi)(a - bi) = a^2 - b^2 i^2 = a^2 + b^2$ is a real number. This property plays an important role in the division of complex numbers, as shown in the next two examples.

Conjugate ▼

The conjugate of $a + bi$ is $a - bi$.

▶ **Self-Check ▼**

Simplify each expression.
1 i^7 **2** i^{80}
3 i^{33} **4** i^{66}
5 $(8 + 3i)(8 - 3i)$

EXAMPLE 5 Find these quotients.

SOLUTION

(a) $\dfrac{16 - 11i}{2 - 3i}$

$\dfrac{16 - 11i}{2 - 3i} = \dfrac{16 - 11i}{2 - 3i} \cdot \dfrac{2 + 3i}{2 + 3i}$

$= \dfrac{32 + 26i - 33i^2}{4 - 9i^2}$

$= \dfrac{65 + 26i}{13}$

$= 5 + 2i$

Multiply the numerator and the denominator by the conjugate of the denominator.

(b) $(-1 + 3i) \div i$

$(-1 + 3i) \div i = \dfrac{-1 + 3i}{i}$

$= \dfrac{-1 + 3i}{i} \cdot \dfrac{-i}{-i}$

$= \dfrac{+i - 3i^2}{-i^2}$

$= \dfrac{3 + i}{1}$

$= 3 + i$

The conjugate of $0 + i$ is $0 - i$.

Replace i^2 with -1.

▶ **Self-Check Answers ▼**

1 $-i$ **2** 1 **3** i **4** -1 **5** 73

Observe that division of complex numbers is similar to rationalizing radicals. Conjugates are used in both cases. In fact, we could have interpreted the division problem $\dfrac{a+bi}{c+di}$ as a problem involving radicals, since $i = \sqrt{-1}$.

We must note that some rules given earlier cannot be extended and are false when used for inappropriate values. For example, if x and y are *both* negative, then $\sqrt{xy} \neq \sqrt{x}\sqrt{y}$. For instance, if $x = -4$ and $y = -9$, then

$$\sqrt{(-4)(-9)} = \sqrt{36} = 6$$

but

$$\sqrt{-4}\sqrt{-9} = (2i)(3i) = 6i^2 = 6(-1) = -6$$

Thus $\sqrt{(-4)(-9)} \neq \sqrt{-4}\sqrt{-9}$. Likewise, $\sqrt{\dfrac{x}{y}} \neq \dfrac{\sqrt{x}}{\sqrt{y}}$ if just y is negative. For instance, if $x = 36$ and $y = -4$, then

$$\sqrt{\dfrac{36}{-4}} = \sqrt{-9} = 3i$$

but

$$\dfrac{\sqrt{36}}{\sqrt{-4}} = \dfrac{6}{2i} = \dfrac{6}{2i} \cdot \dfrac{-2i}{-2i} = \dfrac{-12i}{-4i^2} = \dfrac{-12i}{4} = -3i$$

Thus

$$\sqrt{\dfrac{36}{-4}} \neq \dfrac{\sqrt{36}}{\sqrt{-4}}.$$

▶ **Self-Check** ▼

1 Write the conjugate of $-5 + 9i$.
2 Divide $2i$ by $1 + i$.

EXAMPLE 6 Evaluate $\dfrac{-b + \sqrt{b^2 - 4ac}}{2a}$ for $a = 1$, $b = -3$, and $c = 4$.

SOLUTION
$$\dfrac{-b + \sqrt{b^2 - 4ac}}{2a} = \dfrac{-(-3) + \sqrt{(-3)^2 - 4(1)(4)}}{2(1)}$$

$$= \dfrac{3 + \sqrt{9 - 16}}{2}$$

$$= \dfrac{3 + \sqrt{-7}}{2}$$

$$= \dfrac{3 + i\sqrt{7}}{2}$$

$$= \dfrac{3}{2} + \dfrac{\sqrt{7}}{2}i \qquad \square$$

▶ **Self-Check Answers** ▼

1 $-5 - 9i$ **2** $1 + i$

SECTION 1-8 COMPLEX NUMBERS

EXAMPLE 7 Verify that $1 + 2i$ is a solution of $x^2 - 2x + 5 = 0$.

SOLUTION
$$x^2 - 2x + 5 = 0$$
$$(1 + 2i)^2 - 2(1 + 2i) + 5 \stackrel{?}{=} 0 \quad \text{Substitute } 1 + 2i \text{ for } x \text{ and then simplify.}$$
$$(1 + 4i + 4i^2) - (2 + 4i) + 5 \stackrel{?}{=} 0$$
$$-3 + 4i - 2 - 4i + 5 \stackrel{?}{=} 0$$
$$0 = 0 \text{ checks} \qquad \square$$

Just as the number line is used as a model of the real numbers, a pair of perpendicular lines can be used to create a complex plane that is a model for the complex numbers. The horizontal axis is called the **real axis** and the vertical axis is called the **imaginary axis**.

To graph the complex number $a + bi$, we plot the ordered pair (a, b). Figure 1-7 shows how the complex numbers $3 + 4i$, $3 - 4i$, 4, and $-2i$ are plotted on the complex plane.

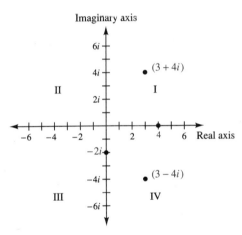

Figure 1-7 Complex numbers plotted in complex plane

#1-43 eoo, 45, 51, 53, 55, 57, 59, 61

EXERCISES 1-8

A

In Exercises 1–10, identify the real values for a and b that will make each statement true.

1. $a + bi = 18 - 5i$
2. $a + bi$ is the conjugate of $-3 + 8i$.
3. $a + bi$ is the conjugate of $7 - 13i$.
4. $7 + 3bi = a - 15i$
5. $a + bi = \sqrt{-16} - \sqrt{-25}$
6. $a + bi = \sqrt{49} - \sqrt{-81}$
7. $a + bi = \frac{2}{3}(3 - 6i)$
8. $4(a + bi) = -12 + 44i$
9. $a + bi = -\sqrt{17}$
10. $a + bi = \sqrt{-17}$

In Exercises 11–44, perform the indicated operations and express the result in standard form.

11. $(1 + 2i) + (8 - 3i)$
12. $(7 + i) - (5 - 2i)$
13. $2(i + 3) - 3(1 - i)$
14. $5(-2 + 7i) - 4(-3 + 8i)$
15. $2\sqrt{-75} + \sqrt{-27}$
16. $3\sqrt{-8} - 5\sqrt{-98}$
17. $\sqrt{4} + \sqrt{-9} - \sqrt{9} - \sqrt{-25}$
18. $\sqrt{64} + \sqrt{-36} - \sqrt{9} - \sqrt{-1}$
19. $\sqrt{-8} - \sqrt[3]{-8}$
20. $-\sqrt{-64} + \sqrt[3]{-64}$
21. $2(5 - 9i) + 7(-2 + 3i) + \frac{1}{2}(6 - 8i)$
22. $5(3 - 7i) + 4(-6 + 2i) + \frac{1}{3}(9 + 6i)$
23. $(2i)(3i) + (-i)^2$
24. $(5i)(7i) + (-4i)(11i)$
25. $(-8i)(-3i) + i(3 - i)$
26. $\left(-\frac{i}{4}\right)(12i) - i(-2 + 3i)$
27. $(2 - 7i)(2 + 7i)$
28. $(6 + 9i)(6 - 9i)$
29. $(5 - 2i)(4 + 7i)$
30. $(6 + i)(3 - 5i)$
31. $(5 + i)^2$
32. $(6 - i)^2$
33. $\dfrac{4}{1 + i}$
34. $\dfrac{6}{1 - i}$
35. $\dfrac{4 - i}{4 + i}$
36. $\dfrac{5 + i}{5 - i}$
37. $\dfrac{85}{7 - 6i}$
38. $\dfrac{-10 + 22i}{2i}$
39. $i^6 - i^8$
40. $i^{15} + i^7$
41. $2i^{64} - 3i^{75}$
42. $4i^{-3} - 6i^{-5}$
43. $i^{7007} + (-i)^{109}$
44. $2i^4 + (-2i)^5$

45. Plot these complex numbers on the complex plane:
    a. $2 - 5i$ b. $2 + 5i$ c. $3i$ d. -4

46. Plot these complex numbers on the complex plane:
    a. $5 + 2i$ b. $5 - 2i$ c. 3 d. $-4i$

B

In Exercises 47–50, perform the indicated operations and express the result in standard form.

47. $(1 - 2i)^3$
48. $(1 + 2i)^3$
49. $\left(\dfrac{-1}{2} + \dfrac{\sqrt{3}}{2}i\right)^3$
50. $\left(\dfrac{-1}{2} - \dfrac{\sqrt{3}}{2}i\right)^3$

51. Evaluate $\sqrt{b^2 - 4ac}$ for $a = 3$, $b = 2$, and $c = 1$.
52. Evaluate $\dfrac{-b - \sqrt{b^2 - 4ac}}{2a}$ for $a = 2$, $b = 1$, and $c = 1$.
53. Evaluate $\dfrac{-b - \sqrt{b^2 - 4ac}}{2a}$ for $a = 3$, $b = -1$, and $c = 1$.
54. Determine if $1 - 2i$ is a solution of $x^2 - 2x + 5 = 0$.
55. Determine if $2 - i$ is a solution of $x^2 - 4x + 5 = 0$.
56. Determine if $2 + i$ is a solution of $x^2 - 4x + 5 = 0$.

C

57. Give two complex numbers whose square is -25.
58. Give two complex numbers whose square is -121.
59. Verify that $(-1 + i\sqrt{3})$ is a cube root of 8.
60. Verify that $(-1 - i\sqrt{3})$ is a cube root of 8.
61. List the three cube roots of 8, and graph these roots on the complex plane.

look at problem 59, 60

KEY CONCEPTS

1. Using $\subseteq$ to indicate a subset, the relationship of some of the subsets of the complex numbers is:
   Natural numbers $\subseteq$ whole numbers $\subseteq$ integers $\subseteq$ rational numbers $\subseteq$ real numbers $\subseteq$ complex numbers.

2. For a real number x and a positive real number a:

   $|x| = a$ is equivalent to $x = -a$ or $x = a$.
   $|x| < a$ is equivalent to $-a < x < a$.
   $|x| > a$ is equivalent to $x < -a$ or $x > a$.

3. $\dfrac{a}{0}$ is undefined; 0^0 is undefined.

4. Order of operations:
   Step 1 Start with the expression within the innermost pair of grouping symbols.
   Step 2 Perform all exponentiations.
   Step 3 Perform all multiplications and divisions, working from left to right.
   Step 4 Perform all additions and subtractions, working from left to right.

5. Properties of exponents:
   For nonzero real numbers x and y,
   Product rule: $x^m x^n = x^{m+n}$
   Quotient rule: $\dfrac{x^m}{x^n} = x^{m-n}$
   Power rule: $(x^m)^n = x^{mn}$
   Product to a power: $(xy)^m = x^m y^m$
   Quotient to a power: $\left(\dfrac{x}{y}\right)^m = \dfrac{x^m}{y^m}$
   $\left(\dfrac{x}{y}\right)^{-n} = \left(\dfrac{y}{x}\right)^n$

6. Factoring special forms:
   1. $a^2 - b^2 = (a+b)(a-b)$
   2. $a^2 + 2ab + b^2 = (a+b)^2$
   3. $a^2 - 2ab + b^2 = (a-b)^2$
   4. $a^3 + b^3 = (a+b)(a^2 - ab + b^2)$
   5. $a^3 - b^3 = (a-b)(a^2 + ab + b^2)$
   6. $a^3 + 3a^2b + 3ab^2 + b^3 = (a+b)^3$
   7. $a^3 - 3a^2b + 3ab^2 - b^3 = (a-b)^3$

7 Rational exponents:
  1. $x^{m/n} = (\sqrt[n]{x})^m = \sqrt[n]{x^m}$ for natural numbers m and n and real numbers x.
  2. $\sqrt{x^2} = |x|$

8 Properties of radicals:
For positive values of a and b,
  1. $\sqrt[n]{ab} = \sqrt[n]{a}\,\sqrt[n]{b}$
  2. $\sqrt[n]{\dfrac{a}{b}} = \dfrac{\sqrt[n]{a}}{\sqrt[n]{b}}$
  3. $\sqrt[n]{a^n} = a$

9 $a + bi$ is a complex number for real numbers a and b and $i = \sqrt{-1}$.

REVIEW EXERCISES FOR CHAPTER 1

In Exercises 1–18 simplify each expression. Assume all denominators and all bases are not zero.

1 $3x^0 - 5y^0 + (3x)^0 - (5y)^0 - (3x - 5y)^0$

2 $144^{1/2} - 169^{1/2} - (169 - 144)^{1/2}$

3 $8^{-2/3} + 32^{2/5}$

4 $(6\sqrt{11} + 5\sqrt{2})(6\sqrt{11} - 5\sqrt{2})$

5 $3(4\sqrt{13} - 2\sqrt{5}) - 4(5\sqrt{13} - 13\sqrt{5})$

6 $3\sqrt{27} + 6\sqrt{45} - 2\sqrt{48} - \sqrt{75}$

7 $|17 - 5(13 - 7)|$

8 $\dfrac{15}{\sqrt{19} - \sqrt{14}}$

9 $(5 - 3i)^2$

10 $\dfrac{26}{3 + 2i}$

11 $2i^{19}$

12 $((\sqrt{2})^{\sqrt{2}})^{\sqrt{2}}$

13 $(3x + 4y)^2 - [(3x)^2 + (4y)^2]$

14 $(v + w)^3 - (v^3 + w^3)$

15 $(2a - b)^4 - (16a^4 + b^4)$

16 $2xy[(x + y)^2 - (x - y)^2]$

17 $\left[\dfrac{(-3vw^4)^2(-2vw^2)^3}{36v^2w^5}\right]^{-1}$

18 $\dfrac{12a^3b^2c^2 - 18a^2b^3c^2 + 30a^2b^2c^5}{6a^2b^2c^2}$

In Exercises 19–31, factor each polynomial completely over the integers. If the polynomial cannot be factored, indicate that it is prime.

19 $30a^3b^2 - 45a^2b^3$

20 $7v^3 - 63v$

21 $125y^3 + 8$

22 $10a^2 - ab - 21b^2$

23 $49m^2 + 140mn + 100n^2$

24 $4ax^3 - 28ax^2 + 24ax$

25 $a(2x + 3y) - b(2x + 3y)$

26 $x^2 - xy - 5x + 5y$

27 $x^3 + 3x^2y + 3xy^2 + y^3$

28 $x^2 - y^2 + 6y - 9$

29 $a^4 + 2a^2 + 9$

30 $(a + 3b)^2 + 6(a + 3b) + 5$

31 $x^{2m} - 25$

In Exercises 32–39, simplify each expression. Assume all denominators and all bases are not zero.

32 $\dfrac{15m - 3n}{9mn^2} \cdot \dfrac{121m^2n}{55m - 11n}$

33 $\dfrac{v^2 - w^2}{vw^2} \cdot \dfrac{v^2}{vw + w^2} \div \dfrac{v^2 - vw}{w^4}$

34 $\dfrac{a}{a^2 - 1} + \dfrac{1}{a^2 - 1}$

35 $\dfrac{9n}{4m^2 - 9n^2} - \dfrac{3}{2m - 3n} - \dfrac{m}{3n^2 + 2mn}$

36 $\dfrac{\dfrac{w}{w - 1} - \dfrac{1}{w + 1}}{\dfrac{w}{w^2 - 1}}$

REVIEW EXERCISES FOR CHAPTER 1

37 $\left(\dfrac{x^{-1}+1}{x-x^{-1}}\right)^{-1}$

38 $\dfrac{2y^2+2y}{y^2+y+1} + \dfrac{2y^4}{y^3-1} \cdot \dfrac{y^2-2y+1}{y^5-y^4}$

39 $\dfrac{ac}{(b-a)(b-c)} + \dfrac{ac}{(a-b)(a-c)} - \dfrac{bc}{(a-c)(b-c)}$

In Exercises 40–46, simplify each expression, assuming the variables are positive.

40 $\sqrt{x^2+2xy+y^2}$

41 $\sqrt{75x^5y^7z^9}$

42 $\sqrt[3]{54x^9y^{12}}$

43 $\dfrac{x^2-y^2}{\sqrt{x}-\sqrt{y}}$

44 $\sqrt{45x} + 3\sqrt{500x}$

45 $(\sqrt[3]{5}+1)(\sqrt[3]{25}-\sqrt[3]{5}+1)$

46 $\left[\dfrac{x^{-1/2}y^{3/7}z^{2/7}}{xy^{5/7}z^{1/7}}\right]^{-14}$

In Exercises 47–52, represent each interval using interval notation.

47 $-3 \leq x < 5$
48 $x < 6$
49 $x \geq -4$
50 $|x| < 3$
51 $|x| \geq \pi$
52 $(-3, 7] \cap [1, 8)$

53 List the elements of the set $S = \{-93, -9.3, -\pi, \sqrt{-3}, -\sqrt{2}, 0, 5i, 2, 2+5i, 1+\sqrt{2}, e, 5.\overline{7}, \frac{117}{19}\}$ that are members of each of the following sets.

   a. natural numbers   **b.** whole numbers   **c.** integers   **d.** rational numbers

   e. irrational numbers   **f.** real numbers   **g.** pure imaginary numbers   **h.** complex numbers

54 List all of the prime numbers between 62 and 72.

55 Give an example of a real number x such that $|x| > x$.

56 Give an example of real numbers x and y such that $|x+y| < |x| + |y|$.

57 Give an example of real numbers x and y such that $(x+y)^2 > x^2 + y^2$.

58 Give an example of a real number x such that $\dfrac{1}{x} > x$.

59 Give an example of a real number x such that $x > x^2$.

In Exercises 60–64, identify the property that justifies each equality.

60 $5(x+2y) + 3z = 5(2y+x) + 3z$

61 $5(x+2y) + 3z = 3z + 5(x+2y)$

62 $5(x+2y) + 3z = (5x+10y) + 3z$

63 $5(x+2y) + 3z = (x+2y)5 + 3z$

64 $(x+y+z)(0) = 0$

65 Are the odd numbers closed under addition?

66 Are the odd numbers closed under multiplication?

In Exercises 67 and 68 use a calculator to evaluate each expression accurately to four significant digits. (See the material on calculators in Appendix A if you have a question on proper calculator usage.)

67 $\sqrt{179.45} + \sqrt{47.897} + \sqrt{179.45 + 47.897}$

68 $(1{,}734{,}260{,}000{,}000)^2 - (457{,}135{,}000)^3$

In Exercises 69 and 70 a student's answer to a calculus problem is given along with the book's answer. Factor the variable with the smallest exponent from the student's answer and simplify to obtain the book's answer. Problem 70 is from Laurence D. Hoffmann, *Calculus for Business, Economics, and the Social and Life Sciences*, fourth edition, New York: McGraw-Hill Book Company, 1989.

	Student's Answer	Book's Answer
69	$-3x^{-2} + 6x^{-3}$	$-\dfrac{3(x-2)}{x^3}$
70	$2 - 8x^{-2}$	$\dfrac{2(x+2)(x-2)}{x^2}$

MASTERY TEST FOR CHAPTER 1

Exercise numbers correspond to Section Objective numbers.

1. **a.** What prime number is between 50 and 58?
   b. Which of the following numbers is a rational number?
   $$1 - \sqrt{3}, \quad \pi, \quad \tfrac{22}{7}, \quad \pi + 1, \quad 1 + \sqrt{3}$$
   c. List each natural number whose absolute value is less than 3.
   d. The temperature in Chicago at noon is recorded to the nearest degree. Would the number of degrees be best described as a natural number, a whole number, an integer, or an even number?

2. Represent these intervals using interval notation.
   a. $-1 < x \leq 3$ **b.** $2 \leq x \leq 7$ **c.** $x < 5$ **d.** $3 < x < 4$ or $x \geq 6$

3. Identify the property that justifies each statement.
   a. $(x + y)z = z(x + y)$
   b. $(x + y)z = xz + yz$
   c. $(x + y)z = (y + x)z$
   d. If $x < 7$ and $7 < y$, then $x < y$.

4. Simplify each expression.
   a. $(7 - 5)^2 + 7^2 - 5^2$
   b. $(7x + 5y)^0 + (7x)^0 + (5y)^0 + 5y^0$
   c. $3^{-1} + 4^{-1} + (3 + 4)^{-1}$
   d. $-1^{50} + 50^{-1}$

5. Simplify each expression.
   a. $(2x^2y^3)^5$
   b. $(3x^{-2}y^4)^{-2}$
   c. $\dfrac{(4x^2y^{-3})^2}{(2x^{-1}y^2)^{-3}}$
   d. $\left(\dfrac{24x^{-3}y^5}{18x^{-4}y^3}\right)^{-2}$

6. Simplify each expression.
   a. $5 + 4 \cdot 3^2 - 12 \div 6$
   b. $\dfrac{5 + 6 \cdot 10}{7 - 2(-3)}$
   c. $8 - 9[5 + 4(2^3 - 3) + 6]$

7. Write each number in scientific notation.
   a. 4,513,000
   b. 0.000176

8. For the polynomial $7x^3y^2 - 18xy^3 + y^4$, determine the following information:
   a. the degree of the third term
   b. the degree of the polynomial
   c. the classification of the polynomial according to the number of its terms
   d. the coefficient of the second term

9. Perform the indicated operations.
   a. $(2v - 13)^2$
   b. $(5v - 7)(5v + 7)$
   c. $2(3a - 4b + c) - 6(7a - 2b - 5c)$
   d. $(2x^2 + 3x - 20)(x + 4)$
   e. $(2x^2 + 3x - 20) \div (x + 4)$
   f. $(a - 2b)^3$

10. Factor each polynomial completely over the integers.
    a. $10ab^2 + 25ab - 15a$
    b. $490bx^2 - 40b$
    c. $28 - 7y^2$
    d. $8w^3 - 1$
    e. $9x^2 - 6x + 1 - 25y^2$
    f. $av^4 + 3av^2 + 4a$

11. Reduce each rational expression to lowest terms.
    a. $\dfrac{3a^2 + 14ab - 5b^2}{3a^2 - 4ab + b^2}$
    b. $\dfrac{12x^2 + 24xy + 12y^2}{16x^2 - 16y^2}$

MASTERY TEST FOR CHAPTER 1

12 Simplify each expression.

a. $\dfrac{x^2 - 3x + 2}{x^2 - 4x + 4} \div \dfrac{x^2 - 2x + 1}{3x^2 - 12}$

b. $\left(\dfrac{25w^2 - wz}{5w - z} - w\right)\left(5 - \dfrac{z^2}{5w^2}\right)$

c. $\dfrac{\dfrac{m^2}{n^2} - \dfrac{n^2}{m^2}}{\dfrac{m}{n} + \dfrac{n}{m}}$

13 Simplify each expression

a. $64^0 + 64^{1/2} + 64^{1/3} + 64^{1/6}$

b. $\sqrt[3]{24}$

c. $(-8)^{2/3} - 27^{2/3}$

14 Simplify each expression.

a. $\sqrt{24}$

b. $\sqrt[6]{8}$

c. $\sqrt{\dfrac{36}{81}}$

d. $\sqrt{\dfrac{25}{3}}$

15 Simplify each expression.

a. $3\sqrt{500} - 7\sqrt{45} + 3\sqrt{605}$

b. $(\sqrt{2} + \sqrt{3})^2 + (\sqrt{2})^2 + (\sqrt{3})^2$

c. $(2\sqrt{5} - 3\sqrt{7})(2\sqrt{5} + 3\sqrt{7})$

d. $\dfrac{75(\sqrt{11} + 1)}{3\sqrt{11} - 7}$

16 Write each expression in the standard form $(a + bi)$.

a. $3(7 + 4i) - 7(4 - 3i)$

b. $(8 - 5i)(8 + 5i)$

c. $\dfrac{12}{4 - 3i}$

d. $i^5 - 5i$

17 Plot the points on the complex plane corresponding to $2 - 4i$ and to its conjugate.

CHAPTER TWO

Equations and Inequalities

Chapter Two Objectives

1. Identify an equation as either an identity, a contradiction, or a conditional equation.
2. Solve linear equations in one variable.
3. Solve an equation for a specified variable.
4. Apply formulas to solve word problems.
5. Translate word problems into equations.
6. Solve any quadratic equation.
7. Use the discriminant to determine the nature of the roots of a quadratic equation.
8. Write a quadratic equation given its solutions.
9. Solve equations of quadratic form.
10. Solve radical and fractional equations.
11. Solve applied problems using linear and quadratic equations.
12. Solve linear and absolute value inequalities.
13. Solve nonlinear inequalities using a sign graph.

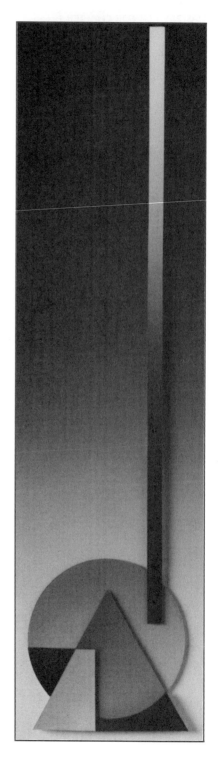

Emmy Noether, 1882–1935

Noether, a German, is known as the most creative woman mathematician in history. After Adolph Hiltler came to power, she left Germany and accepted a position at Bryn Mawr College in the United States. Noether is known for her elegant mathematical formulations for some general relativity concepts and for her work with the formal properties of algebra.

One of the most important objectives of algebra is to solve equations. Many of the problems that you will encounter outside the classroom will not be stated as precisely as those in textbooks. Thus your first task is to analyze both the question and the given information so that you can restate the problem concisely and unambiguously. This chapter will help you systematically develop the skills needed to form the *word equation* and then to translate this word equation into algebraic form. The restatement is called a **mathematical model** of the problem.

This chapter covers the basic techniques for solving linear, quadratic, and absolute value equations and inequalities.

SECTION 2-1

Linear Equations

Section Objectives

1. Identify an equation as an identity, a contradiction, or a conditional equation.
2. Solve linear equations in one variable.
3. Solve an equation for a specified variable.

An equation is a statement that two quantities are equal.* This statement of equality may be either true or false. If the equation contains a variable, it may be true for some values of the variable but false for other values. Such an equation is called a **conditional equation** because its truth depends on the conditions—that is, on the values of the variables. An equation that is always true is an **identity**, and an equation that is always false is a **contradiction**.

Most of our work consists of determining values for variables that will make conditional equations true. These values are said to **satisfy** the equation and are called **roots** or **solutions** of the equation. The set of all solutions is the **solution set**.

---

* The first recorded use of the equality symbol ($=$) was by Robert Recorde in 1557.

EXAMPLE 1 Identify each equation as an identity, a contradiction, or a conditional equation.

SOLUTIONS

(a) $4x = 12$ — There is a solution: $4(3) = 12$. Since 3 is the only solution, this equation is conditional.

(b) $w = 2w$ — The only root of this equation is 0. The solution set of this conditional equation is $\{0\}$.

(c) $2x + 3x = 5x$ — Since this statement is true for each real number that can be substituted for x, this equation is an identity. The solution set is the set of all real numbers.

(d) $w = w + 1$ — No value of w will satisfy this equation. Since there are no solutions, it is a contradiction. The solution set is the empty or null set, denoted by $\emptyset$. □

As the last two parts of Example 1 illustrate, the presence of a variable in the equation does not necessarily mean that the equation is conditional. Some equations containing variables are contradictions, whereas other equations containing variables are identities. Identities containing variables can be used to state important properties. For example, the identity $a(b + c) = ab + ac$ is used to state the distributive property.

An equation in which each variable term is in the first degree is called a **first-degree equation**. First-degree equations in two variables have straight lines as graphs, so it has become customary to refer to all first-degree equations as **linear equations**. Every linear equation in one variable can be rewritten in the form $ax = b$, where x represents the variable and a and b are real constants.

Examples are given below to clarify this terminology:

- $6y = 30$ is a linear equation, since y has an exponent of 1.
- $w^2 = 36$ is not a linear equation, since it is not first degree. The exponent on w is 2.
- $3v - 5 = 18 - 4v$ is a linear equation in one variable. The only variable is v and the degree is 1. We can rewrite this equation as $7v = 23$.
- $13x - 5 = 4y + 7$ is a linear equation that contains two variables, x and y.

The general strategy for solving a linear equation, regardless of its complexity, is to isolate the variable terms on one side of the equation and all other terms on the other side. As we work toward our goal of isolating the variable, we form new—but simpler—equations. These equations should be **equivalent** to the given equation; that is, they should have the same

SECTION 2-1 LINEAR EQUATIONS

solution set. Two properties that are used to solve many equations are given in the following box; they are a direct result of the theorems of equality in Chapter 1.

▶ Equivalent Equations ▼

If a, b, and c are real numbers, then the equation
1. $a = b$ is equivalent to the equation $a + c = b + c$.
2. $a = b$ is equivalent to the equation $ac = bc$ for $c \neq 0$.

▶ Self-Check ▼

Identify each equation as an identity, a contradiction, or a conditional equation.
1. $a + a = a$
2. $a - a = 2$
3. $a + a = 2a$

EXAMPLE 2 Solve $7(w - 1) - 4(2w + 3) = 2(w + 1) - 3(3 - w)$.

SOLUTION
$$7(w - 1) - 4(2w + 3) = 2(w + 1) - 3(3 - w)$$
$$7w - 7 - 8w - 12 = 2w + 2 - 9 + 3w \quad \text{Use the distributive property to simplify both sides.}$$
$$-w - 19 = 5w - 7$$
$$-w - 19 + 19 = 5w - 7 + 19 \quad \text{Next isolate the variable terms.}$$
$$-w = 5w + 12$$
$$-w - 5w = 5w + 12 - 5w$$
$$-6w = 12 \quad \text{Then solve for } w.$$
$$\frac{-6w}{-6} = \frac{12}{-6}$$
$$w = -2$$

Check
$$7(w - 1) - 4(2w + 3) = 2(w + 1) - 3(3 - w)$$
$$7(-2 - 1) - 4[2(-2) + 3] \stackrel{?}{=} 2(-2 + 1) - 3[3 - (-2)]$$
$$7(-3) - 4(-1) \stackrel{?}{=} 2(-1) - 3(5)$$
$$-21 + 4 \stackrel{?}{=} -2 - 15$$
$$-17 = -17 \text{ checks.}$$

Answer The solution is -2; the solution set is $\{-2\}$.

Note: The equations at each step in this solution are equivalent; that is, $w = -2$ is the solution of each equation. The properties produce simpler equations at each step, but the solution set is unchanged. □

▶ Self-Check Answers ▼

1. Conditional equation   2. Contradiction   3. Identity

EXAMPLE 3 Solve $2(x + 1) = 2(x - 1)$.

SOLUTION
$$2(x + 1) = 2(x - 1)$$
$$2x + 2 = 2x - 2$$
$$2 = -2 \text{ is a contradiction.}$$

Answer The equation has no solution. □

If a linear equation contains fractions, then you may first simplify the equation by converting it to an equivalent equation that does not involve fractions.

Solving Equations Containing Fractions ▼

Step 1 Determine the LCD (least common denominator) of all the terms of the equation and list all excluded values.

Step 2 Multiply both sides of the equation by the LCD. (Caution: to accomplish this, multiply each term by the LCD.)

Step 3 Solve the resulting equivalent equation by isolating the variable on one side of the equation.

Step 4 Check for extraneous values.

EXAMPLE 4 Solve $\dfrac{3v - 3}{6} = \dfrac{4v + 1}{15} + 2$ and check the solution.

SOLUTION

[1] First determine the LCD.
$$6 = 2 \cdot 3$$
$$15 = 3 \cdot 5$$
$$\text{LCD} = 2 \cdot 3 \cdot 5 = 30$$

[2] Next multiply both sides by this LCD.
$$30\left(\frac{3v - 3}{6}\right) = 30\left(\frac{4v + 1}{15} + 2\right)$$

[3] Then solve this equation.
$$5(3v - 3) = 2(4v + 1) + 60$$
$$15v - 15 = 8v + 2 + 60$$
$$15v - 15 = 8v + 62$$
$$15v = 8v + 77$$
$$7v = 77$$
$$v = 11$$

SECTION 2-1 LINEAR EQUATIONS

[4] This solution does check, as you can verify.

Answer $v = 11$

If the LCD of a fractional equation contains a variable, then it is possible that the solution process will produce an **extraneous value**—that is, a number that is *not* a solution of the original equation. Extraneous values occur when we produce nonequivalent equations (equations with different solution sets) in the solution process. For example, if $c = 0$, then $a = b$ may not be equivalent to $ac = bc$.

▶ **Self-Check** ▼

Solve these equations.
1. $(4k - 3) - 2(k + 4) = 3(k + 7)$
2. $3z + 9 = 3(z + 3)$
3. $5(w + 1) = 5w + 1$
4. Solve $\dfrac{2w}{7} = 1 - \dfrac{2w + 1}{3}$.

EXAMPLE 5 Solve $\dfrac{2x + 3}{x - 5} = \dfrac{2 - 3x}{5 - x}$.

SOLUTION
LCD: $x - 5$ Note that $5 - x = -(x - 5)$.
Excluded value: $x = 5$ This value cannot be a solution since $x = 5$ causes division by zero.

$(x - 5)\left(\dfrac{2x + 3}{x - 5}\right) = (x - 5)\left(\dfrac{2 - 3x}{5 - x}\right)$ Multiply by the LCD, $x - 5$, for $x \neq 5$.

$2x + 3 = -(2 - 3x)$
$2x + 3 = 3x - 2$
$x = 5$ This is an excluded value.

Answer There is no solution for this equation.   The solution set is the null set.

A value obtained by solving an equation with variables in the denominator must always be checked to determine if this value causes division by zero and is therefore an extraneous value.

Nonequivalent equations can also arise from squaring both sides of an equation. The equation $\sqrt{x} = -2$ has no real solution; however, if we square both sides of this equation, we obtain $x = 4$. Since this equation has the obvious solution 4, we note that squaring both sides of an equation can produce an equation with more solutions than the original equation has.

▶ **Self-Check Answers** ▼

1. $k = -32$ 2. Identity, all real numbers are solutions
3. Contradiction, no solution   4. $w = \dfrac{7}{10}$

Power Rule ▼

> If A and B are algebraic expressions and n is a natural number, then any solution of $A = B$ is also a solution of $A^n = B^n$.
> *Caution*: The equations $A = B$ and $A^n = B^n$ are not always equivalent. The equation $A^n = B^n$ may have a solution that is not a solution of $A = B$.

Since raising both sides of an equation to a power can produce extraneous values, all possible solutions of a radical equation must be checked to determine if they are extraneous.

EXAMPLE 6 Solve these equations and check for extraneous values.

SOLUTIONS

(a) $\sqrt{5 - x} = 3$

$\sqrt{5 - x} = 3$
$5 - x = 9$ Square both sides.
$-x = 4$
$x = -4$

Check $\sqrt{5 - x} = 3$
$\sqrt{5 - (-4)} \stackrel{?}{=} 3$
$\sqrt{9} \stackrel{?}{=} 3$
$3 = 3$ checks.

Answer $x = -4$

(b) $\sqrt{31 - 2x} = -5$

$\sqrt{31 - 2x} = -5$
$31 - 2x = 25$ Square both sides.
$-2x = -6$
$x = 3$

Check $\sqrt{31 - 2x} = -5$
$\sqrt{31 - 2(3)} \stackrel{?}{=} -5$
$\sqrt{25} \stackrel{?}{=} -5$
$+5 = -5$ does not check.

Answer   No solution; the solution set is the null set.   □

The equation $\sqrt{31 - 2x} = -5$ in Example 6(b) can be solved by inspection. Because a principal square root can never be negative, this equation has no solution.

Many well-known formulas such as $I = PRT$ (Interest = Principal · Rate · Time) are commonly used in business and all the sciences. The equation

SECTION 2-1 LINEAR EQUATIONS

$I = PRT$ is particularly convenient if we wish to calculate interest I, given P, R, and T. However, if we wish to calculate the interest rate R, then the form $R = \dfrac{I}{PT}$ is more convenient. The properties of equality can be used to create alternative forms of literal equations that contain more than one variable. We can specify which variable we wish to be the subject of the statement of equality and then transpose the equality to write it in the desired form. This is called **transposing an equation** or **solving for a specified variable**.

Solving Linear Equations for a Specified Variable ▼

Step 1 Simplify each member of the equation. This may include using the distributive property to eliminate fractions and to remove parentheses.

Step 2 Isolate the terms containing the specified variable in one member of the equation.

Step 3 Solve the equation produced in Step 2. This may include factoring out the specified variable and dividing by its coefficient.

EXAMPLE 7 Solve $2(3h - b) = 7(2h + b) - 5$ for h.

SOLUTION

[1] Simplify the equation using the distributive property.

$$2(3h - b) = 7(2h + b) - 5$$
$$6h - 2b = 14h + 7b - 5$$

[2] Isolate h on the left side using the addition-subtraction property.

$$6h - 2b + 2b = 14h + 7b - 5 + 2b$$
$$6h = 14h + 9b - 5$$
$$6h - 14h = 14h + 9b - 5 - 14h$$
$$-8h = 9b - 5$$

[3] Solve for h using the multiplication-division property.

$$\dfrac{-8h}{-8} = \dfrac{9b - 5}{-8}$$
$$h = \dfrac{9b - 5}{-8}$$

Answer $h = -\dfrac{9b - 5}{8}$

EXAMPLE 8 Solve $2yy' = xy' + y$ for y'.

SOLUTION

$2yy' = xy' + y$ Note that y and y' (read "y prime") are distinct variables.

$2yy' - xy' = y$ Put all terms involving y' in the left member.

$y'(2y - x) = y$ Factor out the common factor y'.

$\dfrac{y'(2y - x)}{2y - x} = \dfrac{y}{2y - x}$ Divide both members by $2y - x$ (for $2y - x \neq 0$).

$y' = \dfrac{y}{2y - x}$ For $x \neq 2y$

Answer $y' = \dfrac{y}{2y - x}$ □

Translating from English into algebra is an essential skill for solving applied problems. An intermediate step in this process is to write an equation using both algebraic symbols and words.

EXAMPLE 9 Four less than six times a certain number is ten more than four times the same number. Find this number.

SOLUTION Let $n = $ The number sought Identify the unknown quantity with a variable.

$\underbrace{4 \text{ less than } 6 \text{ times } n}_{6n - 4} \underbrace{\text{ is }}_{=} \underbrace{10 \text{ more than } 4 \text{ times } n}_{4n + 10}$

Write an equation using algebraic symbols and words. Translate this *word equation* into an algebraic equation.

$2n = 14$ Solve the equation.

$n = 7$

Answer The number is 7. Does this number satisfy the given conditions? □

The English language has many key words to indicate the operations of addition, subtraction, multiplication, division, and exponentiation. Sometimes we use the word "variation" to suggest an operation when we are describing how the change in one quantity affects the change in another quantity. Three common types of variation are given in the following box.

Variation ▼

If x, y, and z are variables and k is a constant,

Direct variation $y = kx^n$ means that "y varies directly as the nth power of x" or that "y is directly proportional to the nth power of x."

Inverse variation $y = \dfrac{k}{x^n}$ means that "y varies inversely as the nth power of x" or that "y is inversely proportional to the nth power of x."

Joint variation $z = kxy$ means that "z varies directly as the product of x and y" or that "z varies jointly as x and y."

The following examples illustrate some commonly used statements of variation:

▶ **Self-Check** ▼

Solve $h = k + (n - 2)m$ for n.

- Hooke's law states that the distance d a spring stretches is directly proportional to the force f applied. In algebraic form this is written as $d = kf$. See Figure 2-1.

- The volume V of a sphere varies directly as the cube of the radius r. This could be written $V = kr^3$. In fact, the constant is $\frac{4\pi}{3}$ and the formula is written as $V = \frac{4\pi r^3}{3}$. See Figure 2-2.

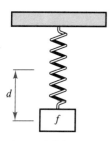

Figure 2-1

Figure 2-2

- The current I in an electrical circuit is inversely proportional to the resistance R. This is written $I = \frac{k}{R}$. See Figure 2-3.

- The volume of a cylinder varies jointly as the square of its radius and its height: $V = kr^2h$. In fact, k is π, and the formula is written as $V = \pi r^2 h$. See Figure 2-4.

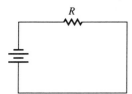

Figure 2-3

Figure 2-4

▶ **Self-Check Answer** ▼

$n = \dfrac{h - k + 2m}{m}$

EXAMPLE 10 A brokerage company interested in increasing its sales volume V claimed that its sales per month varied directly with the time t spent talking to clients on the telephone. For one month its sales were $203,000, with an average of 7 minutes of conversation per client for the month. If this claim is accurate, what would the sales volume be if the time spent averaged 10 minutes on the phone with each client per month?

SOLUTION Let $V =$ Sales volume in thousands of dollars

$t =$ Time in minutes of phone conversation per client

Then

Sales volume varies directly as time

$V = k \cdot t$ "Varies directly" means that the volume V is the product of a constant k and the time t.

$203 = k(7)$ Substitute the given values and solve for the constant k.

$k = 29$

$V = 29t$

$V = 29(10)$ Substitute 10 for the time and solve for the volume V.

$V = 290$

Answer Sales volume of $290,000   □

Additional applications of variation will be given in later exercise sets as we solve different types of equations.

EXERCISES 2-1

A

In Exercises 1–4, solve each linear equation and check the solution.

1 $4x = 20 - x$
2 $2p + 5 = -15 - 2p$
3 $7(m - 2) - 2(3 + m) = 80$
4 $2(3y - 5) = 7(2y + 5) - 5$

In Exercises 5–18, identify each equation as a contradiction, an identity, or a conditional equation. Solve the conditional equations.

5 $3a - 6 = 4a - 6$
6 $5b + 2 = 3b - 8$
7 $x + 3 = x - 3$
8 $28t = 27t$
9 $5x = 3x + 2x$
10 $\frac{5}{7}v = \frac{3}{5}$
11 $\frac{3}{5}v = 15$
12 $4.9m = 2.9m - 4.36$
13 $2.3w = 1.3w - 7.2$
14 $3n - 7 + (n + 1) = 11n + 1$
15 $2 - 6(s + 1) = 4(2 - 3s) + 6$
16 $2(z + 5) - 2(5 - z) - 4z = 8$
17 $3(2w - 5) + (3 - 6w) + 24 = 12$
18 $7[3(1 - 2n) + 4(n + 1)] = 8 + 3(2 - 4n) - 4n$

SECTION 2-1 LINEAR EQUATIONS

In Exercises 19–26, solve each equation.

19 $\dfrac{w}{3} - \dfrac{1}{6} = \dfrac{w+1}{9}$

20 $\dfrac{2(t+1)}{3} = \dfrac{2(t-5)}{5}$

21 $\dfrac{3a+5}{a-2} = \dfrac{2a+7}{a-2}$

22 $\dfrac{8-3b}{b+7} = \dfrac{5b-8}{b+7}$

23 $\dfrac{5(c-6)}{2c-3} = \dfrac{6(c-5)}{3-2c}$

24 $\dfrac{5(c+1)}{3c-2} = \dfrac{2(c+3)+1}{2-3c}$

25 $\dfrac{12}{5-v} - \dfrac{5}{v+3} = \dfrac{15}{v+3}$

26 $\dfrac{1}{(w+1)(w-2)} - \dfrac{1}{w+1} + \dfrac{2}{2-w} = 0$

In Exercises 27–36, solve each equation for the indicated variable.

27 $p = a + b + c$ for b

28 $F = \tfrac{9}{5}C + 32$ for C

29 $A = \tfrac{1}{2}bh$ for b

30 $V = lwh$ for h

31 $2mn + n = m$ for m

32 $2bx = b + 1$ for b

33 $A = \tfrac{1}{2}h(a+b)$ for a

34 $3(x + a) = b$ for x

35 $x^2y' + 2xy = 3xy^2y' + y^2$ for y'

36 $y' = 2x^2yy' + 2xy$ for y'

In Exercises 37–44, solve each equation. (Checking for extraneous roots is part of the solution process.)

37 $\sqrt{x-3} = 6$

38 $\sqrt{x+6} = 4$

39 $\sqrt{x+5} = -3$

40 $\sqrt{7-3x} = -8$

41 $\sqrt{11-x} = \sqrt{7+x}$

42 $\sqrt{3x-2} = \sqrt{5x-8}$

43 $\sqrt{5-x} = \sqrt{11-x}$

44 $\sqrt{2x-3} = \sqrt{6x-29}$

In Exercises 45–54, give the algebraic equation that can be formed from each problem. Then solve this equation.

45 Six more than twice a number is equal to 2 increased by the number. Find the number.

46 Six times a number decreased by five times the number is equal to 10. Find the number.

47 If the sum of a number and 7 is divided by 9, the quotient is 3. What is this number?

48 If the sum of a number and -3 is divided by 10, the quotient is 4. What is this number?

49 y varies directly as x. When x is 12, y is 18. Find y when x is 20.

50 y varies inversely as x. When x is 12, y is 3. Find y when x is 4.

51 z varies jointly as x and y. When x is 10 and y is 8, z is 16. Find z when x is 5 and y is 11.

52 Hooke's law states that the distance a spring stretches is directly proportional to the force applied. If a five-pound weight stretches a spring three inches, what weight is required to stretch the spring five inches?

53 The weight of an object relatively near planet Earth varies inversely as the square of its distance from the center of the Earth. An astronaut weighs approximately 175 pounds on the surface of the Earth. How much will this astronaut weigh when the spacecraft he is in is 400 miles above the surface of the Earth? Assume the radius of the Earth is approximately 4000 miles.

54 The volume of a right pyramid varies jointly as the height and the area of the base. Find the constant of variation of a pyramid with a base of 24 cm², a height of 8 centimeters, and a volume of 64 cm³. (See the figure shown to the right.)

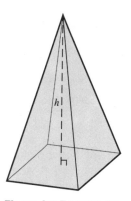

Figure for Exercise 54

B

Exercises 55–58, have two parts. In part (a) solve the equation, and in part (b) perform the indicated operations and simplify the result.

55 a. $\dfrac{2}{x+1} - \dfrac{1}{x-1} = 0$ b. $\dfrac{2}{x+1} - \dfrac{1}{x-1}$ 56 a. $\dfrac{6}{2y-3} + \dfrac{2}{1-y} = 0$ b. $\dfrac{6}{2y-3} + \dfrac{2}{1-y}$

57 a. $\dfrac{t}{2(t+2)} + \dfrac{1}{2} - \dfrac{2t}{t+2} = 0$ b. $\dfrac{t}{2(t+2)} + \dfrac{1}{2} - \dfrac{2t}{t+2}$ 58 a. $\dfrac{u-1}{3u-5} + \dfrac{1}{3} = 0$ b. $\dfrac{u-1}{3u-5} + \dfrac{1}{3}$

59 Give an example of real numbers a, b, and c for which $ac = bc$ but $a \neq b$.

60 Is the equation $x = 5$ equivalent to the equation $x - 2 = 3$?

61 Is the equation $x = 5$ equivalent to the equation $x^2 = 25$?

62 Is the equation $\sqrt{x} = -7$ equivalent to the equation $x = 49$?

63 A manager of a business hypothesizes that sales volume is directly proportional to the advertising budget. For one month the sales volume was $90,000 and the advertising budget was $600. For the next month, the advertising budget was increased to $1000. What must the sales volume be to verify the manager's hypothesis?

64 The pressure P exerted on a submerged object is directly proportional to the depth the object is submerged. If the pressure at 0.6 meter is 5880 pascals, find the pressure at 0.8 meter.

C

In Exercises 65–72, translate each algebraic equation into words.

65 $7a = 14$ 66 $c + 9 = 4$ 67 $2v + 5 = 6v$ 68 $4w = w - 13$

69 $\tfrac{1}{2}x = 9$ 70 $\tfrac{3}{4}y = y + 1$ 71 $2(a+b) = 2a + 2b$ 72 $3(c-d) = 3c - 3d$

In Exercises 73 and 74, use a calculator to solve each equation to the nearest hundredth.

73 $6.72w - 1819.41 = 2376.83 - 5.03w$ 74 $4.17(3.1t + 57) = 2\pi(2.2t - 16)$

SECTION 2-2

Using Formulas and Forming Equations

Section Objectives

4 Apply formulas to solve word problems.

5 Translate word problems into equations.

The first step in solving any word problem is to read through it carefully until you know exactly what you are being asked to find. Often it is helpful to rephrase the problem, condensing it to its essential points. You must understand the equation verbally before you can write it algebraically. We

SECTION 2-2 USING FORMULAS AND FORMING EQUATIONS

will refer to this concise wording of the equation as the *word equation*. In many cases you will recognize that the equation involves a well-known formula. A formula offers two advantages. It not only provides the needed equation, but also provides a standard labeling for the variables. Most formulas use representative letters, such as A for area, V for volume, and I for interest. A list of well-known formulas, including those for the areas of common regions and the volumes of common solids, is given on the inside back cover of this book.

Many problems concerning perimeter, area, and volume involve shapes that can be separated into more basic shapes for which we know the formulas. This basic strategy is illustrated in the next example.

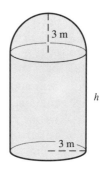

EXAMPLE 1 The total volume of the radar facility shown in the figure to the right is 280 m³. Find the height of this facility to the nearest tenth of a meter.

SOLUTION

$\left(\begin{array}{c}\text{Volume of}\\ \text{cylindrical base}\end{array}\right) + \left(\begin{array}{c}\text{Volume of}\\ \text{hemispherical top}\end{array}\right) = \text{Total volume}$ *Word equation*

$\pi r^2 h + \frac{1}{2}(\frac{4}{3}\pi r^3) = V$ Write the volume using formulas for the volume of a cylinder and a hemisphere. (A hemisphere is half of a sphere.)

$\pi r^2 h = V - \frac{2}{3}\pi r^3$ Isolate the term involving h in the left member.

$h = \dfrac{V - \frac{2}{3}\pi r^3}{\pi r^2}$ Divide both members by the coefficient of h and simplify.

$h = \dfrac{V}{\pi r^2} - \dfrac{2}{3}r$

$h = \dfrac{\boxed{280}}{\pi(\boxed{3})^2} - \dfrac{2}{3}(\boxed{3})$ Substitute the given values and then simplify.

$h = \dfrac{280}{9\pi} - 2$

$h \approx 7.9029742$

Answer   The height is approximately 7.9 meters.   □

EXAMPLE 2 Determine the yearly interest rate for a $7000 investment that earned $507.50 interest in one year.

SOLUTION   Interest = Principal · Rate · Time   *Word equation*

$I = PRT$ Simple interest formula

$R = \dfrac{I}{PT}$ First solve for R.

$R = \dfrac{\boxed{507.50}}{\boxed{7000}(\boxed{1})}$ Substitute the given values for I, P, and T. (The time is 1 year.)

$R = 0.0725$

Answer   The interest rate is 7.25%.   □

If a word problem that you are solving does not fit a familiar formula, you may need to select your own variable to represent the unknown quantity. First read through the problem carefully until you know exactly what you are trying to find and then precisely identify this quantity. Choose a representative variable for this quantity, such as *c* for cost or *p* for profit. Do not get into the habit of always using the variable *x*. Using a representative variable will give you a sharper focus on the precise quantity you are trying to find. This habit will also be helpful in any computer programming you might do.

Forming an equation from a stated problem is generally more difficult than actually solving the equation. Thus it is helpful to rephrase the problem and to condense it into a *word equation*. This concise wording can then be translated into an algebraic equation. A general strategy for solving word problems is given below. A summary of this strategy is given in the box that follows.

▶ **Self-Check** ▼

Determine the radius of the semicircular ends of the track shown below. The perimeter of this track is 400 meters.

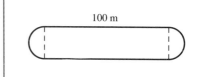

Strategy for Solving Word Problems

Step 1   **a.** *Determine what you are asked to find.* Quickly read through the problem for an overview, and then reread the problem until you understand what you are asked to find. Look for a question mark or the word "how," "what," "find," "determine," or "evaluate."

b. *Clearly identify what you are trying to find with a variable or variables.* It is helpful to use appropriately chosen letters such as *c* for cost and *w* for width. When feasible, make a sketch to clarify the relationship between the variables, or organize the information given into a table.

Step 2   **a.** *Form the word equation.* Determine the key concept or principle on which the equality is based. Rephrase the key idea or concept clearly and concisely. For example,

$$\text{Profit} = \text{Revenue} - \text{Cost}$$

or

$$\text{Area} = \text{Length} \cdot \text{Width}$$

b. *Translate the word equation into an algebraic equation.* Look for key words that suggest operations. Identify the word or phrase that indicates equality.

Step 3   *Solve the equation.* Find the variable identified in Step 1 and any other quantities asked for.

Step 4   *Check the reasonableness of your answer.* For example, mentally check to determine that prices are reasonable, that sizes are realistic, and so on.

▶ **Self-Check Answer** ▼

Total perimeter = Length of sides + Length of ends. The radius is $100/\pi$ m ≈ 31.8 m.

SECTION 2-2 USING FORMULAS AND FORMING EQUATIONS

Summary of Word Problem Strategy ▼

Step 1 a. Determine what you are asked to find.
b. Identify this numeric value with an appropriately chosen variable.
Step 2 a. Form the *word equation*.
b. Translate this word equation into an algebraic equation.
Step 3 Solve this equation and answer the question asked.
Step 4 Check the reasonableness of your answer.

EXAMPLE 3 The sum of two consecutive integers is three times the next consecutive integer. Find these three integers.

SOLUTION

1 Let n = The smallest of these three integers
$n + 1$ = The second of these three integers   Consecutive integers differ by 1.
$n + 2$ = The third of these three integers

2 $\left(\begin{array}{c}\text{First}\\\text{integer}\end{array}\right) + \left(\begin{array}{c}\text{Second}\\\text{integer}\end{array}\right) = \left(\begin{array}{c}\text{Three times the}\\\text{third integer}\end{array}\right)$ Word equation

$\quad\quad n \quad\quad + \quad (n+1) \quad = \quad\quad 3(n+2)$ Algebraic equation

3 $2n + 1 = 3n + 6$ Simplify and solve for the three integers.
$-n = 5$
$n = -5$
$n + 1 = -4$
$n + 2 = -3$

4 Are the integers $-5, -4,$ and -3 reasonable answers?

Answers The consecutive integers are $-5, -4,$ and -3. □

EXAMPLE 4 One of two supplementary angles is twice as large as the other angle. Find the number of degrees in each angle.

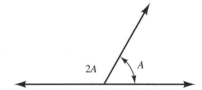

SOLUTION Let A = Number of degrees in the smaller angle
$2A$ = Number of degrees in the larger angle

$\left(\begin{array}{c}\text{Number of degrees}\\\text{in the smaller angle}\end{array}\right) + \left(\begin{array}{c}\text{Number of degrees}\\\text{in the larger angle}\end{array}\right) = 180$

The *word equation* is based on the fact that two angles are supplementary if the sum of their measures is 180°.

$A + 2A = 180$ Algebraic equation
$3A = 180$ Simplify and solve for both angles.
$A = 60$
$2A = 120$

Answers The smaller angle is $60°$, and the larger angle is $120°$.   Are these answers reasonable? □

EXAMPLE 5 A company's fixed overhead (building rental, insurance, etc.) is $10,000 per month. The company produces small windmills that cost $3250 each to produce. How many windmills can they produce if $172,500 is budgeted to cover all their monthly expenses?

SOLUTION   Let w = Number of windmills for the month   *The question is "How many windmills?"*

$\quad\quad$ Total cost = Overhead cost + Production cost   *Word equation*

$\quad\quad\quad$ $172{,}500 = 10{,}000 + 3250w$ *Algebraic equation.*

$\quad\quad\quad$ $162{,}500 = 3250w$ *Simplify and solve for w.*

$\quad\quad\quad\quad\quad$ $50 = w$

Answer   They can produce 50 windmills per month.   *Is this answer reasonable?*
$\quad\quad\quad\quad\quad\quad\quad\quad\quad\quad\quad\quad\quad\quad\quad\quad\quad\quad\quad\quad\quad\quad\quad\quad$ ☐

EXAMPLE 6 An orange paint is a mixture of four parts of yellow paint for every five parts or red paint. Determine the number of liters (L) of each paint to mix in order to obtain a total of 301.5 liters.

SOLUTION   Let y = Number of liters of yellow paint   *There are five parts of red paint for each four parts of yellow paint. (See the Self-Check on page 93.)*

$\quad\quad\quad\quad$ $\dfrac{5}{4}y$ = Number of liters of red paint

$\left(\begin{array}{c}\text{Liters of}\\ \text{yellow paint}\end{array}\right) + \left(\begin{array}{c}\text{Liters of}\\ \text{red paint}\end{array}\right) = \left(\begin{array}{c}\text{Liters of paint}\\ \text{in the mixture}\end{array}\right)$ *Word equation*

$\quad\quad\quad\quad\quad\quad y + \dfrac{5}{4}y = 301.5$ *Algebraic equation*

$\quad\quad\quad\quad\quad\quad\quad \dfrac{9}{4}y = 301.5$ *Simplify and solve for y.*

$\quad\quad\quad\quad\quad\quad\quad\quad y = \dfrac{4}{9}(301.5)$

$\quad\quad\quad\quad\quad\quad\quad\quad y = 134$

$\quad\quad\quad\quad\quad\quad\quad \dfrac{5}{4}y = 167.5$

Answers   Use 134 liters of yellow paint and 167.5 liters of red paint.   *Are these answers reasonable?*
$\quad\quad\quad\quad\quad\quad\quad\quad\quad\quad\quad\quad\quad\quad\quad\quad\quad\quad\quad\quad\quad\quad\quad\quad\quad\quad\quad\quad\quad$ ☐

EXAMPLE 7 A $5000 investment was split into two parts. Part of the principal, p, was invested at 7%, and the rest of the principal was invested at 8%. How much was invested at each rate if the interest for one year was $362.50?

SOLUTION   Let p = Principal invested at 7%

$\quad\quad\quad\quad$ $5000 - p$ = Principal invested at 8%   *If one part of the 5000 is p, the other part is $5000 - p$.*

Interest at 7% + Interest at 8% = Total interest   *Word equation*

SECTION 2-2 USING FORMULAS AND FORMING EQUATIONS

Interest Formula:	$P \cdot R = I$
7% investment	$p \cdot 0.07 = 0.07p$
8% investment	$(5000 - p) \cdot 0.08 = 400 - 0.08p$

$$0.07p + (400 - 0.08p) = 362.50$$
$$-0.01p = -37.50$$
$$p = 3750$$
$$5000 - p = 1250$$

This is the interest formula $I = P \cdot R \cdot T$, with T omitted since the time is one year.

Substitute the values from the table into the word equation. Then simplify and solve for p.

Answers $3750 was invested at 7% and $1250 was invested at 8%.

Are these answers reasonable? ☐

▶ **Self-Check** ▼

Complete the following table for a paint that contains four parts of yellow for every five parts of red.

Liters of Yellow	Liters of Red	Total Liters
4	5	9
8		
	15	
1		
y		

EXAMPLE 8 The base of a right triangle is 12 meters. The hypotenuse is 2 meters more than the height. Find the height.

SOLUTION Let
  h = Height of the triangle in centimeters
  $h + 2$ = Length of the hypotenuse in centimeters
  (Length of base)² + (Height)² = (Length of hypotenuse)²

$$(12)^2 + h^2 = (h + 2)^2$$
$$144 + h^2 = h^2 + 4h + 4$$
$$140 = 4h$$
$$h = 35$$

The *word equation* is based on the Pythagorean theorem.

Algebraic equation

Simplify and solve for h.

Answer The height of the triangle is 35 centimeters.

Is this answer reasonable? ☐

▶ **Self-Check Answer** ▼

Liters of Yellow	Liters of Red	Total Liters	
4	5	9	
8	10	18	← Multiply row one by 2
12	15	27	← Multiply row one by 3
1	$\frac{5}{4}$	$\frac{9}{4}$	← Divide row one by 4
y	$\frac{5}{4}y$	$\frac{9}{4}y$	← Multiply row four by y

EXERCISES 2-2

A

All the formulas needed to work these exercises are included on the inside back cover of this book. Use the π key on your calculator whenever this is appropriate. Assume that all curved portions are either circles or semicircles. For each exercise write the *word equation* and the algebraic equation that you use to solve the problem. In Exercises 1–3, determine the area of each region to the nearest hundredth of a square centimeter.

1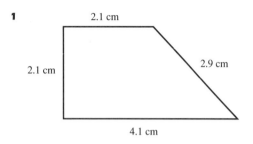

2 Shaded portion only:
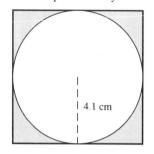

3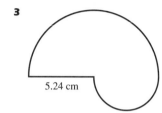

4 Determine the surface area of the top of the steel wall of the pipe in the figure below. The inner radius is 15 centimeters and the thickness of the pipe is 1 centimeter.

5 The volume of the grain bin shown below is 515.2 m³. The radius of the base is 4 meters, and the height of the conical top is 0.75 meter. Determine to the nearest hundredth of a meter the height of the cylindrical portion of this grain bin.

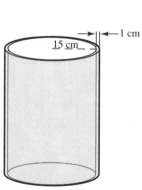

Figure for Exercise 4

Figure for Exercise 5

6 The volume of the container shown at right is 1204.1 cm³. Determine to the nearest hundredth of a centimeter the length of the cylindrical portion of this container.

Figure for Exercise 6

SECTION 2-2 USING FORMULAS AND FORMING EQUATIONS

7 The volume of steel in the pipe (shown shaded in the figure below) is 255.6 cm³. Determine to the nearest tenth of a centimeter the length of this pipe.

8 The volume of insulation in the walls of the heat duct (shown shaded in the figure below) is 512.6 cm³. Determine to the nearest tenth of a centimeter the length of this portion of the heat duct.

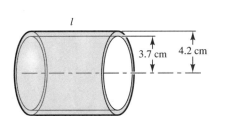

Figure for Exercise 7

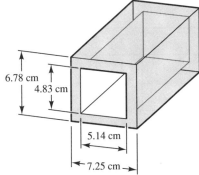

Figure for Exercise 8

9 Determine the principal required to earn $625 interest in one year at an interest rate of 6.75%.

10 What interest rate is necessary for an investment of $4000 to earn $375 interest in one year?

11 One of two supplementary angles is three times as large as the other. Find the number of degrees in each of these angles.

12 One of two complementary angles is 32° larger than the other. Find the number of degrees in each of these angles.

13 The first of three consecutive integers is twice the sum of the next two. Find these integers.

14 The sum of two consecutive integers is two more than seven times the next integer. Find these integers.

15 The sum of three consecutive odd integers is 99. Find these integers.

16 The sum of two consecutive odd integers is four times as large as the next odd integer. Find these integers.

17 The fixed overhead for a small business is $6000 per month. The business makes tabletops that cost $35 each to produce. How many tabletops can it produce if $34,000 is budgeted to cover all its monthly expenses?

18 A company that makes custom chairs has a fixed monthly overhead of $5500. It costs the company $55 to produce and sell each chair. The price received for each chair is $75. How many chairs must the company produce and sell each month to make a profit of $3000?

19 All tickets to a country and western concert were sold for $6.50 apiece. The concert hall charged a $12,000 fee for rental, electricity, security, etc.; the booking agency charged 75¢ per ticket for advertising, insurance, etc. If the performers had a profit of $59,300 after these expenses, how many tickets were sold?

20 Two partners decide to sell their combined interest in a business for $77,000. If one owns one-half of the business and the other owns a one-fifth interest, how much is the entire business worth?

21 A trust fund was established with the plan of investing $3 in secure grade A+ tax-free bonds for each dollar invested in speculative stocks. For a fund of $120,000, how much would be invested in bonds?

22 A mixute of two chemicals requires three parts of chemical A for every four parts of chemical B.

  a. Use this information to complete the table below.

Chemical A	3	6		1	a
Chemical B	4		40		
Mixture	7				

  b. Write the *word equation* for a total mixture of 80.5 liters.

  c. Use parts (a) and (b) of this problem to determine the amount of each chemical needed to produce 80.5 liters of this mixture.

23 A supplier for a construction company prepares a mixture that consists of three parts of sand for every five parts of clay.

  a. Use this information to complete the table below.

Pounds of Sand	3	6		1	s
Pounds of Clay	5		25		
Pounds of Mixture	8				

  b. Write the *word equation* for a total mixture of 3600 pounds.

  c. Use parts (a) and (b) of this problem to determine the amount of sand and of clay needed to produce 3600 pounds of this mixture.

24 a. Complete the following table for pairs of numbers whose sum is 27.

Larger Number	26	16		l
Smaller Number	1		5	
Their Sum	27	27	27	27
Their Difference	25			

  b. Determine the pair of numbers whose sum is 27 and whose difference is 15.

25 a. Complete the following table for pairs of numbers whose difference is 14.

Larger Number	27	20		
Smaller Number	13		5	s
Their Difference	14	14	14	14
Their Sum	40			

  b. Determine the pair of numbers whose difference is 14 and whose sum is 42.

SECTION 2-2 USING FORMULAS AND FORMING EQUATIONS

B

26. The difference between the length and the width of a rectangle is seven centimeters. Find the dimensions of the rectangle if the perimeter is 53 centimeters.

27. The United States Post Office defines the size of a package as the girth plus the length. The girth is the perimeter of the end. The size of the package shown to the right is 48 inches. Its width is twice its height, and its length is three times its width. Determine the height of this package.

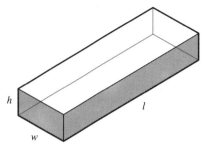

Figure for Exercise 27

28. A homeowner borrowed $60,000 from two sources to finance a home. The interest rates on the two mortages were 9.5% and 10.25%. The total monthly interest charges for the first month were $481.25. Determine the amount of each mortgage. (One month is treated as $\frac{1}{12}$ of a year by these lenders.)

29. A retiree deposited a total of $7500 in two different savings accounts. The interest rates on these two accounts were 6.0% and 6.5%. At the end of the first year, the total interest earned on these two investments was $456.25. Determine the amount deposited in each account.

30. A chemistry experiment requires two parts of chemical A for each part of chemical B and three parts of chemical C for each part of chemical B. If 36 milliliters of the mixture is required, how much of each chemical should be used?

31. A lubricant is formed by mixing two oils from the warehouse. The mixture contains three parts of oil A for every eight parts of oil B. Determine the number of liters of each oil to mix in order to obtain 880 liters of this lubricant.

32. The arc length s of an arc varies directly as the measure of the angle θ. If an angle of $1°$ forms an arc 2 centimeters long, what size angle will form an arc of 7 centimeters?

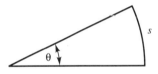

Figure for Exercise 32

33. A manager of a business hypothesizes that sales volume is directly proportional to the advertising budget. For one month the sales volume was $90,000 and the advertising budget was $150. For the next month the advertising budget was increased to $400. What must the sales volume be to verify the manager's hypothesis?

34. The load that a beam of a given depth and width can support varies inversely as the length of the beam. If a 10-foot beam can support a load of 12 tons, what weight can a similar 15-foot beam support? (See figure below.)

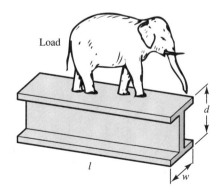

35 The bottom of the ladder shown below is 3.3 meters from the bottom of the wall. If the length of the ladder from the ground to the wall is only 0.9 meter more than the height of the wall, determine the height of the wall.

36 Two airplanes departed from the same airport, one flying due south and the other flying due east. The plane flying east traveled 480 kilometers. The plane flying south traveled only 180 kilometers less than the current distance between the planes. What is the current distance between the planes?

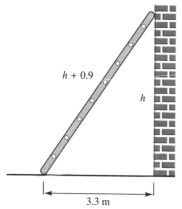

Figure for Exercise 35

Figure for Exercise 36

In Exercises 37–45, select an appropriate variable for the problem, form the *word equation*, and translate this word equation into an algebraic equation. Do *not* solve the equation at this time; only practice setting up these exercises.

37 The difference of the squares of two consecutive even integers is 52. Find the smaller of these integers.

38 The sum of the squares of three consecutive odd integers is 515. Find these integers.

39 The length of a side of the larger square in the figure to the right is two centimeters longer than the length of a side of the smaller square. If the total area of these two squares is 290 cm², find the length of each side of the smaller square.

40 The length of a side of the outer square shown in the figure to the right is two centimeters longer than the length of a side of the inner square. If the shaded area between the squares is 52 cm², find the length of each side of the inner square.

41 The radius of the larger circle shown in the figure to the right is three centimeters less than twice the radius of the smaller circle. Find the radius of the smaller circle if the area between the circles is 57.75 cm².

42 A plumber examining the blueprints for a rectangular room that is 1 foot longer than it is wide determines that a pipe run diagonally across this room will be 29 feet long. What is the width of this room?

43 The hypotenuse of a right triangle is two meters longer than the longest leg and nine meters longer than the shortest leg. Find the length of the shortest leg.

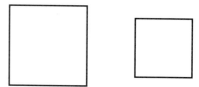

Figures for Exercise 39

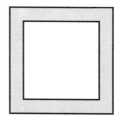

Figure for Exercise 40

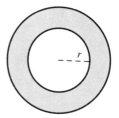

Figure for Exercise 41

44 Sixty meters of string is used to outline a rectangular play area of 189 m². Find the width and the length of this area.

45 The volume of steel in a pipe whose length is 200 centimeters is 15,700 cm³. The thickness of the walls of this pipe is one centimeter. Determine to the nearest tenth of a centimeter the outer radius of this pipe.

Figure for Exercise 45

SECTION 2-3

Quadratic Equations

Section Objectives

6 Solve any quadratic equation.

7 Use the discriminant to determine the nature of the roots of a quadratic equation.

8 Write a quadratic equation given its solutions.

The importance of quadratic equations was recognized by ancient civilizations, which then devised methods for solving them. It is known that the Babylonians were able to solve these equations as early as 2000 B.C.

> **Quadratic Equation ▼**
>
> If a, b, and c are real constants with $a \neq 0$, then
>
> $$\underset{\text{Quadratic term}}{ax^2} + \underset{\text{Linear term}}{bx} + \underset{\text{Constant term}}{c} = 0$$
>
> is a quadratic equation in x.

The form $ax^2 + bx + c = 0$ is called **standard form**. If we allowed the coefficient a to be zero, then the equation would not be quadratic, but instead would be the linear equation $bx + c = 0$. Examples are given below to clarify this terminology:

- $5y - 6 = -y^2$ is a quadratic equation in y and can be written in standard form as $y^2 + 5y - 6 = 0$, with $a = 1$, $b = 5$, and $c = -6$. It is customary to write the equation so that the quadratic coefficient is positive.
- $5v^2 = 3v - 4$ is a quadratic equation in v and can be written in standard form as $5v^2 - 3v + 4 = 0$. The quadratic term is $5v^2$, the linear term is $-3v$, and the constant term is 4.

- $5x = 12$ is a linear equation, not a quadratic equation.
- $w^2 + 7w = w^3 - 5$ is a third-degree equation, not a quadratic equation.
- $x^2 + 3x + 7$ is not an equation, but merely a polynomial expression.

A quadratic equation can have either two **distinct solutions** or one repeated solution, which is referred to as a **double root** or a root of **multiplicity two**. The equation $x^2 = 9$ has two distinct solutions, -3 and 3, whereas $x^2 = 0$ has 0 as a double root. Since $\sqrt{R}$ denotes the principal square root of R, we also need to use $-\sqrt{R}$ to denote the negative square root of R. The symbol $\pm$, read "plus or minus," is used to denote both roots.

If $b = 0$ in the standard form of a quadratic equation, then the equation can be written as $ax^2 + c = 0$. Since this equation can also be written as $x^2 = \dfrac{-c}{a}$, we can solve for x by taking the square roots of both sides. This method of solving the equation is called **extraction of roots**.

▶ **Self-Check** ▼

Identify a, b, and c in these quadratic equations.
1. $x^2 - 4x = -11$
2. $3v = v^2 + 4$
3. $5y^2 = 27$

EXAMPLE 1 Solve these quadratic equations.

SOLUTION

(a) $5x^2 - 95 = 0$

$x^2 = \dfrac{95}{5}$ Solve for x^2.

$x^2 = 19$

$x = \pm\sqrt{19}$ Take the square roots of both sides.

$x = -\sqrt{19}$ or $x = \sqrt{19}$ Note: There are two distinct solutions.

(b) $y^2 + 4 = 0$

$y^2 = -4$ Solve for y^2.

$y = \pm\sqrt{-4}$ Take the square roots of both sides.

$y = \pm 2i$

$y = -2i$ or $y = 2i$ Note: There are two distinct solutions.

(c) $(2v - 5)^2 = 0$

$2v - 5 = \pm\sqrt{0}$ Take the square roots of both sides.

$2v = 5 \pm 0$

$v = \dfrac{5 \pm 0}{2}$

$v = \dfrac{5}{2}$ A root of multiplicity two ☐

The easiest method for solving many quadratic equations is the method of factoring. This method relies upon the zero-factor principle given in Section 1-2.

▶ **Self-Check Answers** ▼

1. $a = 1$, $b = -4$, $c = 11$ 2. $a = 1$, $b = -3$, $c = 4$ 3. $a = 5$, $b = 0$, $c = -27$

SECTION 2-3 QUADRATIC EQUATIONS

Zero-Factor Principle ▼

If A and B are algebraic expressions and if $AB = 0$, then
$$A = 0 \quad \text{or} \quad B = 0$$

▶ **Self-Check** ▼

Solve each quadratic equation by the method of extraction of roots.

1 $s^2 - 49 = 0$
2 $(y - 4)^2 = 25$
3 $(z - 8)^2 = 0$

EXAMPLE 2 Solve $(x - 2)(x + 3) = 0$ using the zero-factor principle.
SOLUTION If $(x - 2)(x + 3) = 0$, then $x - 2 = 0$ or $x + 3 = 0$. Thus $x = 2$ or $x = -3$. ☐

EXAMPLE 3 Solve $x^2 - 7x + 12 = 2$ and check the solution.

SOLUTION
$x^2 - 7x + 12 = 2$	
$x^2 - 7x + 10 = 0$	Write the equation in standard form.
$(x - 2)(x - 5) = 0$	Factor the left side.
$x - 2 = 0 \quad \text{or} \quad x - 5 = 0$	Set each factor equal to zero.
$x = 2 \qquad\qquad x = 5$	Solve each equation.

Check

$x = 2$: $(2)^2 - 7(2) + 12 \stackrel{?}{=} 2$
$\qquad 4 - 14 + 12 \stackrel{?}{=} 2$
$\qquad 2 = 2$ checks.

$x = 5$: $(5)^2 - 7(5) + 12 \stackrel{?}{=} 2$
$\qquad 25 - 35 + 12 \stackrel{?}{=} 2$
$\qquad 2 = 2$ checks. ☐

EXAMPLE 4 Solve $(3z + 1)(2z - 1) = 11$.

SOLUTION
$(3z + 1)(2z - 1) = 11$
$6z^2 - z - 1 = 11$
$6z^2 - z - 12 = 0$
$(3z + 4)(2z - 3) = 0$
$3z + 4 = 0 \quad \text{or} \quad 2z - 3 = 0$
$3z = -4 \qquad\qquad 2z = 3$
$z = -\dfrac{4}{3} \qquad\qquad z = \dfrac{3}{2}$

Caution: First write in the standard form $ax^2 + bx + c = 0$. The zero-factor principle works only if the product is zero.

☐

EXAMPLE 5 Solve $x^2 - 5xy + 4y^2 = 0$ for x.

SOLUTION
$x^2 - 5xy + 4y^2 = 0$
$(x - y)(x - 4y) = 0$
$x - y = 0 \quad \text{or} \quad x - 4y = 0$
$x = y \qquad\qquad x = 4y$ ☐

▶ **Self-Check Answers** ▼

1 $s = -7$ or $s = 7$ **2** $y = -1$ or $y = 9$ **3** $z = 8$ (a double root)

The zero-factor principle can also be used to solve higher-order equations if the left member can be set equal to zero and factored.

EXAMPLE 6 Solve $x^3 - x^2 - 12x = 0$ by factoring.

SOLUTION $x(x^2 - x - 12) = 0$ The GCF is x.

$x(x - 4)(x + 3) = 0$ Factor by trial and error.

$x = 0$ or $x - 4 = 0$ or $x + 3 = 0$ The zero-factor principle

$x = 0$ $x = 4$ $x = -3$ □

By reversing the factoring process, we can write a quadratic equation given the solutions of the equation.

EXAMPLE 7 Write a quadratic equation x with solutions of 3 and $-\frac{2}{5}$.

SOLUTIONS $x = 3$ or $x = -\frac{2}{5}$

$x - 3 = 0$ $5x = -2$

                   $5x + 2 = 0$

Thus $(x - 3)(5x + 2) = 0$.

Answer $5x^2 - 13x - 6 = 0$ has solutions of 3 and $-\frac{2}{5}$.   □

▶ **Self-Check** ▼

Solve each quadratic equation by factoring.

1 $y^2 - y = 12$ **2** $8t^2 = 2t + 3$

Some quadratic equations cannot be solved by factoring over the integers. We shall develop the quadratic formula to solve these problems. The development of the quadratic formula uses the method of completing the square, which is also a useful tool for analyzing conic sections.

If the left side of a quadratic equation is a perfect square, then we can solve the equation by extraction of roots. Note the perfect square shown in Example 8.

EXAMPLE 8 Solve $y^2 - 10y + 25 = 9$.

SOLUTION $(y - 5)^2 = 9$ The left side is a perfect square.

$y - 5 = \pm 3$ Extraction of roots

$y = 5 \pm 3$

$y = 2$ or $y = 8$ □

By using the properties of perfect square trinomials, we can always write the left side of a quadratic equation as a perfect square. This process of completing the square is outlined in the next box.

▶ **Self-Check Answers** ▼

1 $y = 4$ or $y = -3$ **2** $t = -\frac{1}{2}$ or $t = \frac{3}{4}$

SECTION 2-3 QUADRATIC EQUATIONS

Solving Quadratic Equations by Completing the Square ▼

Step 1 Shift the constant term to the right side of the equation.
Step 2 Divide both sides of the equation by the coefficient of x^2.
Step 3 Add the square of half the coefficient of x to both sides of the equation.
Step 4 Write the left side of the equation as a perfect square.
Step 5 Extract the square roots of both sides of the equation.
Step 6 Simplify both solutions of the quadratic equation.

EXAMPLE 9 Solve $2x^2 + 3x - 2 = 0$ by completing the square.

SOLUTION

1 Shift the constant term to the right side.
$$2x^2 + 3x - 2 = 0$$
$$2x^2 + 3x = 2$$

2 Divide by the coefficient of x^2.
$$x^2 + \frac{3}{2}x = 1$$

3 Add the square of $\frac{1}{2}$ the coefficient of x.
$$x^2 + \frac{3}{2}x + \left(\frac{3}{4}\right)^2 = 1 + \left(\frac{3}{4}\right)^2$$

4 Write the left side as a perfect square.
$$\left(x + \frac{3}{4}\right)^2 = \frac{25}{16} \qquad \text{Also simplify the right side.}$$

5 Extract the square roots of both sides.
$$x + \frac{3}{4} = \pm\frac{5}{4}$$

6 Simplify both solutions.
$$x = -\frac{3}{4} \pm \frac{5}{4}$$

$$x = -\frac{3}{4} - \frac{5}{4} \quad \text{or} \quad x = -\frac{3}{4} + \frac{5}{4}$$

$$x = -2 \qquad\qquad\qquad x = \frac{1}{2} \qquad □$$

Any quadratic equation can be solved by completing the square. Rather than actually performing the completing the square procedure each time, it is simpler to do this once and preserve the answer as a formula, which is often easier to apply.

Quadratic Formula ▼

For $ax^2 + bx + c = 0$ and $a \neq 0$,

$$x = \frac{-b \pm \sqrt{b^2 - 4ac}}{2a}$$

▶ Self-Check ▼

Fill in the blanks at each step of the solution $5x^2 - 3x - 2 = 0$.

1. $5x^2 - 3x = $ _____
2. $x^2 + $ _____ $x = \frac{2}{5}$
3. $x^2 - \frac{3}{5}x + \frac{9}{100} = \frac{2}{5} + $ _____
4. $(x - \frac{3}{10})^2 = $ _____
5. $x - \frac{3}{10} = \pm$ _____
6. $x = \frac{3}{10} \pm \frac{7}{10}$
   $x = \frac{3}{10} - \frac{7}{10}$ or $x = \frac{3}{10} + \frac{7}{10}$
   $x = -\frac{2}{5}$ $x = $ _____

Derivation of the Quadratic Formula

$$ax^2 + bx + c = 0$$

$$ax^2 + bx = -c \qquad \text{Shift the constant to the right side.}$$

$$x^2 + \frac{b}{a}x = \frac{-c}{a} \qquad \text{Divide both sides by the coefficient of } x^2.$$

$$x^2 + \frac{b}{a}x + \left(\frac{b}{2a}\right)^2 = \frac{-c}{a} + \left(\frac{b}{2a}\right)^2 \qquad \text{Add the square of half the coefficient of } x.$$

$$\left(x + \frac{b}{2a}\right)^2 = \frac{-4ac}{4a^2} + \frac{b^2}{4a^2} \qquad \text{Write the left side as a perfect square.}$$

$$\left(x + \frac{b}{2a}\right)^2 = \frac{b^2 - 4ac}{4a^2} \qquad \text{Simplify the right side.}$$

$$x + \frac{b}{2a} = \pm\sqrt{\frac{b^2 - 4ac}{4a^2}} \qquad \text{Extract the roots.}$$

$$x = \frac{-b}{2a} \pm \frac{\sqrt{b^2 - 4ac}}{2a} \qquad \text{Simplify}$$

$$x = \frac{-b \pm \sqrt{b^2 - 4ac}}{2a} \qquad \text{The quadratic formula.}$$

▶ **Self-Check Answers** ▼

1 2   **2** $-\frac{3}{5}$ **3** $\frac{9}{100}$ **4** $\frac{49}{100}$ **5** $\frac{7}{10}$ **6** 1

SECTION 2-3 QUADRATIC EQUATIONS

EXAMPLE 10 Solve these equations using the quadratic formula.

SOLUTIONS

(a) $10y^2 = 11y + 6$

$10y^2 = 11y + 6$

$10y^2 - 11y - 6 = 0$

To identify a, b, and c, first write the equation in standard form.

$y = \dfrac{-(-11) \pm \sqrt{(-11)^2 - 4(10)(-6)}}{2(10)}$

Substitute $a = 10$, $b = -11$, $c = -6$ into

$y = \dfrac{11 \pm \sqrt{121 + 240}}{20}$

$y = \dfrac{-b \pm \sqrt{b^2 - 4ac}}{2a}$.

$y = \dfrac{11 \pm \sqrt{361}}{20}$

$y = \dfrac{11 \pm 19}{20}$

$y = \dfrac{11 - 19}{20}$ or $y = \dfrac{11 + 19}{20}$

$y = -\dfrac{2}{5}$ $\qquad$ $y = \dfrac{3}{2}$

(b) $w^2 + 2w = -5$

$w^2 + 2w = -5$

$w^2 + 2w + 5 = 0$

First write in standard form.

$w = \dfrac{-2 \pm \sqrt{(2)^2 - 4(1)(5)}}{2(1)}$

Substitute $a = 1$, $b = 2$, $c = 5$ into

$w = \dfrac{-2 \pm \sqrt{4 - 20}}{2}$

$w = \dfrac{-b \pm \sqrt{b^2 - 4ac}}{2a}$.

$w = \dfrac{-2 \pm \sqrt{-16}}{2}$

$w = \dfrac{-2 \pm 4i}{2}$

$w = -1 \pm 2i$

$w = -1 - 2i$ or $w = -1 + 2i$

Check $w = -1 - 2i$: $(-1 - 2i)^2 + 2(-1 - 2i) \stackrel{?}{=} -5$

$(-3 + 4i) - 2 - 4i \stackrel{?}{=} -5$

$-5 = -5$ checks.

$w = -1 + 2i$: $(-1 + 2i)^2 + 2(-1 + 2i) \stackrel{?}{=} -5$

$(-3 - 4i) - 2 + 4i \stackrel{?}{=} -5$

$-5 = -5$ checks. ∎

The nature of the solutions of a quadratic equation can be determined by examining only the radicand $b^2 - 4ac$ of the quadratic formula

$$x = \frac{-b \pm \sqrt{b^2 - 4ac}}{2a}$$

In Example 10(a), the radicand is positive and there are two distinct real roots. In Example 10(b), the radicand is negative and there are two imaginary roots that are complex conjugates. If the radicand is zero, $\frac{-b \pm 0}{2a}$ will yield a real root of multiplicity two. Because its value can be used to discriminate between the real solutions and the imaginary solutions, $b^2 - 4ac$ is called the **discriminant**.

Nature of the Solutions of a Quadratic Equation ▼

Discriminant, $b^2 - 4ac$	Nature of the Solutions
Positive	Two distinct real roots
Zero	Double real root, a root of multiplicity two
Negative	Two imaginary roots that are complex conjugates

▶ **Self-Check** ▼

Solve $v^2 = 4 - 2v$ using the quadratic formula.

EXAMPLE 11 Determine the nature of the solutions of $3x^2 - 5x + 7 = 0$.

SOLUTION $b^2 - 4ac = (-5)^2 - 4(3)(7)$ Substitute $a = 3$, $b = -5$, and $c = 7$ into the discriminant.
$= 25 - 84$
$= -59$

Since the discriminant is negative, the solutions are imaginary and are complex conjugates. ☐

Many computer languages are not designed to evaluate the square root of a negative number. Thus programs written in these languages must first evaluate the discriminant before computing the square root of $b^2 - 4ac$. A program written without this feature would give an error message for the first equation it encountered with imaginary roots.

▶ **Self-Check Answer** ▼

$v = -1 - \sqrt{5}$ or $v = -1 + \sqrt{5}$

SECTION 2-3 QUADRATIC EQUATIONS

EXAMPLE 12 Determine k such that $4x^2 - 12x + k = 0$ will have a real root of multiplicity two.

SOLUTION
$b^2 - 4ac = 0$ The discriminant equals zero for double roots.
$(-12)^2 - 4(4)(k) = 0$ Substitute $a = 4$, $b = -12$, and $c = k$ into the discriminant.
$144 - 16k = 0$
$144 = 16k$
$k = 9$

▶ **Self-Check** ▼

Use the discriminant to determine the nature of the solutions of each quadratic equation.
1. $4y^2 - 8y + 4 = 0$
2. $4v^2 - 8v + 5 = 0$
3. $4w^2 - 8w - 5 = 0$

Check The resulting quadratic equation is $4x^2 - 12x + 9 = 0$. Solving this equation by factoring gives $(2x - 3)^2 = 0$, which yields $x = \frac{3}{2}$ as a root of multiplicity two. □

The four methods of solving quadratic equations given in this section are extraction of roots, factoring, completing the square, and the quadratic formula. Of these four methods, only completing the square and the quadratic formula can be used with all quadratic equations. However, the methods of factoring and extraction of roots are easier to apply for many quadratic equations and hence are used frequently.

EXERCISES 2-3

A

In Exercises 1–4, write each quadratic equation in standard form and identify a, b, and c.

1. $(4n + 1)(n - 3) = 0$
2. $(5 - 3p)(2 + p) = 8$
3. $3m^2 = 5m^2 - 4$
4. $-8z^2 = 9z$

In Exercises 5–10, solve each quadratic equation by extraction of roots.

5. $x^2 = 64$
6. $y^2 - 4 = 0$
7. $5r^2 + 80 = 0$
8. $7t^2 = -175$
9. $(3w - 4)^2 = 100$
10. $(2x - 1)^2 = 81$

In Exercises 11–18, solve each quadratic equation by factoring.

11. $(5n + 3)(n - 2) = 0$
12. $p^2 - 121 = 0$
13. $m^2 + 6m + 5 = 0$
14. $m^2 + 3m + 2 = 0$
15. $v^2 = 7v$
16. $3w^2 = 17w + 6$
17. $2w^2 - 7w = 15$
18. $r(r + 3) = 10$

In Exercises 19–24, calculate the discriminant and then use it to determine the nature of the solutions of each equation.

19. $x^2 + 8 = 0$
20. $y^2 - 8 = 0$
21. $z^2 = 10z - 25$
22. $z^2 = 22z - 121$
23. $4t^2 + 12t + 15 = 0$
24. $3t^2 - 7t - 4 = 0$

In Exercises 25–30, use the quadratic formula to solve each equation.

25. $9v^2 = 3v + 20$
26. $x^2 + 18 = 10x$
27. $t^2 = 4t + 1$
28. $z^2 + 8 = 0$
29. $y^2 + 2y + 2 = 0$
30. $4v(v - 2) - 1 = 0$

▶ **Self-Check Answers** ▼

1. Double real root
2. Imaginary, complex conjugates
3. Distinct real solutions

In Exercises 31–42, solve each quadratic equation by any method.

31 $r^2 = 50$
32 $3w^2 - 4w = 0$
33 $5w^2 - 2w = 0$
34 $6x^2 = -19x - 10$
35 $-2w^2 + 6w - 5 = 0$
36 $(t - 3)(t + 2) = 1$
37 $(2y - 1)^2 = -49$
38 $(s - 1)^2 + (s + 3)^2 = 0$
39 $(v + 5)(v + 3) = v + 8$
40 $(v - 12)(v + 1) = -40$
41 $\dfrac{s^2}{18} - \dfrac{s}{6} - 1 = 0$
42 $\dfrac{t^2}{20} - \dfrac{t}{4} + \dfrac{1}{5} = 0$

In Exercises 43–48, write a quadratic equation in x with the specified solutions.

43 -3 and 3
44 -5 and 4
45 $\frac{3}{7}$ and -2
46 $-\frac{2}{5}$ and $\frac{5}{2}$
47 $-\sqrt{2}$ and $\sqrt{2}$
48 $1 - \sqrt{3}$ and $1 + \sqrt{3}$

In Exercises 49–52, solve each quadratic equation by completing the square.

49 $v^2 + 2v = 4$
50 $v^2 + 4v - 7 = 0$
51 $4x^2 - 8x + 3 = 0$
52 $3x^2 + 6x + 1 = 0$
53 Solve $x^2 - xy - 2y^2 = 0$ for x.
54 Solve $x^2 + 2xy + y^2 = 0$ for y.

B

In Exercises 55–57, write a quadratic equation in x with the specified solutions.

55 $-2i$ and $2i$
56 $2 - i$ and $2 + i$
57 $1 - i$ and $1 + i$

In Exercises 58–61, determine a value of k such that each equation will have a root of multiplicity two.

58 $x^2 + kx + 25 = 0$
59 $x^2 + kx + 81 = 0$
60 $9y^2 + 30y + k = 0$
61 $kz^2 + (3k + 1)z + 4 = 0$

In Exercises 62–65, solve each equation by factoring.

62 $(t - 5)(t + 3)(2t + 3) = 0$
63 $w(6w^2 + 5w - 6) = 0$
64 $14y^3 = 3y - 19y^2$
65 $v^3 + 5v^2 = 4v + 20$
66 Solve $6x^2 - xy - 2y^2 = 0$ for y.
67 Solve $10x^2 + 13xy - 3y^2 = 0$ for y.

68 Find two consecutive integers the sum of whose squares is 41.
69 Find two consecutive odd integers whose product is 143.
70 The height in feet of a projectile after t seconds of flight is given by $h(t) = -16t^2 + 100t + 120$. Solve $-16t^2 + 100t + 120 = 0$ for the positive value of t at which the projectile is at ground level. (Give your answer to the nearest tenth of a second.)
71 Forty meters of rope is used to fence off a rectangular area of 90 m². Find the length and width of this rectangle to the nearest tenth of a meter.
72 The centripetal force F on an object on a curved path is directly proportional to the square of its velocity V. The force on the object at 10 kilometers per hour is 83 newtons. Determine the force (to the nearest 10 newtons) on the object on this same path at 35 kilometers per hour.
73 The load L that a beam can support is directly proportional to the square of its cross-sectional depth d. A beam that is 8 centimeters deep can support a mass of 93 kilograms. If all other dimensions of the beam remain the same, determine to the nearest tenth of a centimeter the depth of beam needed to support a mass of 127 kilograms.

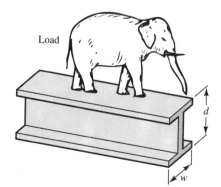

Figure for Exercise 73

C

In Exercises 74–77, solve each equation for the indicated variable. Assume the denominators are nonzero.

74 $A = \pi r^2$ for r ($A, r > 0$)

75 $V = \pi r^2 h$ for r ($V, r, h > 0$)

76 $x^2 + 3xy + y^2 = 0$ for x ($y > 0$)

77 $x^2 + 4xy + y^2 = 0$ for x ($y > 0$)

*78 **a.** Prove that the sum of the two solutions of $ax^2 + bx + c = 0$ is $\dfrac{-b}{a}$.

  b. Prove that the product of the two solutions of $ax^2 + bx + c = 0$ is $\dfrac{c}{a}$.

  c. Solve $6x^2 - 17x + 12 = 0$ and show that the sum of the two solutions is $\frac{17}{6}$ and the product of these solutions is $\frac{12}{6} = 2$.

Use a calculator or computer to solve the equations in Exercises 79 and 80 accurately to four significant digits.

79 $45.78x^2 - 789.9x - 9826 = 0$

80 $0.0004572x^2 + 0.0008764x - 0.0007381 = 0$

SECTION 2-4

Equations That Result in Quadratic Equations

Section Objectives

9 Solve equations of quadratic form.

10 Solve radical and fractional equations.

Many equations that are not quadratic equations can be worked by quadratic methods if an appropriate substitution is made to obtain a quadratic equation.

Quadratic Form ▼

An equation in x is of quadratic form if it can be written as

$$az^2 + bz + c = 0$$

where z is an algebraic expression in x and $a \neq 0$.

---

* You can use the sum and the product of the solutions as a quick check on the accuracy of your calculations.

EXAMPLE 1 Write each equation as a quadratic equation in z and identify which expression z equals.

SOLUTIONS

(a) $x^4 - 13x^2 + 36 = 0$ $z^2 - 13z + 36 = 0$ with $z = x^2$

(b) $(x + 3)^2 - 5(x + 3) + 4 = 0$ $z^2 - 5z + 4 = 0$ with $z = x + 3$

(c) $x - 10\sqrt{x} + 9 = 0$ $z^2 - 10z + 9 = 0$ with $z = \sqrt{x}$

(d) $\left(\dfrac{x-1}{x}\right)^2 - 5\left(\dfrac{x-1}{x}\right) + 6 = 0$ $z^2 - 5z + 6 = 0$ with $z = \dfrac{x-1}{x}$

(e) $x^{-2} - x^{-1} - 12 = 0$ $z^2 - z - 12 = 0$ with $z = \dfrac{1}{x}$

EXAMPLE 2 Solve $x^4 - 13x^2 + 36 = 0$.

SOLUTION
$x^4 - 13x^2 + 36 = 0$
$z^2 - 13z + 36 = 0$ Substitute z into the equation. Let $z = x^2$, then $z^2 = x^4$.
$(z - 4)(z - 9) = 0$
$z = 4$ or $z = 9$
$x^2 = 4$ $x^2 = 9$ Substitute x^2 for z.
$x = \pm 2$ $x = \pm 3$ Solve using extraction of roots.

Answer The solution set is $\{-3, -2, 2, 3\}$.   We leave the check to you.

▶ **Self-Check** ▼

Write each of the following equations as a quadratic equation in z and identify which expression z equals.

1 $x^4 + 5x^2 - 36 = 0$

2 $2x - 9\sqrt{x} + 4 = 0$

3 $x^{-2/3} - 9x^{-1/3} + 8 = 0$

EXAMPLE 3 Solve $(3y + 1)^{-2} - 2(3y + 1)^{-1} - 15 = 0$.

SOLUTION
$(3y + 1)^{-2} - 2(3y + 1)^{-1} - 15 = 0$
$z^2 - 2z - 15 = 0$ Substitute z into the equation. Let $z = (3y + 1)^{-1}$, then $z^2 = (3y + 1)^{-2}$.
$(z + 3)(z - 5) = 0$
$z + 3 = 0$ or $z - 5 = 0$
$z = -3$ $z = 5$
$\dfrac{1}{3y + 1} = -3$ $\dfrac{1}{3y + 1} = 5$ Substitute $(3y + 1)^{-1}$ for z.
$1 = -9y - 3$ $1 = 15y + 5$ Multiply by the LCD, $3y + 1$, and then solve for y.
$9y = -4$ $15y = -4$
$y = -\dfrac{4}{9}$ $y = -\dfrac{4}{15}$

Answers $y = -\dfrac{4}{9}$ or $y = -\dfrac{4}{15}$. (Both solutions check.)

▶ **Self-Check Answers** ▼

1 $z^2 + 5z - 36 = 0$ with $z = x^2$ **2** $2z^2 - 9z + 4 = 0$ with $z = \sqrt{x}$ **3** $z^2 - 9z + 8 = 0$ with $z = x^{-1/3}$

SECTION 2-4 EQUATIONS THAT RESULT IN QUADRATIC EQUATIONS

Fractional equations may simplify to nonequivalent quadratic equations; therefore, we must check for excluded values in order to eliminate any extraneous values from the solution set.

> ▶ **Self-Check** ▼
>
> Solve each equation.
> 1 $(x^2 - x)^2 - 14(x^2 - x) + 24 = 0$
> 2 $x + \sqrt{x} - 6 = 0$

EXAMPLE 4 Solve $1 - \dfrac{14}{(v+2)^2} = \dfrac{7v}{(v+2)^2}$.

SOLUTION LCD: $(v+2)^2$
Excluded value: $v = -2$

$1 - \dfrac{14}{(v+2)^2} = \dfrac{7v}{(v+2)^2}$

$(v+2)^2 - 14 = 7v$ Multiply both sides by the LCD, $(v+2)^2$.

$(v^2 + 4v + 4) - 14 - 7v = 0$ Simplify this quadratic equation and then solve it by factoring.

$v^2 - 3v - 10 = 0$

$(v-5)(v+2) = 0$ Since -2 is an excluded value, this value is extraneous. We leave the check of the value 5 for you.

$v = 5$ or $v = -2$

Answer $v = 5$

The procedure for solving radical equations often produces one or more extraneous values. Thus checking possible solutions is an essential part of the solution process.

EXAMPLE 5 Solve $w = \sqrt{w + 6}$.

SOLUTION

$w = \sqrt{w + 6}$

$w^2 = w + 6$ Square both sides of the equation.

$w^2 - w - 6 = 0$ Write in standard form.

$(w - 3)(w + 2) = 0$ Factor.

$w - 3 = 0$ or $w + 2 = 0$

$w = 3$ $w = -2$

Check $w = 3$: $\;3 \stackrel{?}{=} \sqrt{3 + 6}$ $w = -2$: $\;-2 \stackrel{?}{=} \sqrt{-2 + 6}$

$3 \stackrel{?}{=} \sqrt{9}$ $-2 \stackrel{?}{=} \sqrt{4}$

$3 = 3$ checks. $-2 = 2$ is false. -2 is an extraneous value.

Answer $w = 3$

If more than one radical occurs in an equation, begin by first isolating one of these radical terms on one side of the equation.

▶ **Self-Check Answers** ▼

1 $x = -3, x = -1, x = 2,$ or $x = 4$ 2 $x = 4$

CHAPTER 2 EQUATIONS AND INEQUALITIES

EXAMPLE 6 Solve $\sqrt{2x+1} - \sqrt{x} = 1$.

SOLUTION

$\sqrt{2x+1} - \sqrt{x} = 1$

$\sqrt{2x+1} = \sqrt{x} + 1$ Add $\sqrt{x}$ to both sides to isolate $\sqrt{2x+1}$ on the left side.

$(\sqrt{2x+1})^2 = (\sqrt{x} + 1)^2$ Square both sides.

$2x + 1 = x + 2\sqrt{x} + 1$ Note the middle term, $2\sqrt{x}$.

$x = 2\sqrt{x}$ Simplify and isolate $2\sqrt{x}$.

$x^2 = 4x$ Square both sides.

$x^2 - 4x = 0$

$x(x - 4) = 0$ Factor.

$x = 0$ or $x - 4 = 0$

$x = 0$ $x = 4$

Check $x = 0$: $\sqrt{2(0)+1} - \sqrt{0} \stackrel{?}{=} 1$ $x = 4$: $\sqrt{2(4)+1} - \sqrt{4} \stackrel{?}{=} 1$

$\sqrt{1} \stackrel{?}{=} 1$ $\sqrt{9} - \sqrt{4} \stackrel{?}{=} 1$

$1 = 1$ checks. $3 - 2 \stackrel{?}{=} 1$

                                                                          $1 = 1$ checks.

Answer $x = 0$ or $x = 4$.

EXAMPLE 7 Solve $\sqrt[3]{3t^2 - 2t - 6} = -t$.

SOLUTION

$\sqrt[3]{3t^2 - 2t - 6} = -t$

$3t^2 - 2t - 6 = -t^3$ Cube both sides of the equation.

$t^3 + 3t^2 - 2t - 6 = 0$

$t^2(t + 3) - 2(t + 3) = 0$ Factor by grouping.

$(t + 3)(t^2 - 2) = 0$ Common factor of $t + 3$

$t + 3 = 0$ or $t^2 - 2 = 0$

$t = -3$ $t^2 = 2$

$t = -3$ $t = -\sqrt{2}$ or $t = \sqrt{2}$

Answer   All of these values check, so the solution set is $\{-3, -\sqrt{2}, \sqrt{2}\}$.

> ▶ **Self-Check** ▼
>
> **1** Solve $\sqrt{28 - 3y} = 7$.
>
> **2** Verify that -3, $-\sqrt{2}$, and $\sqrt{2}$ are all roots of $\sqrt[3]{3t^2 - 2t - 6} = -t$.

▶ **Self-Check Answers** ▼

1 $y = -7$ **2** Check for $-\sqrt{2}$ (the other checks are similar):

$\sqrt[3]{3t^2 - 2t - 6} = -t$

$\sqrt[3]{3(-\sqrt{2})^2 - 2(-\sqrt{2}) - 6} \stackrel{?}{=} -(-\sqrt{2})$

$\sqrt[3]{6 + 2\sqrt{2} - 6} \stackrel{?}{=} \sqrt{2}$

$\sqrt[3]{2\sqrt{2}} \stackrel{?}{=} \sqrt{2}$

$\sqrt[3]{\sqrt{8}} \stackrel{?}{=} \sqrt{2}$

$\sqrt{\sqrt[3]{8}} \stackrel{?}{=} \sqrt{2}$

$\sqrt{2} = \sqrt{2}$ checks.

EXERCISES 2-4

A

In Exercises 1–10, write each equation as a quadratic equation in z and identify which expression z represents, but don't solve these equations.

1. $x^4 + 6x^2 + 5 = 0$
2. $3x^4 + 2x^2 - 5 = 0$
3. $y - 5\sqrt{y} - 6 = 0$
4. $y - 3\sqrt{y} - 10 = 0$
5. $\left(\dfrac{v-2}{v}\right)^2 = 2\left(\dfrac{v-2}{v}\right) + 15$
6. $\left(\dfrac{v^2+5}{2v}\right)^2 + 6 = 5\left(\dfrac{v^2+5}{2v}\right)$
7. $\dfrac{1}{w^2} + \dfrac{1}{w} - 2 = 0$
8. $\dfrac{3}{w^2} - \dfrac{1}{w} - 2 = 0$
9. $r^{2/3} - 2r^{1/3} - 35 = 0$
10. $(r-2)^{1/2} - 11(r-2)^{1/4} + 18 = 0$

In Exercises 11–30, solve each equation. Each equation can be written in quadratic form.

11. $x^4 - 10x^2 + 9 = 0$
12. $x^4 - 26x^2 + 25 = 0$
13. $2y^4 - 5y^2 + 2 = 0$
14. $4y^4 - 37y^2 + 9 = 0$
15. $2r + 4 = 9\sqrt{r}$
16. $(r^2 + 7) = \sqrt{r^2 + 7} + 12$
17. $m^{-4} = 4m^{-2}$
18. $x^{-2} + 4 = 5x^{-1}$
19. $4(v^2 - v)^2 + 6 = 11(v^2 - v)$
20. $2\left(\dfrac{1}{v} + 1\right)^2 = 3\left(\dfrac{1}{v} + 1\right) + 20$
21. $w^4 = 16$
22. $w^4 = 81$
23. $k - 5k^{1/2} + 4 = 0$
24. $k - 26k^{1/2} + 25 = 0$
25. $2a^{2/5} + 5a^{1/5} + 2 = 0$
26. $a^{-2/3} + 2a^{-1/3} + 1 = 0$
27. $3(s^2 - 1) + \sqrt{s^2 - 1} = 2$
28. $r - \sqrt{r} + 72$
29. $4(4x - 3)^{-2} + (4x - 3)^{-1} - 3 = 0$
30. $6(3x + 5)^{-2} + (3x + 5)^{-1} - 15 = 0$

In Exercises 31–48, solve each equation.

31. $4 + \dfrac{9}{v} + \dfrac{2}{v^2} = 0$
32. $3 + \dfrac{1}{v} = \dfrac{4}{v^2}$
33. $\dfrac{y-1}{y+2} + \dfrac{y}{y-2} + 1 = \dfrac{8}{y^2 - 4}$
34. $\dfrac{6}{p-1} + 1 = \dfrac{8}{p+3}$
35. $\dfrac{7x}{(x+3)^2} = 2 - \dfrac{21}{(x+3)^2}$
36. $\dfrac{5}{(x-1)^2} = 3 + \dfrac{5x}{(x-1)^2}$
37. $\dfrac{z}{z^2 - z - 2} - \dfrac{z}{z^2 + 4z + 3} = \dfrac{3z}{z^2 + z - 6}$
38. $\dfrac{4}{z^2 - 1} = \dfrac{6-z}{z^2 - z - 2} - \dfrac{8}{z^2 - 3z + 2}$ QUAD. FORM
39. $y = \sqrt{y + 12}$
40. $2v = \sqrt{-9v - 2}$
41. $\sqrt{w^2 - 2w + 1} = 2w$
42. $\sqrt{2w + 1} = w + 1$
43. $4 + \sqrt{4x - x^2} = x$
44. $\sqrt{x - 1} = 7 - x$
45. $\sqrt{2z + 1} + \sqrt{z} = 1$
46. $\sqrt{z + 1} = 1 - \sqrt{z}$
47. $\sqrt{3s + 3} + \sqrt{2 - 3s} = 1$
48. $1 + \sqrt{y - 3} = \sqrt{2y - 5}$

B

49. **a.** Factor $x^4 - 17x^2 + 16$ completely.
    b. Use this factorization to solve $x^4 - 17x^2 + 16 = 0$.
    c. Now solve $x^4 - 17x^2 + 16 = 0$ by substituting z for x^2.

50. **a.** Factor $x^4 - 13x^2 + 36$ completely.
    b. Use this factorization to solve $x^4 - 13x^2 + 36 = 0$.
    c. Now solve $x^4 - 13x^2 + 36 = 0$ by substituting z for x^2.

In Exercises 51–54, solve each equation.

51. $\sqrt[3]{p^2 - 2p - 7} - 2 = 0$
52. $\sqrt[3]{p^2 + 2} - 3 = 0$
53. $-1 + \sqrt{3t + 16} = \sqrt{5t + 21}$
54. $\sqrt{8t + 5} = 3 + \sqrt{2 - 4t}$

In Exercises 55–58, solve each equation for the indicated variable. Assume the denominators are nonzero.

55 $P = 2\pi \sqrt{\dfrac{L}{g}}$ for L

56 $\dfrac{V_a}{V_b} = \sqrt{\dfrac{m_b}{m_a}}$ for m_a

57 $a = \sqrt{c^2 - b^2}$ for b

58 $c = \sqrt{a^2 + b^2}$ for a

59 The number of kilowatts of electricity that can be produced by a windmill varies directly as the cube of the speed of the wind. In a 20 mile per hour wind, the windmill can produce 8 kilowatts. How fast must the wind blow to produce 27 kilowatts?

60 The load that a beam can support varies jointly as its width and the square of its cross-sectional depth, and inversely as the length of the beam. What is the result of doubling the width if the other measurements are unchanged? What is the result of doubling the width, depth, and the length?

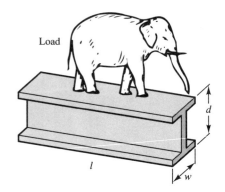

Figure for Exercise 60

C

In Exercises 61 and 62 use a calculator to approximate each solution to the nearest thousandth.

61 $\sqrt{3.09w^2 + 7.248} = 4.021w + 39.041$

62 $\dfrac{10.89}{p^4} = \dfrac{17.048}{p^2} - 0.304$

In Exercises 63–70, solve each equation.

63 $\sqrt{2w + \sqrt{2w + 4}} = 4$

64 $1 + \sqrt{w} = \sqrt{2\sqrt{w} + 3}$

65 $\sqrt{2t + 3} - \sqrt{t + 1} = \sqrt{t - 2}$

66 $\sqrt{t + 5} + \sqrt{8 - t} = \sqrt{23 - 2t}$

67 $a^2 + \sqrt{a^2 - 3a + 3} = 3a - 1$ (*Hint*: Let $z = \sqrt{a^2 - 3a + 3}$)

68 $a^2 + \sqrt{a^2 + 3a - 2} = 22 - 3a$ (*Hint*: Let $z = \sqrt{a^2 + 3a - 2}$)

69 $\sqrt{v + \sqrt{v}} + \sqrt{v - \sqrt{v}} = \sqrt{2}$

70 $\sqrt{1 - 14x} = 7 - x$

SECTION 2-5

Applications of Equations

Section Objective

11 Solve applied problems using linear and quadratic equations.

This section will use the strategy introduced in Section 2-2, which is repeated in the box on page 115. Although the problems will vary considerably, this strategy will focus your efforts so that you learn how to solve problems in general rather than memorizing specific types of problems.

SECTION 2-5 APPLICATIONS OF EQUATIONS

Summary of Word Problem Strategy ▼

Step 1 a. Determine what you are asked to find.
   b. Identify this numeric value with an appropriately chosen variable.

Step 2 a. Form the *word equation*.
   b. Translate this word equation into an algebraic equation.

Step 3 Solve this equation and answer the question asked.

Step 4 Check the reasonableness of your answer.

Practice this strategy even on easy problems; then you will have the skill when you need it for harder problems. Some additional skills pointed out in this section are

- recognizing a specific application of a general principle,
- organizing information into a table so as to clarify the important relationships, and
- drawing figures to illustrate the problem and the key relationships.

Two principles, commonly identified as the mixture principle and the rate principle, are used in a variety of settings. These principles form the basis for the word equation in many problems.

Mixture Principle for Two Ingredients ▼

Amount in first + Amount in second = Amount in mixture

Rate Principle ▼

Amount = Rate · Base

The following lists show that we have already applied these principles in a number of formulas. It also illustrates the power of a general mathematical principle to solve seemingly unrelated problems.

Mixture Principle

Amount from first + Amount from second = Total amount

Applications:
1 Work by craftsman + Work by apprentice = Total work
2 Distance by car + Distance by plane = Total distance

3 Interest on bonds + Interest on CDs = Total interest
4 Value of dimes + Value of quarters = Total value of the coins
5 Medicine in first solution + Medicine in second solution = Total medicine in mixture

Rate Principle

$$\text{Rate} \cdot \text{Base} = \text{Amount}$$

Applications:
1 Rate of work · Time worked = Amount of work ($W = RT$)
2 Rate of travel · Time traveled = Amount of distance ($D = RT$)
3 Principal · Rate of interest · Time = Amount of interest ($I = PRT$)
4 Value per coin · Number of coins = Amount of money
5 Percent of medicine in solution · Quantity of solution = Amount of medicine in solution

EXAMPLE 1 A tollway machine contained $2100 worth of quarters, dimes, and nickels. A typical collection will contain nine times as many dimes as quarters and five times as many nickels as quarters. If this collection is typical, determine how many quarters are in the collection.

SOLUTION Let q = Number of quarters   *The problem is to find the number of quarters.*

$9q$ = Number of dimes

$5q$ = Number of nickels

$$\begin{pmatrix}\text{Value of}\\ \text{quarters}\end{pmatrix} + \begin{pmatrix}\text{Value of}\\ \text{dimes}\end{pmatrix} + \begin{pmatrix}\text{Value of}\\ \text{nickels}\end{pmatrix} = \text{Total value}$$

The word equation is based on the mixture principle.

A table is a convenient means of displaying all of this information in an organized manner.

Rate Principle:	Unit value of Coins	·	Number of Coins	=	Value of Coins
Quarters	0.25	·	q	=	$0.25q$
Dimes	0.10	·	$9q$	=	$0.90q$
Nickels	0.05	·	$5q$	=	$0.25q$

Summarize the given information in a table, forming the headings from the rate principle.

Each expression in the last column is a term in the word equation.

$$0.25q + 0.90q + 0.25q = 2100.00$$
$$1.40q = 2100.00$$
$$q = 1500$$

Substitute the values from the table into the word equation.

Is this answer reasonable?

Check for reasonableness:
$$0.25(1500) = \$\ 375$$
$$0.10[9(1500)] = \$1350$$
$$0.05[5(1500)] = \$\ 375$$
$$\overline{\$2100}$$

Answer There are 1500 quarters in the collection. □

A classic example of the rate principle is the distance formula.

Distance Formula ▼

Distance = Rate of travel · Time traveled
$D = R \cdot T$

Distance problems are often clarified by making a sketch of the relationships involved.

EXAMPLE 2 A cruise missile is 200 kilometers from a target and moving at 600 kilometers per hour when it is detected and an anti-missile missile is fired from the target toward this attacker. The anti-missile missile travels at 1000 kilometers per hour and destroys its objective. How many minutes and seconds is the anti-missile missile in flight before it destroys the cruise missile?

SOLUTION Let t = Time in hours until the explosion   The question is to find the time.

$$\begin{pmatrix}\text{Distance traveled}\\ \text{by cruise missile}\end{pmatrix} + \begin{pmatrix}\text{Distance traveled by}\\ \text{anti-missile missile}\end{pmatrix} = \text{Total distance}$$

The *word equation* is based on the mixture principle.

Rate Principle:	Rate · Time = Distance
Cruise missile	600 · t = 600t
Anti-missile missile	1000 · t = 1000t

Summarize the given information in a table, forming headings from the rate principle.

Each expression in the last column is a term in the word equation.

$$600t + 1000t = 200$$
$$1600t = 200$$
$$t = \frac{200}{1600}$$

Substitute the values from the table into the word equation.

$$t = \frac{1}{8}\ h = \frac{1}{8}\ (60\ \text{min})$$

$$t = 7.5\ \text{min} = 7\ \text{minutes}\ 30\ \text{seconds}$$ Is this answer reasonable?

Answer The anti-missile missile will be in flight for 7 minutes and 30 seconds. □

EXAMPLE 3 A pharmacist must prepare 50 mililiters of a solution that contains 25% of a drug. She does not have this particular solution in stock, but she does have both a 40% solution and a 20% solution on hand. Fill in the table below to determine how much of each solution she should mix to obtain the desired prescription.

SOLUTION Let s = Number of mililiters of 40% solution

$50 - s$ = Number of mililiters of 20% solution

The problem is to determine the number of mL of each solution to use.

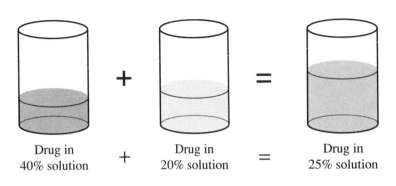

Drug in 40% solution $+$ Drug in 20% solution $=$ Drug in 25% solution

The *word equation* is based on the mixture principle.

Rate Principle:	Percent of Drug	·	mL of Solution	=	mL of Drug
40% solution	40%	·	s	=	$0.40s$
20% solution	20%	·	$(50 - s)$	=	$0.20(50 - s)$
Total mixture	25%	·	50	=	$0.25(50)$

Note that the rate principle forms the headings of this table.

Each expression in the last column is a term in the word equation.

$$0.40s + 0.20(50 - s) = 0.25(50)$$
$$0.40s + 10 - 0.20s = 12.5$$
$$0.20s = 2.5$$
$$s = 12.5$$
$$50 - s = 37.5$$

Substitute values from the table into the word equation.

Simplify and solve for s.

Check for reasonableness. 12.5 mL + 37.5 mL = 50 mL

$(0.40)(12.5) = 5.0$ mL in 40% solution
$+(0.20)(37.5) = 7.5$ mL in 20% solution
$\overline{(0.25)(50.0) = 12.5}$ mL in 25% solution

Answer She should mix 12.5 mililiters of the 40% solution with 37.5 mililiters of the 20% solution. ☐

SECTION 2-5 APPLICATIONS OF EQUATIONS

Another application of the rate principle is the formula for calculating the amount of work accomplished.

Work Formula ▼

$$\text{Work} = \text{Rate of work} \cdot \text{Time worked}$$
$$W = R \cdot T$$

Since $W = RT$, the rate of work, R, is equal to W/T. Following are some examples of rates of work:

- A streetcleaner can clean one sector of a city in 6 hours; therefore the rate of work is $R = \frac{1}{6}$ sector per hour.
- If a delivery van can make 18 deliveries in 5 hours, its rate of work is $R = \frac{18}{5} = 3.6$ deliveries per hour.
- If an outlet pipe takes t hours to drain a swimming pool, its rate of work is $R = \frac{1}{t}$ pool per hour.
- If a computer takes h hours to print 5000 payroll checks, its rate of work is $R = \frac{5000}{h}$ checks per hour.

Rate of Work ▼

If one job can be done in t units of time, then $\frac{1}{t}$ of the job can be done in one unit of time.

▶ **Self-Check** ▼

1. A homeowner purchased a mixture of two types of grass seed: fine perennial bluegrass and a rapid-growing annual rye. The bluegrass cost $1.20 per pound and the rye cost $0.95 per pound. If she purchased 20 pounds of the mixture for $23.00, how many pounds of each type of seed did she buy?
2. An ambulance leaves a town and proceeds to an accident at an average rate of 120 kilometers per hour. A police helicopter leaves town 15 minutes later at an average rate of 200 kilometer per hour. How long will it take the helicopter to overtake the ambulance?

▶ **Self-Check Answer** ▼

1. She bought 4 pounds of rye and 16 pounds of bluegrass.
2. It will take 22.5 minutes.

EXAMPLE 4 Search plane A can search an area for a crash victim in 112 hours. Planes A and B can jointly search the area in 63 hours. How many hours would it take plane B to search the area if it were working alone?

SOLUTION Let $t =$ Time in hours for plane B to search the area alone

$$\begin{pmatrix} \text{Area A} \\ \text{searches} \end{pmatrix} + \begin{pmatrix} \text{Area B} \\ \text{searches} \end{pmatrix} = 1 \text{ complete area searched}$$

The *word equation* is based on the mixture principle.

Rate Principle	$R \cdot T = W$
Plane A	$\dfrac{1}{112} \cdot 63 = \dfrac{63}{112}$
Plane B	$\dfrac{1}{t} \cdot 63 = \dfrac{63}{t}$

Summarize the given information in a table, forming the headings from the rate principle.

Each expression in the last column is a term in the word equation.

$$\frac{63}{112} + \frac{63}{t} = 1$$

Substitute the values from the table into the word equation.

$$63t + 63(112) = 112t$$

Multiply both sides by the LCD, 112t.

$$63(112) = 49t$$

$$t = 144$$

Is this answer reasonable?

Answer Plane B could search the area in 144 hours. □

Not all of the solutions to the algebraic model for a problem will always apply to the real problem. That is, the algebraic model may be more general than the specific application that is being considered. In particular, an algebraic model will often yield both a positive and a negative solution, although negative values may not be applicable to the given problem.

EXAMPLE 5 A chain of hobby stores grossed $2000 on the sales of one type of model airplane in one month. The next month the price was reduced by $1 per plane, and sales of this model increased by 100 units. The chain's gross on this model for the second month was $2700. What was the original price of the model?

SOLUTION Let $p =$ Original price of the model airplane

$p - 1 =$ Reduced price of the model airplane

The problem is to find the original price of the model.

$$\begin{pmatrix} \text{Number of models} \\ \text{sold the first month} \end{pmatrix} + 100 = \begin{pmatrix} \text{Number of models} \\ \text{sold the second month} \end{pmatrix}$$

Word equation

SECTION 2-5 APPLICATIONS OF EQUATIONS

	Amount of Gross Sales	÷	Price Per Model	=	Number Sold
First Month	2000	÷	p	=	$\dfrac{2000}{p}$
Second Month	2700	÷	$(p-1)$	=	$\dfrac{2700}{p-1}$

By the rate principle, Gross sales = (Price per model) · (Number sold).

Thus (Gross sales) ÷ (Price per model) = Number sold.

Each expression in the last column is a term in the word equation.

$$\frac{2000}{p} + 100 = \frac{2700}{p-1}$$

Substitute the values from the table into the word equation.

$$2000(p-1) + 100p(p-1) = 2700p$$

Multiply both sides by the LCD, $p(p-1)$.

$$100p^2 - 800p - 2000 = 0$$

$$p^2 - 8p - 20 = 0$$

Divide both sides by 100.

$$(p-10)(p+2) = 0$$

Factor and solve for p.

$p - 10 = 0$ or $p + 2 = 0$
$p = 10$ $p = -2$

Negative values are not meaningful in this problem.

Answer The original cost of the airplane model was $10. □

Problems based on geometric shapes such as circles, squares, and triangles are often easier to analyze if sketches are made to illustrate the problem.

EXAMPLE 6 A rectangular swimming pool 40 meters by 20 meters is surrounded by a concrete apron of uniform width. Determine the proper width in order to have an apron of area 256 m².

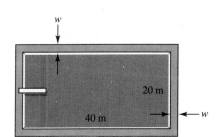

SOLUTION Let w = Width of the apron in meters.

Total area − Area of pool = Area of apron

$(40 + 2w)(20 + 2w) - 40(20) = 256$

$800 + 120w + 4w^2 - 800 = 256$

$4w^2 + 120w - 256 = 0$

$w^2 + 30w - 64 = 0$

$(w - 2)(w + 32) = 0$

$w - 2 = 0$ or $w + 32 = 0$
$w = 2$ $w = -32$

The problem is to find the width of the apron.

The *word equation*

Translate into the algebraic equation.

Simplify the left member.

Divide both sides by 4.

Factor and solve for w.

Negative values are not meaningful in this problem.

Answer The apron should be 2 meters wide. □

EXERCISES 2-5

Solve each of these word problems using the four-step strategy given in this book.

A

1. Each week a shopping mall donates the coins from its fountain to a local charity. Past statistics indicate that the charity should expect six times as many nickels as dimes, and three times as many pennies as nickels. The total of the dimes, nickels, and pennies in one week's collection was $37.70. Complete the following table and then determine the number of dimes in this collection.

	Unit Value of Coins	·	Number of Coins	=	Value of Coins
Dimes					
Nickels					
Pennies					

2. A vending machine takes quarters and dimes but gives change in nickels and dimes. The selections chosen by the customers usually result in four times as many nickels as dimes being used in change. How many of each coin should be placed in the change hoppers by the vendor, if he is putting $45 in change in the machine?

3. An insurance agent receives a 12% fee on each new policy payment and a 3% fee on each renewal payment. If the agent received a $1350 fee on total payments of $26,250, how much were the payments for new policies? Compute the table below to get your answer.

	Fee Rate · Base Payment = Amount of Fee
New Policies	
Old Policies	

4. A school district planned to invest $200,000 of its building funds in two kinds of six-month certificates of deposit. If the rates are 8.5% and 9%, how much should be invested at each rate in order to earn $8862.50 in six months?

5. A community college board decided to invest a portion of its tax receipts in two types of certificates of deposit until these funds were needed to meet monthly expenses. A total of $350,000 was invested in 6.75% and 7.0% six-month certificates and earned $12,093.75 interest. Determine the amount invested at each rate.

6. Two cars start toward each other at the same time from two cities 360 kilometers apart. If one car averages 40 kilometers per hour and the other 50 kilometers per hour, how much time will elapse before they meet?

SECTION 2-5 APPLICATIONS OF EQUATIONS

7 One train is 30 kilometers per hour faster than another. If both depart a terminal at the same time in opposite directions, they will be 800 kilometers apart after 4 hours. How fast is each traveling?

	Rate · Time = Distance
Slower Train	
Faster Train	

8 Two search teams are 45 kilometers apart. They plan to rendezvous at some intermediate point. If one team hikes one kilometer per hour faster than the other and they meet in three hours, what is the rate of each team?

9 Two police cars depart from the same point, traveling in opposite directions to look for a suspect in a stolen truck. One car averages 6 kilometers per hour faster than the other car. Find the speed of each if they are 264 kilometers apart at the end of 55 minutes.

10 One type of paint contains four parts of pigment per liter and another contains six parts of pigment per liter. How much of each should be mixed to fill a 10-liter tank with paint containing 4.7 parts of pigment per liter?

11 A hospital needs 82.5 liters of a 20% disinfectant solution. How many liters of a 60% and a 15% solution could be mixed to obtain this 20% solution?

	Percent of Disinfectant · Liters of Solution = Liters of Disinfectant
60% Solution	
15% Solution	
Total Mixture	

12 A druggist needs 20 mililiters of a 30% solution. To obtain this she mixes an 80% stock of the solution with a dilutant (0% solution). How many milliliters of the stock and how many milliliters of the dilutant should be used?

13 A window washer can wash the windows on one side of a building in 72 working hours. With an assistant he can do the job in 40 hours. How many hours would the assistant take to do the job working alone?

14 One pipe can fill a watering trough in 50 minutes and another can fill it in 75 minutes. If both pipes are turned on, how many minutes will it take to fill the trough?

15 Two conveyor belts can unload grain from a barge in six hours. Working alone, the slower belt would take nine hours more than the faster belt to do the job. How many hours would it take each belt working alone?

16 A gravel pit operator can move a pile of sand in six hours with two end loaders. Working alone, the larger end loader could do the job five hours quicker than the smaller machine. How many hours would it take the larger machine to do the job alone?

17 A firm is test-marketing a product. The sales of this product the first week totaled $10,000. After reducing the price by $5 per item, the firm sold 200 more units the second week than the first week, producing $12,000 in sales for this week. What was the original price per item of this product?

	Amount of Gross Sales ÷ Price per Item = Number Sold
First Week	
Second Week	

18 The sales of a product totaled $21,250 one week. As part of a marketing strategy the price of each item was reduced by $3. The result of this strategy was that the sales for the second week were 200 items more than in the first week and the gross sales were $23,100. What was the price per item for the first week?

19 A picture w inches wide by $w + 2$ inches long was matted so that the mat formed a $1\frac{1}{2}$-inch boundary on all sides of the picture.

   a. Write an algebraic expression for the total area this matted picture would occupy on a wall.

   b. Write an algebraic expression for the area of the exposed mat.

   c. Find w if the area of the exposed mat is 63 in^2.

20 A metal box with an open top can be formed by cutting 4-centimeter squares from each corner of a rectangular sheet of metal and folding up the sides. If the length of the box is 10 centimeters more than the width and its capacity is 384 cm^3, what should the dimensions of the sheet be?

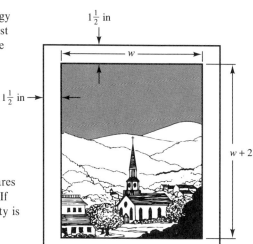

Figure for Exercise 19

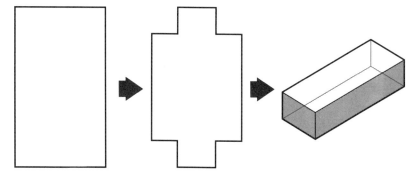

B

21 Thirty meters of fencing encloses a rectangular pen of 54 m^2. Find the length and width of the pen.

22 The diagonal of a rectangle is one more than twice the width, whereas the length is one less than twice the width. Find the dimensions of the rectangle.

23 A theater can seat 500 people. On a sell-out night the theater grossed $1945 from tickets. The adults' tickets were $5 each, and the children's tickets were $2 each. How many adults were there that night?

SECTION 2-5 APPLICATIONS OF EQUATIONS

24 A landscape contractor has two employees plant the shrubbery around a new office building. The older employee could do the job alone in two days, whereas the newer employee would take three days working alone. How many days will it take them working together?

25 The current in a river is three miles per hour. A supply boat must make a delivery 88 miles upstream. The throttle setting on the boat is the same both ways; thus the speed of the boat in still water is the same for the trips upstream and downstream. If it takes an hour and a half longer to go upstream than it does to go downstream, what is the speed of the boat in still water?

26 One pipe can fill a cooling tank with warm water in four hours. A second pipe can drain the cooled water from the tank in six hours. If both pipes are open, how many hours will it take to fill the empty tank?

27 Two barrels contain different mixtures of dog food. The mixture in the first barrel contains 20% protein and the mixture in the second barrel contains 36% protein. Determine how many kilograms of each food to mix in order to obtain 40 kilograms of a mixture that is 30% protein.

28 A retired couple received $300 interest in one year from a money market fund. They also earned $420 interest in one year on a certificate of deposit (CD). The interest rate of the CD was one percentage point higher than the money market rate. What was the interest rate of each investment? The total amount invested in the money market fund and the CD was $11,000.

29 How many liters of water must evaporate from a kiloliter of an 18% salt solution in order to leave a 25% solution?

30 The total resistance in the parallel circuit shown to the right is given by

$$\frac{1}{R_T} = \frac{1}{R_1} + \frac{1}{R_2} + \frac{1}{R_3}$$

The total resistance of this circuit is 2 Ω (ohms). If R_2 is five times R_1, and R_3 is one more than R_1, find each of these values in ohms.

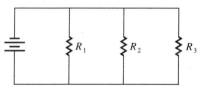

Figure for Exercise 20

C

31 The overhead for marketing and advertising a product is $55 per week. To sell t items the profit per item will have to be adjusted to $16 - t$ dollars. The manager making a decision on this product may want to know the break-even points—that is, the number of items that must be sold in order to pay the overhead expenses. Find the break-even points.

32 An auto dealer makes a profit of $2000 by selling a car at the sticker price. However, the dealer anticipates selling only one car per month at this price. Data suggest that each drop of $100 in the selling price per car will result in the sale of one additional car, increasing total sales for the month by one. (All cars must be sold at the same price.)

  a. Write an expression for profit per car that involves the number of cars sold.

  b. Write an expression for total profit that involves the number of cars sold.

  c. Determine how many cars the dealer would have to sell to generate a monthly profit of $6800.

33 A piece of wire 100 centimeters long is cut into two distinct pieces, which are then bent into squares. The combined area of these squares is 317 cm². Find the length of each piece of wire.

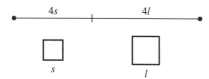

34 A pipe with a three-inch outer diameter has walls $\frac{1}{4}$-in thick.
  a. How much liquid (in cubic inches) can be put into 10 feet of this pipe?
  b. What is the volume of the steel the manufacturer put into the walls of this steel pipe?
  c. What length of pipe is needed to contain 4.12 ft³ of liquid?

35 Two planes depart from the same airport. The first plane flies due north at 400 miles per hour and the second plane flies due east at 300 miles per hour.
  a. How far apart are the planes after one hour?
  b. After how long will the planes be 250 miles apart?

Figure for Exercise 34

SECTION 2-6

Linear and Absolute Value Inequalities

Section Objective

12 Solve linear and absolute value inequalities.

The techniques used to solve inequalities rely on and are similar to the methods that we use to solve equations. As we work toward the goal of isolating the variable on one side of the inequality, we form simpler inequalities that are equivalent to the given inequality. **Equivalent inequalities** are inequalities that have the same solution set. In addition to the trichotomy property and the transitive property given in Chapter 1, we will frequently use the properties given in the following box.

Equivalent Inequalities ▼

If a, b, and c are real numbers,
Order-preserving properties: **1** $a < b$ is equivalent to $a + c < b + c$.
                                   2 $a < b$ is equivalent to $ac < bc$ for $c > 0$.
Order-reversing property: **3** $a < b$ is equivalent to $ac > bc$ for $c < 0$.

SECTION 2-6 LINEAR AND ABSOLUTE VALUE INEQUALITIES

Similar statements can also be made for the order relations $\leq$, $>$, and $\geq$. Statements for subtraction and division were omitted, because all subtraction statements can be rewritten as addition statements, since $a - b = a + (-b)$, and all division statements can be rewritten as multiplication statements, since $a \div b = a \cdot \frac{1}{b}$ for $b \neq 0$.

> **▶ Self-Check ▼**
>
> Complete the following properties of inequalities. If a, b, and c are real numbers:
>
> **1** If $a < b$, then $a - c$ _____ $b - c$.
>
> **2** If $a < b$ and $c > 0$, then $\dfrac{a}{c}$ _____ $\dfrac{b}{c}$.
>
> **3** If $a < b$ and $c < 0$, then $\dfrac{a}{c}$ _____ $\dfrac{b}{c}$.

EXAMPLE 1 Solve $7(v - 1) - 3(v - 3) > 2(v + 1) + 4(2v + 3)$ and graph the solution.

SOLUTION

$$7(v - 1) - 3(v - 3) > 2(v + 1) + 4(2v + 3)$$
$$7v - 7 - 3v + 9 > 2v + 2 + 8v + 12 \quad \text{First simplify both sides of the inequality.}$$
$$4v + 2 > 10v + 14$$
$$4v > 10v + 12 \quad \text{Subtracting 2 preserves the order.}$$
$$-6v > 12 \quad \text{Subtracting } 10v \text{ preserves the order.}$$
$$v < -2 \quad \text{Dividing by } -6 \text{ reverses the order.}$$

Answer $(-\infty, -2)$

☐

The same operation often can be performed on both pairs of inequalities that form a compound inequality. In such cases we can solve the compound inequality directly, as illustrated in the next example.

EXAMPLE 2 Solve $x - 13 \leq 7 - 4x < x + 17$ and graph the solution.

SOLUTION

$$x - 13 \leq 7 - 4x < x + 17$$
$$x - 20 \leq -4x < x + 10 \quad \text{Subtracting 7 preserves the order.}$$
$$-20 \leq -5x < 10 \quad \text{Subtracting } x \text{ preserves the order.}$$
$$4 \geq x > -2 \quad \text{Dividing by } -5 \text{ reverses the order.}$$
or $\quad -2 < x \leq 4 \quad$ Compound inequalities are usually written with the smaller number on the left and the larger number on the right.

Answer $(-2, 4]$

☐

For some compound inequalities the variable cannot be isolated between the inequalities the way it was in the previous example. In such cases the compound inequality must be split into two simple inequalities that can be solved individually. The final solution is then formed by the intersection of these individual sets.

▶ Self-Check Answers ▼

1 $<$ **2** $<$ **3** $>$

EXAMPLE 3 Solve $3(2w - 1) \leq 2(5w + 1) < 4(w + 3) - 6$ and then graph the solution.

SOLUTION
$$3(2w - 1) \leq 2(5w + 1) < 4(w + 3) - 6$$
$$6w - 3 \leq 10w + 2 < 4w + 6 \qquad \text{Simplify each member of the inequality.}$$

This compound inequality is equivalent to

$6w - 3 \leq 10w + 2$	and	$10w + 2 < 4w + 6$	Solve each of these inequalities individually. The word "and" indicates that a solution must satisfy both of these individual inequalities.
$-4w \leq 5$		$6w < 4$	
$w \geq -\dfrac{5}{4}$		$w < \dfrac{2}{3}$	

Answer $\left[-\dfrac{5}{4}, \dfrac{2}{3}\right)$

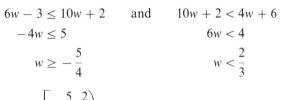

EXAMPLE 4 Solve $3w + 5 \leq 20$ or $w + 7 < 3w - 5$ and graph the solution.

SOLUTION

$3w + 5 \leq 20$	or	$w + 7 < 3w - 5$	The word "or" indicates that a solution must satisfy at least one of the individual inequalities. The solution set is the union of these two intervals.
$3w \leq 15$		$w < 3w - 12$	
$w \leq 5$		$-2w < -12$	
$w \leq 5$		$w > 6$	

Answer $(-\infty, 5] \cup (6, +\infty)$.

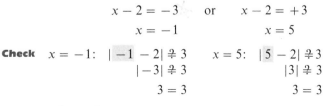

If a is larger than b, then the distance between a and b is given by the difference $a - b$. To indicate that the distance from a to b is always nonnegative, we can denote this distance by $|a - b|$. In particular, this means that $|a| = |a - 0|$ can be interpreted as the distance from a to the origin (as shown in Section 1-1). Likewise, $|a + b| = |a - (-b)|$ equals the distance from a to $-b$. In the next example, $|x - 2| = 3$ can be interpreted as the points x such that the distance between x and 2 is 3 units.

EXAMPLE 5 Solve $|x - 2| = 3$.

SOLUTION This means that the distance between x and 2 is three units. Thus x is either three units to the left of 2 or three units to the right of 2.

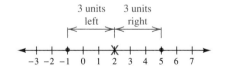

$$x - 2 = -3 \quad \text{or} \quad x - 2 = +3$$
$$x = -1 \qquad\qquad x = 5$$

Check $x = -1$: $|-1 - 2| \stackrel{?}{=} 3$ $\qquad x = 5$: $|5 - 2| \stackrel{?}{=} 3$
$\qquad\qquad\qquad |-3| \stackrel{?}{=} 3 \qquad\qquad\qquad\quad |3| \stackrel{?}{=} 3$
$\qquad\qquad\qquad\quad 3 = 3 \qquad\qquad\qquad\qquad\quad 3 = 3$

Answer $\{-1, 5\}$

SECTION 2-6 LINEAR AND ABSOLUTE VALUE INEQUALITIES

EXAMPLE 6 Interpret each of these absolute value expressions using the geometric concept of distance.

▶ **Self-Check** ▼

Solve $3y + 2 \geq 2y + 3$ and $y - 3 \leq 5 - y$ and graph the intersection.

SOLUTIONS

(a) $|v - 3| = 2$

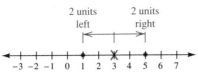

The values two units from 3 are 1 and 5.

Answer $\{1, 5\}$

(b) $|2y + 6| = 8$
$2|y + 3| = 8$
$|y - (-3)| = 4$

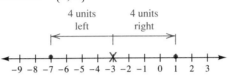

The values four units from -3 are -7 and 1. You can check them in the original expression.

Answer $\{-7, 1\}$

(c) $|x - 1| < 4$

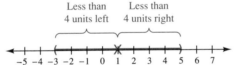

The values less than four units from 1 form the interval from -3 to 5.

Answer $(-3, 5)$

(d) $|x + 4| \geq 3$
$|x - (-4)| \geq 3$

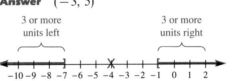

The values three or more units from -4 form a solution composed of two disjoint intervals, $(-\infty, -7]$ and $[-1, +\infty)$.

Answer $(-\infty, -7] \cup [-1, +\infty)$ □

The results of Example 6 are stated algebraically in the following box; these results are a generalization of those given for $|x|$ in Section 1-1.

▶ **Self-Check Answer** ▼

[1, 4]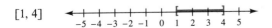

Solving Absolute Value Equations and Inequalities ▼

If E is a real algebraic expression and d is a positive number,

1. $|E| = d$ is equivalent to $E = -d$ or $E = +d$.
2. $|E| < d$ is equivalent to $-d < E < +d$.
3. $|E| > d$ is equivalent to $E < -d$ or $E > +d$.

EXAMPLE 7 Solve each of the following absolute value expressions.

SOLUTIONS

(a) $|5r + 8| = 8$

$|5r + 8| = 8$

$5r + 8 = -8$ or $5r + 8 = 8$ $|E| = d$ is equivalent to $E = -d$ or $E = d$.

$5r = -16$ $5r = 0$

$r = -\dfrac{16}{5}$ $r = 0$

Answer $\left\{ -\dfrac{16}{5}, 0 \right\}$

(b) $|4m - 3| + 5 \le 10$

$|4m - 3| + 5 \le 10$

$|4m - 3| \le 5$ Subtracting 5 preserves the order.

$-5 \le 4m - 3 \le 5$ $|E| \le d$ is equivalent to $-d \le E \le d$.

$-2 \le 4m \le 8$ Adding 3 preserves the order.

$-\dfrac{1}{2} \le m \le 2$ Dividing by $+4$ preserves the order.

Answer $\left[-\dfrac{1}{2}, 2 \right]$

(c) $\left| 3 - \dfrac{z}{2} \right| > 1$

$\left| 3 - \dfrac{z}{2} \right| > 1$

$3 - \dfrac{z}{2} < -1$ or $3 - \dfrac{z}{2} > 1$ $|E| > d$ is equivalent to $E < -d$ or $E > d$.

$\dfrac{-z}{2} < -4$ $\dfrac{-z}{2} > -2$ Subtracting 3 preserves the order.

$z > 8$ $z < 4$ Multiplying by -2 reverses the order.

Answer $(-\infty, 4) \cup (8, +\infty)$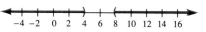

(d) $|7x - 8| < -7$

$|7x - 8| < -7$ has no solution.   Since an absolute value expression is always nonnegative, this expression has no solution.

Answer The solution set is the null set, $\varnothing$.

□

SECTION 2-6 LINEAR AND ABSOLUTE VALUE INEQUALITIES

EXAMPLE 8 Solve $|2n - 3| = |n + 6|$.

SOLUTION If the absolute values of these expressions are equal, then the expressions are either equal or opposite in sign.

$$
\begin{array}{lll}
2n - 3 = n + 6 & \text{or} & 2n - 3 = -(n + 6) \\
2n = n + 9 & & 2n - 3 = -n - 6 \\
n = 9 & & 3n = -3 \\
n = 9 & & n = -1
\end{array}
$$

Answer $\{-1, 9\}$

Figure 2-5 Tolerance interval

Since measurements are not usually exact, applications where measurements are crucial often state the amount of allowable variation from the desired measurement. This acceptable variation is called **tolerance**. For example, the weight of a part in an aircraft engine must be within 1 milligram of 2 grams. (See Figure 2-5.) Parts outside this tolerance interval would cause the engine to fail. The engineer designing this part describes this tolerance algebraically with the absolute value inequality $|m - 2.000| \leq 0.001$.

EXAMPLE 9 A producer of an industrial powder sells bags that are labeled 50 kilograms. The machines doing the filling vary the actual amount in each bag. If too little is placed in the bags, the producer will lose customers. If too much is placed in the bags, the producer will lose money. A testing program was established to ensure that bags would not vary from 50 kilograms by more than 0.5 kilograms. Express the tolerance interval for one bag algebraically. Express the tolerance interval for the combined weight of a 1000-bag shipment.

SOLUTION The tolerance interval for one bag is $49.5 \leq b \leq 50.5$, which can be given by

$$|b - 50.0| \leq 0.5$$

The tolerance interval for a 1000-bag shipment is $49{,}500 \leq s \leq 50{,}500$, which can be given by

$$|s - 50{,}000| \leq 500$$

▶ **Self-Check** ▼

1. Solve $|5 - 2t| = 1$.
2. Solve $|3v - 7| < 4$.

▶ **Self-Check Answers** ▼

1. $\{2, 3\}$ 2. $(1, 3\frac{2}{3})$

132 CHAPTER 2 EQUATIONS AND INEQUALITIES

EXERCISES 2-6

A

In Exercises 1–18, solve each inequality. Express each answer in interval notation and sketch the graph of the solution set.

1 $3w + 4 > 2w + 3$
2 $6b - 11 \geq 3b - 2$
3 $-7s \leq -21$
4 $\dfrac{-6y}{7} > 42$
5 $3(c - 5) < 5(c + 3)$
6 $2(3d + 5) \geq 4(6 - d)$
7 $5(w + 3) > 4(w - 1) + 17$
8 $3(x - 7) - 5(4 - x) - 3 > 4(x - 1)$
9 $4(x - 1) + 3(x + 2) \leq 7(x + 1) - 5$
10 $\dfrac{7q}{8} - \dfrac{2}{3} \geq \dfrac{3q}{4} - \dfrac{1}{2}$
11 $\dfrac{3s}{4} - \dfrac{2}{9} < \dfrac{2s}{3} - \dfrac{7}{18}$
12 $-13 < 13m \leq 26$
13 $-35 < -5w < -20$
14 $2y \leq 6$ and $3y > -12$
15 $2p \leq 8$ or $4p > 20$
16 $|q| \leq 2$
17 $|r| \geq 1.5$
18 $|x - 3| \leq 2$

In Exercises 19–44, solve each equation or inequality. Express each answer in interval notation.

19 $\dfrac{3(t - 1)}{4} \geq \dfrac{4(t - 2)}{6}$
20 $\dfrac{7r}{3} - \dfrac{6}{5} \geq \dfrac{12r}{5} - \dfrac{2}{3}$
21 $-7r \leq 21$ and $-r > 0$
22 $5x \geq -10$ or $7x \geq 14$
23 $x < 2x + 1 \leq x + 3$
24 $2x + 3 \leq 3x + 2 < 2x + 5$
25 $3(y - 6) \geq 5y - 12$ or $-4y + 3 < -y + 6$
26 $7x + 3 \geq 5x - 5$ or $3x - 2 \geq 5x + 8$
27 $-4(2 - r) < 3 - 3(2r + 7)$ and $4(3 - r) \geq 18 + 5(3 - 2r)$
28 $8 - 2d < 3d - 2$ and $4d + 3 \leq 2d + 11$
29 $5(3w - 2) + 1 \leq 6(2w + 3) \leq 4(5w + 7)$
30 $4(2v - 3) - 2 < 3(v - 1) + 4 \leq 5(2v + 3)$
31 $|a - 5| = 3$
32 $|x - 2| = 5$
33 $|v + 2| < 4$
34 $|w - 3| \geq 2$
35 $|2x - 3| \geq 1$
36 $|3x + 2| \leq 4$
37 $|5x - 8| = |x + 4|$
38 $|3 - w| = |3w + 1|$
39 $\left|\dfrac{3h - 5}{2}\right| + 5 > 15$
40 $|3q - 4| - 6 \geq 8$
41 $|x| \geq -3$
42 $|z| \leq -2$
43 $|7x + 5| + 8 \leq 0$
44 $|5x - 7| + 3 > 0$

In Exercises 45–50, write an absolute value inequality to represent each set of points.

45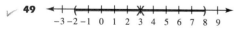
46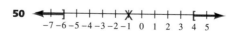
47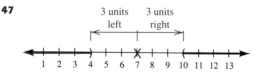
48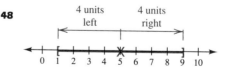
49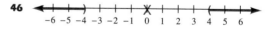
50

B

51 a. Find a value of m for which $\dfrac{4}{m} > -8$ and $4 > -8m$. **b.** Find a value of m for which $\dfrac{4}{m} > -8$ but $4 < -8m$.

SECTION 2-6 LINEAR AND ABSOLUTE VALUE INEQUALITIES

52 Find value of x and y for which
  a. $x < y$ and $\dfrac{1}{x} < \dfrac{1}{y}$
  b. $x < y$ and $\dfrac{1}{x} > \dfrac{1}{y}$

53 Find a value of x for which
  a. $8x > 3x$
  b. $8x < 3x$
  c. $8x = 3x$

54 Find values of x and y for which
  a. $x < y$ and $x^2 < y^2$
  b. $x < y$ and $x^2 > y^2$

In Exercises 55–62, use absolute value notation to describe each interval.

55 $[-4, 4]$
56 $(-7, 7)$
57 $(-\infty, -4) \cup (4, +\infty)$
58 $(-\infty, -1] \cup [1, +\infty)$
59 $[-5, 3]$
60 $(-2, 8)$
61 $(-\infty, -5) \cup (9, +\infty)$
62 $(-\infty, 1] \cup [3, +\infty)$

In Exercises 63–65, use inequalities to solve each problem.

63 Two x minus 5 is greater than or equal to 7 and is less than 15.

64 Five t is less than negative 15 or is greater than 10.

65 If a student has test scores of 85, 73, and 81, how many points must the student receive on the next exam in order to have a total of at least 320 points? Is this possible on a 100-point exam?

C

In Exercises 66–68, express each tolerance interval using absolute value notation.

66 Desired volume is 117.8 mililiters, with a tolerance of ± 4.3 mililiters.

67 Desired length is 160 kilometers, with a tolerance of ± 0.5 kilometers.

68 An order of 100 bags of fertilizer, each at 25 kilograms with a tolerance of ± 0.05 kilograms.

69 A statistician, after collecting data about 60-watt light bulbs for a manufacturer, was 95% sure that a given bulb would burn from 800 to 900 hours. Express this confidence interval using absolute value notation.

70 The cost of t units of a product is $C = 8.92t + 57.77$, and the revenue from selling these t units is $R = 10.01t$. Determine how many units must be sold in order to generate a profit $P > 0$.

71 A thermostat will maintain the temperature of a room between 68° and 77° Fahrenheit. Given that $C = \frac{5}{9}(F - 32)$, what is this temperature range in Celsius?

72 Fill in values of x and y for which
  a. $|x - y| > |x| - |y|$
  b. $|x - y| = |x| - |y|$

73 Fill in values of x and y for which
  a. $|x + y| < |x| + |y|$
  b. $|x + y| = |x| + |y|$

In Exercises 74 and 75, use a calculator to solve each inequality.

74 $4823(83t - 497) > 5009(33t + 41)$

75 $|7.22t - 2.72| - 3.71 \le 7.39$

SECTION 2-7

Nonlinear Inequalities

Section Objective

13 Solve nonlinear inequalities using a sign graph.

The zero-factor principle is an important tool for solving quadratic equations. To solve a quadratic equation by factoring, we first write the equation in standard form so that the right side is zero. A similar procedure can be used to solve many inequalities. We start by rewriting the inequality so that the right side is zero and then use the sign rule for multiplication.

Factor Principle ▼

> For real algebraic expressions A and B,
>
> **1** If $AB < 0$, then A and B have opposite signs.
> **2** If $AB = 0$, then $A = 0$ or $B = 0$.
> **3** If $AB > 0$, then A and B have the same sign.

By observing the sign pattern of each factor, we can solve many inequalities. To facilitate the solution process, we will show these signs in a **sign graph** like the one in Example 1.

EXAMPLE 1 Solve $(x + 3)(x - 2) < 0$.

SOLUTION

Factors	Sign of Factors
$x + 3$	
$x - 2$	
$(x + 3)(x - 2)$	

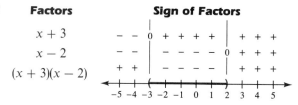

The factor $x + 3$ is negative for values of x less than -3, zero at -3, and positive for values of x greater than -3. The sign graph shows the sign changes at -3.

The sign of $x - 2$ changes at 2.

Use the signs above and the sign rule for multiplication to obtain the signs of the product.

Since the problem $(x + 3)(x - 2) < 0$ calls for negative values, the answer is the interval from -3 to 2.

Answer $(-3, 2)$ ☐

The intervals on the number line where the expression changes from negative to positive must be separated by points where the expression either is zero or is undefined. These points, where the sign of the expression may change, are called **critical values**. Thus the basic strategy for solving nonlinear inequalities hinges on finding all the critical values.

SECTION 2-7 NONLINEAR INEQUALITIES

Solving Nonlinear Inequalities ▼

Step 1 Rewrite the inequality so that the right side is zero, and factor (over the real numbers) both the numerator and the denominator of the left side.

Step 2 Determine the critical values.
   a. Zeros of the numerator only make the expression equal to zero.
   b. Zeros of the denominator cause division by zero.

Step 3 Determine the sign of each factor on the intervals between the critical values and form a sign graph.

Step 4 Determine the sign of the left side of the inequality and select the solution from the intervals formed by the critical values.

As you form the sign graph, remember that each row of the sign graph gives the sign of one of the factors. The columns of the sign graph are separated by the critical values.

▶ **Self-Check ▼**

Solve $(2x + 1)(x - 5) \geq 0$.

EXAMPLE 2 Solve $y^2 \geq 5y + 6$.

SOLUTION

1 Rewrite the inequality so that the right side is 0, and factor the left side.
$$y^2 - 5y - 6 \geq 0$$
$$(y + 1)(y - 6) \geq 0$$

2 Determine the critical values.
$$y + 1 = 0 \quad \text{or} \quad y - 6 = 0$$
$$y = -1 \qquad\qquad y = 6$$

3 **Factors**   **Sign of Factors**

$y + 1$ The factor $y + 1$ changes sign at -1.
$y - 6$ The factor $y - 6$ changes sign at 6.
4 $(y + 1)(y - 6)$ Use the signs of the factors $y + 1$ and $y - 6$ and the sign rule for multiplication to obtain the signs of the product, $(y + 1)(y - 6)$.

Answer $(-\infty, -1] \cup [6, +\infty)$

The solution to $(y + 1)(y - 6) \geq 0$ consists of the intervals where the product of the factors is positive or zero. □

▶ **Self-Check Answer ▼**

$(-\infty, -\frac{1}{2}] \cup [5, +\infty)$

EXAMPLE 3 Solve $w^3 - 48 < 16w - 3w^2$.

SOLUTION

$$w^3 - 48 < 16w - 3w^2$$

$$w^3 + 3w^2 - 16w - 48 < 0 \qquad \text{Rewrite the inequality so that the right side is zero.}$$

$$w^2(w + 3) - 16(w + 3) < 0 \qquad \text{Factor the left side by grouping.}$$

$$(w + 3)(w^2 - 16) < 0$$

$$(w + 3)(w + 4)(w - 4) < 0$$

$$w = -3 \quad \text{or} \quad w = -4 \quad \text{or} \quad w = 4 \qquad \text{Determine the critical values.}$$

Factors	Sign of Factors	
$w + 4$		The factor $w + 4$ changes sign at -4.
$w + 3$		The factor $w + 3$ changes sign at -3.
$w - 4$		The factor $w - 4$ changes sign at 4.
$(w + 4)(w + 3)(w - 4)$		The signs of the factors and the sign rule for multiplication are used to obtain the signs of the product.

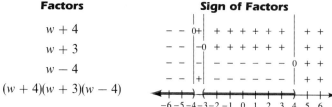

The solution to $(w + 3)(w + 4)(w - 4) < 0$ consists of the intervals where the product of the factors is negative.

Answer $(-\infty, -4) \cup (-3, 4)$ ☐

Inequalities involving rational expressions may have excluded values—that is, values causing division by zero. Since the expression is undefined at an excluded value but is defined on both sides of the point, it is possible that the expression has opposite signs on the intervals formed by this point. Critical values are therefore of two types:

1. *Points of equality*—zeros of the numerator only and thus zeros of the expression.
2. *Excluded values*—zeros of the denominator and thus values for which the expression is undefined.

Hence the strategy, as we outlined earlier, is designed to determine these two types of critical values by examining (1) factors of the numerator and (2) factors of the denominator. *A rational expression can change sign only if either its numerator or its denominator changes sign.*

EXAMPLE 4 Solve $\dfrac{4}{3 - t} \geq t + 3$.

SOLUTION

$$\frac{4}{3 - t} \geq t + 3$$

$$\frac{4}{3 - t} - (t + 3) \geq 0 \qquad \text{Rewrite the inequality so that the right side is zero.}$$

$$\frac{4 - (3 + t)(3 - t)}{3 - t} \geq 0 \qquad \text{Rewrite the left side as a single fraction.}$$

SECTION 2-7 NONLINEAR INEQUALITIES

$$\frac{4-(9-t^2)}{3-t} \geq 0$$

$$\frac{t^2-5}{3-t} \geq 0 \qquad \text{Simplify the numerator.}$$

$$\frac{(t+\sqrt{5})(t-\sqrt{5})}{3-t} \geq 0 \qquad \text{Rewrite the numerator and the denominator in factored form.}$$

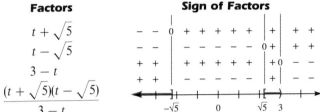

$t+\sqrt{5}$ changes sign at $-\sqrt{5}$. $t-\sqrt{5}$ changes sign at $\sqrt{5}$.

$3-t$ is positive for $t<3$, zero at $t=3$, and negative for $t>3$.

Use the signs above and the sign rules for multiplication and division to obtain the signs of the expression.

Note that 3 cannot be a solution, since the expression is undefined for this value.

Answer $(-\infty, -\sqrt{5}] \cup [\sqrt{5}, 3)$

EXAMPLE 5 Solve $\dfrac{4x+5}{1-x} < x+1$.

SOLUTION

$$\frac{4x+5}{1-x} < x+1$$

$$\frac{4x+5}{1-x} - (x+1) < 0 \qquad \text{Rewrite the inequality so that the right side is zero.}$$

$$\frac{4x+5-(1+x)(1-x)}{1-x} < 0 \qquad \text{Rewrite the left side as a single fraction.}$$

$$\frac{4x+5-(1-x^2)}{1-x} < 0 \qquad \text{Reorder the terms so that the numerator and the denominator can be factored.}$$

$$\frac{x^2+4x+4}{1-x} < 0$$

$$\frac{(x+2)^2}{1-x} < 0$$

$(x+2)^2$ is zero at $x=-2$ and positive for all other values of x. $1-x$ is positive for $x<1$, zero at $x=1$, and negative for $x>1$. The signs above are used to determine the signs of the expression. Note that the sign of the expression did not change at -2, since $(x+2)^2$ is a factor of multiplicity two.

If the inequality had been $\leq$ instead of $<$, then -2, the point of equality, would have been included in the solution set, but 1 could not be included because $x=1$ would cause division by zero.

Answer $(1, +\infty)$

Example 6 uses the fact that $|E| \geq d$ is equivalent to $E \leq -d$ or $E \geq d$ to solve an absolute value inequality.

▶ **Self-Check** ▼

1. Solve $x^3 + 18 \geq 2x^2 + 9x$.
2. Solve $\dfrac{6}{m-3} \geq 2$.

EXAMPLE 6 Solve $\left|\dfrac{2}{n+1}\right| \geq \dfrac{1}{2}$.

SOLUTION $\left|\dfrac{2}{n+1}\right| \geq \dfrac{1}{2}$

$\dfrac{2}{n+1} \leq -\dfrac{1}{2}$ or $\dfrac{2}{n+1} \geq \dfrac{1}{2}$ Rewrite the inequality without absolute value notation.

$\dfrac{2}{n+1} + \dfrac{1}{2} \leq 0$ $\dfrac{2}{n+1} - \dfrac{1}{2} \geq 0$ Rewrite each inequality so that the right side is 0.

$\dfrac{n+5}{2(n+1)} \leq 0$ $\dfrac{3-n}{2(n+1)} \geq 0$ Then use a sign graph to solve each inequality.

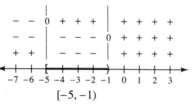

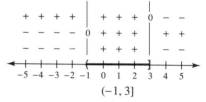

$[-5, -1)$ or $(-1, 3]$

The word "or" indicates that the solution set is the union of these two intervals. The expression is undefined at -1.

Answer $[-5, -1) \cup (-1, 3]$

EXERCISES 2-7

A

In Exercises 1–30, solve each inequality.

1. $(s + 2)(s - 1) < 0$
2. $(t - 4)(t + 6) \geq 0$
3. $(4 - v)(v + 7) \geq 0$
4. $(r - 5)(3 - r) \leq 0$
5. $(m + 2)(m - 1)(m - 4) > 0$
6. $(n + 3)(n + 1)(n - 2) < 0$
7. $(w - 3)^2(w + 5) \geq 0$
8. $(3z - 2)(z + 5)^2 < 0$
9. $4x^2 > 4x + 3$
10. $6z^2 + 17z > 14$
11. $w^2 - 2w \geq 24$
12. $3x^2 < 6 - 17x$
13. $(s - 1)^2 \geq 36$
14. $(t + 5)^2 \leq 25$
15. $\dfrac{4r - 2}{r} > 0$

▶ **Self-Check Answers** ▼

1. $[-3, 2] \cup [3, +\infty)$ 2. $(3, 6]$

SECTION 2-7 NONLINEAR INEQUALITIES

16 $\dfrac{3s+2}{2s-3} \geq 0$

17 $\dfrac{(x-3)(x-1)}{x+4} \geq 0$

18 $\dfrac{(x-2)(x+2)}{x} \leq 0$

19 $\dfrac{6}{5-r} > r$

20 $\dfrac{5}{s-4} \leq s$

21 $\dfrac{t^2-6}{5t} \geq 1$

22 $v \leq \dfrac{2v-1}{v}$

23 $3w+2 \leq \dfrac{5}{2w+1}$

24 $2 - \dfrac{2}{w+3} \leq \dfrac{3}{w+1}$

25 $\left|\dfrac{2r-3}{r}\right| < \dfrac{1}{2}$

26 $\left|\dfrac{5t-2}{t}\right| > 1$

27 $\left|\dfrac{5}{v-1}\right| \geq 3$

28 $\left|\dfrac{w}{1+2w}\right| \leq 3$

29 $\dfrac{c^3-c^2-2c}{c^2-c-6} > 0$

30 $\dfrac{p^3+p^2-20p}{2p^2+5p+2} \leq 0$

B

In Exercises 31 and 32, determine the values of x such that the expression will be a real number.

31 $\sqrt{6x^2-7x-5}$

32 $\sqrt{x^2-7}$

In Exercises 33 and 34, use the discriminant to determine the value(s) of k such that each equation will have real roots.

33 $y^2 - ky + 4 = 0$

34 $w^2 - kw + k = 0$

35 The profit made by selling t units of a product was determined by a company analyst to be $P = -t^2 + 60t - 500$. Determine the values of t that will
  a. generate a profit
  b. be break-evenpoints
  c. generate a loss

36 The cost of t units of a product is $C = t^2 + 5t$, and the revenue from selling these t units is $R = 45t$. Determine the number of units that must be sold to generate a profit. *practice*

37 The number of meters a projectile is above ground level t seconds after firing is given by $s = -4.9t^2 + 100t$. An observer can see the projectile when its height is more than five meters. During what time interval after firing can the observer see the projectile?

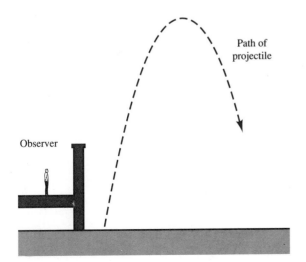

140 CHAPTER 2 EQUATIONS AND INEQUALITIES

C

In Exercises 38–45, solve each inequality.

38 $(t-3)(t+2) \leq 1$

39 $s^3 - 45 > 9s - 5s^2$

40 $x^4 < 5x^2 - 4$

41 $|x^2 + 5x + 5| \leq 1$

42 $\dfrac{m^2 - m - 2}{m^2 - 1} \geq 0$

43 $|x - 1| \leq 3|x - 2|$

44 $\left|\dfrac{u^2 - 2u + 5}{u + 1}\right| > 2$

45 $\dfrac{v^3 + 3v^2 - 4v - 12}{v^3 + 2v^2 - 9v - 18} \geq 0$

46 For some inequalities it may be easier to test a point in each interval formed by the critical points than to form the sign graph. (This technique is illustrated in *Algebra for College Students* by James W. Hall, published by PWS-KENT Publishing Company 1988.)

Solve $\left|\dfrac{2}{n+1}\right| \geq \dfrac{1}{2}$ (Example 6) by testing a value in each interval formed by the critical points $-5, -1,$ and 3. (*Hint*: Use the test values $-6, -2, 0,$ and 4.)

In Exercises 47 and 48, use a calculator to solve each inequality.

47 $\dfrac{3.46}{c - 9.24} < 7.3$

48 $x^3 + 5.43x^2 + 7.41x < 0$

KEY CONCEPTS

1 Equivalent Equations:

$a = b$ is equivalent to the equation $a + c = b + c$.

$a = b$ is equivalent to the equation $ac = bc$ for $c \neq 0$.

2 The steps for solving equations with fractions or radicals must include a check for extraneous values.

3 Summary of Word Problem Strategy:

Step 1 Determine what you are asked to find. Identify this numeric value with an appropriately chosen variable.

Step 2 Form the *word equation*. Translate this word equation into an algebraic equation.

Step 3 Solve this equation and answer the question asked.

Step 4 Check the reasonableness of your answer.

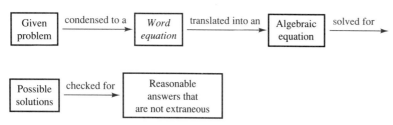

Key Concept 4

4 Quadratic formula:

For $ax^2 + bx + c = 0$ and $a \neq 0$,

$$x = \frac{-b \pm \sqrt{b^2 - 4ac}}{2a}$$

5 Discriminant $b^2 - 4ac$ of $ax^2 + bx + c = 0$:
  1. Positive implies two distinct real roots.
  2. Zero implies a real root of multiplicity two.
  3. Negative implies two imaginary roots that are complex conjugates.

6 Mixture principle:

Amount in first + Amount in second = Amount in mixture

7 Rate principle: Amount = Rate · Base

8 Rate of work:

If one job can be done in t units of time, then $\frac{1}{t}$ of the job can be done in one unit of time.

9 Equivalent inequalities:
  1. $a < b$ is equivalent to $a + c < b + c$.
  2. $a < b$ is equivalent to $ac < bc$ for $c > 0$.
  3. $a < b$ is equivalent to $ac > bc$ for $c < 0$.

10 Solving absolute value equations and inequalities:

If E is a real algebraic expression and d is a positive number.
  1. $|E| = d$ is equivalent to $E = -d$ or $E = +d$.
  2. $|E| < d$ is equivalent to $-d < E < +d$.
  3. $|E| > d$ is equivalent to $E < -d$ or $E > +d$.

11 Factor principle:
  1. If $AB < 0$, then A and B have opposite signs.
  2. If $AB = 0$, then $A = 0$ or $B = 0$.
  3. If $AB > 0$, then A and B have the same sign.

REVIEW EXERCISES FOR CHAPTER 2

In Exercises 1–26, solve each equation.

1 $8n + 11 = 4 - 6n$

2 $17(2t - 3) = 17(2t) - 3$

3 $\dfrac{5x}{6} + 1 = \dfrac{5x}{3} - \dfrac{3x}{4}$

4 $2[2(b - 2) - 2] = 3b - (12 - b)$

5 $2|5z - 10| - 6 = 24$

6 $|3 - 4w| = |6w - 4|$

7 $8t - 4[2t - (5 - t)] = 3[2(t - 2) - 7(t - 8) - 5]$

8 $\dfrac{5r - 3}{2r - 9} = \dfrac{7r - 5}{9 - 2r}$

9 $(3 + 2v)^2 = (3v - 2)^2$
10 $(5z + 1)^2 = -4$
11 $d^2 = 33 - 8d$
12 $(7w - 2)(2w + 7) = -45$
13 $(3w + 1)(w + 5) = 2w(w + 7)$
14 $x^2 - 4x + 1 = 0$
15 $9k^2 - 66k + 121 = 0$
16 $\dfrac{4}{t^2 + 2t - 3} = \dfrac{3}{t + 2} + \dfrac{1}{t - 1}$
17 $\dfrac{3}{z + 2} = \dfrac{4}{z^2 + 7z + 10} + \dfrac{8}{z + 5}$
18 $\sqrt{5z + 4} = 8$
19 $\sqrt[3]{m^2 + 4} = 5$
20 $\sqrt{8t + 5} - \sqrt{2 - 4t} = 3$
21 $\sqrt{13 + \sqrt{y}} = 1 + \sqrt{y}$
22 $21w^{-2} + 11w^{-1} = 2$
23 $2x^4 - 51x^2 + 25 = 0$
24 $\left(\dfrac{r + 5}{r - 1}\right)^2 + 8\left(\dfrac{r + 5}{r - 1}\right) - 20 = 0$
25 $2n(n - 5)(n + 7) = 0$
26 $n^3 + 2n^2 - 9n - 18 = 0$

In Exercises 27–31, solve each equation for the indicated variable.

27 $V = \dfrac{\pi}{3} r^2 h$ for h
28 $V = \dfrac{\pi}{3} r^2 h$ for r
29 $S = 2\pi r(r + h)$ for h
30 $S = 2\pi r(r + h)$ for r
31 $|x - y| = 7$ for x

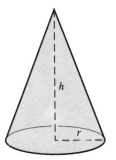

Figure for Exercise 27–28

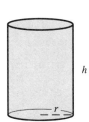

Figure for Exercises 29–30

In Exercises 32–36, write a quadratic equation in x with the given solutions.

32 5 and -7
33 $-\tfrac{3}{5}$ and $\tfrac{5}{3}$
34 $2 - \sqrt{3}$ and $2 + \sqrt{3}$
35 $2 - 3i$ and $2 + 3i$
36 a double root of $\tfrac{4}{7}$
37 Write an equation with solutions -3, 3, and 5.

In Exercises 38–40, use the discriminant to determine the nature of the solutions of each quadratic equation.

38 $4x^2 + 14x + 9 = 0$
39 $4x^2 + 12x + 9 = 0$
40 $x^2 = 7x - 13$

In Exercises 41–49, solve each inequality.

41 $-\tfrac{2}{3}(3w - 2) \geq -12$
42 $-4 < 5z + 1 \leq 6$
43 $|5 - 2a| \leq 11$
44 $\dfrac{1}{v - 3} > 1$
45 $6n^2 + 17n > 14$
46 $(p + 5)(p + 3) < 5p + 25$
47 $p^3 - 2p^2 \geq 8p$
48 $\dfrac{6}{5 - m} > m$
49 $\dfrac{x^2 - 5x - 14}{x^2 - 3x - 10} > 0$

50 A metal sheet 60 centimeters wide is used to form a trough by bending up each side as illustrated in the figure to the right. Determine the height of each side if the cross-sectional area is 450 cm².

51 The length of a room is 5 meters less than twice the width. Find the width of the room if the perimeter is 38 meters.

52 An investment for one year at a rate of 11% resulted in total principal and interest of $1720.50. What principal was invested at the beginning of the year?

53 In 1983, a homeowner obtained two mortgages to finance a new home, one at 15% and the other at 12%. The total of the mortgages was $60,000. The total interest paid the first month ($\frac{1}{12}$ year) was $700. What was the amount of the 15% mortgage?

54 Two brothers depart from home and travel in opposite directions on an expressway. One travels 8 kilometers per hour faster than the other. If the brothers are 380 kilometers apart after $2\frac{1}{2}$ hours, how fast is the faster one traveling?

55 A nurse must administer 18 milliliters of a 20% solution of medicine. In stock are a 25% solution and a 5% solution of this medicine. How many milliliters of 5% solution and 25% solution should he mix to obtain the proper dosage?

56 The diagonal of a rectangle is 8 more than the width, and the length is 7 more than the width. Find the width of the rectangle.

57 It takes the hot-water faucet 15 minutes longer than the cold-water faucet to fill a bathtub. If both faucets are turned on, the tub is filled in 18 minutes. How many minutes does it take the cold-water faucet to fill the tub by itself?

58 Within limits, the revenue for a publisher is directly proportional to the square of the amount spent by the sales force. For one year the company had revenue of $25,000,000 on a sales force budget of $100,000. Determine the approximate sales force budget needed to generate revenue of $36,000,000.

59 A furniture manufacturer produces chairs that are sold to retailers at $80 each. The company's overhead for this production is $7490 per month, and the production cost per chair is $45. How many chairs must be sold each month in order to break even?

60 A shipping container has a square top and bottom of material costing 0.2¢ per cm². The height of the rectangular sides is twice the width, and the material in these sides costs 0.1¢ per cm². If the cost of the container is $50.70, find the dimensions of the container.

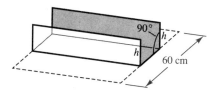

Figure for Exercise 50

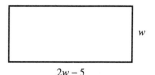

Figure for Exercise 51

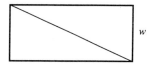

Figure for Exercise 56

OPTIONAL EXERCISES FROM CHAPTERS 1 AND 2 FOR CALCULUS-BOUND STUDENTS

1 Show that $-15x^{-4} + 7x^{-3} + 2x^{-2}$ can be written in the form $\dfrac{(2x-3)(x+5)}{x^4}$.

2 Show that $x^{3/2} - 25x^{-1/2}$ can be written in the form $\dfrac{(x+5)(x-5)}{\sqrt{x}}$.

3 Solve $2xy' - 3yy' + 7 = 0$ for y'.

4 Solve $(y')^2 - 3yy' - 4y^2 = 0$ for y'.

5 Solve $\sqrt{\dfrac{b^2 - ab}{2}} = a$ for b given $b > a > 0$.

6 Explain why $a \div 0$ is undefined for $a \neq 0$. Also explain why $0 \div 0$ is called indeterminant.

7 Factor $7y^4 - 567$ completely, over the integers.

8 Factor $8w^6 - 1$ completely, over the integers.

9 Solve $x^3 - 7x^2 + 6x = 0$.

10 Solve $x^3 + 2x^2 - 9x - 18 = 0$.

11 Use completing the square to factor $x^4 + x^2y^2 + y^4$.

12 Use completing the square to factor $4x^4 - 13x^2y^2 + y^4$.

13 Solve $x^2 + x + 1 = 0$ by completing the square.

14 Simplify $\dfrac{10}{\sqrt[3]{25x}}$.

15 Simplify $\dfrac{h}{\sqrt{x+h} - \sqrt{x}}$.

16 Simplify $\dfrac{1}{\sqrt[3]{x} - \sqrt[3]{y}}$. (*Hint*: $a^3 - b^3 = (a-b)(a^2 + ab + b^2)$).

17 Determine the values of x for which $\sqrt{x^2 - x - 12}$ assumes real values.

18 Given $x^2 + kx + 16 = 0$, determine k so that the roots of this equation are
   a. distinct real roots   **b.** a double real root   **c.** imaginary roots.

19 Use absolute value notation to represent the set of points:

20 Determine the excluded values of $\dfrac{x - \dfrac{1}{x-1}}{2 - \dfrac{1}{x}}$.

21 Solve $3v^2 - \sqrt{3v^2 + 2v + 3} = -(1 + 2v)$. (*Hint*: Let $z = \sqrt{3v^2 + 2v + 3}$.))

22 Solve $|x^2 - x - 13| > 7$.   **23** Solve $\left|\dfrac{x^2 - 3}{x - 3}\right| \leq 1$.

24 A baseball is thrown vertically so that its height in meters after t seconds is $s = -4.9t^2 + 44t$. During what time interval will the ball be higher than 2 meters?

25 Heron's formula for the area A of a triangle with sides a, b, and c is
$A = \sqrt{s(s-a)(s-b)(s-c)}$ where $s = \dfrac{a+b+c}{2}$. Use this formula to compute the area of the triangle shown to the right.

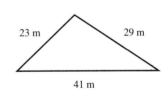

Figure for Exercise 25

***26** The axle of a train is solid. Thus on curves the wheels must travel different distances while turning the same number of times. This is accomplished by using a slightly slanted wheel, so that as the point of contact with the rail varies the turning radius of each wheel will also vary. If the difference in effective radii is $\tfrac{1}{32}$ inch, find the difference in the distances covered with one revolution of each wheel.

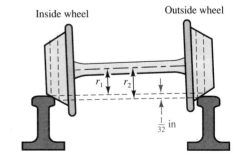

Figure for Exercise 26

* See the May 1982 issue of *The Mathematics Teacher* for an interesting article by Eugene F. Krause.

27 The examination of the rings in a cross-section of a tree reveals much about the history of the tree.

   a. From the log sketched below, determine the amount of cross-sectional area added during the last year of growth.

   b. Given that the log is 1000 centimeters long, calculate the volume of the new wood added during the last year of growth.

   c. What width growth ring would have to be added next year in order to produce an additional 137,000 cm³ of wood?

28 A man two meters tall starts walking away from a light pole eight meters high at the rate of one meter per second.

   a. How far will he be from the pole after five seconds?

   b. How long will the man's shadow be after five seconds?

   c. At what time will the length of the man's shadow be three meters?

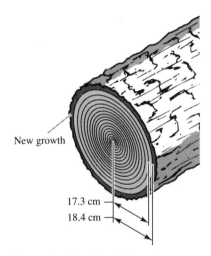

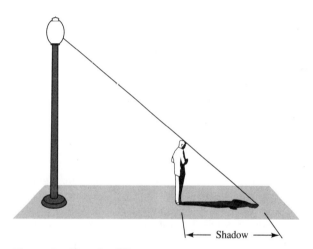

Figure for Exercise 27   **Figure for Exercise 28**

MASTERY TEST FOR CHAPTER 2

Exercise numbers correspond to Section Objective numbers.

1 Identify each equation as an identity, a contradiction, or a conditional equation.

   a. $2x = x + 5$ **b.** $2x = x + x$

   c. $2x = 0$ **d.** $2x = x - (5 - x)$

2 Solve these equations.

   a. $3t - 5[4 + 2(t - 6)] = 11 - 2(2t - 19)$ **b.** $\sqrt{2m - 3} = 5$ **c.** $\dfrac{3a + 7}{a - 4} = \dfrac{27 - 2a}{a - 4}$

3 Solve each equation for the specified variable.

a. $I = PRT$ for T **b.** $s = \dfrac{a+b+c}{2}$ for c **c.** $2y - 2xy' = 2yy'$ for y'

4 Find the volume of the shaded region in the figure to the right.

5 a. The difference between the squares of two consecutive integers is 25. Find these integers.

  b. A total of $4000 was invested, part at 6% and part at 8%. How much was invested at each rate if the total yearly interest on these investments was $270?

6 Solve these quadratic equations.

a. $3v^2 = 10 - v$ **b.** $(w-2)^2 = 7$ **c.** $2x^2 = 8x - 10$

7 Use the discriminant to determine the nature of the solutions of each equation.

a. $2y^2 - 5y + 3 = 0$ **b.** $2y^2 + 5y + 4 = 0$ **c.** $4w^2 - 28w + 49 = 0$

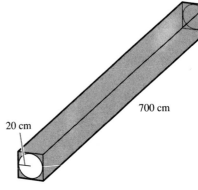

Figure for Exercise 4

8 Write a quadratic equation given these solutions.

a. $\tfrac{2}{7}, -\tfrac{3}{5}$ **b.** a double root of 5 **c.** $-\sqrt{3}, \sqrt{3}$

d. $3+i, 3-i$

9 Solve these equations of quadratic form.

a. $t^4 - 101t^2 + 100 = 0$ **b.** $\dfrac{1}{(x+3)^2} - \dfrac{10}{x+3} + 9 = 0$

c. $a^{-2/5} - 3a^{-1/5} + 2 = 0$

10 Solve these equations.

a. $\dfrac{1}{3y^2 - 10y - 8} + \dfrac{2}{y-4} = \dfrac{1}{3y+2}$ **b.** $\sqrt{3-2x} + \sqrt{7-4x} = 5$

11 a. A woman is driving 55 miles per hour on an interstate highway. Upon discovering that an important document was left behind, her boss sends a messenger after her with the document. If the messenger started 30 minutes after the woman left and drove at 65 miles per hour, how long would it take him to catch up with her?

  b. Two men working together can paint a house in 24 hours. Working alone the faster painter can paint the house in 40 hours. How many hours would it take the slower painter to paint the house working alone?

12 Solve these inequalities.

a. $4(z-2) - 2(z-4) \leq 5(3z-2)$

b. $7w - 6 < 15$ or $4(w-1) \geq 3w + 11$

c. $|8n - 7| + 3 < 20$

13 Solve these inequalities.

a. $2x^3 - 3x^2 < 2x$ **b.** $\dfrac{13}{t^2 - 1} - \dfrac{3t}{t+1} \leq \dfrac{5}{t-1}$

CHAPTER THREE

Functions, Relations, and Graphs

Chapter Three Objectives

1. Determine whether a relation is a function.
2. Evaluate an expression using functional notation.
3. Determine the domain and range of a function.
4. Graph lines.
5. Calculate the slope of a line.
6. Determine if lines are parallel or perpendicular.
7. Use the point-slope form and the slope-intercept form.
8. Use translations and reflections to graph functions.
9. Graph a parabola.
10. Determine the maximum or minimum value of a quadratic function.
11. Determine if the graph of a relation has symmetry to the x-axis, the y-axis, or the origin.
12. Determine the distance and the midpoint between two points.
13. Add, subtract, multiply, and divide two functions.
14. Form the composition of two functions.
15. Determine whether or not a function is a one-to-one function.
16. Determine the inverse of a one-to-one function.
17. Graph the inverse of a function.
18. Write in standard form the equation of a circle or an ellipse.
19. Graph a circle or an ellipse.

*20   Write the equation of a hyperbola in standard form.

*21   Graph a hyperbola.

*22   Identify the type of conic section defined by an equation.

René Descartes, 1596–1650

Descartes was born to a noble French family and was known for his studies in anatomy, astronomy, chemistry, physics, philosophy, as well as mathematics. He is most famous for his development of analytic geometry.

One of the central goals of all areas of science is to develop an understanding of nature by studying the relationship between different quantities. As this understanding becomes clear, numerical quantities are often represented mathematically by variables and their relationship expressed by a formula. Thus mathematics performs a service to many diverse areas of study.

Several branches of mathematics are devoted to studying properties of a special type of relationship called a function. Trigonometry, for example, is the study of a class of functions known as trigonometric functions, and calculus covers many important properties of functions. Therefore it is important to clarify and study in detail mathematical relations and functions.

The graph of a function visually displays the properties of the function and the relationship that exists between the variables. Thus a study of functions must include a careful examination of graphs. Optional material on graphing functions on a graphics calculator is presented in this chapter. This material includes exercises that can be worked using either a graphics calculator or a computer.

SECTION 3-1
Introduction to Functions and Relations
Section Objectives

1   Determine whether a relation is a function.

2   Evaluate an expression using functional notation.

3   Determine the domain and range of a function.

A **function** is a special type of correspondence that associates elements or quantities from one set with those in another set.

* These are optional objectives.

SECTION 3-1 INTRODUCTION TO FUNCTIONS AND RELATIONS

Each of the following correspondences is a function.

- To each wire on a piano $\xrightarrow{\text{there corresponds}}$ a unique note. *positive*
- To each natural number $\xrightarrow{\text{there corresponds}}$ its principal square root.
- To each given weight of club steak $\xrightarrow{\text{there corresponds}}$ exactly one price.
- To each person $\xrightarrow{\text{there corresponds}}$ a birth date.

Function ▼

> A function is a correspondence from a domain set D to a range set R that pairs *each* element in the domain with *exactly one* element in the range.

EXAMPLE 1 Determine if each correspondence is a function with domain D.

mapping notation

SOLUTIONS

(a)
$D \quad R$
$7 \to 15$
$8 \to 17$
$11 \to 23$

This is a function. The domain is $D = \{7, 8, 11\}$, and the range is $R = \{15, 17, 23\}$.

(b)
$D \quad R$
$49 \to 7$
$\searrow -7$
$0 \to 0$

This is *not* a function. The element 49 is not paired with exactly one element in R; it is paired with both 7 and -7.

(c)
$D \quad R$
$4 \to 3$
$0 \to 1$
-1

This is *not* a function because -1 in the domain $\{4, 0, -1\}$ is not paired with any element in the range. Each element in D must be paired with an element in R for this correspondence to be a function with domain D. If the domain was restricted to $\{4, 0\}$, then it would be a function.

(d)
$D \quad R$
$7 \to 49$
$-7 \nearrow$
$10 \to 100$

This is a function. $D = \{7, -7, 10\}$, and $R = \{49, 100\}$. 7 is paired only with 49; -7 is paired only with 49; and 10 is paired only with 100. Thus each element in the domain is paired with exactly one element in the range. □

The mapping or arrow notation used above is helpful in illustrating that a function is just a correspondence between two sets. However, **ordered-pair notation** can convey this same information more concisely.

EXAMPLE 2   Rewrite the function as a set of ordered pairs.

SOLUTION

D	R	As ordered pairs:
−1 →	−2	(−1, −2)
0 →	1	(0, 1)
1 →	4	(1, 4)
2 →	7	(2, 7)
3 →	10	(3, 10)

Both notations indicate a function with $D = \{-1, 0, 1, 2, 3\}$ and $R = \{-2, 1, 4, 7, 10\}$. Both the arrows and the parentheses indicate how the elements are paired. This set of ordered pairs could also be written horizontally as $\{(-1, -2), (0, 1), (1, 4), (2, 7), (3, 10)\}$. □

▶ **Self-Check** ▼

Determine if each correspondence is a function with domain D.

1 D   R
      0 → 1
      1 → 0
      −1 ↗

2 D   R
      7 →   7
          ↘ −7
      4 →   4
          ↘ −4

3 D   R
      7 → 7
      −7
      4 → 4
      −4 ↗

4 D   R
      2 → $\frac{1}{2}$
      1 → 1
      $\frac{1}{2}$ → 2
      0 ↗

Note that in ordered-pair notation the domain of a function is the set of first coordinates of the ordered pairs and the range is the set of second coordinates.

We can state an alternative definition of a function using ordered pairs. This restatement does not change the fact that a function is a correspondence.

Function (Ordered-Pair Definition) ▼

A function is a set of ordered pairs such that no two distinct ordered pairs have the same first coordinate. The domain is the set of first coordinates, and the range is the set of second coordinates.

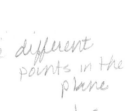
locate different points in the plane

Although functions are extremely important sets of ordered pairs, some other sets of ordered pairs are also useful. A more general concept is that of a relation; a **relation** is any set of ordered pairs.

EXAMPLE 3   Determine if the relation $\{(4, 2), (4, -2), (9, 3), (0, 0)\}$ is a function.

SOLUTION

D	R
4 →	2
	−2
9 →	3
0 →	0

This relation is not a function because the domain element 4 results in the distinct ordered pairs (4, 2) and (4, −2), which have the same first coordinate. □

If the domain set is small, it may be feasible to actually list all ordered pairs of the relation or to illustrate this correspondence using the mapping

▶ **Self-Check Answers** ▼

1 Function   **2** Not a function   **3** Not a function   **4** Function

notation. However, if the domain is infinite, we cannot possibly list all the ordered pairs. Thus we will look for a pattern or formula that describes how all the elements in the domain are paired with those in the range.

EXAMPLE 4 Write a formula that describes how the elements in the domain are paired with the elements in the range.

SOLUTION

D	R
-2	$\to 4$
-1	$\to 1$
0	$\to 0$
1	$\to 1$
2	$\to 4$
3	$\to 9$

In this function each element of the domain is squared to obtain the corresponding value in the range. We can denote this by $x \to x^2$ or (x, x^2). The ordered pairs (x, y) are also described by $y = x^2$.

▶ **Self-Check** ▼

Rewrite each relation using the mapping notation, and determine if it is a function.

1 $\{(5, 4), (6, 4), (7, 4)\}$

2 $\{(4, 5), (4, 6), (4, 7)\}$

Functional notation could also be used to describe Example 4. The domain is $\{-2, -1, 0, 1, 2, 3\}$, and the range is $\{0, 1, 4, 9\}$. The formula f that describes how the x's are paired with the y's can be denoted by $y = x^2$ or by $f(x) = x^2$.

Functional Notation ▼

$f(x)$ is read "f of x" or "the value of f at x." The variable x is an element of the domain, and $f(x)$ is the corresponding element in the range.

The x-coordinate of (x, y) is called the **independent variable**, and the y-coordinate, which equals $f(x)$, is called the **dependent variable**. Once a value of x in the domain has been selected, the formula f determines a unique functional value of y in the range. A value of x from the domain of f is sometimes called the **argument of the function** and the resulting y in the range the **value of the function** for this argument. For example, for the function $f(x) = x^2$,

$$f(-2) = (-2)^2 = 4$$

is read "f of -2 equals 4" or "the value of f at the argument -2 is 4." The independent value is -2, and the dependent value is 4.

▶ **Self-Check Answers ▼**

1 $D \to R$
$5 \to 4$
$6 \nearrow$
$7 \nearrow$
Function

2 $D \quad R$
$4 \to 5$
$\searrow 6$
$\searrow 7$
Relation, but not a function

EXAMPLE 5 Given $f(x) = \dfrac{2x^2 - x + 5}{x + 3}$, evaluate each functional value.

SOLUTIONS

(a) $f(-1)$

$$f(\ ) = \frac{2(\ )^2 - (\ ) + 5}{(\ ) + 3}$$

$$f(-1) = \frac{2(-1)^2 - (-1) + 5}{(-1) + 3}$$

$$= \frac{2 + 1 + 5}{2}$$

$$= 4$$

The use of parentheses, as illustrated in this example, is strongly recommended since this format will help prevent careless order-of-operation errors when the independent variable is substituted into the formula.

(b) $f(a)$

$$f(a) = \frac{2(a)^2 - (a) + 5}{(a) + 3}$$

$$= \frac{2a^2 - a + 5}{a + 3}$$

EXAMPLE 6 Given $g(x) = x^2 + 3x - 5$, evaluate each functional value.

SOLUTIONS

(a) $g(-1)$

$$g(-1) = (-1)^2 + 3(-1) - 5$$
$$= 1 - 3 - 5$$
$$= -7$$

(b) $g(-x)$

$g(-x) = (-x)^2 + 3(-x) - 5 = x^2 - 3x - 5$

(c) $g(2w)$

$g(2w) = (2w)^2 + 3(2w) - 5 = 4w^2 + 6w - 5$

(d) $g(h + 1)$

$g(h+1) = (h+1)^2 + 3(h+1) - 5$
$= h^2 + 2h + 1 + 3h + 3 - 5$
$= h^2 + 5h - 1$

(e) $\dfrac{g(x+h) - g(x)}{h}$

$$\frac{g(x+h) - g(x)}{h}$$

$$= \frac{[(\ )^2 + 3(\ ) - 5] - [x^2 + 3x - 5]}{h}$$

$$= \frac{[(x+h)^2 + 3(x+h) - 5] - [x^2 + 3x - 5]}{h}$$

$$= \frac{x^2 + 2xh + h^2 + 3x + 3h - 5 - x^2 - 3x + 5}{h}$$

$$= \frac{2xh + h^2 + 3h}{h}$$

$$= 2x + 3 + h, \quad \text{for } h \neq 0$$

▶ **Self-Check** ▼

Given $f(x) = 3x^2 - 7x + 2$, find these functional values.

1 $f(4)$ **2** $f(0)$ **3** $f(-4)$

Use parentheses to set up the form and then substitute $x + h$ into the parentheses.

This particular computation is used in calculus.

▶ **Self-Check Answers** ▼

1 22   **2** 2   **3** 78

SECTION 3-1 INTRODUCTION TO FUNCTIONS AND RELATIONS

Diagrams are frequently used to illustrate functions. Figure 3-1 illustrates a correspondence from a domain set D to a range set R. To each element in D there corresponds exactly one element in R.

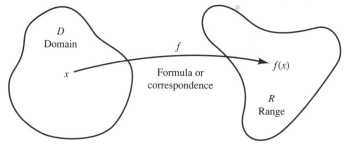

Figure 3-1 Correspondence from domain D to range R

We can produce or calculate the range element of a function, given the domain element. Some teachers emphasize this nature of functions with an illustration of a "function machine." (See Figure 3-2.) The raw material input into the machine must be an element of the domain, the manipulation on this raw material is performed according to the specified formula, and the product output by the machine is a value in the range. In the past, teachers used these imaginary machines as devices for instructional purposes. Today, however, most students have access to genuine function machines in the form of computers and calculators.

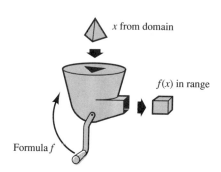

Figure 3-2 Function machine

Table 3-1 illustrates the evaluation of $f(4)$ by a calculator, given $f(x) = x^2$.

Table 3-1 Keystrokes to evaluate $f(4)$

Input (from domian)	Function Machine (a formula)		Output (into range)	
S: 4	x^2	→	16	Keystrokes for a typical scientific calculator.
G: 4	x^2 ENTER	→	16	Keystrokes for a TI-81.

EXAMPLE 7 Illustrate the correct key sequence to use with a calculator in order to evaluate $f(x)$ for the squaring function given in Example 5. (See the material on calculators in Appendix A if you have any questions on proper calculator usage.)

SOLUTIONS

(a) $f(3)$ S: 3 x^2 → 9
 G: 3 x^2 ENTER → 9

(b) $f(-2)$ S: 2 +/− x^2 → 4
 G: ((−) 2) x^2 ENTER → 4 □

Functions are frequently named by the letters f, g, and h and are denoted by giving the formula that describes how the elements in the domain are paired with those in the range. Rather than actually list the domain and the range, it is more convenient to have these sets implied by the formula. It is usually understood that the domain is the set of all possible real numbers and that the range also consists of real numbers. Certain values may be excluded from the domain because of division by zero, and other values may be excluded from the domain to avoid imaginary numbers in the range.

Domain of $y = f(x)$ ▼

> The domain of $y = f(x)$ is understood to be the set of all real numbers for which the formula is defined and that yield real values in the range. In particular this means that we exclude from the domain values that
>
> 1 cause division by zero, or
> 2 result in imaginary numbers.

EXAMPLE 8 State the domain of each function.

SOLUTIONS

(a) $f(x) = 3x - 1$ $D = \mathbb{R}$

The rule is to "triple x and then subtract 1." There are no excluded values because of division by zero or square roots of negative values. Thus the domain is the set of all real numbers.

(b) $f(x) = \dfrac{1}{(x-3)(x+2)}$ $D = \mathbb{R} \sim \{-2, 3\}$

The domain consists of all real numbers except 3 and -2; this is denoted by $D = \mathbb{R} \sim \{-2, 3\}$. These two values are excluded because they result in division by zero.

(c) $f(x) = \sqrt{x+3}$

$x + 3 \geq 0$
$x \geq -3$
$D = [-3, +\infty)$

The radicand must be nonnegative to obtain real numbers in the range. Thus $x \geq -3$. Values of x less than -3 are excluded because they would produce imaginary numbers in the range. □

A graph is a pictorial means of presenting the relationship between two variables. Often the visual impact of a graph will provide more insight into a relationship than a full page of values on a computer printout. Thus it seems natural for us to use a graphical approach to examine the concept of functions and relations. We can thank René Descartes for making this natural but powerful observation. In 1619 he gave birth to the idea of analytic geometry—the application of algebra to geometry. Prior to Descartes, algebra was concerned with numbers and calculations and geometry was concerned with figures and shapes. By associating points in the plane with ordered pairs of numbers, Descartes merged the power of these areas into analytic geometry. Geometry gives us pictorial intuition and physical insight, and algebra provides us with a concise and powerful notation for calculating results. According to John Stuart Mill, analytic geometry "constitutes the greatest single step ever made in the progress of the exact sciences."

SECTION 3-1 INTRODUCTION TO FUNCTIONS AND RELATIONS

In honor of Descartes, the **rectangular coordinate system**, shown in Figure 3-3, is frequently called the **Cartesian coordinate system**. Since each point in this plane can be uniquely identified by specifying its horizontal and vertical location, each ordered pair of a function can be represented by a point in the plane. Customarily, the *x*-coordinate, or **abscissa**, represents the domain element, and the *y*-coordinate, or **ordinate**, represents the range element.

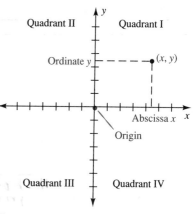

Figure 3-3 Cartesian coordinate system

EXAMPLE 9 Determine the domain and the range of the functions defined by these graphs.

SOLUTIONS

(a)

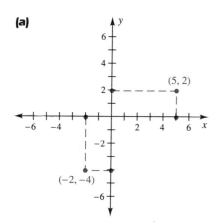

This is the graph of the set of ordered pairs
$$\{(-2, -4), (5, 2)\}$$
$D = \{-2, 5\}$;
$R = \{-4, 2\}$

Note that the domain is shown as the projection of the points onto the *x*-axis. Likewise, the range is the projection of the points onto the *y*-axis.

(b)

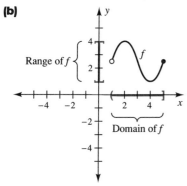

$D = (1, 5]$;
$R = [1, 4]$

The domain, which is the set of all *x*-coordinates, does not include the endpoint 1 but does include the endpoint 5. Note the open circle and closed dot illustrating this. The range is the projection of the graph onto the *y*-axis.

(c)

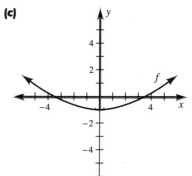

$D = \mathbb{R}$;
$R = [-1, +\infty)$

Since there is a point on the graph for each *x* along the *x*-axis, the domain is the set of all real numbers. The projection of the graph onto the *y*-axis yields all values greater than or equal to -1.

The graph of a function displays more than just the domain and the range of the function; it displays the relationship between the variables. Thus the more we can determine by looking at the graph, the more we know about the relationship between the variables. In particular, we can determine whether the function is increasing, decreasing, or constant on an interval, as defined in the box below.

Increasing, Decreasing, and Constant Functions ▼

A function $y = f(x)$ defined on an interval that includes x_1 and x_2 is

think of the projection on the x-axis

Increasing　　　　　　　　　　　　　**Decreasing**

if $f(x_2) > f(x_1)$ whenever $x_2 > x_1$　　if $f(x_2) < f(x_1)$ whenever $x_2 > x_1$

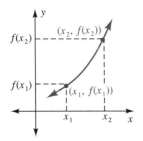 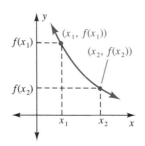

Constant

if $f(x_1) = f(x_2)$ for every x_1 and x_2

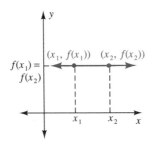

▶ Self-Check ▼

State the domain of these functions, and use a calculator to evaluate $f(-128)$, $f(0)$, and $f(39.0625)$.

1 $f(x) = \dfrac{1}{x}$　　**2** $f(x) = \sqrt{x}$

✱ Note that a function is increasing if its graph rises from left to right and decreasing if its graph falls from left to right. It is constant if it is horizontal.

▶ Self-Check Answers ▼

1 $D = \mathbb{R} \sim \{0\}$; -0.0078125, Error (0 is not in the domain), 0.0256

2 $D = [0, +\infty)$; Error (-128 is not in the domain), 0, 6.25

SECTION 3-1 INTRODUCTION TO FUNCTIONS AND RELATIONS

EXAMPLE 10 Determine the intervals for which each function $y = f(x)$ is increasing, decreasing, or constant.

▶ **Self-Check** ▼

Determine the domain and the range of these functions.

(a)
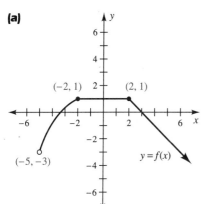

SOLUTIONS
Increasing on $(-5, -2]$
Constant on $[-2, 2]$
Decreasing on $[2, +\infty)$

1
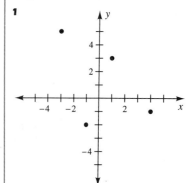

(b)
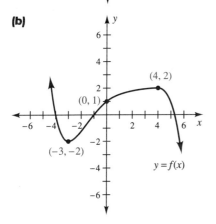

Decreasing on $(-\infty, -3]$
Increasing on $[-3, 4]$
Decreasing on $[4, +\infty)$

2
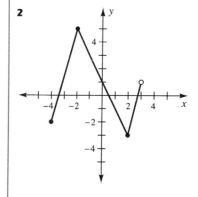

3
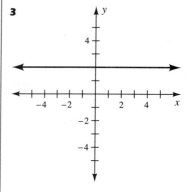

Not all graphs represent functions. Some useful relations assign the same x value to more than one y value. Hence there will be more than one point on the graph for such an x value. We can visually determine this by using the **vertical line test** to inspect the graph, since any two points on the same vertical line must have the same x value.

Vertical Line Test ▼

Imagine a vertical line placed anywhere through the x-axis. If at any position this vertical line intersects the graph at more than one point, then the graph does *not* represent a function.

▶ **Self-Check Answers** ▼

1 $D = \{-3, -1, 1, 4\}$; $R = \{-2, -1, 3, 5\}$ **2** $D = [-4, 3)$; $R = [-3, 5]$ **3** $D = \mathbb{R}$; $R = \{2\}$

EXAMPLE 11 Use the vertical line test to determine whether or not each relation is a function.

SOLUTIONS

(a) 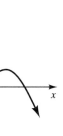 Function

(b) 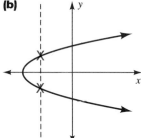 Not a function

(c) 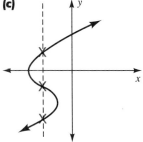 Not a function

(d) 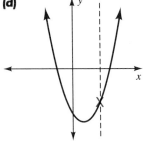 Function

▶ **Self-Check** ▼

Determine the intervals for which $y = f(x)$ is increasing, decreasing, or constant.

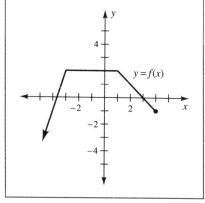

▶ **Self-Check Answers** ▼

Increasing on $(-\infty, -3]$, constant on $[-3, 1]$, decreasing on $[1, 4]$

EXERCISES 3-1

A

In Exercises 1–4, determine which relations are functions with domain D.

1. **a.**
$D \quad R$
$1 \to -1$
$ 1$
$2 \to 2$
$3 \to 3$

   b.
$D \quad R$
$-1 \to 1$
$1 \nearrow$
$2 \to 2$
$-3 \to 3$

   c.
$D \quad R$
$3 \to \frac{1}{2}$
$2 \to 1$
1
$0 \to 0$

   d.

x	y
-3	-1
-2	0
-1	1
0	2
1	3

2. **a.**
$D \quad R$
$3 \to -5$
$ 0$
$ 4$

   b.
$D \quad R$
$-5 \to 3$
$0 \nearrow$
4

   c.
$D \quad R$
$4 \to 2$
$1 \to 1$
$0 \to 0$
-1

   d.

x	y
-2	-8
-1	-1
0	0
1	1
2	8

3. **a.** $\{(1, 1), (-1, 1), (2, 0)\}$ **b.** $\{(1, 1), (1, -1), (0, 2)\}$

4. **a.** $\{(2, 3), (2, -3), (0, 13), (13, 0)\}$ **b.** $\{(3, 2), (-3, 2), (0, 13), (13, 0)\}$

In Exercises 5 and 6, use the vertical line test to determine which graphs represent functions.

5. **a.** 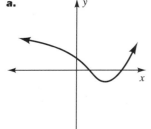 **b.** 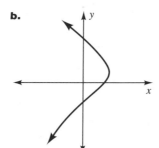 **c.** 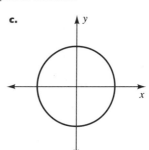 **d.**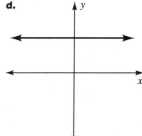

e. 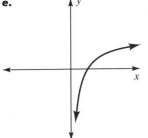 **f.** 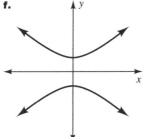 **g.**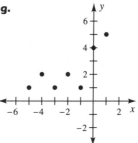

6 a. 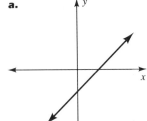 b. 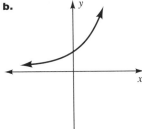 c. 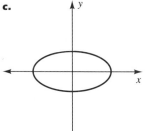 d.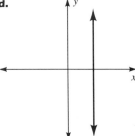

e. 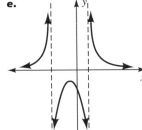 f. 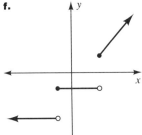 g.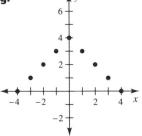

In Exercise 7–10, determine the domain and the range of each function.

7
D	R
3 →	7
−9 →	11
5 →	−2

8

x	$f(x)$
−7	4
0	4
3	4
4	4

9 $\{(11, 4), (6, -1), (17, 0)\}$

10 $\{(18, -3), (2, 8), (-7, 4)\}$

In Exercises 11 and 12, (a) determine the range of the function, (b) express the function using the mapping notation, (c) express the function using the ordered-pair notation, and (d) illustrate the function by making a graph of its points.

11 $f(x) = \sqrt{x}$ with $D = \{\frac{1}{4}, 4, 9\}$ **12** $g(x) = 3x - 5$ with $D = \{1, 2, 3\}$

In Exercises 13–16, (a) express the function shown or defined using ordered-pair notation, (b) express the function using the mapping notation, (c) determine the domain of the function, and (d) determine the range of the function.

13 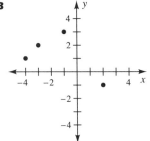   **14**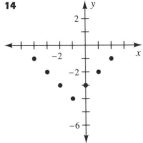

SECTION 3-1 INTRODUCTION TO FUNCTIONS AND RELATIONS

15.

x	$f(x)$
-2	-7
-1	-4
0	-1
1	2
2	5

16.

x	$f(x)$
-5	0
-4	3
-3	4
0	5
3	4
4	3
5	0

In Exercises 17 and 18, find the functional values given $f(x) = 4x - 5$.

17 a. $f(-4)$ b. $f(-x)$ c. $f(h+3)$ d. $f(h) + 3$
     e. $f(2h+1)$ f. $2f(h) + 1$ g. $f(x+h) - f(x)$

18 a. $f(0)$ b. $f(h+5)$ c. $f(h) + 5$ d. $f(3h-1)$
     e. $3f(x) - 1$ f. $f(x+h)$ g. $\dfrac{f(x+h) - f(x)}{h}$

19 Find each of these functional values for $h(x) = \dfrac{5x-4}{4x+5}$.

   a. $h(0)$ b. $h(-1)$ c. $h(\tfrac{3}{8})$ d. $h(\tfrac{-5}{4})$

20 Find each of these functional values for $g(x) = \sqrt{5x-1}$.

   a. $g(0)$ b. $g(1)$ c. $g(\tfrac{1}{5})$ d. $g(\tfrac{2}{5})$

In Exercises 21–24, determine the domain and the range of the function represented by each graph.

21
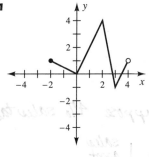

22
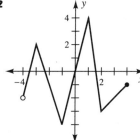

23
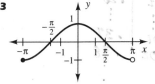

24
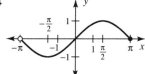

In Exercises 25–30, determine the domain and the range of each function $y = f(x)$ and also determine the intervals for which the function is increasing, decreasing, or constant.

25
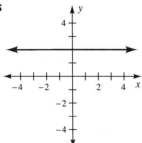

26
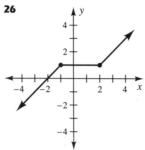

27
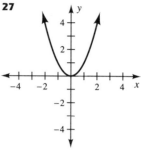

28
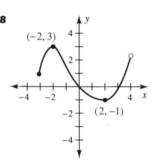

29
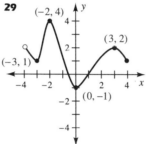

30
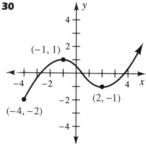

In Exercises 31–40, determine the domain of each function.

31 $g(x) = 5x + 3$

32 $h(x) = |2x - 7|$

33 $f(x) = 20$

34 $g(x) = \dfrac{x^2 - 9}{2x - 1}$

35 $h(x) = \dfrac{x^2 - 9}{3x + 1}$

36 $f(x) = -5$

37 $g(x) = \dfrac{x + 1}{x^2 - 9}$

38 $h(x) = \dfrac{x - 1}{x^2 - 16}$

39 $h(x) = \sqrt{x + 1}$

40 $g(x) = \sqrt{x - 1}$

B

41 Name five different representations used to denote functions.

42 If the equation of a straight line is not a function, then this line must be a _____ line.

43 The temperature corresponding to each time of the day is a function. Explain why the time corresponding to a given temperature is not a function.

44 The sales tax corresponding to a purchase amount is a function. Explain why the sales amount corresponding to a given tax is not a function.

In Exercises 45–48, calculate $\dfrac{f(x + h) - f(x)}{h}$.

45 $f(x) = 5x + 4$

46 $f(x) = 3x - 2$

47 $f(x) = x^2 - x + 4$

48 $f(x) = x^2 + 3x - 5$

SECTION 3-2 LINES

C

In Exercises 49–52, use a calculator to determine the functional values, given $f(x) = \sqrt{\dfrac{1}{x}}$. If an error message results, explain why.

49 $f(0.0016)$ **50** $f(15625)$ **51** $f(-25)$ **52** $f(0)$

In Exercises 53 and 54, calculate $\dfrac{f(x+h) - f(x)}{h}$.

53 $f(x) = x^3$ **54** $f(x) = (x-4)^2$

In Exercises 55–58, determine the domain of each function.

55 $g(x) = \sqrt{x^2 - 5x - 6}$ practice   **56** $h(x) = \sqrt{x^2 - 5x + 6}$

57 $f(x) = \sqrt{\dfrac{x-1}{x+3}}$ **58** $g(x) = \sqrt{\dfrac{x+3}{x-1}}$

 59 Use a computer or a graphics or programmable calculator to evaluate $f(x) = 3x^2 - 7x + 11$ for integral values from -10 to 10.

SECTION 3-2

Lines

PIECEWISE FUNCTIONS

Section Objectives

4 Graph lines.

5 Calculate the slope of a line.

6 Determine if lines are parallel or perpendicular.

7 Use the point-slope form and the slope-intercept form.

The visual impact of graphs is so useful that we use the Cartesian coordinate system and graphs as fundamental tools for studying relations and functions. Some functions occur frequently, and it is helpful to study them in detail and to be able to sketch their graphs by plotting only a few points. In order to do this, we must be able to recognize the basic shape of a graph from its defining equation. Fortunately, nature is very cooperative—many of the functions that we use the most are among the least complicated. For example, the graph of a **linear function** is a straight line, and the equation defining this function can be written in the **general form** $Ax + By + C = 0$.

Exactly two pieces of information are needed to write the equation of a linear function or to draw its graph. Two pieces of information of particular importance that can be used are the points where the line intercepts the axes. The **x-intercept** $(a, 0)$ and the **y-intercept** $(0, b)$ are often denoted by just a and b, respectively.

EXAMPLE 1 Find the intercepts of $2x - 3y = 6$, and then graph this function.

SOLUTION
$$2x - 3y = 6$$
$2x - 3(0) = 6$ Set $y = 0$ to find the x-intercept.
$x = 3$ $(3, 0)$ is the x-intercept.
$2(0) - 3y = 6$ Set $x = 0$ to find the y-intercept.
$y = -2$ $(0, -2)$ is the y-intercept.

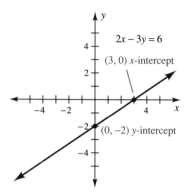

The two intercepts are plotted, and a line is drawn through the two points. □

One property of a line that is extremely important is its slope. The **slope** of a line is a number that measures the steepness, or slant, of the line. This steepness can be calculated by choosing any two points on the line and forming the ratio of the change in y to the change in x.

Slope ▼

Slope $m = \dfrac{\text{Change in } y}{\text{Change in } x}$

$m = \dfrac{y_2 - y_1}{x_2 - x_1}$ for $x_1 \neq x_2$

$P_2(x_2, y_2)$

$y_2 - y_1 = $ change in y

$P_1(x_1, y_1)$

$x_2 - x_1 = $ change in x

EXAMPLE 2 Calculate the slope of the line passing through the points:

SOLUTIONS

(a) $(-4, 5)$ and $(6, -3)$ $m = \dfrac{y_2 - y_1}{x_2 - x_1} = \dfrac{-3 - 5}{6 - (-4)} = \dfrac{-8}{10} = \dfrac{-4}{5}$

(b) $(a, 0)$ and $(0, b)$ $m = \dfrac{y_2 - y_1}{x_2 - x_1} = \dfrac{b - 0}{0 - a} = -\dfrac{b}{a}$ $m = \dfrac{-b}{a}$ where a and b are the x and y intercepts. □

If the slope of a line is positive, the linear function is increasing; that is, $f(x_2) > f(x_1)$ for $x_2 > x_1$. If the slope is negative, the linear function is decreasing; that is, $f(x_2) < f(x_1)$ for $x_2 > x_1$. It is important to be able to visualize the inclination of the line given the slope.

SECTION 3-2 LINES

Classifying Lines by Their Slope ▼

Slope	Description of Line	Graph of Line
Positive	Slopes upward to the right, an increasing function.	
Negative	Slopes downward to the right, a decreasing function.	
Zero	Horizontal line, no change in the constant function.	
Undefined	Vertical line, not a function.	

The ability to picture lines with a slope of $+1$ or -1 will help you to develop the skill of estimating the slope of a line by visual inspection. Observe the lines with $m = 1$ and $m = -1$ in Figure 3-4, and note that all steeper lines have slopes of magnitude greater than 1.

Parallel lines have the same slope, since they rise or fall at the same rate. (See Figure 3-5.) Vertical lines are parallel to each other even though the slope of a vertical line is undefined.

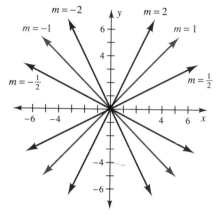

Figure 3-4 Slope of $+1$ and -1

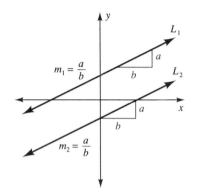

Figure 3-5 Parallel lines

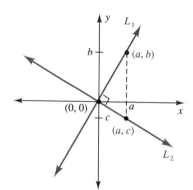

Figure 3-6 Perpendicular lines

The concept of slope can also be used to determine if two lines are perpendicular. The slopes of the perpendicular lines L_1 and L_2 shown in Figure 3-6, are negative reciprocals, $m_1 = -\dfrac{1}{m_2}$. We shall verify this using m_1 and m_2 to designate the slopes of L_1 and L_2, respectively. By the definition of slope,

$$m_1 = \frac{b-0}{a-0} = \frac{b}{a} \quad \text{and} \quad m_2 = \frac{c-0}{a-0} = \frac{c}{a}$$

Using the Pythagorean Theorem and the distance formula (page 197), we have

$$[(a-0)^2 + (b-0)^2] + [(a-0)^2 + (c-0)^2] = (a-a)^2 + (b-c)^2$$

Thus
$$a^2 + b^2 + a^2 + c^2 = b^2 - 2bc + c^2$$
$$2a^2 = -2bc$$
$$bc = -a^2$$
$$\frac{b}{a} = \frac{-a}{c}$$
$$m_1 = -\frac{1}{m_2} \qquad m_1 = \frac{b}{a} \text{ and } m_2 = \frac{c}{a}$$

SECTION 3-2 LINES

Parallel and Perpendicular Lines ▼

If L_1 and L_2 are distinct nonvertical lines with slopes m_1 and m_2, respectively, then

Parallel lines: L_1 and L_2 are parallel if and only if $m_1 = m_2$.

Perpendicular lines: L_1 and L_2 are perpendicular to each other if and only if $m_1 = -\dfrac{1}{m_2}$.

Vertical lines: The slope of a vertical line is undefined. All vertical lines are parallel to each other and are perpendicular to all horizontal lines.

▶ Self-Check ▼

Calculate the slope of the line through each pair of points, and describe the line.

1 (5, 6) and (6, 4)
2 (−3, −2) and (5, 2)
3 (−4, −5) and (4, −5)
4 (3, 2) and (3, −2)

EXAMPLE 3 Use slopes to determine whether a triangle with vertices $A(-4, 8)$, $B(2, 6)$, and $C(0, 4)$ is a right triangle.

SOLUTION ABC is a right triangle if two sides are perpendicular. To determine this, we calculate the slope of each side.

$$m(\overline{AB}) = \frac{6 - 8}{2 - (-4)} = \frac{-2}{6} = -\frac{1}{3}$$

$$m(\overline{AC}) = \frac{4 - 8}{0 - (-4)} = \frac{-4}{4} = -1$$

$$m(\overline{BC}) = \frac{4 - 6}{0 - 2} = \frac{-2}{-2} = 1$$

$\overline{AC}$ is perpendicular to $\overline{BC}$ because their slopes are negative reciprocals.

Answer ABC is a right triangle with the 90° angle at C. □

Suppose we are asked to find the equation of the line through (x_1, y_1) with slope m. One equation relating these values is the definition of slope. Applying this definition to points (x_1, y_1) and (x, y), we have $m = \dfrac{y - y_1}{x - x_1}$, which simplifies to $y - y_1 = m(x - x_1)$. This form of a linear equation is called the **point-slope form**.

▶ Self-Check Answers ▼

1 $m = -2$, a decreasing function
2 $m = \frac{1}{2}$, an increasing function
3 $m = 0$, a horizontal line
4 undefined, a vertical line

EXAMPLE 4 Find the equation of a line passing through $(2, -3)$ with slope -2.

SOLUTION
$$y - y_1 = m(x - x_1) \quad \text{Point-slope form}$$
$$y - (-3) = -2(x - 2) \quad \text{Substitute in the coordinates and the slope.}$$
$$y + 3 = -2x + 4 \quad \text{Simplify.}$$
$$2x + y - 1 = 0 \quad \text{General form}$$

EXAMPLE 5 Find the equation of a line passing through the points $(-5, 4)$ and $(6, 2)$.

SOLUTION $m = \dfrac{y_2 - y_1}{x_2 - x_1} = \dfrac{2 - 4}{6 - (-5)} = \dfrac{-2}{11}$ First calculate m so that the point-slope form can be used.

$$y - 2 = -\frac{2}{11}(x - 6) \quad \text{Substitute the point } (6, 2) \text{ and the slope } -\tfrac{2}{11} \text{ into the point-slope form, } y - y_1 = m(x - x_1).$$
$$11y - 22 = -2(x - 6)$$
$$11y - 22 = -2x + 12$$
$$2x + 11y - 34 = 0$$

Note: If the other point is used, an equivalent equation will result.

The point-slope form and other forms of linear equations readily give specific information about a line. One of the most useful forms is the slope-intercept form, which we will now develop. Applying the point-slope form, $y - y_1 = m(x - x_1)$, to a line with slope m and y-intercept $(0, b)$, we obtain $y - b = m(x - 0)$, which simplifies to $y = mx + b$. The slope and the y-intercept are immediately available from the **slope-intercept form** $y = mx + b$. This is also the functional form of a linear equation and is often written as $f(x) = mx + b$.

EXAMPLE 6 Determine the slope m and the y-intercept b of $10x - 5y = 15$.

SOLUTION
$$10x - 5y = 15 \quad \text{Solve this equation for } y \text{ to express it in slope-intercept form.}$$
$$-5y = -10x + 15$$
$$y = 2x - 3$$

Answer $m = 2$ and $b = -3$

▶ **Self-Check** ▼

Quickly *estimate* the slope of each line without doing any calculations.

1
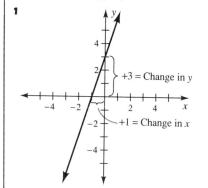
$+3 =$ Change in y
$+1 =$ Change in x

2
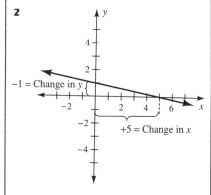
$-1 =$ Change in y
$+5 =$ Change in x

3
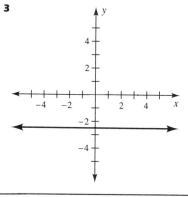

▶ **Self-Check Answers** ▼

1 $m = 3$ **2** $m = -\tfrac{1}{5}$ **3** $m = 0$

SECTION 3-2 LINES

EXAMPLE 7 Graph $y = \frac{7}{4}x - 2$ using its slope and y-intercept.

SOLUTION

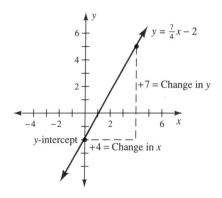

1 Find and plot the y-intercept.

$$y = \frac{7}{4}x - 2$$

Slope $m = \frac{7}{4}$, y-intercept is $(0, -2)$

2 Use the relative changes of x and y as given by the slope to plot a second point.

$$m = \frac{+7}{+4} = \frac{\text{Change in } y}{\text{Change in } x}$$

3 Draw the line through these points. □

Answers to problems involving linear equations are commonly written in the general form $Ax + By + C = 0$. If A, B, and C are rational numbers, this form can be written free of fractions. A summary of the most frequently used forms of linear equations is given in the box.

Forms of Linear Equations ▼

General form:	$Ax + By + C = 0$
Point-slope form:	$y - y_1 = m(x - x_1)$
Slope-intercept form:	$y = mx + b$
Horizontal line:	$y = k$ for any constant k
Vertical line:	$x = h$ for any constant h

▶ Self-Check ▼

1 Is the line L_1 through $(-4, 2)$ and $(6, -3)$ parallel to the line L_2 through $(3, -3)$ and $(1, 0)$?

2 Is the line through $(18, 1)$ and $(-4, -10)$ perpendicular to the line through $(18, 1)$ and $(8, 21)$?

3 By inspection, determine the slope and one point on the line $y - 4 = 3(x + 2)$.

▶ Self-Check Answers ▼

1 Since $m_1 = -\frac{1}{2}$ and $m_2 = -\frac{3}{2}$, the slopes are not equal and the lines are not parallel.

2 Since $m_1 = \frac{1}{2}$ and $m_2 = -2$, $m_1 = -\frac{1}{m_2}$ and the lines are perpendicular

3 $m = 3$ and $(x_1, y_1) = (-2, 4)$

CHAPTER 3 FUNCTIONS, RELATIONS, AND GRAPHS

EXAMPLE 8 Write the equation of a line perpendicular to $2x + 3y = 12$ and passing through the point $(4, -1)$.

SOLUTION

$2x + 3y = 12$	Find the slope of the given line from the slope-intercept form,
$3y = -2x + 12$	$-y = mx + b$.
$y = -\dfrac{2}{3}x + 4$	
$m_1 = -\dfrac{2}{3}$ and $m_2 = \dfrac{3}{2}$	The slopes of perpendicular lines are negative reciprocals.
$y - y_1 = m(x - x_1)$	Point-slope form.
$y - (\boxed{-1}) = \dfrac{3}{2}(x - \boxed{4})$	Substitute the given point and slope.
$2(y + 1) = 3(x - 4)$	Simplify.
$3x - 2y - 14 = 0$	Write in the general form, $Ax + By + C = 0$.

▶ **Self-Check** ▼

1. Write the equation of a line through $(2, 4)$ and $(-5, 4)$.
2. Write the equation of a line through $(2, 4)$ and $(2, -5)$.
3. Write in general form the equation of a line with slope $\frac{2}{7}$ and y-intercept $-\frac{1}{2}$.
4. Find the intercepts of $2x + 3y = 12$.

EXAMPLE 9 A company uses straight-line depreciation to depreciate a lathe purchased new for $15,000. At the end of eight years, the estimated trade-in or salvage value is $2000. Determine the linear function that gives the value V after n years.

SOLUTION Use the slope-intercept form to determine the linear function. Since $15,000 is the value after zero years and $2000 is the value after eight years, two points on the line are $(0, 15{,}000)$ and $(8, 2000)$.

$V(n) = mn + b$	In this problem the number of years, n, is the independent variable, and the value, V, is the dependent variable. Thus the slope-intercept form $y = mx + b$ is replaced by $V(n) = mn + b$.
$m = \dfrac{15{,}000 - 2000}{0 - 8}$	Calculate the slope (the change in value divided by the change in years).
$m = -1625$	
$V(n) = \boxed{-1625}\, n + \boxed{15{,}000}$	Substitute the slope, -1625, and the intercept, $(0, 15{,}000)$, into the equation.

Answer The linear function that gives the value after n years is $V(n) = -1625n + 15{,}000$, and the domain for n is $\{0, 1, \ldots, 8\}$.

▶ **Self-Check Answers** ▼

1 $y = 4$ **2** $x = 2$ **3** $4x - 14y - 7 = 0$ **4** x-intercept $(6, 0)$; y-intercept $(0, 4)$

SECTION 3-2 LINES

EXAMPLE 10 A balloon rises vertically at the rate of 15 meters per second and is blown horizontally by the wind at the rate of 8 meters per second. Express the distance d between the balloon and the release point as a function of the time t in seconds.

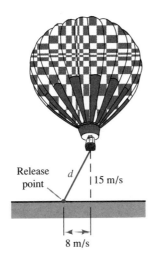

SOLUTION

$$\left(\begin{array}{c}\text{Distance between balloon} \\ \text{and release point}\end{array}\right)^2 = \left(\begin{array}{c}\text{Horizontal} \\ \text{distance}\end{array}\right)^2 + \left(\begin{array}{c}\text{Vertical} \\ \text{distance}\end{array}\right)^2$$

The *word equation* is based on the Pythagorean Theorem.

Rate Principle:	Rate	·	Time	=	Distance
Horizontal	8	·	t	=	$8t$
Vertical	15	·	t	=	$15t$

$$[d(t)]^2 = (8t)^2 + (15t)^2$$
$$[d(t)]^2 = 64t^2 + 225t^2$$
$$[d(t)]^2 = 289t^2$$
$$d(t) = 17t \quad \text{or} \quad d(t) = -17t$$

— Negative values are not appropriate for distance.

Answer $d(t) = 17t$

Many graphs are best analyzed by partitioning the domain and then describing the behavior of the graph in each of these separate pieces. An example of a **piecewise function** is given in Example 11, where each piece of the function is defined by a different linear formula.

172 CHAPTER 3 FUNCTIONS, RELATIONS, AND GRAPHS

EXAMPLE 11 Graph

$$y = \begin{cases} -2 & \text{if } x < -1 \\ 2x + 1 & \text{if } -1 \leq x \leq 2 \\ 5 & \text{if } x > 2 \end{cases}$$

SOLUTION Each piece of this piecewise function is linear. The three pieces are shown separately below.

If $x < -1$, $y = -2$. This piece of the graph is a portion of a horizontal line.

If $-1 \leq x \leq 2$, $y = 2x + 1$. This piece of the graph is a portion of the line with slope 2 and y-intercept 1.

If $x > 2$, $y = 5$. This piece of the graph is a portion of a horizontal line.

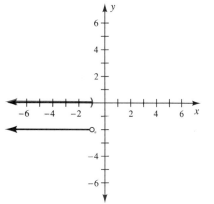 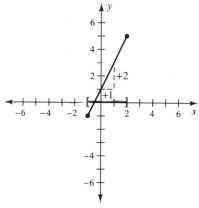 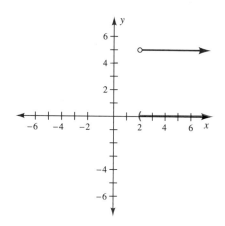

The domain of each individual piece is in color.
  To obtain the graph of the entire function, we combine these individual pieces in the figure below.

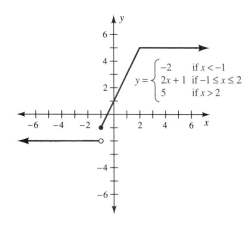

SECTION 3-2 LINES

EXERCISES 3-2

A

In Exercises 1 and 2, calculate the slope of the line through each pair of points.
1. **a.** (2, 7) and (4, 11) **b.** (−2, 6) and (2, 6) **c.** (−11, 7) and (−11, 3)
2. **a.** (−12, 6) and (8, 1) **b.** (2, −6) and (2, 6) **c.** (2, −3) and (14, −3)

In Exercises 3–6, determine the slope of each line. If the slope is defined, describe the linear function as increasing, decreasing, or constant.
3. **a.** $y = 3x - 2$ **b.** $y = -\frac{2}{3}x + \frac{4}{7}$ **c.** $y = 5$
4. **a.** $y = -2x + 3$ **b.** $y = \frac{4}{7}x - \frac{2}{3}$ **c.** $y = 7$
5. **a.** $x = 5$ **b.** $2x + 3y = 12$ **c.** $6x + 2y + 4 = 0$
6. **a.** $x = 7$ **b.** $3x - 2y = 12$ **c.** $6x + 3y - 4 = 0$

In Exercises 7–22, graph the lines described.
7. A line through (5, 3) and (−1, 2)
8. A horizontal line through (1, −2)
9. A vertical line through (1, −2)
10. A line with x-intercept 2 and y-intercept −3
11. A line through (−3, −2) with $m = \frac{5}{4}$
12. A line through (−3, 2) with $m = -\frac{2}{5}$
13. $2x - 3y = 12$
14. $4x + 5y = 20$
15. $y = -2$
16. $x = 3$
17. $y = \frac{5}{3}x - 4$
18. $y = \frac{3}{4}x + 1$
19. $y - 2 = \frac{2}{5}(x + 1)$
20. $y + 3 = -\frac{4}{3}(x - 2)$
21. $f(x) = 2x$
22. $f(x) = -x + 3$

In Exercises 23–28, determine if the line through the first pair of points is parallel or perpendicular to the line through the second pair of points.
23. (2, 0) and (13, 4)
    (1, −2) and (−3, 9)
24. (−2, 8) and (3, −2)
    (4, 1) and (7, −5)
25. (0, 5) and (0, 7)
    (5, 0) and (5, 7)
26. (−3, 2) and (4, −1)
    (4, −5) and (7, −11)
27. (−4, −7) and (0, 5)
    (−4, −10) and (6, 10)
28. (0, −2) and (2, 4)
    (0, 0) and (−3, 1)

In Exercises 29–48, write the general form of a line satisfying the given conditions.
29. $m = 2, b = -5$
30. $m = -5, b = 2$
31. $m = \frac{1}{2}, b = -\frac{3}{4}$
32. $m = -\frac{3}{7}, b = 2$
33. $m = 0$ through (−6, −6)
34. $m = 7, b = 0$
35. A vertical line through (−4, 7)
36. $m = \frac{4}{7}$ through (8, −1)
37. A horizontal line through (4, −7)
38. A vertical line through (4, −7)
39. A line through (−5, 6) and (6, −5)
40. $m = 0$ through (6, 6)
41. A line through (0, 5) and perpendicular to the line segment from (−2, 7) to (2, 3)
42. A line through (4, 5) and (−4, −5)
43. A line through (2, 3) and parallel to $y = 2x + 3$
44. A line through (3, 2) and parallel to $y = 3x + 2$
45. A line through (−1, 4) and perpendicular to $y = 5x - 4$
46. A line through (1, 3) and perpendicular to the line segment from (−4, −1) to (6, 7)
47. A line through (−2, 5) and parallel to $5x + 7y = 35$
48. A line through (5, −2) and perpendicular to $3x - 4y = 12$

In Exercises 49–52, use the y-intercept and the slope to graph each equation.

49 $y = -\frac{2}{5}x + 3$ **50** $y = \frac{5}{2}x - 3$ **51** $y = \frac{7}{3}x - 2$ **52** $y = \frac{3}{7}x + 2$

In Exercises 53 and 54, estimate the x-intercept, the y-intercept, and the slope of each line.

53 a. 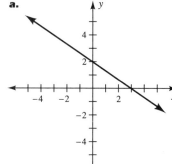 **b.** 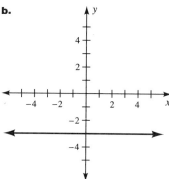 **c.**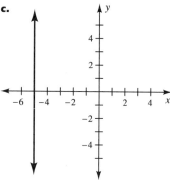

54 a. 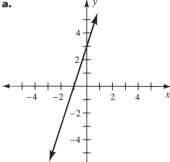 **b.** 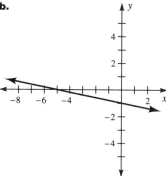 **c.**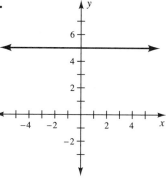

B

55 Determine if the points (3, 9), (5, 7), and (6, 8) form a right triangle.

56 In the quadrilateral ABCD with A(0, 0), B(3, 4), C(3, 8), D(0, 4), are the diagonals perpendicular?

In Exercises 57–58, an item is depreciated using straight-line depreciation. The purchase price and the scrap value are given. Determine the linear function that gives the value V after n years.

57 Cost is $4000, and scrap value after ten years is $800.

58 Cost is $7500, and scrap value after eight years is $300.

59 The cost C of x units of production is $C(x) = 10x + 250$. The marginal cost is the change in cost for one additional unit of production.

  a. Find C(5).
  b. What is the slope of the cost function?
  c. Is the cost an increasing or a decreasing function?
  d. What is the marginal cost?

SECTION 3-2 LINES

60 The perimeter of a rectangle is 60 meters. Express its length as $l(w)$, a function of the width, w.

61 A rope of length x is laid so that it is in the shape of a circle. Express the radius of this circle as $r(x)$, a function of x.

62 A business has fixed monthly costs of $3500 and variable costs of $23 per computer copy stand that they ship. Express the total, T, of these fixed and variable costs as a function of the number, n, of copy stands shipped.

In Exercises 63–66, graph each function and determine the intervals for which the function is increasing, decreasing, or constant.

63 $y = \begin{cases} x + 2 & \text{if } x \leq 0 \\ 2 & \text{if } x > 0 \end{cases}$

64 $y = \begin{cases} 3x & \text{if } x \leq 1 \\ 2x + 1 & \text{if } x > 1 \end{cases}$

65 $y = \begin{cases} -x + 1 & \text{if } x < -3 \\ 2x & \text{if } -3 \leq x \leq 2 \\ 3x - 1 & \text{if } x > 2 \end{cases}$

66 $y = \begin{cases} -1 & \text{if } x < -1 \\ x & \text{if } -1 \leq x \leq 1 \\ 1 & \text{if } x > 1 \end{cases}$

graph by hand first
(note discontinuities)

$f(x) = \begin{cases} x - 1 & \text{if } x < -1 \\ 4 - x^2 & \text{if } -1 \leq x \leq 3 \\ \frac{x}{4} & \text{if } x > 3 \end{cases}$

THEN GRAPH ON TI-81
{DOT} MODE

C

67 A homeowner has 24 meters of fencing to enclose a rectangular playing area for his children. One side of the playing area will be formed by the wall of an existing building.

  a. Express the length, l, of the rectangle as a function of the width, w.

  b. Express the area, A, of the rectangle as a function of the width, w.

Figure for Exercise 67

68 The Norman window shown below will be designed so that the rectangular portion has a height of two meters and a width w.

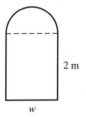

Figure for Exercise 68

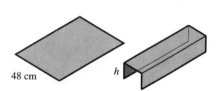

Figure for Exercise 69

  a. Express the perimeter, P, of this window as a function of the width, w.

  b. Express the area, A, of this window as a function of the width, w.

69 A piece of steel sheet 48 centimeters wide is bent to form the frame for a storm drain by turning down the two ends perpendicular to the piece of steel.

  a. Express the width of frame as a function of the height, h, of this frame.

  b. Express the area of the opening of this frame as a function of the height, h. (Consider the thickness of the steel as negligible.)

70 Two vertical radio towers are located 95 kilometers apart. Are these towers parallel?

SECTION 3-3

Translations of Graphs and Parabolas

Section Objectives

8 Use translations and reflections to graph functions.

9 Graph a parabola.

10 Determine the maximum or minimum value of a quadratic function.

One of the main goals of college algebra is to study the basic properties and characteristics of some of the more common functions. Whole families of functions have graphs with the same basic shape. Thus it is efficient to be able to recognize the basic shape of the graph of a function from its defining equation. Then we can determine what modifications to this shape are necessary to produce the specific member from this family of graphs. In this section we will examine the effects to a known graph produced by translations, by reflections, and by shrinking or stretching. In particular, we will examine these effects on the families of functions to which $y = |x|$ and $y = x^2$ belong.

The vertical component of the point $(x, f(x))$ on the graph of $y = f(x)$ is $f(x)$. For $k > 0$, the graph of $y = f(x) + k$ is shifted vertically k units above the graph of $y = f(x)$ and the graph of $y = f(x) - k$ is shifted k units below the graph of $y + f(x)$. For $k > 0$, $(x, y + k)$ is k units above (x, y) and $(x, y - k)$ is k units below (x, y). Figure 3-7 illustrates vertical translations of the graph of $y = |x|$.

The graph of $y = |x|$ can be obtained by plotting points from a table of values such as that given below.

x	y
−4	4
−3	3
−2	2
−1	1
0	0
1	1
2	2
3	3
4	4

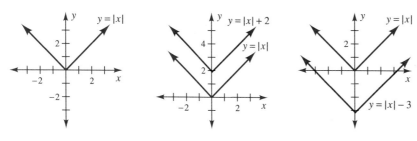

Figure 3-7 Vertical translations

We will now examine horizontal translations of the graph of $y = f(x)$. The horizontal component of $(x, f(x))$ is x. For $h > 0$, the graph of $y = f(x + h)$ is shifted horizontally h units to the left of the graph of $y = f(x)$ and the graph of $y = f(x - h)$ is shifted h units to the right of the graph of $y = f(x)$. Figure 3-8 illustrates horizontal translations of the graph of $y = |x|$.

Inside the parentheses the constant h affects the domain before the function f is applied to this domain element. Adding h inside the parentheses causes the action produced by f to happen h units sooner, that is, h units to the left.

SECTION 3-3 TRANSLATIONS OF GRAPHS AND PARABOLAS

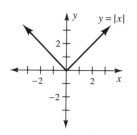 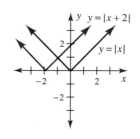 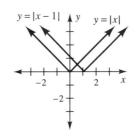

The vertex of $y = |x + 2|$ occurs where $y = 0$. Solving $|x + 2| = 0$ yields $x = -2$. The vertex of $y = |x + 2|$ occurs at $(-2, 0)$. Likewise solving $|x - 1| = 0$ yields $x = 1$. The vertex of $y = |x - 1|$ occurs at $(1, 0)$.

Figure 3-8 Horizontal translations

Horizontal and vertical translations can be combined as described in the following box.

Translations ▼

For positive constants h and k: $f(x-h)$ where h is negative represents a horizontal shift to the left

Horizontal: The graph of $y = f(x + h)$ is obtained by shifting the graph of $y = f(x)$ to the left h units.
The graph of $y = f(x - h)$ is obtained by shifting the graph of $y = f(x)$ to the right h units.

Vertical: The graph of $y = f(x) + k$ is obtained by shifting the graph of $y = f(x)$ up k units.
The graph of $y = f(x) - k$ is obtained by shifting the graph of $y = f(x)$ down k units.

Combined: The graph of $y = f(x - h) - k$ is obtained by shifting the graph of $y = f(x)$ to the right h units and down k units. Other combinations behave similarly.

EXAMPLE 1 Given the graph of $y = f(x)$ shown to the right, graph $y = f(x - 1) + 3$.

SOLUTION The graph of $y = f(x - 1) + 3$ is obtained by shifting the graph of f one unit to the right and three units up. (See figure below.)

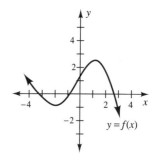

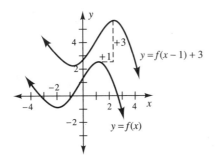

Another useful curve-sketching aid is reflection. The graph of $y = -f(x)$ is a reflection about the x-axis of the graph of $y = f(x)$. The **reflection** of the point (x, y) about the x-axis is the point $(x, -y)$ (see Figure 3-9). Figure 3-10 illustrates the reflection of $y = |x|$ about the x-axis.

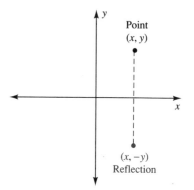

Figure 3-9 Reflection about the x-axis

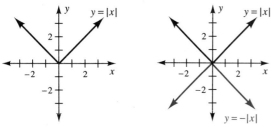

Figure 3-10 Reflection of $y = |x|$ about the x-axis

EXAMPLE 2 Given the graph of $y = f(x)$ shown to the right, graph $y = -f(x)$.

SOLUTION The new function is a reflection of $y = f(x)$ about the x-axis. (Shown in the figure below.)

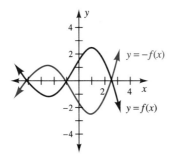

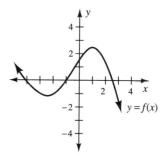

▶ **Self-Check** ▼

Given the graph of $y = x^2$ shown below, graph

$$y = (x + 2)^2 - 1.$$

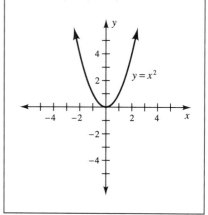

▶ **Self-Check Answer** ▼

Shift the graph of $y = x^2$ left two units and down one unit.

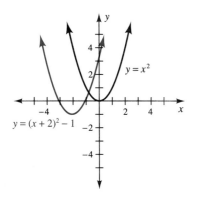

SECTION 3-3 TRANSLATIONS OF GRAPHS AND PARABOLAS

Vertical translations, horizontal translations, and reflections are all rigid transformations that cause no distortion in the graph that is transformed. If these transformations are illustrated with a clear transparency on an overhead projector, then the new graph can be obtained by simply moving the original graph while keeping the axis system fixed. We will now examine stretching and shrinking transformations which distort the original shape somewhat. These transformations cannot be illustrated by simply moving the original graph. Instead we must either change the scale of the axis system or *stretch* or *shrink* the original graph.

Stretchings, Shrinkings, and Reflections ▼

$$y = af(x)$$

$a > 1$: We shall refer to $y = af(x)$ as a *stretching* of $y = f(x)$ if $a > 1$.

$0 < a < 1$: We shall refer to $y = af(x)$ as a *shrinking* of $y = f(x)$ if $0 < a < 1$.

$a = -1$: $y = -f(x)$ is a reflection of $y = f(x)$ about the x-axis.

Figure 3-11 illustrates the stretching and the shrinking of the graph of $y = |x|$.

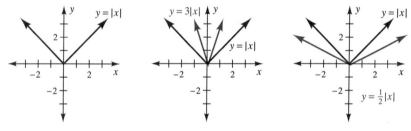

Figure 3-11 Stretching and shrinking

EXAMPLE 3 Given the graph of $y = x^2$ shown below, graph $y = -\frac{1}{2}x^2$.

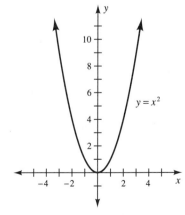

The graph of the parabola $y = x^2$ can be obtained by plotting points from a table of values such as that given below.

x	y
-3	9
-2	4
-1	1
$-\frac{1}{2}$	$\frac{1}{4}$
0	0
$\frac{1}{2}$	$\frac{1}{4}$
1	1
2	4
3	9

SOLUTION The graph of $y = -\frac{1}{2}x^2$ can be obtained from the graph of $y = x^2$ by first using a shrinking factor of $\frac{1}{2}$ and then reflecting this graph about the x-axis.

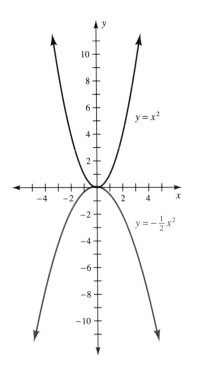

It may be wise to test one or two points in the transformed graph to confirm that the transformation is correct. In this case, note that the point $(2, -2)$ satisfies $y = -\frac{1}{2}x^2$.

▶ **Self-Check** ▼

Given the graph of $y = x^2$ shown below, graph $y = 2x^2$.

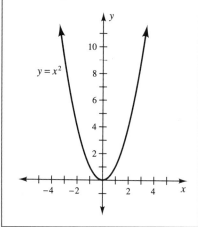

▶ **Self-Check Answer** ▼

A stretching factor of 2 is applied to the graph of $y = x^2$.

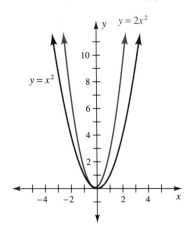

The Parabola

The parabolic shape occurs frequently in nature. (See Figure 3-12.) The path of a thrown ball is parabolic (assuming air resistance is ignored). Some cables supporting bridges hang as parabolas. Parabolic reflectors are used for spotlights and for picking up sound at sporting events. The up-and-down trend of profit and cost functions sometimes has a parabolic form.

The graph of $f(x) = ax^2 + bx + c$ $(a \neq 0)$ is always a parabola that is either concave upward or concave downward (see Figure 3-13). Each graph from this family of parabolas can be obtained by performing transformations to the parabola defined by $y = x^2$. Some of the interesting properties shared by all these parabolas are that each is symmetric about a vertical line and each has a vertex that is either the lowest or the highest point on the curve. For profit or cost functions, these vertices are especially important, for they can indicate levels of least cost or of highest profit.

Figure 3-12 A parabolic stream of water

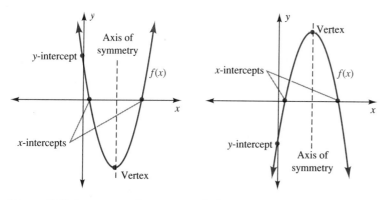

Figure 3-13 Important points on a parabola

The graph of $f(x) = x^2$ is concave upward (see Figure 3-14) and the graph of $f(x) = -x^2$ is concave downward. In general, the coefficient of the second-degree term in the parabola's equation determines the concavity of the parabola. Thus, the parabola $f(x) = ax^2 + bx + c$ is concave upward if $a > 0$ and concave downward if $a < 0$.

The vertex of $f(x) = (x - 1)^2 + 2$ can be obtained from the vertex of $f(x) = x^2$ by shifting that vertex one unit to the right and two units up. That is, the vertex of $f(x) = (x - 1)^2 + 2$ is (1, 2). This analysis can be generalized to the parabola $f(x) = (x - h)^2 + k$ to show that its vertex is (h, k).

The Parabola $f(x) = (x - h)^2 + k$ ▼

This parabola is concave upward with vertex (h, k).

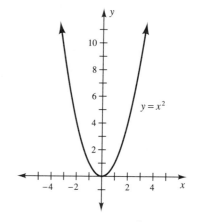

The parabola $y = x^2$
Concave upward with vertex (0, 0)

Figure 3-14

EXAMPLE 4 Determine the vertex and the concavity of $f(x) = (x - 5)^2 + 7$.

SOLUTION $f(x) = (x - \boxed{5})^2 + \boxed{7}$ The vertex of $f(x) = (x - h)^2 + k$ is (h, k).

This parabola is concave upward with vertex (5, 7).   In expanded form $f(x) = (x - 5)^2 + 7$ can be written as $f(x) = x^2 - 10x + 32$. Since the coefficient of the second-degree term is positive, this parabola is concave upward. □

In the next example we will use the process of completing the square to rewrite the equation of the parabola in the form $f(x) = (x - h)^2 + k$. Then we can determine the vertex of the parabola by inspection.

EXAMPLE 5 Determine the vertex of $f(x) = x^2 + 4x + 1$.

SOLUTION $f(x) = x^2 + 4x + 1$

$f(x) = (x^2 + 4x + \boxed{4}) + 1 - \boxed{4}$ Add and subtract 4 to complete the square.

$f(x) = (x + 2)^2 - 3$ Write in perfect-square form.

$f(x) = [x - (\boxed{-2})]^2 + (\boxed{-3})$ Write the equation in the form $f(x) = (x - h)^2 + k$.

Answer The vertex is $(-2, -3)$.   □

Remember that the quadratic formula was developed by using completing the square. This formula can then be applied directly rather than going through the steps of completing the square. Similarly, completing the square can be used to develop a formula for the vertex of a parabola. Sometimes it may be preferable to use this formula as illustrated in the following examples.

▶ **Self-Check** ▼

Determine the concavity of each of these parabolas.

1 $y = -2x^2 + 3x - 7$
2 $y = 2x^2 - 3x - 11$
3 $y + x^2 = 8$

Derivation of the Coordinates of the Vertex of $f(x) = ax^2 + bx + c$

$f(x) = ax^2 + bx + c$

$f(x) = a\left(x^2 + \dfrac{b}{a}x + \dfrac{c}{a}\right)$ Factor out a.

$f(x) = a\left[x^2 + \dfrac{b}{a}x + \boxed{\dfrac{b^2}{4a^2}} + \dfrac{c}{a} - \boxed{\dfrac{b^2}{4a^2}}\right]$ Complete the square.

$f(x) = a\left[\left(x + \dfrac{b}{2a}\right)^2 + \dfrac{4ac - b^2}{4a^2}\right]$ Write in perfect-square form.

$f(x) = a\left(x + \dfrac{b}{2a}\right)^2 + \dfrac{4ac - b^2}{4a}$ Use the distributive property and multiply each term by a.

▶ **Self-Check Answers** ▼

1 Concave downward   **2** Concave upward   **3** Concave downward

SECTION 3-3 TRANSLATIONS OF GRAPHS AND PARABOLAS

Thus the vertex occurs at the point $\left(\dfrac{-b}{2a}, \dfrac{4ac - b^2}{4a}\right)$.

Although this formula can be used to find both coordinates, we recommend memorizing only the x-coordinate, $\dfrac{-b}{2a}$, and using the formula $f(x)$ to calculate the y-coordinate, $f\left(\dfrac{-b}{2a}\right)$. This formula is given in the following box, which summarizes a procedure for graphing the parabola defined by $f(x) = ax^2 + bx + c$. This procedure is an alternative to using transformations, which may include stretchings and shrinkings, of $y = x^2$.

Graphing Parabolas ▼

To graph the parabola defined by $f(x) = ax^2 + bx + c$, follow this procedure.

Step 1 Determine whether the parabola opens upward or downward.
  a. If $a > 0$, it opens upward.
  b. If $a < 0$, it opens downward.

Step 2 Plot the vertex, $\left(\dfrac{-b}{2a}, f\left(\dfrac{-b}{2a}\right)\right)$.

Step 3 Plot the y-intercept, the point $(0, c)$. (Set $x = 0$.)

Step 4 a. Determine and plot the x-intercepts (if any) by solving $ax^2 + bx + c = 0$. (Set $y = 0$.)
  b. If there are no x-intercepts, plot two or three other points.

Step 5 Draw a smooth curve through these points.

There is always exactly one y-intercept of the parabola defined by $f(x) = ax^2 + bx + c$ because this function pairs each x with exactly one y. However, a parabola may have two, one, or no x-intercepts, depending on whether $ax^2 + bx + c = 0$ has two real roots, one double root, or two imaginary roots. (See Figure 3-15.)

Two x-intercepts, x_1 and x_2, are the real roots of $ax^2 + bx + c = 0$.

One x-intercept, x, is the double root of $ax^2 + bx + c = 0$.

There are no x-intercepts; the roots of $ax^2 + bx + c = 0$ are imaginary.

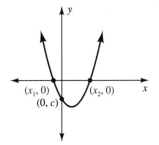
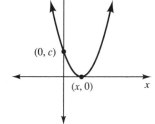
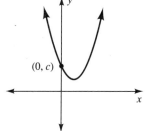

Figure 3-15 Possible x-intercepts

EXAMPLE 6 Graph $f(x) = 3x^2 + 5x - 2$.

SOLUTION
1. Since $3 > 0$, the parabola opens upward.
2. The x-coordinate of the vertex is

$$\frac{-b}{2a} = \frac{-5}{2(3)} = -\frac{5}{6}$$

For $f(x) = 3x^2 + 5x - 2$, $a = 3$ and $b = 5$.

The y-coordinate of the vertex is

$$f\left(\frac{-b}{2a}\right) = f\left(\frac{-5}{6}\right) = 3\left(\frac{-5}{6}\right)^2 + 5\left(\frac{-5}{6}\right) - 2$$

Calculate the y-coordinate using the equation $f(x) = 3x^2 + 5x - 2$.

$$= 3\left(\frac{25}{36}\right) - \frac{25}{6} - 2$$

$$= \frac{75 - 150}{36} - 2$$

$$= \frac{-75}{36} - 2$$

$$= -4\frac{1}{12}$$

Thus the vertex is $\left(-\frac{5}{6}, -4\frac{1}{12}\right)$.

3. Set $x = 0$ to find the y-intercept $(0, -2)$.
4. Set $y = 0$ and solve $3x^2 + 5x - 2 = 0$ to find the x-intercepts.

$$3x^2 + 5x - 2 = 0$$
$$(3x - 1)(x + 2) = 0$$
$$3x - 1 = 0 \quad \text{or} \quad x + 2 = 0$$
$$x = \frac{1}{3} \qquad\qquad x = -2$$

Thus the intercepts are $(\frac{1}{3}, 0)$ and $(-2, 0)$.
5. The graph is shown to the right.

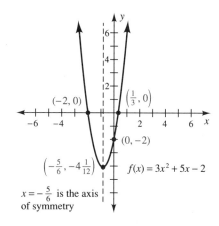

▶ **Self-Check** ▼

Determine the vertices of the parabolas defined by these equations.
1 $f(x) = x^2 - 5$ **2** $f(x) = (x - 5)^2$
3 $f(x) = (x + 10)^2 - 3$

Note that if there are x-intercepts, they are always symmetric about the axis of symmetry. This fact can be used as a quick check on the accuracy of some of your work. This symmetry is actually the basis for a handy way to

$x = \dfrac{-b}{2a}$ is the average of the x-coordinates $\dfrac{-b - \sqrt{b^2 - 4ac}}{2a}$ and $\dfrac{-b + \sqrt{b^2 - 4ac}}{2a}$.

▶ **Self-Check Answers** ▼

1 $(0, -5)$ **2** $(5, 0)$ **3** $(-10, -3)$

SECTION 3-3 TRANSLATIONS OF GRAPHS AND PARABOLAS

remember the formula for the vertex, $\left(\dfrac{-b}{2a}, f\left(\dfrac{-b}{2a}\right)\right)$. The x-intercepts can be found by using the quadratic formula to solve $ax^2 + bx + c = 0$. Thus $\dfrac{-b}{2a}$ plays a key role in both the vertex and the quadratic formula

$$x = \dfrac{-b \pm \sqrt{b^2 - 4ac}}{2a}$$

This fact facilitates memorizing the formula for the vertex.

EXAMPLE 7 Graph $y = -x^2 + 4x - 7$.

SOLUTION
1. Since the coefficient of x^2 is -1, the parabola opens downward.
2. The x-coordinate of the vertex is

$$\dfrac{-b}{2a} = \dfrac{-4}{2(-1)} = 2$$

The y-coordinate of the vertex is

$$f(2) = -(2)^2 + 4(2) - 7 = -3$$

Thus the vertex is $(2, -3)$.

3. Set $x = 0$ to find the y-intercept $(0, -7)$.
4. **(a)** Set $y = 0$ and solve $-x^2 + 4x - 7 = 0$ to find the x-intercepts.

$$-x^2 + 4x - 7 = 0$$

$$x = \dfrac{-4 \pm \sqrt{4^2 - 4(-1)(-7)}}{2(-1)}$$

$$= \dfrac{-4 \pm \sqrt{16 - 28}}{-2}$$

$$x = \dfrac{-4 \pm \sqrt{-12}}{-2} = 2 \pm i\sqrt{3}$$

The calculations are shown in Step 4a to illustrate what happens if we try to find x-intercepts when none exist. However, it is more efficient to note that since the vertex is below the x-axis and the parabola opens downward, there can be no x-intercepts.

There are no x-intercepts, since these roots are imaginary.

(b) Determine three other points to plot. For $x = 1$.

$$f(1) = -1^2 + 4(1) - 7 = -4$$

Thus $(1, -4)$ is on the parabola. The axis of symmetry $x = 2$ passes through the vertex $(2, -3)$. The point $(4, -7)$ is the symmetric image of $(0, -7)$ about the axis of symmetry. The symmetric image of $(1, -4)$ is $(3, -4)$.

5. The graph is shown to the right.

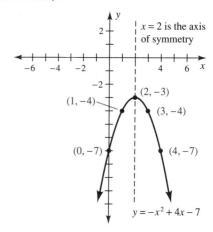

Since the value of $f(x)$ at the vertex is either the greatest value of the function or the least value of the function, quadratic functions can be used to solve some maximum and minimum problems.

By finding the vertex of the parabolas in Examples 8 and 9, we can determine the maximum height in Example 8 and the minimum cost in Example 9.

▶ **Self-Check** ▼

Graph $f(x) = x^2 - 4x + 4$ by determining its most important features.

EXAMPLE 8 The height of a projectile in meters is a function of the time in seconds since the projectile was released. Find the maximum height achieved by the projectile whose height in meters after t seconds is given by $h(t) = 147t - 4.9t^2$. For what intervals is the projectile rising? For what intervals is the projectile falling?

SOLUTION The graph of $h(t) = 147t - 4.9t^2$ is a parabola that is concave downward. Thus the maximum height will be at the y-coordinate of the vertex.

$$t = \frac{-b}{2a} = \frac{-147}{2(-4.9)} = 15$$

$$h(15) = 147(15) - 4.9(15)^2 = 1102.5$$

The vertex is (15, 1102.5)

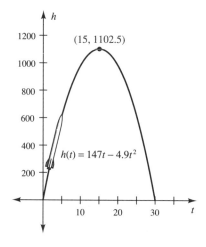

From the moment the projectile is released, the tug of gravity causes the flight path to be concave downward.

Answer The maximum height of 1102.5 meters occurs after 15 seconds. The projectile is rising from 0 to 15 seconds and falling from 15 to 30 seconds.

▶ **Self-Check Answer** ▼

The parabola opens upward. The vertex is (2, 0). The y-intercept is (0, 4). The only x-intercept is (2, 0), which is also the vertex and a point of tangency to the x-axis. (4, 4) is the symmetric image of (0, 4) about the axis of symmetry $x = 2$.

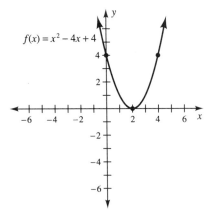

SECTION 3-3 TRANSLATIONS OF GRAPHS AND PARABOLAS

The average cost C of manufacturing t units is often a parabolic function whose graph is concave upward. (See Figure 3-16.) Average cost starts high because the overhead is spread over very few units. As production increases, the overhead cost is spread over more units and thus the average cost drops. Eventually the law of diminishing returns sets in (cost of overtime, limits of machine capacity, etc.), and average cost begins to increase.

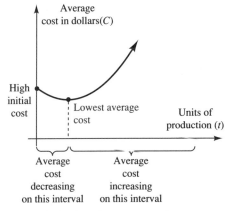

Figure 3-16 Manufacturing costs

EXAMPLE 9 A manufacturer of calculators has gathered data suggesting that the average cost in dollars of producing t units per hour of a new model in its product line will be $C(t) = t^2 - 16t + 69$ for $1 \leq t \leq 100$. (The formula is only valid for 1 to 100 units of production.) Determine how many calculators should be produced per hour in order to minimize the average cost. What will this cost be?

SOLUTION The graph of $C(t) = t^2 - 16t + 69$ is a parabola that opens upward. Thus the minimum cost will be at the y-coordinate of the vertex.

$$t = \frac{-b}{2a} = \frac{-(-16)}{2(1)} = 8$$

$$C(8) = 8^2 - 16(8) + 69 = 64 - 128 + 69 = 5$$

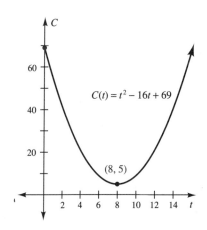

Answer The average cost can be minimized by producing eight calculators per hour. The minimum average cost per calculator is $5. □

Equations of the form $x = ay^2 + by + c (a \neq 0)$ define parabolas that open either to the left or to the right instead of up or down. The key information used to graph these parabolas can be obtained by the same method used to graph $y = ax^2 + bx + c$ except that the roles of x and y are interchanged.

EXAMPLE 10 Graph the relation $x = y^2$.

SOLUTION This parabola opens to the right because $a > 0$. In the form $x = ay^2 + by + c$, $a = 1$ in $x = y^2$.

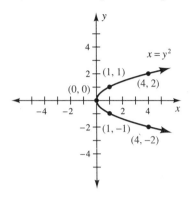

y	$x = y^2$
-2	4
-1	1
0	0
1	1
2	4

Note by the vertical line test that $x = y^2$ is not a function with x as the independent variable. However, $x = y^2$ is a function with y as the independent variable and x as the dependent variable. We customarily graph relationships so that the independent variable is along the horizontal axis. Thus we use the form $y = ax^2 + bx + c$ more often than we use the form $x = ay^2 + by + c$.

EXERCISES 3-3

A

In Exercises 1 and 2 use translations or reflections and the graph of $y = f(x)$ shown to the right to sketch the graph of each function.

1 a. $y = f(x - 2)$ b. $y = f(x) - 2$ c. $y = f(x - 3) + 1$ d. $y = -f(x)$
2 a. $y = f(x + 3)$ b. $y = f(x) + 3$ c. $y = f(x - 1) + 4$ d. $y = -f(x) + 3$

In Exercises 3 and 4, the point (2, 3) lies on the graph of $y = f(x)$. Determine the point that must lie on the graph of each function.

3 a. $y = f(x) + 3$ b. $y = f(x + 5)$ c. $y = f(x - 6) - 9$ d. $y = -f(x) + 1$
4 a. $y = f(x) - 3$ b. $y = f(x - 7)$ c. $y = f(x + 4) + 8$ d. $y = -f(x)$
5 Graph the point $(-3, 5)$ and its reflection about the x-axis.
6 Graph the point $(2, -4)$ and its reflection about the x-axis.

In Exercises 7–10, sketch the graph of $y = |x|$ and the graph of the given function on the same coordinate system.

7 a. $y = |x| - 4$ b. $y = |x - 4|$ c. $y = |x + 1| + 2$ d. $y = -|x| - 3$
8 a. $y = |x| + 4$ b. $y = |x + 4|$ c. $y = |x - 1| - 3$ d. $y = -|x| + 2$

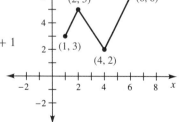

Figure for Exercises 1–2

SECTION 3-3 TRANSLATIONS OF GRAPHS AND PARABOLAS

9 a. $y = 4|x|$ b. $y = \frac{1}{4}|x|$ c. $y = -3|x|$
10 a. $y = 3|x|$ b. $y = \frac{1}{3}|x|$ c. $y = -2|x|$

In Exercises 11 and 12, sketch the graph of $y = x^2$ and the graph of the given function on the same coordinate system.

11 a. $y = (x - 1)^2 + 2$ b. $y = -x^2 + 3$ c. $y = \frac{1}{3}x^2$
12 a. $y = (x + 1)^2 - 3$ b. $y = -x^2 - 1$ c. $y = -\frac{1}{4}x^2$

In Exercises 13–18, (a) determine whether the parabola defined by each equation opens upward or downward, (b) determine the y-intercept of the parabola, and (c) find the x-intercepts, if there are any.

13 $y = x^2 + 3x - 10$ **14** $f(x) = -4x^2 + 20x - 25$ **15** $f(x) = x^2 - 6x + 4$
16 $y = 2x^2 - 3x - 2$ **17** $y = x^2 - 4x + 6$ **18** $f(x) = 2x^2 - 3x + 2$

In Exercises 19 and 20, find the vertex of the parabola defined by each equation.

19 a. $f(x) = x^2 + 2$ b. $y = (x - 3)^2 + 5$ c. $f(x) = -2x^2 + 8x - 15$
20 a. $y = (x + 3)^2$ b. $f(x) = (x - 1)^2 - 3$ c. $y = 3x^2 - 6x + 11$

In Exercises 21–26, graph the parabolas satisfying the given conditions.

	Opens	Vertex	y-intercept	x-intercept	Other Points
21	Upward	(4, −1)	(0, 15)	(3, 0), (5, 0)	(1, 8)
22	Downward	(0, 0)	(0, 0)	(0, 0)	(1, −1), (−1, −1), (2, −4), (−2, −4)
23	Downward	(0, 4)	(0, 4)	(2, 0), (−2, 0)	(1, 3), (−1, 3)
24	Upward	(2, −4)	(0, 0)	(0, 0), (4, 0)	(−1, 5), (1, −3)
25	Upward	(0, −4)	(0, −4)	(2, 0), (−2, 0)	(1, −3), (−1, −3)
26	Downward	(4, 1)	(0, −15)	(3, 0), (5, 0)	(1, −8)

In Exercises 27–34 graph the parabola defined by each equation by determining its most important features. If there are no x-intercepts, plot some additional points.

27 $f(x) = x^2 + 2x - 3$ **28** $f(x) = -x^2 + 6x - 9$ **29** $y = 4x^2 + 12x + 9$ **30** $y = -3x^2 + 3x + 6$
31 $f(x) = -x^2 + 6x - 13$ **32** $y = x^2 + 7$ **33** $y = -x^2 + 6x$ **34** $f(x) = x^2 - 9$

B

In Exercises 35–38, each graph is a translation of the graph of $y = x^2$. Write the equation of each of these functions.

35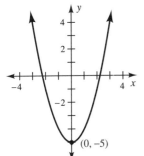

36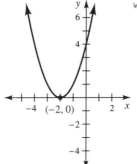

37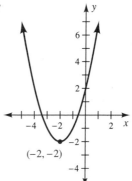

38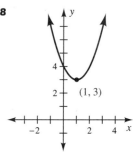

39 Suppose the manager of a small windmill manufacturer projects that the profit in dollars from making t windmills per day will be $P(t) = -t^2 + 70t - 600$.
  a. How many windmills should be produced each day in order to maximize the profit?
  b. What is the maximum profit?
  c. For what interval of values of t will the profit be increasing?
  d. For what interval of values of t will the profit be decreasing?

40 A manufacturer of water pumps has gathered data suggesting that the average cost in dollars of producing t units per hour of a new style pump will be $C(t) = t^2 - 22t + 166$.
  a. Determine how many pumps per hour to produce in order to minimize the average cost.
  b. What will this cost be?
  c. For what interval of values of t will be average cost be increasing?
  d. For what interval of values of t will the average cost be decreasing?

41 The height in meters of a ball released from a ramp is given by the function $h(t) = -4.9t^2 + 29.4t + 34.3$, where t represents the time in seconds since the ball was released from the end of the ramp.
  a. Determine the maximum height of this ball.
  b. Determine the time interval that the ball is rising.
  c. Determine the time interval that the ball is falling.

42 A piece of steel sheet 48 centimeters wide is bent to form the frame for a storm drain by turning down the two ends perpendicular to the piece of steel. How many centimeters should be turned down to maximize the area of the opening?

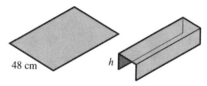

Figure for Exercise 42

43 A homeowner has 14 meters of fencing to enclose a rectangular playing area for his children. One side of the playing area will be formed by the wall of an existing building, and the other three sides will be formed by the fencing. What is the maximum area that can be enclosed in this manner?

44 Determine the maximum rectangular area that can be enclosed by 60 meters of fencing.

C

In Exercises 45–48, graph each function and determine the intervals for which the function is increasing, decreasing, or constant.

45 $y = \begin{cases} 1 & \text{if } x < -1 \\ x^2 & \text{if } -1 \leq x \leq 2 \\ 4 & \text{if } x > 2 \end{cases}$

46 $y = \begin{cases} x^2 & \text{if } x \leq -1 \\ 1 & \text{if } -1 < x \leq 1 \\ x & \text{if } x > 1 \end{cases}$

47 $y = \begin{cases} -x^2 & \text{if } x \leq -1 \\ -1 & \text{if } -1 < x \leq 1 \\ (x-1)^2 & \text{if } x > 1 \end{cases}$

48 $y = \begin{cases} -x^2 + 4 & \text{if } x < 0 \\ 4 - x & \text{if } 0 \leq x < 4 \\ (x-4)^2 & \text{if } x \geq 4 \end{cases}$

49 Determine whether the parabola defined by each equation opens to the left or to the right.

   a. $x = -y^2 + 7y + 9$ **b.** $x = 5y^2 - 7y + 9$

50 Each graph is a translation of the graph of $x = -y^2$. Write the equation of each of these relations.

a. 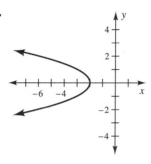   **b.**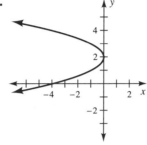

51 Find the value of k such that the graph of $f(x) = x^2 + kx + 16$ will have exactly one x-intercept.

52 Find the value of k such that the graph of $f(x) = x^2 + 5x + k$ will have x-intercepts of $(-3, 0)$ and $(-2, 0)$.

53 Verify that the y-coordinate of the vertex is $\dfrac{4ac - b^2}{4a}$ by evaluating $f(x) = ax^2 + bx + c$ for $x = \dfrac{-b}{2a}$.

54 Find the values of k such that the graph of $f(x) = x^2 + k$ will not cross the x-axis.

55 Find the two natural numbers with a sum of 16 whose product is a maximum.

SECTION 3-4

Symmetry and the Distance and Midpoint Formulas

Section Objectives

11 Determine if the graph of a relation has symmetry to the x-axis, the y-axis, or the origin.

12 Determine the distance and the midpoint between two points.

Every graph of a relation is just a picture of its points. Thus when the shape of a graph is not obvious from its defining equation, one way to obtain the graph is to plot many points and then connect them. This crude approach to graphing can be time-consuming, may fail to reveal all the "interesting" features of the graph, and can be in error if the points are connected improperly. Thus it is important to learn to recognize shapes from the form of the defining equation. It is also important to recognize geometric properties, such as symmetry, which can aid the curve-sketching process by reducing the number of points that must be plotted in order to yield a representative graph of a relation. We will begin this section by examining symmetry geometrically and then we will give a precise algebraic definition of symmetry.

A graph is **symmetric with respect to the y-axis** if the portion of the graph to the left of the y-axis is a mirror image of the portion to the right of the y-axis. A graph is **symmetric with respect to the x-axis** if the portion of the graph below the x-axis is a mirror image of the portion above the x-axis. A graph is **symmetric with respect to the origin** if each point (x, y) on the graph has an image $(-x, -y)$ directly across the origin in the opposite quadrant. (See Figure 3-17.)

▶ **Self-Check** ▼

Sketch another portion of this graph so that it will be symmetric to the

1 y-axis **2** x-axis **3** origin

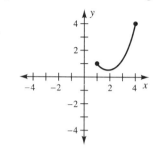

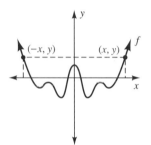

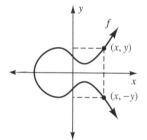

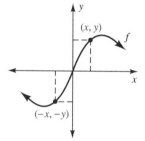

(a) $(-x, y)$ is a mirror image of (x, y) about the y-axis.

(b) $(x, -y)$ is a mirror image of (x, y) about the x-axis.

(c) $(-x, -y)$ is an image of (x, y) across the origin.

Figure 3-17 Symmetry with respect to the (a) y-axis, (b) x-axis, and (c) origin.

▶ **Self-Check Answers** ▼

1 Symmetry to y-axis:

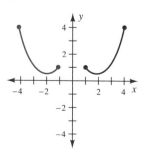

2 Symmetry to x-axis:

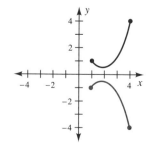

3 Symmetry to origin:

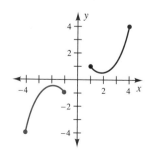

SECTION 3-4 SYMMETRY AND THE DISTANCE AND MIDPOINT FORMULAS

Each of these types of symmetry occurs frequently in nature and in man-made objects, as shown in Figure 3-18.

Figure 3-18 A butterfly's wings display natural symmetry

EXAMPLE 1 Given the point $P(-2, 4)$, graph the following points: A, symmetric to P about the y-axis; B, symmetric to P about the x-axis; and C, symmetric to P about the origin.

SOLUTION

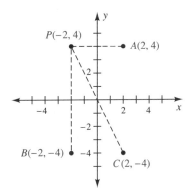

The graph of a relation defined by an equation can be tested for symmetry without actually graphing the relation. Any symmetry can be determined directly from the equation by testing an arbitrary point (x, y) in the equation. The geometric logic supporting these tests is analogous to that shown in Figure 3-19 for a point in quadrant I. We urge you to make your own sketches and analyze each of the cases in the following box.

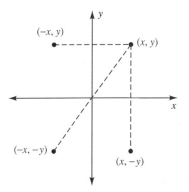

Figure 3-19 Testing an arbitrary point for symmetry

Tests for Symmetry ▼

> To test the graph of a relation for symmetry with respect to the **y-axis**, substitute $-x$ for x in the equation. If this results in an equivalent equation, the graph is symmetric to the y-axis.
>
> To test the graph of a relation for symmetry with respect to the **x-axis**, substitute $-y$ for y in the equation. If this results in an equivalent equation, the graph is symmetric to the x-axis.
>
> To test the graph of a relation for symmetry with respect to the **origin**, substitute $-x$ for x and $-y$ for y in the equation. If this results in an equivalent equation, the graph is symmetric about the origin.

EXAMPLE 2 Test the graph of $y = \frac{1}{5}x^4 - x^2 + 1$ for symmetry about both axes and about the origin.

SOLUTION

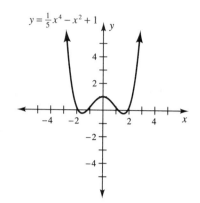

(a) Test for y-symmetry:

$y = \frac{1}{5}(-x)^4 - (-x)^2 + 1$ Substitute $-x$ for x to test for y-symmetry.

$y = \frac{1}{5}x^4 - x^2 + 1$ This equation is equivalent to the original equation.

Has y-symmetry.

(b) Test for x-symmetry:

$(-y) = \frac{1}{5}x^4 - x^2 + 1$ Substitute $-y$ for y to test for x-symmetry.

$-y = \frac{1}{5}x^4 - x^2 + 1$

$y = -\frac{1}{5}x^4 + x^2 - 1$ This equation is not equivalent to the original equation.

Does not have x-symmetry.

(c) Test for symmetry about the origin:

$(-y) = \frac{1}{5}(-x)^4 - (-x)^2 + 1$ Substitute $-x$ for x and $-y$ for y to test for symmetry about the origin.

$-y = \frac{1}{5}x^4 - x^2 + 1$

$y = -\frac{1}{5}x^4 + x^2 - 1$ This equation is not equivalent to the original equation.

Does not have symmetry about the origin.

Answer This graph is symmetric about the y-axis. □

Observe that any relation whose graph is symmetric about the x-axis cannot be a function, since both (x, y) and $(x, -y)$ are points on its graph. Hence functional notation is not used in the description of a relation whose graph is symmetric about the x-axis.

Since $(-x)^n = x^n$ for each even integer n, a polynomial function with only even-degree terms will have $f(-x) = f(x)$, and therefore its graph will be symmetric about the y-axis. In fact, any function with the property that $f(-x) = f(x)$ is called an **even function**. In this case, $(x, f(x))$ and $(-x, f(-x))$ are alternative notations for (x, y) and $(-x, y)$. *Thus the graphs of even functions are symmetric about the y-axis.* Observe in Example 2 that each term of $\frac{1}{5}x^4 - x^2 + 1$ is of even degree (consider $1 = 1x^0$). We could note by inspection that $y = \frac{1}{5}x^4 - x^2 + 1$ is an even function, and therefore its graph is symmetric about the y-axis.

Following this same logic, we define an **odd function** as a function with

SECTION 3-4 SYMMETRY AND THE DISTANCE AND MIDPOINT FORMULAS

$f(-x) = -f(x)$. *Graphs of odd functions are symmetric about the origin.* A function does not have to be a polynomial to be even or odd; there are examples of both even functions and odd functions that are not polynomials.

EXAMPLE 3 Determine if the graph of $y = 4x^3 - 8x$ is symmetric about the origin.

SOLUTION
$$y = 4x^3 - 8x$$
$$(-y) = 4(-x)^3 - 8(-x) \quad \text{Substitute } -y \text{ for } y \text{ and } -x \text{ for } x.$$
$$-y = -4x^3 + 8x$$
$$y = 4x^3 - 8x \quad \text{This is equivalent to the orginal equation.}$$

The graph of this function is symmetric about the origin.

Alternative Solution Note that each term of $4x^3 - 8x$ is of odd degree. Therefore the polynomial function is odd, and its graph must be symmetric about the origin. □

EXAMPLE 4 Test the graph of $y = 5x - 11$ for symmetry about both axes and about the origin.

SOLUTION $y = 5x - 11$ can be written as $y = 5x^1 - 11x^0$. $5x^1$ is an odd term and $-11x^0$ is an even term. Therefore this function is neither an even relation nor an odd relation, and so its graph is not symmetric about either the y-axis or the origin. To test for symmetry about the x-axis, we substitute $-y$ for y, obtaining $-y = 5x - 11$. Since this is not equivalent to $y = 5x - 11$, the graph does not have symmetry about the x-axis. □

EXAMPLE 5 Use symmetry as an aid to graphing $y = x^3$.

SOLUTION $y = x^3$ is an odd function, and its graph is therefore symmetric about the origin. For each point (x, y) in quadrant I from the table below, $(-x, -y)$ will also be a point on the graph.

		By Symmetry	
x	y	$-x$	$-y$
0	0	0	0
$\frac{1}{2}$	$\frac{1}{8}$	$-\frac{1}{2}$	$-\frac{1}{8}$
1	1	-1	-1
2	8	-2	-8

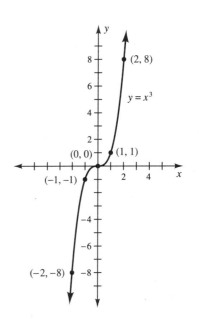

□

The concept of distance between two points is an important tool for studying the graphs of functions and relations. In Section 3-7 we will use the distance formula to develop the equation of a circle. We begin the development of the distance formula by first considering the horizontal and vertical changes between the points P and Q. The distance between P and Q is denoted by $\overline{PQ}$.

The Distance from P to Q, $\overline{PQ}$

Special Case
Calculate the distance between $P(2, 2)$ and $Q(5, 6)$

General Case
Calculate the distance between $P(x_1, y_1)$ and $Q(x_2, y_2)$

▶ **Self-Check** ▼

Test the graph of $y = x^2$ for symmetry about both axes and about the origin.

Step 1 Find the horizontal change from P to Q.

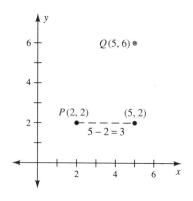
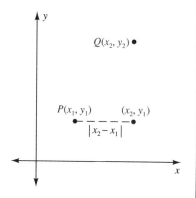

Horizontal distance $= 5 - 2 = 3$

Horizontal distance $= |x_2 - x_1|$

Step 2 Find the vertical change from P to Q.

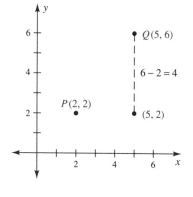
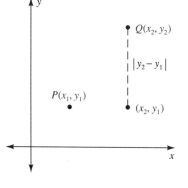

Vertical distance $= 6 - 2 = 4$

Vertical distance $= |y_2 - y_1|$

▶ **Self-Check Answer** ▼

The graph of this even function is a familar parabola that is symmetric about the y-axis.

SECTION 3-4 SYMMETRY AND THE DISTANCE AND MIDPOINT FORMULAS

Step 3 Use the Pythagorean Theorem to find the length of the hypotenuse PQ.

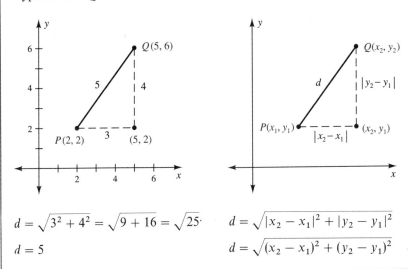

$d = \sqrt{3^2 + 4^2} = \sqrt{9 + 16} = \sqrt{25}$
$d = 5$

$d = \sqrt{|x_2 - x_1|^2 + |y_2 - y_1|^2}$
$d = \sqrt{(x_2 - x_1)^2 + (y_2 - y_1)^2}$

Absolute value notation is not needed in the distance formula because the squares are always nonnegative. The formula is applicable in all cases even if P and Q are on the same horizontal or vertical line.

Distance Formula ▼

The distance d from (x_1, y_1) to (x_2, y_2) is given by
$$d = \sqrt{(x_2 - x_1)^2 + (y_2 - y_1)^2}$$

EXAMPLE 6 Find the distance between $(3, -2)$ and $(-2, 10)$.

SOLUTION $d = \sqrt{(-2 - 3)^2 + [10 - (-2)]^2}$
$d = \sqrt{(-5)^2 + (12)^2}$
$d = \sqrt{25 + 144}$
$d = \sqrt{169}$
$d = 13$

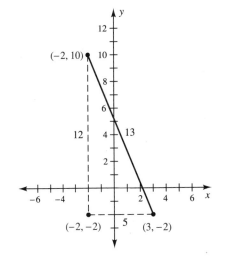

EXAMPLE 7 Determine if the vertices $A(-4, 8)$, $B(2, 6)$, and $C(0, 4)$ form a right triangle.

SOLUTION ABC is a right triangle if the sum of the squares of the shorter two sides equals the square of the longer side.

$$a = \overline{BC} = \sqrt{(0-2)^2 + (4-6)^2} = \sqrt{8}$$
$$b = \overline{AC} = \sqrt{[0-(-4)]^2 + (4-8)^2} = \sqrt{32}$$
$$c = \overline{AB} = \sqrt{[2-(-4)]^2 + (6-8)^2} = \sqrt{40}$$

Now check to see if $a^2 + b^2 = c^2$:

$$(\sqrt{8})^2 + (\sqrt{32})^2 \stackrel{?}{=} (\sqrt{40})^2$$
$$8 + 32 \stackrel{?}{=} 40$$
$$40 = 40 \text{ checks.}$$

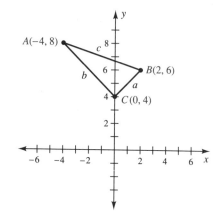

Answer ABC is a right triangle with angle C the 90° angle. □

A student's average grade for two tests is the sum of the two scores divided by two. Thus the average grade is a score midway between the two grades. Likewise, the midpoint of the line segment from P to Q can be found by averaging the coordinates of these points.

Midpoint Formula ▼

The midpoint (x, y) between $P(x_1, y_1)$ and $Q(x_2, y_2)$ is

$$(x, y) = \left(\frac{x_1 + x_2}{2}, \frac{y_1 + y_2}{2}\right)$$

EXAMPLE 8 A line intersects a circle at the points $(-5, 12)$ and $(12, 5)$. Find the midpoint between these points of intersection.

SOLUTION $(x, y) = \left(\dfrac{x_1 + x_2}{2}, \dfrac{y_1 + y_2}{2}\right)$

$$(x, y) = \left(\frac{-5 + 12}{2}, \frac{12 + 5}{2}\right)$$

$$(x, y) = \left(\frac{7}{2}, \frac{17}{2}\right) \quad □$$

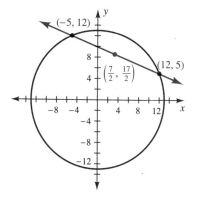

Some important functions do not exhibit symmetry, yet they occur frequently enough to have earned a special name and notation. For example, the **greatest integer function**, denoted by $f(x) = [\![x]\!]$, pairs each real number x with the greatest integer less than or equal to x. Note from the definition that $[\![2.6]\!] = 2$, $[\![2.999]\!] = 2$, and $[\![3]\!] = 3$. Be careful with negative values of x; $[\![-1.7]\!] = -2$, $[\![-1.999]\!] = -2$, $[\![-2]\!] = -2$, and $[\![-2.0001]\!] = -3$.

SECTION 3-4 SYMMETRY AND THE DISTANCE AND MIDPOINT FORMULAS

A function called **INT** appears on some calculators and in some computer languages. Unfortunately, in some instances this function is not the greatest integer function. Thus one should test both positive and negative values in this **INT** function before relying on it for an important calculation. This **INT** function could be used as in Example 9(b) to find that 2 is the remainder when 47 is divided by 5. (See Exercises 45–48 at the end of this section.)

▶ **Self-Check** ▼

Given $P(-2, -3)$ and $Q(12, 1)$, find
1. $\overline{PQ}$
2. the midpoint between P and Q

EXAMPLE 9 Evaluate each of these expressions.

SOLUTION

(a) $\left[\!\left[\dfrac{47}{5}\right]\!\right]$ $\left[\!\left[\dfrac{47}{5}\right]\!\right] = [\![9.4]\!]$
$= 9$

(b) $5\left[\!\left[\dfrac{47}{5}\right]\!\right]$ $5\left[\!\left[\dfrac{47}{5}\right]\!\right] = 5[\![9.4]\!]$
$= 5(9)$
$= 45$

(c) $[\![-3.7]\!]$ $[\![-3.7]\!] = -4$ As illustrated on this number line, -4 is the greatest integer less than -3.7.

EXAMPLE 10 Graph $f(x) = [\![x]\!]$ by plotting points.

SOLUTION

x	$f(x) = [\![x]\!]$
-2	-2
-1.9	-2
-1.5	-2
-1	-1
-0.5	-1
0	0
0.5	0
0.9	0
0.99	0
1	1
1.5	1
2	2

It should be clear from the graph in Example 10 why $f(x) = [\![x]\!]$ is described as a step function. Try to observe patterns about the "length of the steps" and the "rise between the steps."

▶ **Self-Check Answers** ▼

1. $2\sqrt{53}$ 2. $(5, -1)$

Graphics Calculators*

The sketch of the graph of a function should be sufficiently detailed to illustrate the most significant features of the graph. Plotting enough points by hand to reveal all of these features can sometimes be a tedious chore. Fortunately, the graphs of many functions can now be sketched by graphics calculators and computer graphing programs. The material that follows illustrates some of the features of the TI-81, a graphics calculator produced by Texas Instruments.

The window of the graph shown on the display of a calculator or computer is limited along both the *x*- and *y*-axes. (See Figure 3-20.) Thus we must be careful to select a window that accurately portrays the behavior of the entire graph. The graph of the function in the next example is viewed through two different windows.

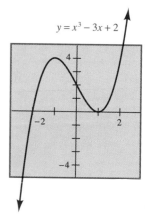

Figure 3-20 The window displays the graph of $y + x^3 - 3x + 2$ within the domain $-3 \leq x \leq 3$ and within the range $-5 \leq y \leq 5$

EXAMPLE 11 Graph $y = -x^3 - x^2 + 2x$ on the TI-81, using first the standard viewing rectangle and then zooming in around the origin.

SOLUTION

(a) Keystrokes:

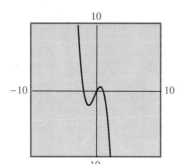

If it is necessary to clear out functions already stored in the calculator, press `CLEAR` immediately after pressing `Y=`.

To be sure the viewing window is set for the standard viewing rectangle with -10 to 10 for x and -10 to 10 for y, use these keystrokes: `ZOOM` `6`.

(b) To magnify the viewing rectangle around the origin use these keystrokes:

`ZOOM` `2` `ENTER`.

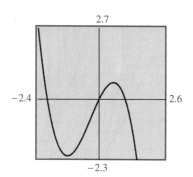

The zooming factors are assummed to be the default values of 4.

* This is an optional section.

SECTION 3-4 SYMMETRY AND THE DISTANCE AND MIDPOINT FORMULAS

The obvious advantages and power of graphics calculators will make them popular tools. However, they also have disadvantages, particularly if the graphs shown on the displays are taken at face value rather than interpreted by someone with some knowledge of the functions being graphed. One pitfall is to graph such a limited portion of the domain that the display window does not include key features of the graph. On the other hand, it is also possible to graph such a large portion of the domain that the large scale obliterates small but important details of the graph. Thus the limitations of screen resolution and size force the user to make many choices and compromises to balance these potential pitfalls.

To overcome the limitations of screen resolution and size, it may be necessary to vary your display window and to make more than one sketch of the graph of a function.

▶ **Self-Check** ▼

Evaluate each of these expressions.
1. $[\![-13.6]\!]$ 2. $[\![-13]\!]$ 3. $[\![13]\!]$
4. $[\![13.6]\!]$ 5. $[\![\frac{17}{5}]\!]$

EXERCISES 3-4

A

In Exercises 1–4, determine whether each graph is symmetric about either axis and/or about the origin.

1
a.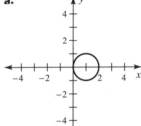
b.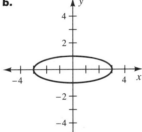
c.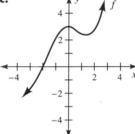

2
a.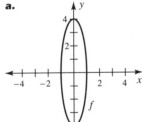
b.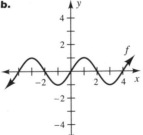
c.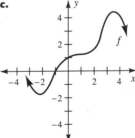

▶ **Self-Check Answers** ▼

1. -14 2. -13 3. 13 4. 13 5. 3

3 a. 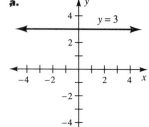 **b.** 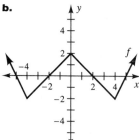 **c.**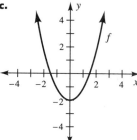

4 a. 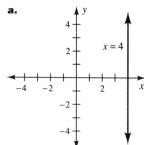 **b.** 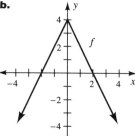 **c.**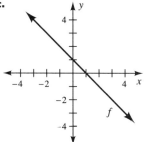

In Exercises 5–8, complete each graph so that it will be symmetric to the (a) x-axis (b) y-axis (c) origin.

5 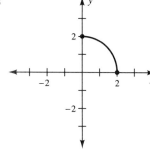 **6** 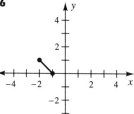 **7** 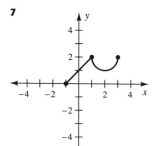 **8**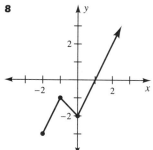

In Exercises 9–12, plot the given point P, a point A that is symmetric to P with respect to the y-axis, and a point B that is symmetric to P with respect to the x-axis, and a point C that is symmetric to P with respect to the origin.

9 $P(3, 4)$ **10** $P(-2, -5)$

11 $P(0, -2)$ **12** $P(4, 0)$

In Exercises 13 and 14, determine by inspection whether each function is an even function, an odd function, or neither odd nor even.

13 a. $f(x) = 5x^{12} - 12x^5 - 3$ **b.** $f(x) = 5x^{12} - 12$ **c.** $g(x) = 6x^7 - 12x^5$

14 a. $g(x) = -12x^5 + 5$ **b.** $h(x) = 3x^4 - 2x^2 + 7$ **c.** $h(x) = 4x^3 - x$

SECTION 3-4 SYMMETRY AND THE DISTANCE AND MIDPOINT FORMULAS

In Exercises 15–18, determine the symmetry of the graph of each relation and use this symmetry and selected points to sketch the graph.

15 $y = \dfrac{1}{x}$ **16** $y = \sqrt[3]{x}$ **17** $y = \dfrac{x^4}{4}$ **18** $x^2 + y^2 = 25$

In Exercises 19–22, calculate the distance between the points and the midpoint of the line segment connecting these points.

19 $(-3, 2)$ and $(1, -1)$
20 $(-6, -2)$ and $(6, 3)$
21 $(a + 1, b)$ and $(a - 3, b + 3)$
22 $(3v, 4v)$ and $(-2v, -4v)$

In Exercises 23 and 24, use the distance formula to calculate the perimeter of the figures described.

23 The triangle formed by connecting the points $(-3, -3)$, $(2, -3)$, and $(2, 9)$.

24 The rectangle $ABCD$ formed by connecting $A(4, 2)$, $B(-2, 2)$, $C(-2, -2)$, and $D(4, -2)$.

In Exercises 25 and 26, use the distance formula to determine if the given vertices form a right triangle.

25 $(3, 9)$, $(5, 7)$, and $(6, 8)$ **26** $(18, 1)$, $(-4, -10)$, and $(8, 21)$

The points A, B, C, are collinear if $\overline{AC} = \overline{AB} + \overline{BC}$. If the points are not collinear, then ABC is a triangle and any one side is less than the sum of the other two sides. In Exercises 27 and 28 determine if the points are collinear.

27 $(6, 5)$, $(4, 3)$, and $(1, 0)$ **28** $(-1, 8)$, $(4, 5)$, and $(8, 1)$

29 Find the distance between $(-2, 3)$ and the midpoint of the line segment with endpoints $(-5, -6)$ and $(-3, 8)$.

30 The endpoints of a diameter of a circle are $(2, -1)$ and $(-5, -3)$. Find the center of the circle and its radius.

In Exercises 31 and 32 evaluate each of these expressions involving the greatest integer function.

31   **a.** $[\![9.79]\!]$ **b.** $[\![-7.32]\!]$ **c.** $[\![\tfrac{13}{2}]\!]$ **d.** $[\![-\tfrac{13}{5}]\!]$

32   **a.** $[\![7.83]\!]$ **b.** $[\![-7.83]\!]$ **c.** $[\![\tfrac{13}{10}]\!]$ **d.** $[\![-\tfrac{13}{4}]\!]$

In Exercises 33–36, sketch the graph of each of these greatest integer functions.

33 $y = [\![x + 2]\!]$ **34** $y = [\![2x]\!]$ **35** $y = 2[\![x]\!]$ **36** $y = \tfrac{1}{2}[\![x]\!]$

B

37 On the graph shown to the right:
   a. complete the graph so that it is symmetric to the y-axis.
   b. complete the graph so that it is symmetric to the x-axis.
   c. complete the graph so that it is symmetric to the origin.
   d. If this is the graph of $y = f(x)$, graph $y = f(x - 2) + 3$.

38 **a.** If a graph is symmetric with respect to both axes, is it symmetric with respect to the origin? Explain why or why not.
   b. If a graph is symmetric with respect to the origin and to the x-axis, is it symmetric with respect to the y-axis?
   c. If a graph is symmetric with respect to the origin and to the y-axis, is it symmetric with respect to the x-axis?
   d. What can you conclude from parts a, b, and c?

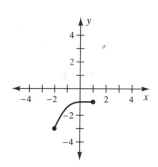

Figure for Exercise 37

39 If f is an even function and $f(-2) = -7.389$, what is the value of $f(2)$?

40 If f is an odd function and $f(-2) = -7.389$, what is the value of $f(2)$?

41 Use the distance formula to verify that $(18, 13)$, $(26, -2)$, $(11, -10)$, and $(3, 5)$ are the vertices of a square.

42 Verify that the vertices $(1, -2)$, $(6, -2)$, $(9, 2)$, and $(4, 2)$ form a rhombus. (A rhombus has four sides of equal length.)

43 Find all values of y for which the distance from $(-2, -3)$ to $(-2, y)$ is 7 units.

44 Find all values of x for which the distance from $(-2, -3)$ to $(x, -3)$ is 7 units.

C

The calculations given in Exercises 45–48 can be used to extract specific digits from a credit card number. Evaluate each of these expressions for $x = 37282695$.

45 $x - 10\left[\dfrac{x}{10}\right]$

46 $\left[\dfrac{x}{10}\right] - 10\left[\dfrac{x}{100}\right]$

47 $\left[\dfrac{x}{100}\right] - 10\left[\dfrac{x}{1000}\right]$

48 $\left[\dfrac{x}{1000}\right] - 10\left[\dfrac{x}{10,000}\right]$

In Exercises 49–50, use a calculator to determine the distance between the given points. Give the distances accurately to six significant digits.

49 $(\pi, \sqrt{2})$ and $(\sqrt{2}, \pi)$

50 $(108, 412)$ and $(-109, 204)$

In Exercises 51–54, use a graphics calculator to obtain the graph of each function. Use the standard range of values $-10 \le x \le 10$ and $-10 \le y \le 10$ and superimpose all three graphs on the same viewing rectangle. Then zoom in on these graphs.

51 **a.** $y = x^2$ **b.** $y = x^2 + 5$ **c.** $y = x^2 - 5$

52 **a.** $y = |5x|$ **b.** $y = |5x| - 5$ **c.** $y = |5x| - 10$

53 **a.** $y = x^3$ **b.** $y = (x + 2)^3$ **c.** $y = (x - 2)^3$

54 **a.** $y = 5^x$ **b.** $y = 5^{x+1}$ **c.** $y = 5^{x-1}$

SECTION 3-5

Combining Functions

Section Objectives

13 Add, subtract, multiply, and divide two functions.

14 Form the composition of two functions.

Problems—such as those in business—are often broken down into simpler components for analysis. For example, in order to determine the profit made by producing and selling an item, both the revenue and the cost must be known. Separate divisions of a business might be asked to find the revenue

SECTION 3-5 COMBINING FUNCTIONS

function and the cost function; the profit function would then be found by properly combining these two functions. We shall examine five ways to combine functions: sum, difference, product, quotient, and composition of functions.

The **sum of two functions** f and g, denoted by $f + g$, is defined as

$$(f + g)(x) = f(x) + g(x)$$

for all values of x that are in the domain of both f and g. Note that if either $f(x)$ or $g(x)$ is undefined, then $(f + g)(x)$ is also undefined.

EXAMPLE 1 Given $f(x) = x^2 - 9x + 3$ and $g(x) = 2x^2 + 7x - 4$, find the following.

SOLUTIONS

(a) $(f + g)(x)$

$(f + g)(x) = f(x) + g(x)$
$= (x^2 - 9x + 3) + (2x^2 + 7x - 4)$
$= 3x^2 - 2x - 1$

(b) the domain of $f + g$

Both f and g are defined for all real numbers; therefore, $f + g$ is also defined for all real numbers.

(c) $(f + g)(2)$

Method I: First we find that

$f(2) = (2)^2 - 9(2) + 3$
$= -11$
$g(2) = 2(2)^2 + 7(2) - 4$
$= 18$

Thus

$(f + g)(2) = f(2) + g(2) = -11 + 18$
$= 7$

Method II: From part (a),

$(f + g)(x) = 3x^2 - 2x - 1$

Thus $(f + g)(2) = 3(2)^2 - 2(2) - 1$
$= 7$

Note the results from Methods I and II are the same. Method I may be preferable if the sum is only evaluated for one argument. Otherwise, Method II is usually more efficient.

The **difference of two functions** f and g, denoted by $f - g$, is defined as

$$(f - g)(x) = f(x) - g(x)$$

for all values of x that are in the domain of both f and g.

EXAMPLE 2 Suppose that the weekly revenue function for u units of a product sold is $R(u) = 5u^2 - 7u$ dollars and the cost function for u units is $C(u) = 8u + 23$. Assume zero is the least number of units that can be produced and 100 is the greatest number that can be marketed. Find the profit function P, and find $P(4)$, the profit made by selling four units.

SOLUTION Profit = Revenue − Cost   *Word equation*

$$P(u) = R(u) - C(u)$$
$$P(u) = (5u^2 - 7u) - (8u + 23)$$
$$P(u) = 5u^2 - 7u - 8u - 23$$

The domain is $0 \leq u \leq 100$. (For some types of units, u may have to be an integer.)

The profit function is
$$P(u) = 5u^2 - 15u - 23$$

The profit on four units is
$$P(4) = 5(4)^2 - 15(4) - 23 = -3$$

Thus $3 will be lost if only four units are sold. □

The operations of multiplication and division are defined similarly to $f + g$ except that $\left(\dfrac{f}{g}\right)$ is not defined if $g(x) = 0$. For functions f and g with domains D_f and D_g, these definitions are summarized in the following table.

Operations on Functions ▼

	Notation	Definition	Domain
Sum	$f + g$	$(f + g)(x) = f(x) + g(x)$	$D_f \cap D_g$
Difference	$f - g$	$(f - g)(x) = f(x) - g(x)$	$D_f \cap D_g$
Product	$f \cdot g$	$(f \cdot g)(x) = f(x)g(x)$	$D_f \cap D_g$
Quotient	$\dfrac{f}{g}$	$\dfrac{f}{g}(x) = \dfrac{f(x)}{g(x)}$	Values of $D_f \cap D_g$ for which $g(x) \neq 0$

EXAMPLE 3 Given $f(x) = x^2 + 5x$ and $g(x) = \dfrac{x+5}{x}$, evaluate

(a) $f + g$ (b) $f - g$ (c) $f \cdot g$ (d) $\dfrac{f}{g}$

SOLUTIONS First note that the domains of f and g are understood to be $D_f = \mathbb{R}$ and $D_g = \mathbb{R} \sim \{0\}$. Also note that $g(-5) = 0$, which is important when we consider $\dfrac{f}{g}$.

(a) $(f + g)(x) = f(x) + g(x) = x^2 + 5x + \dfrac{x+5}{x}$; $D_{f+g} = \mathbb{R} \sim \{0\}$

(b) $(f - g)(x) = f(x) - g(x) = x^2 + 5x - \dfrac{x+5}{x}$; $D_{f-g} = \mathbb{R} \sim \{0\}$

SECTION 3-5 COMBINING FUNCTIONS

(c) $(f \cdot g)(x) = f(x) \cdot g(x) = (x^2 + 5x)\left(\dfrac{x+5}{x}\right) = (x+5)^2; \quad D_{f \cdot g} = \mathbb{R} \sim \{0\}$

(d) $\dfrac{f}{g}(x) = \dfrac{f(x)}{g(x)} = \dfrac{x^2 + 5x}{\dfrac{x+5}{x}} = \dfrac{x^2 + 5x}{1} \cdot \dfrac{x}{x+5} = x^2 \qquad D_{f/g} = \mathbb{R} \sim \{-5, 0\}$

The domains of $f + g$ and $f - g$ in Example 3 are obvious from the formulas that define these functions. However, it is not obvious from $(f \cdot g)(x) = (x + 5)^2$ that this function is not defined at $x = 0$. Likewise, the formula $\left(\dfrac{f}{g}\right)(x) = x^2$ does not show that this function is undefined at $x = -5$ and $x = 0$. These restrictions are necessary, for without them we could not have performed the operations that resulted in these simplified formulas.

Two functions f and g are **equal** if the domain of f equals the domain of g and $f(x) = g(x)$ for each x in their common domain.

▶ **Self-Check** ▼

Given $f(x) = x^3 + 3$ and $g(x) = 5 - x^2$, evaluate these expressions.

1 $(f + g)(2)$ **2** $(f - g)(4)$

3 $(f \cdot g)(-1)$ **4** $\left(\dfrac{f}{g}\right)(1)$

EXAMPLE 4   Given $f(x) = x$, $g(x) = \dfrac{x^3 - 4x}{x^2 - 4}$, and $h(x) = \dfrac{x^3 + 4x}{x^2 + 4}$, determine whether

SOLUTIONS

(a) $f = g$ $f(x) = x$ for all real x

$g(x) = \dfrac{x^3 - 4x}{x^2 - 4}$ for $x \neq \pm 2$

$ = \dfrac{x(x^2 - 4)}{x^2 - 4}$ for $x \neq \pm 2$

$ = x$ for $x \neq \pm 2$

Since $D_f = \mathbb{R}$ and $D_g = \mathbb{R} \sim \{-2, 2\}$, $f \neq g$.

(b) $f = h$ $f(h) = x$ for all real x

$h(x) = \dfrac{x^3 + 4x}{x^2 + 4}$ for all real x There are no real numbers for which the denominator of h is zero.

$ = \dfrac{x(x^2 + 4)}{x^2 + 4}$ for all real x

$ = x$ for all real x

Since $D_f = D_h = \mathbb{R}$ and $f(x) = h(x)$ for each x in $\mathbb{R}$, $f = h$.

▶ **Self-Check Answers** ▼

1 12   **2** 78   **3** 8   **4** 1

The graphs of the three functions f, g, and h in Example 4 are nearly identical. The only difference is that the graph of g has "holes" at $x = -2$ and $x = 2$ because these values are not in its domain.

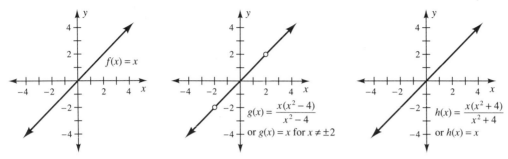

Figure 3-21 Graphs of functions f, g, and h

Functions, especially in formula form, are a powerful means of describing the relationship between two quantities. We can further amplify this power by "chaining" two functions together. This way of combining two functions is called **composition** (see Figure 3-21).

▶ **Self-Check** ▼

Determine if the functions $f(x) = x$ and $g(x) = \sqrt{x^2}$ are equal.

Composite Function ▼

The composition of the function f with the function g, denoted by $f \circ g$, is defined by

$$(f \circ g)(x) = f[g(x)]$$

The domain of $f \circ g$ is the set of x values from the domain of g for which $g(x)$ is in the domain of f.

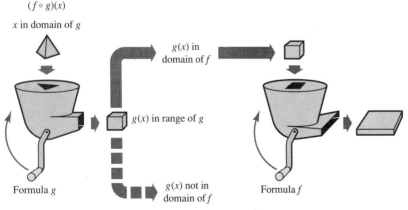

Figure 3-22 Composition of functions

▶ **Self-Check Answer** ▼

$f \neq g$. $g(x) = |x|$ since it is the principal square root. This is different from $f(x) = x$ for negative values of x.

SECTION 3-5 COMBINING FUNCTIONS

EXAMPLE 5 Given

$f = \{(2, 6), (3, 9), (4, 12), (6, 18)\}$ and $g = \{(6, 3), (10, 5), (12, 6), (18, 9)\}$ find the following.

SOLUTIONS

(a) $f \circ g$

$6 \xrightarrow{g} 3 \xrightarrow{f} 9$

$10 \to 5 \to$ undefined

$12 \to 6 \to 18$

$18 \to 9 \to$ undefined

$f \circ g = \{(6, 9), (12, 18)\}$

Note that $f \circ g$ starts with the elements 6, 10, 12, and 18 from the domain of g. The y-values associated with these x-values must be found in the domain of f to complete the composition. Otherwise the composition is undefined.

(b) $g \circ f$

$2 \xrightarrow{f} 6 \xrightarrow{g} 3$

$3 \to 9 \to$ undefined

$4 \to 12 \to 6$

$6 \to 18 \to 9$

$g \circ f = \{(2, 3), (4, 6), (6, 9)\}$

For $g \circ f$ start with the elements 2, 3, 4, and 6 from the domain of f. The y-values associated with these x-values must be found in the domain of g to complete the composition. Otherwise the composition is undefined.

Observe that the domain of $f \circ g$ is $\{6, 12\}$, that the domain of $g \circ f$ is $\{2, 4, 6\}$, and that $f \circ g \neq g \circ f$. □

EXAMPLE 6 Given $f(x) = x^2$ and $g(x) = 3x$, evaluate these expressions.

SOLUTIONS

(a) $(f \circ g)(4)$

$(f \circ g)(4) = f[g(4)] = f[3(\ )]$

$= f[3(4)]$

$= f(12)$

$= (\ )^2$

$= (12)^2$

$= 144$

Use parentheses to set up the form for the function g. Then evaluate g at 4.

Use parentheses to set up the form for the function f. Then evaluate f at 12.

$4 \xrightarrow{g} 12 \xrightarrow{f} 144 \; f \circ g$

(b) $(g \circ f)(4)$

$(g \circ f)(4) = g[f(4)] = g[4^2]$

$= g(16)$

$= 3(16)$

$= 48$

$4 \xrightarrow{f} 16 \xrightarrow{g} 48$

$g \circ f$

(c) $(f \circ g)(x)$

$(f \circ g)(x) = f[g(x)] = f(3x)$

$= (3x)^2$

$= 9x^2$

$x \xrightarrow{g} 3x \xrightarrow{f} 9x^2$

$f \circ g$

(d) $(g \circ f)(x)$

$(g \circ f)(x) = g[f(x)] = g(x^2)$

$= 3(x^2)$

$= 3x^2$

$x \xrightarrow{f} x^2 \xrightarrow{g} 3x^2$

$g \circ f$

The domain of both $f \circ g$ and $g \circ f$ is the set of all real numbers. The formulas are different, however, and so $f \circ g \neq g \circ f$. □

EXAMPLE 7 Given $f(x) = \sqrt{x}$ and $g(x) = \dfrac{1}{x}$, evaluate these expressions.

SOLUTIONS

(a) $(f \circ g)(4)$

$(f \circ g)(4) = f[g(4)] = f\left(\dfrac{1}{4}\right)$ $4 \xrightarrow{g} \dfrac{1}{4} \xrightarrow{f} \dfrac{1}{2}$

$\qquad = \sqrt{\dfrac{1}{4}}$

$\qquad = \dfrac{1}{2}$

(b) $(f \circ g)(0)$

$(f \circ g)(0) = f[g(0)] = f(\text{undefined})$ $\dfrac{1}{0}$ is undefined

$\qquad = \text{undefined}$

(c) $(f \circ g)(-4)$

$(f \circ g)(-4) = f[g(-4)]$ $-4 \xrightarrow{g} -\dfrac{1}{4} \xrightarrow{f}$ imaginary number

$\qquad = f\left(-\dfrac{1}{4}\right)$

$\qquad = \sqrt{-\dfrac{1}{4}}$ is not a real number $(f \circ g)(-4)$ is not a real number and thus cannot be evaluated if the range is restricted to real numbers.

(d) $(f \circ g)(x)$

$(f \circ g)(x) = f[g(x)]$

$g(x) = \dfrac{1}{x}$ for $x \neq 0$ $D_g = \mathbb{R} \sim \{0\}$ to avoid division by zero.

$f[g(x)] = f\left(\dfrac{1}{x}\right)$ for $x \neq 0$ and $\dfrac{1}{x} \geq 0$ $D_f = [0, +\infty)$ to avoid imaginary numbers in the range.

$\qquad = \sqrt{\dfrac{1}{x}}$ for $x > 0$ Thus $x > 0$ is necessary for $\dfrac{1}{x}$ to be in the domain of f.

Thus $(f \circ g)(x) = \sqrt{\dfrac{1}{x}}$ for x in the interval $(0, +\infty)$.

In particular, 0 and -4 are not in this domain.

▶ **Self-Check** ▼

Given $f(x) = 5 - x^2$ and $g(x) = \sqrt{1 - x}$, find and state the domains of

1 $f \circ g$ **2** $g \circ f$

EXAMPLE 8 The quantity of items a factory can produce weekly is a function of the number of hours it operates. For one company this is given by $q(t) = 40t$ for $0 \leq t \leq 168$. The dollar cost of manufacturing these items is a function of the quantity produced; in this case, $C(q) = q^2 - 40q + 750$ for $q \geq 0$. Evaluate and interpret the following expressions.

▶ **Self-Check Answers** ▼

1 $(f \circ g)(x) = x + 4;\ D = (-\infty, 1]$ **2** $(g \circ f)(x) = \sqrt{x^2 - 4};\ D = (-\infty, -2] \cup [2, +\infty)$

SECTION 3-5 COMBINING FUNCTIONS

SOLUTIONS

(a) $q(8)$
$$q(t) = 40t$$
$$q(\boxed{8}) = 40(\boxed{8}) = 320$$
320 units can be produced in 8 hours.

(b) $C(320)$
$$C(q) = q^2 - 40q + 750$$
$$C(\boxed{320}) = (\boxed{320})^2 - 40(\boxed{320}) + 750 = 90{,}350$$
$90{,}350 is the cost of 320 units.

(c) $(C \circ q)(8)$
$$(C \circ q)(\boxed{8}) = C[q(\boxed{8})]$$
$$= C(\boxed{320}) \qquad \text{Substitute from part (a).}$$
$$= \$90{,}350 \qquad \text{Substitute from part (b).}$$
$90{,}350 is the cost of 8 hours of production.

(d) $(C \circ q)(t)$
$$(C \circ q)(t) = C[q(t)]$$
$$= C[\boxed{40t}] \qquad \text{Substitute for } q(t).$$
$$= (\boxed{40t})^2 - 40(\boxed{40t}) + 750$$
$$= 1600t^2 - 1600t + 750$$
This is the cost of t hours of production. □

EXAMPLE 9 A piece of wire 20 meters long is cut into two pieces. The length of the shorter piece is s meters, and the length of the longer piece is L meters. The longer piece is then bent into the shape of a square of area A m² (square meters).

(a) Express the length L as a function of s.
(b) Express the area A as a function of L.
(c) Express the area A as a function of s.

SOLUTIONS

(a) $\begin{pmatrix} \text{Length of the} \\ \text{longer piece} \end{pmatrix} = \begin{pmatrix} \text{Total} \\ \text{length} \end{pmatrix} - \begin{pmatrix} \text{Length of the} \\ \text{shorter piece} \end{pmatrix}$ Word equation

$\qquad L(s) \quad = \quad 20 \quad - \quad s$ This equation expresses the length as a function of s.

(b) Area = Square of the length of one side   The length of one side of the square is one-fourth the perimeter.

$$A(L) = \left(\frac{L}{4}\right)^2$$

(c) $(A \circ L)(s) = A[L(s)]$ The composition $A \circ L$ expresses A as a function of s.
$$= A(20 - s) \qquad \text{Substitute } 20 - s \text{ for } L(s) \text{ from part (a). Evaluate } A \text{ using the formula from part (b).}$$
$$= \left(\frac{20 - s}{4}\right)^2$$

□

An important skill in calculus is the ability to take a given function and decompose it into simpler components.

EXAMPLE 10 Express each of these functions in terms of $f(x) = 2x - 3$ and $g(x) = \sqrt{x}$.

SOLUTIONS

(a) $h(x) = \sqrt{2x - 3}$ $h(x) = \sqrt{2x - 3} = \sqrt{f(x)}$ First substitute $f(x)$ for $2x - 3$. Then replace the
$= g(f(x))$ square root function with the g function.

(b) $h(x) = 2\sqrt{x} - 3$ $h(x) = 2\sqrt{x} - 3 = 2g(x) - 3$ First substitute $g(x)$ for $\sqrt{x}$. Then rewrite the
$= f(g(x))$ expression using the definition of the f function.

EXERCISES 3-5

A

In Exercises 1–4, evaluate each expression, given $f(x) = x^2 - 1$ and $g(x) = 2x + 5$.

1. a. $(f + g)(2)$ b. $(f - g)(2)$ c. $(f \cdot g)(2)$ d. $\left(\dfrac{f}{g}\right)(2)$

2. a. $(f + g)(-3)$ b. $(f - g)(-3)$ c. $(f \cdot g)(-3)$ d. $\left(\dfrac{f}{g}\right)(-3)$

3. a. $(f \circ g)(2)$ b. $(g \circ f)(2)$ c. $(f \circ f)(2)$ d. $(g \circ g)(2)$

4. a. $(f \circ g)(-3)$ b. $(g \circ f)(-3)$ c. $(f \circ f)(-3)$ d. $(g \circ g)(-3)$

In Exercises 5–8, find $f \circ g$ and $g \circ f$.

5. $f = \{(-1, 4), (0, 0), (1, 4), (-3, 12)\}$ and $g = \{(0, -3), (4, 1), (8, 5)\}$

6. $f = \{(2, 3), (3, 5), (5, 9), (9, 2)\}$ and $g = \{(2, 5), (3, 9), (5, 2), (9, 3)\}$

7. a. Graph of f b. Graph of g

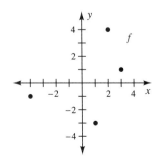
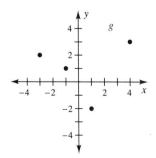

SECTION 3-5 COMBINING FUNCTIONS

8 a. Graph of f **b.** Graph of g

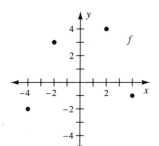
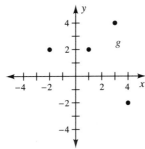

In Exercises 9–12, determine $f + g, f - g, f \cdot g, \dfrac{f}{g}$, and $f \circ g$ for the given functions. State the domain of each function.

9 $f(x) = 2x^2 - x - 3$
$g(x) = 2x - 3$

10 $f(x) = 6x^2 - x - 15$
$g(x) = 3x - 5$

11 $f(x) = \dfrac{x}{x + 1}$
$g(x) = \dfrac{1}{x}$

12 $f(x) = 5x - 7$
$g(x) = 3$

In Exercises 13–18, determine if the functions f and g are equal. If $f \ne g$, state the reason.

13 $f(x) = x - 3$
$g(x) = \dfrac{x^2 - 5x + 6}{x - 2}$

14 $f(x) = 1 - \dfrac{1}{x}$
$g(x) = \dfrac{x - 1}{x}$

15 $f = \{(-1, 1), (0, 0), (1, 1), (2, 4)\}$
$g(x) = x^2$

16 $f = \{(-7, 7), (0, 0), (8, 8)\}$
$g(x) = |x|$

17 $f(x) = x$
$g(x) = \dfrac{x^3 + x}{x^2 + 1}$

18 $f(x) = 5$
$g(x) = \dfrac{5x^2 + 15}{x^2 + 3}$

In Exercises 19–22, graph $f + g$, given the graphs of f and g.

19 a.
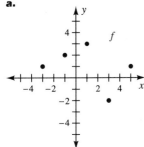

b.
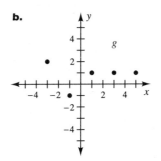

20 a. 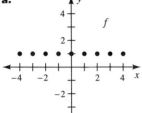 **b.**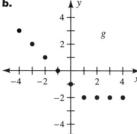

21 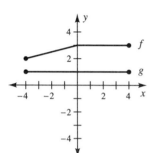 **22**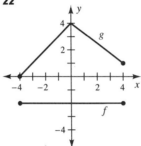

23 The fixed monthly cost F (rent, insurance, etc.) of a manufacturer is $5000. The variable cost (labor, materials, etc.) for producing u units is given by $V(u) = u^2 + 5u$ for $0 \le u \le 12{,}000$. The total cost of producing u units is $C(u) = F(u) + V(u)$. Determine
  a. $F(5000)$ **b.** $V(5000)$ **c.** $C(5000)$
  d. the average cost of producing 5000 units
  e. a formula for $C(u)$
  f. a formula for $A(u)$, the average cost of producing u units

24 The number of sofas a factory can produce weekly is a function of the number of hours t it operates. This function is $S(t) = 5t$ for $0 \le t \le 168$. The cost of manufacturing s sofas is given by $C(s) = s^2 - 6s + 500$ for $s \ge 0$. Evaluate and interpret
  a. $S(10)$ **b.** $C(50)$ **c.** $(C \circ S)(t)$
  d. $(C \circ S)(10)$ **e.** $(C \circ S)(40)$ **f.** $(C \circ S)(100)$

B

In Exercises 25–30 find $f \circ g$ and $g \circ f$. State the domain of each composite function.

25 $f(x) = 3x - 4$
    $g(x) = 4x + 3$

26 $f(x) = x^2 - 5x + 3$
    $g(x) = 4x - 2$

27 $f(x) = |x|$
    $g(x) = x - 8$

28 $f(x) = \sqrt{x}$
    $g(x) = x + 5$

29 $f(x) = \dfrac{1}{x}$
    $g(x) = x^3 + 1$

30 $f(x) = \dfrac{1}{x+2}$
    $g(x) = x^2 - 4$

SECTION 3-5 COMBINING FUNCTIONS

31 A manufacturer produces circuit boards for the electronics industry. The fixed cost F associated with this production is $3000 per week, and the variable cost V is $10 per board. The circuit boards produce a revenue of $12 each. Determine

  a. $V(b)$, the variable cost of producing b boards
  b. $F(b)$, the fixed cost of producing b boards
  c. $C(b)$, the total cost of producing b boards
  d. $A(b)$, the average cost of producing b boards
  e. $R(b)$, the revenue from selling b boards
  f. $P(b)$, the profit from selling b boards
  g. $P(1000)$
  h. $P(1500)$
  i. $P(2000)$
  j. the break-even point

32 The number of units demanded by consumers is a function of the number of months the product has been advertised. The price per item is varied each month as part of the marketing strategy. The number demanded during the mth month is $N(m) = 36m - m^2$, and the price per item during the mth month is $P(m) = 5m + 45$. The revenue for the mth month $R(m)$ is the product of the price per item times the number of items demanded. Determine

  a. $N(7)$ **b.** $P(7)$ **c.** $R(7)$ **d.** a formula for $R(m)$

33 The weekly cost C of making d doses of a vaccine is $C(d) = 0.30d + 400$. The company charges 150% of cost to its wholesaler for this drug; that is, $R(c) = 1.5C$. Evaluate and interpret

  a. $C(5000)$ **b.** $R(1900)$ **c.** $(R \circ C)(d)$ **d.** $(R \circ C)(5000)$

34 Determine which of these functions are equal:
$$f(x) = \sqrt{x^2 + 2x + 1}, \ g(x) = x + 1, \text{ and } h(x) = |x + 1|.$$

35 For $f(x) = 2x + 3$ and $g(x) = \dfrac{x - 3}{2}$, determine $f \circ g$ and its domain. Then graph $y = f(x)$, $y = g(x)$, and $y = (f \circ g)(x)$ on the same coordinate system.

In Exercises 36–39, express each function in terms of $f(x) = x^2 + 1$ and $g(x) = \sqrt{x}$.

36 $h(x) = x^2 + \sqrt{x} + 1$

37 $h(x) = x + 1, \ x \geq 0$

38 $h(x) = \sqrt{x^2 + 1}$

39 $h(x) = \dfrac{\sqrt{x}}{x^2 + 1}$

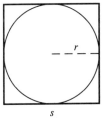

Figure for Exercise 40

40 A circular concrete pad was poured to serve as the base for a grain bin. This pad was inscribed in the square plot shown to the right.

  a. Express the radius r of this circle as a function of the length s of a side of the square.
  b. Express the area A of this circle as a function of the radius r.
  c. Compute $(A \circ r)(s)$ and interpret this result.

41 A border of uniform width x is trimmed from all sides of the square posterboard shown to the right.

  a. Express the area A of the border as a function of the remaining width w.
  b. Express the remaining width w as a function of x.
  c. Compute $(A \circ w)(x)$ and interpret this function.

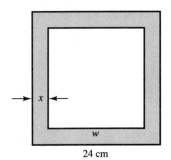

Figure for Exercise 41

42 A metal box with an open top can be formed by cutting squares of sides x centimeters from each corner of a square piece of sheet metal of width 44 centimeters.

    a. Express the area of the base of this box as a function of its width w.
    b. Express w as a function of x.
    c. Compute $(A \circ w)(x)$ and interpret this function.
    d. Express the volume V of the box as a function of x.

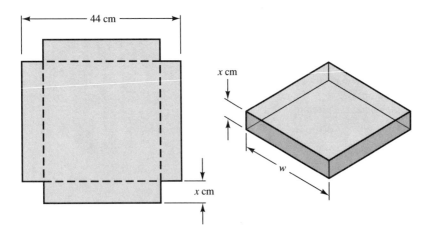

In Exercises 43–46 find $f \circ g$ and $g \circ f$. State the domain of each composite function.

43 $f(x) = \dfrac{1}{x^2 - 5x - 6}$
    $g(x) = 2x$

44 $f(x) = 4x - 5$
    $g(x) = 2$

45 $f(x) = \sqrt[3]{x}$
    $g(x) = x^3 + 8$

46 $f(x) = \sqrt{x^2 - x - 6}$
    $g(x) = \dfrac{1}{x}$

In Exercises 47–50, decompose each function into functions f and g such that $h(x) = (f \circ g)(x)$.

47 $h(x) = \sqrt[3]{x^2 + 4}$

48 $h(x) = \dfrac{1}{3x^2 - 7x + 9}$

49 $h(x) = (x + 2)^2 + 3(x + 2) + 5$

50 $h(x) = x^3 - 2 + \dfrac{1}{x^3 - 2}$

In Exercises 51–54, use a calculator to determine each functional value, accurate to six significant digits. Let $f(x) = \sqrt{x}$ and $g(x) = \dfrac{1}{x}$.

51 $f(129.678)$

52 $g(129.678)$

53 $(f \circ g)(129.678)$

54 $(g \circ f)(129.678)$

In Exercises 55–56, use a graphics calculator to obtain the graph of $y = (f \circ g)(x)$.

55 $f(x) = \sqrt{x},\ g(x) = 2x + 3$

56 $f(x) = 3x - 1,\ g(x) = x^2$

SECTION 3-6

Inverse Functions

Section Objectives

15 Determine whether or not a function is a one-to-one function.

16 Determine the inverse of a one-to-one function.

17 Graph the inverse of a function.

Since a relation f is a set of ordered pairs (x, y), we can think of f as pairing each x-value with a y-value. For example, consider the exponential function for computing continuous compound interest (See Figure 3-23.) For a given principal and interest rate, each period of time is paired with a unique amount yielded by this investment. We can also view this relationship as a pairing in the reverse direction. In this case, each total amount is associated with a period of time that will yield that amount.

The new perspective gained by reversing the original ordered pairs (x, y) and examining the new pairs (y, x) provides additional insight into many relationships. To gain this insight, we will examine **inverse relations**. (The inverses of exponential functions, called logarithmic functions, are particularly important; we will examine these functions in Chapter 5.)

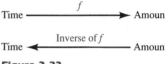

Figure 3-23

Inverse Relation f^{-1} ▼

If f is a set of ordered pairs (x, y), then f^{-1} is the set of ordered pairs (y, x) formed by reversing the coordinates of the ordered pairs of f.

The notation f^{-1} is read "the inverse of f"; this is not exponential notation and does not mean $\dfrac{1}{f}$. The context of this notation should make the meaning clear, as illustrated by the following examples:

- If $f = \{(100, 27), (500, 135), (1000, 271)\}$, then
  $f^{-1} = \{(27, 100), (135, 500), (271, 1000)\}$.
- If $f = \{(2, 4), (3, 4), (7, 10)\}$, then $f^{-1} = \{(4, 2), (4, 3), (10, 7)\}$.

Note that in the second example, f is a function, but the inverse relation f^{-1} is *not* a function. Since f has two ordered pairs—(2, 4) and (3, 4)—with the same second component its inverse has two ordered pairs—(4, 2) and (4, 3)—with the same first component.

If distinct ordered pairs of a function f have distinct second components, then f^{-1} will be a function. In this case, we call f a **one-to-one function** and f^{-1} an **inverse function**.

One-to-One Function ▼

A function is one-to-one if distinct ordered pairs have distinct second components.

▶ **Self-Check** ▼

Give the inverse of each relation.
1 $f = \{(2, 8), (-3, 4), (5, 1)\}$
2 $g = \{(1, 7), (-3, 7), (0, 4)\}$

The following examples illustrate this definition.

- If $f = \{(1, 3), (2, 7), (5, 11), (4, 9)\}$, then f is one-to-one and thus the inverse $f^{-1} = \{(3, 1), (7, 2), (11, 5), (9, 4)\}$ is an inverse function.
- If $f = \{(1, 5), (2, 5), (8, 3)\}$, then f is not one-to-one because both 1 and 2 are paired with 5. Thus the inverse relation $f^{-1} = \{(5, 1), (5, 2), (3, 8)\}$ is not a function.

The vertical line test covered in Section 3-1 is a quick method for determining if a graph represents a function. The test allows us to visually determine if each x in the domain is paired with a unique y element in the range. Similarly, the horizontal line test is a quick method for determining if the graph represents a one-to-one function.

Horizontal Line Test ▼

Imagine a horizontal line placed anywhere through the y-axis. If at any position this horizontal line intersects the graph at more than one point, then the graph does *not* represent a one-to-one function.

EXAMPLE 1 Use the horizontal line test to determine whether or not each graph represents a one-to-one function.

SOLUTIONS

(a) Not a one-to-one function

▶ **Self-Check Answers** ▼

1 $f^{-1} = \{(8, 2), (4, -3), (1, 5)\}$ **2** $g^{-1} = \{(7, 1), (7, -3), (4, 0)\}$

SECTION 3-6 INVERSE FUNCTIONS

(b) 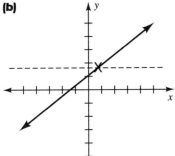 A one-to-one function

(c) 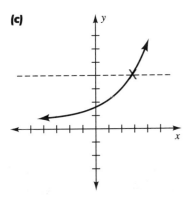 A one-to-one function

(d) 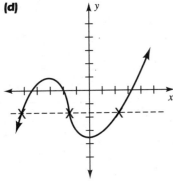 Not a one-to-one function

□

If f is a one-to-one function given by an equation, then the inverse function f^{-1} is also indirectly given by this equation. For example, if f is the doubling function given by $y = 2x$, then this implies that the inverse function f^{-1} is the halving function given by $x = \dfrac{y}{2}$. Note in this example that $x = 5$ is paired with $y = 10$ by the equation $y = 2x$ and then $y = 10$ is paired with $x = 5$ by the equation $x = \dfrac{y}{2}$.

The choice of letters or variables used to describe a function is arbitrary, and thus some texts refer to these symbols as "dummy variables." For example, $f(x) = x^2$, $g(y) = y^2$, and $h(z) = z^2$ all represent the squaring function. Although these notations are distinct, they represent the same set of ordered pairs and would have identical graphs. Thus the inverse of $f(x) = 2x$ can be represented by either $g(y) = \dfrac{y}{2}$ or $f^{-1}(x) = \dfrac{x}{2}$. Although we are using the same symbol x in both $f(x) = 2x$ and $f^{-1}(x) = \dfrac{x}{2}$, the elements represented by these x's come from different domains. Remember that the domain of $f^{-1}(x)$ is the same set as the range of $f(x)$. To obtain the formula for f^{-1}, we can interchange x and y and then solve for y, as illustrated in Example 2. Remember that f must be a one-to-one function for f^{-1} to be an inverse function.

▶ **Self-Check** ▼

Determine whether or not each of the following functions or graphs represents a one-to-one function.

1 $h = \{(-6, 3), (7, 5)\}$

2 $g = \{(0, 0), (-1, 1), (1, 1)\}$

3

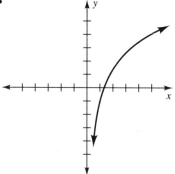

4

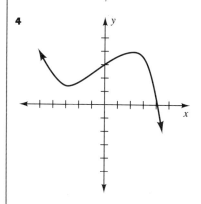

Inverse of a Function ▼

To find the inverse of a function $y = f(x)$:

Step 1 Replace each x with y and each y with x.

Step 2 Solve the resulting equation for y, if possible. This inverse can then be written in functional notation by replacing y with $f^{-1}(x)$.

EXAMPLE 2 Given $f(x) = 2x$, find $f^{-1}(x)$.

SOLUTION

$f(x) = 2x$

$y = 2x$ First replace $f(x)$ with y to express the function in x-y notation.

$x = 2y$ Then form the inverse by interchanging the variables x and y.

$y = \dfrac{x}{2}$ Solve the inverse explicitly for y.

$f^{-1}(x) = \dfrac{x}{2}$ Rewrite the inverse in functional notation by replacing y with $f^{-1}(x)$.

▶ **Self-Check Answers** ▼

1 One-to-one function   **2** Not a one-to-one function   **3** One-to-one function   **4** Not a one-to-one function

SECTION 3-6 INVERSE FUNCTIONS

The domain of a function can often be determined by inspection. (Values that cause division by zero or negative values under a square root are excluded from real-valued functions.) However it is often not possible to determine the range of a function as easily as this. One way to determine the range of a one-to-one function is to determine the domain of its inverse. This is illustrated in the next example.

EXAMPLE 3 Given $f(x) = \dfrac{2x}{x+3}$, find $f^{-1}(x)$ and the domain and range of both functions.

SOLUTION $f(x) = \dfrac{2x}{x+3}$

$y = \dfrac{2x}{x+3}$ Write f in x-y notation.

$x = \dfrac{2y}{y+3}$ Interchange x and y to form the inverse function.

$x(y+3) = 2y$ Multiply both sides by $y+3$.

$xy + 3x = 2y$ Then solve this equation for y.

$xy - 2y = -3x$

$y(x-2) = -3x$

$y = \dfrac{-3x}{x-2}$

$f^{-1}(x) = \dfrac{-3x}{x-2}$ Rewrite the inverse in functional notation.

For f the domain is $\mathbb{R} \sim \{-3\}$.   The value -3 would cause division by zero in
Thus the range of f^{-1} is $\mathbb{R} \sim \{-3\}$. $f(x) = \dfrac{2x}{x+3}$.

The domain of f^{-1} is $\mathbb{R} \sim \{2\}$.   The value 2 would cause division by zero in
Thus the range of f is $\mathbb{R} \sim \{2\}$. $f^{-1}(x) = \dfrac{-3x}{x-2}$. □

In Example 3, observe that $f(1) = \dfrac{2(1)}{1+3} = \dfrac{1}{2}$ and $f^{-1}\left(\dfrac{1}{2}\right) = \dfrac{-3(\frac{1}{2})}{\frac{1}{2} - 2} = 1.$

Thus f maps 1 to $\dfrac{1}{2}$ and f^{-1} reverses this ordered pair to map $\dfrac{1}{2}$ to 1. Checking one or two pairs of values like this does not prove that f^{-1} reverses all ordered pairs, but it can catch careless errors.

EXAMPLE 4 Given $g(x) = \sqrt{x-3}$, find $g^{-1}(x)$.

SOLUTION $g(x) = \sqrt{x-3}$ This is a one-to-one function, since the principal root is always nonnegative.

$y = \sqrt{x-3}$ Write g in $x - y$ notation.

$x = \sqrt{y-3}$ Form the inverse by interchanging x and y. Note that $x \geq 0$.

$x^2 = y - 3$ Solve for y.

$y = x^2 + 3$

$g^{-1}(x) = x^2 + 3$ for $x \geq 0$ Rewrite in functional notation.

Note: The domain of g is $D = [3, +\infty)$, and the range of g is $R = [0, +\infty)$. By interchanging the ordered pairs to form g^{-1}, we interchange the domain and the range so that $D = [0, +\infty)$ and $R = [3, +\infty)$ for g^{-1}.

▶ **Self-Check** ▼

Find the inverse of each function.
1 $f(x) = 5x$
2 $g(x) = \frac{1}{2}x + 3$
3 $h(x) = x^2$ for $x \geq 0$

EXAMPLE 5 Graph $f(x) = 2x + 1$ and its inverse $f^{-1}(x) = \dfrac{x-1}{2}$ on the same coordinate system.

SOLUTION Reverse the ordered pairs of f to obtain the ordered pairs of f^{-1}.

x	$f(x)$
-2	-3
-1	-1
0	1
1	3

x	$f^{-1}(x)$
-3	-2
-1	-1
1	0
3	1

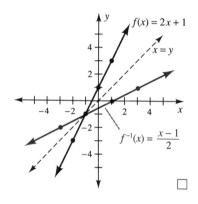

Note in Example 5 that the graphs of f and f^{-1} are symmetric about the line $x = y$. One way to observe this symmetry is to make identical copies of $f(x)$ on separate sheets of clear plastic. To graph $f^{-1}(x)$, lift the second copy, and flip this sheet in a manner that will interchange the x-axis and the y-axis.

▶ **Self-Check Answers** ▼

1 $f^{-1}(x) = \dfrac{x}{5}$ 2 $g^{-1}(x) = 2x - 6$ 3 $h^{-1}(x) = \sqrt{x}$

SECTION 3-6 INVERSE FUNCTIONS

(See Figure 3-24). This symmetry of the graphs of f and f^{-1} can also be observed by graphing both f and f^{-1} on a graphics calculator.

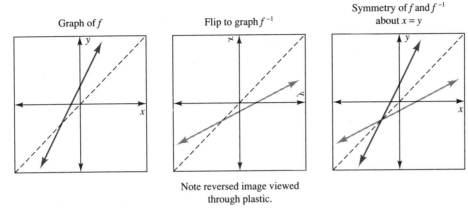

Figure 3-24 Symmetry of the graphs of f and f^{-1}

The graphs of f and its inverse f^{-1} are always symmetric about the line $x = y$ because the points (a, b) and (b, a) are mirror images about the graph of $x = y$. We can verify this by showing that the line $x = y$ is the perpendicular bisector of the line segment connecting (a, b) and (b, a). First, note that the slope of the line $x = y$ is 1 and the slope of the line segment is

$$\frac{b - a}{a - b} = -1$$

Since these numbers are negative reciprocals, the lines are perpendicular. Second, note that the midpoint of (a, b) and (b, a) is

$$\left(\frac{a + b}{2}, \frac{a + b}{2}\right)$$

The coordinates of this midpoint are equal, so it must be on the line $x = y$. This verfies that $x = y$ also bisects the line segment (see Figure 3-25).

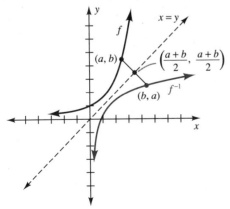

Figure 3-25 Symmetry of f and f^{-1} about the line $x + y$

EXAMPLE 6 Use symmetry and the graph of f to sketch f^{-1}.

SOLUTION

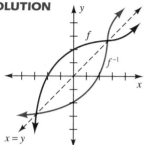

For each point (x, y) on the graph of f, move directly perpendicular to the line $x = y$. Continue in the same direction an equal distance on the other side of $x = y$ to locate the point (y, x) on the graph of f^{-1}.

□

If f^{-1} is the inverse function of f, then reversing the mapping means that $(f^{-1} \circ f)(x) = x$ and also that $(f \circ f^{-1})(x) = x$. This is shown pictorially in Figure 3-26. (Composition of functions was first covered in Section 3-5.)

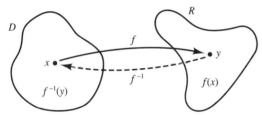

Figure 3-26 Composition of f^{-1} with f

This characterization of the inverse function is actually another way to define the inverse function.

Inverse Function ▼

> If f is a one-to-one function with domain D and range R, and if g is a function with domain R and range D, then g is the inverse function of f if
>
> **1** $(g \circ f)(x) = x$ for each x in D and
> **2** $(f \circ g)(x) = x$ for each x in R.

EXAMPLE 7 Verify that $f(x) = \dfrac{x}{x-7}$ (domain $\mathbb{R} \sim \{7\}$ is understood) and $g(x) = \dfrac{7x}{x-1}$ (domain $\mathbb{R} \sim \{1\}$) are inverses of each other.

SOLUTION First evaluate $(g \circ f)(x)$.

$$(g \circ f)(x) = g\left[\frac{x}{x-7}\right] \qquad \text{Evaluate } f \text{ at } x \text{ for all } x \neq 7.$$

SECTION 3-6 INVERSE FUNCTIONS

$$= \frac{7(\ )}{(\ ) - 1}$$ Set up the form for evaluating the function g.

$$= \frac{7\left(\dfrac{x}{x-7}\right)}{\left(\dfrac{x}{x-7}\right) - 1}$$ Evaluate g at $\dfrac{x}{x-7}$.

$$= \frac{7x}{x - (x - 7)}$$ Multiply the numerator and the denominator by the LCD $x - 7$.

$$= \frac{7x}{7}$$

$$= x$$

Second evaluate $(f \circ g)(x)$.

$$(f \circ g)(x) = f\left(\frac{7x}{x-1}\right)$$ Evaluate g at x for all $x \neq 1$.

$$= \frac{(\ )}{(\ ) - 7}$$ Set up the form for evaluating the function f.

$$= \frac{\dfrac{7x}{x-1}}{\left(\dfrac{7x}{x-1}\right) - 7}$$ Evaluate f at $\dfrac{7x}{x-1}$.

$$= \frac{7x}{7x - 7(x - 1)}$$ Multiply the numerator and the denominator by the LCD $x - 1$.

$$= \frac{7x}{7}$$

$$= x$$

Thus g is the inverse of f.

▶ **Self-Check** ▼

Use symmetry and the graph of $f(x) = 2x + 3$ to graph $f^{-1}(x)$.

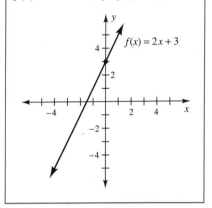

▶ **Self-Check Answer** ▼

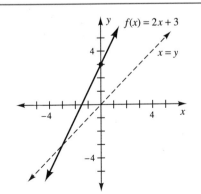

All one-to-one functions have inverse functions. This does not mean that it is easy or even possible to find a known formula that explicitly expresses the inverse. Thus for some functions we must either leave the inverse function implied by the original function or invent a new notation to represent the inverse function. The inverse of an exponential function such as $y = 3^x$ cannot be given using the basic operations. The inverse of $f(x) = 3^x$ is denoted by $f^{-1}(x) = \log_3 x$ and is read "the logarithm of x base 3." Since $f(2) = 3^2 = 9$, it must be true that $f^{-1}(9) = \log_3 9 = 2$. We will study these logarithmic functions in Chapter 5.

EXERCISES 3-6

A

In Exercises 1 and 2, determine whether or not each function is a one-to-one function.

1. **a.** $\{(2, 7), (5, \pi), (9, e)\}$ **b.** $\{(2, e), (5, \pi), (9, e)\}$
   c. $\{(-3, 9), (-2, 4), (-1, 1), (0, 0), (1, 1), (2, 4), (3, 9)\}$
2. **a.** $\{(-1, 8), (0, 8), (1, 1), (7, 3)\}$ **b.** $\{(1, 8), (0, 0), (3, 1), (8, 4)\}$
   c. $\{(-3, -27), (-2, -8), (-1, -1), (0, 0), (1, 1), (2, 8), (3, 27)\}$

In Exercises 3 and 4, determine whether or not each graph represents a one-to-one function.

3. **a.** 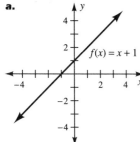 **b.** 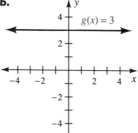 **c.** 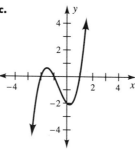 **d.**

4. **a.** 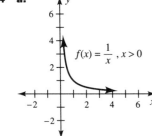 **b.** 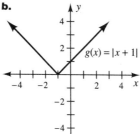 **c.** 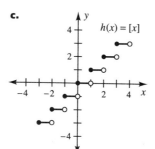 **d.**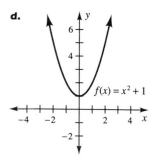

SECTION 3-6 INVERSE FUNCTIONS

In Exercises 5 and 6, complete each expression.

5 If $f(5) = 11$, then $f^{-1}(11) = ?$

6 If $f(3) = -8$, then $f^{-1}(-8) = ?$

In Exercises 7–26, determine the inverse of each function. State the domain of both the function and the inverse function.

7 $\{(0, 1), (1, 4), (2, 7)\}$

8 $\{(0, 0), (1, 1), (4, 2)\}$

9 $\{(1, 0), (2, 1), (4, 2), (8, 3), (16, 4), (32, 5)\}$

10 $\{(1, 0), (3, 1), (9, 2), (27, 3), (81, 4)\}$

11 $f(x) = 5x + 3$

12 $g(x) = 9x - 2$

13 $h(x) = \frac{1}{3}x - 8$

14 $f(x) = \frac{1}{7}x + 4$

15 $4x + 5y = 20$

16 $3x - 7y = 21$

17 $y = \dfrac{3}{x - 1}$

18 $y = \dfrac{5}{x + 2}$

19 $y = \dfrac{x}{x + 5}$

20 $y = \dfrac{x}{x - 3}$

21 $f(x) = \dfrac{2x - 3}{4x + 5}$

22 $g(x) = \dfrac{3x + 2}{5x - 4}$

23 $y = x^2$ for $x \geq 0$

24 $y = x^2$ for $x \leq 0$

25 $y = \sqrt[3]{x}$

26 $y = x^5$

In Exercises 27–34, sketch the inverse of each function graphed.

27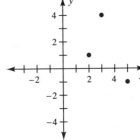

28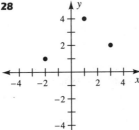

29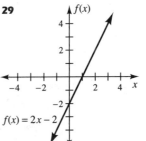

30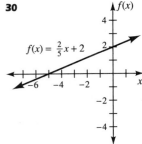

31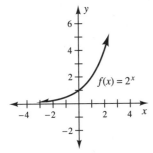

32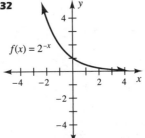

33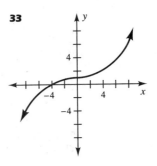

34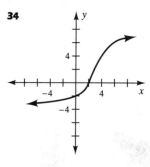

In Exercises 35 and 36, determine whether or not the graphs of f and g represent two functions that are inverses of each other.

35 a. 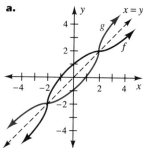 **b.** 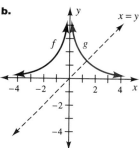 **c.**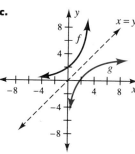

36 a. 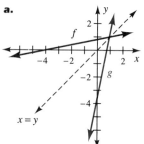 **b.** 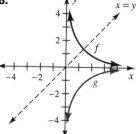 **c.**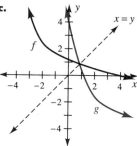

In Exercises 37–40, graph f and f^{-1} on the same coordinate system.

37 $f(x) = \dfrac{x}{2} + \dfrac{1}{3}$

38 $f(x) = \dfrac{x-4}{2}$

39 $f(x) = x^2$ for $x \leq 0$

40 $f(x) = x^2$ for $x \geq 0$

B

In Exercises 41–46, graph each function and determine whether or not it is a one-to-one function.

41 $f(x) = \sqrt{x}$ **42** $f(x) = \sqrt[3]{x}$ **43** $g(x) = x^4$ **44** $g(x) = x^3$ **45** $h(x) = 3^x$ **46** $h(x) = \left(\dfrac{1}{3}\right)^x$

In Exercises 47–50, verify that f and g are inverses by computing $(g \circ f)(x)$ and $(f \circ g)(x)$.

47 $f(x) = 2x + 3$
$g(x) = \dfrac{x-3}{2}$

48 $f(x) = x^2 + 1$ for $x \geq 0$
$g(x) = \sqrt{x-1}$ for $x \geq 1$

49 $f(x) = x^2 - 1$ for $x \leq 0$

$g(x) = -\sqrt{x+1}$ for $x \geq -1$

50 $f(x) = \dfrac{3}{x-7}$ for $x \neq 7$

$g(x) = \dfrac{7x+3}{x}$ for $x \neq 0$

C

In Exercises 51–54, use a graphics calculator to graph, f, g, and $y = x$ on the same coordinate system. Use these graphs to determine whether f and g are inverses of each other.

51 $f(x) = x + 1$
$g(x) = x - 1$

52 $f(x) = \dfrac{1}{x}$
$g(x) = x^2$

53 $f(x) = x^5$
$g(x) = x^{0.2}$

54 $f(x) = x^{-2}$
$g(x) = -2x$

55 Determine the inverse of $y = x^n$ for n an odd natural number.
56 Determine the inverse of $y = mx + b$.   practice
57 Determine the inverse of $y = \sqrt[n]{x}$, $x \geq 0$, for n an even natural number.
58 Give a function f for which $f = f^{-1}$.

SECTION 3-7

Circles and Ellipses

Section Objectives

18 Write in standard form the equation of a circle or an ellipse.

19 Graph a circle or an ellipse.

A **circle** is the set of all points in a plane that are a constant distance from a fixed point. The fixed point is called the **center** of the circle, and the distance from the center to the points on the circle is called the **radius**. (See Figure 3-27.)

Using the distance formula, we will now develop the equation of a circle with center (h, k) and radius r. The distance r from any point (x, y) on the circle to the center (h, k) is given by

$$r = \sqrt{(x - h)^2 + (y - k)^2}$$

Squaring both sides of this equation gives an equation satisfied by the points on the circle.

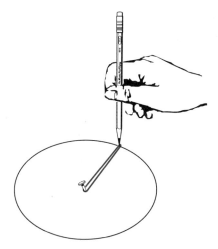

Figure 3-27 To draw a circle, fix a loop of string with a tack and draw the circle as illustrated

Standard Form of the Equation of a Circle ▼

The equation of a circle with center (h, k) and radius r is

$$(x - h)^2 + (y - k)^2 = r^2$$

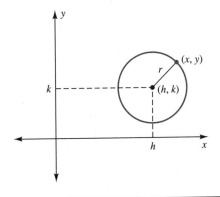

EXAMPLE 1 Write the equation of a circle with the center $(-1, 3)$ and the radius 5.

SOLUTION $[x - (-1)]^2 + (y - 3)^2 = 5^2$ Substitute the given values into the standard
$(x + 1)^2 + (y - 3)^2 = 25$ form $(x - h)^2 + (y - k)^2 = r^2$.

EXAMPLE 2 Determine the equation of the circle shown in the graph.

SOLUTION From the graph we determine that the center is $(4, -3)$ and the radius is 2. Thus the equation is

$$(x - 4)^2 + (y + 3)^2 = 4$$

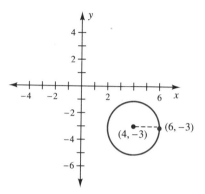

The standard form of the circle $(x - 4)^2 + (y + 3)^2 = 4$ can be expanded to give the general form $x^2 + y^2 - 8x + 6y + 21 = 0$. If we are given the equation of a circle in general form, we can use the process of completing the square to rewrite the equation in standard form so that the center and radius will be obvious.

EXAMPLE 3 Determine the center and the radius of the circle defined by the equation

$$3x^2 + 3y^2 + 30x - 66y + 330 = 0$$

SOLUTION $3x^2 + 3y^2 + 30x - 66y + 330 = 0$
$x^2 + y^2 + 10x - 22y + 110 = 0$ Divide both sides by 3.
$(x^2 + 10x) + (y^2 - 22y) = -110$ Regroup terms.
$(x^2 + 10x + 25) + (y^2 - 22y + 121) = -110 + 25 + 121$ Complete the square.
$(x + 5)^2 + (y - 11)^2 = 6^2$ Write in standard form.

Answer The circle has center $(-5, 11)$ and radius 6.

Equations of the form $x^2 + y^2 = c$, for $c > 0$, define circles centered at the origin. The following illustration also describes the cases where $c = 0$ or $c < 0$.

Equations of the Form $x^2 + y^2 = c$
As shown in the figure to the right,

1. $x^2 + y^2 = 4$ is a circle with center $(0, 0)$ and radius 2.
2. $x^2 + y^2 = 1$ is a circle with center $(0, 0)$ and radius 1.
3. $x^2 + y^2 = \frac{1}{4}$ is a circle with center $(0, 0)$ and radius $\frac{1}{2}$.
4. $x^2 + y^2 = 0$ is the degenerate case of a circle—a single point, $(0, 0)$.
5. $x^2 + y^2 = -1$ has that no real solutions and thus no points to graph. Both x^2 and y^2 are nonnegative so $x^2 + y^2$ must also be nonnegative.

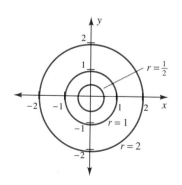

SECTION 3-7 CIRCLES AND ELLIPSES

Many of the uses of circles can be attributed to their symmetry. From the definitions of symmetry given in Section 3-4 we can note that a circle centered at the origin is symmetric with respect to the x-axis, the y-axis, and the origin. Therefore we can plot the intercepts and then use symmetry to complete the circle. According to the vertical line test, the circle $x^2 + y^2 = 16$ is not the graph of a function. However, if we consider only the upper semicircle, then each x is paired with a unique y; thus this semicircle is the graph of a function.

▶ **Self-Check** ▼

Classify the graph of each of these equations.
1. $x^2 + y^2 - 8x + 2y + 8 = 0$
2. $x^2 + y^2 + 10x - 6y + 34 = 0$
3. $x^2 + y^2 - 6x - 14y + 68 = 0$

EXAMPLE 4 Graph the function $y = \sqrt{16 - x^2}$.

SOLUTION Since y is the principal square root, it is never negative. By squaring both sides of the equation, we obtain $y^2 = 16 - x^2$, or $x^2 + y^2 = 16$. This is the equation of a circle with center $(0, 0)$ and radius 4. Because y is never negative, the graph is only the upper semicircle. Note that the domain of this function, determined by solving $16 - x^2 \geq 0$, is $-4 \leq x \leq 4$. □

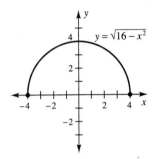

Figure for Example 4

An **ellipse** is the set of all points in a plane, the sum of whose distances from two fixed points is constant. (See Figure 3-28.)

The two fixed points $F_1(-c, 0)$ and $F_2(c, 0)$ are called **foci**. The **major axis** of the ellipse passes through the foci. The **minor axis** is shorter than the major axis and is perpendicular to it at the center. The ends of the major axis are called the **vertices**, and the ends of the minor axis are called the **covertices**. If the ellipse is centered at the origin, then it is symmetric to the x-axis, the y-axis, and the origin. The equation of the ellipse shown in Figure 3-29 is

$$\frac{x^2}{a^2} + \frac{y^2}{b^2} = 1$$

This formula is derived on the next page.

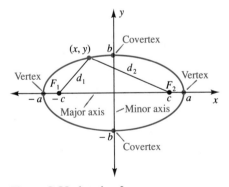

Figure 3-29 $d_1 + d_2 = 2a$

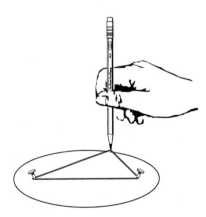

Figure 3-28 To draw an ellipse, loop a piece of string around 2 tacks and draw the ellipse as illustrated

▶ **Self-Check Answers** ▼

1. Circle with center $(4, -1)$ and radius 3
2. The point $(-5, 3)$
3. No graph

Derivation of the Formula for an Ellipse with Center at the Origin and a Horizontal Major Axis

$$d_1 + d_2 = 2a$$ The sum of the distances from (x, y) to the foci is a constant we denote by $2a$.

$$\sqrt{(x + c)^2 + (y - 0)^2} + \sqrt{(x - c)^2 + (y - 0)^2} = 2a$$ Apply the distance formula.

$$\sqrt{(x + c)^2 + y^2} = 2a - \sqrt{(x - c)^2 + y^2}$$ Isolate a radical term.

$$(x + c)^2 + y^2 = 4a^2 - 4a\sqrt{(x - c)^2 + y^2} + (x - c)^2 + y^2$$ Square both sides.

$$x^2 + 2cx + c^2 = 4a^2 - 4a\sqrt{(x - c)^2 + y^2} + x^2 - 2cx + c^2$$ Simplify, and then isolate the radical term.

$$4a\sqrt{(x - c)^2 + y^2} = 4a^2 - 4cx$$

$$a\sqrt{(x - c)^2 + y^2} = a^2 - cx$$ Divide by 4.

$$a^2[(x - c)^2 + y^2] = (a^2 - cx)^2$$ Square both sides.

$$a^2x^2 - 2a^2cx + a^2c^2 + a^2y^2 = a^4 - 2a^2cx + c^2x^2$$ Simplify, and then reorder the terms.

$$a^2x^2 - c^2x^2 + a^2y^2 = a^4 - a^2c^2$$

$$(a^2 - c^2)x^2 + a^2y^2 = a^2(a^2 - c^2)$$

$$b^2x^2 + a^2y^2 = a^2b^2$$ Let $b^2 = a^2 - c^2$.

$$\frac{x^2}{a^2} + \frac{y^2}{b^2} = 1$$ Divide by a^2b^2.

Letting $y = 0$ in this formula, we obtain $\frac{x^2}{a^2} = 1$. Thus the x-intercepts are $(-a, 0)$ and $(a, 0)$. Likewise, letting $x = 0$, we obtain the y-intercepts $(0, -b)$ and $(0, b)$. Note that $a > b$ since the major axis is the horizontal axis.

EXAMPLE 5   Graph $\dfrac{x^2}{49} + \dfrac{y^2}{16} = 1$.

SOLUTION   This is the equation of an ellipse centered at the origin, with $a = 7$ and $b = 4$. Thus the x-intercepts are $(-7, 0)$ and $(7, 0)$, and the y-intercepts are $(0, -4)$ and $(0, 4)$. Plot these intercepts, and then use the known shape and symmetry to complete the ellipse.

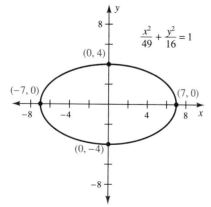

The major axis is horizontal since the denominator of the x^2 term is larger than the denominator of the y^2 term.

$$\frac{x^2}{49} + \frac{y^2}{16} = 1$$

$a^2 = 49$ $b^2 = 16$

$a = 7$ $b = 4$

□

SECTION 3-7 CIRCLES AND ELLIPSES

EXAMPLE 6 Graph $\dfrac{(x-2)^2}{49} + \dfrac{(y+3)^2}{16} = 1$.

SOLUTION This is the graph from Example 1 translated to the right two units and down three units. Thus this is the equation of an ellipse with center $(2, -3)$ and $a = 7$ and $b = 4$.
The vertices are:

$(2 - 7, -3) = (-5, -3)$ Since $a = 7$, the vertices are located 7 units to
$(2 + 7, -3) = (9, -3)$ the left and right of the center, $(2, -3)$.

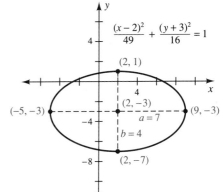

The covertices are:

$(2, -3 - 4) = (2, -7)$ Since $b = 4$, the covertices are located 4 units
$(2, -3 + 4) = (2, 1)$ below and above the center, $(2, -3)$.

The ellipse is completed by connecting these points using the known shape and symmetry. □

In general the graph of $\dfrac{(x-h)^2}{a^2} + \dfrac{(y-k)^2}{b^2} = 1$ is the same as the graph of 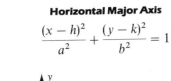 $= 1$ except that the center has been translated from $(0, 0)$ to (h, k).

Standard Form of the Equation of an Ellipse ▼

The equation of an ellipse with center (h, k), major axis of length $2a$, and minor axis of length $2b$ is as follows:

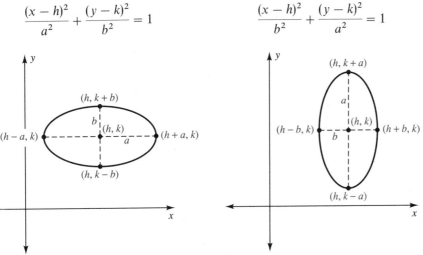

Note: $a > b > 0$.

EXAMPLE 7 Graph $36x^2 + 9y^2 = 324$.

SOLUTION $36x^2 + 9y^2 = 324$ Divide both sides by 324 to obtain 1 on the right side and to express the equation in standard form.

$$\frac{x^2}{9} + \frac{y^2}{36} = 1$$

$$\frac{x^2}{\underset{\underset{b=3}{b^2=9}}{9}} + \frac{y^2}{\underset{\underset{a=6}{a^2=36}}{36}} = 1$$

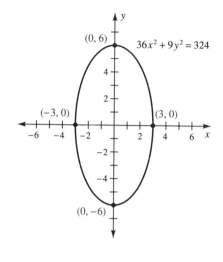

This is the equation of an ellipse centered at the origin with its major axis along the y-axis and its minor axis along the x-axis. Since $a = 6$ and $b = 3$, the y-intercepts are $(0, -6)$ and $(0, 6)$ and the x-intercepts are $(-3, 0)$ and $(3, 0)$. ☐

EXAMPLE 8 Graph $\dfrac{(y-5)^2}{36} + \dfrac{(x+4)^2}{16} = 1$.

SOLUTION The major axis of this ellipse is vertical. The center is $(-4, 5)$, with $a = 6$ and $b = 4$. The vertices are

$(-4, 5 - 6) = (-4, -1)$ The major axis is vertical since the denominator of the y^2 term is larger than the denominator of the x^2 term.
$(-4, 5 + 6) = (-4, 11)$

The covertices are

$$\frac{(x+4)^2}{\underset{\underset{b=4}{b^2=16}}{16}} + \frac{(y-5)^2}{\underset{\underset{a=6}{a^2=16}}{36}} = 1$$

$(-4 - 4, 5) = (-8, 5)$
$(-4 + 4, 5) = (0, 5)$

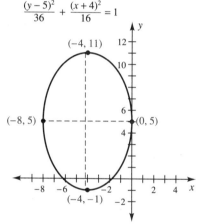

☐

EXAMPLE 9 Write the standard form of the equation of the ellipse whose vertices are $(-1, 1)$ and $(7, 1)$ and whose covertices are $(3, -2)$ and $(3, 4)$. This ellipse is shown to the right for visual reference.

SOLUTION The center $(h, k) = (3, 1)$.   The center of an ellipse is located at the intersection of its major and minor axes.
Thus $h = 3$ and $k = 1$.

$2a = 7 - (-1) = 8$ The length of the horizontal axis is $2a$.
$a = 4$

$2b = 4 - (-2) = 6$ The length of the vertical axis is $2b$.
$b = 3$

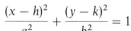

$\dfrac{(x-h)^2}{a^2} + \dfrac{(y-k)^2}{b^2} = 1$ The horizontal axis is the major axis; thus we use the standard form of an ellipse with a horizontal major axis.

$\dfrac{(x-3)^2}{4^2} + \dfrac{(y-1)^2}{3^2} = 1$ Substitute a, b, h, and k into the standard form.

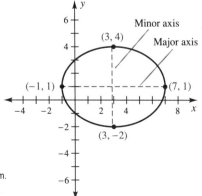

SECTION 3-7 CIRCLES AND ELLIPSES

Answer $\dfrac{(x-3)^2}{16} + \dfrac{(y-1)^2}{9} = 1$

EXAMPLE 10 Write $25x^2 + 4y^2 + 150x + 16y + 141 = 0$ in standard form, and then sketch its graph.

SOLUTION
$$25x^2 + 4y^2 + 150x + 16y + 141 = 0$$
$$(25x^2 + 150x) + (4y^2 + 16y) = -141 \qquad \text{Regroup terms.}$$
$$25(x^2 + 6x) + 4(y^2 + 4y) = -141$$
$$25(x^2 + 6x + 9) + 4(y^2 + 4y + 4) = -141 + 9(25) + 4(4) \qquad \text{Factor out the coefficients of } x^2 \text{ and } y^2 \text{ and then complete the squares.}$$
$$25(x + 3)^2 + 4(y + 2)^2 = 100$$
$$\dfrac{(x + 3)^2}{4} + \dfrac{(y + 2)^2}{25} = 1 \qquad \text{Divide by 100 to obtain 1 on the right side.}$$

This is the equation of an ellipse whose major axis is vertical. The center is $(-3, -2)$, with $a = 5$ and $b = 2$.

▶ **Self-Check** ▼

1 Graph $9x^2 + 36y = 324$.

2 Write $9x^2 + 16y^2 - 72x - 96y + 144 = 0$ in standard form, and then sketch its graph.

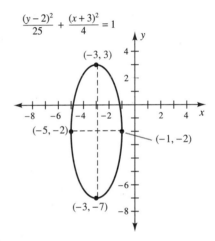

Figure for Example 10

▶ **Self-Check Answer** ▼

1

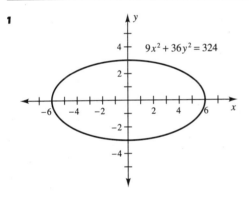

2 $\dfrac{(x-4)^2}{16} + \dfrac{(y-3)^2}{9} = 1$

EXERCISES 3-7

A

In Exercises 1–10, write in standard form and in general form with integer coefficients the equation of the circle satisfying the given conditions.

1. Center (0, 0), radius 10
2. Center (0, 0), radius 7
3. Center (5, −4), radius 4
4. Center(−2, −9), radius 3
5. Center (2, 6), radius $\sqrt{2}$
6. Center (−3, 7), radius $\sqrt{5}$
7. Center (−3, 8), radius $\frac{1}{5}$
8. Center (1, 1), radius 0.1
9. Center (0, $\frac{1}{2}$), radius $\frac{1}{2}$
10. Center ($\frac{1}{3}$, 0), radius 1

In Exercises 11–20, determine the center and the radius of the circle defined by each equation, and then graph the circle.

11. $x^2 + y^2 = 144$
12. $x^2 + y^2 = 169$
13. $(x + 5)^2 + (y - 4)^2 = 64$
14. $(x - 4)^2 + (y + 5)^2 = 36$
15. $x^2 + y^2 - 6x = 0$
16. $x^2 + y^2 + 12y - 13 = 0$
17. $x^2 + y^2 - 2x + 10y + 22 = 0$
18. $x^2 + y^2 - 6x - 18y + 86 = 0$
19. $4x^2 + 4y^2 - 8x - 40y + 103 = 0$
20. $9x^2 + 9y^2 + 54x - 72y + 224 = 0$

In Exercises 21–30, determine the center and the lengths of the major and minor axes of the ellipse defined by each equation, and then graph the ellipse.

21. $\dfrac{x^2}{36} + \dfrac{y^2}{16} = 1$
22. $\dfrac{x^2}{25} + \dfrac{y^2}{9} = 1$
23. $\dfrac{(x-1)^2}{9} + \dfrac{(y+2)^2}{49} = 1$
24. $\dfrac{(y-3)^2}{25} + \dfrac{(x+4)^2}{16} = 1$
25. $\dfrac{(y+3)^2}{9} + \dfrac{4(x+5)^2}{1} = 1$
26. $\dfrac{(y-5)^2}{16} + \dfrac{9(x+2)^2}{25} = 1$
27. $36x^2 + 49y^2 = 1764$
28. $9x^2 + 25y^2 = 225$
29. $16x^2 + 9y^2 + 32x + 54y - 47 = 0$
30. $4x^2 + 25y^2 - 40x + 300y + 900 = 0$

In Exercises 31–40, write in standard form the equation of the ellipse satisfying the given conditions.

31. Center (0, 0), x-intercepts (−9, 0) and (9, 0), y-intercepts (0, −5) and (0, 5)
32. Center (0, 0), x-intercepts (−5, 0) and (5, 0), y-intercepts (0, −3) and (0, 3)
33. Center (3, 4), horizontal major axis of length 4, vertical minor axis of length 2
34. Center (−5, 2), horizontal major axis of length 6, vertical minor axis of length 2
35. Center (6, −2), vertical major axis of length 10, horizontal minor axis of length 6
36. Center (−3, −4), vertical major axis of length 8, horizontal minor axis of length $\frac{1}{2}$
37. Center (5, 0), endpoints of major axis (1, 0) and (9, 0), endpoints of minor axis (5, −2) and (5, 2)
38. Center (0, −3), endpoints of major axis (0, 3) and (0, −9), endpoints of minor axis (−1, −3) and (1, −3)
39. Center (−3, −4), $a = 6$, $b = 2$, major axis vertical
40. Center (2, 5), $a = 8$, $b = 4$, major axis horizontal

SECTION 3-7 CIRCLES AND ELLIPSES

B

In Exercises 41–46, sketch the graph of each equation.

41 $\dfrac{x^2}{25} + \dfrac{y^2}{25} = 1$

42 $9x^2 + 9y^2 = 1$

43 $x^2 + y^2 - 6x + 8y + 25 = 0$

44 $x^2 + y^2 + 10x - 22y + 147 = 0$

45 $9x^2 + 4y^2 + 36 = 0$

46 $\dfrac{(x-5)^2}{9} + \dfrac{(y+2)^2}{4} = 0$

In Exercises 47–50, write an equation for the graph of each relation.

47
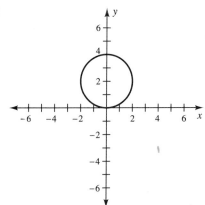

48
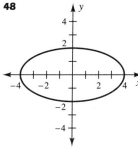

49
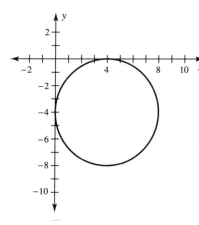

50
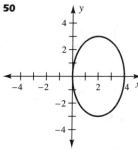

C

51 Write the equation of a circle if the endpoints of a diameter are $(-1, -7)$ and $(5, 1)$.

52 Write the equation of a circle tangent to the x-axis at $(3, 0)$ and to the y-axis at $(0, 3)$.

53 Write the equation of a circle with radius 4, located in the third quadrant and tangent to both the x- and y-axes.

54 Write the equation of a circle with center $(3, -8)$ and passing through the point $(-2, 4)$.

In Exercises 55–57, use the formula $A = \pi ab$ to find the area of the ellipse satisfying the given conditions.

55 Center (0, 0), $a = 5$, $b = 4$

56 Vertices $(-6, 3)$ and $(2, 3)$, covertices $(-2, 4)$ and $(-2, 2)$

57 $\dfrac{(x-4)^2}{100} + \dfrac{(y+3)^2}{64} = 1$

In Exercises 58–60, graph each pair of relations using a graphics calculator.

58 $y_1 = \sqrt{36 - x^2}$
$y_2 = -y_1$

59 $y_1 = -\sqrt{25 - x^2}$
$y_2 = -y_1$

60 $y_1 = \frac{2}{3}\sqrt{9 - x^2}$
$y_2 = -y_1$

SECTION 3-8*

Hyperbolas and a Summary of Conic Sections

Section Objectives

20 Write the equation of a hyperbola in standard form.

21 Graph a hyperbola.

22 Identify the type of conic section defined by an equation.

A **hyperbola** is the set of all points in a plane whose distances from two fixed points have a constant difference. (See Figure 3-30.) The fixed points are the **foci** of the hyperbola. The dvelopment of the formula $\dfrac{x^2}{a^2} - \dfrac{y^2}{b^2} = 1$ for the hyperbola centered at the origin with x-intercepts $(-a, 0)$ and $(a, 0)$ is similar to the development of the equation for an ellipse.

The hyperbola defined by

$$\dfrac{x^2}{a^2} - \dfrac{y^2}{b^2} = 1$$

is symmetric about the origin and about both the x-axis and the y-axis. It also approaches the lines $y = \dfrac{-b}{a} x$ and $y = \dfrac{b}{x} x$ as $|x|$ becomes large. A line is an **asymptote** of a graph if the graph gets closer and closer to the line for points farther and farther from the origin. Asymptotes can simplify the sketching of a hyperbola, for they allow us to quickly picture a portion of this complicated curve. In the following paragraph, we verify that $y = \dfrac{-b}{a} x$ and $y = \dfrac{b}{a} x$ are asymptotes of the hyperbola defined by $\dfrac{x^2}{a^2} - \dfrac{y^2}{b^2} = 1$.

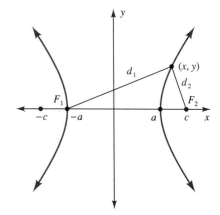

Figure 3-30 $|d_1 - d_2| = 2a$

*This is an optional section.

The asymptotic behavior of the hyperbola defined by $\dfrac{x^2}{a^2} - \dfrac{y^2}{b^2} = 1$ can be understood by examining y as $|x|$ becomes very large. First, solve this equation for y.

$$\frac{y^2}{b^2} = \frac{x^2}{a^2} - 1$$

$$y^2 = \frac{b^2}{a^2} x^2 - b^2 \qquad \text{Multiply by } b^2$$

$$= \frac{b^2}{a^2} x^2 \left(1 - \frac{a^2}{x^2}\right) \qquad \text{Factor out } \frac{b^2}{a^2} x^2.$$

$$y = \pm \frac{b}{a} x \sqrt{1 - \frac{a^2}{x^2}} \qquad \text{Extract the roots.}$$

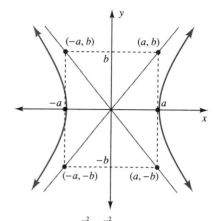

Figure 3-31 $\dfrac{x^2}{a^2} - \dfrac{y^2}{b^2} = 1$

As $|x|$ becomes very large, the fraction $\dfrac{a^2}{x^2}$ approaches 0, the radical approaches 1, and y approaches $\pm \dfrac{b}{a} x$. Thus the hyperbola approaches the lines $y = \dfrac{-b}{a} x$ and $y = \dfrac{b}{a} x$. Note that the asymptotes $y = \dfrac{-b}{a} x$ and $y = \dfrac{b}{a} x$ pass through the origin, which is the center of this hyperbola. The asymptotes also pass through the corners of the rectangle formed by (a, b), $(-a, b)$, $(-a, -b)$, and $(a, -b)$. This rectangle, which is shown in Figure 3-31, is called the **fundamental rectangle** and is used to sketch quickly the linear asymptotes.

EXAMPLE 1 Graph $\dfrac{x^2}{16} - \dfrac{y^2}{9} = 1$.

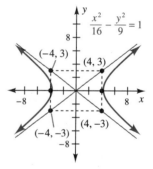

Figure for Example 1

SOLUTION Set $y = 0$ to determine the x-intercepts $(-4, 0)$ and $(4, 0)$. Since $a = 4$ and $b = 3$, the corners of the fundamental rectangle are $(4, 3)$, $(-4, 3)$, $(-4, -3)$, and $(4, -3)$. Draw the asymptotes through the corners of this rectangle and then sketch the hyperbola, using the asymptotes as guidelines. □

Hyperbolas of the form $\dfrac{x^2}{a^2} - \dfrac{y^2}{b^2} = 1$ open horizontally, as illustrated by Example 1. We will now examine a hyperbola of the form $\dfrac{y^2}{a^2} - \dfrac{x^2}{b^2} = 1$. These hyperbolas open vertically, with foci and intercepts on the y-axis. There can be no x-intercepts for this hyperbola, because setting $y = 0$ yields $\dfrac{-x^2}{b^2} = 1$, which has no real solution. Note in Figure 3-32 that the corners of the fundamental rectangle are (b, a), $(-b, a)$, $(-b, -a)$, and $(b, -a)$.

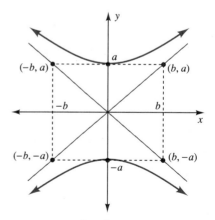

Figure 3-32 $\dfrac{y^2}{a^2} - \dfrac{x^2}{b^2} = 1$

EXAMPLE 2 Graph $36y^2 - 25x^2 = 900$.

SOLUTION

$$36y^2 - 25x^2 = 900$$ First divide both sides of the equation by 900.

$$\frac{y^2}{25} - \frac{x^2}{36} = 1$$

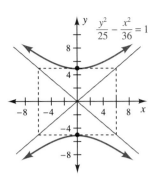

This hyperbola opens vertically. The y-intercepts are $(0, -5)$ and $(0, 5)$. Since $a = 5$ and $b = 6$, the corners of the fundamental rectangle are $(6, 5)$, $(-6, 5)$, $(-6, -5)$, and $(6, -5)$. Draw the asymptotes through the corners of this rectangle and then sketch the hyperbola using the asymptotes as guidelines. ☐

If the center of the hyperbola $\dfrac{x^2}{a^2} - \dfrac{y^2}{b^2} = 1$ is translated to the point (h, k), then the standard form of the equation of this hyperbola is

$$\frac{(x-h)^2}{a^2} - \frac{(y-k)^2}{b^2} = 1$$

Standard Form of the Equation of a Hyperbola ▼

The equation of a hyperbola with center (h, k) is as follows:

Opening Horizontally

$$\frac{(x-h)^2}{a^2} - \frac{(y-k)^2}{b^2} = 1$$

Opening Vertically

$$\frac{(y-k)^2}{a^2} - \frac{(x-h)^2}{b^2} = 1$$

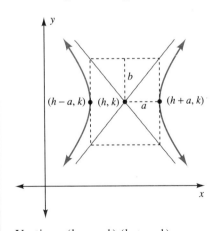
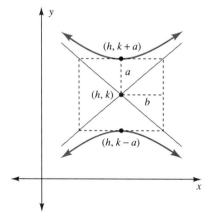

Vertices: $(h - a, k)$ $(h + a, k)$.
The fundamental rectangle has base $2a$ and height $2b$.

Vertices: $(h, k - a)$, $(h, k + a)$.
The fundamental rectangle has base $2b$ and height $2a$.

SECTION 3-8 (OPTIONAL) HYPERBOLAS AND A SUMMARY OF CONIC SECTIONS

EXAMPLE 3 Graph $\dfrac{(x-4)^2}{64} - \dfrac{(y+2)^2}{25} = 1$.

SOLUTION From the standard form, note that the x term is positive and so the hyperbola opens horizontally, with center $(4, -2)$ and

$$a = 8 \qquad \underset{\underset{a^2 = 64}{\uparrow}}{\dfrac{(x-4)^2}{64}} - \underset{\underset{b^2 = 25}{\nearrow}}{\dfrac{(y+2)^2}{25}} = 1$$
$$b = 5$$
$$a = 8 \quad b = 5$$

▶ **Self-Check** ▼

Graph $4x^2 - 25y^2 = 100$.

1 Plot the vertices

$$(4 - 8, -2) = (-4, -2)$$
$$(4 + 8, -2) = (12, -2)$$

2 Sketch the rectangle with corners

$$(4 + 8, \ -2 + 5) = (12, 3)$$
$$(4 - 8, -2 + 5) = (-4, 3)$$
$$(4 - 8, -2 - 5) = (-4, -7)$$
$$(4 + 8, -2 - 5) = (12, -7)$$

3 Sketch the asymptotes passing through the corners of this rectangle.
4 Sketch the hyperbola opening horizontally through the vertices and approaching the asymptotes. □

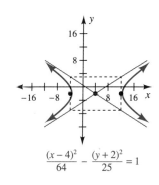

$\dfrac{(x-4)^2}{64} - \dfrac{(y+2)^2}{25} = 1$

▶ **Self-Check Answer** ▼

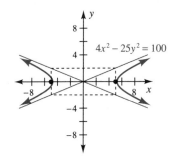

$4x^2 - 25y^2 = 100$

EXAMPLE 4 Graph $4x^2 + 24x + 47 = y^2 + 10y$.

SOLUTION

$4x^2 + 24x + 47 = y^2 + 10y$

$4(x^2 + 6x) + 47 = y^2 + 10y$

To obtain the key information for this hyperbola, first rewrite the equation in standard form. Begin by completing the square.

$4(x^2 + 6x + 9) + 47 + 25 = (y^2 + 10y + 25) + 4(9)$

$4(x + 3)^2 + 72 = (y + 5)^2 + 36$

$(y + 5)^2 - 4(x + 3)^2 = 36$

$$\frac{(y + 5)^2}{36} - \frac{(x + 3)^2}{9} = 1$$

Rewrite in standard form with 1 on the right side of the equation.

From the standard form, we know that the hyperbola has center $(-3, -5)$, $a = 6$, and $b = 3$ and opens vertically, since the y term is positive.

[1] Plot the vertices

$(-3, -5 - 6) = (-3, -11)$

$(-3, -5 + 6) = (-3, 1)$

[2] Sketch the rectangle with corners

$(0, 1), \quad (-6, 1), \quad (-6, -11), \quad (0, -11)$

[3] Sketch the asymptotes passing through the corners of this rectangle.

[4] Sketch the hyperbola opening vertically through the vertices and approaching the asymptotes.

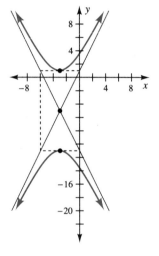

$$\frac{(y+5)^2}{36} - \frac{(x+3)^2}{9} = 1$$

A Summary of Conic Sections

Parabolas, circles, ellipses, and hyperbolas can all be formed by cutting a cone with a plane. Therefore these figures are collectively referred to as conic sections. Conic sections were originally studied from a geometric viewpoint. (See Figure 3-33.)

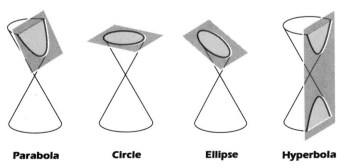

Figure 3-33 Conic sections

SECTION 3-8 (OPTIONAL) HYPERBOLAS AND A SUMMARY OF CONIC SECTIONS

The Geometric Definition of a Parabola ▼

A parabola is the set of all points equidistant from a fixed line L (the **directrix**) and a fixed point F (the **focus**).

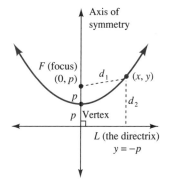

Figure 3-34 $d_1 = d_2$

The vertex of the parabola is on the axis of symmetry midway between the focus and the directrix, and the axis of symmetry is perpendicular to the directrix.

Using the geometric definition of a parabola, we develop the standard form of the equation of a parabola. (See Figure 3-34.)

$$d_1 = d_2$$

$$\sqrt{(x-0)^2 + (y-p)^2} = |y + p|$$

$x^2 + y^2 - 2py + p^2 = y^2 + 2py + p^2$ Square both sides and then combine
$ x^2 = 4py$ like terms.

If the vertex is translated from the origin to the point (h, k), the formula becomes $(x - h)^2 = 4p(y - k)$, which is the standard form of the equation of a parabola that opens up or opens down. The parabola opens up if $p > 0$ and opens down if $p < 0$. Note that $|p|$ is the distance from the focus to the vertex and also the distance from the vertex to the directrix. Thus the distance from the focus to the directrix is $2|p|$.

EXAMPLE 5 Write the standard form of the equation of the parabola with focus $(-2, 4)$ and directrix $y = -2$.

SOLUTION The parabola opens up since the focus is above the directrix.

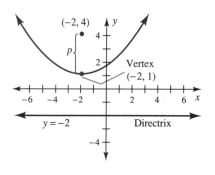

$(x - h)^2 = 4p(y - k)$	The standard form of the equation of a parabola. $p > 0$ since it opens up; thus $\|p\| = p$.
$2p = 4 - (-2) = 6$	The distance from the focus to the directrix is $2p$.
$p = 3$	
$h = -2$	Since the axis of symmetry is vertical, the x-coordinates of the vertex and the focus are the same.
$k = 1$	The vertex is $p = 3$ units below the focus; thus $k = 4 - 3 = 1$.
Answer $(x + 2)^2 = 12(y - 1)$	Substitute the values of h, k, and p into the standard form.

The Geometric Definition of an Ellipse ▼

> An ellipse is the set of all points in a plane, the sum of whose distances from two fixed points (the foci) is constant.

$\dfrac{(x-h)^2}{a^2} + \dfrac{(y-k)^2}{b^2} = 1$ is the standard form of the equation of an ellipse with a horizontal major axis, center (h, k), foci $(h \pm c, k)$, vertices $(h \pm a, k)$, and covertices $(h, k \pm b)$, where a is the distance from the center to a vertex, c is the distance from the center to a focal point, and b is the distance from the center to a covertex. $b^2 = a^2 - c^2$ was defined in the development of this standard form in the previous section. (See Figure 3-35.)

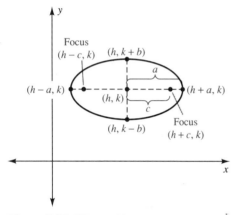

Figure 3-35 Ellipse with horizontal major axis

EXAMPLE 6 Write the standard form of the equation of an ellipse with foci $(-2, 4)$, $(6, 4)$ and a vertex $(7, 4)$.

SOLUTION

$\dfrac{(x-h)^2}{a^2} + \dfrac{(y-k)^2}{b^2} = 1$ The major axis is horizontal since the foci and the vertex are on the same horizontal line.

Center $= \left(\dfrac{6+(-2)}{2}, \dfrac{4+4}{2}\right)$ The center is the midpoint of the segment connecting the foci.

$(h, k) = (2, 4)$

$c = 6 - 2 = 4$ c is the distance from a focal point $(6, 4)$ to the center $(2, 4)$.

$a = 7 - 2 = 5$ a is the distance from a vertex $(7, 4)$ to the center $(2, 4)$.

$b^2 = 25 - 16 = 9$ $b^2 = a^2 - c^2$

$b = 3$

Answer $\dfrac{(x-2)^2}{25} + \dfrac{(y-4)^2}{9} = 1$ Substitute the values of h, k, a, and b into the standard form. ☐

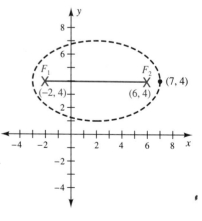

The Geometric Definition of a Hyperbola ▼

A hyperbola is the set of all points in a plane whose distances from two fixed points (the foci) have a constant difference.

▶ **Self-Check ▼**

Write the standard form of the equation of the parabola with focus $(2, -4)$ and directrix $y = 2$.

$\dfrac{(x-h)^2}{a^2} - \dfrac{(y-k)^2}{b^2} = 1$ is the standard form of the equation of a hyperbola opening horizontally with center (h, k), foci $(h \pm c, k)$, and vertices $(h \pm a, k)$, where a is the distance from the center to a vertex, c is the distance from the center to a focal point, and $2b$ is the height of the fundamental rectangle. $b^2 = c^2 - a^2$ was defined in the development of this standard form. (See Figure 3-36.)

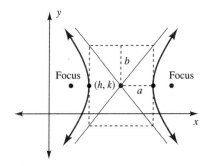

Figure 3-36 Hyperbola opening horizontally

EXAMPLE 7 Write an equation in standard form of the hyperbola with foci $(1, -3), (11, -3)$ and a vertex $(9, -3)$.

SOLUTION

$\dfrac{(x-h)^2}{a^2} - \dfrac{(y-k)^2}{b^2} = 1$ The hyperbola opens horizontally, since the foci and the vertices are on a horizontal line.

$\text{Center} = \left(\dfrac{1+11}{2}, \dfrac{-3+(-3)}{2}\right)$ The center is the midpoint of the line segment joining the foci.

$(h, k) = (6, -3)$

$c = 11 - 6 = 5$ c is the distance from a focal point $(11, -3)$ to the center $(6, -3)$.

$a = 9 - 6 = 3$ a is the distance from a vertex $(9, -3)$ to the center $(6, -3)$.

$b^2 = 25 - 9 = 16$

$b = 4$ $b^2 = c^2 - a^2$

Answer $\dfrac{(x-6)^2}{9} - \dfrac{(y+3)^2}{16} = 1$ Substitute the values of h, k, a, and b into the standard form. □

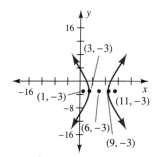

Figure for Example 7

A hyperbola is *not* a pair of parabolas; a parabola has no asymptotes, whereas a hyperbola has a pair of asymptotes crossing at its center.

All conic sections with a vertical or horizontal axis have equations of the **general form** $Ax^2 + Cy^2 + Dx + Ey + F = 0$. An equation in general form can be converted to standard form by completing the square, and thus the key features of its graph can be determined. However, it is useful to be able to at least identify the type of conic section from an equation in general form. Fortunately this can be done by just examining the coefficients of $Ax^2 + Cy^2 + Dx + Ey + F = 0$.

▶ **Self-Check Answer ▼**

$(x-2)^2 = -12(y+1)$

Conic Sections ▼

General Form: $Ax^2 + Cy^2 + Dx + Ey + F = 0$

	General Form	Standard Form	Features
Parabola:	$Ax^2 + Dx + Ey + F = 0$ ($C = 0, A \neq 0, E \neq 0$)	$(x - h)^2 = 4p(y - k)$	Vertex: (h, k) $p > 0$: opens upward $p < 0$: opens downward
	$Cy^2 + Dx + Ey + F = 0$ ($A = 0, C \neq 0, D \neq 0$)	$(y - k)^2 = 4p(x - h)$	$p > 0$: opens to right $p < 0$: opens to left
Circle:	$Ax^2 + Ay^2 + Dx + Ey + F = 0$ ($A = C \neq 0$)	$(x - h)^2 + (y - k)^2 = r^2$	Center: (h, k) Radius: r $(r > 0)$
Ellipse:	$Ax^2 + Cy^2 + Dx + Ey + F = 0$ ($AC > 0$)	$\dfrac{(x - h)^2}{a^2} + \dfrac{(y - k)^2}{b^2} = 1$	Center: (h, k) Length of major axis: $2a$
		$\dfrac{(y - k)^2}{a^2} + \dfrac{(x - h)^2}{b^2} = 1$	Length of minor axis: $2b$ $(a > b > 0)$
Hyperbola:	$Ax^2 + Cy^2 + Dx + Ey + F = 0$ ($AC < 0$)	$\dfrac{(x - h)^2}{a^2} - \dfrac{(y - k)^2}{b^2} = 1$	Center: (h, k) Opens horizontally through vertices $(h \pm a, k)$ Fundamental rectangle has base $2a$ and height $2b$
		$\dfrac{(y - k)^2}{a^2} - \dfrac{(x - h)^2}{b^2} = 1$	Opens vertically through vertices $(h, k \pm a)$ Fundamental rectangle has base $2b$ and height $2a$

SECTION 3-8 (OPTIONAL) HYPERBOLAS AND A SUMMARY OF CONIC SECTIONS

EXAMPLE 8 Identify the type of conic sections described by each of the following equations.

(a) $9x^2 - y^2 = 4$

(b) $9x^2 - 18x + 9 = 24y - 4y^2$

(c) $x^2 + y^2 + 8x - 14y + 65 = 0$

SOLUTIONS

This is the equation of a hyperbola, since the coefficients of x^2 and y^2 have opposite signs.

First rewrite this equation in general form as

$$9x^2 + 4y^2 - 18x - 24y + 9 = 0$$

This is the equation of an ellipse, since the coefficients of x^2 and y^2 have the same sign but different magnitudes.

This equation has the form of a circle, since the coefficients of x^2 and y^2 are equal. However, the standard form of the equation is $(x + 4)^2 + (y - 7)^2 = 0$. Thus the graph is a degenerate case of a circle—the point $(-4, 7)$.

▶ **Self-Check** ▼

Identify the type of conic section defined by these equations.

1 $7x^2 - 9x - y + 11 = 0$

2 $5x^2 + y^2 - 20x + 6y + 34 = 0$

If an axis of a conic section is not vertical or horizontal, the general form of its equation is $Ax^2 + Bxy + Cy^2 + Dx + Ey + F = 0$ with $B \neq 0$. It can be shown that the sign of $B^2 - 4AC$ indicates the type of conic section (see Table 3-2). The proof of this result is given in more advanced mathematical texts.

Table 3-2 Sign of $B^2 - 4AC$

Sign	Type of Conic Section
Negative	Ellipse
Zero	Parabola
Positive	Hyperbola

▶ **Self-Check Answers** ▼

1 Parabola opening upward

2 No graph (right side of standard form of an ellipse cannot be negative)

EXAMPLE 9 Identify the type of conic section represented by these equations in the general form $Ax^2 + Bxy + Cy^2 + Dx + Ey + F = 0$.

SOLUTIONS

(a) $xy + y - 1 = 0$

$A = 0, B = 1, C = 0$
$B^2 - 4AC = 1^2 - 4 \cdot 0 \cdot 0 > 0$
Thus the equation represents a hyperbola.

(b) $2x^2 + xy + y^2 - x - 60 = 0$

$A = 2, B = 1, C = 1$
$B^2 - 4AC = 1^2 - 8 < 0$
Thus the equation represents an ellipse.

(c) $x^2 + 2xy + y^2 - x + y - 9 = 0$

$A = 1, B = 2, C = 1$
$B^2 - 4AC = 2^2 - 4 = 0$
Thus the equation represents a parabola. □

When one body is under the gravitational influence of another, its relative orbit must be one of the conic sections, as illustrated in Figure 3-37. All planets, satellites, and asteroids describe elliptical orbits; many comets follow essentially parabolic orbits. A few comets have hyperbolic orbits; after one perihelion passage, such a comet leaves the solar system forever. Artificial space probes have been launched into hyperbolic orbits with respect to the Earth, but they are nearly always captured into elliptical orbits about the Sun. Pioneer 10 is the first space craft with an orbit that, when perturbed by Jupiter, will lead to escape from the solar system.

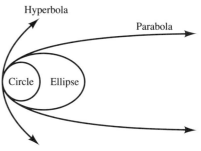

Figure 3-37 Orbits of heavenly bodies

EXERCISES 3-8

A

In Exercises 1–10, determine the center, the values of a and b, and the direction the hyperbola opens, and then sketch the hyperbola.

1 $\dfrac{x^2}{25} - \dfrac{y^2}{81} = 1$

2 $\dfrac{y^2}{36} - \dfrac{x^2}{16} = 1$

3 $25x^2 = 100y^2 + 100$

4 $16x^2 + 64 = 4y^2$

5 $\dfrac{(x-4)^2}{49} - \dfrac{(y-6)^2}{9} = 1$

6 $\dfrac{(y-6)^2}{9} - \dfrac{(x-4)^2}{49} = 1$

7 $16x^2 + 96x = 9y^2 - 126y + 153$

8 $25x^2 + 150x = y^2 + 1375$

9 $36x^2 - 36x = 36y^2 + 48y + 43$

10 $100x^2 - 300x + 289 = 100y^2 - 120y$

In Exercises 11–16, write the standard form of the equation of the hyperbola satisfying the given conditions.

11 The center is (0, 0), the hyperbola opens vertically, and the fundamental rectangle has height 10 and width 8.

12 The center is (0, 0), the hyperbola opens horizontally, and the fundamental rectangle has height 6 and width 12.

SECTION 3-8 (OPTIONAL) HYPERBOLAS AND A SUMMARY OF CONIC SECTIONS

13 The hyperbola has vertices $(-3, 0)$ and $(3, 0)$, and the height of the fundamental rectangle is 14.

14 The hyperbola has vertices $(0, -5)$ and $(0, 5)$, and the width of the fundamental rectangle is 16.

15 The hyperbola has vertices at $(2, -4)$ and $(2, 14)$, and the width of the fundamental rectangle is 30.

16 The hyperbola has vertices $(3, 5)$ and $(11, 5)$, and the height of the fundamental rectangle is 8.

In Exercises 17–24, match each graph with its correct relation.

a. 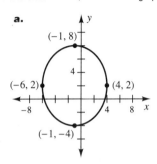 **b.** 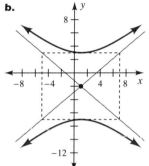 **c.** 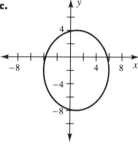 **d.**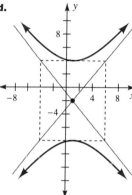

e. 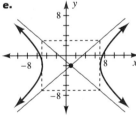 **f.** 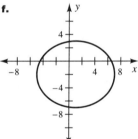 **g.** 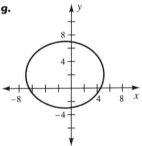 **h.**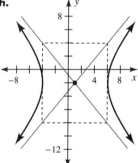

17 $\dfrac{(x-1)^2}{25} + \dfrac{(y+2)^2}{36} = 1$ **18** $\dfrac{(x-1)^2}{36} + \dfrac{(y+2)^2}{25} = 1$ **19** $\dfrac{(x+1)^2}{25} + \dfrac{(y-2)^2}{36} = 1$ **20** $\dfrac{(x+1)^2}{36} + \dfrac{(y-2)^2}{25} = 1$

21 $\dfrac{(x-1)^2}{25} - \dfrac{(y+2)^2}{36} = 1$ **22** $\dfrac{(x-1)^2}{36} - \dfrac{(y+2)^2}{25} = 1$ **23** $\dfrac{(y+2)^2}{25} - \dfrac{(x-1)^2}{36} = 1$ **24** $\dfrac{(y+2)^2}{36} - \dfrac{(x-1)^2}{25} = 1$

B

In Exercises 25–35, identify the type of conic section defined by each equation.

25 $\dfrac{(x-5)^2}{9} + \dfrac{(y+3)^2}{11} = 1$ **26** $\dfrac{(x-5)^2}{9} - \dfrac{(y+3)^2}{11} = 1$ **27** $y + 6 = 5(x-4)^2$

28 $(x+8)^2 + (y-6)^2 = 13$
29 $x + 3 = 2(y-4)^2$
30 $x^2 - y^2 = 0$
31 $x^2 + y^2 = -1$
32 $x^2 + y^2 + 22x - 10y + 110 = 0$
33 $3x^2 - 4y^2 + 18x + 16y - 1 = 0$
34 $xy = 1$
35 $2x^2 - 4xy + 2y^2 + x - y + 5 = 0$

C

In Exercises 36–38, write the standard form of the equation of the conic section described by the given information.

36 A parabola with focus $(-2, 8)$ and directrix $y = 2$
37 An ellipse with foci $(0, 6)$, $(0, 0)$ and vertex $(0, 7)$
38 A hyperbola with foci $(2, 5)$, $(8, 5)$ and vertex $(7, 5)$

In Exercises 39–44, graph the conic section defined by each equation.

39 $y^2 - x - 6y + 7 = 0$
40 $5x^2 - 7y^2 + 30x + 28y + 52 = 0$
41 $x^2 + y^2 + 4x + 16y + 19 = 0$
42 $9x^2 + 4y^2 + 144x - 8y + 579 = 0$
43 $x^2 + 14x - y + 50 = 0$
44 $x^2 - 9y^2 = 0$

KEY CONCEPTS

1 Methods of denoting functions:

   Mapping notation
   Table of values
   Ordered-pair notation
   Graph
   Functional notation

2 $(f \circ g)(x) = f[g(x)]$ for the set of x values in the domain of g for which $g(x)$ is in the domain of f.

3 If a vertical line placed through the x-axis intersects the graph at more than one point, the graph is *not* a function.

4 A function $y = f(x)$ defined on an interval that includes x_1 and x_2 is:
   increasing if $f(x_2) > f(x_1)$ whenever $x_2 > x_1$.
   decreasing if $f(x_2) < f(x_1)$ whenever $x_2 > x_1$.
   constant if $f(x_1) = f(x_2)$ for every x_1 and x_2.

5 Forms of Linear equations:

   General form: $Ax + By + C = 0$
   Point-slope form: $y - y_1 = m(x - x_1)$
   Slope-intercept form: $y = mx + b$
   Horizontal line: $y = k$ for any constant k
   Vertical line: $x = h$ for any constant h

6 Lines with the same slope are parallel. Lines whose slopes are negative reciprocals of each other are perpendicular.

7 Formulas:

   1. Slope of a line: $m = \dfrac{y_2 - y_1}{x_2 - x_1}$

KEY CONCEPTS

2. Midpoint between two points: $\left(\dfrac{x_1 + x_2}{2}, \dfrac{y_1 + y_2}{2}\right)$

3. Distance between two points: $d = \sqrt{(x_2 - x_1)^2 + (y_2 - y_1)^2}$

8 Parabolas:
   a. Opening vertically: $f(x) = ax^2 + bx + c \ (a \neq 0)$
      $a > 0$ The parabola opens upward.
      $a < 0$ The parabola opens downward.
      Vertex of the parabola: $\left(\dfrac{-b}{2a}, f\left(\dfrac{-b}{2a}\right)\right)$
   b. Opening horizontally: $x = ay^2 + by + c \ (a \neq 0)$
      $a > 0$ The parabola opens to the right.
      $a < 0$ The parabola opens to the left.

9 Translations:
   For $h > 0$ and $k > 0$, the graph of $y = f(x - h) - k$ is obtained by shifting the graph of $y = f(x)$ to the right h units and down k units.

10 Stretching:
   We shall refer to $y = af(x)$ as a stretching of $y = f(x)$ if $a > 1$.

   Shrinking:
   We shall refer to $y = af(x)$ as a shrinking of $y = f(x)$ if $0 < a < 1$.

   Reflection:
   $y = -f(x)$ is a reflection of $y = f(x)$ about the x-axis.

11 Symmetry:
   A graph is symmetric to the y-axis if the portion of the graph to the left of the y-axis is a mirror image of the portion to the right of the y-axis.
   A graph is symmetric to the x-axis if the portion of the graph below the x-axis is a mirror image of the portion above the x-axis.
   A graph is symmetric to the origin if each point (x, y) on the graph has an image $(-x, -y)$ directly across the origin in the opposite quadrant.

12 The horizontal line test can be used to determine if a graph represents a one-to-one function.

13 A function and its inverse are symmetric about the line $y = x$.

14 The range of a one-to-one function can be determined by finding the domain of its inverse.

15 Standard form of the equations of the conic sections:
   a. **Circle:** $(x - h)^2 + (y - k)^2 = r^2$
   b. **Ellipse** (a horizontal major axis): $\dfrac{(x - h)^2}{a^2} + \dfrac{(y - k)^2}{b^2} = 1$
   c. **Hyperbola** (opening horizontally): $\dfrac{(x - h)^2}{a^2} - \dfrac{(y - k)^2}{b^2} = 1$
   d. **Parabola (opening vertically):** $(x - h)^2 = 4p(y - k)$

16 Whole families of relations have graphs with the same basic shape. Thus an important part of curve sketching is the recognition of these basic shapes from the defining equations. Specific members of these families can often be determined by using translations, reflections, or stretchings or shrinkings. This approach to curve sketching is generally faster than plotting points and yields a better overall understanding of the properties of the graph. Basic shapes covered in this chapter include: lines, parabolas, circles, ellipses, hyperbolas, absolute value functions (V-shaped), and greatest integer functions (step functions).

REVIEW EXERCISES FOR CHAPTER 3

1 Find the midpoint between $(-2, -4)$ and $(5, 20)$, the distance between these points, and the slope of the line connecting these points.

In Exercises 2–9, evaluate each expression, given $f(x) = 6x^2 + 11x - 35$ and $g(x) = 2x + 7$.

2 $f(3)$ **3** $g(-3)$ **4** $(f - g)(2)$ **5** $(f \cdot g)(-1)$

6 $(f \circ g)(-4)$ **7** $f(-x)$ **8** $f(x + h)$ **9** $\dfrac{g(x + h) - g(x)}{h}$

In Exercises 10–12, determine which of the relations are functions.

10 $\{(5, 6), (-3, 7), (4, 2), (2, 6)\}$

11
```
D    R
6 → 3
5 → 2
4 ↗
```

12
```
D    R
6 → 3
5 → 2
4 → 1
  ↘ 7
```

In Exercises 13 and 14, determine whether each graph represents a function.

13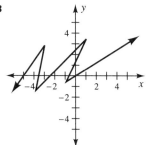

14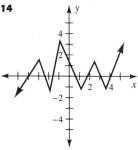

In Exercises 15–26, write the general form of the equation of each line described.

15 A line through $(-4, 5)$ with slope $\tfrac{2}{3}$

16 A line through $(3, -2)$ and $(-5, 4)$

17 Slope $-\tfrac{3}{7}$, y-intercept 6

18 Slope $-\tfrac{3}{7}$, x-intercept 6

19 A vertical line through $(2, -5)$

20 A horizontal line through $(2, -5)$

REVIEW EXERCISES FOR CHAPTER 3

21 A line through $(2, -1)$ and parallel to $2x - y - 1 = 0$
22 A line through $(2, -1)$ and perpendicular to $2x - y - 1 = 0$
23 A line through $(2, -1)$ and perpendicular to $x = 7$
24 A line through $(2, -1)$ and parallel to $y = 5$
25 x-intercept $\frac{1}{2}$, y-intercept $-\frac{1}{3}$
26 x and y coordinates equal

In Exercises 27–32, write the standard form of the equation of each circle or ellipse.

27 Circle with center $(0, 0)$ and radius $\frac{1}{2}$
28 Circle with center $(0, 0)$ and x-intercepts $(-4, 0)$ and $(4, 0)$
29 Ellipse with center $(0, 0)$, x-intercepts $(-4, 0)$ and $(4, 0)$, and y-intercepts $(0, -3)$ and $(0, 3)$
30 Ellipse with center $(3, 1)$, vertical major axis of length 4, and horizontal minor axis of length 1.
31 $x^2 + y^2 - 4x + 4y - 17 = 0$
32 $4x^2 + 25y^2 - 8x + 150y + 129 = 0$

In Exercises 33–35, determine the domain of each function. In Exercise 33 also determine the range.

33 $\{(3, 11), (5, 9), (7, -2)\}$
34 $y = \dfrac{5x - 35}{x^2 - 9x + 14}$
35 $y = \sqrt{x + 2}$

36 Determine the domain and the range of each function $y = f(x)$ and also determine the intervals for which the function is increasing, decreasing, or constant.

a.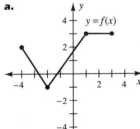

b.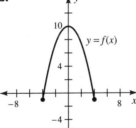

c.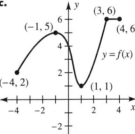

37 Determine whether each relation is even, odd, or neither. Also determine whether the graph of each relation is symmetric about the x-axis, the y-axis, or the origin.
   a. $y = 3x^4 - 7x^2 - 8$
   b. $y = |x| + 2$
   c. $y = 4x^3 - 7x - 8$
   d. $f(x) = \dfrac{1}{x^3 - x}$

38 Complete the drawing of the relation so that its graph will be symmetric about the
   a. x-axis
   b. y-axis
   c. origin

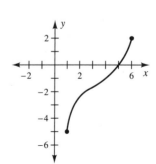

Figure for Exercise 38

39 Sketch the graph of the functions described, given the graph of $y = f(x)$ sketched to the right.
  a. $y = f(x - 2)$ b. $y = f(x) - 2$
  c. $y = -f(x)$ d. $y = f(x + 1) + 3$

40 Write the equation for each function and state its domain, given $f(x) = 2x^2 - 5x - 12$ and $g(x) = x - 4$.
  a. $(f + g)(x)$ b. $\left(\dfrac{g}{f}\right)(x)$
  c. $(f \circ g)(x)$ d. $f(-x)$

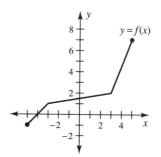

Figure for Exercise 39

In Exercises 41–55, match each relation with its graph.

a. 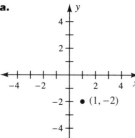 b. 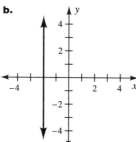 c. 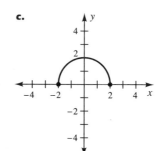 d.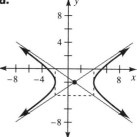

e. 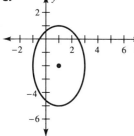 f. 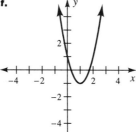 g. 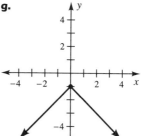 h.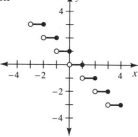

i. 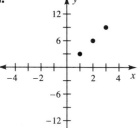 j. 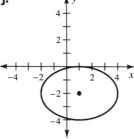 k. 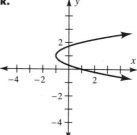 l.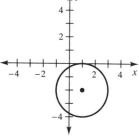

REVIEW EXERCISES FOR CHAPTER 3

m. 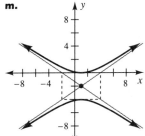   **n.** 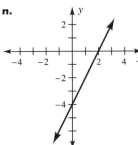   **o.**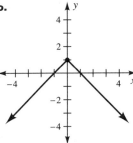

41 $y = 2x^2 - 4x + 1$
42 $y = 2x - 4$
43 $x = -2$
44 $y = -|x| + 1$
45 $f(x) = 3x$, domain $= \{1, 2, 3\}$
46 $(x - 1)^2 + (y + 2)^2 = 0$
47 $(x - 1)^2 + (y + 2)^2 = 4$
48 $x = 2y^2 - 4y + 1$
49 $\dfrac{(x - 1)^2}{4} + \dfrac{(y + 2)^2}{9} = 1$
50 $\dfrac{(x - 1)^2}{9} + \dfrac{(y + 2)^2}{4} = 1$
51 $\dfrac{(x - 1)^2}{9} - \dfrac{(y + 2)^2}{4} = 1$
52 $\dfrac{(y + 2)^2}{4} - \dfrac{(x - 1)^2}{9} = 1$
53 $y = \sqrt{4 - x^2}$
54 $y = [\![1 - x]\!]$
55 $y = \begin{cases} x - 1 & \text{if } x \leq 0 \\ -x - 1 & \text{if } x > 0 \end{cases}$

56 Determine whether the functions $f(x) = \dfrac{x^2 - 2x}{x^2 - 4}$ and $g(x) = \dfrac{x}{x + 2}$ are equal. State why or why not.

57 Graph $y = \begin{cases} 1 - x & \text{if } x \leq 1 \\ x - 1 & \text{if } x > 1 \end{cases}$

In Exercises 58 and 59, determine whether or not each graph represents a one-to-one function. If the function is one-to-one, give its inverse and the domain of the inverse.

58 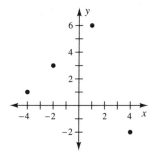   **59**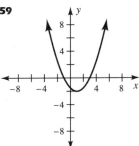

In Exercises 60 and 61 determine whether or not each function is a one-to-one function. If the function is one-to-one, give its inverse and the domain of the inverse.

60 $\{(7, 8), (8, 9), (9, 7)\}$
61 $f(x) = \dfrac{x - 4}{5}$

62 Graph $y = 3x - 2$ and its inverse on the same coordinate system.

63 The height h in feet of an artillery shell t sec after firing is given by the equation $h(t) = -16t^2 + 720t$. Determine how many seconds will elapse before the shell reaches its highest point.

64 a. Select the two graphs below that illustrate that y varies directly as x. What can you say about the constant of variation in each case?

b. Select the graph below that illustrates that y varies inversely as x.

c. Select the two graphs below that illustrate that y varies directly as x^2. What can you say about the constant of variation in each case?

a.
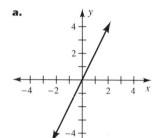

b.
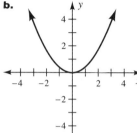

c.
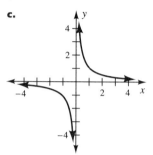

d.
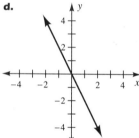

e.
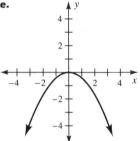

In Exercises 65–75, graph each relation.

65 $y = -\frac{2}{5}x + 3$
66 $y - 1 = 2(x + 2)$
67 $x = 3$
68 $y = -3$
69 $f(x) = |x - 3| + 2$
70 $y = -x^2 + 1$
71 $y = 4x^2 - 4x + 1$
72 $y = [\![x - 1]\!]$
73 $x^2 + y^2 = 25$
74 $4(x - 2)^2 + 9(y + 3)^2 = 36$
75 $x^2 - y^2 = 4$

OPTIONAL EXERCISES FOR CALCULUS-BOUND STUDENTS

1 Calculate $\dfrac{f(x + h) - f(x)}{h}$ for $f(x) = 2x^2 - 5x - 7$.

2 a. What is the slope of the line through the intercepts $(a, 0)$ and $(0, b)$?

b. Verify that the equation of this line can be written as $\dfrac{x}{a} + \dfrac{y}{b} = 1$.

OPTIONAL EXERCISES FOR CALCULUS-BOUND STUDENTS

3 The endpoints of a chord of the circle $x^2 + y^2 = 169$ are (5, 12) and (12, 5). Find the equation of a line perpendicular to this chord and through their midpoint (8.5, 8.5). Does this line pass through the origin?

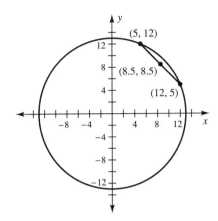

In Exercises 4–7, express each function in terms of $f(x) = 3x^2 - 5$ and $g(x) = 2x$.

4 $h(x) = 3x^2 + 2x - 5$

5 $h(x) = \dfrac{3x^2 - 5}{2x}$

6 $h(x) = 2(3x^2 - 5)$

7 $h(x) = 3(2x)^2 - 5$

In Exercises 8 and 9, decompose each function into functions f and g such that $h(x) = (f \circ g)(x)$.

8 $h(x) = \sqrt{x^2 - 3x - 1}$

9 $h(x) = \dfrac{1}{|3x + 5|}$

In Exercises 10 and 11, graph each piecewise function and give each interval for which the function is increasing, decreasing, and constant.

10 $y = \begin{cases} 2x + 1 & \text{if } x \leq -1 \\ 2 & \text{if } x > -1 \end{cases}$

11 $y = \begin{cases} -x - 1 & \text{if } x < -1 \\ 1 - x^2 & \text{if } -1 \leq x \leq 1 \\ x - 1 & \text{if } x > 1 \end{cases}$

12 Graph $y = x^4$ and $y = (x - 1)^4 - 5$ on the same coordinate system.

13 Determine the inverse of $f(x) = (x - 1)^2$ for $x \geq 1$.

14 If (y, x) is on the graph whenever (x, y) is on the graph, what is the line of symmetry?

15 If a graph is symmetric with respect to the x-axis, determine a line of symmetry for the graph obtained by shifting each point (x, y) to the new location $(x + h, y + k)$.

16 If the graph of $y = f(x)$ is symmetric with respect to the y-axis, determine a line of symmetry for the graph of $y = f(x - 3)$.

17 Determine whether the points (2, 15), (18, 3), and (18, 47) form a right triangle.

18 Find k such that (3, 5) is the midpoint between (8, 2) and $(k, 8)$.

19 Find k such that $\overline{AB}$ is perpendicular to $\overline{BC}$ given that $A = (1, 1)$, $B = (k, k)$, and $C = (4, 2)$.

20 Write the equation of the set of all points (x, y) that are five units from the point $(3, -4)$.

21 Find all values of k such that $(5, k)$ is five units from the point $(2, 1)$.

22 Find the perimeter of the quadrilateral with vertices at $(2, 0), (0, 2), (-2, 0)$, and $(0, -2)$. Is it a square? Explain why or why not.

23 Find k such that the line through the origin and $(-2, k)$ is parallel to the line through $(4, 1)$ and $(6, 3)$.

24 An electrician plans to run a wire from a switch on the wall to a light on the ceiling. Determine the shortest path along the walls and ceiling from the switch to the ceiling fixture. Also determine the point b where the path goes from the wall to the ceiling. (*Hint*: Mentally fold the ceiling so that the wall and the ceiling are in the same plane. The shortest distance from a to c is a straight line in this plane.)

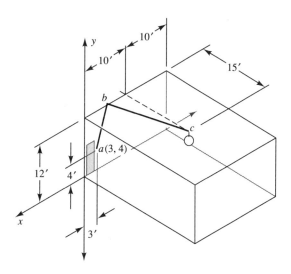

25 A person 6 feet tall is walking away from a streetlight that is 12 feet high. Express the length, s, of the person's shadow in terms of the distance, d, from the base of the light.

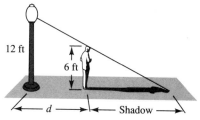

Figure for Exercise 25

Exercises 26 and 27 are from *Calculus for Business, Economics, and the Social and Life Sciences*, fourth edition, 1989, by Laurence D. Hoffmann, New York: McGraw-Hill Book Company.

26 A manufacturer can produce bookcases at a cost of $10 apiece. Sales figures indicate that if the bookcases are sold for x dollars apiece, approximately $50 - x$ will be sold each month. Express the manufacturer's monthly profit as a function of the selling price x, draw the graph, and estimate the optimal selling price.

27 A retailer can obtain cameras from the manufacturer at a cost of $50 apiece. The retailer has been selling the cameras at the price of $80 apiece, and, at this price, consumers have been buying 40 cameras a month. The retailer is planning to lower the price to stimulate sales and estimates that for each $5 reduction in the price, 10 more cameras will be sold each month. Express the retailer's monthly profit from the sale of the cameras as a function of the selling price. Draw the graph and estimate the optimal selling price.

MASTERY TEST FOR CHAPTER 3

Exercise numbers correspond to Section Objective numbers.

1. Determine which of the relations defined or graphed below are functions.

   a. $\{(1, 1), (2, 8), (3, 27), (4, 64)\}$ b. $\{(1, 1), (1, -1), (4, 2)\}$

   c. 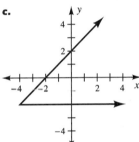   d.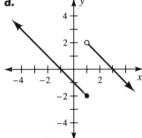

   e. $y = x^2$ f. $x = y^2$

2. Evaluate each of these functional values, given $f(x) = x^3 - 2$ and $g(x) = \sqrt{x}$.

   a. $f(2)$ b. $\dfrac{f}{g}(4)$ c. $f(0) + g(4)$

   d. $(g \circ f)(3)$ e. $f(x + h)$ f. $f(-x)$

3. Determine the domain and range of the functions defined or graphed below. In parts f and g, determine the domain only.

   a. $f = \{(2, 5), (7, -11), (-3, 8)\}$ b. $f(x) = x^2$ c.

x	4	3	5	0
y	6	7	-2	9

   d. 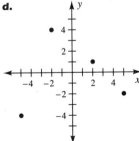   e.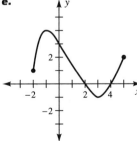

   f. $f(x) = \dfrac{2}{3x^2 - 12}$ g. $g(x) = \sqrt{x + 5}$

4 Graph the lines described below.
   a. $3x - 2y = 6$
   b. x-intercept -2, slope $\frac{2}{5}$
   c. $y = -\frac{3}{7}x + 5$
   d. $x = 3$

5 Find the slopes of the lines defined by these equations.
   a. $y = 7x - 9$
   b. $2x - 3y = 5$
   c. $x = 4$
   d. $y = 4$
   e. the line through (6, 0) and (2, −4)

6 Determine if the lines defined by $2x + 3y = 5$ and $3x - 2y = 5$ are
   a. parallel
   b. perpendicular
   c. neither parallel nor perpendicular

7 Write the equation of a line
   a. with slope 7 and passing through (2, 3)
   b. vertical and passing through (2, 3)
   c. parallel to $y = 2x - 4$ and with y-intercept (0, 2)
   d. horizontal and passing through (0, 2)

8 Use the graph of $y = x^3$ shown to the right to graph:
   a. $y = x^3 + 1$
   b. $y = (x + 1)^3$
   c. $y = (x - 2)^3 - 4$

9 Graph
   a. $y = (x - 2)^2$
   b. $y = -x^2 + 4x + 5$

10 A manufacturer of a new type of railroad coupling has data indicating that the average cost in dollars of producing t units per day is

$$C(t) = t^2 - 20t + 300$$

Determine how many couplings should be produced per day in order to minimize the average cost.

11 Given $f(x) = x^2 + 2$, $g(x) = x^3 - x$, and $h(x) = x^2 - x$, determine which (if any) of these relations have graphs that are
   a. symmetric about the x-axis
   b. symmetric about the y-axis
   c. symmetric about the origin

12 a. Find the distance from (3, −4) to (−2, 8).
   b. Find the midpoint between (3, −4) and (−2, 8).

13 Given $f(x) = x^2 - x - 6$ and $g(x) = x - 3$, find the function and state its domain.
   a. $(f + g)(x)$
   b. $(f - g)(x)$
   c. $(f \cdot g)(x)$
   d. $\left(\dfrac{f}{g}\right)(x)$

14 Given $f(x) = \sqrt{3x - 1}$ and $g(x) = x^2 + 1$, find the function and state its domain.
   a. $(f \circ g)(x)$
   b. $(g \circ f)(x)$

15 Determine whether or not each of the functions defined or graphed below is a one-to-one function.
   a. $\{(7, 4), (4, 7), (3, 3)\}$
   b. $\{(1, 9), (2, 9), (3, 1)\}$
   c. $f(x) = x^4$
   d. $f(x) = x^5$

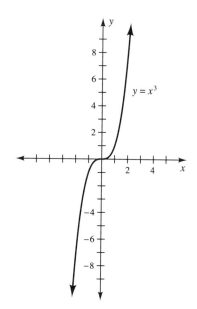

Figure for Exercise 8

MASTERY TEST FOR CHAPTER 3

e. 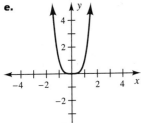   f.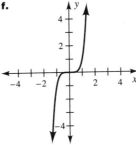

16 Determine the inverse of each function.
   a. $\{(10, 1), (100, 2), (1000, 3)\}$ **b.** $f(x) = 4x - 3$
   c. $g(x) = \log x$ **d.** $y = e^x$

17 Graph the inverse of the functions graphed in (a) and (b).

a. 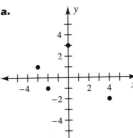   b.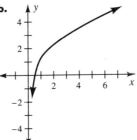

18 Write the standard form of the equation of each of the following circles and ellipses.
   a. circle with center $(-3, 6)$ and radius 4
   b. $x^2 + y^2 - 10x - 4y + 13 = 0$
   c. Ellipse with center $(0, 0)$, horizontal major axis of length 8, and vertical minor axis of length 6
   d. $1024x^2 + y^2 + 6144x + 8y + 9168 = 0$

19 Graph the circles and ellipses defined by the following equations.
   a. $(x - 1)^2 + (y + 2)^2 = 9$ **b.** $x^2 + y^2 + 6y - 16 = 0$
   c. $\dfrac{x^2}{9} + \dfrac{y^2}{4} = 1$ **d.** $9x^2 + 4y^2 - 18x + 16y - 11 = 0$

20 Write the standard form of the equation of each of the following hyperbolas:
   a. hyperbola with vertices $(\pm 4, 2)$ and a fundamental rectangle of height 12
   b. $25x^2 - y^2 - 100x + 10y - 25 = 0$

21 Graph the hyperbolas defined by the following equations:
   a. $\dfrac{x^2}{9} - \dfrac{y^2}{36} = 1$ **b.** $16x^2 - 4y^2 + 32x + 24y + 44 = 0$

22 Identify the type of conic section defined by each of these equations.
   a. $3x^2 + 3y^2 = 11$ **b.** $3x^2 + 11y = 3$
   c. $3x^2 + 11y^2 = 3$ **d.** $3x^2 - 3y^2 = 11$

CHAPTER FOUR

Polynomial and Rational Relations

Chapter Four Objectives

1. Sketch and interpret the graph of a polynomial function given in factored form.

2. Use synthetic division.

3. Evaluate $P(a)$ for a polynomial $P(x)$ and a constant a by using synthetic division.

4. Find a real polynomial with given zeros.

5. Find rational zeros of a polynomial function with integer coefficients.

* 6. Use the location theorem to isolate a real zero between two numbers.

* 7. Apply Descartes's rule of signs to find the number of possible positive and negative zeros of a polynomial.

* 8. Find upper and lower bounds for the real zeros of a polynomial.

* 9. Combine the rational zeros theorem, bounds on zeros, Descartes's rule of signs, and the location theorem to find zeros of a polynomial.

*10. Approximate the real zeros of a polynomial.

11. Determine the horizontal and vertical asymptotes of the graph of a rational function.

12. Sketch the graph of a rational function.

* These are optional objectives.

Niels Henrik Abel, 1802-1829

Niels Abel, a Norwegian, spent most of his life in poverty. In 1823, in an attempt to gain recognition, he published at his own expense a paper showing the impossibility of solving general fifth-degree polynomial equations. A few years later, heavily in debt, he died of tuberculosis.

Historically, polynomial equations have received the attention of many great mathematicians. The quadratic formula was one of the first formulas that could solve a class of polynomial equations. Techniques for the solution of cubic equations evolved over a period of many years. Heron of Alexandria solved some cubics in approximately A.D. 100. About 1400 years later, the general solution of cubic equations with one real solution or with three real solutions, two of which coincide, was independently discovered by two Italian mathematicians, Master Scipione del Ferro and Niccolo Tartaglia. Tartaglia gave the formula to Girolamo Cardano only after Cardano promised not to reveal the formula to anyone. Cardano could not resist the temptation, however, and published the results in 1545. Although Tartaglia received some credit, the formula was named after Cardano. The solution of the cubic equation with three distinct real roots was discovered by Vieta about 1615.

Ludovico Ferrari, a student of Cardano, derived a formula for the solution of quartic (fourth-degree) equations about 1545. The formulas for solving quadratic, cubic, and quartic equations all involve radicals. Many mathematicians attempted to derive formulas for higher-degree equations until Niels Henrik Abel proved that radical formulas could not solve general equations of degree greater than four.

One of the main themes of this chapter is the relationship between the factors of a polynomial $P(x)$, the zeros of the polynomial $P(x)$, the solutions of the polynomial equation $y = P(x)$, and the x-intercepts of the graph of $y = P(x)$. The popular approach has long been to carefully study the algebraic questions concerning this relationship and then to use this algebraic information to generate the graph of the corresponding polynomial function. However, this book will start with the geometric approach, use the graphs to develop an understanding of the behavior of the function, and then use this understanding as a basis for answering the corresponding algebraic questions. This is a more intuitive approach and one that is quite practical, considering the relative ease with which graphs of functions can be obtained using graphics calculators.

A graphics calculator is an excellent tool for examining polynomial functions and for solving polynomial equations. However, anyone using a graphics calculator must have an overall understanding of the function being examined. Since the viewing window on a calculator presents a limited view of a graph, it is crucial that the user select a window that both gives a representative view of the whole graph and provides sufficient detail to answer any specific question. Thus the user of a graphics calculator should be able to anticipate the general behavior of a function and then rely on the calculator to perform the more tedious details. This chapter will not only describe how one can easily predict some of the general characteristics of a polynomial function but will also show how to determine some of the most important details of these functions.

Techniques for solving some higher-degree polynomial equations will also be developed in this chapter. These procedures can be used to locate all rational roots and to approximate irrational roots. The last section of this chapter builds on the earlier sections to analyze rational functions.

SECTION 4-1

Graphs of Polynomial Functions

Section Objective

1 Sketch and interpret the graph of a polynomial function given in factored form.

The graph of any polynomial function

$$y = a_n x^n + a_{n-1} x^{n-1} + a_{n-2} x^{n-2} + \cdots + a_2 x^2 + a_1 x + a_0$$

is a smooth continuous curve over the entire domain of real numbers. Typical graphs are shown in Figure 4-1. To sketch a graph such as these by plotting points would be a tedious task requiring many points. This may be appropriate for a computer or a graphics calculator, but it is not appropriate for students who need a quick procedure for sketching these graphs or for anticipating what general shape to expect from a graphics calculator.

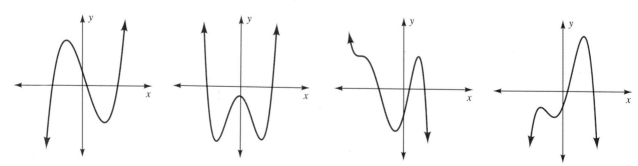

Figure 4-1 Typical graphs of polynomial functions

Our main goal in this section is to learn to predict the general shape of a polynomial function from the defining equation—or, equivalently, to use the general shape of the curve to predict information about the defining

SECTION 4-1 GRAPHS OF POLYNOMIAL FUNCTIONS

equation. The information we are concentrating on can often be obtained with relatively little effort; it is not the detailed type of information that is generally reserved for calculus. This section will concentrate on the relationship between the factors of $P(x)$ and the x-intercepts of the polynomial function $y = P(x)$.

A value of x that makes the polynomial $P(x)$ equal 0 is called a **zero of** $P(x)$. For example, the zeros of the polynomial $P(x) = (x + 1)(x - 2)$ are -1 and 2 since $P(-1) = 0$ and $P(2) = 0$. These zeros correspond directly to $(-1, 0)$ and $(2, 0)$, the x-intercepts of the graph of $y = P(x)$, which is given in Example 1. Both the zeros of $P(x)$ and the x-intercepts of $y = P(x)$ can be readily determined from the factored form of $P(x)$.

A sign graph of the factors of $P(x)$ can be used to analyze where the graph of $y = P(x)$ is above the x-axis and where the graph is below the x-axis. The graph will be above the x-axis in intervals for which $y > 0$ and below the x-axis in intervals for which $y < 0$. This is illustrated in the next example for $P(x) = (x + 1)(x - 2)$.

EXAMPLE 1 Compare the sign graph of $(x + 1)(x - 2)$ and the graph of $y = (x + 1)(x - 2)$.

SOLUTION

First make a sign graph for y that equals $(x + 1)(x - 2)$.

$x + 1$ changes sign at $x = -1$.
$x - 2$ changes sign at $x = 2$.
The sign of y indicates whether the graph of $y = (x + 1)(x - 2)$ is above or below the x-axis.

The zeros of the polynomial, -1 and 2, are also the x-intercepts of the graph of the polynomial function. The graph crosses the x-axis at these x-intercepts because the sign of y changes at these values.

The sketch is made by drawing a smooth curve through these x-intercepts and through the y-intercept $(0, -2)$.

The next example illustrates that the sign of $y = P(x)$ does not necessarily change at each zero of $P(x)$. Thus the graph of $y = P(x)$ does not necessarily cross the x-axis at each x-intercept.

EXAMPLE 2 Compare the sign graph of $(x + 1)^2(x - 2)$ and the graph of $y = (x + 1)^2(x - 2)$.

SOLUTION

Factors | Sign of Factors

$(x + 1)^2$ + + + + 0 + + | + + +
$x - 2$ - - - -   - - 0 + + +
y - - - -   - -   + + +

   -5 -4 -3 -2   0   1   3   4   5

y below x-axis   y below x-axis   y above x-axis

First make a sign graph for y that equals $(x + 1)^2(x - 2)$.

The square $(x + 1)^2$ does *not* change sign at $x = -1$; it is positive for values on both sides of $x = -1$.

$x - 2$ changes sign at $x = 2$.

The sign of y indicates whether the graph of $y = (x + 1)^2(x - 2)$ is above or below the x-axis.

The zeros of the polynomial, -1 and 2, are also the x-intercepts of the graph of the polynomial function. The graph crosses the x-axis at $x = 2$ because the sign of y changes at $x = 2$, but it is tangent to the x-axis at $x = -1$ since the sign of y does not change at $x = -1$.

The sketch is made by drawing a smooth curve through these x-intercepts and through the y-intercept $(0, -2)$.

□

If n is the largest integer such that $(x - a)^n$ is a factor of the polynomial of $P(x)$, then a is a **zero of multiplicity n** of this polynomial. For $P(x) = (x - 4)^2(x + 5)^3$, 4 is a zero of multiplicity 2 and -5 is a zero of multiplicity 3. A generalization of the results of Examples 1 and 2 is that the graph of a polynomial function will cross the x-axis at x-intercepts corresponding to factors of odd multiplicity. The graph will be tangent to the x-axis at x-intercepts corresponding to factors of even multiplicity.

x-intercepts of Even and Odd Multiplicity ▼

Even multiplicity x-intercepts:	The graph of $y = P(x)$ is tangent to the x-axis at x-intercepts of even multiplicity.
Odd multiplicity x-intercepts:	The graph of $y = P(x)$ crosses the x-axis at x-intercepts of odd multiplicity.

SECTION 4-1 GRAPHS OF POLYNOMIAL FUNCTIONS

EXAMPLE 3 From the given graph of $y = -x^4 + 5x^3 - x^2 - 21x + 18$, determine each x-intercept and whether this x-intercept is of even or odd multiplicity.

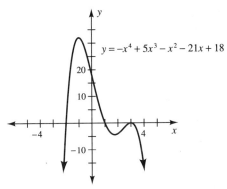

SOLUTION $(-2, 0)$ is an x-intercept of odd multiplicity.   The graph crosses the x-axis at $(-2, 0)$.

$(1, 0)$ is an x-intercept of odd multiplicity.   The graph crosses the x-axis at $(1, 0)$.

$(3, 0)$ is an x-intercept of even multiplicity.   The graph is tangent to the x-axis at $(3, 0)$.

□

Knowledge of each x-intercept of $y = P(x)$ and its multiplicity enables us to rapidly sketch the general shape of the graph. The method outlined in the following box is not intended to supply the exact curve or the turning points of the graph, just the intercepts and the general shape.

Rapid Sketching of Graphs of Polynomial Functions ▼

To sketch the graph of $y = P(x)$:

Step 1 Plot the y-intercept $(0, P(0))$.

Step 2 Plot the x-intercepts, and determine whether each is of even or odd multiplicity.

Step 3 Determine whether the graph is above or below the x-axis to the right of all the x-intercepts and illustrate this behavior on the graph.

Step 4 Starting from the right, complete the sketch by drawing a smooth curve to connect the portion to the right with the portion to the left. Draw the curve through all intercepts. The graph crosses the x-axis at each x-intercept of odd multiplicity and is tangent to the x-axis at each x-intercept of even multiplicity.

EXAMPLE 4 Sketch the graph of $y = (x + 1)(x - 1)^2(x - 3)$.

SOLUTION

1 y-intercept: $(0, -3)$

2 x-intercepts: $(-1, 0)$, multiplicity 1 (odd)
$(1, 0)$, multiplicity 2 (even)
$(3, 0)$, multiplicity 1 (odd)

3 $y > 0$ for $x > 3$

$y(0) = (1)(-1)^2(-3) = -3$.
From the factor $x + 1$.
From the factor $(x - 1)^2$.
From the factor $x - 3$.

For $x > 3$, all factors of $y = (x + 1)(x - 1)^2(x - 3)$ are positive and so $y > 0$ in this interval. This means that the graph of $y = (x + 1)(x - 1)^2(x - 3)$ is above the x-axis for $x > 3$.

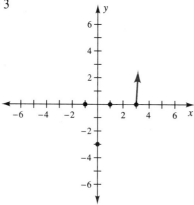

4

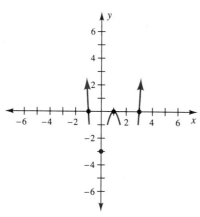

The graph crosses the x-axis at $(3, 0)$ because of the odd multiplicity.
The graph is tangent to the x-axis at $(1, 0)$ because of the even multiplicity.
The graph crosses the x-axis at $(-1, 0)$ because of the odd multiplicity.
Each of these features is sketched to the left.

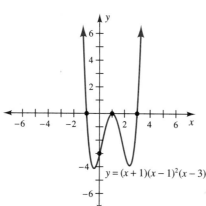

Complete the sketch by connecting these portions with a smooth curve.
Note that the y-intercept plotted in step 1 can be used as a check to verify that the sketch of the graph is correctly positioned above or below the x-axis at this intercept.

SECTION 4-1 GRAPHS OF POLYNOMIAL FUNCTIONS

EXAMPLE 5 Sketch the graph of $y = (x - 1)^2(x + 2)^3$.

SOLUTION

y-intercept: (0, 8)
x-intercept: (1, 0) of multiplicity 2, even
x-intercept: (−2, 0) of multiplicity 3, odd
The graph is above the x-axis for $x > 1$.

$y(0) = (-1)^2(2)^3 = 8$
From the factor $(x - 1)^2$
From the factor $(x + 2)^3$
For $x > 1$, $(x - 1)^2(x + 2)^3 > 0$.

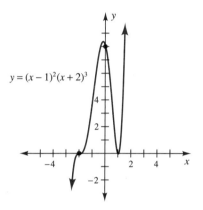

Starting from the right, sketch the curve tangent to the x-axis at (1, 0), through the y-intercept (0, 8), and crossing the x-axis at (−2, 0).

EXAMPLE 6 Use the graph of $y = x^2 - 5x - 6$ given to the right to solve the inequality $x^2 - 5x - 6 > 0$.

SOLUTION

$x^2 - 5x - 6 > 0$ in the intervals $(-\infty, -1)$ and $(6, +\infty)$.

If the graph of $y = P(x)$ is above the x-axis, then $y > 0$. Also note that in factored form $y = (x + 1)(x - 6)$.

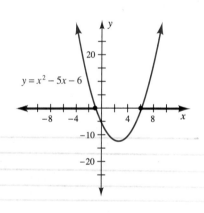

The up or down behavior of the graph of $y = P(x)$ to the far left and far right can easily be determined by examining the factors of $P(x)$ and working from left to right as described in the rapid sketching of graphs of polynomial functions. Another quick method for determining this behavior of the graph of $y = P(x)$ is based on inspection of the leading term, $a_n x^n$, of

$$P(x) = a_n x^n + a_{n-1} x^{n-1} + \cdots + a_1 x + a_0$$

Table 4-1 on page 270 shows how the leading term $a_n x^n$ can be used to determine whether the graph of $y = P(x)$ points up or down to the far left [that is, left of all zeros of $P(x)$] and to the far right [that is, right of all zeros of $P(x)$]. We can determine the behavior to the far left by examining $x < 0$ (negative values of x) and the behavior to the far right by examining $x > 0$ (positive values of x). If $a_n x^n > 0$ (positive), then the graph points up. If $a_n x^n < 0$ (negative), then the graph points down.

▶ **Self-Check** ▼

Classify the multiplicity of each x-intercept of the graph of $y = (x - 3)^2(x + 4)^3$ as either even or odd.

▶ **Self-Check Answer** ▼

(−4, 0) is an x-intercept of odd multiplicity; (3, 0) is an x-intercept of even multiplicity.

Table 4-1 The behavior of the graph of $y = P(x)$

Degree n of $a_n x^n$	Leading Coefficient of a_n	Graph of $y = P(x)$	
n even	a_n positive	Points up both to the far right and far left	n even: same
	a_n negative	Point down both to the far right and far left	
n odd	a_n positive	Points up on the far right and down on the far left	n odd: opposite
	a_n negative	Points down on to the far right and up on the far left	

EXAMPLE 7 Determine whether the graph of each function points up or down to the far left and the far right

SOLUTIONS

		Degree, n	Sign of the Coefficient a_n	Graph of $y = P(x)$
(a)	$y = x^3 - 2x^2 - 5x + 6$	$n = 3$, odd	1, the coefficient of x^3 is positive	Points up to the far right Points down to the far left
(b)	$y = -4x^5 + 36x^3$	$n = 5$, odd	-4 is negative	Points down to the far right Points up to the far left
(c)	$y = -2x^4 + 26x^2 - 72$	$n = 4$, even	-2 is negative	Points down to the far right and far left
(d)	$y = x^6 - 7x^4 + 3x + 11$	$n = 6$, even	1, the coefficient of x^6 is positive	Points up to the far right and far left

The polynomial functions $y = P(x)$ in this section have mostly been given in factored form since this form readily yields information about the x-intercepts of the graph of $y = P(x)$. If the functions are not already in factored form, then we often start by factoring the polynomial. For some polynomials we can use the factoring skills reviewed in Section 1-5. For more difficult polynomials we will use techniques developed later in this chapter.

EXAMPLE 8 Sketch the graph of $y = -(x + 3)(4x^2 - 4x + 1)$.

SOLUTION $y = -(x + 3)(4x^2 - 4x + 1)$ First factor the trinomial, which is a perfect square.
$y = -(x + 3)(2x - 1)^2$

y-intercept: $(0, -3)$ $y(0) = -(3)(-1)^2 = -3$

x-intercept: $(\frac{1}{2}, 0)$ of multiplicity 2, even   From the factor $(2x - 1)^2$

x-intercept: $(-3, 0)$ of multiplicity 1, odd   From the factor $(x + 3)$

▶ **Self-Check** ▼

Determine from the given graph of $y = P(x)$ whether the leading coefficient is positive or negative.

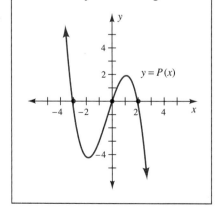

▶ **Self-Check Answer** ▼

The leading coefficient is negative.

SECTION 4-1 GRAPHS OF POLYNOMIAL FUNCTIONS

The graph is below the x-axis for $x > \frac{1}{2}$.

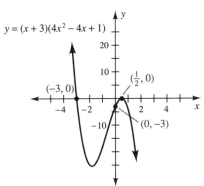

Figure for Example 8

For $x > \frac{1}{2}$, $-(x + 3)(2x - 1)^2 < 0$.

Starting from the right, sketch the curve tangent to the x-axis at $(\frac{1}{2}, 0)$, through the y-intercept $(0, -3)$, and crossing the x-axis at $(-3, 0)$.

▶ **Self-Check** ▼

Use the given graph of $y = x^3 + 2x^2 - 15x - 36$ to factor $x^3 + 2x^2 - 15x - 36$.

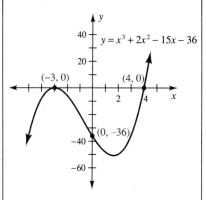

EXERCISES 4-1

A

For the graphs of the polynomial functions defined in Exercises 1 and 2, list each x-intercept and its multiplicity. Determine whether the graph of this function crosses the x-axis or is tangent to the x-axis at each x-intercept.

1 **a.** $y = (x + 2)(x - 3)^2(x - 5)$ **b.** $y = -x^2(2x + 5)(5x - 2)$ **c.** $y = (x + 2)(x^2 - x - 6)$

2 **a.** $y = (x + 2)^3(x - 3)(x - 5)^2$ **b.** $y = x^3(3x - 1)(x + 3)^3$ **c.** $y = (x - 2)(x^2 - x - 2)$

For each of the graphs of $y = P(x)$ in Exercises 3 and 4, list each real number that is a zero of $P(x)$. State whether the multiplicity of each of these zeros is even or odd.

3 **a.**

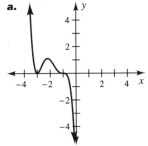

b.

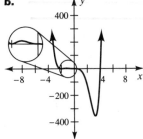

c.

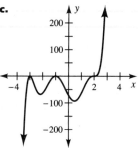

▶ **Self-Check Answer** ▼

$x^3 + 2x^2 - 15x - 36 = (x + 3)^2(x - 4)$

4 a. 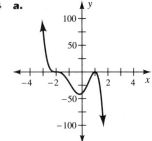 b. c.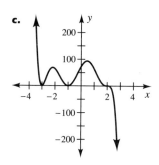

In Exercises 5 and 6, use the graphs of $y = P(x)$ below to determine whether $P(x)$ is positive or negative in each of the given intervals.

5 $y = x^3 - 3x^2 - x + 3$
  a. $(-\infty, -1)$ **b.** $(-1, 1)$ **c.** $(1, 3)$ **d.** $(3, +\infty)$

6 $y = x^3 - 2x^2 - 4x + 8$
  a. $(-\infty, -2)$ **b.** $(-2, 2)$ **c.** $(2, \infty)$

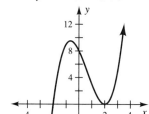

Figure for Exercise 5   **Figure for Exercise 6**

In Exercises 7–10, determine whether the graph of $y = P(x)$ is above or below the x-axis in each interval formed by the x-intercepts.

7 $P(x) = (x - 3)^4(x - 1)^5$
  a. $(3, +\infty)$ **b.** $(1, 3)$ **c.** $(-\infty, 1)$

8 $P(x) = (x - 2)^7(x + 1)^8$
  a. $(2, +\infty)$ **b.** $(-1, 2)$ **c.** $(-\infty, -1)$

9 $P(x) = (x + 2)^3(x - 3)^2(x + 6)^5$
  a. $(3, +\infty)$ **b.** $(-2, 3)$ **c.** $(-6, -2)$ **d.** $(-\infty, -6)$

10 $P(x) = (x + 7)^2(x + 3)^4(x + 5)^3$
  a. $(-3, +\infty)$ **b.** $(-5, -3)$ **c.** $(-7, -5)$ **d.** $(-\infty, -7)$

In Exercises 11 and 12, determine whether the graph of $y = P(x)$ points up or down to the far left and to the far right.

11 **a.** $y = 3x^5 + 9x^4 - 7x + 2$ **b.** $y = -2x^4 - x^3 + x^2 + 17$ **c.** $y = -(2x + 5)(x - 4)^2$
12 **a.** $y = 7x^4 - 9x^3 + 6x^2 - 2x + 1$ **b.** $y = -4x^3 + 7x^2 - 8x + 5$ **c.** $y = (2x - 5)^3(x + 4)^2$

SECTION 4-1 GRAPHS OF POLYNOMIAL FUNCTIONS

In Exercises 13 and 14, determine from the graph of $y = P(x)$ whether $P(x)$ is of even degree or of odd degree. Also determine whether the leading coefficient is positive or negative.

13 a. 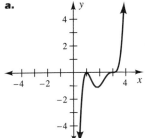   b. 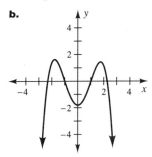   c.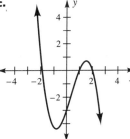

14 a. 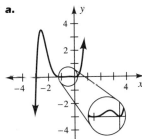   b. 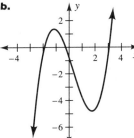   c.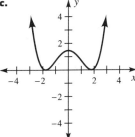

The polynomial functions in Exercises 15–18 were graphed on a graphics calculator. (The scale has been added here on the axes to facilitate your examination of each graph.) Determine by inspection the graphical properties of each polynomial function and then select the graph that best matches the function.

15 $y = (x - 3)^4(x - 1)^5$

a. 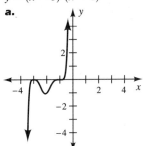   b. 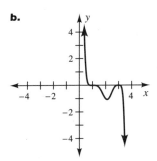   c.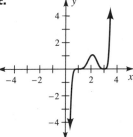

16 $y = (x - 2)^3(x + 1)^6$

a. 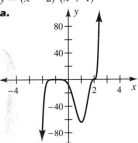   b. 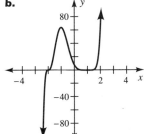   c.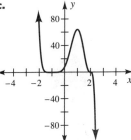

17 $y = (x - 2)^2(x + 1)^3(x + 4)$

a.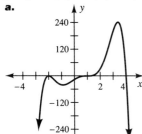
b.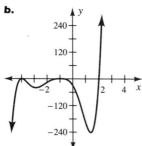
c.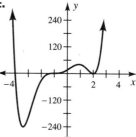

18 $y = (x - 3)^2(x - 1)^2(x + 2)^3$

a.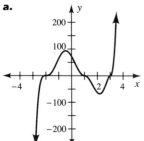
b.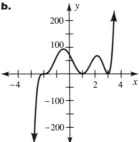
c.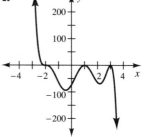

In Exercises 19–22, sketch the graph of a polynomial function with the given features.

19 x-intercept: (2, 0), multiplicity two, even
y-intercept: (0, 4)
x-intercept: (−1, 0), multiplicity one, odd
The graph points up on the far right.

20 x-intercept: (1, 0), multiplicity one, odd
y-intercept: (0, −10)
x-intercept: (−4, 0), multiplicity two, even
The graph points up on the far right.

21 y-intercept: (0, −540)
x-intercept: (6, 0), multiplicity odd
(3, 0), multiplicity even
(−5, 0), multiplicity odd
The graph points up to the far right.

22 y-intercept: (0, −30)
x-intercepts: (10, 0), multiplicity odd
(−2, 0), mutiplicity even
(−5, 0), multiplicity even
The graph points up to the far right.

In Exercises 23–32, sketch the graph of each polynomial function.

23 $y = (x - 2)(x + 1)(x - 3)$
24 $y = (x + 2)(x - 3)(x + 1)$
25 $y = (x - 1)^2(x + 4)$
26 $y = (x + 2)(x - 3)^2$
27 $y = (x - 1)^3$
28 $y = (x + 2)^3$
29 $y = (x + 6)(x - 4)(x + 2)(2 - x)$
30 $y = (x + 5)(x - 5)(x - 3)(1 - x)$
31 $y = (x - 4)^2(x + 6)^2$
32 $y = (x + 6)^2(x + 2)^2$

SECTION 4-1 GRAPHS OF POLYNOMIAL FUNCTIONS

In Exercises 33 and 34, use the given graph of $y = P(x)$ to solve each inequality.

33 **a.** $x^3 - 9x > 0$ **b.** $x^3 - 9x \leq 0$

34 **a.** $2x^3 + x^2 - 13x + 6 < 0$ **b.** $2x^3 + x^2 - 13x + 6 \geq 0$

B

In Exercises 35–40, write a polynomial $y = P(x)$ of lowest degree with the given features.

35 The graph of the function has an even x-intercept at $(3, 0)$, an odd x-intercept at $(-1, 0)$, and a y-intercept at $(0, 9)$.

36 The graph of the function has an odd x-intercept at $(2, 0)$, an even x-intercept at $(1, 0)$, and a y-intercept at $(0, -2)$.

37 The graph of the function has an odd x-intercept at $(2, 0)$, an even x-intercept at $(-3, 0)$, and a y-intercept at $(0, 18)$.

38 The graph of the function has an even x-intercept at $(3, 0)$, an odd x-intercept at $(-2, 0)$, and a y-intercept at $(0, -18)$.

39 The graph of the function has an even x-intercept at $(4, 0)$ and odd x-intercepts at $(-1, 0)$ and $(2, 0)$. The graph points down on both the far right and the far left.

40 The graph of the function has an even x-intercept at $(-3, 0)$ and an even x-intercept at $(2, 0)$. The graph points down on the far right.

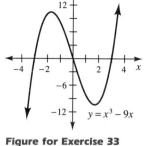

Figure for Exercise 33

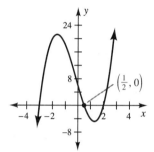

Figure for Exercise 34

The functions in Exercises 41–44, were graphed on a graphics calculator. (The scale has been added here on the axes to facilitate your examination of each graph.) Select the polynomial function that matches the graph.

41 **a.** $y = (x - 5)^4(x - 2)(x + 2)^2$ **b.** $y = (x + 5)^2(x - 2)^2(x + 2)$
c. $y = (x + 5)^2(x - 2)(x + 2)^2$ **d.** $y = -(x + 5)^2(x - 2)(x + 2)^2$
e. $y = (x + 5)^3(x - 2)(x + 2)^2$

42 **a.** $y = -(x + 4)^2(x + 1)(x + 3)$ **b.** $y = -(x + 4)(x + 1)^3(x + 3)$
c. $y = (x + 4)(x + 1)^2(x + 3)^4$ **d.** $y = -(x - 4)^3(x + 1)^4(x + 3)$
e. $y = (x + 4)^2(x - 1)^3(x - 3)$

43 **a.** $y = (x - 3)(5x + 3)(x^2 - 5)$ **b.** $y = (x + 3)(5x + 3)(x^2 - 5)$
c. $y = -(x + 3)(5x + 3)(x^2 - 5)$ **d.** $y = (x - 3)(5x - 3)(x^2 - 5)$
e. $y = (x - 3)(5x + 3)(x - 5)^2$

44 **a.** $y = -x(3x + 2)(x^2 - 10)$ **b.** $y = -x(3x - 2)(x^2 - 10)$
c. $y = x^2(3x - 2)(x - 10)^2$ **d.** $y = x^2(3x - 2)(x^2 - 10)$
e. $y = x^2(3x + 2)(x^2 - 10)$

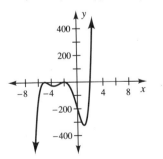

Figure for Exercise 41

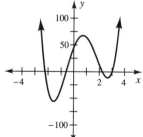

Figure for Exercise 42

Figure for Exercise 43

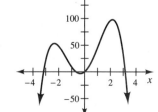

Figure for Exercise 44

C

In Exercises 45 and 46, graph each polynomial function.

45 $y = (x + 3)^2(x + 1)(x - 1)^2$

46 $y = (x - 3)^3(x - 1)(x + 1)^2$

In Exercises 47 and 48, use your factoring skills from Section 1-5 to factor $P(x)$ and then graph $y = P(x)$.

47 $y = x^3 + 2x^2 - x - 2$

48 $y = x^3 - 3x^2 + 3x - 1$

In Exercises 49 and 50, graph each polynomial function $y = P(x)$ using a graphics calculator. Then use this graph to factor the fourth-degree polynomial $P(x)$.

49 $y = 4x^4 - 5x^2 + 1$

50 $y = 2x^4 - 3x^3 - 10x^2 + 12x + 8$

SECTION 4-2

Synthetic Division and the Factor Theorem

Section Objectives

2 Use synthetic division.

3 Evaluate $P(a)$ for a plynomial $P(x)$ and a constant a by using synthetic division.

The solutions of the polynomial equation

$$a_n x^n + a_{n-1} x^{n-1} + \cdots + a_1 x + a_0 = 0$$

and the factors of

$$P(x) = a_n x^n + a_{n-1} x^{n-1} + \cdots + a_1 x + a_0$$

and the values for which $P(x) = 0$, and the x-intercepts of the graph of $y = P(x)$ are all related. The relationship between the factors of $P(x)$ and the solutions of $P(x) = 0$ is used each time we solve a quadratic equation by factoring. For example, -3 and 2 are solutions of $(x + 3)(x - 2) = 0$, as well as zeros of $P(x) = (x + 3)(x - 2) = x^2 + x - 6$, and $(-3, 0)$ and $(2, 0)$ are the x-intercepts of the graph of $y = x^2 + x - 6$.

The search for the linear factors of $P(x)$ may require many trial divisions. Synthetic division, a simplification of the long division process, can be used to save time when one is looking for linear factors. We begin by considering division of the polynomial $P(x)$ by the divisor $D(x)$.

The Division Algorithm

For arithmetic: Let p, q, r, and d be whole numbers with $d \neq 0$. If q is the quotient and r is the remainder obtained by dividing p by d, then p may be written *uniquely* as $p = qd + r$, where $r < d$.

SECTION 4-2 SYNTHETIC DIVISION AND THE FACTOR THEOREM

For polynomials: Let $P(x)$, $Q(x)$, $R(x)$, and $D(x)$ be polynomials. If $Q(x)$ is the quotient and $R(x)$ is the remainder obtained by dividing $P(x)$ by $D(x)$, then $P(x)$ may be written *uniquely* in the form $P(x) = Q(x)D(x) + R(x)$, where either $R(x) = 0$ or the degree of $R(x)$ is less than the degree of $D(x)$.

Each of these algorithms is illustrated below.

$$\begin{array}{r} 3 \\ 6\overline{)20} \\ \underline{18} \\ 2 \end{array} \text{ means that } \quad \underset{\underset{p}{\text{Dividend}}}{20} = \underset{\underset{q}{\text{Quotient}}}{3} \cdot \underset{\underset{d}{\text{Divisor}}}{6} + \underset{\underset{r}{\text{Remainder}}}{2}$$

$$\begin{array}{r} 2x^2 + 6x - 2 \\ x-3\overline{)2x^3 + 0x^2 - 20x + 3} \\ \underline{2x^3 - 6x^2} \\ 6x^2 - 20x \\ \underline{6x^2 - 18x} \\ -2x + 3 \\ \underline{-2x + 6} \\ -3 \end{array} \text{ means that}$$

$$\underbrace{2x^3 - 20x + 3}_{\substack{\text{Dividend} \\ P(x)}} = \underbrace{(2x^2 + 6x - 2)}_{\substack{\text{Quotient} \\ Q(x)}} \underbrace{(x - 3)}_{\substack{\text{Divisor} \\ D(x)}} + \underbrace{(-3)}_{\substack{\text{Remainder} \\ R(x)}}$$

Synthetic division is a shortened form of long division. We will demonstrate this method by eliminating the redundant information in the previous problem, where we used long division to divide $2x^3 - 20x + 3$ by $x - 3$.

Step 1 Suppress the variable x.

$$\begin{array}{r} \boxed{2 \quad 6 \quad -2} \\ 1 - 3\overline{)2 \quad 0 \quad -20 \quad 3} \\ \underline{\boxed{2} \quad -6} \\ 6 \quad -20 \\ \underline{\boxed{6} \quad -18} \\ \boxed{-2} \quad 3 \\ \underline{\boxed{-2} \quad 6} \\ -3 \end{array}$$

The dividend and the divisor are written in standard form, with a coefficient of zero for each missing term. The variable x can be deleted, since the position of each term indicates the power of x.

Step 2 Condense the notation.

$$\begin{array}{r|rrrr} -3 & 2 & -0 & -20 & 3 \\ & & (-6) & (-18) & 6 \\ \hline & 2 & 6 & -2 & -3 \end{array}$$

Many numbers, including the coefficients of the quotient, may be deleted because they are repeated within the division. Also eliminate the 1 in the divisor and bring down the 2. The -6, -18, and 6 in the second row may be obtained by multiplying the bottom row numbers, 2, 6, and -2 by -3. Then subtract to get the numbers in the bottom row.

Step 3 Change the sign on the divisor to convert from subtraction to addition.

Divisor $x - 3$

$$\begin{array}{r|rrrr} 3 & 2 & 0 & -20 & 3 \\ & & +6 & +18 & +(-6) \\ \hline & 2 & 6 & -2 & -3 \end{array}$$

← Dividend $2x^3 + 0x^2 - 20x + 3$

Quotient $2x^2 + 6x - 2$

Remainder -3

Answer $2x^3 - 20x + 3 = (2x^2 + 6x - 2)(x - 3) - 3$

Synthetic Division ▼

Divide the polynomial $P(x)$ by $x - a$.

Step 1 Arrange $P(x)$ in the standard form

$$a_n x^n + a_{n-1} x^{n-1} + \cdots + a_2 x^2 + a_1 x + a_0$$

including coefficients of 0 for missing terms.

Step 2 Record the coefficients of $P(x)$ in a row and place the constant a of the divisor $x - a$ to the left.

Step 3 Bring down the first coefficient, a_n; multiply a_n by a; and add the result to the second coefficient, a_{n-1}.

Step 4 Repeat the zigzag process until all columns have been completed. The remainder is the last number in the bottom row. The other numbers in the bottom row are the coefficients of the quotient.

SECTION 4-2 SYNTHETIC DIVISION AND THE FACTOR THEOREM

EXAMPLE 1 Use synthetic division to divide $12x^2 + 6x^3 + 8$ by $x + 3$. Express the answer in the form $P(x) = Q(x)(x - a) + R(x)$.

SOLUTION

[1] Arrange $P(x)$ in standard form.

$$P(x) = 6x^3 + 12x^2 + 0x + 8$$
$$x + 3 = x - (-3)$$

[2] Record the coefficients of $P(x)$ in a row, and place the constant -3 of the divisor $x - (-3)$ to the left.

$$-3 \,|\; 6 \quad 12 \quad 0 \quad 8$$

[3] Bring down the first coefficient, 6; multiply 6 by -3, and add the product -18 to 12.

$$\begin{array}{r|rrrr} -3 & 6 & 12 & 0 & 8 \\ & & -18 & & \\ \hline & 6 & -6 & & \end{array}$$

[4] Repeat the zigzag process until all columns have been completed.

$$\begin{array}{r|rrrr} -3 & 6 & 12 & 0 & 8 \\ & & -18 & 18 & -54 \\ \hline & 6 & -6 & 18 & -46 \end{array}$$

Answer $6x^3 + 12x^2 + 8 = (6x^2 - 6x + 18)(x + 3) + (-46)$

The quotient $6x^2 - 6x + 18$ has degree 2, which is one less than the degree of the dividend, $6x^3 + 12x^2 + 8$. The remainder is -46.

□

EXAMPLE 2 Use synthetic division to divide $-2x^4 + x^3 - 3x + 10$ by $x + 2$. Express the answer in the form $P(x) = Q(x)(x - a) + R(x)$.

SOLUTION
$$\begin{array}{r|rrrrr} -2 & -2 & 1 & 0 & -3 & 10 \\ & & 4 & -10 & 20 & -34 \\ \hline & -2 & 5 & -10 & 17 & -24 \end{array}$$

$P(x) = -2x^4 + x^3 + 0x^2 - 3x + 10$
$x + 2 = x - (-2)$; thus $a = -2$.

Answer $-2x^4 + x^3 - 3x + 10 = (-2x^3 + 5x^2 - 10x + 17)(x + 2) + (-24)$

□

Synthetic division can be used to divide by $x - a$ for any complex number a, even when a is an imaginary number. This is illustrated in the next example.

EXAMPLE 3 Use synthetic division to divide $x^3 - 3x^2 + 4x - 2$ by $x - 1 - i$. Express the answer in the form $P(x) = Q(x)(x - a) + R(x)$.

SOLUTION

$$\begin{array}{r|rrrr} 1+i & 1 & -3 & 4 & -2 \\ & & 1+i & -3-i & 2 \\ \hline & 1 & -2+i & 1-i & 0 \end{array}$$

$P(x) = x^3 - 3x^2 + 4x - 2$
$x - 1 - i = x - (1 + i)$; thus $a = 1 + i$.

Answer $x^3 - 3x^2 + 4x - 2 = [x^2 + (-2 + i)x + (1 - i)](x - 1 - i) + 0$ □

EXAMPLE 4 Given $P(x) = x^3 - 2x^2 + 3x - 5$,

SOLUTIONS

(a) Evaluate $P(4)$.

$P(x) = x^3 - 2x^2 + 3x - 5$
$P(4) = 4^3 - 2 \cdot 4^2 + 3 \cdot 4 - 5$
$= 64 - 2 \cdot 16 + 12 - 5$
$= 64 - 32 + 12 - 5$
$P(4) = 39$

(b) Find the remainder for $P(x) \div (x - 4)$.

$$\begin{array}{r|rrrr} 4 & 1 & -2 & 3 & -5 \\ & & 4 & 8 & 44 \\ \hline & 1 & 2 & 11 & 39 \end{array}$$

The remainder is 39. □

In Example 4, $P(x) = x^3 - 2x^2 + 3x - 5 = (x^2 + 2x + 11)(x - 4) + 39$. Thus

$$P(4) = [4^2 + 2(4) + 11](4 - 4) + 39$$
$$= [4^2 + 2(4) + 11](0) + 39$$
$$= 0 + 39$$
$$= 39$$

Therefore, the value of $P(4)$ in part (a) is equal to the remainder obtained by dividing $P(x)$ by $x - 4$. This result is generalized in the remainder theorem.

Remainder Theorem ▼

If $P(x)$ is a polynomial, a is a complex number, and $P(x)$ is divided by $x - a$, then the remainder equals $P(a)$.

Proof When the polynomial $P(x)$ is written in the form $P(x) = Q(x)(x - a) + r$, the remainder r must be a constant. Substituting a for x, we obtain $P(a) = Q(a)(a - a) + r$. This simplifies to $P(a) = 0 + r$. Thus $P(a) = r$, the remainder.

SECTION 4-2 SYNTHETIC DIVISION AND THE FACTOR THEOREM

EXAMPLE 5 Use synthetic division and the remainder theorem to evaluate $P(2)$ for

$$P(x) = x^5 - 3x^4 + 7x - 9$$

SOLUTION

$$\begin{array}{r|rrrrrr} 2 & 1 & -3 & 0 & 0 & 7 & -9 \\ & & 2 & -2 & -4 & -8 & -2 \\ \hline & 1 & -1 & -2 & -4 & -1 & -11 \end{array}$$

Note the zero coefficients for x^3 and x^2.

$P(2) = -11$, which is the remainder when $P(x)$ is divided by $x - 2$.

Answer $P(2) = -11$

A divisor is an exact divisor or a factor of the dividend if the remainder is zero. If $x - a$ is a divisor of $P(x)$, then the degree of the quotient must be exactly one less than the degree of $P(x)$. Therefore, this quotient polynomial is sometimes called a **depressed factor**.

Recall from Section 4-1 that a value of x that makes the polynomial $P(x) = 0$ is called a zero of $P(x)$. A direct consequence of the remainder theorem is the factor theorem, which states that $x - a$ is a factor of $P(x)$ if and only if a is a zero of $P(x)$.

▶ **Self-Check** ▼

Use synthetic division to find:
1 $(2x^3 - 3x^2 - 2x - 10) \div (x - 3)$
2 $(x^5 - 3x^3 + 2x^2 - 5) \div (x + 2)$

Factor Theorem ▼

For any polynomial $P(x)$ and complex number a,
1 If $P(a) = 0$, then $x - a$ is a factor of $P(x)$.
2 If $x - a$ is a factor of $P(x)$, then $P(a) = 0$.

Proof of Part 1 Let $P(a) = 0$. By the remainder theorem, $P(a) = r$; thus $r = 0$. Therefore $P(x) = Q(x)(x - a)$, and $x - a$ is a factor of $P(x)$.

Proof of Part 2 Suppose $x - a$ is a factor of $P(x)$. Then $P(x) = Q(x)(x - a)$, where $Q(x)$ is a polynomial. The uniqueness part of the division algorithm states that there is exactly one way $P(x)$ can be written in the form $P(x) = Q(x)(x - a) + r$. Therefore $r = 0$ and $P(a) = 0$, since $P(a) = r$.

▶ **Self-Check Answers** ▼

1 $2x^3 - 3x^2 - 2x - 10 = (2x^2 + 3x + 7)(x - 3) + 11$
2 $x^5 - 3x^3 + 2x^2 - 5 = (x^4 - 2x^3 + x^2)(x + 2) + (-5)$

EXAMPLE 6 (a) Use synthetic division to show that $x - 1$ is a factor of $P(x) = x^3 - 3x^2 + 4x - 2$. (b) Use the factor theorem to show that $x - 1$ is a factor of $P(x) = x^3 - 3x^2 + 4x - 2$.

SOLUTIONS

(a)
$$\begin{array}{r|rrrr} 1 & 1 & -3 & 4 & -2 \\ & & 1 & -2 & 2 \\ \hline & 1 & -2 & 2 & 0 \end{array}$$

Since the remainder is zero, $x - 1$ is a factor of $P(x)$. The depressed polynomial factor is $x^2 - 2x + 2$.

$$P(x) = x^3 - 3x^2 + 4x - 2 = (x^2 - 2x + 2)(x - 1)$$

(b) $P(x) = x^3 - 3x^2 + 4x - 2$

$P(1) = 1^3 - 3 \cdot 1^2 + 4 \cdot 1 - 2 = 0$ Substitute 1 for x in $P(x)$. The remainder r equals $P(1)$; thus $r = 0$.

$x - 1$ is a factor of $P(x)$, since $P(1) = 0$.

EXAMPLE 7 Show that $x - 5$ is not a factor of
$$P(x) = x^5 - 4x^4 - 23x^2 - 11x + 9$$

SOLUTION
$$\begin{array}{r|rrrrrr} 5 & 1 & -4 & 0 & -23 & -11 & 9 \\ & & 5 & 5 & 25 & 10 & -5 \\ \hline & 1 & 1 & 5 & 2 & -1 & 4 \end{array}$$
Use synthetic division to determine the remainder r.

Answer $x - 5$ is not a factor of $x^5 - 4x^4 - 23x^2 - 11x + 9$ because the remainder is 4, not zero.

The factors of a polynomial, the zeros of a polynomial, the solutions of a polynomial equation, and the x-intercepts of the graph of the corresponding polynomial equation are related as summarized in the following box.

Equivalent Statements About Polynomials ▼

1. $x - a$ is a factor of the polynomial $P(x)$.
2. a is a zero of $P(x)$; that is, $P(a) = 0$.
3. a is a solution of the polynomial equation $P(x) = 0$.
4. If a is a real number, $(a, 0)$ is an x-intercept of $y = P(x)$.

The equivalency of statements 1 and 2 is a restatement of the factor theorem.

SECTION 4-2 SYNTHETIC DIVISION AND THE FACTOR THEOREM

EXAMPLE 8 Given $P(x) = x^3 + 3x^2 - 10x - 24$ and -4 is a zero of $P(x)$:
- (a) Completely factor $P(x)$.
- (b) Find all zeros of $P(x)$.
- (c) Find the solution set of the equation $P(x) = 0$.
- (d) Find the x-intercepts of $y = P(x)$.

SOLUTIONS

$$\begin{array}{r|rrrr} -4 & 1 & 3 & -10 & -24 \\ & & -4 & 4 & 24 \\ \hline & 1 & -1 & -6 & 0 \end{array}$$

Use synthetic division to divide $P(x)$ by $x - (-4)$.

The remainder is zero, and the depressed factor is $x^2 - x - 6$.

$P(x) = (x^2 - x - 6)(x + 4)$

Since $x + 4$ is a factor of $P(x)$, -4 is a zero of $P(x)$. The other zeros of $P(x)$ are determined by examining the depressed factor $x^2 - x - 6$.

$ = (x - 3)(x + 2)(x + 4)$

Factor $x^2 - x - 6$.

- (a) $P(x) = (x - 3)(x + 2)(x + 4)$
- (b) The zeros of $P(x)$ are 3, -2, and -4.
- (c) The solution set of $P(x) = 0$ is $\{3, -2, -4\}$.
- (d) The x-intercepts of $y = P(x)$ are $(3, 0)$, $(-2, 0)$, and $(-4, 0)$.

The graph of $y = x^3 + 3x^2 - 10x - 24$ is shown below for your reference.

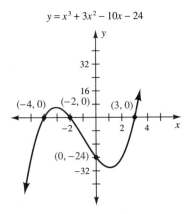

All the zeros of $P(x)$ are zeros of a factor of $P(x)$. Thus for $P(x) = (x - a)Q(x)$, all zeros of $P(x)$ other than a must be zeros of the depressed factor $Q(x)$. Therefore, once we find one zero of $P(x)$ by synthetic division, we can examine the depressed factor $Q(x)$, which is one degree less than $P(x)$. This approach is illustrated in the next example.

▶ **Self-Check ▼**

If $P(x) = x^4 - 2x^3 - 4x^2 - 2x + 10$, find $P(5)$ first by using synthetic division and then by substitution.

▶ **Self-Check Answer ▼**

$P(5) = 275$

EXAMPLE 9 Use the graph of $y = P(x)$ shown to the right to determine the real numbers that are zeros of the fourth-degree polynomial $P(x)$. Then use synthetic division to determine the other zeros of $P(x) = 2x^4 + 3x^3 + 3x - 2$.

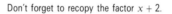

SOLUTION By inspection, $(-2, 0)$ and $(\frac{1}{2}, 0)$ are x-intercepts of the graph of $y = P(x)$. Thus the real numbers that are zeros of $P(x)$ are -2 and $\frac{1}{2}$.

$$\underline{-2 |} \quad 2 \quad 3 \quad 0 \quad 3 \quad -2$$
$$\phantom{\underline{-2 |} \quad 2} \quad -4 \quad 2 \quad -4 \quad 2$$
$$\phantom{\underline{-2 |}} \quad \overline{2 \quad -1 \quad 2 \quad -1 \quad 0}$$

Use synthetic division to divide $P(x)$ by $x - (-2)$.

$P(x) = (x + 2)(2x^3 - x^2 + 2x - 1)$

The depressed factor is $2x^3 - x^2 + 2x - 1$.

$$\underline{\tfrac{1}{2} |} \quad 2 \quad -1 \quad 2 \quad -1$$
$$\phantom{\underline{\tfrac{1}{2} |} \quad 2} \quad 1 \quad 0 \quad 1$$
$$\phantom{\underline{\tfrac{1}{2} |}} \quad \overline{2 \quad 0 \quad 2 \quad 0}$$

The other zeros of $P(x)$ must be zeros of this depressed factor, which is now divided by $x - \frac{1}{2}$.

$P(x) = (x + 2)(x - \frac{1}{2})(2x^2 + 2)$

Don't forget to recopy the factor $x + 2$.

$P(x) = (x + 2)(x - \frac{1}{2})(2)(x^2 + 1)$

Factor 2 out of the depressed factor and then multiply it times the factor $x - \frac{1}{2}$.

$P(x) = (x + 2)(2x - 1)(x^2 + 1)$

The other two zeros of $P(x)$ are zeros of $x^2 + 1$.

The zeros of $x^2 + 1$ are the same as the solutions of $x^2 + 1 = 0$, which can be determined by extraction of roots.

$x^2 + 1 = 0$
$x^2 = -1$
$x = \pm i$

Answer The zeros of $2x^4 + 3x^3 + 3x - 2$ are $-2, \frac{1}{2}, -i$, and i. ☐

Following is an example of a problem for which the memory feature of your calculator is helpful. The keystrokes vary considerably from model to model; refer to Appendix A if you have a question concerning the use of the memory on your calculator.

EXAMPLE 10 Use synthetic division and a calculator to divide

$$P(x) = 2x^4 + 6.48x^3 - 28.8384x^2 - 0.6016x + 32.7168$$

by $x + 5.68$. Express the answer in the form $P(x) = Q(x)(x - a) + R(x)$.

SOLUTION
$$\underline{-5.68 |} \quad 2 \quad 6.48 \quad -28.8384 \quad -0.6016 \quad 32.7168$$
$$\phantom{\underline{-5.68 |} \quad 2} \quad -11.36 \quad 27.7184 \quad 6.3616 \quad -32.7168$$
$$\phantom{\underline{-5.68 |}} \quad \overline{2 \quad -4.88 \quad -1.12 \quad 5.76 \quad 0}$$

Store -5.68 in the memory of your calculator so that it can be recalled at each step of the process. The individual steps are shown in Appendix A.

Answer $P(x) = (2x^3 - 4.88x^2 - 1.12x + 5.76)(x + 5.68)$

The quotient is $2x^3 - 4.88x^2 - 1.12x + 5.76$ with remainder 0. Thus -5.68 is a zero of $P(x)$.

☐

SECTION 4-2 SYNTHETIC DIVISION AND THE FACTOR THEOREM

EXERCISES 4-2

A

In Exercises 1–10 divide using synthetic division. Express your answers in the form $P(x) = Q(x)(x - a) + R(x)$.

1. $(x^3 - 6x^2 + 5x - 3) \div (x - 5)$
2. $(x^3 + 5x^2 - 7x + 11) \div (x - 3)$
3. $(3a^5 - a^3 + 3a - 8) \div (a + 1)$
4. $(2b^6 + b^4 - b^2 + b + 3) \div (b - 1)$
5. $(6p^3 - 2p + 5) \div (p - \frac{1}{3})$
6. $(2q^3 + 4q^2 - q + 6) \div (q + \frac{1}{2})$
7. $(x^3 + 8) \div (x + 2)$
8. $(x^3 - 27) \div (x - 3)$
9. $(x^3 + x^2 + x - 5) \div [x - (1 + i)]$
10. $(x^3 + 2x^2 - x + 3) \div [x - (2 - i)]$

In Exercises 11–18, use synthetic division to find $P(a)$ for the polynomial $P(x)$ and the complex number a. If a is a zero of $P(x)$, write $P(x)$ in the form $Q(x)(x - a)$.

11. $P(x) = x^3 - 8x^2 + 18x - 11$, $a = 1$
12. $P(x) = x^3 + 8x^2 + 22x + 21$, $a = -3$
13. $P(x) = 3x^4 - 5x^2 + 4x + 6$, $a = -4$
14. $P(x) = 2x^3 - 4x^2 + 8x - 12$, $a = 4$
15. $P(x) = x^5 - x^3 + 2x - 8$, $a = -1$
16. $P(x) = x^6 + x^4 - x^2 + x + 3$, $a = 1$
17. $P(x) = x^3 + 8$, $a = -2$
18. $P(x) = x^3 - 27$, $a = 3$
19. Show that $v + 4$ is a factor of $v^3 + 5v^2 - 16v - 80$, and determine the depressed factor.
20. Show that $y - 3$ is a factor of $y^3 + 4y^2 - 9y - 36$, and determine the depressed factor.
21. Show that $r - 5$ is *not* a factor of $r^4 - 5r^2 - 6r - 7$.
22. Show that $s + 6$ is *not* a factor of $s^4 + 3s^3 - 10s^2 - 5s + 6$.
23. Factor $t^3 - 31t + 30$, given that $t - 1$ is a factor.
24. Factor $w^3 - w^2 - 14w + 24$, given that $w - 3$ is a factor.
25. Solve the equation $2x^3 - 3x^2 - 32x + 48 = 0$, given that 4 is a solution of the equation.
26. Solve the equation $2x^3 - x^2 - 13x - 6 = 0$, given that -2 is a solution of the equation.

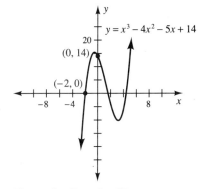

Figure for Exercise 27

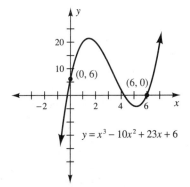

Figure for Exercise 28

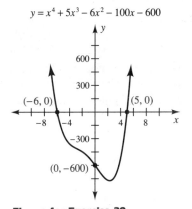

Figure for Exercise 29

B

27. Use the graph of $y = x^3 - 4x^2 - 5x + 14$ shown to the right as an aid to finding the exact solutions of $x^3 - 4x^2 - 5x + 14 = 0$.
28. Use the graph of $y = x^3 - 10x^2 + 23x + 6$ shown to the right as an aid to finding the exact solutions of $x^3 - 10x^2 + 23x + 6 = 0$.
29. Use the graph of $y = x^4 + 5x^3 - 6x^2 - 100x - 600$ shown to the right as an aid to finding all the zeros of $P(x) = x^4 + 5x^3 - 6x^2 - 100x - 600$.

30 Show that $x - 1$ is a factor of $P(x) = x^{100} - 51x^{87} + 43x^{43} - 10x^{21} + 17$.

31 Show that $x + 1$ is a factor of $P(x) = x^{100} + 32x^{81} + 16x^{52} + 5x^6 + 10$.

In Exercises 32–34, use synthetic division to find the quotient and the remainder.

32 $(y^3 - 3y^2 + 6y - 4) \div (y - 1 + i\sqrt{3})$

33 $(v^3 - v^2 - 3v - 1) \div (v - 1 - \sqrt{2})$

34 $(w^5 - a^5) \div (w - a)$ (*Hint*: Both a and a^5 are constants.)

35 Find the value of k such that $x - 2$ is a factor of $4x^3 - 3x^2 + kx - 8$.

36 Find all values of k such that $x + 3$ is a factor of $x^3 + k^2x^2 + kx + 9$.

C

In Exercises 37–39, use synthetic division to find $P(a)$ for the polynomial $P(x)$ and the complex number a. If $P(a) = 0$, write $P(x)$ in the form $Q(x)(x - a)$, where $Q(x)$ is the quotient.

37 $P(x) = x^3 + x^2 + x - 5$, $a = 1 - i$

38 $P(x) = x^3 - x^2 - 3x - 1$, $a = 1 + \sqrt{2}$

39 $P(x) = x^3 + bx^2 + b^3$, $a = -b$

40 Show that $x + a$ is *not* a factor of $x^4 + a^4$, if $a \neq 0$.

41 Show that $x + a$ is a factor of $x^5 + a^5$.

42 If the sum of the coefficients of $P(x)$ is zero, show that $x - 1$ is a factor of $P(x)$.

43 Show that $-x^5 + x^4 - 7x^3 + 11x^2 + 19$ cannot have a negative zero.

44 Show that $x^6 + 3x^4 + 7x^2 + 2$ can have no factor $x - k$ where k is a real number.

45 Show that $-x^8 - 9x^4 - 17$ can have no real zero.

46 Show that $x - a$ is a factor of $x^n - a^n$ for any positive integer n.

47 Show that $x + a$ is a factor of $x^n + a^n$ when n is an odd positive integer.

48 Give an example of a polynomial function that has no zeros.

Use a calculator to work Exercises 49–53.

49 Use synthetic division to show that 2.21 is a zero of the polynomial $50v^3 + 6.8v^2 - 0.933v - 570.843$ and find the depressed factor.

50 Use synthetic division to find the quotient polynomial and the remainder polynomial when $2.2w^3 - 10.06w^2 - 40.136w + 128.984$ is divded by $w + 3.9$.

51 Use synthetic division to evaluate $P(5.5)$ for

$$P(x) = 6.8x^5 + 7.4x^4 - 12.44x^3 + 142.28x^2 + 60x - 59.065$$

52 Completely factor $P(x) = 20x^3 + 2299x^2 - 4777x + 2340$ over the integers, given that $P(-117) = 0$.

53 Completely factor $P(x) = 10x^3 - 339x^2 - 2701x + 1134$ over the integers, given that $P(40.5) = 0$.

SECTION 4-3

Complex Zeros of Polynomials

Section Objective

4 Find a real polynomial with given zeros.

A **real polynomial** is a polynomial with real number coefficients. The quadratic formula has already shown us that a real quadratic polynomial with complex zero $a + bi$ must also have its conjugate $a - bi$ as a zero. The conjugate zeros theorem is a generalization of this property.

If the discriminant of $x = \dfrac{-b \pm \sqrt{b^2 - 4ac}}{2a}$ is negative, then x will assume two values that are complex conjugates.

Conjugate Zeros Theorem ▼

> The imaginary zeros of a real polynomial occur in conjugate pairs.

For example, if $2 + i$ and $3 - i\sqrt{2}$ are zeros of a real polynomial, then their conjugates $2 - i$ and $3 + i\sqrt{2}$ must also be zeros of this polynomial.

EXAMPLE 1 Find a real polynomial of smallest degree with zeros 1, -2 and $3 - i$.

▶ **Self-Check ▼**

If $3 - 2i$ and $4 + 3i$ are zeros of a real polynomial, find two other zeros. *of smallest degree*

SOLUTION

Zero	Corresponding Factor
1	$z - 1$
-2	$z + 2$
$3 - i$	$z - (3 - i)$
$3 + i$	$z - (3 + i)$

By the factor theorem a is a zero of $P(z)$ if and only if $z - a$ is a factor of $P(z)$.

The conjugate of $3 - i$ must also be a zero of $P(z)$.

Therefore,

$P(z) = (z - 1)(z + 2)[z - (3 - i)][z - (3 + i)]$ $P(z)$ is the product of its factors.

$= (z^2 + z - 2)[(z - 3) + i][(z - 3) - i]$ Regroup terms, and multiply the factors arising from the conjugates. This will enable you to take advantage of the products of the special forms.

$= (z^2 + z - 2)[(z - 3)^2 - i^2]$

$= (z^2 + z - 2)[(z^2 - 6z + 9) - (-1)]$

$= (z^2 + z - 2)(z^2 - 6z + 10)$

Answer $P(z) = z^4 - 5z^3 + 2z^2 + 22z - 20$ $P(z)$ does have real coefficients. □

▶ **Self-Check Answer ▼**

$3 + 2i$, $4 - 3i$

The polynomial $kP(x)$ for $k \neq 0$ will have the same zeros as $P(x)$. For example, the polynomials $(x - 2)(x + 3)$ and $5(x - 2)(x + 3)$ have the same zeros, 2 and -3. Thus there are an infinite number of polynomials that all have exactly the same zeros; only by specifying some other condition can we single out one polynomial from a family of these polynomials. This fact is used in the next example.

EXAMPLE 2 Find a real polynomial $P(z)$ of smallest degree with zeros 1, -2, and $3 - i$ and $P(-1) = -2$.

SOLUTION From Example 1, we know that

$$P(z) = z^4 - 5z^3 + 2z^2 + 22z - 20$$

has the desired zeros. Thus, so will

$$P(z) = k(z^4 - 5z^3 + 2z^2 + 22z - 20)$$

for each nonzero real number k.

$$P(-1) = k[(-1)^4 - 5(-1)^3 + 2(-1)^2 + 22(-1) - 20] \quad \text{Substitute } -1 \text{ for } z.$$
$$= k[1 + 5 + 2 - 22 - 20]$$
$$= -34k$$

Thus $\quad -34k = -2 \qquad\qquad\qquad\qquad\qquad P(-1) = -2, \text{ and } P(-1) = -34k.$

$$k = \frac{1}{17}$$

and $\quad P(z) = \frac{1}{17}(z^4 - 5z^3 + 2z^2 + 22z - 20)$

Answer $\quad P(z) = \frac{1}{17}z^4 - \frac{5}{17}z^3 + \frac{2}{17}z^2 + \frac{22}{17}z - \frac{20}{17}$ ☐

The polynomial in the next example will be factored over the complex numbers. Earlier in the text we factored polynomials using only integer coefficients and constants. For example, we noted in Section 1-5 that $w^2 + 1$ cannot be factored using only integers. Thus we say that $w^2 + 1$ is prime over the integers. However, $w^2 + 1$ can be factored over the complex numbers as $w^2 + 1 = (w + i)(w - i)$.

EXAMPLE 3 Verify that i, $1 + i$, and their conjugates are zeros of

$$P(x) = x^4 - 2x^3 + 3x^2 - 2x + 2$$

and give the linear factors of $P(x)$.

SECTION 4-3 COMPLEX ZEROS OF POLYNOMIALS

SOLUTION

$$\begin{array}{r|rrrrr} i & 1 & -2 & 3 & -2 & 2 \\ & & i & -1-2i & 2+2i & -2 \\ \hline & 1 & -2+i & 2-2i & 2i & 0 \end{array}$$

$P(x)$ is of degree 4. The 0 remainder indicates that i is a zero of $P(x)$.

$Q_1(x) = x^3 + (-2+i)x^2 + (2-2i)x + 2i$

The quotient $Q_1(x)$ is a depressed polynomial factor of degree 3.

$$\begin{array}{r|rrrr} -i & 1 & -2+i & 2-2i & 2i \\ & & -i & 2i & -2i \\ \hline & 1 & -2 & 2 & 0 \end{array}$$

The conjugate of i, $-i$, is also a zero of $Q_1(x)$ and thus of $P(x)$.

$Q_2(x) = x^2 - 2x + 2$

$Q_2(x)$ is a depressed polynomial factor of degree 2.

$$\begin{array}{r|rrr} 1+i & 1 & -2 & 2 \\ & & 1+i & -2 \\ \hline & 1 & -1+i & 0 \end{array}$$

$1+i$ is a zero of $Q_2(x)$ and thus of $P(x)$.

$Q_3(x) = x + (-1+i)$

$Q_3(x)$ is a depressed polynomial factor of degree 1.

$$\begin{array}{r|rr} 1-i & 1 & -1+i \\ & & 1-i \\ \hline & 1 & 0 \end{array}$$

$1-i$ is a zero of $Q_3(x)$ and thus of $P(x)$.

Answer
$P(x) = x^4 - 2x^3 + 3x^2 - 2x + 2$
$= (x-i)(x+i)[x-(1+i)][x-(1-i)]$ □

A polynomial $P(x)$ with a zero a of multiplicity n has a repeated factor $(x-a)^n$. In the next example we use the given information to produce the complete factorization of $P(x)$ and to obtain all the zeros of $P(x)$.

▶ **Self-Check** ▼

Find a real polynomial $P(x)$ with smallest degree that has zeros 2, -3, and $1+i$ and $P(-1) = -60$.

EXAMPLE 4 Find all zeros and state their multiplicities for the polynomial

$$P(x) = x^4 - 6x^3 + x^2 + 24x + 16$$

given that 4 is a zero.

SOLUTION

$$\begin{array}{r|rrrrr} 4 & 1 & -6 & 1 & 24 & 16 \\ & & 4 & -8 & -28 & -16 \\ \hline & 1 & -2 & -7 & -4 & 0 \end{array}$$

Use synthetic division to find the depressed factor.

4 is a zero.

$Q_1(x) = 1x^3 - 2x^2 - 7x - 4$

The depressed factor is $Q_1(x)$.

$$\begin{array}{r|rrrr} 4 & 1 & -2 & -7 & -4 \\ & & 4 & 8 & 4 \\ \hline & 1 & 2 & 1 & 0 \end{array}$$

Divide $Q_1(x)$ by $x-4$ to see if 4 is a multiple zero.

4 is a multiple zero.

$Q_2(x) = 1x^2 + 2x + 1 = (x+1)^2$

The depressed factor is $Q_2(x)$.

Thus $P(x) = (x-4)^2(x+1)^2$.

Answer 4 and -1 are each zeros of multiplicity 2. □

▶ **Self-Check Answer** ▼

$P(x) = 2x^4 - 2x^3 - 12x^2 + 28x - 24$

We know that each quadratic equation has two roots, counting multiplicity, and so each second-degree polynomial has two linear factors over the complex numbers. A direct consequence of the fundamental theorem of algebra is that an nth-degree polynomial has n linear factors. This result was proven by the 22-year-old mathematician Carl Friedrich Gauss in 1799. The proof is beyond the scope of college algebra.

Fundamental Theorem of Algebra ▼

Each polynomial of degree $n \geq 1$ has a linear factor $x - a$, where a is a complex number.

Complete Factorization Theorem ▼

Each nth-degree polynomial can be factored into n linear factors over the complex numbers.

Since there is a zero of a polynomial corresponding to each linear factor, the complete factorization theorem implies that the sum of the multiplicities of the zeros of a polynomial is equal to the degree of the polynomial.

EXAMPLE 5 What is the smallest degree possible for a real polynomial with zeros -2 (multiplicity 3) and $2 + 3i$? Write a polynomial with these zeros.

SOLUTION To be a real polynomial, $P(x)$ must have zeros -2 (multiplicity 3), $2 + 3i$, and $2 - 3i$. Therefore $P(x)$ must be at least fifth degree.

The conjugate zeros theorem implies that $2 - 3i$ must also be a zero.

Zero	Corresponding Factor	
-2	$(x + 2)^3$	-2 is a zero of multiplicity 3.
$2 + 3i$	$x - (2 + 3i)$	
$2 - 3i$	$x - (2 - 3i)$	$2 - 3i$ is the conjugate of $2 + 3i$.

Thus

$$P(x) = (x + 2)^3[x - (2 + 3i)][x - (2 - 3i)]$$

$P(x)$ is the product of its factors.

$$= (x^3 + 6x^2 + 12x + 8)[(x - 2) - 3i][(x - 2) + 3i]$$

Cube the first factor and regroup the other factors.

$$= (x^3 + 6x^2 + 12x + 8)[(x - 2)^2 - 9i^2]$$

Multiply and then simplify.

$$= (x^3 + 6x^2 + 12x + 8)(x^2 - 4x + 13)$$

Answer $P(x) = x^5 + 2x^4 + x^3 + 38x^2 + 124x + 104$

$P(x)$ is fifth degree.

SECTION 4-3 COMPLEX ZEROS OF POLYNOMIALS

Each nth-degree polynomial has n linear factors. If just one factor can be found, then we can use synthetic division to determine the depressed factor of degree $n - 1$. Thus we can simplify the search for other factors by examining the depressed factor at each step. Once a second-degree depressed polynomial factor is obtained, the last two factors can be found by determining the zeros of the corresponding quadratic polynomial.

▶ **Self-Check** ▼

1. Find all zeros and state their multiplicity for the polynomial
$$P(x) = x^4 + x^3 - 18x^2 - 52x - 40$$
given that -2 is a zero.

2. What is the smallest degree possible for a real polynomial with zeros i (multiplicity 2), $1 + \sqrt{2}$, $1 - \sqrt{2}$, and 7 (multiplicity 4)?

EXAMPLE 6 Find the zeros of the polynomial
$$P(z) = z^4 - 9z^3 + 27z^2 + 23z - 150$$
given that $4 - 3i$ is a zero.

SOLUTION

$$\begin{array}{r|rrrrr} 4-3i & 1 & -9 & 27 & 23 & -150 \\ & & 4-3i & -29+3i & 1+18i & 150 \\ \hline & 1 & -5-3i & -2+3i & 24+18i & 0 \end{array}$$

Divide by $z - (4 - 3i)$ to obtain the depressed factor $Q_1(z)$.

$Q_1(z) = z^3 + (-5 - 3i)z^2 + (-2 + 3i)z + (24 + 18i)$

$$\begin{array}{r|rrrr} 4+3i & 1 & -5-3i & -2+3i & 24+18i \\ & & 4+3i & -4-3i & -24-18i \\ \hline & 1 & -1 & -6 & 0 \end{array}$$

$4 + 3i$, the conjugate of $4 - 3i$, must be a zero of $Q_1(z)$.

$Q_2(z) = z^2 - z - 6$

The depressed factor is $Q_2(z)$.

$Q_2(z) = (z - 3)(z + 2)$

Factor the quadratic polynomial $Q_2(z)$ to find the other two zeros.

3 and -2 are zeros of $Q_2(z)$.

Answer The zeros of $P(z)$ are $4 - 3i$, $4 + 3i$, 3, and -2.

Note that this fourth-degree polynomial has four zeros.

□

The graph of a polynomial function $y = P(x)$ shows all the real zeros of $P(x)$ as the x-intercepts of the graph, but it does not show the imaginary zeros of $P(x)$. However, the graph of $y = P(x)$ can indicate that $P(x)$ has imaginary zeros. For example, the parabola defined by $y = x^2 + 2x + 2$ is concave upward with no x-intercepts. This is a graphical indication that $x^2 + 2x + 2 = 0$ has only imaginary solutions. The parabola defined by $y = x^2 + 2x - 3 = (x + 3)(x - 1)$, which can be obtained by shifting the parabola $y = x^2 + 2x + 2$ down 5 units, has x-intercepts of $(-3, 0)$ and $(1, 0)$. Any graph of a polynomial function that can be translated vertically to produce more x-intercepts will have an imaginary zero corresponding to each of these potential new x-intercepts. This is illustrated in the next example.

▶ **Self-Check Answers** ▼

1. -2 is a zero of multiplicity 3, and 5 is a zero of multiplicity 1

2. Tenth degree

EXAMPLE 7 Use the given graphs of $y = x^3 + 2x^2 - x - 2$ and $y = x^3 + 2x^2 - x - 14$ to factor each of the corresponding polynomials.

The graph of $y = x^3 + 2x^2 - x - 14$ can be obtained by shifting the graph of $y = x^3 + 2x^2 - x - 2$ down 12 units.

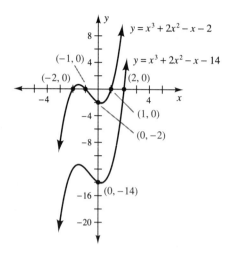

SOLUTION

$y = x^3 + 2x^2 - x - 2$
$y = (x + 2)(x + 1)(x - 1)$

The x-intercepts of the graph of this function are $(-2, 0)$, $(-1, 0)$, and $(1, 0)$.

$y = x^3 + 2x^2 - x - 14$

$$\underline{2|} \begin{array}{cccc} 1 & 2 & -1 & -14 \\ & 2 & 8 & 14 \\ \hline 1 & 4 & 7 & 0 \end{array}$$

The only x-intercept of the graph of this function is $(2, 0)$. The corresponding zero of 2 is used to produce the depressed factor of $x^2 + 4x + 7$.

$y = (x - 2)(x^2 + 4x + 7)$
$y = (x - 2)(x + 2 - i\sqrt{3})(x + 2 + i\sqrt{3})$

The zeros of $x^2 + 4x + 7$ can be found by using the quadratic formula to solve $x^2 + 4x + 7 = 0$. The solutions of this equation are $-2 \pm i\sqrt{3}$. □

EXERCISES 4-3

A

Exercises 1 and 2 list some zeros of a real polynomial $P(x)$. Use the conjugate zeros theorem to determine other zeros of $P(x)$. Also determine the smallest degree possible for $P(x)$.

1. **a.** $2i$ and $2 + 7i$
   b. 2, 3, and $2 + i\sqrt{3}$ (multiplicity 2)
   c. 8 (multiplicity 2), $3 + \sqrt{2}$, $3 - \sqrt{2}$, $5 + 6i$, and i (multiplicity 3)
2. **a.** $3 - 2i$ and $-5i$
   b. -7, $-i\sqrt{2}$, and $3 + 2i$ (multiplicity 2)
   c. -6 (multiplicity 3), $2 - \sqrt{5}$, $2 + \sqrt{5}$, and $8 - 10i$ (multiplicity 2)

SECTION 4-3 COMPLEX ZEROS OF POLYNOMIALS

For the polynomials in Exercises 3–6, list each zero and its multiplicity.

3. a. $P(x) = (x - 2)^5(x + 3)^2$ b. $P(x) = x^3(2x - 1)^2(x + 2)^2$ c. $P(t) = t^2 + 4t + 4$
4. a. $P(x) = (x - 6)^4(x + 8)^3$ b. $P(x) = x^4(3x + 1)^2(x + 3)^5$ c. $P(t) = t^2 - 6t + 9$
5. a. $P(z) = (z - 5)^3(z^2 + 4z + 3)$ b. $P(x) = (x - 1)(x^2 + x + 1)$ c. $P(x) = x^3 + 8$
6. a. $P(z) = z(z + 5)^2(z^2 - 4z - 5)$ b. $P(x) = (x - 2)(x^2 + 2x + 4)$ c. $P(x) = x^3 - 27$

In Exercises 7–16, find a real polynomial $P(x)$ of smallest degree that has the zeros listed.

7. $2i$
8. $-3i$
9. $1 + i$
10. $1 - i$
11. $2, 2 - i$
12. $-2, 2 + i$
13. $-3, 4, -3 + 2i$
14. $3, -4, -2 - 3i$
15. $0, -4i, 1 - 2i$
16. $0, -2i, 2 - i$

17. Find $P(x)$, a real polynomial with zeros 3 and $4 - i$ and $P(2) = -15$.
18. Find $P(x)$, a real polynomial with zeros $1 + \sqrt{2}, 1 - \sqrt{2}$, and i and $P(1) = -2$.
19. Find $P(x)$, a real polynomial with zeros $1 - \sqrt{3}, 1 + \sqrt{3}$, and $1 + i$ and $P(-1) = 50$.
20. Find $P(x)$, a real polynomial with zeros $-3, -2, -1, 0$, and i and $P(1) = 96$.

Exercises 21–28 each list a polynomial and one of its zeros. Find the other zeros.

21. $P(y) = y^3 + 3y^2 + y + 3; -3$
22. $P(y) = y^3 + 2y^2 + 3y + 6; -2$
23. $P(w) = 6w^4 - w^3 + 12w^2 - 4w - 48; 2i$
24. $P(x) = 6x^4 + x^3 + 52x^2 + 9x - 18; 3i$
25. $P(v) = v^4 - v^3 - 6v^2 + 14v - 12; 1 + i$
26. $P(u) = u^4 + 3u^3 + 5u^2 + u - 10; -1 + 2i$
27. $P(t) = t^4 + 4t^3 + 9t^2 + 20t + 20; -2$ of multiplicity 2
28. $P(t) = t^4 - 8t^3 + 17t^2 + 6t - 36; 3$ of multiplicity 2

In Exercises 29 and 30, factor the polynomials and determine their zeros.

29. $P(x) = 2x(x^2 - 5x - 6) - 3(x^2 - 5x - 6)$
30. $P(y) = y(y^2 - 9) - 7(y^2 - 9)$

B

31. Use the graph of $y = x^4 + x^3 - 19x^2 + x - 20$ shown below as an aid to determining all the zeros of $x^4 + x^3 - 19x^2 + x - 20$.
32. Use the graph of $y = x^4 - 3x^3 - 6x^2 - 12x - 40$ shown below as an aid to determining all the zeros of $x^4 - 3x^3 - 6x^2 - 12x - 40$.

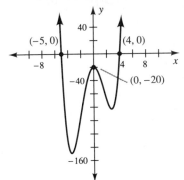

Figure for Exercise 31

Figure for Exercise 32

In Exercises 33–35, factor the polynomials and determine their zeros.

33 $P(u) = u^3 - 3u^2 + 3u - 1$
34 $P(y) = y^3 + 3y^2 + 3y + 1$
35 $P(w) = w^4 - 26w^2 + 25$

36 Why is it not possible to find a real polynomial of degree 3 with zeros of 1, -2, and $1 + 2i$?

37 Why does a real polynomial of odd degree have at least one real zero?

38 $-2i$ is a zero of $z^2 + 2iz$, but $2i$ is not a zero. Explain the relationship of this result to the conjugate zeros theorem. What is the other zero of this polynomial?

39 Give a polynomial with only these zeros: 1, 2, and $1 + 2i$.

40 What is the degree of a polynomial that has no zeros?

C

41 Since 2 is a zero of $P(x) = x^3 - 8$, 2 is one cube root of 8. Find the other two cube roots of 8.

42 The velocity at time t of a particle moving along a line is given by $V(t) = t^3 - 7t^2 + 16t - 12$. If the velocity is zero at $t = 3$ seconds, find the other time(s) that the velocity is zero.

43 Does $P(x) = x^2 + x + 1$ have a linear factor over the complex numbers? If so, find all linear factors.

44 Write a polynomial $P(x)$ of smallest degree that has zeros of i (multiplicity 2) and 0 (multiplicity 3) and with $P(1) = 100$.

45 Write a polynomial $P(x)$ of smallest degree that has zeros of $2i$ (multiplicity 2) and 0 (multiplicity 4) and with $P(1) = 100$.

Use a calculator to work Exercises 46 and 47.

46 Completely factor $P(x) = 36x^4 - 396x^3 + 1085x^2 + 44x - 121$ over the integers, given that 5.5 is a zero of $P(x)$ of multiplicity 2.

47 Given that $P(7.1) = 0$ and $P(4.5) = 0$, find the other zeros of

$$P(x) = 100x^4 - 1160x^3 + 3295x^2 - 1160x + 3195$$

In Exercises 48 and 49, graph each polynomial function $y = P(x)$ using a graphics calculator. Then use this graph to determine all the zeros of the fourth-degree polynomial $P(x)$.

48 $y = 2x^4 - 5x^3 + 15x^2 - 45x - 27$
49 $y = x^4 - x^3 - 7x^2 - 5x - 60$

SECTION 4-4

Rational Zeros of Polynomials

Section Objective

5 Find rational zeros of a polynomial function with integer coefficients.

One method for determining the rational zeros of a polynomial function is to find a set of all possible rational zeros. We then test the members of this set, using synthetic division, until all rational zeros of the polynomial have

SECTION 4-4 RATIONAL ZEROS OF POLYNOMIALS

been found. Our experience with quadratic polynomials suggests that the set of possible rational zeros can be determined by examining only two of the coefficients of the polynomial. For example, the quadratic polynomial $12x^2 + 17x - 40$ factors as $(4x - 5)(3x + 8)$. Thus its zeros are $\frac{5}{4}$ and $-\frac{8}{3}$. Note that the numerators 5 and -8 both divide the constant term, -40, and the denominators 4 and 3 both divide the leading coefficient, 12.

The logic used to form the set of possible rational zeros for a quadratic polynomial is similar to that used to factor trinomials by trial and error. this logic is examined again in the paragraph following the rational zeros theorem.

Rational Zeros Theorem ▼

If $P(x) = a_n x^n + a_{n-1} x^{n-1} + \cdots + a_1 x + a_0$, $a_n \neq 0$ is a polynomial with integer coeffcients and $\frac{p}{q}$ is a rational zero reduced to lowest terms, then the numerator p must divide the constant, a_0, and the denominator q must divide the leading coefficient, a_n.

Rather than prove this theorem, we will review the logic further in the following special case. Suppose that $\frac{3}{2}$ is a zero of a third-degree polynomial $P(x)$ with integer coefficients. Then $2x - 3$ is a factor of $P(x)$, and the other factor can be written as $ax^2 + bx + c$. That is,

$$P(x) = (2x - 3)(ax^2 + bx + c) = 2ax^3 + (2b - 3a)x^2 + (2c - 3b)x - 3c$$

Note that 3, the numerator of $\frac{3}{2}$, divides the constant, $-3c$; and 2, the denominator, divides $2a$, the coefficient of x^3.

EXAMPLE 1 Use the rational zeros theorem to list the possible rational zeros of

$$P(x) = 2x^3 + 5x^2 - 4x + 4$$

SOLUTION Possible numerators must be integer divisors of the constant term 4:

$$\pm 1 \quad \pm 4$$
$$\pm 2$$

Possible denominators are integer divisors of the leading coefficient 2:

$$\pm 1 \quad \pm 2$$

Possible rational zeros are $-4, -2, -1, -\frac{1}{2}, \frac{1}{2}, 1, 2,$ and 4. □

The possible rational zeros $\frac{\pm 1}{\pm 1}, \frac{\pm 2}{\pm 1}, \frac{\pm 4}{\pm 1}, \frac{\pm 1}{\pm 2},$ $\frac{\pm 2}{\pm 2},$ and $\frac{\pm 4}{\pm 2}$ are listed to the left in increasing order.

Note that duplicates such as $\frac{2}{1}, \frac{-2}{-1}, \frac{4}{2},$ and $\frac{-4}{-2}$ are listed only once in simplified form.

The degree of $P(x)$ is three; thus a maximum of three of these eight possible rational zeros are actually zeros.

CHAPTER 4 · POLYNOMIAL AND RATIONAL RELATIONS

EXAMPLE 2 Use the rational zeros theorem to list the possible rational zeros of

$$P(x) = 10x^4 + 7x^3 - 64x^2 + 16x + 6$$

SOLUTION Possible numerators must be integer divisors of 6:

$$\pm 1 \quad \pm 6$$
$$\pm 2 \quad \pm 3$$

Possible denominators are integer divisors of 10:

$$\pm 1 \quad \pm 10$$
$$\pm 2 \quad \pm 5$$

Possible rational zeros are $-6, -3, -2, -\dfrac{3}{2}, -\dfrac{6}{5}, -1, -\dfrac{3}{5}, -\dfrac{1}{2}, -\dfrac{2}{5},$ $-\dfrac{3}{10}, -\dfrac{1}{5}, -\dfrac{1}{10}, \dfrac{1}{10}, \dfrac{1}{5}, \dfrac{3}{10}, \dfrac{2}{5}, \dfrac{1}{2}, \dfrac{3}{5}, 1, \dfrac{6}{5}, \dfrac{3}{2}, 2, 3,$ and 6. □

It is often convenient to pair the factors so that each pair yields the desired product. This pairing makes it less likely that one factor will inadvertently be omitted.

The degree of $P(x)$ is four; thus a maximum of four of these 24 possible rational zeros are actually zeros.

The search for rational zeros should be a systematic and orderly procedure, such as that outlined in the following box.

Determining Rational Zeros of a Polynomial Function ▼

Step 1 List all possible rational zeros.
Step 2 Test the possible rational zeros one by one. (After each zero is found, use the depressed polynomial factor and delete any value from the original list that cannot be a zero of this polynomial. Be sure to test for possible multiple zeros.)
Step 3 If a depressed factor of degree two is obtained, find the remaining zeros by factoring or by using the quadratic formula.

▶ Self-Check ▼

List all possible rational zeros
$$P(x) = 3x^3 + 2x^2 - 5x + 6$$

All rational zeros can be found using this procedure. In fact, if a depressed polynomial of degree two can be obtained, then all zeros can be found.

▶ Self-Check Answer ▼

$-6, -3, -2, -1, -\dfrac{2}{3}, -\dfrac{1}{3}, \dfrac{1}{3}, \dfrac{2}{3}, 1, 2, 3, 6$

SECTION 4-4 RATIONAL ZEROS OF POLYNOMIALS

EXAMPLE 3 Factor and find all zeros of

$$P(x) = 2x^4 + x^3 - 9x^2 - 4x + 4$$

SOLUTION

1 Possible numerators are ± 1, ± 2, and ± 4. Possible denominators are ± 1 and ± 2. Possible rational zeros are -4, -2, -1, $-\frac{1}{2}$, $\frac{1}{2}$, 1, 2, and 4.

2 Test the possible rational zeros.

```
 1 | 2   1   -9   -4    4
   |     2    3   -6  -10
   -----------------------
     2   3   -6  -10   -6
```
Start by testing integral possibilities. Test 1 first. Examine each remainder r to determine if the test value is a zero of $P(x)$.

Since $r = -6$, 1 is not a zero of $P(x)$.

```
 2 | 2   1   -9   -4    4
   |     4   10    2   -4
   -----------------------
     2   5    1   -2    0
```
Since 1 is not a zero, test 2, the next possible integral zero.

Since $r = 0$, 2 is a zero of $P(x)$.

$P(x) = (2x^3 + 5x^2 + x - 2)(x - 2)$

The depressed factor is $Q_1(x) = 2x^3 + 5x^2 + x - 2$. The possible rational zeros of this depressed factor are ± 1, ± 2, and $\pm \frac{1}{2}$. Thus only these possibilities from the original list need to be considered now.

```
 2 | 2   5    1   -2
   |     4   18   38
   ------------------
     2   9   19   36
```
Use $Q_1(x)$ to find the next zero. Is 2 a multiple zero?

Since $r = 36$, 2 is not a multiple zero.

```
-1 | 2   5    1   -2
   |    -2   -3    2
   ------------------
     2   3   -2    0
```
Now test -1.

Since $r = 0$, -1 is a zero of $P(x)$.

$P(x) = (2x^2 + 3x - 2)(x + 1)(x - 2)$ The depressed factor is $Q_2(x) = 2x^2 + 3x - 2$.

3 $P(x) = (2x - 1)(x + 2)(x + 1)(x - 2)$ Factor $2x^2 + 3x - 2$. Set each factor of $P(x)$ equal to zero.

Answer $P(x) = (2x - 1)(x + 2)(x + 1)(x - 2)$; the zeros of $P(x)$ are $\frac{1}{2}$, -2, -1, and 2. □

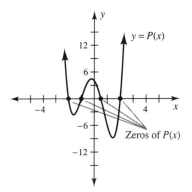

Figure 4-2 Graph of $y = 2x^4 + x^3 - 9x^2 - 4x + 4$

An alternate approach to Example 3 is to use a graphics calculator to produce the graph of $y = 2x^4 + x^3 - 9x^2 - 4x + 4$ shown to the right and then use the estimated x-intercepts as an aid to producing the factored form $y = (2x - 1)(x + 2)(x + 1)(x - 2)$.

EXAMPLE 4 Factor $P(x) = 30x^3 - 41x^2 + 17x - 2$ over the integers, and find the x-intercepts of the graph of this function.

SOLUTION Possible numerators are ± 1 and ± 2. Possible denominators are ± 1, ± 2, ± 3, ± 5, ± 6, ± 10, ± 15, and ± 30. Possible rational zeros are ± 1, ± 2, $\pm \frac{1}{2}$, $\pm \frac{1}{3}$, $\pm \frac{2}{3}$, $\pm \frac{1}{5}$, $\pm \frac{2}{5}$, $\pm \frac{1}{6}$, $\pm \frac{1}{10}$, $\pm \frac{1}{15}$, $\pm \frac{2}{15}$, and $\pm \frac{1}{30}$.

$$\underline{1|}\;\begin{array}{rrrr} 30 & -41 & 17 & -2 \\ & 30 & -11 & 6 \\ \hline 30 & -11 & 6 & \boxed{4} \end{array}$$

Start the test for a zero by examining the positive integers.

Since $r = 4$, 1 is not a zero of $P(x)$.

$$\underline{2|}\;\begin{array}{rrrr} 30 & -41 & 17 & -2 \\ & 60 & 38 & 110 \\ \hline 30 & 19 & 55 & \boxed{108} \end{array}$$

Since $r = 108$, 2 is not a zero of $P(x)$.

$$\underline{\tfrac{1}{2}|}\;\begin{array}{rrrr} 30 & -41 & 17 & -2 \\ & 15 & -13 & 2 \\ \hline 30 & -26 & 4 & 0 \end{array}$$

Now start testing the fractions.

Since $r = 0$, $\frac{1}{2}$ is a zero of $P(x)$.

$P(x) = (x - \tfrac{1}{2})(30x^2 - 26x + 4)$ Express $P(x)$ in factored form.

$ = (x - \tfrac{1}{2})(\boxed{2})(15x^2 - 13x + 2)$ Pull a factor of 2 out of the trinomial factor and multiply it times $\left(x - \dfrac{1}{2}\right)$, thus obtaining integral coefficients.

$ = (\boxed{2x - 1})(15x^2 - 13x + 2)$ Complete the factorization by factoring the trinomial by trial and error.

$ = (2x - 1)(5x - 1)(3x - 2)$

The x-intercepts are

$\left(\dfrac{1}{5}, 0\right)$, $\left(\dfrac{1}{2}, 0\right)$, and $\left(\dfrac{2}{3}, 0\right)$ The zeros of the function are the x-intercepts of the graph of $y = P(x)$. □

EXAMPLE 5 Find all zeros of $P(x) = x^3 - 4x + 3$.

SOLUTION Possible numerators are ± 1 and ± 3. Possible denominators are ± 1. Possible rational zeros are -3, -1, 1, and 3.

$$\underline{1|}\;\begin{array}{rrrr} 1 & 0 & -4 & 3 \\ & 1 & 1 & -3 \\ \hline 1 & 1 & -3 & \boxed{0} \end{array}$$

Start by testing 1.

Since $r = 0$, 1, is a zero of $P(x)$.

$P(x) = (x - 1)(x^2 + x - 3)$

$x = \dfrac{-1 \pm \sqrt{1 - 4(1)(-3)}}{2}$ The depressed factor, $x^2 + x - 3$, does not factor over the integers, so the quadratic formula is used to determine its zeros.

$ = -\dfrac{1}{2} \pm \dfrac{\sqrt{13}}{2}$

SECTION 4-4 RATIONAL ZEROS OF POLYNOMIALS

Answer The zeros of $P(x)$ are 1, $-\dfrac{1}{2} + \dfrac{\sqrt{13}}{2} \approx 1.3$, and $-\dfrac{1}{2} - \dfrac{\sqrt{13}}{2} \approx -2.3$. □

In Example 5 all three zeros of $P(x) = x^3 - 4x + 3$ are real numbers. However, only one of these real zeros is among the possible rational zeros listed. The other two zeros are irrational numbers. The graph of $y = x^3 - 4x + 3$ shown below could serve as an aid to approximating these zeros, but just by inspecting this graph we could not determine the exact values of the zeros $-\dfrac{1}{2} \pm \dfrac{\sqrt{13}}{2}$.

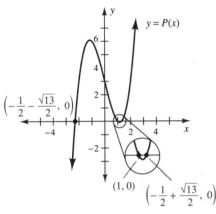

Figure 4-3 Graph of $y = x^3 - 4x + 3$

Inequalities involving polynomials can be solved using the procedure introduced in Section 2-7. We can also approximate the solution of inequalities like $P(x) > 0$ directly from the graph of $y = P(x)$ as we illustrated in Section 4-1. Recall that a polynomial function crosses the x-axis at x-intercepts of odd multiplicity and is tangent to the x-axis at x-intercepts of even multiplicity. This information is restated here in terms of the zeros of $P(x)$.

Sign Changes of a Polynomial $P(x)$ ▼

> $P(x)$ has opposite signs on both sides of zeros of odd multiplicity.
> $P(x)$ has the same sign on both sides of zeros of even multiplicity.

This information can be used to solve polynomial inequalities rapidly, as illustrated in the next example.

EXAMPLE 6 Solve $(x + 2)^3(x - 1)(x - 3)^2 > 0$.

SOLUTION Let $P(x) = (x + 2)^3(x - 1)(x - 3)^2$.

The zeros of $P(x)$ are $-2, 1$, and 3.

The zeros of $P(x)$ are the critical values for this inequality. These critical values can be determined by inspection, since $P(x)$ is in factored form.

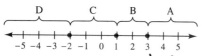

Plot these zeros on a number line. ↳ leave these zeros open circles

A: $P(x) > 0$ in the interval $(3, +\infty)$

Determine the sign of $P(x)$ in each interval formed by these zeros, starting with the interval to the far right, interval A. All factors of $P(x)$ are positive in this interval.

B: $P(x) > 0$ in the interval $(1, 3)$

3 is a zero of even multiplicity; thus the sign of $P(x)$ does not change.

C: $P(x) < 0$ in the interval $(-2, 1)$

1 is a zero of odd multiplicity, thus the sign of $P(x)$ changes.

D: $P(x) > 0$ in the interval $(-\infty, -2)$

-2 is a zero of odd multiplicity; thus the sign of $P(x)$ changes.

Graph the solution set for $P(x) > 0$.

Answer $(-\infty, -2) \cup (1, 3) \cup (3, +\infty)$

Note that 3 is not a solution of this inequality, since $P(3) = 0$. □

The strategy used in the previous example uses the factored form of the polynomial. To solve polynomial inequalities that are not in factored form, the first step is to factor the polynomial. This is illustrated in the next example, which uses the rational zeros theorem to assist in factoring the given polynomial.

▶ **Self-Check** ▼

Find all zeros of

$P(x) = 6x^4 - 7x^3 - 10x^2 + 17x - 6$

EXAMPLE 7 Solve the inequality $2x^3 + 5x^2 + x - 3 < 0$. over the real numbers

SOLUTION First factor $P(x) = 2x^3 + 5x^2 + x - 3$.

Possible rational zeros of $P(x)$: $\pm 1, \pm 3, \pm\frac{1}{2}, \pm\frac{3}{2}$

$$\begin{array}{r|rrrr} -\frac{3}{2} & 2 & 5 & 1 & -3 \\ & & -3 & -3 & 3 \\ \hline & 2 & 2 & -2 & 0 \end{array}$$

One of the possible rational zeros, $-\frac{3}{2}$, is checked using synthetic division.

$P(x) = \left(x + \dfrac{3}{2}\right)(2x^2 + 2x - 2)$

The depressed factor is $2x^2 + 2x - 2$.

▶ **Self-Check Answer** ▼

$-\frac{3}{2}, \frac{2}{3}, 1$ (multiplicity two)

SECTION 4-4 RATIONAL ZEROS OF POLYNOMIALS

$$P(x) = \left(x + \frac{3}{2}\right)(2)(x^2 + x - 1)$$

Factor 2 out of the trinomial and then multiply this constant times the binomial factor.

$$P(x) = (2x + 3)(x^2 + x - 1)$$

$$x = \frac{-1 \pm \sqrt{1^2 - 4(-1)}}{2}$$

The zeros of $x^2 + x - 1$ are determined by using the quadratic formula to solve $x^2 + x - 1 = 0$.

$$x = \frac{-1 \pm \sqrt{5}}{2}$$

$$P(x) = (2x + 3)\left(x + \frac{1}{2} + \frac{\sqrt{5}}{2}\right)\left(x + \frac{1}{2} - \frac{\sqrt{5}}{2}\right)$$

Note that each of these factors is of odd multiplicity.

has zeros at $-\frac{1}{2} - \frac{\sqrt{5}}{2} \approx -1.6$, $-\frac{3}{2} = -1.5$, and $-\frac{1}{2} + \frac{\sqrt{5}}{2} \approx 0.6$

$P(x) > 0$ in the interval $\left(-\frac{1}{2} + \frac{\sqrt{5}}{2}, \infty\right)$

All three factors of $P(x)$ are positive for x greater than $-\frac{1}{2} + \frac{\sqrt{5}}{2}$.

$P(x) < 0$ in the interval $\left(-1.5, -\frac{1}{2} + \frac{\sqrt{5}}{2}\right)$

$P(x)$ changes sign at zeros of odd multiplicity.

$P(x) > 0$ in the interval $\left(-\frac{1}{2} - \frac{\sqrt{5}}{2}, -1.5\right)$

$P(x)$ changes sign at zeros of odd multiplicity.

$P(x) < 0$ in the interval $\left(-\infty, -\frac{1}{2} - \frac{\sqrt{5}}{2}\right)$

$P(x)$ changes sign at zeros of odd multiplicity.

Answer $\left(-\infty, -\frac{1}{2} - \frac{\sqrt{5}}{2}\right) \cup \left(-1.5, -\frac{1}{2} + \frac{\sqrt{5}}{2}\right)$ ☐

EXERCISES 4-4

A

In Exercises 1–6, use the rational zeros theorem to list all possible rational zeros of each polynomial.

1 $P(x) = x^3 - 2x^2 - 5x + 6$
2 $P(x) = x^3 - 4x^2 + 5x - 8$
3 $P(w) = 2w^4 - w + 10$
4 $P(w) = 3w^5 - w^2 + 15$
5 $P(v) = 6v^4 - v^3 + v^2 + 5v - 20$
6 $P(v) = 15v^4 - 10v^3 + 12v^2 - v + 45$

In Exercises 7–10, use the rational zeros theorem to select the number that *cannot* possibly be a zero of the given polynomial $P(x)$.

7 $P(x) = 2x^4 - 9x^3 + 7x^2 + 9x - 9$
   a. 1   **b.** 2   **c.** 3   **d.** $\frac{3}{2}$

8 $P(x) = 9x^4 - 27x^3 - 7x^2 + 23x - 6$
   a. $-\frac{1}{2}$ **b.** $\frac{1}{3}$ **c.** $\frac{2}{3}$ **d.** 3

9 $P(x) = 16x^3 - 40x^2 + 13x + 6$
   a. 2   **b.** $-\frac{1}{4}$ **c.** $\frac{3}{4}$ **d.** $\frac{2}{3}$

10 $P(x) = 6x^3 - 31x^2 + 23x + 20$
   a. $-\frac{1}{2}$ **b.** $\frac{5}{3}$ **c.** 6   **d.** 4

In Exercises 11–14, (a) use the rational zeros theorem to list all possible rational zeros of the given polynomial $P(x)$; (b) use synthetic division to show that the given number is a zero of $P(x)$ and to determine the depressed factor of $P(x)$; (c) then use the rational zeros theorem on this depressed factor to shorten the list of possibilities from part (a).

11. $P(x) = 4x^3 - 48x^2 - x + 12,\ a = 12$
12. $P(x) = 9x^3 - 90x^2 - x + 10,\ a = 10$
13. $P(x) = 2x^3 - 31x^2 + 14x + 15,\ a = 15$
14. $P(x) = 3x^3 - 19x^2 - 38x - 16,\ a = 8$

In Exercises 15–24, find each zero of the polynomial function.

15. $P(x) = x^3 - 2x^2 - 4x + 8$
16. $P(x) = x^4 - x^3 - x^2 + x$
17. $P(t) = t^3 + 2t^2 - 9t - 18$
18. $P(t) = t^3 - t^2 - 14t + 24$
19. $P(y) = 12y^3 - 4y^2 - 3y + 1$
20. $P(y) = 36y^3 + 9y^2 - 4y - 1$
21. $P(z) = z^3 + 3z^2 + z - 2$
22. $P(z) = z^3 + 3z^2 - 2z - 4$
23. $P(v) = 3v^4 - 14v^3 + 11v^2 + 16v - 12$
24. $P(v) = 3v^4 - 2v^3 - 17v^2 + 8v + 20$

In Exercises 25–28, determine the sign of $P(x)$ in each interval formed by its zeros.

25. $P(x) = (x - 2)^5(x + 5)^6$
    a. $(2, +\infty)$ b. $(-5, 2)$ c. $(-\infty, -5)$
26. $P(x) = (x + 1)^8(x + 3)^9$
    a. $(-1, +\infty)$ b. $(-3, -1)$ c. $(-\infty, -3)$
27. $P(x) = (x - 3)^2(x - 6)^3(x - 8)^4$
    a. $(8, +\infty)$ b. $(6, 8)$ c. $(3, 6)$ d. $(-\infty, 3)$
28. $P(x) = (x - 5)^5(x + 3)^6(x - 9)^7$
    a. $(9, +\infty)$ b. $(5, 9)$ c. $(-3, 5)$ d. $(-\infty, -3)$

In Exercises 29–36, solve each inequality.

29. $(x + 1)(x - 2)(x - 4) > 0$
30. $x^2(x + 1)(x - 2) \geq 0$
31. $(x + 3)^2(x - 1)(x + 1)^2 > 0$
32. $x^3(x - 3)^2(x + 3)^3 < 0$
33. $x^3 + 2x^2 - 9x - 18 < 0$
34. $x^3 - x^2 - 14x + 24 \leq 0$
35. $12x^3 - 4x^2 - 3x + 1 \geq 0$
36. $36x^3 + 9x^2 - 4x - 1 > 0$

B

In Exercises 37 and 38, use the graphs of $y = P(x)$ given below as aids to selecting the exact rational zeros of $P(x)$.

37. $P(x) = 24x^3 - 22x^2 - 41x + 30$
38. $P(x) = 40x^3 - 22x^2 - 83x + 60$

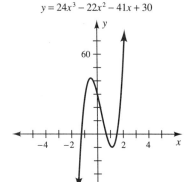

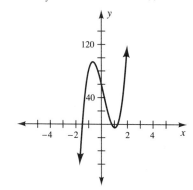

SECTION 4-5 (OPTIONAL) SHORTCUTS FOR FINDING ZEROS

In Exercises 39 and 40, find the exact x-intercepts of the graph of $y = P(x)$ and then graph this polynomial function.

39 $P(x) = 2x^3 + 3x^2 - 4x - 6$

40 $P(x) = 3x^3 - 2x^2 - 9x + 6$

41 Use $P(x) = x^2 - 2$ to show that $\sqrt{2}$ is irrational.

42 Use $P(x) = x^3 - 27$ to find all three cube roots of 27.

C

In Exercises 43–46, find all zeros of each polynomial.

43 $P(x) = 3x^4 - 7x^3 - 6x^2 + 12x + 8$

44 $P(u) = 3u^4 + 8u^3 - 16u^2 - 44u - 15$

45 $P(x) = 2x^5 + 9x^4 - x^3 - 42x^2 - 28x + 24$

46 $P(z) = 2z^5 - 13z^4 + 15z^3 - 3z^2 + 13z + 10$

47 Show that $P(x) = x^4 + x^3 + 3x^2 + 2x + 2$ has no rational zeros.

48 Find the dimensions of a box if the width is 1 centimeter longer than the length, the height exceeds the length by 2 centimeters, and the volume is 6 cm^3.

49 A slice 1 centimeter wide is cut off the end of a cube, leaving a solid of volume 48 cm^3. Find the length of each side of the cube.

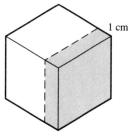

Figure for Exercise 49

Use a graphics calculator to graph the polynomial functions in Exercises 50 and 51. Then use the graph to answer the question.

50 Graph $y = 2x^3 + 3x^2 - 29x - 60$: For which interval(s) is $2x^3 + 3x^2 - 29x - 60 > 0$?

51 Graph $y = 2x^3 - 7x^2 - 8x + 28$. For which interval(s) is $2x^3 - 7x^2 - 8x + 28 > 0$?

SECTION 4-5*

Shortcuts for Finding Zeros

Section Objectives

6 Use the location theorem to isolate a real zero between two numbers.

7 Apply Descartes's rule of signs to find the number of possible positive and negative zeros of a polynomial.

8 Find upper and lower bounds for the real zeros of a polynomial.

9 Combine the rational zeros theorem, bounds on zeros, Descartes's rule of signs, and the location theorem to find zeros of a polynomial.

A polynomial may have many possible rational zeros. This section will discuss three shortcuts that can be used to eliminate some of these possible rational zeros. The first shortcut examined will be the location theorem. One

---

* This is an optional section.

way to use this theorem is to proceed as usual, checking possible rational zeros of $P(x)$, but taking advantage of information that can pinpoint a zero between two other real numbers. The logic of the location theorem is outlined below.

The graph of a polynomial function $y = P(x)$ is one continuous curve. If $P(a)$ and $P(b)$ are opposite in sign, then the points $(a, P(a))$ and $(b, P(b))$ are on opposite sides of the x-axis. The curve must cross the x-axis at some x-intercept $(c, 0)$ in order to connect the points $(a, P(a))$ and $(b, P(b))$. Thus $P(c) = 0$, and c is a zero of $P(x)$ between a and b. This is illustrated in Figure 4-4.

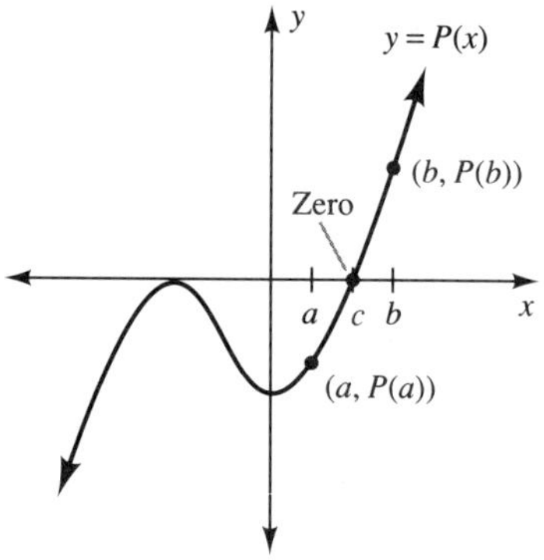

Figure 4-4 A real zero between a and b

Location Theorem ▼

If $P(x)$ is a polynomial and a and b are real numbers such that $P(a)$ and $P(b)$ are opposite in sign, then there is a real number c between a and b such that c is a zero of $P(x)$.

EXAMPLE 1 Use the location theorem to show that the polynomial

$$P(x) = 4x^3 - 4x^2 - 5x + 3$$

has a zero between 1 and 2.

SOLUTION $P(x) = 4x^3 - 4x^2 - 5x + 3$

$P(1) = 4 - 4 - 5 + 3 = -2$

The polynomial $P(x)$ can be evaluated at $x = a$ by either substituting a into the polynomial or by using synthetic division. For $a = 1$, substitution is quicker.

$$\underline{2|4 -4 -5 3}$$
$$\phantom{2|4}886$$
$$\phantom{2|}443\boxed{9}$$

Synthetic division and the remainder theorem are used to calculate $P(2)$.

$P(2) = 9$

Since $P(1) < 0$ and $P(2) > 0$, there is a zero of $P(x)$ between 1 and 2.

$$\underline{\tfrac{3}{2}|4 -4 -5 3}$$
$$\phantom{\tfrac{3}{2}|4}63-3$$
$$\phantom{\tfrac{3}{2}|}42-20$$

The possible rational zeros of $P(x)$ are: ± 1, ± 3, $\pm \frac{1}{2}$, $\pm \frac{3}{2}$, $\pm \frac{1}{4}$, and $\pm \frac{3}{4}$. Since $\frac{3}{2}$ is the only possible rational zero between 1 and 2, it is tested and is shown to be a zero of $P(x)$.

The real zero indicated between 1 and 2 is the rational number $\frac{3}{2}$. □

A second shortcut that can sometimes be used to reduce the number of possisble zeros that must be examined is Descartes's rule of signs. One advantage of this shortcut is that it can signal the end of the search for either the positive or the negative zeros of a polynomial.

SECTION 4-5 (OPTIONAL) SHORTCUTS FOR FINDING ZEROS

The number of possible positive and negative real zeros can be determined by calculating the number of variations of sign of $P(x)$ and of $P(-x)$. A polynomial written in standard form has a variation in sign when the signs of two consecutive nonzero terms differ. For example,

$$-3x^5 + 2x^4 + 7x^3 - 8x - 4$$

has two variations in sign.

Descartes's Rule of Signs

Assume $P(x)$ is a polynomial of degree greater than or equal to 1:
1. If $P(x)$ has k variations in sign, then the number of positive real zeros of $P(x)$ is either k or $k - 2m$, where m is a positive integer.
2. If $P(-x)$ has j variations in sign, then the number of negative real zeros is either j or $j - 2n$, where n is a positive integer.

▶ Self-Check

Use the location theorem to show that

$$P(x) = 2x^4 + x^3 - 11x^2 - 5x + 5$$

has a zero between 0 and 1.

Both m and n are positive integers, so $2m$ and $2n$ are even integers. Therefore, if k and j are even (odd) integers, $k - 2m$ and $j - 2n$ are also even (odd) integers.

EXAMPLE 2 Use Descartes's rule of signs to examine the number of positive and negative zeros of these polynomials.

SOLUTIONS

(a) $P(x) = x^3 - 2x + 1$

Two changes in sign indicate two or no positive zeros.

$P(-x) = -x^3 + 2x + 1$

One change in sign indicates one negative zero.

(b) $P(x) = x^7 - 4x^6 + 2x^4 + 11x^3 + 10x^2 - 12x - 8$

Three changes in sign indicate three or one positive zero(s).

$P(-x) = -x^7 - 4x^6 + 2x^4 - 11x^3 + 10x^2 + 12x - 8$

Four changes in sign indicate four, two, or no negative zeros.

▶ Self-Check Answer

$P(0) = 5 > 0$ and $P(1) = -8 < 0$

In the next example Descartes's rule of signs is used to eliminate all the positive real zeros, thus eliminating half the possibilities from the list of possible rational zeros.

EXAMPLE 3 Find all zeros of $P(x) = 6x^3 + 19x^2 + 19x + 6$.

SOLUTION The possible rational zeros of $P(x)$ are ± 1, ± 2, ± 3, ± 6, $\pm\frac{1}{2}$, $\pm\frac{3}{2}$, $\pm\frac{1}{3}$, $\pm\frac{2}{3}$, and $\pm\frac{1}{6}$.

$P(x) = 6x^3 + 19x^2 + 19x + 6$ has no sign changes.

$P(-x) = -6x^3 + 19x^2 - 19x + 6$ has three sign changes.

Descartes's rule of signs enables us to quickly eliminate all the positive numbers from the list of rational possibilities.

Thus $P(x)$ has no positive real zeros and either three or one negative real zeros. Since the real zeros of $P(x)$ must be negative, we will start testing the negative integral possibilities.

$$\begin{array}{r|rrrr} -1 & 6 & 19 & 19 & 6 \\ & & -6 & -13 & -6 \\ \hline & 6 & 13 & 6 & \boxed{0} \end{array}$$

Start by testing -1.

Since $r = 0$, $(x + 1)$ is a factor of $P(x)$.

$P(x) = (x + 1)(6x^2 + 13x + 6)$ Factor the depressed factor $6x^2 + 13x + 6$.
$ = (x + 1)(3x + 2)(2x + 3)$

Thus, -1, $-\frac{2}{3}$, and $-\frac{3}{2}$ are the zeros of $P(x)$. □

The third shortcut uses the upper and lower bounds theorem. This theorem can sometimes significantly reduce the number of possible zeros of a polynomial. It is important to keep in mind as you read this material that you would usually not assume as your goal the establishment of bounds on the zeros of $P(x)$. Rather, you would begin your search for zeros as normal and then take advantage of any bounds or other shortcuts that you notice as you perform your usual work. We will begin by defining upper and lower bounds and illustrating these definitions with some examples.

Bounds ▼

The real number b is an **upper bound (lower bound)** on the real zeros of a polynomial $P(x)$ if b is greater (less) than or equal to each real zero of $P(x)$.

Suppose $P(x) = x^3 + 3x^2 - 4x - 12 = (x + 3)(x + 2)(x - 2)$; then -3, -2, and 2 are zeros of $P(x)$. Bounds on these zeros are illustrated in Figure 4-5.

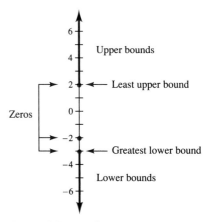

Figure 4-5 Bounds on zeros

- 5 is an upper bound on the zeros of $P(x)$, since 5 is greater than or equal to -3, -2, and 2.
- Similarly, 2, π, 2.5, or any other number greater than or equal to 2 is an upper bound on the zeros of $P(x)$. Since 2 is the smallest number that is an upper bound on the zeros, it is called the least upper bound.
- -5 is a lower bound on the zeros of $P(x)$, since -5 is less than or equal to -3, -2, and 2.
- Similarly, -3, $-\sqrt{11}$, -8.7, or any other number less than or equal to -3 is a lower bound on the zeros of $P(x)$. Since -3 is the largest number that is a lower bound on the zeros, it is called the greatest lower bound.

EXAMPLE 4 List all zeros of $P(x) = (x + 5)(x - 2)(x^2 - 6)$ and determine the least upper bound and the greatest lower bound of the real zeros of $P(x)$.

SOLUTION The zeros of $P(x)$ are -5, $-\sqrt{6}$, 2, and $\sqrt{6}$. Thus the least upper bound is $\sqrt{6}$, and the greatest lower bound is -5. □

The factor $x^2 - 6$ yields the irrational zeros $\pm\sqrt{6}$.

The next theorem describes a criterion for determining upper and lower bounds of the real zeros even if these zeros are unknown.

Upper and Lower Bounds Theorem ▼

If a real polynomial

$$P(x) = a_n x^n + a_{n-1} x^{n-1} + \cdots + a_1 x + a_0, \quad a_n > 0$$

is divided by $x - c$, where c is a real number.

1. If $c \geq 0$ and the numbers in the last row of the synthetic division are nonnegative, then c is an upper bound on the real zeros of $P(x)$.
2. If $c \leq 0$ and the numbers in the last row of the synthetic division are alternating nonnegative and nonpositive numbers, then c is a lower bound on the real zeros of $P(x)$.

Proof of Part 1 $P(x) = Q(x)(x - c) + P(c)$ by the division algorithm. The nonnegative numbers in the last row of the synthetic division represent the coefficients of the quotient polynomial $Q(x)$ and the remainder $P(c)$. Therefore $P(c) \geq 0$ and the coefficients of $Q(x)$ are nonnegative. This means that $Q(b) > 0$ for $c \geq 0$ and $b > c$. Thus, $P(b) = Q(b)(b - c) + P(c)$ is greater than zero for $b > c$, and b cannot be a zero of $P(x)$. Hence c is an upper bound of the zeros of $P(x)$.

The proof of part 2 is similar, but more complex. It involves the absolute values of $Q(b)$, $b - c$, and $P(c)$, if $b < c$.

EXAMPLE 5 Verify that 3 is an upper bound and that -6 is a lower bound on the real zeros of $P(x) = 6x^3 + 17x^2 - 59x + 30$.

SOLUTION

$$\begin{array}{r|rrrr} 3 & 6 & 17 & -59 & 30 \\ & & 18 & 105 & 138 \\ \hline & 6 & 35 & 46 & 168 \end{array}$$

Since $3 \geq 0$ and the numbers in the last row of this synthetic division are nonnegative, 3 is an upper bound on the real zeros of $P(x)$.

$$\begin{array}{r|rrrr} -6 & 6 & 17 & -59 & 30 \\ & & -36 & 114 & -330 \\ \hline & 6 & -19 & 55 & -300 \end{array}$$

Since $-6 \leq 0$ and the numbers in the last row of this synthetic division are alternating nonnegative and nonpositive, -6 is a lower bound on the real zeros of $P(x)$.

Using the knowledge of the upper and lower bounds established in Example 5, we can reduce the list of 36 possible rational zeros of the given polynomial to 23 possibilities, as shown in Example 6.

EXAMPLE 6 Given the polynomial $P(x) = 6x^3 + 17x^2 - 59x + 30$, list all possible rational zeros of $P(x)$ and then use the information in Example 5 to reduce this list by eliminating values.

SOLUTION Possible numerators are $\pm 1, \pm 2, \pm 3, \pm 5, \pm 6, \pm 10, \pm 15$, and ± 30. Possible denominators are $\pm 1, \pm 2, \pm 3,$ and ± 6. Possible rational roots (thirty-six possibilities) are $-30, -15, -10, -\frac{15}{2}, -6, -5, -\frac{10}{3}, -3,$ $-\frac{5}{2}, -2, -\frac{5}{3}, -\frac{3}{2}, -1, -\frac{5}{6}, -\frac{2}{3}, -\frac{1}{2}, -\frac{1}{3}, -\frac{1}{6}, \frac{1}{6}, \frac{1}{3}, \frac{1}{2}, \frac{2}{3}, \frac{5}{6}, 1, \frac{3}{2}, \frac{5}{3}, 2, \frac{5}{2}, 3,$ $\frac{10}{3}, 5, 6, \frac{15}{2}, 10, 15,$ and 30.
The reduced list (23 possibilities) is $-5, -\frac{10}{3}, -3, -\frac{5}{2}, -2, -\frac{5}{3}, -\frac{3}{2}, -1,$ $-\frac{5}{6}, -\frac{2}{3}, -\frac{1}{2}, -\frac{1}{3}, -\frac{1}{6}, \frac{1}{6}, \frac{1}{3}, \frac{1}{2}, \frac{2}{3}, \frac{5}{6}, 1, \frac{3}{2}, \frac{5}{3}, 2,$ and $\frac{5}{2}$.

▶ **Self-Check** ▼

Examine the number of positive and negative zeros of each polynomial.

1 $P(x) = 3x^3 - 2x^2 - x + 1$
2 $P(x) = 2x^3 - x - 2$

All the values that are in color can be eliminated from the list because they are less than the lower bound -6, or greater than the upper bound 3. The values -6 and 3 can also be eliminated because the work in Example 5 showed that these values are not zeros.

The three shortcuts just covered in this section are not used in every problem, but when they are applicable they can significantly reduce the effort required to find the zeros of a polynomial. The following box describes how the procedure for determining rational zeros can be amended to include these shortcuts.

▶ **Self-Check Answers** ▼

1 Two or no positive zeros and one negative zero
2 One positive zero and two or no negative zeros

SECTION 4-5 (OPTIONAL) SHORTCUTS FOR FINDING ZEROS

Determining Rational Zeros of a Polynomial Function ▼

To determine rational zeros of a polynomial with integer coefficients:

Step 1 List all possible rational zeros (rational zeros theorem).

Step 2 Examine the number of possible positive and negative zeros (Descartes's rule of signs).

Step 3 Test the possible integer zeros listed in Step 1 for
  a. zeros and multiple zeros (after each zero is found, use the depressed factor to delete values from the original list and to test the other possible zeros).
  b. a change in sign of remainders that indicates a real zero between the integers (the location theorem).
  c. an upper bound indicated by all nonnegative numbers on the last row of the synthetic division, and a lower bound indicated by alternating nonpositive and nonnegative numbers on the last row of the synthetic division (upper and lower bounds theorem).

Step 4 Test the possible rational zeros in the intervals indicated by the location theorem in Step 3b.

Step 5 If the depressed polynomial has not been reduced to degree two, test for fractional zeros that have not been eliminated by the previous steps.

EXAMPLE 7 Find all zeros of $P(x) = 3x^5 + 10x^4 - 2x^3 - 28x^2 - 8x + 16$.

SOLUTION

[1] Possible rational zeros are $-16, -8, -\frac{16}{3}, -4, -\frac{8}{3}, -2, -\frac{4}{3}, -1, -\frac{2}{3}, -\frac{1}{3}, \frac{1}{3}, \frac{2}{3}, 1, \frac{4}{3}, 2, \frac{8}{3}, 4, \frac{16}{3}, 8,$ and 16.

[2] $P(x) = 3x^5 + 10x^4 - 2x^3 - 28x^2 - 8x + 16$ Two sign changes indicate two or no positive zeros.

$P(-x) = -3x^5 + 10x^4 + 2x^3 - 28x^2 + 8x + 16$ Three sign changes indicate three or one negative zero(s).

[3] $P(0) = 16$ $P(0) > 0$ and $P(1) < 0$ indicate a zero between 0 and 1.

$P(1) = 3 + 10 - 2 - 28 - 8 + 16 = -9$

```
2 | 3   10   -2   -28   -8    16
        6    32    60    64   112
    ─────────────────────────────
    3   16   30    32    56   128
```

$P(1) < 0$ and $P(2) > 0$ indicates a zero between 1 and 2.

2 is an upper bound because the numbers in the bottom row are positive.

The two positive zeros are not integers since they are between 0 and 1 and between 1 and 2, respectively, so we will now investigate the possible negative integer zeros.

$P(-1) = -3 + 10 + 2 - 28 + 8 + 16 = 5$ -1 is not a zero.

$$\underline{-2\,|}\begin{array}{rrrrrr} 3 & 10 & -2 & -28 & -8 & 16 \\ & -6 & -8 & 20 & 16 & -16 \\ \hline 3 & 4 & -10 & -8 & 8 & 0 \end{array}$$

-2 is a zero.

$$\underline{-2\,|}\begin{array}{rrrrr} 3 & 4 & -10 & -8 & 8 \\ & -6 & 4 & 12 & -8 \\ \hline 3 & -2 & -6 & 4 & 0 \end{array}$$

Use the depressed polynomial to test for a multiple zero.

-2 is a zero of multiplicity 2.

Any other zeros of $P(x)$ must be zeros of the depressed factor

$3x^3 - 2x^2 - 6x + 4$

Descartes's rule of signs predicts three or one negative zero(s). We have already found two negative zeros. Therefore there must be another negative zero.

$$\underline{-2\,|}\begin{array}{rrrr} 3 & -2 & -6 & 4 \\ & -6 & 16 & -20 \\ \hline 3 & -8 & 10 & -16 \end{array}$$

-2 is not a zero of multiplicity three; however, -2 is a lower bound on the zeros of the depressed factor, because the numbers in the bottom row alternate signs.

4 We have found all integer zeros, so we will now examine the fractional values. The only possible rational zero between 1 and 2 is $\frac{4}{3}$.

$$\underline{\tfrac{4}{3}\,|}\begin{array}{rrrr} 3 & -2 & -6 & 4 \\ & 4 & \tfrac{8}{3} & -\tfrac{40}{9} \\ \hline 3 & 2 & -\tfrac{10}{3} & -\tfrac{4}{9} \end{array}$$

$\frac{4}{3}$ is not a zero. Thus the zero between 1 and 2 must be an irrational number.

The only possible rational zeros between 0 and 1 are $\frac{1}{3}$ and $\frac{2}{3}$.

$$\underline{\tfrac{1}{3}\,|}\begin{array}{rrrr} 3 & -2 & -6 & 4 \\ & 1 & -\tfrac{1}{3} & -\tfrac{19}{9} \\ \hline 3 & -1 & -\tfrac{19}{3} & \tfrac{17}{9} \end{array}$$

$\frac{1}{3}$ is not a zero.

$$\underline{\tfrac{2}{3}\,|}\begin{array}{rrrr} 3 & -2 & -6 & 4 \\ & 2 & 0 & -4 \\ \hline 3 & 0 & -6 & 0 \end{array}$$

$\frac{2}{3}$ is a zero.

5 $3x^2 - 6 = 3(x^2 - 2)$ Factor the depressed polynomial of degree two.

$ = 3(x - \sqrt{2})(x + \sqrt{2})$ $\sqrt{2}$ is the zero indicated between 1 and 2.

Answer The zeros of $P(x)$ are -2 (multiplicity 2), $\frac{2}{3}$, $-\sqrt{2}$, and $\sqrt{2}$. There are two positive zeros and three negative zeros, as suggested by Descartes's rule of signs. □

SECTION 4-5 (OPTIONAL) SHORTCUTS FOR FINDING ZEROS

All three of the shortcuts given in this section relate to information about the real zeros of a polynomial $P(x)$. Thus one should be careful when examining the list of possible rational roots to remember that the real roots can also be irrational. In addition, a real polynomial can also include imaginary zeros that will occur in conjugate pairs.

EXERCISES 4-5

A

In Exercises 1 and 2, use the location theorem to show that the polynomial $P(x)$ has a zero between the numbers a and b.

1. **a.** $P(x) = 6x^3 + 31x^2 + 3x - 10$, $a = 0$, $b = 1$ **b.** $P(x) = 2x^3 + 5x^2 - 2x - 5$, $a = -2$, $b = -3$
2. **a.** $P(x) = 9x^3 + 15x^2 - 8x - 4$, $a = 0$, $b = 1$ **b.** $P(x) = 3x^3 + 7x^2 - 28x - 60$, $a = -4$, $b = -3$

In Exercises 3 and 4, use Descartes's rule of signs to examine the number of possible positive and negative real zeros of each polynomial.

3. **a.** $P(x) = 5x^3 + 2x^2 + x - 5$ **b.** $P(y) = y^3 - 2y^2 - y + 5$
   c. $P(t) = 4t^3 + t^2 + 3t + 12$ **d.** $P(u) = 6u^3 - u^2 + 5u - 12$
4. **a.** $P(x) = x^3 - x^2 - x - 1$ **b.** $P(y) = y^3 + 4y^2 - 2y + 8$
   c. $P(t) = 5t^4 + 6t^3 + 8t^2 + 8t + 16$ **d.** $P(u) = 7u^4 - 5u^3 + 6u^2 - 7u + 8$

5. Use the upper and lower bounds theorem to show that 7 is an upper bound on the real zeros of $P(x) = x^4 - 2x^3 - 19x^2 + 14x + 84$.

6. Use the upper and lower bounds theorem to show that -7 is a lower bound on the real zeros of $P(x) = x^4 - 2x^3 - 19x^2 + 14x + 84$.

7. Use the upper and lower bounds theorem to show that -6 is a lower bound on the real zeros of $P(x) = x^4 + x^3 - 25x^2 - 19x + 144$.

8. Use the upper and lower bounds theorem to show that 6 is an upper bound on the real zeros of $P(x) = x^4 + x^3 - 25x^2 - 19x + 144$.

9. Given the polynomial $P(x) = 2x^3 - 39x^2 + 187x - 84$, answer each part of this exercise.
   a. List all possible rational zeros of $P(x)$ in order from smallest to largest.
   b. Use Descartes's rule of signs to eliminate one-half of the possible zeros listed in part a.
   c. Use the location theorem to verify that there is a zero of $P(x)$ between 0 and 1.
   d. Find the zero between 0 and 1 and then use the depressed factor to determine the other zeros.

10. Given the polynomial $P(x) = 3x^3 + 49x^2 + 196x + 60$, answer each part of the exercise.
    a. List all possible rational zeros of $P(x)$ in order from smallest to largest.
    b. Use Descartes's rule of signs to eliminate one-half of the possible zeros listed in part a.
    c. Use the location theorem to verify that there is a zero of $P(x)$ between -1 and 0.
    d. Find the zero between -1 and 0 and then use the depressed factor to determine the other zeros.

11 Given the polynomial $P(x) = x^3 + 6x^2 - 35x - 210$, answer each part of the exercise.
   a. List all possible rational zeros of $P(x)$ in order from smallest to largest.
   b. Test the positive possibilities in increasing order until an upper bound is determined.
   c. Explain why the work in parts (a) and (b) shows that there are no positive rational zeros.
   d. Find a rational zero of this polynomial and then use the depressed factor to determine the other zeros.

12 Given the polynomial $P(x) = x^3 - 10x^2 - 21x + 210$, answer each part of the exercise.
   a. List all possible rational zeros of $P(x)$ in order from smallest to largest.
   b. Test the negative possibilities in decreasing order until a lower bound is determined.
   c. Explain why the work in parts a and b shows that there are no negative rational zeros.
   d. Find a rational zero of this polynomial and then use the depressed factor to determine the other zeros.

13 Use the rational zeros theorem, the location theorem, and synthetic division to show that there is an irrational zero of $P(x) = x^4 - x^2 - 20$ between 2 and 3.

14 Use the rational zeros theorem, the location theorem, and synthetic division to show that there is an irrational zero of $P(x) = 2x^4 + x^3 - x^2 - 2x - 6$ between 1 and 2.

15 Use Descartes's rule of signs to show that $15x^5 + 13x^3 + 2x^2 + 7$ must have four imaginary roots.

16 Use Descartes's rule of signs to show that $x^6 + 3x^4 + 7x^2 + 2$ can have no real zeros.

For the polynomials in Exercises 17–20, locate each of the zeros between two consecutive integers.

17 $P(t) = 8t^3 - 20t^2 - 2t + 5$
18 $P(x) = 27x^3 - 72x^2 - 93x + 22$
19 $P(s) = 12s^3 - 68s^2 + 97s - 33$
20 $P(z) = 12z^3 + 88z^2 + 205z + 150$

21 Use the upper and lower bounds theorem to show, without graphing, that the graph of $y = x^4 - x^3 - 18x^2 + 52x - 40$ does not have an x-intercept for $x > 5$.

22 Use the upper and lower bounds theorem to show, without graphing, that the graph of $y = x^4 + 4x^3 - 48x^2 + 112x - 80$ does not have an x-intercept for $x < -10$.

B

23 Given the polynomial $P(x) = 8x^3 - 14x^2 - 7x + 15$, answer each part of the exercise.
   a. Determine $P(1)$ and $P(2)$.
   b. Show that $\frac{3}{2}$ and $\frac{5}{4}$ are zeros of $P(x)$.
   c. Explain why the location theorem does not detect the zeros between 1 and 2.

24 Show, without graphing, that the graph of $y = x^5 + 2x^3 + 5x$ has exactly one x-intercept. What is this x-intercept?

In Exercises 25–30, find all zeros of each polynomial. Also factor each polynomial and solve the equation generated by setting the polynomial equal to zero.

25 $P(t) = 2t^3 - 19t^2 + 57t - 54$
26 $P(t) = 3t^3 - 19t^2 + 36t - 20$
27 $P(x) = 9x^3 + 18x^2 - 4x - 8$
28 $P(x) = 9x^3 + 44x^2 + 49x - 6$
29 $P(t) = 3t^3 - t^2 - 15t + 5$
30 $P(t) = 3t^3 - 2t^2 - 27t + 18$

In Exercises 31–35, determine all zeros of each polynomial.

31 $P(u) = 8u^4 + 48u^3 + 80u^2 + 39u + 14$
32 $P(y) = 2y^4 + y^3 - 16y^2 - 3y + 30$
33 $P(x) = 8x^5 - 18x^4 - 3x^3 + 37x^2 - 33x + 9$
34 $P(t) = 6t^6 + 7t^5 - 82t^4 - 101t^3 + 114t^2 + 204t + 72$
35 $P(y) = 6y^{12} + 7y^{10} - 82y^8 - 101y^6 + 114y^4 + 204y^2 + 72$ (*Hint:* Let $t = y^2$.)

In Exercises 36–38, sketch the graph of each polynomial function.

36 $y = x^4 + x^3 - 6x^2$
37 $y = 2x^3 + 5x^2 - 6x - 15$
38 $y = 2x^5 - 3x^4 - 10x^3 + 15x^2$

SECTION 4-6*

Approximation of Zeros

Section Objective

10 Approximate the real zeros of a polynomial.

Because there is no formula or procedure for exactly solving all polynomial equations, it is important to be able to approximate the real roots of these equations. The roots of $P(x) = 0$ are the zeros of the polynomial $P(x)$. This section will develop two techniques for approximating the real zeros of a polynomial function. Both techniques rely on the same basic strategy, which will be described below. The first technique is algebraic in nature and primarily uses synthetic division, whereas the second technique is graphic in nature and uses the capabilities available on graphics calculators.

These techniques can be used to approximate any real zero of a polynomial, rational or irrational. Since the procedure developed for determining the rational zeros of a polynomial may not yield all of the irrational zeros, these techniques are particularly useful for approximating irrational zeros.

The strategy behind both of these techniques is based on the fact that a real zero of $P(x)$ is also an x-intercept of the graph of $y = P(x)$ (see Figure 4-6). We can often make a rough estimate of a zero or an x-intercept, either algebraically or by looking at the graph. What we will concentrate on in this section is refining this rough estimate in order to obtain the desired accuracy.

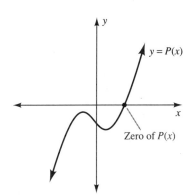

Figure 4-6 Real zero of $P(x)$ is an x-intercept of $y = P(x)$

* This is an optional section.

Algebraic Approximations

The algebraic procedure for determining the rational zeros of a polynomial often yields important information even when no zeros are found. For example, we can use synthetic division to place upper and lower bounds on the zeros (Section 4-5). We also can often use the location theorem (Section 4-5) to determine that a zero is in an interval (a, b) when $P(x)$ changes sign in this interval (see Figure 4-7). This zero can then be more accurately approximated by testing the midpoint of this interval. By continuing to use this procedure, we can gradually approximate the zero to the desired accuracy.

In working the next example, it is expedient to use a calculator that has a memory feature to perform the synthetic division (see Section 4-2).

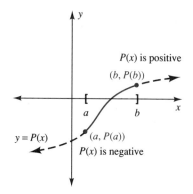

Figure 4-7 $P(x)$ changes sign

EXAMPLE 1 Given that $P(x) = x^4 + x^3 - 5x^2 - 6x - 6$, $P(2.4) = -2.1948$, and $P(2.5) = 2.4375$, approximate the zero between 2.4 and 2.5 to the nearest tenth.

SOLUTION $P(2.4) < 0$ and $P(2.5) > 0$, so the zero is between 2.4 and 2.5 by the location theorem. Test the midpoint 2.45 to approximate the zero to the nearest tenth.

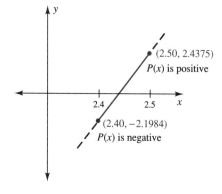

```
2.45 | 1   1.0000   -5.0000   -6.000000   -6.00000000
           2.4500    8.4525    8.458625    6.02363125
     ─────────────────────────────────────────────────
       1   3.4500    3.4525    2.458625    0.02363125
```

Since $P(2.40) < 0$ and $P(2.45) > 0$, the zero is between 2.40 and 2.45. Thus to the nearest tenth the zero is 2.4.

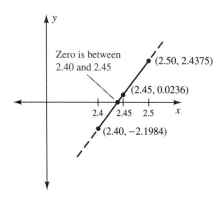

The polynomial $P(x) = x^4 - x^3 - 2x^2 - 3x - 1$ has one change in sign. Thus by Descartes's rule of signs this polynomial must have exactly one positive real zero. The location theorem is first used in the next example to locate this positive zero of $P(x)$ between two integers. This zero can then be approximated further to any desired degree of accuracy.

SECTION 4-6 (OPTIONAL)   APPROXIMATION OF ZEROS

EXAMPLE 2 Approximate to the nearest tenth the only positive real zero, a, of
$$P(x) = x^4 - x^3 - 2x^2 - 3x - 1$$

SOLUTION The strategy is to approximate the zero, first to the nearest integer, and then to the nearest tenth.

Approximate a to the nearest integer.
$P(0) = -1$
$P(1) = 1 - 1 - 2 - 3 - 1 = -6$

Apply the location theorem to isolate a zero between two consecutive integers.

```
2 | 1   -1   -2   -3    -1
         2    2    0    -6
    1    1    0   -3    -7
```
$P(2) = -7 < 0$

```
3 | 1   -1   -2   -3    -1
         3    6   12    27
    1    2    4    9    26
```
$P(3) = 26 > 0$, so $2 < a < 3$.

```
2.5 | 1  -1.0  -2.00  -3.000  -1.0000
            2.5   3.75   4.375   3.4375
       1   1.5   1.75   1.375   2.4375
```
Then test the midpoint 2.5 to approximate a to the nearest integer. $P(2) < 0$ and $P(2.5) = 2.4375 > 0$, so $2.0 < a < 2.5$.

Thus $a \approx 2$. Now approximate a to the nearest tenth.

a is in one of these intervals between 2.0 and 2.5.

```
  2.0   2.1   2.2   2.3   2.4   2.5
```

```
2.3 | 1  -1.0  -2.00  -3.000  -1.0000
            2.3   2.99   2.277  -1.6629
       1   1.3   0.99  -0.723  -2.6629
```
Locate the interval containing a by starting with a middle value of 2.2 or 2.3. $P(2.5) > 0$ and $P(2.3) < 0$, so $2.3 < a < 2.5$.

```
2.4 | 1  -1.0  -2.00  -3.000  -1.0000
            2.4   3.36   3.264   0.6336
       1   1.4   1.36   0.264  -0.3664
```
$P(2.5) > 0$ and $P(2.4) < 0$, so $2.4 < a < 2.5$.

```
2.45 | 1  -1.00  -2.0000  -3.000000  -1.00000000
             2.45   3.5525    3.803625    1.96888125
        1   1.45   1.5525    0.803625    0.96888125
```
Test the midpoint 2.45. $P(2.4) < 0$ and $P(2.45) > 0$, so $2.40 < a < 2.45$; thus $a \approx 2.4$.

Answer $a \approx 2.4$

Geometric Approximations

The geometric procedure for approximating the real zeros of an elementary function (in particular, a polynomial function) is based on visually approximating the x-intercepts of the graph of the function. To refine the accuracy of an approximation, we can then zoom in on a specific portion of the graph

▶ **Self-Check** ▼

1. If $P(x) = x^4 + x^3 - 5x^2 - 6x - 6$, $P(-2.4) = -1.0464$, and $P(-2.5) = 1.1875$, approximate the zero between -2.4 and -2.5 to the nearest tenth.
2. Approximate the zero in Example 2 to the nearest hundredth.

▶ **Self-Check Answers** ▼

1. -2.4 2. $a \approx 2.41$

316 CHAPTER 4 POLYNOMIAL AND RATIONAL RELATIONS

and examine this magnified portion of the graph. The original graph and the magnified portions can all be obtained by plotting points by hand. Hand plotting these functions, however, is quite-time consuming. Fortunately, these tasks can all be done automatically on a computer or a graphics calculator.

Approximating Zeros with a Graphics Calculator ▼

Step 1 Enter the formula for the function $y = f(x)$.

Step 2 Select an appropriate viewing window for the first view and graph the function. (The standard viewing rectangle is appropriate for many functions.)

Step 3 Zoom in on one of the x-intercepts. (Repeat this step to obtain the desired accuracy.)

Step 4 Use the trace option to approximate this x-intercept.

The following example illustrates the capabilities of a TI-81 graphics calculator. Although other graphics calculators have similar capabilities, the keystrokes vary from model to model.

For the polynomial $P(x) = x^4 - 2x^3 - x^2 + 4x - 2$, $P(-x)$ has one change in sign. Thus by Descartes's rule of signs, this polynomial must have exactly one negative real zero. In the next example we will graphically approximate this zero to the nearest tenth.

EXAMPLE 3 Approximate to the nearest tenth the only negative real zero of

$$P(x) = x^4 - 2x^3 - x^2 + 4x - 2$$

SOLUTION

1 [Y=] [X|T] [^] [4] [−] [2] [X|T] [^] [3]   Enter the formula for the function into y_1.

[−] [X|T] [x^2] [+] [4] [X|T] [−] [2]

2 [ZOOM] [Standard (option 6)]   Select option 6, which is the standard viewing rectangle with $-10 \le x \le 10$ and $-10 \le y \le 10$.

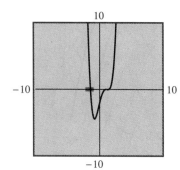

Zoom in on the region that is shaded.

SECTION 4-6 (OPTIONAL) APPROXIMATION OF ZEROS

3 ZOOM Box (option 1)

Use the arrow keys to select approximately $(-0.74, -0.48)$ as one corner of the box and then press ENTER. Then move the cursor to approximately $(-2.00, 0.79)$ and press ENTER to fix the diagonal corner of the box.

Select option 1, which is the box option. Then fix the corners of the new viewing rectangle and zoom in on this rectangle.

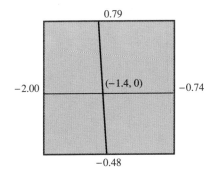

4 TRACE

Use the arrow keys to move the cursor over this x-intercept and determine the approximate value of this point.
To the nearest tenth this negative zero is -1.4.

Answer $x \approx -1.4$

The accuracy of this approximation can be observed by how much the coordinates change as the cursor moves from one side of the point to the other side of this point.

▶ **Self-Check** ▼

Zoom in on the zero in Example 3 in order to approximate this zero to the nearest hundredth.

EXERCISES 4-6

A

$P(x) = x^4 - 4x^3 - 2x^2 + 12x - 3$ has four irrational zeros. In Exercises 1–4, approximate to the nearest tenth the zero of $P(x)$ between a and b, given the values of $P(a)$ and $P(b)$.

1 $P(3.7) = -1.1759, P(3.8) = 2.7456$
2 $P(0.2) = -0.7104, P(0.3) = 0.3201$
3 $P(1.7) = 0.3201, P(1.8) = -0.7104$
4 $P(-1.8) = 2.7456, P(-1.7) = -1.1759$

$P(x) = x^4 - 6x^3 + 2x^2 + 20x + 8$ has four irrational zeros. In Exercises 5–8, approximate to the nearest tenth the zero of $P(x)$ between a and b, given the values of $P(a)$ and $P(b)$.

5 $P(3.0) = 5, P(3.5) = -4.6875$
6 $P(4.0) = -8, P(4.5) = 1.8125$
7 $P(-0.5) = -0.6875, P(0) = 8$
8 $P(-1.5) = 7.8125, P(-1.0) = -3$

▶ **Self-Check Answer** ▼

$x \approx -1.41$

$P(x) = x^4 - 14x^3 + 64x^2 - 102x + 39$ has four irrational zeros. In Exercises 9–12, approximate to the nearest hundredth the zero of $P(x)$ between a and b, given the values of $P(a)$ and $P(b)$.

9 $P(5.40) = 0.2496$, $P(5.45) = -0.002244$
10 $P(0.55) = 0.02226$, $P(0.60) = -2.0544$
11 $P(2.2) = -1.2864$, $P(2.3) = 0.6061$
12 $P(5.7) = -0.1419$, $P(5.8) = 0.4416$

In Exercises 13–16, determine to the nearest tenth the zero of $P(x)$ in each interval, given $P(x) = x^4 + 4x^3 - 5x^2 - 24x - 6$.

13 $(-1, 0)$ **14** $(-3, -2)$ **15** $(2, 3)$ **16** $(-4, -3)$

In Exercises 17–20, graph each function on a graphics calculator and estimate each real zero to the nearest integer. For each graph, first set the window to the values given to the right.

17 $y = x^3 - 3x^2 + 2x - 6$
18 $y = x^3 + 2x^2 + 3x + 6$
19 $y = x^3 + 6x^2 - x - 6$
20 $y = x^3 + 2x^2 - 5x - 6$

```
RANGE
Xmin = -6
Xmax = 6
Xscl = 1
Ymin = -50
Ymax = 50
Yscl = 10
Xres = 1
```

Figure for Exercises 17–20

The functions in Exercises 21–24 each have exactly one real zero. Use a graphics calculator or a computer graphing program to approximate this real zero to the nearest hundredth.

21 $y = 3x^3 + 4x^2 + 18x + 24$
22 $y = 7x^3 + 3x^2 + 35x + 15$
23 $y = 10x^3 + 3x^2 - 8x - 12$
24 $y = 6x^3 + 11x^2 - 29x + 21$

In Exercises 25 and 26, use either the algebraic approach or the geometric approach to approximate to the nearest tenth each positive zero of the polynomial function.

25 $P(y) = y^4 + 6y^3 + y^2 - 22y + 12$
26 $P(x) = x^4 - 4x^3 - 8x^2 + 24x + 11$

In Exercises 27 and 28, use a graphics calculator or a computer to approximate to the nearest tenth each zero of the polynomial function.

27 $y = -x^4 + 13x^2 - 35$
28 $y = x^4 - 16x^2 + 54$

In Exercises 29 and 30, use a graphics calculator or a computer to approximate to the nearest hundredth each zero of the polynomial function.

29 $P(v) = 2v^4 - 16v^3 + 37v^2 - 24v - 7$
30 $P(x) = 4x^4 + 16x^3 - 51x^2 - 26x + 12$

SECTION 4-7

Rational Functions

Section Objectives

11 Determine the horizontal and vertical asymptotes of the graph of a rational function.

12 Sketch the graph of a rational function.

A function $f(x) = \dfrac{P(x)}{Q(x)}$ with $Q(x) \neq 0$ is a **rational function** if $P(x)$ and $Q(x)$ are polynomials. Because these functions can contain variables in the

SECTION 4-7 RATIONAL FUNCTIONS

denominator, some real values may be exluded from the domain to prevent division by zero. To graph these functions, we must not only exclude certain values from the domain, but also determine the behavior of the function "near" any excluded values.

EXAMPLE 1 Determine which of these functions are rational functions. For those that are rational, state the domain.

SOLUTIONS

(a) $f(x) = \dfrac{2x + 7}{x^2 - 9}$

This is a rational function whose domain is $\mathbb{R} \sim \{-3, 3\}$, because $x^2 - 9 = 0$ for $x = \pm 3$.

Recall that $\mathbb{R} \sim \{-3, 3\}$ denotes the set of all real numbers except -3 and 3.

(b) $g(x) = \dfrac{3x^2 - 7x + 19}{2x^2 - 5x - 3}$

This is a rational function whose domain is $\mathbb{R} \sim \left\{-\dfrac{1}{2}, 3\right\}$, because $2x^2 - 5x - 3 = (2x + 1)(x - 3) = 0$ for $x = -\dfrac{1}{2}$ or $x = 3$.

(c) $h(x) = \dfrac{\sqrt{x^2 + 3}}{5x + 9}$

This is *not* a rational function, because the numerator $\sqrt{x^2 + 3}$ is not a polynomial. ☐

The denominator of a rational function produces some interesting graphical features that are not found in polynomial functions. Two of these features are holes and asymptotes. We shall now examine a rational function with a hole.

The rational function $f(x) = \dfrac{x^2 + 4x + 3}{x + 1}$ has a factor of $x + 1$ in both the numerator and the denominator. Thus $f(x) = \dfrac{(x + 1)(x + 3)}{x + 1}$ and $g(x) = x + 3$ are nearly identical functions; the only difference is that f is not defined for $x = -1$, whereas g is defined for all real numbers. As the following example shows, the graph of f has a hole at $x = -1$; otherwise, it is the same as the graph of g.

▶ **Self-Check** ▼

Determine which of these functions are rational functions. For those that are rational, state the domain.

1 $f(x) = \dfrac{5}{x - 7}$

2 $g(x) = \dfrac{3x}{x^{1/2}}$

3 $h(x) = \dfrac{4x + 9}{x^2 + 1}$

▶ **Self-Check Answers** ▼

1 Rational with domain $\mathbb{R} \sim \{7\}$ **2** Not rational, $x^{1/2}$ is not a polynomial.
3 Rational with domain $\mathbb{R}$; the denominator is never zero.

EXAMPLE 2 Graph the following functions.

SOLUTIONS

(a) $g(x) = x + 3$

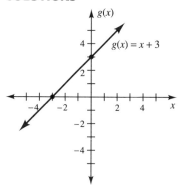

Domain of $g = \mathbb{R}$

(b) $f(x) = \dfrac{x^2 + 4x + 3}{x + 1}$

$= \dfrac{(x + 1)(x + 3)}{x + 1}$

$= x + 3 \quad \text{for } x \neq -1$

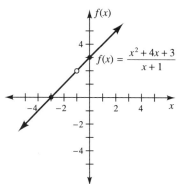

This graph is identical to the graph in part (a) except for the hole at $x = -1$.

Domain of $f = \mathbb{R} \sim \{-1\}$.

□

The graph of $f(x) = \dfrac{P(x)}{Q(x)}$ will have a hole at $x = c$ if $(x - c)^n$ divides $P(x)$ whenever $(x - c)^n$ divides $Q(x)$. That is, c is a zero of the original numerator and denominator of $f(x)$, but it is not a zero of the denominator of the reduced form of $f(x)$. We will now examine the case where $x - c$ is a factor of $Q(x)$ but not of $P(x)$; that is, c is a zero of the denominator only. Thus the function is undefined at $x = c$, and the behavior near c is described as asymptotic in nature. A line is an **asymptote** of a graph if this graph gets closer and closer to the line for points farther and farther from the origin. Asymptotes can simplify curve sketching, since they allow us to picture quickly a portion of a more complicated curve. The graph in the next example has both a vertical and a horizontal asymptote.

EXAMPLE 3 Graph $f(x) = \dfrac{1}{x}$.

SOLUTION The domain of f is $\mathbb{R} \sim \{0\}$. Since $f(-x) = -f(x)$, this is an odd function. Thus its graph is symmetric about the origin. The values in

SECTION 4-7 RATIONAL FUNCTIONS

the table show the behavior of $f(x)$ for positive values of x. The behavior of $f(x)$ for negative values of x can be obtained by symmetry.

x	$f(x)$
10	0.1
5	0.2
2	0.5
1	1
0.5	2
0.1	10
0.01	100
0.001	1000

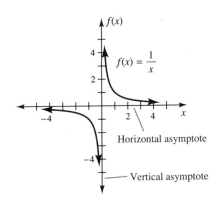

The line $x = 0$ (the y-axis) is a **vertical asymptote** of the graph of $f(x) = \dfrac{1}{x}$, because this graph approaches this vertical line even though it does not touch it. Algebraically, the absolute value of $f(x)$ becomes very large as x approaches 0. This graph also has a **horizontal asymptote**, because it approaches the line $y = 0$ (the x-axis) as the absolute value of x becomes large. We will use asymptotes rather than a large table of values to sketch the graphs of rational functions, as the asymptotes provide the needed guidelines more quickly and more accurately.

a vertical asymptote occurs at a number x where the deno. is 0 and the numerator is not 0.

Vertical Asymptote ▼

If $f(x) = \dfrac{P(x)}{Q(x)}$ and $Q(c) = 0$, then

1. $x = c$ is a vertical asymptote of the graph of $f(x)$ if $P(c) \neq 0$.
2. There is a hole in the graph of $f(x)$ at $x = c$ if c is not a zero of the denominator of the reduced form of $f(x)$.

The even and odd multiplicities of vertical asymptotes will be used the same way we used the even and odd multiplicities of x-intercepts to determine the sign of a rational function. The exponents on the linear factors of $Q(x)$ in $f(x) = \dfrac{P(x)}{Q(x)}$ will determine whether each vertical asymptote is of even or odd multiplicity.

EXAMPLE 4 Determine the vertical asymptotes of the graph of

$$f(x) = \frac{x^2 - 4}{x^4 + 3x^3 - 10x^2}$$

SOLUTION $f(x) = \dfrac{(x+2)(x-2)}{x^2(x+5)(x-2)}$ \hspace{1em} Factor the numerator and the denominator.

$f(x) = \dfrac{x+2}{x^2(x+5)}$ for $x \neq 2$ \hspace{1em} This is the reduced form of $f(x)$.

Vertical asymptotes: $x = 0$, even multiplicity \hspace{1em} The exponent on x^2 is even (2).
$x = -5$, odd multiplicity \hspace{1em} The exponent on $x + 5$ is odd (1).
There is a hole at $x = 2$. \hspace{1em} 2 is a zero of the original numerator and denominator, but it is not a zero of the denominator of the reduced form.

The domain of f is $\mathbb{R} \sim \{-5, 0, 2\}$.

□

A vertical asymptote describes the behavior of a graph very near a value of $x = c$, which is not in the domain at the function, whereas a horizontal asymptote describes the behavior of the graph as the magnitude of the x values in the domain gets larger and larger. The graph of $f(x) = \dfrac{1}{x}$ illustrates that the fraction $\dfrac{1}{x}$ gets closer and closer to zero as the denominator x gets larger and larger.

The graph of the rational function $f(x) = \dfrac{P(x)}{Q(x)}$ has $y = 0$ as a horizontal asymptote if the degree of $P(x)$ is less than the degree of $Q(x)$. The logic supporting this claim is illustrated by examining $\dfrac{x^2 - 4}{x^3 + 3x^2 - 10x}$, which can be rewritten as

$$\frac{\dfrac{1}{x} - \dfrac{4}{x^3}}{1 + \dfrac{3}{x} - \dfrac{10}{x^2}}$$

by dividing both the numerator and denominator by x^3. As the absolute value of x becomes larger, all the fractions $\dfrac{1}{x}, -\dfrac{4}{x^3}, \dfrac{3}{x}$, and $-\dfrac{10}{x^2}$ approach zero.

Thus $\dfrac{\dfrac{1}{x} - \dfrac{4}{x^3}}{1 + \dfrac{3}{x} - \dfrac{10}{x^2}}$ approaches $\dfrac{0 - 0}{1 + 0 - 0} = \dfrac{0}{1} = 0$

SECTION 4-7 RATIONAL FUNCTIONS

Therefore, $y = 0$ is a horizontal asymptote of the graph of

$$f(x) = \frac{x^2 - 4}{x^3 + 3x^2 - 10x}$$

EXAMPLE 5 Determine all asymptotes of the graph of $f(x) = \dfrac{3}{x - 2}$.

SOLUTION The vertical asymptote is $x = 2$ Only the denominator is zero for $x = 2$.
The domain of f is $\mathbb{R} \sim \{2\}$.

The horizontal asymptote is $y = 0$ The degree of the numerator is less than the degree of the denominator. ☐

> **Self-Check ▼**
>
> Determine the behavior of the graph of
> $$f(x) = \frac{2x - 6}{(x + 3)(x - 3)(x - 5)}$$
> at
> **1** $x = -3$ **2** $x = 3$ **3** $x = 5$
>
> **4** Determine all asymptotes of the graph of
> $$y = \frac{x - 1}{x^2 + 5x - 6}$$

We will now examine $f(x) = \dfrac{6x - 12}{3x + 15}$ to illustrate the behavior of the graph of $f(x) = \dfrac{P(x)}{Q(x)}$ when the degree of $P(x)$ equals the degree of $Q(x)$. First, divide both the numerator and the denominator by x, so that

$$f(x) = \frac{6 - \dfrac{12}{x}}{3 + \dfrac{15}{x}}$$

As the absolute value of x becomes larger,

$$\frac{6 - \dfrac{12}{x}}{3 + \dfrac{15}{x}} \text{ approaches } \frac{6 - 0}{3 + 0} = \frac{6}{3} = 2$$

Thus $y = 2$ is a horizontal asymptote of the graph of $f(x) = \dfrac{6x - 12}{3x + 15}$. In general if $P(x)$ and $Q(x)$ are of the same degree, then $y = c$ will be a horizontal asymptote, where c is the ratio of the leading coefficients of $P(x)$ and $Q(x)$.

▶ **Self-Check Answers ▼**

1 Vertical asymptote   **2** Hole   **3** Vertical asymptote
4 The vertical asymptote is $x = -6$, the hole is at $x = 1$, and the horizontal asymptote is $y = 0$.

Horizontal Asymptote ▼

[margin note: there is always at most 1 horz. asymp]

The graph of the rational function $f(x) = \dfrac{P(x)}{Q(x)}$ has

1. $y = 0$ as a horizontal asymptote if the degree of the numerator, $P(x)$, is less than the degree of the denominator, $Q(x)$.
2. $y = c$ as a horizontal asymptote if $P(x)$ and $Q(x)$ have the same degree and $c = \dfrac{a_n}{b_n}$, where
$$f(x) = \frac{a_n x^n + a_{n-1} x^{n-1} + \cdots + a_0}{b_n x^n + b_{n-1} x^{n-1} + \cdots + b_0}$$
3. No horizontal asymptote if the degree of the numerator, $P(x)$, is greater than the degree of the denominator, $Q(x)$. (Oblique asymptotes, which are neither vertical nor horizontal, are covered in the **C** Exercises at the end of this section.)

[margin notes:
$\dfrac{ax^m + \cdots}{bx^n + \cdots}$

If $m = n$, $y = \dfrac{a}{b}$ is the horizontal asymptote

If $m < n$, $y = 0$, i.e. the x-axis is the horizontal asymptote

If $m > n$, no horizontal asymptote.

note if $m = n+1$ then there's an oblique asymptote]

EXAMPLE 6 Determine the intercepts and asymptotes of the graph of
$$y = \frac{2x^3 - 4x^2 - 8x + 16}{3x^3 - 3x^2 - 3x + 3}$$

SOLUTION
$$y = \frac{2x^3 - 4x^2 - 8x + 16}{3x^3 - 3x^2 - 3x + 3}$$

$$= \frac{2(x^3 - 2x^2 - 4x + 8)}{3(x^3 - x^2 - x + 1)} \quad \text{Factor the numerator and the denominator.}$$

$$= \frac{2[x^2(x-2) - 4(x-2)]}{3[x^2(x-1) - (x-1)]}$$

$$= \frac{2(x^2 - 4)(x - 2)}{3(x^2 - 1)(x - 1)}$$

$$= \frac{2(x+2)(x-2)^2}{3(x+1)(x-1)^2}$$

y-intercept: $\left(0, \dfrac{16}{3}\right)$
　　Set $x = 0$ in the original expression.

Horizontal asymptote: $y = \dfrac{2}{3}$
　　The numerator and denominator are of the same degree; $\dfrac{2}{3}$ is the ratio of the leading coefficients.

x-intercepts: $(-2, 0)$, multiplicity 1 (odd)
　　　　　　$(2, 0)$, multiplicity 2 (even)
　　Set $y = 0$, which means that the numerator must equal zero. The x-intercepts can be determined directly from the factors of the numerator.

SECTION 4-7 RATIONAL FUNCTIONS

Vertical asymptotes: $x = -1$, multiplicity 1 (odd)
$x = 1$, multiplicity 2 (even)

Set the denominator equal to zero. The vertical asymptotes can be determined directly from the factors of the denominator.

The domain of the function is $\mathbb{R} \sim \{-1, 1\}$.

□

EXAMPLE 7 Determine the intercepts and asymptotes of the graph of

$$y = \frac{7x^2 + 7x - 140}{5x^4 - 5x^3 - 150x^2}$$

SOLUTION $y = \dfrac{7x^2 + 7x - 140}{5x^4 - 5x^3 - 150x^2}$

$= \dfrac{7(x^2 + x - 20)}{5x^2(x^2 - x - 30)}$ Factor the numerator and the denominator.

$= \dfrac{7(x + 5)(x - 4)}{5x^2(x + 5)(x - 6)}$

$= \dfrac{7(x - 4)}{5x^2(x - 6)}$ for $x \neq -5$

y-intercept: none y is undefined for $x = 0$.

Horizontal asymptote: $y = 0$ The degree of the numerator is less than the degree of the denominator.

x-intercept: (4, 0), multiplicity 1 (odd)   Set $y = 0$ and then determine values that make the numerator 0 without making the denominator 0.

Hole: $\left(-5, \dfrac{63}{1375}\right)$ Substitute -5 into the reduced form of the function to determine the y-coordinate.

Vertical asymptotes: $x = 0$, multiplicity 2 (even)
$x = 6$, multiplicity 1 (odd)

Set the denominator of the reduced form equal to zero.

The domain of the function is $\mathbb{R} \sim \{-5, 0, 6\}$.

□

The strategy for graphing rational functions is keyed to determining the asymptotes and intercepts. The main features of the graph can be determined by analyzing the asymptotic behavior together with the multiplicities of the x-intercepts and vertical asymptotes. This is an extension of the strategy used to graph polynomials in Section 4-1. This rapid curve-sketching technique will provide an important foundation for those who need to graph these functions in more detail in later courses. This study also gives an important perspective to those using graphics calculators.

Graphing Rational Functions ▼

To sketch the graph of a rational function $y = \dfrac{P(x)}{Q(x)}$:

Step 1 Plot the y-intercept (if any) and sketch the horizontal asymptote (if any).

Step 2 Plot the x-intercepts and the holes, and sketch the vertical asymptotes. Determine the multiplicity of each of these intercepts and asymptotes.

Step 3 Determine the behavior of the graph to the right of all x-intercepts and vertical asymptotes. Then illustrate this behavior of the graph.

Step 4 Complete the sketch by using the multiplicity of each x-intercept and each vertical asymptote.

The graph of a rational function switches to the opposite side of the x-axis across odd vertical asymptotes and odd x-intercepts. The graph is on the same side of the x-axis across even vertical asymptotes and even x-intercepts.

EXAMPLE 8 Graph $y = \dfrac{2x + 2}{x - 2}$.

SOLUTION

1

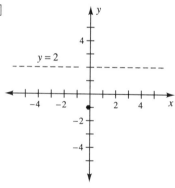

For $x = 0$, $y = \dfrac{2(0) + 2}{0 - 2} = -1$.

y-intercept: $(0, -1)$.

$y = 2$ is the horizontal asymptote because the numerator and the denominator are of the same degree and 2 is the quotient of the leading coefficients.

2

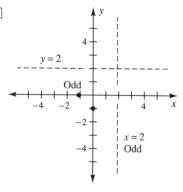

$y = \dfrac{2(x + 1)}{x - 2}$.

x-intercept: $(-1, 0)$ (odd); -1 makes the numerator 0 and the exponent on the factor $(x + 1)$ is odd.

Vertical asymptote: $x = 2$ (odd); 2 makes the denominator 0 and the exponent on the factor $(x - 2)$ is odd.

The domain of the function is $\mathbb{R} \sim \{2\}$.

SECTION 4-7 RATIONAL FUNCTIONS

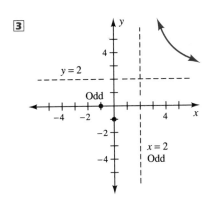

For $x > 2$ all factors in the numerator and the denominator are positive. Thus the graph approaches the vertical asymptote $x = 2$ going up and then approaches the horizontal asymptote $y = 2$. The graph cannot approach the vertical asymptote going down because this would force the graph to cross the x-axis as it approaches the horizontal asymptote $y = 2$. It does not have an x-intercept for $x > 2$.

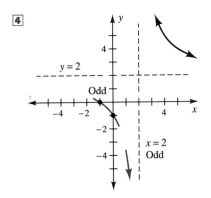

The vertical asymptote $x = 2$ is odd, so mark a curve segment below the x-axis left of the asymptote. The curve segment on the y-axis indicates the y-intercept. The curve segment crossing the x-axis denotes the odd x-intercept $(-1, 0)$.

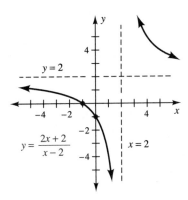

Complete the sketch by connecting the curve segments and letting the curve approach the horizontal asymptote $y = 2$.

□

In the region to the right of all the vertical asymptotes, the graph either approaches the vertical asymptote going up or going down. Sometimes the easiest way to determine the correct choice is to eliminate one of the possibilities as impossible. In the previous example it was impossible for the graph to cross the x-axis for $x > 2$, so the graph could not approach $x = 2$ going downward for $x > 2$. The other portions of the graph depend on this first portion of the graph being correct. Thus it is crucial that the approach to the vertical asymptote be accurately selected for this portion of the graph. Just remember to start by considering both possibilities.

328 CHAPTER 4 POLYNOMIAL AND RATIONAL RELATIONS

EXAMPLE 9 Graph $y = \dfrac{(x-1)(x-5)}{(x-3)^2}$.

SOLUTION $y = \dfrac{(x-1)(x-5)}{(x-3)^2}$

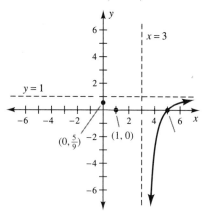

y-intercept: $\left(0, \dfrac{5}{9}\right)$

Horizontal asymptote: $y = 1$

x-intercepts: $(5, 0)$ multiplicity 1 (odd)
$(1, 0)$ multiplicity 1 (odd)

Vertical asymptotes: $x = 3$ multiplicity 2 (even)
The domain of the function is $\mathbb{R} \sim \{3\}$.

For $x > 5$ all factors in the numerator and the denominator are positive, thus $y > 0$ in this interval. Since $(5, 0)$ is an odd intercept, the graph crosses the x-axis at $(5, 0)$ and approaches the vertical asymptote $x = 3$ going downward.

Since $x = 3$ is an even vertical asymptote, the graph must approach $x = 3$ going downward on both sides of $x = 3$. Then the graph crosses the x-axis at $(1, 0)$ and the y-axis at $\left(0, \dfrac{5}{9}\right)$.

Complete the graph by sketching it asymptotic to the horizontal line $y = 1$. □

EXAMPLE 10 Graph $y = \dfrac{x(x-1)^2}{x(x-2)(x+2)^2}$.

SOLUTION

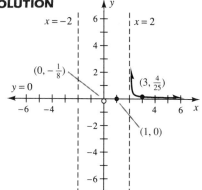

y-intercepts: none
Horizontal asymptote: $y = 0$
Hole: $\left(0, -\dfrac{1}{8}\right)$

x-intercept: $(1, 0)$, multiplicity 2 (even)
Vertical asymptotes:
$x = -2$, multiplicity 2 (even)
$x = 2$, multiplicity 1 (odd)
The domain of the function is $\mathbb{R} \sim \{-2, 0, 2\}$.
For $x > 2$, all factors in the numerator and the denominator are positive. Thus $y > 0$. This portion of the graph is shown passing through $\left(3, \dfrac{4}{25}\right)$ and asymptotic to $x = 2$ and $y = 0$.

SECTION 4-7 RATIONAL FUNCTIONS

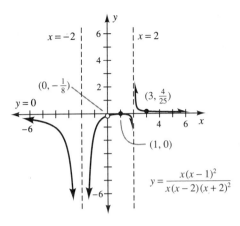

$$y = \frac{x(x-1)^2}{x(x-2)(x+2)^2}$$

Since $x = 2$ is a vertical asymptote of odd multiplicity, the graph must approach $x = 2$ through negative values for $x < 2$.
Since $(1, 0)$ is an x-intercept of even multiplicity, the graph is tangent to the x-axis at this point but then continues to have points with negative y-values.
The graph approaches $x = -2$ asymptotically. Since this asymptote is of even multiplicity, the graph also approaches $x = -2$ through negative values for $x < -2$.
It then approaches the horizontal asymptote $y = 0$.

☐

EXERCISES 4-7

A

In Exercises 1 and 2, give the domain of each relation.

1 a. $f(x) = \dfrac{7x - 9}{3x + 5}$ b. $y = \dfrac{5x^2 - 8}{6x^2 - 5x - 4}$

2 a. $g(x) = \dfrac{2x + 3}{5x - 4}$ b. $h(x) = \dfrac{2x^2 + 3x - 35}{5x^2 - 7x - 6}$

In Exercises 3 and 4, determine the value of x for which the graph of $f(x)$ has a hole.

3 a. $f(x) = \dfrac{(x-2)(x+3)}{(x+3)(x+5)}$ b. $f(x) = \dfrac{x^2 - x - 12}{2x^2 - 13x + 20}$

4 a. $f(x) = \dfrac{(2x - 1)(x + 9)}{(2x - 1)(9x + 1)}$ b. $f(x) = \dfrac{15x^2 - 7x - 2}{10x^2 - 13x - 3}$

In Exercises 5–8, determine the horizontal asymptote and the vertical asymptotes of the graph of each function. State whether each vertical asymptote is of even or odd multiplicity.

5 a. $g(x) = \dfrac{5}{2x + 7}$ b. $h(x) = \dfrac{3x - 4}{4x + 3}$

6 a. $y = \dfrac{6}{3x - 5}$ b. $f(x) = \dfrac{5 - 2x}{4x + 5}$

7 a. $y = \dfrac{2x^2 - 5}{x^2 + 7x + 12}$ b. $y = \dfrac{2x^2 + 7x - 15}{x^3 - 10x^2 + 25x}$

8 a. $y = \dfrac{3x^2 + 7}{x^2 - 4x + 4}$ b. $h(x) = \dfrac{3x^2 + 5}{5x^2 + 9}$

330 CHAPTER 4 POLYNOMIAL AND RATIONAL RELATIONS

In Exercises 9–16, use the step-by-step approach described to graph the rational function.

9 The function $y = f(x)$ has a horizontal asymptote $y = 1$ and a vertical asymptote $x = 2$.
  a. Sketch the two asymptotes.
  b. Sketch the graph of $y = f(x)$ to the right of the vertical asymptote, given that there is an odd intercept at (3, 0).
  c. Sketch the remainder of the graph, given that $x = 2$ is an odd vertical asymptote and $(0, \frac{3}{2})$ is the y-intercept.

10 The function $y = f(x)$ has a horizontal asymptote $y = -2$ and a vertical asymptote $x = 1$.
  a. Sketch the two asymptotes.
  b. Sketch the graph of $y = f(x)$ to the right of the vertical asymptote, given that there is an odd x-intercept at (2, 0).
  c. Sketch the remainder of the graph, given that $x = 1$ is an odd vertical asymptote and $(0, -4)$ is a y-intercept.

11 The function $y = f(x)$ has a horizontal asymptote $y = 1$ and a vertical asymptote $x = 2$.
  a. Sketch the two asymptotes.
  b. Sketch the graph of $y = f(x)$ to the right of the vertical asymptote, given that there are no x-intercepts to the right of $x = 2$. (*Hint*: This means that the graph cannot touch the x-axis to the right of the vertical asymptote.)
  c. Sketch the remainder of the graph, given that $x = 2$ is an odd vertical asymptote, (1, 0) is an odd x-intercept, and (0, 0.5) is the y-intercept.

12 The function $y = f(x)$ has a horizontal asymptote $y = -2$ and a vertical asymptote $x = 1$.
  a. Sketch the two asymptotes.
  b. Sketch the graph of $y = f(x)$ to the right of the vertical asymptote, given that there are no x-intercepts to the right of $x = 1$.
  c. Sketch the remainder of the graph, given that $x = 1$ is an odd vertical asymptote, $(-1, 0)$ is an odd x-intercept, and (0, 2) is the y-intercept.

13 The function $y = f(x)$ has a horizontal asymptote $y = 1$ and a vertical asymptote $x = 2$. *graph is incorrectly sketched*
  a. Sketch the two asymptotes.
  b. Sketch the graph of $y = f(x)$ to the right of the vertical asymptote, given that there is an odd x-intercept at (3, 0).
  c. Sketch the remainder of the graph, given that $x = 2$ is an even vertical asymptote, (1, 0) is an odd x-intercept, and (0, 0.75) is the y-intercept.

14 The function $y = f(x)$ has a horizontal asymptote $y = -2$ and a vertical asymptote $x = 1$.
  a. Sketch the two asymptotes.
  b. Sketch the graph of $y = f(x)$ to the right of the vertical asymptote, given that there is an odd x-intercept at (2, 0).
  c. Sketch the remainder of the graph, given that $x = 1$ is an even vertical asymptote, (0.5, 0) is an odd x-intercept, and $(0, -2)$ is the y-intercept.

SECTION 4-7 RATIONAL FUNCTIONS

15 The function $y = f(x)$ has a horizontal asymptote $y = 0$ and a vertical asymptote $x = 2$.
  a. Sketch the two asymptotes.
  b. Sketch the graph of $y = f(x)$ to the right of the vertical asymptote, given that there are no x-intercepts.
  c. Sketch the remainder of the graph, given that $x = 2$ is an even vertical asymptote and $(0, 2)$ is the y-intercept.

16 The function $y = f(x)$ has a horizontal asymptote $y = 0$ and a vertical asymptote $x = 1$.
  a. Sketch the two asymptotes.
  b. Sketch the graph of $y = f(x)$ to the right of the vertical asymptote, given that there are no x-intercepts.
  c. Sketch the remainder of the graph, given that $x = 1$ is an even vertical asymptote and $(0, -3)$ is the y-intercept.

In Exercises 17–22, sketch a rational function with the given properties.

17 y-intercept: $(0, -2)$
Horizontal asymptote: $y = 0$
No x-intercepts
Vertical asymptote: $x = -1$ (odd)

18 y-intercept: $(0, 5)$
Horizontal asymptote: $y = 0$
No x-intercepts
Vertical asymptote: $x = 2$ (even)

19 y-intercept: $(0, -5.25)$
Horizontal asymptote: $y = 1$
x-intercepts: $(3, 0)$ (odd), $(-7, 0)$ (odd)
Vertical asymptote: $x = 2$ (even)

graph is incorrectly sketched in that there's not enough info to assume that there are points above the horizontal asymptote

20 y-intercept: $(0, 2)$
Horizontal asymptote: $y = -3$
x-intercept: $(2, 0)$ (odd)
Vertical asymptote: $x = -3$ (odd)

21 y-intercept: $(0, 9)$
Horizontal asymptote: $y = 2$
x-intercepts: $(-1, 0)$ (odd), $(-3, 0)$ (even)
vertical asymptotes: $x = 1$ (even), $x = -2$ (odd)

22 y-intercept: $(0, -2)$
Horizontal asymptote: $y = 2$
x-intercepts: $(-2, 0)$ (odd), $(1, 0)$ (even)
Vertical asymptotes: $x = -1$ (even), $x = 2$ (odd)

In Exercises 23–30, graph each rational function.

23 $y = \dfrac{1}{x + 1}$

24 $y = \dfrac{2}{x - 3}$

25 $y = \dfrac{4}{x^2}$

26 $y = \dfrac{1}{(x + 2)^2}$

27 $f(x) = \dfrac{2x + 1}{x - 4}$

28 $g(x) = \dfrac{-x - 1}{2x + 1}$

29 $y = \dfrac{3x^2 - x - 10}{3x + 5}$

30 $g(x) = \dfrac{9}{2x - 10}$

B

In Exercises 31–34, write an equation that defines a function with the stated properties.

31 Odd vertical asymptote: $x = 3$
Odd x-intercept: $(-2, 0)$
Horizontal asymptote: $y = 1$

32 Odd vertical asymptote: $x = -2$
Odd x-intercept: $(-4, 0)$
Horizontal asymptote: $y = 1$

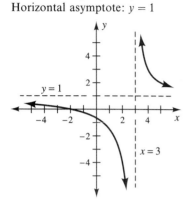
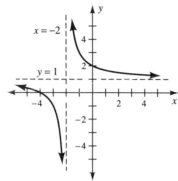

33 Even vertical asymptote: $x = -2$
Odd x-intercept: $(4, 0)$
Horizontal asymptote: $y = 0$

34 Even vertical asymptote: $x = 2$
Odd x-intercept: $(3, 0)$
Horizontal asymptote: $y = 0$

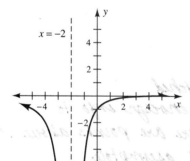
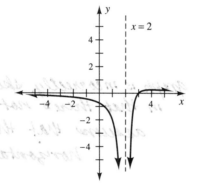

The functions in Exercises 35–44 were graphed on a graphics calculator. (The scale has been added here on the axes to facilitate your examination of each graph.) Match each function with the appropriate graph.

35

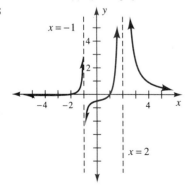

36

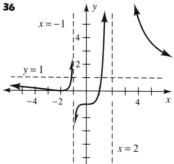

37

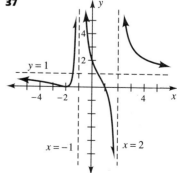

SECTION 4-7 RATIONAL FUNCTIONS

38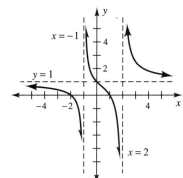

39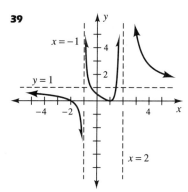

a. $y = \dfrac{(x-1)(x+2)}{(x+1)(x-2)^2}$ b. $y = \dfrac{(x-1)^2(x+2)}{(x+1)(x-2)^2}$ c. $y = \dfrac{(x-1)(x+2)}{(x+1)(x-2)}$

d. $y = \dfrac{(x-1)(x+2)^2}{(x+1)^2(x-2)}$ e. $y = \dfrac{(x-1)(x+2)^2}{(x+1)(x-2)^2}$

40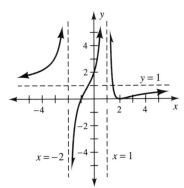

41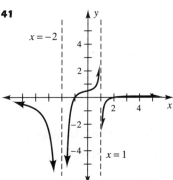

42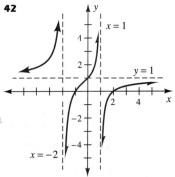

43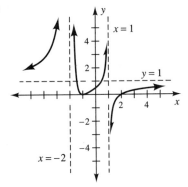

44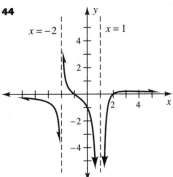

a. $y = \dfrac{(x+1)(x-2)}{(x-1)(x+2)}$ b. $y = \dfrac{(x+1)^2(x-2)}{(x-1)(x+2)^2}$ c. $y = \dfrac{(x+1)(x-2)}{(x-1)^2(x+2)}$

d. $y = \dfrac{(x+1)(x-2)^2}{(x-1)^2(x+2)}$ e. $y = \dfrac{(x+1)(x-2)}{(x-1)(x+2)^2}$

45 Can the graph of a rational function have both a vertical and a horizontal asymptote? Give an example or explain why not.

46 Can the graph of a rational function have three vertical asymptotes? Give an example or explain why not.

47 Can the graph of a rational function have three horizontal asymptotes? Give an example or explain why not.

48 Can the graph of a rational function ever cross a horizontal asymptote? Give an example or explain why not. (*Hint*: See Example 9.)

49 Can the graph of a rational function ever cross a vertical asymptote? Give an example or explain why not.

50 Is it possible to have a rational function whose graph has no *x*-intercepts? Give an example or explain why not.

51 Is it possible to have a rational function whose graph has no *y*-intercept? Give an example or explain why not.

52 Is it possible to have a rational function whose graph has no vertical asymptotes? Give an example or explain why not.

53 Explain why all polynomial functions are also rational functions.

C

In Exercises 54–58, graph each rational function.

54 $f(x) = \dfrac{10x^2 - 33x - 7}{2x - 7}$

55 $y = \dfrac{2x^2 + 5x - 3}{1 - 2x - 3x^2}$

56 $g(x) = \dfrac{24}{x^3 - x^2}$

57 $y = \dfrac{x^3 - 2x^2 - 3x}{x^3 + 4x^2 + 4x}$

58 $y = \dfrac{2x^3 - x^2 - 25x - 12}{x^3 - x^2 - 5x + 5}$

59 If a factory operates t hours per day ($0 \le t \le 24$), it can produce $N(t) = 48t - t^2$ unit as a total dollar cost of $C(t) = 1000t + 1200$. Express the average cost as a function of t and graph this function for $0 < t \le 24$.

60 Write the equation of a function identical to $f(x) = \dfrac{3x}{x^2 - 1}$ except that its graph has a hole at $x = 4$.

61 Write the equation of a function identical to $y = \dfrac{2x + 1}{x - 2}$ except that its graph has a hole at $x = 3$.

A line that is neither horizontal nor vertical is **oblique**. The graph of

$$f(x) = \dfrac{x^2 - 5}{x - 2} = x + 2 - \dfrac{1}{x - 2}$$

has an oblique asymptote $y = x + 2$, since the term $\dfrac{1}{x - 2}$ approaches zero as the magnitude of x gets larger. In Exercises 62–65, determine the oblique asymptote of the graph of each function. Then sketch this graph.

62 $y = \dfrac{x^2 + 6}{x - 3}$

63 $f(x) = \dfrac{2x^2 - x - 5}{2x}$

64 $f(x) = \dfrac{(4x - 1)^2}{2x + 1}$

65 $f(x) = \dfrac{4x^2 - 16x + 16}{2x - 3}$

KEY CONCEPTS

66 State a rule involving the degrees of $P(x)$ and $Q(x)$ that will determine when the graph of $f(x) = \dfrac{P(x)}{Q(x)}$ has an oblique linear asymptote.

67 Can the graph of a rational function have both a horizontal and an oblique asymptote? Give an example or explain why not.

68 Can the graph of a rational function have both a vertical and an oblique linear asymptote? Give an example or explain why not.

69 Write the equation of a rational function whose graph has an oblique asymptote of $y = x + 2$ and vertical asymptote of $x = 1$.

In Exercises 70 and 71 use a graphics calculator or a computer graphing program to graph each of the following rational functions.

70 $y = \dfrac{2x^2 - 5}{x^2 - 6}$

71 $y = \dfrac{2x^3 - 5x}{x^2 - 6}$

In Exercises 64 and 65, use a graphics calculator or a computer graphing program to approximate all zeros of each function to the nearest tenth.

72 $y = \dfrac{x^2 + 4x + 2}{x^2 + x - 2}$

73 $y = \dfrac{4x^2 - 4x - 1}{x^2 - 9}$

KEY CONCEPTS

1 Equivalent statements about polynomials:
   1. $x - a$ is a factor of the polynomial $P(x)$.
   2. a is a zero of $P(x)$; that is, $P(a) = 0$.
   3. a is a solution of the equation $P(x) = 0$.
   4. If a is a real number, $(a, 0)$ is an x-intercept of the graph of $y = P(x)$.

2 The y-coordinate of the graph of a rational function changes signs at x-intercepts and vertical asymptotes of odd multiplicity but does not change sign at x-intercepts and vertical asympotes of even multiplicity.

3 Remainder theorem:
   If the polynomial $P(x)$ is divided by $x - a$, the remainder equals $P(a)$.

4 Conjugate zeros theorem:
   The complex zeros of a real polynomial occur in conjugate pairs.

5 Complete factorization theorem:
   Each nth-degree polynomial can be factored into n linear factors over the complex numbers.

6 Rational zeros theorem:
   If $\quad P(x) = a_n x^n + a_{n-1} x^{n-1} + \cdots + a_1 x + a_0, \quad a_n \neq 0$

   is a polynomial with integer coefficients and $\dfrac{p}{q}$ is a rational zero in lowest terms, then p must divide a_0 and q must divide a_n.

7 Location theorem:
If $P(a)$ and $P(b)$ are opposite in sign, then the polynomial $P(x)$ has a zero between a and b.

8 Descartes's rule of signs:
If $P(x)$ has k variations in sign, then the number of positive real zeros of $P(x)$ is either k or $k - 2m$, where m is a positive integer. If $P(-x)$ has j variations in sign, then the number of negative real zeros is either j or $j - 2n$, where n is a positive integer.

9 The rational function $y = \dfrac{P(x)}{Q(x)}$ has

  1. a vertical asymptote $x = c$ if $Q(c) = 0$ and $P(c) \neq 0$.
  2. a hole at $x = c$ if $Q(c) = 0$ but c is not a zero of the denominator of the reduced form of y.
  3. a horizontal asymptote $y = 0$ if the degree of $P(x)$ is less than the degree of $Q(x)$.
  4. a horizontal asymptote $y = c = \dfrac{a_n}{b_n}$ where a_n and b_n are the leading coefficients of $P(x)$ and $Q(x)$, respectively, if the degrees of $P(x)$ and $Q(x)$ are the same.
  5. no horizontal asymptote if the degree of the numerator, $P(x)$, is greater than the denominator, $Q(x)$.

REVIEW EXERCISES FOR CHAPTER 4

In Exercises 1 and 2, use synthetic division to perform the divisions.

1 $(3y^4 - 2y^2 + 1) \div (y + 2)$
2 $(t^4 - 3t^3 + 10t^2 + 9t + 13) \div (t - 2 + 3i)$

In Exercises 3 and 4, use synthetic division to find $P(a)$ for the polynomial $P(x)$ and the complex number a. If a is a zero, write $P(x)$ in the form $Q(x)(x - a)$.

3 $P(x) = x^5 - 5x^4 - 2x^3 + 11x^2 - 10x + 25$, $a = 5$
4 $P(x) = 4x^3 + 5bx^2 - 4b^2x + b^3$, $a = -2b$
5 Find the value of k such that $x + 5$ is a factor of $x^3 + 7x^2 + 7x + k$.

In Exercises 6–8, find a real polynomial $P(x)$ of smallest degree that has the zeros listed and the given value for $P(3)$.

6 $2, -3, 4; P(3) = 12$
7 $-1, \sqrt{3}, -\sqrt{3}, 5; P(3) = 48$
8 $2, 3 - i; P(3) = -1$

For the polynomials in Exercises 9–11, list each zero and its multiplicity.

9 $P(x) = x^2(x - 1)^4(x + 5)^3$
10 $P(t) = 4t^4 - 4t^3 + t^2$
11 $P(u) = 4u^3 + 32u^2 + 69u + 45$
12 Determine the zeros of $P(x) = x^4 - 6x^3 + 12x^2 - 6x - 5$ given that $2 - i$ is a zero.

REVIEW EXERCISES FOR CHAPTER 4

13 Use the rational zeros theorem to list all possible rational zeros of $P(t) = 6t^3 - 5t^2 + t + 30$.

14 Show that 5 is an upper bound and -3 is a lower bound on the real zeros of $P(x) = x^4 - 2x^3 - 10x^2 + 14x - 3$.

15 Show that $P(x) = 2x^4 - x^3 + 9x^2 + 8x + 10$ has no rational zeros. (*Hint*: Use the upper and lower bounds theorem to limit the possibilities that must be examined.)

16 Use $P(x) = x^2 - 7$ and the rational zeros theorem to show that $\sqrt{7}$ is irrational.

In Exercises 17 and 18, use Descartes's rule of signs to determine the number of possible positive and negative zeros of each polynomial.

17 $P(y) = 5y^5 - y - 1$

18 $P(t) = 6t^4 - 5t^3 + 4t^2 - 5t + 8$

In Exercises 19 and 20, use the location theorem to show that the graph of each polynomial function has an x-intercept between a and b.

19 $P(x) = 3x^4 + 13x^3 + 5x^2 + 2x - 8,\ a = 0,\ b = 1$

20 $P(x) = x^4 - 6x^3 + 12x^2 - 6x - 5,\ a = 2,\ b = 3$

In Exercises 21–24, determine all zeros of each polynomial.

21 $P(u) = u^3 + 6u^2 + 3u - 10$

22 $P(t) = 2t^4 + 5t^3 - 39t^2 - 62t + 40$

23 $P(z) = 6z^4 + 17z^3 + 7z^2 + z - 10$

24 $P(y) = 3y^5 + 8y^4 - 16y^3 - 48y^2 + 16y + 64$

25 Solve $12t^3 + 41t^2 - 38t - 40 = 0$.

26 Solve $2x^3 + x^2 - 43x - 60 > 0$.

27 Solve $2x^3 + 17x^2 + 48x + 45 \leq 0$.

28 $P(x) = x^4 - 6x^3 - x^2 + 12x - 2$ has four irrational zeros. Use the given information to approximate to the nearest tenth the zero between 0.1 and 0.2.

$$P(0.1) = -0.8159,\ P(0.2) = 0.3136$$

In Exercises 29–31, give the domain of each function.

29 $f(t) = t^2 + t - 1$

30 $g(y) = \dfrac{5 - 2y}{3y - 2}$

31 $h(x) = \dfrac{3x^2 - 2x + 5}{8x^2 + 22x - 21}$

In Exercises 32–35, determine the horizontal and vertical asymptotes of the graph of each function.

32 $h(s) = \dfrac{1}{s - 1}$

33 $y = \dfrac{2 - x}{x - 3}$

34 $S(t) = \dfrac{t^2 - t - 6}{4t^2 - 4t - 3}$

35 $R(t) = \dfrac{2t^2 - 18}{t^2 + 5}$

338 CHAPTER 4 POLYNOMIAL AND RATIONAL RELATIONS

The functions in Exercises 36–46 were graphed with a graphics calculator. (The scale has been added here on the axes to facilitate your examination of each graph.) Match each function with the appropriate graph.

36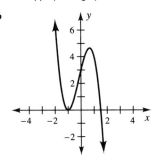

37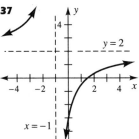

38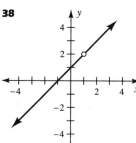

39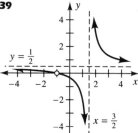

40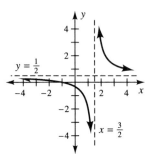

41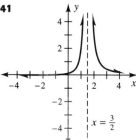

42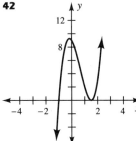

43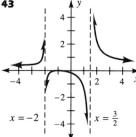

44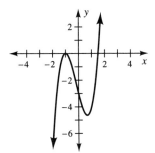

45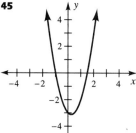

46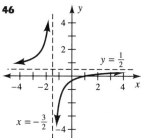

a. $y = (x + 1)(2x - 3)$

b. $y = (x + 1)(2x - 3)^2$

c. $y = (x + 1)^2(2x - 3)$

d. $y = (x + 1)^2(3 - 2x)$

e. $y = \dfrac{x + 1}{2x - 3}$

f. $y = \dfrac{2x - 3}{x + 1}$

g. $y = \dfrac{x^2 + 2x + 1}{2x^2 - x - 3}$

h. $y = \dfrac{x + 1}{4x^2 - 6x + 9}$

i. $y = \dfrac{x - 1}{2x + 3}$

j. $y = \dfrac{x^2 + 2x + 1}{2x^2 + x - 6}$

k. $y = \dfrac{2x^2 - x - 3}{2x - 3}$

47 Graph $y = \dfrac{2}{(x - 1)^2}$ and determine the intervals for which the function is increasing, decreasing, or constant.

48 Graph $y = \dfrac{x + 2}{x - 2}$ and use this graph to solve $\dfrac{x + 2}{x - 2} > 0$.

In Exercises 39–44, graph the relations defined by each equation.

49 $y = (x - 1)(x + 2)(x - 4)$

50 $R(t) = t(t - 1)^2(t + 3)^3$

51 $C(t) = 3t^5 - t^4 - 24t^3 + 8t^2 + 48t - 16$

52 $y = \dfrac{5}{x - 4}$

53 $g(r) = \dfrac{r^2 - r - 6}{r^2 - r - 20}$

54 $f(x) = \dfrac{x^2 - 2x + 1}{x^2 - 1}$

55 Write an equation of a function whose graph has a horizontal asymptote $y = 0$, vertical asymptotes $x = -1$ and $x = 2$, and an x-intercept $(\tfrac{1}{2}, 0)$.

OPTIONAL EXERCISES FOR CALCULUS-BOUND STUDENTS

In Exercises 1–4 graph each function.

1 $y = (2x + 3)(2x - 1)(x + 2)(x - 5)$

2 $y = x^3 - x^2 - 12x$

3 $y = 2x^4 + 9x^3 + 6x^2 - 11x - 6$

4 $y = \dfrac{2x^3 - x^2 - 25x - 12}{x^3 - x^2 - 5x + 5}$

5 Use the graph of $y = \dfrac{2x - 1}{2x^2 - x - 1}$ to solve $\dfrac{2x - 1}{2x^2 - x - 1} > 0$.

6 Show that $x + a$ is a factor of $x^{17} + a^{17}$.

7 Find all zeros of $x^5 - 4x^4 - 30x^3 + 148x^2 - 184x + 48$.

8 Determine the intervals for which $y = \dfrac{3x - 1}{2x + 4}$ is increasing.

9 Determine the oblique asymptote of the graph of $y = \dfrac{2x^2 + 4}{2x}$, and sketch this graph.

10 Determine the oblique asymptote of the graph of $y = \dfrac{(x + 1)^2(x - 2)}{(x - 1)(x + 2)}$, and sketch this graph.

In Exercises 11 and 12, use a graphics calculator or a computer graphing program to approximate all the real zeros of each function to the nearest tenth.

11 $y = 7x^3 + 9x^2 + 4x - 10$

12 $y = \dfrac{x^2 - 2x - 4}{x^2 + 3x + 2}$

MASTERY TEST FOR CHAPTER 4

Exercise numbers correspond to Section Objective numbers.

1. **a.** Sketch the graph of the polynomial function $P(x) = x(x - 2)(x + 2)(x - 4)$.
   b. Write the equation of the third-degree polynomial function whose graph is given to the right.
2. Use synthetic division to divide $2x^4 - 5x^2 + x - 5$ by $x + 2$.
3. $P(x) = 2x^4 - 5x^2 + x - 5$; use synthetic division to find $P(3)$.
4. Find a real polynomial of smallest degree with zeros -1 (multiplicity 2) and $2 + i$.
5. List the possible rational zeros of $P(x) = 6x^3 + 4x^2 - 5x - 15$.
6. Show that $P(x) = x^4 - 2x^3 - 10x^2 + 14x - 3$ has a zero between 3 and 4.
7. Find the number of possible positive and negative zeros for $P(x) = x^5 - x^4 + 3x^3 + 9x^2 - 12x + 5$.
8. Find upper and lower bounds for the zeros of $P(x) = x^4 - 2x^3 - 10x^2 + 14x - 3$.
9. Determine the zeros of $P(x) = 3x^4 - 14x^3 + x^2 + 38x + 12$.
10. Approximate to the nearest tenth the zero of $P(x) = x^4 - 2x^3 - 10x^2 + 14x - 3$ that is between 3 and 4.
11. Determine the horizontal and vertical asymptotes of the graph of $y = \dfrac{4x^2 - 16}{2x^2 + 3x - 9}$.
12. Graph the rational functions
    a. $y = \dfrac{2x - 1}{x + 2}$ **b.** $y = \dfrac{x - 1}{x^2 - x - 2}$

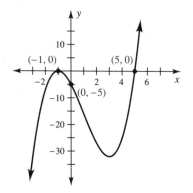

Figure for Example 1b

CHAPTER FIVE

Exponential and Logarithmic Functions

Chapter Five Objectives

1. Graph an exponential function.

2. Solve word problems involving growth and decay.

3. Interpret and use logarithmic notation.

4. Simplify logarithmic expressions using the basic identities.

5. Graph a logarithmic function.

6. Evaluate common and natural logarithms.

7. Use the change of base formula for logarithms.

8. Use the properties of logarithms.

9. Solve exponential and logarithmic equations.

John Napier, 1550–1617

Napier, from Scotland, is most known for his invention of logarithms. He also invented "Napier's Bones," a mechanical calculating device using logarithms that is an ancestor of the slide rule, and a hydraulic screw and revolving axle for controlling the water level in coal pits. At one time, he worked on plans for using mirrors to burn and destroy enemy ships.

Growth and decay problems are typical applications of exponential and logarithmic

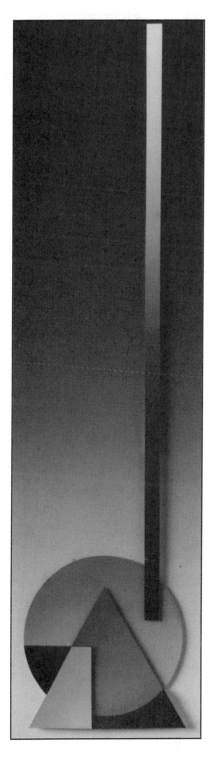

functions. Biologists use exponential functions to describe the growth of bacteria cultures for a given period of time. Bankers use similar functions to compute compound interest. Archaeologists can determine the age of ancient objects from the rate of decay of carbon-14. We will examine these problems in the exercises of this chapter.

SECTION 5-1

Exponential Functions

Section Objectives

1. Graph an exponential function.
2. Solve word problems involving growth and decay.

An **exponential function** has a variable rather than a constant in the exponent. For example, $f(x) = 2^x$ is an exponential functions, but $f(x) = x^2$ is not an exponential function, since the exponent is the constant 2.

Exponential Function ▼

> If $f(x) = b^x$, for a real constant $b > 0$ and $b \neq 1$, then f is an exponential function with base b. The domain of f is the set of all real numbers.

Examples illustrating the details of this definition are as follows.

- $y = (\frac{3}{7})^x$ is an exponential function with base $\frac{3}{7}$.
- $y = 1^x$ simplifies to $y = 1$. This is a constant function, not an exponential function.
- $f(x) = x^3$ has a constant exponent and is a cubic function, not an exponential function.
- $g(x) = (-7)^x$ has a negative base and therefore does not satisfy the conditions of the definition of an exponential function.
- $h(x) = x^x$ has a variable base and thus is not an exponential function.

In an exponential function $f(x) = b^x$, the exponent x can be any real number. This means that we must be able to interpret expressions with

SECTION 5-1 EXPONENTIAL FUNCTIONS

irrational exponents, such as $2^{\sqrt{2}}$. The irrational number $\sqrt{2}$ can be approximated by the rational numbers, 1, 1.4, 1.41, 1.414, 1.4142, etc. As these numbers "approach" $\sqrt{2}$, the values 2^1, $2^{1.4}$, $2^{1.41}$, $2^{1.414}$, and $2^{1.4142}$ "approach" $2^{\sqrt{2}}$.

This intuitive idea of defining irrational exponents through approximations is covered formally in more advanced mathematics texts. The important point is that the exponents in an exponential expression can be any real number—even an irrational number. The rules for exponents given in Section 1-3 hold for all real exponents, provided we make the restrictions given in the following box.

Restrictions on Real Exponents ▼

If b is a real base, $b > 0$ and $b \neq 1$, and if x is a real number, then
1. b^x is a unique real number.
2. If $0 < b < 1$ and if $r < x < s$ for rational numbers r and s, then $b^r > b^x > b^s$.
3. If $b > 1$ and if $r < x < s$ for rational numbers r and s, then $b^r < b^x < b^s$.

EXAMPLE 1 Approximate $3^{\sqrt{2}}$ accurate to six significant digits.

SOLUTION The key sequence on a calculator is

S: [3] [y^x] [2] [$\sqrt{}$] [=] → 4.7288044

G: [3] [^] [$\sqrt{}$] [2] [ENTER] → 4.728804388

Answer $3^{\sqrt{2}} \approx 4.72880$ (See the material on calculators in Appendix A if you have a question on proper calculator usage.) □

To graph exponential functions, we will begin by plotting points, and then we will examine the characteristics our graphs exhibit. We will choose most values for convenience, but we will use calculator values whenever appropriate.

▶ Self-Check ▼

1 Fill in the first two blanks with the correct inequality symbol and the last two blanks with either "increases" or "decreases."

For $b = \frac{1}{2}$, b^2 _____ b^3.

For $b = 2$, b^2 _____ b^3.

For $0 < b < 1$, if x increases, then b^x _____.

For $b > 1$, if x increases, then b^x _____.

2 Approximate $2^{\sqrt{3}}$ accurate to six significant digits.

▶ Self-Check Answers ▼

1 >; <; decreases; increases   2 3.32200

EXAMPLE 2 Graph the following functions:

SOLUTIONS

(a) $f(x) = 2^x$

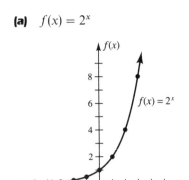

x	y
-3	$\frac{1}{8}$
-2	$\frac{1}{4}$
-1	$\frac{1}{2}$
0	1
1	2
2	4
3	8

(b) $f(x) = (\frac{1}{2})^x$

x	y
-3	8
-2	4
-1	2
0	1
1	$\frac{1}{2}$
2	$\frac{1}{4}$
3	$\frac{1}{8}$

□

Note that both graphs in Example 2 pass through the point (0, 1). All exponential functions $f(x) = b^x$ pass through (0, 1), since $b^0 = 1$. Also note that the graph of $f(x) = 2^x$ approaches the negative x-axis asymptotically. The graph rises rapidly to the right of the y-axis. These characteristics are common to all exponential functions with $b > 1$. An exponential function $f(x) = b^x$ with $b > 1$ is an increasing function that is described as an **exponential growth function**. Similarly, if $b < 1$, $f(x) = b^x$ is a decreasing function described as an **exponential decay function**. The decay function $f(x) = (\frac{1}{2})^x$ approaches the positive x-axis asymptotically. Note that $f(x) = (\frac{1}{2})^x$ can also be written in the form $f(x) = 2^{-x}$.

The domain of an exponential function $f(x) = b^x$ is the set of all real numbers, and the range is the set of all positive real numbers.

$f(x) = (\frac{1}{2})^x$
$f(x) = (2^{-1})^x$
$f(x) = 2^{-x}$

SECTION 5-1 EXPONENTIAL FUNCTIONS

Growth and Decay ▼

If $f(x) = b^x$ for real x, then for

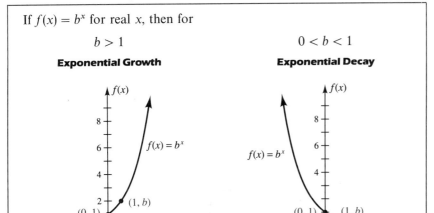

EXAMPLE 3 Suppose the number of units of bacteria in a culture after t minutes is modeled by $N(t) = 2^{0.5t}$.

SOLUTIONS

(a) How many units are present originally?

$N(t) = 2^{0.5t}$

$N(0) = 2^{0.5(0)} = 2^0 = 1$ $t = 0$ is the original time.

(b) How many units are present after 15 minutes?

$N(15) = 2^{0.5(15)}$ Calculator value for $2^{7.5}$ has been rounded to two significant digits.

$= 2^{7.5}$

≈ 180 units   ☐

The equations in the next example are all solved without the use of a calculator because they involve familiar powers of 2, 3, 4, or other small integers. The principles we will use are given in the following box.

Properties of Exponential Functions ▼

For real exponents x and y and bases $a > 0$, $b > 0$:

1. For $b \neq 1$, $b^x = b^y$ if and only if $x = y$.
2. For $x \neq 0$, $a^x = b^x$ if and only if $a = b$.

EXAMPLE 4 Solve these equations without using a calculator.

SOLUTIONS

(a) $2^{y+1} = 32$

$2^{y+1} = 32$
$2^{y+1} = 2^5$ The key to solving this problem by inspection is to express both sides of the equation in terms of the same base. So substitute 2^5 for 32.
$y + 1 = 5$ Since the bases are the same, the exponents must be equal.
$y = 4$

(b) $8^x = \dfrac{1}{16}$

$8^x = \dfrac{1}{16}$
$(2^3)^x = 2^{-4}$ Express 8 and $\dfrac{1}{16}$ in terms of the common base 2.
$2^{3x} = 2^{-4}$ Use the power rule for exponents.
$3x = -4$ The exponents are equal, since the bases are the same.
$x = -\dfrac{4}{3}$

(c) $(b-1)^3 = 64$

$(b-1)^3 = 64$
$(b-1)^3 = 4^3$ Substitute 4^3 for 64.
$b - 1 = 4$ The bases are equal, since the exponents are equal.
$b = 5$ □

The irrational number e, which arises in many problems, is called the **natural base** for exponential functions.* One common usage of this value is in computing compound interest. This concept is developed in the following material.

The formula for simple interest is $I = PRT$. The amount accumulated in one year is therefore

$$A = P + I = P + PR(1) = P(1 + R)$$ Substitute 1 for T, since the time is one year.

If this amount is compounded for t years, we have

Year t	Amount A	
0	$A = P$	The original amount is the principal P.
1	$A = P(1 + r)$	Amount at end of first year.
2	$A = [P(1 + r)](1 + r)$	
	$ = P(1 + r)^2$	Amount at end of second year.
3	$A = P(1 + r)^3$	
⋮	⋮	
t	$A = P(1 + r)^t$	Amount at end of tth year.

---

* The Euler number e was used by Léonhard Euler (1707–1783) as early as 1727.

SECTION 5-1 EXPONENTIAL FUNCTIONS

The general formula for the periodic compounding of a principal is given in the following box.

Periodic Compounding Formula ▼

$$A = P\left(1 + \frac{r}{n}\right)^{nt}$$

is the formula for determining the amount A when a principal P is invested at an annual rate r and compounded n times a year for t years.

EXAMPLE 5 A total of $1000 is invested at 8.5% annual interest. If this amount is left to compound quarterly for five years, what amount will result?

SOLUTION $A = P\left(1 + \frac{r}{n}\right)^{nt}$ Substitute the given values into the interest formula: $t = 5$ years, $n = 4$ times a year, and $r = 0.085$.

$= 1000\left(1 + \frac{0.085}{4}\right)^{4(5)}$

$= 1000(1.02125)^{20}$

S: [1][0][0][0][×][1][.][0][2][1][2][5][y^x][2][0][=] → 1522.7948
G: [1][0][0][0][×][1][.][0][2][1][2][5][^][2][0][ENTER] → 1522.79482

Answer $A \approx \$1522.79$ Calculator value has been rounded to the nearest penny. □

To develop the number e, consider the hypothetical case of an investment of $1 at 100% annual interest. Table 5-1 shows various compounding periods and the amount to which $1 accumulates in each situation after one year. The interest formula, with $P = 1$ and $r = 1.00$, is

$$A = 1\left(1 + \frac{1}{n}\right)^{n(1)} = \left(1 + \frac{1}{n}\right)^n$$

Note that compounding monthly produces a big gain over simple interest. However, further gains are realtively small, and as n increases the expression $\left(1 + \frac{1}{n}\right)^n$ approaches the irrational number e. The first eleven digits of e are 2.7182818284 (you should remember that $e \approx 2.718$).

▶ **Self-Check** ▼

Solve these equations without using a calculator.

1 $3^y = 81$
2 $9^x = 3$
3 $(2a + 4)^3 = 8$
4 A total of $500 is invested at 9% annual interest. If this amount is compounded monthly for two years, what amount will result?

▶ **Self-Check Answers** ▼

1 $y = 4$ **2** $x = \frac{1}{2}$ **3** $a = -1$ **4** $598.21

Table 5-1 Periodic compounding

Compounded	n	$\left(1+\dfrac{1}{n}\right)^n$ (to six places)	A (to nearest penny)
annually	1	2.000000	$2.00
semiannually	2	2.250000	2.25
quarterly	4	2.441406	2.44
monthly	12	2.613035	2.61
weekly	52	2.692597	2.69
daily	365	2.714567	2.71
hourly	8,760	2.718127	2.72
every minute	525,600	2.718279	2.72
every second	31,536,000	2.718282	2.72
⋮	⋮	⋮	
continuously	infinite	e	

EXAMPLE 6 Use a calculator to evaluate these expressions accurate to six significant digits.

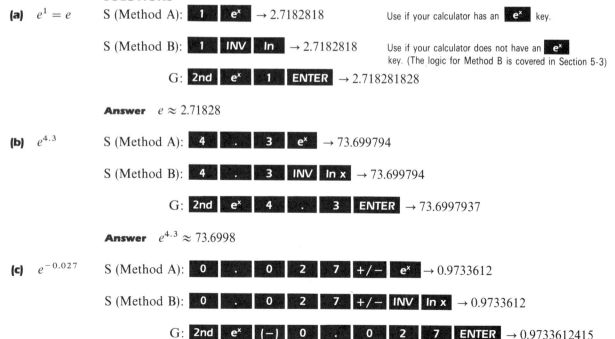

SOLUTIONS

(a) $e^1 = e$

S (Method A): [1] [e^x] → 2.7182818 Use if your calculator has an [e^x] key.

S (Method B): [1] [INV] [ln] → 2.7182818 Use if your calculator does not have an [e^x] key. (The logic for Method B is covered in Section 5-3)

G: [2nd] [e^x] [1] [ENTER] → 2.718281828

Answer $e \approx 2.71828$

(b) $e^{4.3}$

S (Method A): [4] [.] [3] [e^x] → 73.699794

S (Method B): [4] [.] [3] [INV] [ln x] → 73.699794

G: [2nd] [e^x] [4] [.] [3] [ENTER] → 73.6997937

Answer $e^{4.3} \approx 73.6998$

(c) $e^{-0.027}$

S (Method A): [0] [.] [0] [2] [7] [+/−] [e^x] → 0.9733612

S (Method B): [0] [.] [0] [2] [7] [+/−] [INV] [ln x] → 0.9733612

G: [2nd] [e^x] [(−)] [0] [.] [0] [2] [7] [ENTER] → 0.9733612415

Answer $e^{-0.027} \approx 0.973361$

SECTION 5-1 EXPONENTIAL FUNCTIONS

(d) 3^e

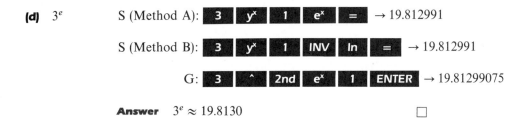

Answer $3^e \approx 19.8130$

The general formula for the continuous compounding of a principal is given in the following box. If $r = 100\%$, as in the hypothetical case in Table 5-1, then $A = Pe^{1.00t}$, or $A = Pe^t$.

Continuous Compounding Formula ▼

$A = Pe^{rt}$ is the formula for determining the amount A when a principal P is invested at an annual rate r and compounded continuously for t years.

EXAMPLE 7   A total of $100 is invested at 7% annual interest. If this amount is compounded continuously for three years, what amount will result?

SOLUTION $A = Pe^{rt}$
$= 100e^{0.07(3)}$ Substitute the given values into the continuous interest formula.
$= 100e^{0.21}$

Answer $A \approx \$123.37$ Calculator value has been rounded to the nearest cent.

▶ **Self-Check ▼**

Approximate these values accurate to six significant digits.

1 e^2 **2** $\dfrac{1}{e}$ **3** $e^{\sqrt{3}}$

4 What amount results if $500 is invested at 7.25% annual interest compounded continuously for two years?

The function $A = Pe^{rt}$ applies to all problems for which the rate of change of a quantity varies directly with the amount of the quantity present. Specifically, this formula is used in continuous growth and decay problems, with positive r denoting growth and negative r denoting decay.

EXAMPLE 8   The amount of radioactive carbon-14 remaining after t years is given by $A = Pe^{-0.0001245t}$, where P is the original amount present. How much of a 1000-gram sample remains a century later?

SOLUTION $A = Pe^{-0.0001245t}$
$= 1000e^{-0.0001245(100)}$ Substitute the given values into the formula.
$= 1000e^{-0.01245}$

Answer $A \approx 987.6$ grams   Calculator value has been rounded to four significant digits.

▶ **Self-Check Answers ▼**

1 7.38906   **2** 0.367879   **3** 5.65223   **4** $578.02

EXERCISES 5-1

A

In Exercises 1–8, solve each equation without using a calculator.

1. a. $2^x = 64$ b. $8^v = 64$ c. $16^w = 64$ d. $\left(\frac{3}{11}\right)^z = \frac{121}{9}$
2. a. $4^x = 64$ b. $64^v = 64$ c. $32^w = 64$ d. $\left(\frac{3}{5}\right)^z = \frac{125}{27}$
3. a. $9^y = 27$ b. $10^x = 0.001$ c. $8^z = 0.125$ d. $2^m = \sqrt[3]{2}$
4. a. $125^y = 25$ b. $5^x = 0.04$ c. $(0.5)^z = 4$ d. $3^m = \sqrt[5]{3}$
5. a. $16^{2x+1} = 32^{4x}$ b. $3^{x^2} = 81$ c. $3^{x^2+2x} = \frac{1}{3}$
6. a. $4^{2y} = 2^{1-y}$ b. $5^{x^2-1} = 1$ c. $3^{x^2+4x} = \frac{1}{81}$
7. a. $x^5 = 32$ b. $z^{-3} = \frac{125}{64}$ c. $(b+3)^2 = 25$
8. a. $y^3 = \frac{1}{27}$ b. $w^{-7} = \frac{1}{128}$ c. $(2b-1)^3 = 1000$

In Exercises 9 and 10, use a calculator to determine each value accurate to six significant digits, given that $f(x) = 2^x$ and $g(y) = e^y$.

9. a. $g(3)$ b. $g(-3)$ c. $g(\pi)$ d. $(g \circ f)(3)$
10. a. $g(4)$ b. $g(-4)$ c. $f(e)$ d. $(g \circ f)(-2)$

In Exercises 11–14, graph each pair of functions on the same coordinate system.

11. $f(x) = 3^x$ $g(x) = 3^{-x}$
12. $f(x) = 4^x$ $g(x) = 4^{-x}$
13. $f(x) = \left(\frac{2}{5}\right)^x$ $g(x) = \left(\frac{5}{2}\right)^x$
14. $f(x) = \left(\frac{4}{3}\right)^x$ $g(x) = \left(\frac{3}{4}\right)^x$

In Exercises 15–18, use the formula $A = P\left(1 + \dfrac{r}{n}\right)^{nt}$ to determine the amount resulting from the given investment.

15. $600 invested at 7% per year compounded daily for five years.
16. $5000 invested at 7.5% per year compounded monthly for two years.
17. $1275 invested at 6.75% per year compounded monthly for $3\frac{1}{2}$ years.
18. $1000 invested at 8.25% per year compounded semiannually for four years.

In Exercises 19–24, use the formula $A = Pe^{rt}$ to determine the answer.

19. Find the amount resulting from a $1000 investment at 8.5% per year compounded continuously for two years.
20. Find the amount resulting from a $5000 investment at 8% per year compounded continuously for three years.
21. The number of bacteria in a culture is growing continuously. After t hours, the number present is given by $A = Pe^{0.2t}$. If $P = 100$, find the number present after one day (accurate to three significant digits).
22. The number of locusts invading a crop area is growing continuously. After t days, the number present is given by $A = Pe^{1.5t}$. If the initial number of locusts is $P = 1000$, find the number present after one week (accurate to three significant digits).
23. The strontium-90 in a nuclear reactor decays continuously. If 100 milligrams is present initially, the amount present after t years is given by $A = 100e^{-0.0248t}$. Find the number of milligrams present after 10 years (accurate to three significant digits).

SECTION 5-1 EXPONENTIAL FUNCTIONS

24 The amount of carbon-14 remaining after t years is given by $A = Pe^{-0.0001245t}$, where P is the original amount present. How much of a 100-gram sample remains after 1000 years?

In Exercises 25–28, mentally estimate the value of each expression and select the choice that best matches your answer.

25 e^{-1}
   a. $e^{-1} < \frac{1}{3}$ **b.** $\frac{1}{3} < e^{-1} < \frac{1}{2}$ **c.** $e^{-1} > \frac{1}{2}$

26 e^2
   a. $e^2 < 4$ **b.** $e^2 > 9$ **c.** $4 < e^2 < 9$

27 $\sqrt{e}$
   a. $\sqrt{e} < 1$ **b.** $1 < \sqrt{e} < 1.5$ **c.** $1.5 < \sqrt{e} < 2$ **d.** $2 < \sqrt{e} < 3$

28 2^e
   a. $2^e < 4$ **b.** $4 < 2^e < 5$ **c.** $5 < 2^e < 8$ **d.** $2^e > 8$

In Exercises 29–30, use the graph of $y = e^x$ shown to the right to approximate each expression accurate to the nearest tenth.

29 $e^{1.5}$ **30** $e^{1.2}$

Figure for Exercises 29–30

B

In Exercises 31–34, each graph is a translation or reflection of the graph of the exponential growth function $y = 2^x$. Match each function with the appropriate graph.

31 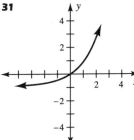   **32** 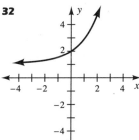   **33** 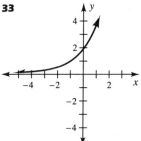   **34**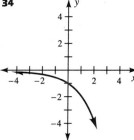

a. $y = 2^{x+1}$ **b.** $y = 2^x + 1$ **c.** $y = -2^x$ **d.** $y = 2^x - 1$

In Exercises 35–38, each graph is a translation or reflection of the graph of the exponential decay function $y = (\frac{1}{2})^x$. Match each function with the appropriate graph.

35 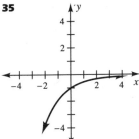   **36** 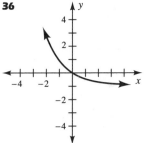   **37** 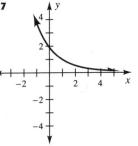   **38**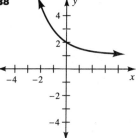

a. $y = (\frac{1}{2})^x - 1$ **b.** $y = (\frac{1}{2})^{x-1}$ **c.** $y = -(\frac{1}{2})^x$ **d.** $y = (\frac{1}{2})^x + 1$

39 If $f(x) = 2^x$, solve each equation for x.
  a. $f(x) = 8$ **b.** $f(x) = 32$ **c.** $f(x) = \frac{1}{4}$ **d.** $f(x) = \sqrt[3]{2}$

40 If $f(x) = 3^x$, solve each equation for x.
  a. $f(x) = 9$ **b.** $f(x) = \frac{1}{9}$ **c.** $f(x) = 1$ **d.** $f(x) = \sqrt{3}$

In Exercises 41 and 42, identify the shape of each function without sketching the graph.

41 **a.** $y = 2x$ **b.** $y = x^2$ **c.** $y = 2^x$
42 **a.** $y = -2x$ **b.** $y = -x^2$ **c.** $y = 2^{-x}$

43 How much more would be earned by investing $1,000 at 7% compounded continuously for a year instead of investing this same amount at 7% compounded monthly?

C

44 Give an example of real values of b, x, and y for which $b^x = b^y$ but $x \neq y$.

45 Give an example of real values of a, b, and x for which $a^x = b^x$ but $a \neq b$.

In Exercises 46–49, use a graphics calculator to graph each pair of functions on the same coordinate system.

46 $f(x) = e^x$
   $g(x) = e^{-x}$

47 $f(x) = 3^{x+1}$
   $g(x) = 3^x + 1$

48 $f(x) = e^{2x}$, $g(x) = \dfrac{1}{e^{2x}}$

49 $f(x) = 2^x$, $g(x) = 2^{|x|}$

In Exercises 50–52, use a graphics calculator to approximate each solution of the following equations to the nearest tenth.

50 Solve $e^x = 2$ by finding the x-intercept of $y = e^x - 2$.

51 Solve $e^x = 5$ by finding the x-intercept of $y = e^x - 5$.

52 Solve $e^{x-1} = 7$ by finding the x-intercept of $y = e^{x-1} - 7$.

SECTION 5-2

Logarithmic Functions

Section Objectives

3 Interpret and use logarithmic notation.

4 Simplify logarithmic expressions using the basic identities.

5 Graph a logarithmic function.

From the graph of an exponential function and the horizontal line test, we know that $f(x) = b^x$ is a one-to-one function for $b > 0$ and $b \neq 1$. Thus each exponential function has an inverse function, which is called a **logarithmic function**. The inverse of $f(x) = b^x$ is written $f^{-1}(x) = \log_b x$. The notation $y = \log_b x$ is read "y equals the log of x base b." The inverses of two exponential functions are shown in Figure 5-1. Note the domain and the

SECTION 5-2 LOGARITHMIC FUNCTIONS

range of the logarithmic function corresponding to the growth function with $b > 1$ and the decay function with $0 < b < 1$.

We observe from these graphs that logarithms are defined only for positive real numbers. That is, the range of an exponential function—and thus the domain of its inverse—is the set of positive real numbers. Likewise, we note that the range of a logarithmic function is the set of all real numbers. The graph of a logarithmic function is asymptotic to the negative y-axis if $b > 1$ and to the positive y-axis if $0 < b < 1$.

Logarithmic Function ▼

For $x > 0$, $b > 0$, and $b \neq 1$,

$$y = \log_b x \quad \text{if and only if} \quad b^y = x$$

- Base: b
- Logarithm or exponent: y

Think: Logarithms are exponents.

(a) $b > 1$

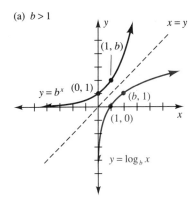

(b) $0 < b < 1$

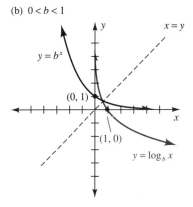

Figure 5-1 Logarithmic functions are inverses of exponential functions

This relationship between logarithmic form and exponential form is illustrated in Table 5-2.

Table 5-2 Logarithmic and exponential forms

Logarithmic Form	Verbal Form	Exponential Form
$\log_7 49 = 2$	The log of 49 base 7 is 2.	$7^2 = 49$
$\log_5(\frac{1}{25}) = -2$	The log of $\frac{1}{25}$ base 5 is -2.	$5^{-2} = \frac{1}{25}$
$\log_3 \sqrt{3} = \frac{1}{2}$	The log of $\sqrt{3}$ base 3 is $\frac{1}{2}$.	$3^{1/2} = \sqrt{3}$

To evaluate $\log_b x$, think "what exponent on b is needed to obtain x?" For problems that involve familiar powers of small integers, it may be possible to find the answer without the use of a calculator. For more complicated values, we will assume the use of a calculator.

EXAMPLE 1 Determine these logarithmic values without using a calculator.

SOLUTIONS

(a) $\log_2 32$ $\log_2 32 = 5$, since $2^5 = 32$

(b) $\log_{2/5}\left(\dfrac{25}{4}\right)$ $\log_{2/5}\left(\dfrac{25}{4}\right) = -2$, since $\left(\dfrac{2}{5}\right)^{-2} = \dfrac{25}{4}$

(c) $\log_7 1$ $\log_7 1 = 0$, since $7^0 = 1$

(d) $\log_2(-8)$ $\log_2(-8)$ is undefined, since the argument -8 is negative.

(e) $\log_5 0$ $\log_5 0$ is undefined, since the argument cannot be zero.

□

The following examples are first rewritten in the more familiar exponential form and then solved by inspection. With practice you should be able to work with the logarithmic form as comfortably as with the exponential form.

▶ **Self-Check** ▼

Write each logarithmic equation in exponential form and each exponential equation in logarithmic form.

1 $\log_2 8 = 3$ **2** $\log_6 \sqrt[5]{6} = \dfrac{1}{5}$

3 $\log_3\left(\dfrac{1}{81}\right) = -4$ **4** $9^2 = 81$

5 $2^{-3} = \dfrac{1}{8}$ **6** $5^0 = 1$

EXAMPLE 2 Solve $\log_2(x - 7) = 3$ for x.

SOLUTION $2^3 = x - 7$ Rewrite the equation in exponential form.
$\ \ \ 8 = x - 7$
$\ \ \ x = 15$ □

EXAMPLE 3 Solve $\log_b 81 = -4$ for b.

SOLUTION $b^{-4} = 81$ Rewrite the equation in exponential form.
$\phantom{SOLUTION\ \ b^{-4}}= 3^4$
$\phantom{SOLUTION\ \ b^{-4}}= \left(\dfrac{1}{3}\right)^{-4}$ Express each side of the equation in terms of the same exponent.
$b = \dfrac{1}{3}$ The bases are equal, since exponents are the same.   □

EXAMPLE 4 Solve $\log_8 16 = x$ for x.

SOLUTION $8^x = 16$ Rewrite the equation in exponential form.
$(2^3)^x = 2^4$ Express each side of the equation in terms of the same base.
$2^{3x} = 2^4$ Use the power rule for exponents.
$3x = 4$ The exponents are equal, since the bases are the same.
$x = \dfrac{4}{3}$ □

▶ **Self-Check Answers** ▼

1 $2^3 = 8$ **2** $6^{1/5} = \sqrt[5]{6}$ **3** $3^{-4} = \dfrac{1}{81}$ **4** $\log_9 81 = 2$

5 $\log_2\left(\dfrac{1}{8}\right) = -3$ **6** $\log_5 1 = 0$

SECTION 5-2 LOGARITHMIC FUNCTIONS

The following identities are used so frequently in logarithmic problems that we have put them in a box for future reference. Note that some of these statements follow directly from the fact that logarithms are exponents on some base b.

Logarithmic Identities ▼

For $b > 0$, $b \neq 1$, and $x > 0$,

1. $\log_b 1 = 0$ since $b^0 = 1$
2. $\log_b b = 1$ since $b^1 = b$
3. $\log_b \dfrac{1}{b} = -1$ since $b^{-1} = \dfrac{1}{b}$
4. $\log_b y = y$ since $b^y = b^y$
5. $b^{\log_b x} = x$ since $\log_b x = \log_b x$
6. If $\log_b x_1 = \log_b x_2$, then $x_1 = x_2$ since a log function is a one-to-one function
7. If $\log_{b_1} x = \log_{b_2} x \neq 0$, then $b_1 = b_2$ since $b_1 = b_2$ if $b_1^y = b_2^y$ and $y \neq 0$

▶ Self-Check ▼

Solve these equations without using a calculator.

1. $\log_4 16 = x$
2. $\log_3 y = -2$
3. $\log_b 125 = 3$

EXAMPLE 5 Solve these equations by inspection.

	SOLUTIONS	
(a) $\log_{183} 1 = y$	$y = 0$	Each of these results follows directly from the properties in the preceding box.
(b) $\log_{17} 17 = y$	$y = 1$	
(c) $\log_5 5^7 = y$	$y = 7$	
(d) $8^{\log_8 64} = y$	$y = 64$	The log of 64 base 8 is the exponent that we put on 8 to get 64. So if we put this exponent on 8 we must get 64.
(e) $\log_7(x + 3) = \log_7 83$	$x + 3 = 83$	
	$x = 80$	
(f) $\log_4 17.3 = \log_b 17.3$	$b = 4$	□

The graphs earlier in this section have already demonstrated the characteristic shape of a logarithmic function. Thus we shall plot only a few points in order to sketch the functions in the following examples. Note that $(1, 0)$ and $(b, 1)$ are always on these graphs, since $\log_b 1 = 0$ and $\log_b b = 1$.

▶ Self-Check Answers ▼

1. $x = 2$ 2. $y = \dfrac{1}{9}$ 3. $b = 5$

EXAMPLE 6 Graph $f(x) = \log_3 x$.

SOLUTION First write $y = \log_3 x$ in the exponential form $x = 3^y$. Then form a table of values to plot.

x	y
$\frac{1}{3}$	-1
1	0
3	1
9	2

$3^{-1} = \frac{1}{3}$
$3^0 = 1$
$3^1 = 3$
$3^2 = 9$

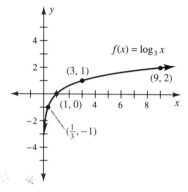

The domain of f is the set of positive real numbers. The range of f is the set of all real numbers.

EXAMPLE 7 Graph $f(x) = \log_3 x$ and $g(x) = 2 + \log_3 x$ on the same coordinate system.

SOLUTION Since $g(x) = f(x) + 2$, the graph of $f(x) = \log_3 x$ from Example 6 is translated up 2 units.

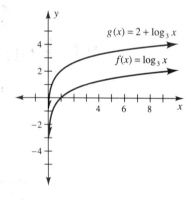

▶ **Self-Check** ▼

Solve these logarithmic equations by inspection.

1 $\log_{3.2} 3.2 = y$ **2** $\log_{0.91} 1 = y$
3 $11^{\log_{11} 71} = x$ **4** $\log_{2/3}(\frac{3}{2}) = y$
5 Graph

$f(x) = \log_2 x$ and
$g(x) = \log_2(x - 3)$

on the same coordinate system.

▶ **Self-Check Answers** ▼

1 $y = 1$ **2** $y = 0$ **3** $x = 71$ **4** $y = -1$ **5**

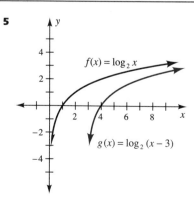

SECTION 5-2 LOGARITHMIC FUNCTIONS

EXAMPLE 8 Write the inverse of $f(x) = \log_2 x$, and graph both of these functions on the same coordinate system.

SOLUTION

$y = \log_2 x$	Write f in x-y notation.
$x = \log_2 y$	Interchange x and y to form the inverse function.
$y = 2^x$	Rewrite the equation in exponential form.
$f^{-1}(x) = 2^x$	Express this inverse in functional notation.

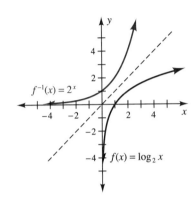

EXERCISES 5-2

A

In Exercises 1 and 2, write each logarithmic equation in exponential form. Assume that $b > 0$, $b \neq 1$, and $x > 0$.

1. **a.** $\log_8 64 = 2$ **b.** $\log_{23} 1 = 0$ **c.** $\log_{7/5}(\frac{25}{49}) = -2$ **d.** $\log_7 x = 2.3$
2. **a.** $\log_2 64 = 6$ **b.** $\log_{17} 1 = 0$ **c.** $\log_8(\frac{1}{2}) = -\frac{1}{3}$ **d.** $\log_b 73 = 2.3$

In Exercises 3 and 4, write each exponential equation in logarithmic form.

3. **a.** $27^{1/3} = 3$ **b.** $(\frac{3}{7})^{-2} = \frac{49}{9}$ **c.** $6^0 = 1$ **d.** $b^{11} = x$
4. **a.** $(\frac{1}{2})^{-3} = 8$ **b.** $9^{3/2} = 27$ **c.** $7^0 = 1$ **d.** $b^x = 11$

In Exercises 5–20, evaluate each logarithmic expression.

5. $\log_{13} 169$
6. $\log_3 81$
7. $\log_{10} 100{,}000$
8. $\log_{10} 10{,}000$
9. $\log_6\left(\dfrac{1}{36}\right)$
10. $\log_{3/5}\left(\dfrac{125}{27}\right)$
11. $\log_{11}\sqrt{11}$
12. $\log_7\sqrt[3]{7}$
13. $\log_8 0$
14. $\log_{11}(-3)$
15. $\log_3(-11)$
16. $\log_4 0$
17. $\log_2\left(\dfrac{16}{8}\right) - \dfrac{\log_2 16}{\log_2 8}$
18. $\log_3\left(\dfrac{81}{27}\right) - \dfrac{\log_3 81}{\log_3 27}$
19. $\log_5\left(\dfrac{125}{5}\right) - (\log_5 125 - \log_5 5)$
20. $\log_2\left(\dfrac{1}{2}\right) - (\log_2 1 - \log_2 2)$

In Exercises 21–46, solve each equation for x.

21 $\log_5(2x - 4) = 2$
22 $\log_3(x + 7) = 3$
23 $\log_x 4 = -2$
24 $\log_x 7 = \dfrac{1}{2}$

25 $\log_{10} 100 = 3x - 4$
26 $\log_{10} 1000 = 2x - 11$
27 $\log_{10} \sqrt[5]{10} = \dfrac{x}{10}$
28 $\log_{10} 0.01 = x - 7$

29 $\log_{43} 1 = x$
30 $\log_{16} 8 = x$
31 $\log_{25} 125 = x$
32 $\log_{27} 81 = x$

33 $\log_8 32 = x$
34 $\log_{19} 19 = x$
35 $\log_{11} 11^5 = x$
36 $\log_b b^8 = x$

37 $\log_b\left(\dfrac{1}{b^2}\right) = x$
38 $\log_b \sqrt{b} = x$
39 $\log_b x = 0$
40 $\log_b x = 1$

41 $b^{\log_b 51} = x$
42 $b^{\log_b 37} = x$
43 $\log_x 17 = \log_{11} 17$
44 $\log_3 29 = \log_x 29$

45 $\log_b 47 = \log_b x$
46 $\log_b 9.3 = \log_b x$

In Exercises 47–50, graph each logarithmic function.

47 $y = \log_5 x$
48 $y = \log_{10} x$
49 $y = \log_{1/5} x$
50 $y = \log_{0.1} x$

B

In Exercises 51 and 52, use the graph of $y = \log_2 x$ shown below to approximate each expression accurate to the nearest tenth.

51 $\log_2 3$
52 $\log_2 5$

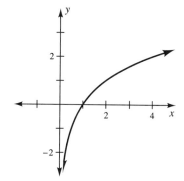

In Exercises 53–58, mentally estimate the value of each expression. Then select the pair of consecutive integers between which this value falls.

53 $\log_2 73$
    a. 1, 2 **b.** 2, 3 **c.** 3, 4 **d.** 5, 6 **e.** 6, 7

54 $\log_3 73$
    a. 1, 2 **b.** 2, 3 **c.** 3, 4 **d.** 5, 6 **e.** 6, 7

55 $\log_5 0.125$
    a. $-3, -2$ **b.** $-2, -1$ **c.** $-1, 0$ **d.** 0, 1 **e.** 1, 2

56 $\log_e 3$
    a. $-3, -2$ **b.** $-2, -1$ **c.** $-1, 0$ **d.** 0. 1 **e.** 1, 2

57 $\log_e 2$
    a. $-3, -2$ **b.** $-2, -1$ **c.** $-1, 0$ **d.** 0, 1 **e.** 1, 2

58 $\log_e 0.5$
    a. $-3, -2$ **b.** $-2, -1$ **c.** $-1, 0$ **d.** 0, 1 **e.** 1, 2

SECTION 5-3 COMMON AND NATURAL LOGARITHMS

In Exercises 59–62, each graph is a translation or reflection of the graph of the function $y = \log_2 x$. Match each function with the appropriate graph.

59 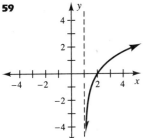　60 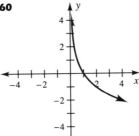　61 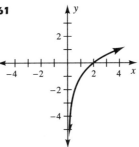　62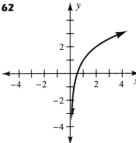

a. $y = \log_2(x - 1)$　**b.** $y = \log_2 x - 1$　**c.** $y = -\log_2 x$　**d.** $y = \log_2 x + 1$

C

In Exercises 63 and 64, use a graphics calculator to approximate each solution of the following equations to the nearest tenth.

63 Solve $\ln x = 2$ by finding the x-intercept of $y = \ln x - 2$. ($\ln x$ denotes $\log_e x$).

64 Solve $\ln n = -1$ by finding the x-intercept of $y - \ln x + 1$.

65 The graph of the logarithmic function $y = \log_b x$ passes through the points $(1, 0)$ and $(5, 1)$. What is the base of this function?

In Exercises 66–68, evaluate each expression, given $f(x) = 10^x$ and $g(x) = \log_{10} x$.

66 $(f \circ g)(100)$　　67 $(g \circ f)(3)$　　68 $f(1) - g(1)$

69 Graph $y = \log_{0.5} x$ and its inverse on the same coordinate system.

70 Graph $y = \log_{1/3} x$ and its inverse on the same coordinate system.

SECTION 5-3

Common and Natural Logarithms

Section Objectives

6 Evaluate common and natural logarithms.

7 Use the change of base formula for logarithms.

8 Use the properties of logarithms.

Although $\log_b N$ is defined for any base $b > 0$ and $b \neq 1$, only two bases are commonly used. We often use logarithms base 10, since our number system is based on powers of 10. Logarithms base 10 are called **common logarithms**, and $\log_{10} x$ is denoted by the abbreviated form $\log x$. Logarithms to the natural base e are used for many growth and decay applications. Logarithms base e are called **natural logarithms**, and $\log_e x$ is denoted by $\ln x$. Natural logarithms are used extensively by mathematicians because many formulas are easier to state in base e. Common and natural logarithmic functions are graphed as shown in Figure 5-2.

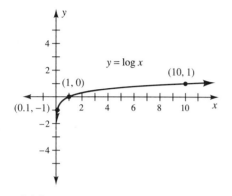

(a) Common logarithmic function

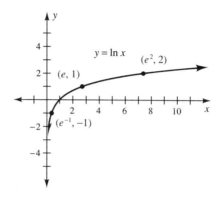

(b) Natural logarithmic function

Figure 5-2

Common and Natural Logarithms ▼

Common logarithms:	$\log x$ means $\log_{10} x$
Natural logarithms:	$\ln x$ means $\log_e x$

Examples of common and natural logarithms are shown in Table 5-3.

Table 5-3 Common and natural logarithms

Logarithmic Form	Verbal Form	Exponential Form
$\log 10{,}000 = 4$	The common log of 10,000 is 4.	$10^4 = 10{,}000$
$\log 0.001 = -3$	The common log of 0.001 is -3.	$10^{-3} = 0.001$
$\ln e^3 = 3$	The natural log of e^3 is 3.	$e^3 = e^3$
$\ln e^{-2} = -2$	The natural log of e^{-2} is -2.	$e^{-2} = e^{-2}$

Only a relatively few logarithms, such as $\log 100 = 2$, can be determined by inspection. Thus common and natural logs have historically been determined from tables, slide rules, and other devices. Today calculators are usually used to determine these values. A typical key sequence is illustrated in the next example; the actual labeling of the keys and the number of significant digits will vary from model to model.

SECTION 5-3 COMMON AND NATURAL LOGARITHMS

EXAMPLE 1 Use a calculator to determine these logarithmic values accurate to five significant digits.

SOLUTIONS

(a) log 76.2

Key sequence:

S: [7] [6] [.] [2] [log] → 1.8819550

G: [log] [7] [6] [.] [2] [ENTER] → 1.881954971

Answer log 76.2 ≈ 1.8820

(b) log 0.00459

Key sequence:

S: [0] [.] [0] [0] [4] [5] [9] [log] → −2.3381873

G: [log] [0] [.] [0] [0] [4] [5] [9] [ENTER] → −2.338187314

Answer log 0.00459 ≈ −2.3382

(c) $\log(3.4097 \times 10^{-7})$

Key sequence:

S: [3] [.] [4] [0] [9] [7] [EE] [7] [+/−] [log] → −6.4672838

G: [log] [3] [.] [4] [0] [9] [7] [EE] [(−)] [7] [ENTER] → −6.46728383

Answer $\log(3.4097 \times 10^{-7}) \approx -6.4673$

(d) ln 42.46

Key sequence:

S: [4] [2] [.] [4] [6] [ln] → 3.7485625

G: [ln] [4] [2] [.] [4] [6] [ENTER] → 3.748562456

Answer ln 42.46 ≈ 3.7486

(e) ln 0.0123

Key sequence:

S: [0] [.] [0] [1] [2] [3] [ln] → −4.398156

G: [ln] [0] [.] [0] [1] [2] [3] [ENTER] → −4.398156017

Answer ln 0.0123 ≈ −4.3982 ◻

Warning for Computer Users ▼

Some computer languages provide only the natural logarithmic function, in which case this function may be called LOG instead of ln even though the base is e. If you are unsure of the meaning of LOG on a particular computer, test a value such as LOG 10 to determine which base is being used.

► Self-Check ▼

Determine these values accurate to six significant digits.

1 log 192.7
2 ln 192.7
3 $\ln(1.927 \times 10^8)$

Since logarithms and exponential functions are inverses of each other, many calculators incorporate both ln x and e^x into one key, in which case some other key such as an **INV** key (inverse key) is used to indicate which of the two functions is being selected. This same logic allows dual usage of a key for both log x and 10^x.

EXAMPLE 2 Use a calculator to evaluate these expressions accurate to six significant digits

SOLUTIONS

(a) $e^{2.378}$

S (Method A): 2 . 3 7 8 e^x → 10.783315

S (Method B): 2 . 3 7 8 INV ln → 10.783315

G: 2nd e^x 2 . 3 7 8 ENTER → 10.78331465

Answer $e^{2.378} \approx 10.7833$

(b) $10^{-1.0047}$

S (Method A): 1 . 0 0 4 7 +/− 10^x → 0.0989236

S (Method B): 1 . 0 0 4 7 +/− INV log → 0.0989236

G: 2nd 10^x (−) 1 . 0 0 4 7 ENTER → 0.0989236199

Answer $10^{-1.0047} \approx 0.0989236$ □

Before the advent of calculators, logarithms were used extensively for computations involving multiplication, division, and exponentiation. Logarithmic tables can be used in two ways: to find logs of given values and, in reverse, to find numbers whose logs are known. The term **antilogarithm** or **antilog** is used to denote a number whose logarithm is known. For example, if log $x = 3.4097$, then $x =$ antilog 3.4097. Thus

$$x = \text{antilog } 3.4097 = \text{inverse log } 3.4097 = 10^{3.4097}$$

► Self-Check Answers ▼

1 2.28488   **2** 5.26113   **3** 19.0766

SECTION 5-3 COMMON AND NATURAL LOGARITHMS

EXAMPLE 3 Use a calculator to find antilog 3.4097 accurate to five significant digits.

SOLUTION antilog $3.4097 = 10^{3.4097}$ Rewrite in exponential form.

S (Method A): `3` `.` `4` `0` `9` `7` `10^x` → 2568.6208

S (Method B): `3` `.` `4` `0` `9` `7` `INV` `log` → 2568.6208

G: `2nd` `10^x` `3` `.` `4` `0` `9` `7` `ENTER` → 2568.620829

Answer antilog $3.4097 \approx 2568.6$

The properties of logarithms facilitate the solving of exponential equations and provide a means of simplifying many algebraic expressions.

Properties of Logarithms ▼

For $x, y > 0$, $b > 0$, and $b \neq 1$:

Product rule: $\log_b xy = \log_b x + \log_b y$ The log of a product is the sum of the logs.

Quotient rule: $\log_b \dfrac{x}{y} = \log_b x - \log_b y$ The log of a quotient is the difference of the logs.

Power rule: $\log_b x^p = p \log_b x$ The log of the pth power of x is p times the log of x.

▶ Self-Check ▼

Determine these values accurate to six significant digits.

1 e^{-2}

2 $10^{0.30103}$

3 antilog 0.17609

Since logarithms are exponents, these logarithmic properties are actually just restatements of the exponential properties. For example,

Product Rule for Logarithms

$\log_b \underbrace{xy}_{\text{Product}} = \underbrace{\log_b x + \log_b y}_{\text{Add logarithms}}$

Product Rule for Exponents

$\underbrace{b^x b^y}_{\text{Product}} = \underbrace{b^{x+y}}_{\text{Add exponents}}$

— Logarithms are exponents. —

▶ Self-Check Answers ▼

1 0.135335   **2** 2.00000   **3** 1.50000

CHAPTER 5 EXPONENTIAL AND LOGARITHMIC FUNCTIONS

Proof of the Product Rule Let $x = b^m$ and $y = b^n$, and thus $\log_b x = m$ and $\log_b y = n$. Then

$$xy = b^m b^n$$
$$xy = b^{m+n}$$ Use the product rule for exponents.
$$\log_b xy = \log_b b^{m+n}$$ Take logs of both sides.
$$\log_b xy = m + n$$ Evaluate the right side.
$$\log_b xy = \log_b x + \log_b y$$ Substitute for m and n.

EXAMPLE 4 Use the properties of logarithms to write these expressions in terms of logarithms of simpler expressions.

SOLUTIONS

(a) $\log x^2(y-5)^3$

$\log x^2(y-5)^3 = \log x^2 + \log(y-5)^3$ Product rule
$\quad\quad\quad\quad\quad\quad = 2 \log x + 3 \log(y-5)$ Power rule

(b) $\ln \sqrt{\dfrac{x+5}{x-9}}$

$\ln \sqrt{\dfrac{x+5}{x-9}} = \ln\left(\dfrac{x+5}{x-9}\right)^{1/2}$

$\quad\quad\quad\quad = \dfrac{1}{2} \ln \dfrac{x+5}{x-9}$ Power rule

$\quad\quad\quad\quad = \dfrac{1}{2} [\ln(x+5) - \ln(x-9)]$ Quotient rule

(c) $\log \dfrac{x\sqrt[3]{y^2}}{z^4}$

$\log \dfrac{x\sqrt[3]{y^2}}{z^4} = \log xy^{2/3} - \log z^4$ Quotient rule

$\quad\quad\quad\quad = \log x + \log y^{2/3} - \log z^4$ Product rule

$\quad\quad\quad\quad = \log x + \dfrac{2}{3} \log y - 4 \log z$ Power rule

EXAMPLE 5 Combine these logarithms into a single logarithmic expression with a coefficient of 1.

(a) $2 \ln x - 3 \ln z$

(b) $\dfrac{1}{2} \log(x-2) + \log y - \dfrac{1}{3} \log z$

SOLUTIONS

(a) $2 \ln x - 3 \ln z = \ln x^2 - \ln z^3$ Power rule

$\quad\quad\quad\quad = \ln \dfrac{x^2}{z^3}$ Quotient rule

SECTION 5-3 COMMON AND NATURAL LOGARITHMS

(b) $\dfrac{1}{2}\log(x-2) + \log y - \dfrac{1}{3}\log z = \log(x-2)^{1/2} + \log y - \log(z)^{1/3}$ Power rule

$= \log y\sqrt{x-2} - \log \sqrt[3]{z}$ Product rule

$= \log \dfrac{y\sqrt{x-2}}{\sqrt[3]{z}}$ Quotient rule

Certain problems result in logarithmic expressions that are neither base 10 nor base e. To evaluate these logarithms with a calculator, we must convert the expressions to common logs or natural logs.

Change of Base Formulas ▼

For $a, b > 0$ and $a, b \neq 1$:

a. $\log_a x = \dfrac{\log_b x}{\log_b a}$ for $x > 0$

b. $a^x = b^{x \log_b a}$

▶ Self-Check ▼

1. Express $\log \dfrac{x(y-9)^3}{z^2}$ in terms of logarithms of simpler expressions.
2. Rewrite $3 \ln v - \dfrac{2}{5} \ln w$ as a single logarithmic expression.

Proof of Change of Base Formula for Logarithms   Let $\log_a x = y$. Then

$x = a^y$ Exponential form

$\log_b x = \log_b a^y$

$\log_b x = y \log_b a$ Power rule

$y = \dfrac{\log_b x}{\log_b a}$ Solve for y.

$\log_a x = \dfrac{\log_b x}{\log_b a}$ Substitute for y.

The most useful forms of this identity are

$\log_a x = \dfrac{\log x}{\log a}$ and $\log_a x = \dfrac{\ln x}{\ln a}$

The proof of the change of base formula for exponents is similar to the proof given for logarithms.

▶ Self-Check Answers ▼

1. $\log x + 3 \log(y-9) - 2 \log z$ 2. $\ln \dfrac{v^3}{\sqrt[5]{w^2}}$

EXAMPLE 6 Evaluate $\log_{8.3} 47.3$ accurate to six significant digits.

SOLUTION Using Method A, we have $\log_{8.3} 47.3 = \dfrac{\log 47.3}{\log 8.3}$:

S: [4][7][.][3][log][÷][8][.][3][log][=] → 1.8223273

G: [log][4][7][.][3][÷][log][8][.][3][ENTER] → 1.822327346

Using Method B, we have $\log_{8.3} 47.3 = \dfrac{\ln 47.3}{\ln 8.3}$:

S: [4][7][.][3][ln][÷][8][.][3][ln][=] → 1.8223273

G: [ln][4][7][.][3][÷][ln][8][.][3][ENTER] → 1.822327346

Answer $\log_{8.3} 47.3 \approx 1.82233$

EXAMPLE 7 Use the [y^x] key to verify that $\log_{8.3} 47.3 \approx 1.8223273$.

SOLUTION $\log_{8.3} 47.3 \approx 1.8223273$

$8.3^{1.8223273} \approx 47.3$ Rewrite this expression in exponential form.

S: [8][.][3][y^x][1][.][8][2][2][3][2][7][3][=] → 47.299995

G: [8][.][3][^][1][.][8][2][2][3][2][7][3][ENTER] → 47.29999541

Answer $8.3^{1.8223273} \approx 47.299995 \approx 47.3$ (The relatively small difference is due to round-off errors.)

EXAMPLE 8 Convert $y = 2^x$ to an exponential function base e.

SOLUTION $a^x = b^{x \log_b a}$ Substitute into the change of base formula with $a = 2$ and $b = e$.

$y = 2^x = e^{x \ln 2}$

Answer $y = e^{x \ln 2}$

Example 8 illustrates how we can rewrite any exponential growth or decay function in terms of base e. This is a timely topic, since some calculators and microcoputers use only base e. Also, as a student you may find the properties of logarithms useful for transforming answers into a more desirable form.

▶ **Self-Check** ▼

1 Determine the values of $\log_2 32768$ and $\log_\pi 10.94$ accurate to five significant digits, then verify these values using the y^x key.

2 Convert $y = (\tfrac{1}{2})^x$ to base e.

▶ **Self-Check Answers** ▼

1 15.000, 2.0899   2 $y = e^{-x \ln 2}$

SECTION 5-3 COMMON AND NATURAL LOGARITHMS

EXAMPLE 9 Verify these identities.

SOLUTIONS

(a) $e^{2\ln x} = x^2$

$e^{2\ln x} = e^{\ln x^2}$ Power rule for logarithms

$= x^2$ $b^{\log_b y} = y$

(b) $x \log_2 e = \dfrac{x}{\ln 2}$

$x \log_2 e = x \dfrac{\ln e}{\ln 2}$ Change of base formula

$= \dfrac{x}{\ln 2}$ $\ln e = 1$

(c) $\log_4 12^x - \log_4 3^x = x$

$\log_4 12^x - \log_4 3^x = x \log_4 12 - x \log_4 3$ Power rule

$= x(\log_4 12 - \log_4 3)$ Factor out x.

$= x \log_4 \dfrac{12}{3}$ Quotient rule

$= x \log_4 4$ $\log_4 4 = 1$

$= x$

EXERCISES 5-3

A

In Exercises 1 and 2, fill in the missing entries in the table.

	Logarithmic Form	Verbal Form	Exponential Form
1 a.	$\log w = z$	_____	_____
b.	_____	The natural log of v is w.	_____
c.	_____	_____	$e^m = n$
2 a.	$\ln x = 4$	_____	_____
b.	_____	The common log of m is t.	_____
c.	_____	_____	$10^{-3} = 0.001$

In Exercises 3 and 4, determine the value of each logarithm by inspection.

3 a. $\log 100{,}000$ **b.** $\ln e^7$ **c.** $\ln \dfrac{1}{e^3}$ **d.** $\log 10^{-7}$

4 a. $\log 10{,}000$ **b.** $\log 0.0001$ **c.** $\log 10^{11}$ **d.** $\ln 1$

In Exercises 5–8, use a calculator to determine the value of each logarithm accurate to six significant digits.

5 a. $\log 8.29$ **b.** $\ln 8.29$ **c.** $\log 928$ **d.** $\ln 7^2 - (\ln 7)^2$

6 a. $\log 0.073$ **b.** $\ln 0.073$ **c.** $\ln(7.4 \times 10^7)$ **d.** $\ln \dfrac{3}{7} - \dfrac{\ln 3}{\ln 7}$

7 a. $\log(1.45 \times 10^{-8})$ **b.** $\ln(1.45 \times 10^{-8})$ **c.** $\ln(22 \cdot 13)$ **d.** $(\ln 22)(\ln 13)$

8 a. $\ln 1{,}234{,}567$ **b.** $\ln(9 \cdot 11) - (\ln 9)(\ln 11)$ **c.** $\ln 10$ **d.** $\log e$

368 CHAPTER 5 EXPONENTIAL AND LOGARITHMIC FUNCTIONS

In Exercises 9 and 10, use a calculator to determine the value of x accurate to six significant digits.

9 **a.** $x = $ antilog 2.70341   **b.** $\log x = 0.123456$ **c.** $\ln x = 2.58$ **d.** $\ln x = -1.00098$

10 **a.** $x = $ antilog 1.97458   **b.** $\log x = 2.04572$ **c.** $\ln x = 3.09806$ **d.** $\ln x = -2.08913$

In Exercises 11–20, express each logarithm in terms of logarithms of simpler expressions. Assume that the arguments of the logarithms are all positive.

11 $\log x^3 y$ **12** $\log x^2 y^5$ **13** $\log \dfrac{x+3}{y-3}$ **14** $\log \dfrac{y+3}{x-2}$ **15** $\ln \dfrac{(x-9)^3}{\sqrt{y+1}}$ **16** $\ln \dfrac{\sqrt{2x-7}}{(y+3)^2}$

17 $\ln \dfrac{\sqrt[5]{x^2 y^3}}{z^4}$ **18** $\ln \sqrt[4]{\dfrac{x^3 y^4}{z^2}}$ **19** $\ln\left(\dfrac{x(y+5)^2}{(z-3)^4}\right)^3$ **20** $\ln\left(\dfrac{(x^2-5)(x-5)^2}{(y+3)^4}\right)^2$

In Exercises 21–28, rewrite each logarithmic expression as a single logarithmic expression. Assume that the arguments of the logarithms are all positive.

21 $3 \log x + 2 \log y$ **22** $5 \log x + 3 \log y$ **23** $3 \ln(x+5) - 4 \ln y$

24 $7 \ln x - 4 \ln(y+2)$ **25** $2 \log x + \dfrac{3}{7} \log y$ **26** $\dfrac{4}{5} \log x - 3 \log(y+4)$

27 $\dfrac{1}{2}[\ln(5x+2) - \ln(x-7)]$ **28** $\dfrac{1}{3}[\ln(4x-3) + \ln(y+1) - \ln(z+5)]$

In Exercises 29–32, use the change of base formula to compute each logarithm accurate to four significant digits.

29 $\log_7 92.5$ **30** $\log_{13} 1469$ **31** $\log_{3.1} 0.00687$ **32** $\log_\pi e$

B

In Exercises 33–38, mentally estimate the value of each expression and then select the answer that is closest to your mental estimate.

33 $\log 143$
   a. 0.1553   **b.** 1.1553   **c.** 2.1553   **d.** 3.1553   **e.** 4.1553

34 $\log 45.7$
   a. 0.6599   **b.** 1.6599   **c.** 2.6599   **d.** 3.6599   **e.** 4.6599

35 $\log 0.0729$
   a. -1.1373 **b.** -0.1373 **c.** 0.1373   **d.** 1.1373   **e.** 2.7313

36 $\log 0.956$
   a. -1.0195 **b.** -0.0195 **c.** 0.0195   **d.** 1.0195   **e.** 2.0195

37 $\ln 2.8$ 2.718
   a. 0.9853   **b.** 1.9853   **c.** 0.0296   **d.** 1.0296   **e.** 2.9853

38 $\ln 10$
   a. 0.3026   **b.** 1.3026   **c.** 2.3026   **d.** 3.3026   **e.** 4.3026

In Exercises 39–42, use the change of base formula to convert each exponential function to base e.

39 $y = 3^x$ **40** $y = \left(\dfrac{1}{3}\right)^x$ **41** $y = \left(\dfrac{1}{5}\right)^x$ **42** $y = 5^x$

C

In Exercises 43–46, given $\log_5 2 \approx 0.43068$ and $\log_5 3 \approx 0.68261$, use the properties of logarithms to calculate the approximate value of each expression to the nearest ten-thousandth. Do not use the change of base formula to work these exercises.

43 $\log_5 6$ **44** $\log_5 8$ **45** $\log_5 \dfrac{2}{3}$ **46** $\log_5 10$

47 For what value of x does $\log x = \ln x$?   **48** Give a value of x for which $\log x^2 \neq (\log x)^2$.

In Exercises 49–54, verify each identity.

49 $e^{-\ln x} = \dfrac{1}{x}$

50 $e^{3 \ln x} = x^3$

51 $\log_7 x \approx 0.5138983 \ln x$

52 $\left(\dfrac{1}{2}\right)^x = e^{-x \ln 2}$

53 $\log 24^x - \log 60^x - \log 0.2^{2x} = x$

54 $x \log e = \dfrac{x}{\ln 10}$

In Exercises 55 and 56, use a graphics caluclator and the change of base formula to graph each function.

55 $y = \log_\pi x$

56 $y = \log_{\sqrt{2}} x$

SECTION 5-4

Exponential and Logarithmic Equations

Section Objective

9 Solve exponential and logarithmic equations.

An **exponential equation** has a variable in at least one exponent. Only trivial exponential equations such as $3^x = 9$ can be solved by inspection. We will use logarithms and a calculator to solve nontrivial problems.

Either natural or common logs can be used to solve exponential equations. Example 1 illustrates the use of natural logs, and Example 2 illustrates the use of common logs. If the natural base e is involved, we usually use natural logs, as in Example 3. Likewise, we use common logs for problems involving base 10.

EXAMPLE 1 Solve $7^w = 21$.

SOLUTION

$7^w = 21$

$\ln 7^w = \ln 21$ Take the natural log of both sides.

$w \ln 7 = \ln 21$ Power rule

$w = \dfrac{\ln 21}{\ln 7}$ The exact answer

Answer $w \approx 1.564575$ Calculator value is rounded to seven significant digits.

Check $7^{1.564575} \approx 21$ checks Use the power key on your calculator to check this answer. ☐

EXAMPLE 2 Solve $5^v = 4^{3v+2}$.

SOLUTION

$5^v = 4^{3v+2}$

$\log 5^v = \log 4^{3v+2}$ — Take the common log of both sides.

$v \log 5 = (3v + 2) \log 4$ — Power rule

$v \log 5 = 3v \log 4 + 2 \log 4$

$v \log 5 - 3v \log 4 = 2 \log 4$ — Collect all the terms involving v on the left side of the equation.

$v(\log 5 - 3 \log 4) = 2 \log 4$ — Factor out v.

$v = \dfrac{2 \log 4}{\log 5 - 3 \log 4}$ — The exact value of v

Answer $v \approx -1.0875263$ — Calculator value is accurate to eight significant digits.

Check $5^{-1.0875263} \approx 4^{3(-1.0875263)+2}$

$5^{-1.0875263} \approx 4^{-1.2625789}$

$0.1737208 = 0.1737208$ checks. — Values found using the power key are equal.

EXAMPLE 3 Solve $2e^{x^2} = 17.890$ accurate to five significant digits.

SOLUTION $2e^{x^2} = 17.89$

$e^{x^2} = 8.945$ — Divide by 2.

$x^2 = \ln 8.945$ — Take ln of both sides; $\ln e^{x^2} = x^2$.

$x^2 \approx 2.1910947$ — Calculator approximation

$x \approx \pm 1.4802347$ — Calculator approximation

Answer $x = -1.4802$ or $x = 1.4802$ — Round to five-place accuracy.

▶ **Self-Check** ▼

Solve $3^{2z+4} = 13^z$ for z accurate to eight significant digits.

A **logarithmic equation** has a variable in the argument of at least one logarithmic expression. In the following examples, note the use of the properties of logarithms to combine the logarithmic expressions into a single expression. This equation is then written as an equivalent equation free of logarithms. *Be sure to check for possible extraneous values, since a logarithmic expression cannot have a negative argument.*

An important property of logarithms is that they are one-to-one functions. Thus in Section 5-3 we noted that $\log_b x_1 = \log_b x_2$ if and only if $x_1 = x_2$ and $x_1, x_2 > 0$. The importance of this result in solving equations is stressed in the box on page 371.

▶ **Self-Check Answer** ▼

$z \approx 11.950375$

SECTION 5-4 EXPONENTIAL AND LOGARITHMIC EQUATIONS

Logarithmic Functions Are One-to-One ▼

If $b > 0$, $b \neq 1$, and $x_1 \, x_2 > 0$, then

$$x_1 = x_2 \quad \text{if and only if} \quad \log_b x_1 = \log_b x_2$$

That is, these equations are equivalent.

EXAMPLE 4 Solve $\log(3x - 2) = 1$.

SOLUTION $\log(3x - 2) = 1$

$\qquad\qquad 10^1 = 3x - 2 \qquad$ Express in exponential form.

$\qquad\qquad 12 = 3x$

$\qquad\qquad x = 4$

Check $\log[3(4) - 2] \stackrel{?}{=} 1$

$\qquad\qquad \log 10 = 1$ checks.

Answer $x = 4$

EXAMPLE 5 Solve $\log(3 - x) + \log(1 - x) = \log(11 - 6x)$.

SOLUTION $\log(3 - x) + \log(1 - x) = \log(11 - 6x)$

$\qquad\qquad \log[(3 - x)(1 - x)] = \log(11 - 6x) \qquad$ Use the product rule to rewrite the left member as a single term.

$\qquad\qquad (3 - x)(1 - x) = 11 - 6x \qquad$ The arguments are equal.

$\qquad\qquad 3 - 4x + x^2 = 11 - 6x$

$\qquad\qquad x^2 + 2x - 8 = 0$

$\qquad\qquad (x + 4)(x - 2) = 0$

$\qquad x + 4 = 0 \qquad \text{or} \qquad x - 2 = 0$

$\qquad\qquad x = -4 \qquad\qquad\qquad x = 2$

Check $x = -4$: $\log[3 - (-4)] + \log[1 - (-4)] \stackrel{?}{=} \log[11 - 6(-4)]$

$\qquad\qquad\qquad\qquad \log 7 + \log 5 \stackrel{?}{=} \log 35$

$\qquad\qquad\qquad\qquad \log 35 = \log 35$ checks.

$\qquad x = 2$: $\log(3 - 2) + \log(1 - 2) \stackrel{?}{=} \log[11 - 6(2)]$

$\qquad\qquad\qquad\qquad\qquad$ —Undefined—

Answer $x = -4$ (2 is an extraneous value.)

EXAMPLE 6 Solve $\ln(z^2 - 1) - \ln(z - 1) = -1$.

SOLUTION $\ln(z^2 - 1) - \ln(z - 1) = -1$

$$\ln\left(\frac{z^2 - 1}{z - 1}\right) = -1 \qquad \text{Use the quotient rule to rewrite the left number as a single term.}$$

$$\ln(z + 1) = -1 \qquad \text{Reduce the fraction.}$$

$$e^{-1} = z + 1 \qquad \text{Express in exponential form and then solve for } z.$$

$$z = \frac{1}{e} - 1 \qquad \text{The exact value of } z.$$

$$z \approx -0.63212 \qquad \text{Calculator approximation is rounded to five significant digits.}$$

Check $\ln(-0.63212)^2 - 1] + \ln(-0.63212 - 1) \stackrel{?}{=} -1$

Both of these logarithmic expressions are undefined for these negative values. Therefore this value is extraneous.

Answer No solution

As Examples 1–3 show, we often take logarithms of both members of an exponential equation in order to solve the equation. Similarly, we often rewrite a logarithmic equation in exponential form in order to solve it. This symmetrical usage of exponents and logarithms should not be surprising, since exponential and logarithmic functions are inverses of each other.

The next examples use the formulas from Section 5-1 for periodic growth (such as compound interest) and for continuous growth and decay. The formula for continuous growth is also a good model for periodic growth when the number n is large. (This was suggested by the development of e in Section 5-1.)

▶ **Self-Check** ▼

Solve $\log(y - 3) + \log y = 1$.

▶ **Self-Check Answer** ▼

$y = 5$ (-2 is an extraneous value.)

SECTION 5-4 EXPONENTIAL AND LOGARITHMIC EQUATIONS

Growth and Decay Functions

Periodic growth:

$$A = P\left(1 + \frac{r}{n}\right)^{nt}$$

An orginal amount P grows at an annual rate r with periodic compounding n times a year for t years, yielding an amount A.

Continuous growth and decay:

$$A = Pe^{rt}$$

An original amount P grows (decays) continuously at a rate r for a time t, yielding an amount A. For $r > 0$ there is growth, and for $r < 0$ there is decay.

EXAMPLE 7 Determine the number of years it will take a $10,000 certificate of deposit to double in value if it is invested at 8.75% interest compounded quarterly.

SOLUTION Let t = Number of years for the investment to double in value.

Then
$$A = P\left(1 + \frac{r}{n}\right)^{nt}$$
Compound interest formula; growth every three months indicates periodic growth.

$$20{,}000 = 10{,}000\left(1 + \frac{0.0875}{4}\right)^{4t}$$
Substitute in the given values: A = $20,000 (double the original amount), P = $10,000, r = 0.0875, and n = four times a year.

$$2 = (1.021875)^{4t}$$
Simplify.

$$\log 2 = \log(1.021875)^{4t}$$
Take the common log of both sides.

$$\log 2 = 4t \log(1.021875)$$
Power rule

$$t = \frac{\log 2}{4 \log(1.021875)}$$
Solve for t.

$$t \approx 8.0080$$
Calculator approximation

Answer The investment will double in approximately eight years. □

EXAMPLE 8 The population of a species of whales is estimated to be 20,000 and is decreasing continuously at 5% per year ($r = -0.05$). If this rate of decrease continues, in how many years will the population have declined to 4000?

SOLUTION Let t = Number of years for this decline to occur.

Then
$$A = Pe^{rt}$$ Continuous decay formula

$$4000 = 20{,}000e^{-0.05t}$$ Substitute in the given values: $A = 4000$ (population at the end of the term), $P = 20{,}000$ (original population), and $r = -0.05$.

$$0.2 = e^{-0.05t}$$ Simplify.

$$\ln 0.2 = -0.05t$$ Take ln of both sides; $\ln e^{-0.05t} = -0.05t$.

$$t = \frac{\ln 0.2}{-0.05}$$ Solve for t.

$$t \approx 32.189$$ Calculator approximation

Answer $t \approx 32$ years (rounded to the nearest year)

▶ **Self-Check** ▼

Determine the number of years needed for the investment in Example 7 to triple in value.

Seismologists use the Richter scale to measure the magnitude of earthquakes. The equation $R = \log \dfrac{A}{a}$ compares the amplitude A of the shock wave of an earthquake to the amplitude a of a reference shock wave of minimal intensity.

EXAMPLE 9 The amplitude of the Alaskan earthquake of 1964 was $10^{8.4}$ times the reference amplitude. Calculate the measurement on the Richter scale.

SOLUTION
$$R = \log \frac{A}{a}$$ Given formula for R

$$= \log \frac{10^{8.4}a}{a}$$ Substitute in the given value of A.

$$= \log 10^{8.4}$$

$$R = 8.4$$

Answer The Alaskan earthquake measured 8.4 on the Richter scale.

▶ **Self-Check Answer** ▼

$t = 12.7$ years

SECTION 5-4 EXPONENTIAL AND LOGARITHMIC EQUATIONS

Chemists use pH to measure the hydrogen potential of a solution. Water has a pH of 7, acids have a pH of less than 7, and alkalines have a pH of greater than 7. The formula for the pH of a solution is $pH = -\log H^+$, where H^+ measures the number of hydrogen ions in moles per liter.

EXAMPLE 10 Determine the pH of a beer if its H^+ has been measured at 6.3×10^{-5} mole/liter.

SOLUTION $pH = -\log H^+$ Given formula for pH
$\qquad\quad = -\log(6.3 \times 10^{-5})$ Substitute in the given value of H^+.

Answer $pH \approx 4.2$

EXERCISES 5-4

A

In Exercises 1–10, solve each equation without using a calculator.

1 $7^x = \dfrac{1}{49}$ **2** $3^y = 81$ **3** $\left(\dfrac{2}{5}\right)^n = \dfrac{625}{16}$ **4** $2^t = \sqrt[3]{2}$ **5** $\log_4 64 = y$ **6** $\log_8 64 = z$

7 $\log(7x + 1) = \log(6x + 3)$ **8** $\log(8x + 31) = \log(3x + 16)$

9 $\ln(5x + 13) = \ln(4x + 11)$ **10** $\ln(4x + 1) = \ln(6x - 9)$

In Exercises 11–28, solve each equation. Give answers accurate to six significant digits.

11 $8^{w+1} = 47.93$ **12** $11^{2v-3} = 8154$ **13** $3.41^{-2z} = 0.0728$ **14** $0.06472^{-3t+5} = 4.1078$

15 $6^{4v+1} = 11^{v-2}$ **16** $13^{3-2v} = 5^{7-3v}$ **17** $e^{5x} = 139.45$ **18** $e^{7z+2} = 3.0836$

19 $10^{3x-1} = 8011.12$ **20** $10^{x^2+2} = 219.83$ **21** $72^{y^2-11} = 139.42$ **22** $4.5^{3-s^2} = 21.403$

23 $\left(\dfrac{3}{5}\right)^{2-x} = 11$ **24** $\left(\dfrac{2}{7}\right)^{3-2x} = 8.5$ **25** $\left(1 + \dfrac{0.10}{12}\right)^{12t} = 2$ **26** $\left(1 + \dfrac{0.145}{365}\right)^{365t} = 3$

27 $e^{-0.07t} = \dfrac{1}{3}$ **28** $e^{-0.05t} = \dfrac{1}{2}$

In Exercises 29–44, solve each eqaution.

29 $6^{t^2-4t+3} = 1$ (Use base 6 logs.)   **30** $7^{x^2-3x-10} = 1$ (Use base 7 logs.)   **31** $\log(t + 2) = 1$

32 $\log(5s + 80) = 2$ **33** $\log(3v + 2) - \log(v - 4) = 1$ **34** $\log(w + 3) + \log(w - 1) = \log 5$

35 $\ln(5x + 1) + \ln(x - 1) = \ln(2 - 6x)$ **36** $\log(y - 5) + \log 2 = \log(3y + 7)$

37 $\log(z^2 + 2z + 15) - \log(z + 2) = \log 6$ **38** $\ln(-4v - 4) - \ln(1 - v) = \ln(6v + 11)$

39 $\ln x + \ln(x + 21) = \ln 100$ **40** $\ln(x^2 + 3x + 4) - \ln 2 = 0$

41 $\log \sqrt{\dfrac{5x - 12}{x}} = 0$ **42** $\log_3 \sqrt{3x^2} = \dfrac{3}{2}$

43 $\ln(5 - x) + \ln(x + 2) = \ln(3 - 3x)$ **44** $\log 2(7 - x) = \log(4 - x) + \log(x + 5)$

B

In Exercises 45–52, solve each equation.

45 $\log(\log y) = 2$
46 $\log(\log x) = 1$
47 $(\log x)^2 = \log x^2$
48 $\log \sqrt{x} = \sqrt{\log x}$
49 $\log \log \log x = 1$
50 $\log x^3 = (\log x)^3$ (Give the answer accurate to six significant digits.)
51 $\ln |x| = 5$ (Give the answer accurate to six significant digits.)
52 $|\ln x| = 7$ (Give the answer accurate to six significant digits.)

53 How many years will it take an investment to triple in value if it is compounded monthly at 8.25%?

54 How many years will it take an investment to double in value if it is compounded continuously at 8%?

55 How much more interest would a $1000 investment earn in one year at 8% compounded continuously than at 8% compounded monthly?

56 If prices would double in eight years at the current rate of inflation, what is the rate of inflation? Assume that the effect of inflation is continuous.

57 If a radioactive material decays continuously at the rate of 0.5% per year, determine the half-life of this material; that is, find the time for each gram to decay to one-half of a gram.

58 Carbon-14, which has a half-life of approximately 5568 years, is used by archaeologists to date objects. The half-life is the time it takes for each gram to decay to one-half of a gram. Determine the annual rate of decay r accurate to four significant digits.

59 The growth rate of bacteria in foods can be used to determine the safe shelf-life of products. Once the bacteria count reaches a certain level, the products are no longer safe. If the growth rate of bacteria in cheese is 5% per day, determine the number of days before the bacteria count is 1000 times the current count.

60 The rat population of a city is estimated to be 50,000 and is increasing at 15% per year. At this rate, what will the rat population be in six years?

61 The population of a species of whales is estimated to be 20,000 and is decreasing at 4% per year. How many years will it be before the population has declined to 10,000?

62 The radioactive material used to power a satellite decays at a rate that decreases the power available by 0.02% per day. When the power supply reaches $\frac{1}{100}$ of its original power, it is no longer functional. How many days should the power supply last?

63 After t years of depreciation, the value of an industrial machine is $A = 15{,}000 e^{-0.2t} + 1000$. How many years will it be before the value depreciates to $5000?

Use the formula $R = \log \dfrac{A}{a}$ to answer Exercises 64 and 65.

64 The amplitude of the 1906 earthquake in San Francisco was $10^{8.3}$ times that of the reference amplitude. Calculate the measurement on the Richter scale.

65 How many times larger was the amplitude of the 1906 San Francisco earthquake (see Exercise 64) than that of the October 1989 San Francisco earthquake, which measured 6.9 on the Richter scale?

KEY CONCEPTS

A type of measurement similar to the Richter scale is the decibel, which measures the intensity of sound. $D = 10 \log \dfrac{I}{I_0}$ gives the number of decibels D of a sound of intensity I as compared to the reference intensity I_0, which is at the threshold of hearing ($I_0 \approx 10^{-16}$ watts per cm^2). Use this information to answer Exercises 66–68.

66 Find the number of decibels of background music whose intensity is 10^{-14} watts per cm^2.

67 Find the number of decibels of city traffic noise whose intensity is a billion times the reference intensity.

68 The noise in a bar is 80 decibels. Calculate the intensity of this noise.

Use the formula pH $= -\log H^+$ to answer Exercises 69 and 70.

69 Determine the pH of a type of coffee whose H^+ is measured at 3.98×10^{-7} moles per liter.

70 Determine the hydrogren ion concentration (in moles per liter) for a soap whose pH is 7.563.

C

71 Solve $3^{x+y} = 9$ for x in terms of y.

72 Solve $\ln(x + y) = 2$ for x in terms of y.

73 Solve $\log(x + y) = 2$ for x in terms of y.

74 Solve $5^{3x-2y} = 125$ for x in terms of y.

75 Solve $e^{x^3+x-1} = 5$ to the nearest tenth by using a graphics calculator to find the x-intercept of $y = e^{x^3+x-1} - 5$.

76 Solve $e^{-x^3+x^2-1} = 9$ to the nearest tenth by using a graphics calculator to find the x-intercept of $y = e^{-x^3+x^2-1} - 9$.

77 Solve $\ln(x+1) + \ln(x+2) + \ln(x+3) = 1$ to the nearest tenth by using a graphics calculator to find the x-intercept of $y = \ln(x+1) + \ln(x+2) + \ln(x+3) - 1$.

78 Solve $\log(x-1) + \log(x+3) + \log(x-5) = 2$ to the nearest tenth by using a graphics calculator to find the x-intercept of $y = \log(x-1) + \log(x+3) + \log(x-5) - 2$.

KEY CONCEPTS

1 The exponential function $f(x) = b^x$ is called a growth function for $b > 1$ and a decay function for $0 < b < 1$.

2 Periodic compounding formula: $A = P\left(1 + \dfrac{r}{n}\right)^{nt}$

Continuous compounding formula: $A = Pe^{rt}$

For r positive, this function represents growth. But for r negative, this function represents decay.

3 Logarithms are exponents:
For $b > 0$, $b \neq 1$, and $x > 0$, $y = \log_b x$ if and only if $b^y = x$.

4 Logarithmic identities: For $b > 0$, $b \neq 1$, and $x > 0$,

    a. $\log_b 1 = 0$ **b.** $\log_b b = 1$ **c.** $\log_b \dfrac{1}{b} = -1$

    d. $\log_b b^y = y$ **e.** $b^{\log_b x} = x$

5 Properties of logarithms: For $x, y > 0$ and $b > 0$, $b \neq 1$,

    1. Product rule: $\log_b xy = \log_b x + \log_b y$

    2. Quotient rule: $\log_b \dfrac{x}{y} = \log_b x - \log_b y$

    3. Power rule: $\log_b x^p = p \log_b x$

6 Change of base formulas: for $a, b > 0$ and $a, b \neq 1$,

    a. $\log_a x = \dfrac{\log_b x}{\log_b a}$, for $x > 0$

    b. $a^x = b^{x \log_b a}$

7 The steps for solving a logarithmic equation must include a check for extraneous values. In particular, the logarithm of a negative number is undefined.

REVIEW EXERCISES FOR CHAPTER 5

In Exercises 1 and 2 rewrite each expression in exponential form.

1 $\log_x 10 = y$ **2** $\log y = z$

In Exercise 3 and 4, rewrite each expression in logarithmic form.

3 $7^{x+3} = y$ **4** $e^z = w$

In Exercises 5–28, solve each equation without using a calculator.

5 $\left(\dfrac{2}{7}\right)^x = \dfrac{343}{8}$ **6** $64^x = 1$

7 $64^x = 64$ **8** $64^x = 8$

9 $64^x = 16$ **10** $9^{x+5} = 27^{x-2}$

11 $5^{x^2-1} = 25^{x^2-5}$ **12** $(\ln x)^2 = \ln x^2$

13 $y^{3/5} = 8$ **14** $y^{-2/3} = \dfrac{25}{16}$

15 $z = \log_7 \dfrac{1}{49}$ **16** $z = \log_5 \sqrt{5}$

17 $\log_8 x = 0$ **18** $\log_{289} 289 = t$

19 $\log_x 16 = 2$ **20** $\log_w 5 = -\dfrac{1}{2}$

21 $\log_v 47 = \log_5 47$ **22** $\log_8 113 = \log_8 v$

23 $\log_2(x - 3) = 4$ **24** $\log_9 81 = x^2 - 34$

25 $19^{\log_{19} 7} = v$ **26** $\log_3 3^x = 11$

27 $\log_3 10 - \log_3 5 = \log_3 z$ **28** $\log_5(x^2 + 9) = 2$

REVIEW EXERCISES FOR CHAPTER 5

In Exercises 29–36, use a calculator to solve each equation accurate to six significant digits.

29 $\log 283.49 = r$

30 $\ln 0.0039 = v$

31 $\ln 8.014 \times 10^{-6} = x$

32 $y = 2.89^{4.76}$

33 $a = e^{-0.6931472}$

34 $\log c = 0.1139434$

35 $z = \text{antilog } 0.9542425$

36 $x = \log_8 19$

In Exercises 37 and 38, rewrite each expression in terms of logarithms of simpler expressions.

37 $\log \dfrac{(x-3)(x+4)^2}{x-7}$

38 $\ln \dfrac{\sqrt{x+y}}{z}$

In Exercises 39 and 40, combine each expression into a single logarithmic expression.

39 $3 \ln x - \ln(x+5) - \ln(x-7)$

40 $3 \log x - 0.5 \log y$

In Exercises 41–49, solve each equation accurate to six significant digits.

41 $5^{v+3} = 713.89$

42 $17^{2-7w} = 13^{4w-2}$

43 $e^{11x} = 193.21$

44 $13^{y^2-7y-8} = 1$

45 $(1 + \frac{0.085}{12})^{12t} = 2$

46 $\ln(m-2) + \ln(m-3) = \ln 2$

47 $\ln(z^2 - 4z + 7) - \ln(z - 5) = \ln 14$

48 $\ln(x-2) + \ln(x-7) = \ln(4-16x)$ *no solution*

49 $\log(x - 15) + \log x = 2$

In Exercises 50 and 51, determine whether or not each function is a one-to-one function. If the function is one-to-one, give its inverse and the domain of the inverse.

50 $f(x) = e^{x+2}$

51 $f(x) = \log_3(2x - 1)$

52 Graph $y = (\frac{3}{2})^x$ and its inverse on the same coordinate system.

53 The graph of every exponential function $y = b^x$ passes through the point _____.

54 The graph of every logarithmic function $y = \log_b x$ passes through the point _____.

55 Which of these expressions are undefined?

a. $\log_7 0$ b. $\log_7 7$ c. $\log_{-7} 7$ d. $\log_7(-7)$
e. $\log_7(\frac{1}{7})$ f. $\log_7 1$ g. $\log_1 7$ h. $\log e$

In Exercises 56 and 57, verify each identity.

56 $e^{-x \ln 3} = \left(\dfrac{1}{3}\right)^x$

57 $x \log_3 e = \dfrac{x}{\ln 3}$

58 How many years will it take an investment to double in value if it is compounded monthly at 9.5%?

59 The radioactive material used to power a satellite decays at a rate that decreases the power supply by 0.015% per day. When the power supply has diminished to 5% of its original power, it is no longer functional. How many days should the power supply last?

380 CHAPTER 5 EXPONENTIAL AND LOGARITHMIC FUNCTIONS

In Exercises 60–75, match each graph with the appropriate relation.

a. $y = 2x + 3$

b. $y = \dfrac{2x^2 + 5x + 3}{x + 1}$

c. $y = 2$

d. $x = -3$

e. $y = x^2 + 3$

f. $y = x^2 - 4x - 5$

g. $\dfrac{x^2}{4} + \dfrac{y^2}{9} = 1$

h. $\dfrac{(x-1)^2}{4} + \dfrac{(y+2)^2}{9} = 1$

i. $\dfrac{x^2}{4} - \dfrac{y^2}{9} = 1$

j. $(x - 1)^2 + (y + 2)^2 = 9$

k. $y = (x-1)(x-2)(x-3)$

l. $y = (x+1)^2(x+3)$

m. $y = \dfrac{x-1}{x+2}$

n. $y = \dfrac{x-1}{(x-3)(x+2)}$

o. $y + 2^x$

p. $y = \log_2 x$

60.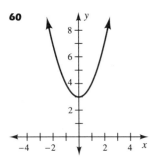

61.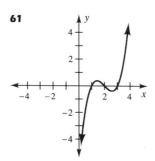

62.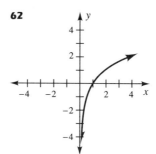

63.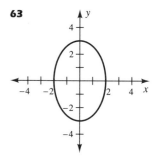

64.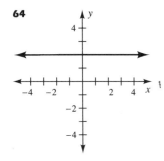

65.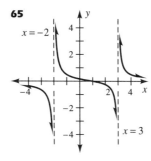

66.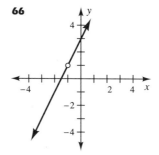

67.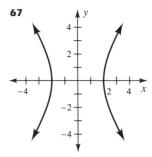

68.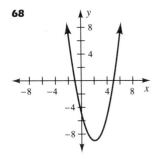

69.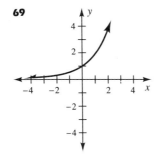

70.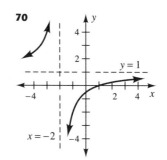

71.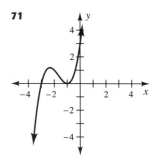

OPTIONAL EXERCISES FOR CALCULUS-BOUND STUDENTS

72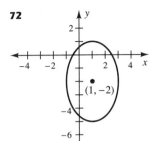
73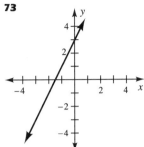
74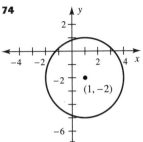
75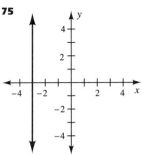

OPTIONAL EXERCISES FOR CALCULUS-BOUND STUDENTS

1. Verify that $f(x) = 3x - 2$ and $g(x) = \dfrac{x+2}{3}$ are inverses of each other by computing $(g \circ f)(x)$ and $(f \circ g)(x)$. State the domain and range of both of these functions.

2. Determine the inverse of $f(x) = \dfrac{3x - 5}{2x + 7}$.

3. Solve $\ln(v - 5) + \ln(v + 4) = \ln(v + 13)$.

4. Solve $\log(x - 3y) = 2$ for x in terms of y.

5. Solve $2^{x - 3y} = 5$ for x in terms of y.

6. Solve $(\log x)^2 + \log x - 2 = 0$.

7. Solve $2^{2x} - 12(2^x) + 32 = 0$.

8. Give two values of x for which $(\log x)^2$ and $\log x^2$ are *not* equal and then list all values of x for which these expressions are equal.

9. The graph of an exponential function $y = b^x$ passes through the points $(0, 1)$ and $(1, 4)$. What is the base b of this exponential function?

10. The graph of a logarithmic function $y = \log_b x$ passes through the points $(1, 0)$ and $(6, 1)$. What is the base b of this logarithmic function?

11. The vertical asymptote of the graph of $y = \ln(x + 3)$ is _____.

12. The horizontal asymptote of the graph of $y = e^x + 2$ is _____.

13. Verify that $(0.5)^x = e^{-x \ln 2}$ is an identity.

14. Verify that $\log 45^x - \log 9^x + \log 0.5^{-x} = x$ is an identity.

15. Verify the quotient rule for logarithms.

16. Verify the change of base formula for exponents.

17. The hyperbolic sine function, denoted by $\sinh x$, and the hyperbolic cosine function, denoted by $\cosh x$, are defined by $\sinh x = \dfrac{e^x - e^{-x}}{2}$ and $\cosh x = \dfrac{e^x + e^{-x}}{2}$. Verify that $\sinh x + \cosh x = e^x$.

18. Solve $y = \sinh x$ for x in terms of y. (See the definition of $\sinh x$ in Exercise 14.)

19. Graph $y = e^x$ and $y = e^{x+2}$ on the same coordinate system.

20. Graph $y = \ln x$ and $y = \ln x + 2$ on the same coordinate system.

21. Graph $y = \ln|x|$.

22 Approximate to the nearest thousandth the solution(s) of $e^{-x^2} = 0.5$ by using a graphics calculator to find the x-intercept(s) of $y = e^{-x^2} - 0.5$.

23 Approximate to the nearest thousandth the solution(s) of $\ln(-x^2 + 5) = 1$ by using a graphics calculator to find the x-intercept(s) of $y = \ln(-x^2 + 5) - 1$.

24 An important property used to solve logarithmic equations is

$$\text{If } \log x_1 = \log x_2, \text{ then } x_1 = x_2.$$

Thus we could state

$$\text{If } f(x) = \log x \text{ and } f(x_1) = f(x_2), \text{ then } x_1 = x_2.$$

Give an example of a function $g(x)$ and values x_1 and x_2 such that $g(x_1) = g(x_2)$ but $x_1 \neq x_2$.

25 Newton's cooling law is a formula for calculating how much time it takes a hot object to cool when surrounded by a medium at some constant temperature. This formula can be applied to a hot steel ingot at a steel factory, to a heat source used to warm air, or to a murder victim at the scene of a crime. The formula is

$$t = k \ln \frac{b - r}{b_0 - r}$$

where
k = constant
t = time
r = room temperature
b = body temperature at end of time period
b_0 = original body temperature at beginning of time period

A coroner at a murder scene found the room temperature to be 22°Celsius and the temperature of the victim to be 29°Celsius. Assume that normal body temperature is 37°Celsius and that the constant k for a body of the nature of the victim is -3.8065. Determine the number of hours the victim had been dead when these temperatures were taken.

MASTERY TEST FOR CHAPTER 5

Exercise numbers correspond to Section Objective numbers.

1 Graph each function.
  a. $y = 4^x$ b. $y = 4^{-x}$ c. $y = -4^x$ d. $y = 4^x - 3$

2 a. Determine the amount that results from a $250 investment at 12.5% interest per year compounded quarterly for five years.

  b. Determine the amount that results from a $250 investment at 12.5% interest per year compounded continuously for five years.

  c. The purchasing power of $250 stored in a safety deposit box declines continuously at 12.5% per year for five years. In terms of the dollar's present value, what is the purchasing power of this amount after five years?

MASTERY TEST FOR CHAPTER 5

3 Determine the value of each logarithm without using a calculator.

a. $\log_7 49$ **b.** $\log_6 \dfrac{1}{\sqrt{6}}$ **c.** $\log_5\left(\dfrac{1}{625}\right)$ **d.** $\log_4(-4)$

e. $\log_{27} 9$ **f.** $\log_{1/4} 8$

4 Solve each equation for x without using a calculator.

a. $\log_{93} 93 = x$ **b.** $\log_5 5^7 = x$ **c.** $\log_7 9 = \log_7 x$
d. $\log_9 1 = x$ **e.** $\log_{2/3}(\tfrac{3}{2}) = x$ **f.** $22^{\log_{22} 33} = x$

5 Graph each function.

a. $y = \log_4 x$ **b.** $y = \log_{1/4} x$ **c.** $y = (\log_4 x) + 3$

6 Use a calculator to determine the value of each logarithm accurate to five significant digits.

a. $\log 943.7$ **b.** $\log(3.479 \times 10^{-8})$ **c.** $\ln 73.49$
d. $\ln(4.973 \times 10^5)$

7 Use a calculator to determine the value of each logarithm accurate to five significant digits.

a. $\log_5 7$ **b.** $\log_2 80$ **c.** $\log_\pi 2\pi$

8 In parts (a) and (b), express the logarithms in terms of logarithms of simpler expressions.

a. $\log \dfrac{x^3 y^2}{z}$ **b.** $\log \dfrac{\sqrt{xy}}{z}$

In parts (c) and (d), combine the logarithms into single logarithmic expressions.

c. $3 \log x - \tfrac{1}{2} \log y$ **d.** $2 \log(x + y) + 3 \log z$

9 a. Solve $11^{3-x} = 15^{2x+1}$ for x.

b. Solve $\ln(2x + 1) + \ln(11 - 3x) = \ln(5x - 1)$ for x.

c. If an investment is compounded monthly at 9.5% annual interest, how many years will pass before the investment triples in value?

d. The population of a small Pacific island is currently 850. The population is decreasing continuously at an annual rate of 4%. How many years will it be before the population decreases to 700?

CHAPTER SIX

Systems of Equations and Inequalities

Chapter Six Objectives

1. Solve a system of linear equations using the substitution method.

2. Solve a system of linear equations using the elimination method.

3. Determine the number of solutions for a system of two linear equations in two variables by comparing the ratios of the coefficients.

4. Solve a system of linear equations using an augmented matrix.

5. Solve word problems using systems of equations.

6. Use the substitution method to solve a nonlinear system of equations.

7. Use the elimination method to solve a nonlinear system of equations.

8. Graph an inequality in two variables.

9. Graph a system of inequalities.

10. Solve a linear programming problem using a graphing strategy.

Srinivas Ramanujan, 1887–1920

The Indian mathematical genius Ramanujan was renowned for his unusual memory and powers of calculation. He could recite whole tables of values and values of π and $\sqrt{2}$ to any number of decimal places. He often "dreamed" new formulas and upon awakening would write them down and verify them. To him almost every natural number was interesting in some special way. For example, when a friend remarked that a taxi had a dull number, 1729, he replied "No, it is a very interesting number; it is the smallest number expressible as a sum of two cubes in two different ways."

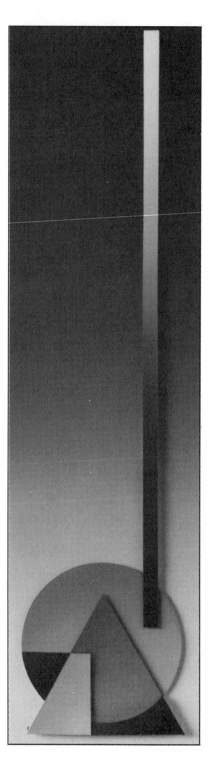

SECTION 6-1 THE SUBSTITUTION AND ELIMINATION METHODS

This chapter will develop methods appropriate for solving systems of equations containing two or three variables. We will also consider matrix methods that make it practical to solve systems of equations containing several variables. The availability of computers and calculators that can manipulate arrays or matrices makes it feasible to solve systems of equations that would involve an unwieldy number of hand calculations.

Two types of problems that lead to systems of equations are included in this chapter. Linear programming, an important managerial tool, is introduced in Section 6-6.

SECTION 6-1

The Substitution and Elimination Methods

Section Objectives

1 Solve a system of linear equations using the substitution method.
2 Solve a system of linear equations using the elimination method.
3 Determine the number of solutions for a system of two linear equations in two variables by comparing the ratios of the coefficients.

Two or more equations that are considered simultaneously are referred to as a **system of equations**. A system containing only linear equations is called a **linear system**. A solution of a linear system of two equations with two unknowns is an ordered pair of numbers that satisfies both equations. Thus the graph of the solution set is the intersection of the graphs of the two equations.

The following systems of equations illustrate certain basic concepts about linear systems:

- $\begin{cases} 3x - 4y = -18 \\ 5x + 3y = -1 \end{cases}$ This is a 2 × 2 (read "two-by-two") linear system. The name reflects the fact that there are two equations in two variables.

- $\begin{cases} 2x + 3y - 4z = 22 \\ 3x - 5y + 2z = -7 \\ 8x + y - 7z = 43 \end{cases}$ This is a 3 × 3 linear system.

- $\begin{cases} x^2 + xy + y^2 = 1 \\ 3x + 4y = -1 \end{cases}$ This system of equations in two variables is *not* a linear system. The first equation contains second-degree terms and is therefore not a linear equation.

One method that can be used to solve many systems of equations with two variables is the graphical method. To solve a system using this method, we graph each equation and then estimate the coordinates of each point of intersection. Section 6-4 will show how to do this using a graphics calculator. Whether using hand plotting or a graphics calculator, serious errors of estimation can occur. Thus it is important to check an estimated solution by substituting it into each equation in the system.

EXAMPLE 1 Solve the system $\begin{cases} x - y = -1 \\ x + y = 5 \end{cases}$ graphically.

SOLUTION Both equations are graphed on the same coordinate system by plotting their intercepts. Then the point of intersection is visually determined to be (2, 3).

Check Substitute (2, 3) into both equations.

$$x - y = -1 \qquad x + y = 5$$
$$2 - 3 \stackrel{?}{=} -1 \qquad 2 + 3 \stackrel{?}{=} 5$$
$$-1 = -1 \qquad 5 = 5$$

Answer The ordered pair (2, 3) is the solution of this system. □

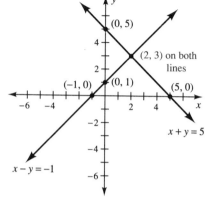

There are three ways in which two lines in a plane can be related. Thus there are three situations that can occur when we solve a linear system of equations in two variables. The two lines can:

1 intersect in a single point (exactly one solution),
2 be parallel with no point of intersection (no solution), or
3 coincide (infinitely many solutions).

If two equations have no common solution, they are **inconsistent**; otherwise, the system is **consistent**. If the graphs of two equations coincide, the system is **dependent**; otherwise, the equations have different graphs and the system is **independent**.

We can determine what type of solution a system of equations has by examining the coefficients and constants in the system. To accomplish this, we first transform the equations from their general form into slope-intercept form:

General Form | Slope-Intercept Form for $b_1, b_2 \neq 0$

$$\begin{cases} a_1 x + b_1 y = c_1 \\ a_2 x + b_2 y = c_2 \end{cases} \qquad \begin{cases} y = \dfrac{-a_1}{b_1} x + \dfrac{c_1}{b_1} \\ y = \dfrac{-a_2}{b_2} x + \dfrac{c_2}{b_2} \end{cases} \qquad \text{Slope } \dfrac{-a_1}{b_1} \text{ and } y\text{-intercept } \dfrac{c_1}{b_1}$$

$$\text{Slope } \dfrac{-a_2}{b_2} \text{ and } y\text{-intercept } \dfrac{c_2}{b_2}$$

SECTION 6-1 THE SUBSTITUTION AND ELIMINATION METHODS

We can use the slope and y-intercept to analyze the three types of solution sets for linear systems.

1. **One solution (a consistent and independent system):** The lines intersect in a single point. Since the lines are not parallel, the slopes are not equal. Therefore,

$$\frac{-a_1}{b_1} \neq \frac{-a_2}{b_2}, \quad \frac{a_1}{b_1} \neq \frac{a_2}{b_2} \quad \text{and so} \quad \frac{a_1}{a_2} \neq \frac{b_1}{b_2}$$

> **▶ Self-Check ▼**
>
> Solve the system $\begin{cases} 2x - y = -5 \\ x + 2y = 0 \end{cases}$ graphically.

2. **No solution (an inconsistent system):** The lines are parallel; thus the slopes are equal, but the y-intercepts are not equal.

$$\frac{-a_1}{b_1} = \frac{-a_2}{b_2} \quad \text{and so} \quad \frac{a_1}{a_2} = \frac{b_1}{b_2}$$

However,

$$\frac{c_1}{b_1} \neq \frac{c_2}{b_2} \quad \text{and so} \quad \frac{b_1}{b_2} \neq \frac{c_1}{c_2}$$

Thus

$$\frac{a_1}{a_2} = \frac{b_1}{b_2} \neq \frac{c_1}{c_2}$$

3. **Infinitely many solutions (a dependent system):** The lines coincide; thus the slopes and the y-intercepts are the same.

$$\frac{-a_1}{b_1} = \frac{-a_2}{b_2} \quad \text{and so} \quad \frac{a_1}{a_2} = \frac{b_1}{b_2}$$

and

$$\frac{c_1}{b_1} = \frac{c_2}{b_2} \quad \text{and so} \quad \frac{b_1}{b_2} = \frac{c_1}{c_2}$$

Thus

$$\frac{a_1}{a_2} = \frac{b_1}{b_2} = \frac{c_1}{c_2}$$

▶ Self-Check Answer ▼

$(-2, 1)$

Types of Solution Sets ▼

For linear systems of the form

$$\begin{cases} a_1x + b_1y = c_1 \\ a_2x + b_2y = c_2 \end{cases}$$

there may be

One Solution

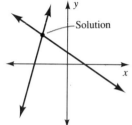

Independent;
consistent

$$\frac{a_1}{a_2} \neq \frac{b_1}{b_2}$$

No Solution

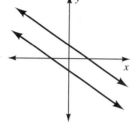

Inconsistent;
parallel lines

$$\frac{a_1}{a_2} = \frac{b_1}{b_2} \neq \frac{c_1}{c_2}$$

Infinite Number of Solutions

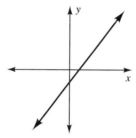

Dependent;
coincident lines

$$\frac{a_1}{a_2} = \frac{b_1}{b_2} = \frac{c_1}{c_2}$$

EXAMPLE 2 Determine the number of solutions for each system of equations.

(a) $2x + 3y = 6$
$3x + 2y = 5$

(b) $3x = 2y + 5$
$6x - 4y = 7$

(c) $2y = 3x - 5$
$6x - 4y = 10$

SOLUTION

1️⃣ Put each system into the form $ax + by = c$.

(a) $2x + 3y = 6$
$3x + 2y = 5$

(b) $3x - 2y = 5$
$6x - 4y = 7$

(c) $-3x + 2y = -5$
$6x - 4y = 10$

2️⃣ Consider the ratios of the corresponding coefficients and constants.

(a) $\dfrac{2}{3} \neq \dfrac{3}{2}$

(b) $\dfrac{3}{6} = \dfrac{-2}{-4} \neq \dfrac{5}{7}$

(c) $\dfrac{-3}{6} = \dfrac{2}{-4} = \dfrac{-5}{10}$

Answers (a) Exactly one solution; (b) no solution; (c) infinitely many solutions.

SECTION 6-1 THE SUBSTITUTION AND ELIMINATION METHODS

Graphs are useful for obtaining an intuitive visualization of the solution of a system of equations or an approximation of a solution. However, algebraic methods efficiently produce the exact solution. The axiom used in the substitution method is that a quantity may be substituted for its equal.

The Substitution Method ▼

Step 1 Solve one of the equations for one variable in terms of the other variable.

Step 2 Substitute the expression obtained in Step 1 into the other equation (eliminating one of the variables), and solve the resulting equation.

Step 3 Substitute the value obtained in Step 2 into the equation produced in Step 1 to find the value of the other variable.

▶ Self-Check ▼

Determine the number of solutions for each system of equations.

1. $2x = 4y - 6$
   $5x + 15 = 10y$
2. $4x - y = 2$
   $12x - 3y = 4$
3. $5x + 2y - 4 = 0$
   $4x + 3y = 5$

EXAMPLE 3 Solve $\begin{cases} 3x - 4y = 1 \\ x + 6y = 4 \end{cases}$ using the substitution method.

SOLUTION

$\boxed{1}$ $x = 4 - 6y.$ Since the coefficient of x is 1, we will solve $x + 6y = 4$ for x.

$\boxed{2}$ $3(4 - 6y) - 4y = 1$ Substitute for x in $3x - 4y = 1$.

$12 - 18y - 4y = 1$ Solve the resulting equation.

$-22y = -11$

$y = \dfrac{1}{2}$

$\boxed{3}$ $x = 4 - 6\left(\dfrac{1}{2}\right)$ Substitute $\dfrac{1}{2}$ for y in $x = 4 - 6y$.

$= 4 - 3$

$= 1$

Answer The solution is $\left(1, \dfrac{1}{2}\right)$. Does this solution check?   □

The algebraic solution of an independent and consistent system will result in a unique value for each variable and thus a unique ordered pair satisfying the system. The algebraic solution of an inconsistent system will result in a contradiction and thus no solution. The algebraic solution of a dependent system will result in an identity and therefore an infinite number of solutions.

▶ Self-Check Answers ▼

1 Infinitely many solutions   **2** No solution   **3** One solution

EXAMPLE 4 Solve $\begin{cases} 2x - y = 3 \\ 4x - 2y = -6 \end{cases}$ using the substitution method.

SOLUTION
$\begin{cases} 2x - y = 3 \\ 4x - 2y = -6 \end{cases} \to y = 2x - 3$ Solve for y.
$\phantom{\begin{cases} 2x - y = 3 \\ 4x - 2y = -6 \end{cases}} \to 4x - 2(2x - 3) = -6$ Substitute for y.
$ 4x - 4x + 6 = -6$ Simplify.
$ 6 = -6$ This equation is a contradiction.

Answer An inconsistent system with no solution.   □

EXAMPLE 5 Solve $\begin{cases} x - 5y = 8 \\ -3x + 15y = -24 \end{cases}$ using the substitution method.

SOLUTION
$\begin{cases} x - 5y = 8 \\ -3x + 15y = -24 \end{cases} \to x = 5y + 8$ Solve for x.
$\phantom{\begin{cases} x - 5y = 8 \\ -3x + 15y = -24 \end{cases}} \to -3(5y + 8) + 15y = -24$ Substitute for x.
$ -15y - 24 + 15y = -24$ Simplify.
$ 0 = 0$ This is an identity.

Answer A dependent system with an infinite number of solutions.   □

The **general solution** of a dependent system describes all solutions of the system and is given by indicating the relationship between the coordinates of the solutions. The **particular solutions** obtained from the general solution contain only constant coordinates.

EXAMPLE 6 Give the general solution and three particular solutions of the dependent system from Example 5.

SOLUTIONS All solutions lie on the line $x - 5y = 8$. Solve this equation for x to obtain $x = 5y + 8$. For any value of y denoted by a, the value of x is $5a + 8$. We call $(5a + 8, a)$ the general solution. Substitute arbitrary values for a (such as -2, -1, and 0) into the general solution to obtain three particular solutions.

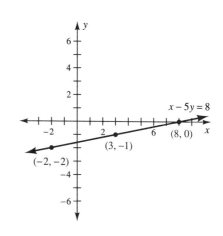

$x = 5a + 8$	$y = a$
-2	-2
3	-1
8	0

Answer The general solution is $(5a + 8, a)$. The ordered pairs $(-2, -2)$, $(8, 0)$, and $(18, 2)$ are three of an infinite number of particular solutions.   □

SECTION 6-1 THE SUBSTITUTION AND ELIMINATION METHODS

The substitution method is well suited to situations in which at least one of the equations in the system has a coefficient of 1 or −1. Otherwise, it may be easier to use transformations to eliminate a variable from one of the equations. **Equivalent systems** of equations have the same solution set. The strategy for solving a linear system is to transform the system into an equivalent system composed of simpler equations. The properties in Chapter 1 can be used to verify that the following transformations can be applied to a system of equations to produce an equivalent system.

Transformations Resulting in Equivalent Systems ▼

1. Any two equations in a system may be interchanged.
2. Both sides of any equation in a system may be multiplied by a nonzero constant.
3. Any equation in a system may be replaced by the sum of itself and a constant multiple of another equation in the system.

▶ Self-Check ▼

Solve each system using the substitution method.

1. $\begin{cases} 3x - y = 3 \\ 6x + 4y = -6 \end{cases}$

2. $\begin{cases} 3x - y = 3 \\ 6x - 2y = 6 \end{cases}$

3. $\begin{cases} 3x - y = 3 \\ 6x - 2y = -6 \end{cases}$

Since this alternative solution method involves the use of transformations to eliminate a variable from one of the equations in the equivalent system, it is called the **elimination method**. Sometimes this method is also known as the **addition method** because a multiple of one equation in the system is added to another equation in the system.

The Elimination Method ▼

Step 1 Multiply each equation by a nonzero constant so that the coefficients of one of the variables will be the same except for their signs.

Step 2 Add the resulting equations to eliminate a variable, and solve the resulting equation.

Step 3 Substitute this value back into one of the original equations (back-substitution), and solve for the other variable.

▶ Self-Check Answers ▼

1 $(\frac{1}{3}, -2)$ **2** Dependent system; general solution is $(a, 3a - 3)$.
3 Inconsistent system; no solution

EXAMPLE 7 Solve $\begin{cases} 3x - 4y = 7 \\ 2x + 3y = -1 \end{cases}$ using the elimination method.

SOLUTION

$\begin{cases} 3x - 4y = 7 \\ 2x + 3y = -1 \end{cases} \xrightarrow[r_2' = -3r_2]{r_1' = 2r_1} \begin{aligned} 6x - 8y &= 14 \\ -6x - 9y &= 3 \\ \hline -17y &= 17 \end{aligned}$

The notation $r_1' = 2r_1$ indicates that the new first row equation is obtained by multiplying the original first row equation by 2. Now that the coefficients of x are the same except for their signs (6 and -6), we can add these new equations to eliminate x.

$y = -1$ Solve for y.

$3x - 4y = 7$ Back-substitute the y-value into the first equation in the original system.

$3x - 4(-1) = 7$

$3x + 4 = 7$ Solve for x.

$3x = 3$

$x = 1$

Answer The solution is $(1, -1)$. Does this solution check?

EXAMPLE 8 Solve $\begin{cases} 2x - 6y = 4 \\ 3x - 9y = 5 \end{cases}$ using the elimination method.

SOLUTION

$\begin{cases} 2x - 6y = 4 \\ 3x - 9y = 5 \end{cases} \xrightarrow[r_2' = -2r_2]{r_1' = 3r_1} \begin{aligned} 6x - 18y &= 12 \\ -6x + 18y &= -10 \\ \hline 0x + 0y &= 2 \\ 0 &= 2 \end{aligned}$

Multiply the first equation by 3.

Multiply the second equation by -2.

Add the new equations.

This is a contradiction.

Answer Inconsistent system, no solution.

Sometimes it is convenient to use both the substitution method and the elimination method to solve a system of three or more equations.

EXAMPLE 9 Solve the system $\begin{cases} x + y = 1 \\ 2x - y + z = 10 \\ 3x + 2y - 3z = -1 \end{cases}$.

SOLUTION

$\begin{cases} x + y = 1 \\ 2x - y + z = 10 \\ 3x + 2y - 3z = -1 \end{cases} \longrightarrow \begin{cases} y = 1 - x \\ 2x - (1 - x) + z = 10 \\ 3x + 2(1 - x) - 3z = -1 \end{cases}$

Solve the first equation for y, and substitute for y in the second and third equations.

$\longrightarrow \begin{cases} y = 1 - x \\ 3x + z = 11 \\ x - 3z = -3 \end{cases}$

Simplify the last two equations.

$\xrightarrow{r_3' = -3r_3} \begin{cases} y = 1 - x \\ 3x + z = 11 \\ -3x + 9z = 9 \end{cases}$

Multiply the third equation by -3 in order to eliminate x from the last pair of equations.

SECTION 6-1 THE SUBSTITUTION AND ELIMINATION METHODS

$10z = 20$ Add the second and third equations.
$z = 2$ Solve for z.
$3x + 2 = 11$ Back-substitute for z, and solve for x in the
$3x = 9$ second equation.
$x = 3$
$y = 1 - 3$ Back-substitute for x in the first equation.
$y = -2$

Answer The ordered triple $(3, -2, 2)$ is the solution. Does this solution check? ☐

EXAMPLE 10 Some grocery orders from 1950 were found in a desk drawer. One order listed three gallons of milk and two loaves of bread with a total price of $2.70. A second order listed two gallons of milk and three loaves of bread with a total price of $2.05. Individual prices were not lised. Determine the price of each item.

SOLUTION Let m = Price of a gallon of milk
b = Price of a loaf of bread

Word Equation:	Price of Milk	+	Price of Bread	=	Total Price
First order	$3m$	+	$2b$	=	2.70
Second order	$2m$	+	$3b$	=	2.05

The *word equation* is based on the mixture principle.

The cost of each item is determined by the rate principle.

$3m + 2b = 2.70 \xrightarrow{r_1' = 2r_1} 6m + 4b = 5.40$ Multiply the first equation by 2.
$2m + 3b = 2.05 \xrightarrow{r_2' = -3r_2} -6m - 9b = -6.15$ Multiply the second equation by -3.
$\phantom{2m + 3b = 2.05 \xrightarrow{r_2' = -3r_2}} -5b = -0.75$ Add the new equations.
$\phantom{2m + 3b = 2.05 \xrightarrow{r_2' = -3r_2}} b = 0.15$
$3m + 2(0.15) = 2.70$ Substitute for b in the first equation.
$3m = 2.40$
$m = 0.80$

Answer The price of milk was $0.80 a gallon, and the price of bread was $0.15 a loaf. ☐

▶ **Self-Check** ▼

Solve each system.

1 $\begin{cases} 8x - 6y = 6 \\ 4x + 9y = -5 \end{cases}$

2 $\begin{cases} 8x - 2y = 6 \\ 12x - 3y = 9 \end{cases}$

3 A total of $8000 is to be deposited in a savings account and a money market fund for one year. The annual interest rates are 6% and 8%, respectively. How much should be put in each in order to earn interest of exactly $540?

▶ **Self-Check Answers** ▼

1 $(\frac{1}{4}, -\frac{2}{3})$ 2 $(a, 4a - 3)$; dependent system
3 $5000 in the savings account and $3000 in the money market fund.

EXERCISES 6-1

A

1 Verify that $(-6, 5)$ is a solution of $\begin{cases} 4x - 3y = -39 \\ 2x + 5y = 13 \end{cases}$.

2 Verify that $\left(\dfrac{3}{2}, \dfrac{5}{3}\right)$ is a solution of $\begin{cases} 2x + 3y = 8 \\ 6x - 9y = -6 \end{cases}$.

In Exercises 3–6, use graphs to approximate the solution of each system of equations.

3 $x - y = 2$
   $x + 3y = 6$

4 $2x + y = 6$
   $2x - y = 6$

5 $5x - 10y = 15$
   $2x - 4y = 6$

6 $x - y = 3$
   $2x - 2y = 4$

In Exercise 7–14, determine the number of solutions for each system of equations.

7 $7x + 3y = 10$
   $6x + 5y = 8$

8 $2u + 6t = 10$
   $9t = 15 - 3u$

9 $3s = 12t + 21$
   $5s - 20t = 30$

10 $6a = 15 + 15b$
   $2a = 6 - 5b$

11 $8t + 18u = 30$
   $4t + 9u = 15$

12 $x = 5y - 6$
   $4x - 20y + 24 = 0$

13 $5x = 4y$
   $5x = 6y$

14 $2a + 6b = 5$
   $4a = 8b$

In Exercises 15–20, solve each system of equations using the substitution method. Give the general solution and three particular solutions for each dependent system.

15 $x = y + 1$
   $2x - 3y = -2$

16 $y = x - 5$
   $4x + y = 5$

17 $2x + 4y = 2$
   $x + 2y = 1$

18 $x - 4y = 1$
   $2x - 5y = -1$

19 $x = 5 - 2y$
   $3x + 6y = 4$

20 $y = 3x + 2$
   $6x - 2y = -4$

In Exercises 21–28, solve each system of equations using the elimination method. Give the general solution and three particular solutions for each dependent system.

21 $2x - 3y = 1$
   $4x + 3y = -7$

22 $2x + 5y = 11$
   $2x - 3y = -13$

23 $2x - 7y = -8$
   $4x + 5y = 60$

24 $11x - 3y = 46$
   $5x + 6y = 43$

25 $4x - 3y = -38$
   $9x + 2y = -33$

26 $4x - 9y = 48$
   $5x + 13y = -37$

27 $8x - 10y = 4$
   $12x - 15y = 6$

28 $20x - 70y = 30$
   $4x - 14y = 20$

In Exercises 29–42 use the substitution method or the elimination method to solve each system of equations. Give the general solution and three particular solutions for each dependent system.

29 $6x + 6y = -1$
   $4x - 3y = -3$

30 $5x + 4y = -1$
   $10x - 8y = 18$

31 $x = 5y + 3$
   $2x - 10y = 6$

32 $3y = 2 - 2x$
   $6x = 5 - 9y$

33 $3x = 7y$
   $11x = 5y$

34 $12x = 5y$
   $9x = 2y$

35 $\dfrac{1}{2}x - \dfrac{1}{3}y = \dfrac{1}{6}$
   $15x - 10y = 2$

36 $x - \dfrac{1}{4}y = \dfrac{3}{4}$
   $12x - 3y = 9$

37 $\dfrac{1}{x} + \dfrac{1}{y} = 0$
   $\dfrac{2}{x} + \dfrac{5}{y} = -6$
   $\left(\text{Hint: Let } v = \dfrac{1}{x} \text{ and } w = \dfrac{1}{y}.\right)$

38 $\dfrac{1}{x} - \dfrac{2}{y} = 6$
   $\dfrac{3}{x} + \dfrac{5}{y} = -4$

39 $x + y = 2$
   $-y + z = 2$
   $x - z = -1$

40 $2x + y = 7$
   $y - z = 2$
   $x + z = 2$

41 $x + y + z = 3$
   $x - y + 2z = 7$
   $2x + 4y + 3z = 4$

42 $2x + y + 2z = -3$
   $3x - 2y - 3z = -3$
   $2x + 3y + 2z = 3$

SECTION 6-1 THE SUBSTITUTION AND ELIMINATION METHODS

B

Solve Exercises 43–48 by first forming a system of equations.

43 One number is five more than three times a second number. Find these numbers if their sum is 77.

44 The perimeter of a rectangular lot is 500 feet. The length of the lot is 20 feet less than twice the width. Find the dimensions.

45 A group of 6 adults and 12 children purchased round-trip tickets on Amtrak to Disney World at a total cost of $900. One month later, another group of 4 adults and 16 children purchased round-trip tickets at the same rate for a total cost of $900. How much did each adult and each child have to pay for a round-trip ticket?

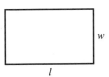

Figure for Exercise 44

46 A music concert was given on two consecutive nights at a high school gymnasium. On Monday, 200 section-A tickets and 350 section-B tickets were sold for a total of $7200. On Tuesday, 250 section-A tickets and 500 section-B tickets were sold for a total of $9750. What was the price of each type of ticket?

47 A 15% solution and an 8% solution of an acid are mixed to produce 40 liters of a 10.8% solution. How many liters of each are used?

48 An investment of $4000 resulted in an annual income of $365. Part of the investment was at 8%, and the rest was at 11%. Find the amount invested at each rate.

In Exercises 49 and 50, solve each system of equations.

49 $\dfrac{2}{x+3} - \dfrac{3}{y-2} = -\dfrac{1}{10}$

$\dfrac{5}{x+3} - \dfrac{4}{y-2} = \dfrac{9}{20}$

50 $x - 2y + 3z = -3$
$x + 2y - z = 3$
$3x + 2y + 3z = 1$

C

In Exercises 51–53, solve each system of equations.

51 $\dfrac{2}{x} - \dfrac{3}{y} + \dfrac{4}{z} = -10$

$\dfrac{3}{x} - \dfrac{4}{y} + \dfrac{2}{z} = -6$

$\dfrac{4}{x} + \dfrac{3}{y} - \dfrac{2}{z} = 18$

52 $\dfrac{3}{x-1} + \dfrac{2}{y+1} + \dfrac{1}{z-3} = 8$

$\dfrac{2}{x-1} + \dfrac{1}{y+1} + \dfrac{3}{z-3} = 7$

$\dfrac{1}{x-1} + \dfrac{3}{y+1} + \dfrac{2}{z-3} = 9$

53 $2x^2 - 5y^2 = -17$
$4x^2 - y^2 = 29$

(*Hint*: Let $v = x^2$ and $w = y^2$.)

In Exercises 54–56, solve each system of equations with the aid of a calculator. Your answers should be accurate to the nearest hundredth.

54 $1.23x - 4.8y = 0.3864$
$4.07x + 1.7y = 0.2236$

55 $806x - 711y = -5587$
$639x - 477y = -702$

56 $\sqrt{2}x - \sqrt{3}y = 5$
$\sqrt{5}x + \sqrt{7}y = -1.42$

In Exercises 57 and 58, find the general solution of each system of equations and list three particular solutions of each dependent system.

57 $2x - 5y - 11z = 0$
$7x - 6y - 4z = 23$

58 $x + y + z = 8$
$x + 3y + 4z = 6$
$x - y - 2z = 10$

SECTION 6-2

Solving Linear Systems Using Augmented Matrices

Section Objective

4 Solve a system of linear equations using an augmented matrix.

The substitution method and the elimination method are appropriate for solving linear systems of only two or three equations. For larger systems, these methods may require several steps, which can be difficult to keep organized. The augmented matrix method is very systematic and simplifies the solution of larger systems of linear equations. This method is also well suited to solution by computer.

A **matrix** is a rectangular array of numbers. The entries in an **augmented matrix** for a system of linear equations consist of the coefficients and constants in the equations. To form the augmented matrix for a system, we first align the similar variables on the left side and the constants on the right side. Then we form each row in the matrix from the coefficients and the constant of the corresponding equation. A zero should be written in any position that corresponds to a missing variable in an equation.

System of Equations

$$\begin{cases} 2x - 3y + 3z = 2 \\ 4x + 2z = -1 \\ -2x + 4y - 3z = -2 \end{cases}$$

Augmented Matrix

$$\begin{bmatrix} 2 & -3 & 3 & | & 2 \\ 4 & 0 & 2 & | & -1 \\ -2 & 4 & -3 & | & -2 \end{bmatrix}$$

Coefficients of x, Coefficients of y, Coefficients of z, Constants

The augmented matrix method for the solution of a linear system of equations is simply a more compact notation for the elimination method. Only the coefficients and constants in the system are used in the augmented matrix method; the variables are denoted by the positions of the coefficients. Each transformation that yields an equivalent system of equations in the elimination method corresponds to an elementary row operation that can be applied to the augmented matrix.

Elementary Row Operations on Matrices ▼

1. Any two rows in a matrix may be interchanged.
2. Any row in a matrix may be multiplied by a nonzero constant.
3. Any row in a matrix may be replaced by the sum of itself and a constant multiple of another row.

SECTION 6-2 SOLVING LINEAR SYSTEMS USING AUGMENTED MATRICES

We use the elementary row operations on the augmented matrix just as if the rows were the equations they represent. This is illustrated by the parallel development in Example 1.

EXAMPLE 1 Solve the system $\begin{cases} 3x + 2y = 6 \\ x - 3y = 4 \end{cases}$.

SOLUTIONS

Elimination Method

$\begin{cases} 3x + 2y = 6 \\ x - 3y = 4 \end{cases}$

$\downarrow$

$\begin{cases} x - 3y = 4 \\ 3x + 2y = 6 \end{cases}$

$\downarrow$

$\begin{cases} x - 3y = 4 \\ 11y = -6 \end{cases}$

$\downarrow$

$\begin{cases} x - 3y = 4 \\ y = -\frac{6}{11} \end{cases}$

$\downarrow$

$\begin{cases} x = \frac{26}{11} \\ y = -\frac{6}{11} \end{cases}$

Answer $\left(\frac{26}{11}, -\frac{6}{11}\right)$

Augmented Matrix Method

$\begin{bmatrix} 3 & 2 & | & 6 \\ 1 & -3 & | & 4 \end{bmatrix}$

$\downarrow r_1 \leftrightarrow r_2$ This notation means that the first row is interchanged with the second row.

$\begin{bmatrix} 1 & -3 & | & 4 \\ 3 & 2 & | & 6 \end{bmatrix}$

$\downarrow r'_2 = r_2 - 3r_1$ Replace the second row with itself minus 3 times the first row.

$\begin{bmatrix} 1 & -3 & | & 4 \\ 0 & 11 & | & -6 \end{bmatrix}$

$\downarrow r'_2 = \frac{1}{11} r_2$ Multiply the second row by $\frac{1}{11}$.

$\begin{bmatrix} 1 & -3 & | & 4 \\ 0 & 1 & | & -\frac{6}{11} \end{bmatrix}$

$\downarrow r'_1 = r_1 + 3r_2$ Replace the first row with itself plus 3 times the second row.

$\begin{bmatrix} 1 & 0 & | & \frac{26}{11} \\ 0 & 1 & | & -\frac{6}{11} \end{bmatrix}$ This form gives the answer.

▶ Self-Check ▼

1 Form the augmented matrix for the system

$\begin{cases} x + 2y + z = -2 \\ 2x + 4y - 3z = 0 \\ -x + 2y = -3 \end{cases}$

2 Write the system of equations represented by the augmented matrix

$\begin{bmatrix} 1 & -1 & 0 & | & 2 \\ 2 & 3 & -1 & | & 4 \\ 0 & 3 & -4 & | & 2 \end{bmatrix}$

▶ Self-Check Answers ▼

1 $\begin{bmatrix} 1 & 2 & 1 & | & -2 \\ 2 & 4 & -3 & | & 0 \\ -1 & 2 & 0 & | & -3 \end{bmatrix}$ **2** $\begin{cases} x - y = 2 \\ 2x + 3y - z = 4 \\ 3y - 4z = 2 \end{cases}$

The final form of the augmented matrix produced in Example 1 can be written as

$$\begin{bmatrix} 1 & 0 & | & k_1 \\ 0 & 1 & | & k_2 \end{bmatrix}$$

where k_1 and k_2 are the values of x and y, respectively. Likewise, the final form for the augmented matrix associated with an independent and consistent 3×3 system of linear equations is

$$\begin{bmatrix} 1 & 0 & 0 & | & k_1 \\ 0 & 1 & 0 & | & k_2 \\ 0 & 0 & 1 & | & k_3 \end{bmatrix}$$

The solution of this system is (k_1, k_2, k_3). Similar forms can be obtained for dependent or inconsistent systems. The reduced form defined here is also called **echelon form**.

Properties of the Reduced Form of a Matrix ▼

1. The first nonzero entry in a row is a 1. All other entries in the column containing the leading 1 are zeros.
2. All nonzero rows are above any rows containing only zeros.
3. The first nonzero entry in a row is to the left of the first nonzero entry in the following row.

The following cases show how to identify a matrix in reduced form:

- Matrices in reduced form:

$$\begin{bmatrix} 1 & 0 & 0 & | & 2 \\ 0 & 1 & 0 & | & 3 \\ 0 & 0 & 1 & | & 4 \end{bmatrix} \quad \begin{bmatrix} 1 & 0 & 0 & | & 1 \\ 0 & 1 & 0 & | & -1 \\ 0 & 0 & 0 & | & 0 \end{bmatrix} \quad \begin{bmatrix} 1 & 5 & 0 & 0 & | & 6 \\ 0 & 0 & 1 & 2 & | & 4 \end{bmatrix}$$

- Matrices not in reduced form:

$$\begin{bmatrix} 1 & 0 & 0 & | & -6 \\ 0 & 2 & 0 & | & 5 \\ 0 & 0 & 1 & | & 4 \end{bmatrix}$$ Violates Property 1 since the first nonzero entry in the second row is not a 1

$$\begin{bmatrix} 1 & 0 & 0 & | & 2 \\ 0 & 1 & -1 & | & 5 \\ 0 & 0 & 1 & | & 6 \end{bmatrix}$$ Violates Property 1 since the third column contains a nonzero value above the 1 in the third row

$$\begin{bmatrix} 1 & 0 & 0 & | & -8 \\ 0 & 0 & 0 & | & 0 \\ 0 & 1 & 5 & | & 3 \end{bmatrix}$$ Violates Property 2. Rows containing only zeros must be below all nonzero rows.

SECTION 6-2 SOLVING LINEAR SYSTEMS USING AUGMENTED MATRICES

$$\begin{bmatrix} 0 & 1 & 0 & | & 3 \\ 1 & 0 & 0 & | & -2 \\ 0 & 0 & 1 & | & 5 \end{bmatrix}$$ Violates Property 3. If the first and second rows were interchanged, this would be in reduced form.

The solution of a system of linear equations can be found by using the elementary row operations to transform the augmented matrix to its reduced form. We will now develop a strategy for efficiently producing this reduced form. Example 2 demonstrates that some elementary row operations may be easier to apply than others.

▶ **Self-Check** ▼

Which of these matrices are in reduced form?

1. $\begin{bmatrix} 1 & 1 & 0 & | & 3 \\ 0 & 1 & 0 & | & -2 \\ 0 & 0 & 1 & | & 2 \end{bmatrix}$

2. $\begin{bmatrix} 1 & 0 & 0 & | & 0 \\ 0 & 1 & 0 & | & 0 \\ 0 & 0 & 1 & | & 0 \end{bmatrix}$

3. $\begin{bmatrix} 1 & 0 & 0 & | & -5 \\ 0 & 1 & 0 & | & 4 \\ 0 & 0 & 0 & | & 0 \end{bmatrix}$

4. $\begin{bmatrix} 0 & 0 & 0 & | & 0 \\ 1 & 0 & 0 & | & 5 \\ 0 & 0 & 1 & | & -4 \end{bmatrix}$

EXAMPLE 2 Use the elementary row operations to transform the matrix

$$\begin{bmatrix} 5 & 6 & -4 & | & -8 \\ 2 & 4 & -1 & | & 1 \\ 1 & 1 & -3 & | & 0 \end{bmatrix} \quad \text{into the form} \quad \begin{bmatrix} 1 & \_ & \_ & | & \_ \\ \_ & \_ & \_ & | & \_ \\ \_ & \_ & \_ & | & \_ \end{bmatrix}.$$

SOLUTION

Method A:

$$\begin{bmatrix} 5 & 6 & -4 & | & -8 \\ 2 & 4 & -1 & | & 1 \\ 1 & 1 & -3 & | & 0 \end{bmatrix} \xrightarrow{r_1 \leftrightarrow r_3} \begin{bmatrix} 1 & 1 & -3 & | & 0 \\ 2 & 4 & -1 & | & 1 \\ 5 & 6 & -4 & | & -8 \end{bmatrix}$$ Interchange the first and third rows.

Method B:

$$\begin{bmatrix} 5 & 6 & -4 & | & -8 \\ 2 & 4 & -1 & | & 1 \\ 1 & 1 & -3 & | & 0 \end{bmatrix} \xrightarrow{r_1' = r_1 - 2r_2} \begin{bmatrix} 1 & -2 & -2 & | & -10 \\ 2 & 4 & -1 & | & 1 \\ 1 & 1 & -3 & | & 0 \end{bmatrix}$$ Replace the first row with the sum of itself and -2 times the second row. Note that $5 - 2(2) = 1$ is the new entry in the first row and the first column.

Method C:

$$\begin{bmatrix} 5 & 6 & -4 & | & -8 \\ 2 & 4 & -1 & | & 1 \\ 1 & 1 & -3 & | & 0 \end{bmatrix} \xrightarrow{r_1' = \frac{1}{5}r_1} \begin{bmatrix} 1 & \frac{6}{5} & -\frac{4}{5} & | & -\frac{8}{5} \\ 2 & 4 & -1 & | & 1 \\ 1 & 1 & -3 & | & 0 \end{bmatrix}$$ Multiply the first row by $\frac{1}{5}$ (divide by 5).

Method A is the easiest to apply but is applicable only when there is a coefficient of 1 to shift into the upper-left position. Method B is often used to avoid the fractions that may result with Method C. ☐

▶ **Self-Check Answers** ▼

1 Violates Property 1   **2** Reduced   **3** Reduced   **4** Violates Property 2

EXAMPLE 3 Use the elementary row operations to transform the matrix in Example 2 into the form

$$\begin{bmatrix} 1 & - & - & | & - \\ 0 & - & - & | & - \\ 0 & - & - & | & - \end{bmatrix}$$

The TI-81 keystrokes for performing these matrix row operations are given in the Graphics Calculator Supplement.

SOLUTION

$$\begin{bmatrix} 5 & 6 & -4 & | & -8 \\ 2 & 4 & -1 & | & 1 \\ 1 & 1 & -3 & | & 0 \end{bmatrix} \xrightarrow{r_1 \leftrightarrow r_3} \begin{bmatrix} 1 & 1 & -3 & | & 0 \\ 2 & 4 & -1 & | & 1 \\ 5 & 6 & -4 & | & -8 \end{bmatrix}$$

Place a one in the upper-left position by interchanging the first and third rows.

$$\xrightarrow{r_2' = r_2 - 2r_1} \begin{bmatrix} 1 & 1 & -3 & | & 0 \\ 0 & 2 & 5 & | & 1 \\ 5 & 6 & -4 & | & -8 \end{bmatrix}$$

With practice, you should be able to combine the last two steps, which produce zeros in the first column.

$$\xrightarrow{r_3' = r_3 - 5r_1} \begin{bmatrix} 1 & 1 & -3 & | & 0 \\ 0 & 2 & 5 & | & 1 \\ 0 & 1 & 11 & | & -8 \end{bmatrix} \quad \square$$

EXAMPLE 4 Solve the 3×3 system $\begin{cases} 2a + 3b - 2c = -8 \\ a - b + 2c = 25 \\ 4a + 6b - c = -7 \end{cases}$.

SOLUTION The first step is to form the augmented matrix:

$$\begin{bmatrix} 2 & 3 & -2 & | & -8 \\ 1 & -1 & 2 & | & 25 \\ 4 & 6 & -1 & | & -7 \end{bmatrix} \xrightarrow{r_1 \leftrightarrow r_2} \begin{bmatrix} 1 & -1 & 2 & | & 25 \\ 2 & 3 & -2 & | & -8 \\ 4 & 6 & -1 & | & -7 \end{bmatrix}$$

$$\xrightarrow[r_3' = r_3 - 4r_1]{r_2' = r_2 - 2r_1} \begin{bmatrix} 1 & -1 & 2 & | & 25 \\ 0 & 5 & -6 & | & -58 \\ 0 & 10 & -9 & | & -107 \end{bmatrix}$$

Transform the first column into the form

$$\begin{bmatrix} 1 & - & - & | & - \\ 0 & - & - & | & - \\ 0 & - & - & | & - \end{bmatrix}$$

$$\xrightarrow{r_2' = \frac{1}{5} r_2} \begin{bmatrix} 1 & -1 & 2 & | & 25 \\ 0 & 1 & -\frac{6}{5} & | & -\frac{58}{5} \\ 0 & 10 & -9 & | & -107 \end{bmatrix}$$

Transform the second column into the form

$$\begin{bmatrix} 1 & 0 & - & | & - \\ 0 & 1 & - & | & - \\ 0 & 0 & - & | & - \end{bmatrix}$$

$$\xrightarrow[r_3' = r_3 - 10r_2]{r_1' = r_1 + r_2} \begin{bmatrix} 1 & 0 & \frac{4}{5} & | & \frac{67}{5} \\ 0 & 1 & -\frac{6}{5} & | & -\frac{58}{5} \\ 0 & 0 & 3 & | & 9 \end{bmatrix}$$

$$\xrightarrow{r_3' = \frac{1}{3} r_3} \begin{bmatrix} 1 & 0 & \frac{4}{5} & | & \frac{67}{5} \\ 0 & 1 & -\frac{6}{5} & | & -\frac{58}{5} \\ 0 & 0 & 1 & | & 3 \end{bmatrix}$$

Transform the third column into the form

$$\begin{bmatrix} 1 & 0 & 0 & | & - \\ 0 & 1 & 0 & | & - \\ 0 & 0 & 1 & | & - \end{bmatrix}$$

SECTION 6-2 SOLVING LINEAR SYSTEMS USING AUGMENTED MATRICES

$$\xrightarrow[r'_2 = r_2 + \frac{6}{5}r_3]{r'_1 = r_1 - \frac{4}{5}r_3} \begin{bmatrix} 1 & 0 & 0 & | & 11 \\ 0 & 1 & 0 & | & -8 \\ 0 & 0 & 1 & | & 3 \end{bmatrix}$$

Answer $(11, -8, 3)$ □

The overall strategy used in the previous example can be summarized as "work from left to right and produce the 'leading ones' before you produce the zeros." This strategy is formalized in the following box.

Transforming an Augmented Matrix into Reduced Form ▼

Step 1 $\begin{bmatrix} 1 & \cdots \\ 0 & \\ 0 & \\ \vdots & \vdots \\ 0 & \cdots \end{bmatrix}$ **Transform the first column** into this form by using the elementary row operations to
  a. produce a 1 in the top position and
  b. use the 1 in row 1 to produce zeros in the other positions of column 1.

Step 2 $\begin{bmatrix} 1 & 0 & \cdots \\ 0 & 1 & \\ 0 & 0 & \\ \vdots & \vdots & \vdots \\ 0 & 0 & \cdots \end{bmatrix}$ **Transform the next column**, if possible, into this form by using the elementary operations to
  a. produce a 1 in the next row and
  b. use the 1 in this row to produce zeros in the other positions of this column.
  If it is not possible to produce a 1 in the next row, proceed to next column.

Step 3 Repeat Step 2 column by column, always producing the 1 in the next row, until you arrive at the reduced form.

▶ Self-Check ▼

Produce a 1 in row two, column two of the matrix

$$\begin{bmatrix} 1 & 1 & -3 & | & 0 \\ 0 & 2 & 5 & | & 1 \\ 0 & 1 & 11 & | & -8 \end{bmatrix}$$

by interchanging the second and third rows. Then introduce two zeros into the second column of this matrix.

For emphasis, remember your goal is to produce the "leading 1's". You may use shortcuts in this process whenever they are appropriate. This procedure is also appropriate for dependent and inconsistent systems.

▶ Self-Check Answers ▼

$$\begin{bmatrix} 1 & 1 & -3 & | & 0 \\ 0 & 2 & 5 & | & 1 \\ 0 & 1 & 11 & | & -8 \end{bmatrix} \xrightarrow{r_2 \leftrightarrow r_3} \begin{bmatrix} 1 & 1 & -3 & | & 0 \\ 0 & 1 & 11 & | & -8 \\ 0 & 2 & 5 & | & 1 \end{bmatrix} \xrightarrow[r'_3 = r_3 - 2r_2]{r'_1 = r_1 - r_2} \begin{bmatrix} 1 & 0 & -14 & | & 8 \\ 0 & 1 & 11 & | & -8 \\ 0 & 0 & -17 & | & 17 \end{bmatrix}$$

EXAMPLE 5 Solve the 3×3 system $\begin{cases} r + 8s + 2t = 20 \\ 11s + t = 28 \\ -22s - 2t = -55 \end{cases}$.

SOLUTION

$$\begin{bmatrix} 1 & 8 & 2 & | & 20 \\ 0 & 11 & 1 & | & 28 \\ 0 & -22 & -2 & | & -55 \end{bmatrix} \xrightarrow{r_3' = r_3 + 2r_2} \begin{bmatrix} 1 & 8 & 2 & | & 20 \\ 0 & 11 & 1 & | & 28 \\ 0 & 0 & 0 & | & 1 \end{bmatrix}$$

Although this matrix is not in reduced form, the last row indicates that the system is inconsistent with no solution.

Answer This last row of the matrix represents the equation

$$0r + 0s + 0t = 1$$

This statement is a contradiction, so there is no solution. □

When one row of a matrix has the form $[0 \ 0 \ 0 \ \cdots \ 0 \ k]$ where $k \neq 0$, the system is inconsistent and has no solution. Thus it is not necessary to transform the matrix to its reduced form.

EXAMPLE 6 Solve the 3×3 system $\begin{cases} x_1 + 4x_2 + 5x_3 = -2 \\ x_2 + 2x_3 = -1 \\ -5x_2 - 10x_3 = 5 \end{cases}$.

SOLUTION

$$\begin{bmatrix} 1 & 4 & 5 & | & -2 \\ 0 & 1 & 2 & | & -1 \\ 0 & -5 & -10 & | & 5 \end{bmatrix} \xrightarrow[r_3' = r_3 + 5r_2]{r_1' = r_1 - 4r_2} \begin{bmatrix} 1 & 0 & -3 & | & 2 \\ 0 & 1 & 2 & | & -1 \\ 0 & 0 & 0 & | & 0 \end{bmatrix}$$

This matrix is in reduced form. A row of zeros in an $n \times n$ consistent system indicates a dependent system with infinitely many solutions.

$$\longrightarrow \begin{cases} x_1 + 0x_2 - 3x_3 = 2 \\ 0x_1 + x_2 + 2x_3 = -1 \\ 0x_1 + 0x_2 + 0x_3 = 0 \end{cases}$$

This is the system represented by the reduced matrix. The system is dependent, since the last equation is an identity satisfied by all values of x_1, x_2, and x_3.

Thus

$$x_1 = 3x_3 + 2$$
$$x_2 = -2x_3 - 1$$

The general solution is acquired by solving the first two equations for x_1 and x_2 in terms of x_3 and then replacing x_3 by a, which can be any real number.

Answer The general solution is $(3a + 2, -2a - 1, a)$; three particular solutions are $(8, -5, 2)$, $(2, -1, 0)$, and $(-7, 5, -3)$. □

These particular solutions were found by letting a be 2, 0, and -3, respectively.

Forms That Indicate a Dependent or Inconsistent $n \times n$ System of Equations ▼

Let A be the augmented matrix of an $n \times n$ system of equations.

1. If the reduced form of A has a row of the form $[0\ 0\ 0\ \cdots\ 0\ k]$, where $k \neq 0$, then the system is inconsistent and has no solution.
2. If the system is consistent and the reduced form of A has a row of zeros, then the system is dependent and has infinitely many solutions.

▶ **Self-Check** ▼

Write the solution for each system of linear equations represented by the following reduced augmented matrices.

$$1\quad \begin{bmatrix} 1 & 0 & 0 & | & -1 \\ 0 & 1 & 0 & | & 1 \\ 0 & 0 & 1 & | & 3 \end{bmatrix}$$

$$2\quad \begin{bmatrix} 1 & 0 & 1 & | & 5 \\ 0 & 1 & -1 & | & 2 \\ 0 & 0 & 0 & | & 0 \end{bmatrix}$$

$$3\quad \begin{bmatrix} 1 & 0 & 0 & | & -2 \\ 0 & 1 & 2 & | & 1 \\ 0 & 0 & 0 & | & 1 \end{bmatrix}$$

The matrix method is also a convenient method for solving systems of linear equations that do not have the same number of variables as equations.

EXAMPLE 7 Find a general solution and three particular solutions for the 2×3 system $\begin{cases} 2x - 3y + 17z = 12 \\ 8x + 2y - 2z = 20 \end{cases}$.

SOLUTION

$$\begin{bmatrix} 2 & -3 & 17 & | & 12 \\ 8 & 2 & -2 & | & 20 \end{bmatrix} \xrightarrow{r_1' = \frac{1}{2}r_1} \begin{bmatrix} 1 & -\frac{3}{2} & \frac{17}{2} & | & 6 \\ 8 & 2 & -2 & | & 20 \end{bmatrix}$$

$$\xrightarrow{r_2' = r_2 - 8r_1} \begin{bmatrix} 1 & -\frac{3}{2} & \frac{17}{2} & | & 6 \\ 0 & 14 & -70 & | & -28 \end{bmatrix}$$

$$\xrightarrow{r_2' = \frac{1}{14}r_2} \begin{bmatrix} 1 & -\frac{3}{2} & \frac{17}{2} & | & 6 \\ 0 & 1 & -5 & | & -2 \end{bmatrix}$$

$$\xrightarrow{r_1' = r_1 + \frac{3}{2}r_2} \begin{bmatrix} 1 & 0 & 1 & | & 3 \\ 0 & 1 & -5 & | & -2 \end{bmatrix}$$

$$\begin{cases} x + z = 3 \\ y - 5z = -2 \end{cases}$$

This is the system represented by the reduced matrix.

$$x = 3 - z$$
$$y = 5z - 2$$

The general solution is obtained by solving these equations for x and y in terms of z and then replacing z by a, which can be any real number.

Answer The general solution is $(3 - a, 5a - 2, a)$; particular solutions are $(3, -2, 0)$, $(4, -7, -1)$, and $(1, 8, 2)$. ☐

The particular solutions were found by letting a be 0, -1, and 2, respectively.

The matrix method can be modified to save time when you are solving several systems of linear equations that share the same coefficients. This is illustrated with two similar systems in Example 8.

▶ **Self-Check Answers** ▼

1 $(-1, 1, 3)$ **2** General solution $(5 - a, 2 + a, a)$ **3** No solution

EXAMPLE 8 Solve the 2 × 2 systems of linear equations

$$\begin{cases} 2x + y = 1 \\ x + 3y = 0 \end{cases} \text{ and } \begin{cases} 2x + y = 0 \\ x + 3y = 1 \end{cases}$$

SOLUTION The augmented matrices are $\begin{bmatrix} 2 & 1 & | & 1 \\ 1 & 3 & | & 0 \end{bmatrix}$ and $\begin{bmatrix} 2 & 1 & | & 0 \\ 1 & 3 & | & 1 \end{bmatrix}$, respectively. The first two columns are identical, so we can combine them into one matrix.

$$\begin{bmatrix} 2 & 1 & | & 1 & 0 \\ 1 & 3 & | & 0 & 1 \end{bmatrix} \xrightarrow{r_1 \leftrightarrow r_2} \begin{bmatrix} 1 & 3 & | & 0 & 1 \\ 2 & 1 & | & 1 & 0 \end{bmatrix} \xrightarrow{r_2' = r_2 - 2r_1} \begin{bmatrix} 1 & 3 & | & 0 & 1 \\ 0 & -5 & | & 1 & -2 \end{bmatrix}$$

$$\xrightarrow{r_2' = -\frac{1}{5}r_2} \begin{bmatrix} 1 & 3 & | & 0 & 1 \\ 0 & 1 & | & -\frac{1}{5} & \frac{2}{5} \end{bmatrix} \xrightarrow{r_1' = r_1 - 3r_2} \begin{bmatrix} 1 & 0 & | & \frac{3}{5} & -\frac{1}{5} \\ 0 & 1 & | & -\frac{1}{5} & \frac{2}{5} \end{bmatrix}$$

Where the constants come from: first column of constants — first equation; second column of constants — second equation. Solution to first system ↑ Solution to second system.

Answer The solutions are $(\frac{3}{5}, -\frac{1}{5})$ and $(-\frac{1}{5}, \frac{2}{5})$ for the first and second systems, respectively. □

Sections 6-1 and 6-2 have covered three methods for solving systems of linear equations: the substitution method, the elimination method, and the augmented matrix method. Two other methods will be covered in Chapter 7: inverse matrices in Section 7-2 and Cramer's rule in Section 7-3. For many systems of equations the easiest way to solve the system utilizing a TI-81 calculator would be the inverse matrix method.

EXERCISES 6-2

A

In Exercises 1–3, determine the augmented matrix for each of the linear systems.

1. $2x + 3y = 5$
   $6x - 4y = 2$

2. $6x - 3y = 2$
   $3y + 2z = 1$
   $2x - z = 5$

3. $x_1 + x_2 + x_3 + x_4 = 1$
   $2x_1 - x_3 = 0$
   $x_2 = 0$

In Exercises 4–6, write the system of equations represented by each augmented matrix.

4. $\begin{bmatrix} 4 & 5 & | & 6 \\ 9 & 2 & | & 8 \end{bmatrix}$

5. $\begin{bmatrix} 1 & 0 & 3 & | & 7 \\ 0 & 4 & -2 & | & 9 \end{bmatrix}$

6. $\begin{bmatrix} 1 & -2 & 5 & | & 0 \\ 2 & 4 & -3 & | & 8 \\ 3 & 5 & 7 & | & 11 \end{bmatrix}$

SECTION 6-2 SOLVING LINEAR SYSTEMS USING AUGMENTED MATRICES

In Exercises 7–10, perform the indicated row operations to complete the missing entries in each matrix.

7. $\begin{bmatrix} 2 & 4 & 8 & | & 6 \\ 3 & 5 & 7 & | & 1 \\ 4 & 9 & 2 & | & 8 \end{bmatrix} \xrightarrow{x_1' = \frac{1}{2}x_1} \begin{bmatrix} & & & | & \\ 3 & 5 & 7 & | & 1 \\ 4 & 9 & 2 & | & 8 \end{bmatrix}$

8. $\begin{bmatrix} 1 & 2 & 5 & | & 11 \\ 0 & 5 & 6 & | & 4 \\ 0 & 8 & 2 & | & 5 \end{bmatrix} \xrightarrow{x_2' = \frac{1}{5}x_2} \begin{bmatrix} 1 & 2 & 5 & | & 11 \\ & & & | & \\ 0 & 8 & 2 & | & 5 \end{bmatrix}$

9. $\begin{bmatrix} 1 & 3 & 5 & | & 11 \\ 2 & 7 & 9 & | & 13 \\ 4 & 8 & 3 & | & 7 \end{bmatrix} \xrightarrow{x_2' = x_2 - 2x_1} \begin{bmatrix} 1 & 3 & 5 & | & 11 \\ & & & | & \\ 4 & 8 & 3 & | & 7 \end{bmatrix}$

10. $\begin{bmatrix} 1 & 3 & 5 & | & 11 \\ 2 & 7 & 9 & | & 13 \\ 4 & 8 & 3 & | & 7 \end{bmatrix} \xrightarrow{x_3' = x_3 - 4x_1} \begin{bmatrix} 1 & 3 & 5 & | & 11 \\ 2 & 7 & 9 & | & 13 \\ & & & | & \end{bmatrix}$

Which of the matrices in Exercises 11–20 are in reduced form? If the matrix is not in reduced form, use an elementary row operation to transform the matrix to reduced form.

11. $\begin{bmatrix} 1 & 0 & 0 & | & 2 \\ 0 & 1 & 0 & | & 3 \\ 0 & 0 & 1 & | & 4 \end{bmatrix}$

12. $\begin{bmatrix} 1 & 0 & 1 & | & 2 \\ 0 & 3 & 2 & | & 3 \\ 0 & 0 & 0 & | & 0 \end{bmatrix}$

13. $\begin{bmatrix} 1 & 0 & 8 & -2 & 0 & | & 1 \\ 0 & 0 & 0 & 0 & 0 & | & 0 \\ 0 & 1 & 0 & 0 & 0 & | & 0 \end{bmatrix}$

14. $\begin{bmatrix} 1 & 2 & 0 & 0 & 1 & 2 & | & 0 \\ 0 & 1 & 0 & 2 & 3 & 0 & | & 0 \\ 0 & 0 & 1 & 0 & 0 & 0 & | & 0 \\ 0 & 0 & 0 & 0 & 0 & 1 & | & 1 \end{bmatrix}$

15. $\begin{bmatrix} 1 & 2 & 0 & 0 & | & 3 \\ 0 & 1 & 1 & 0 & | & 1 \\ 0 & 0 & 0 & 1 & | & 0 \end{bmatrix}$

16. $\begin{bmatrix} 0 & 1 & 0 & 0 & | & 1 \\ 1 & 0 & 0 & 0 & | & 0 \end{bmatrix}$

17. $\begin{bmatrix} 1 & 2 & 0 & 0 & | & 0 \\ 0 & 0 & 2 & 0 & | & 0 \\ 0 & 0 & 0 & 1 & | & 1 \end{bmatrix}$

18. $\begin{bmatrix} 0 & 1 & 0 & 0 & | & 0 \\ 1 & 0 & 1 & 0 & | & 0 \\ 0 & 0 & 0 & 1 & | & 1 \end{bmatrix}$

19. $\begin{bmatrix} 1 & 0 & 0 & 2 & 3 & 0 & 0 & 0 & | & 1 \\ 0 & 0 & 1 & 4 & -3 & 0 & 0 & 0 & | & 2 \\ 0 & 0 & 0 & 0 & 0 & 0 & 0 & 1 & | & 0 \end{bmatrix}$

20. $\begin{bmatrix} 1 & 2 & 3 & 0 & 0 & | & 6 \\ 0 & 0 & 0 & 0 & 1 & | & 6 \end{bmatrix}$

In Exercises 21 and 22, give the solution for the system of equations represented by each augmented matrix.

21. **a.** $\begin{bmatrix} 1 & 0 & 0 & | & 2 \\ 0 & 1 & 0 & | & 5 \\ 0 & 0 & 1 & | & -3 \end{bmatrix}$ **b.** $\begin{bmatrix} 1 & 0 & 0 & | & 2 \\ 0 & 1 & 0 & | & 5 \\ 0 & 0 & 0 & | & -3 \end{bmatrix}$ **c.** $\begin{bmatrix} 1 & 0 & 4 & | & 2 \\ 0 & 1 & 2 & | & 5 \\ 0 & 0 & 0 & | & 0 \end{bmatrix}$

22. **a.** $\begin{bmatrix} 1 & 0 & -1 & | & 2 \\ 0 & 1 & -3 & | & 5 \\ 0 & 0 & 0 & | & -3 \end{bmatrix}$ **b.** $\begin{bmatrix} 1 & 0 & -1 & | & 2 \\ 0 & 1 & -3 & | & 5 \\ 0 & 0 & 0 & | & 0 \end{bmatrix}$ **c.** $\begin{bmatrix} 1 & 0 & 0 & | & 7 \\ 0 & 1 & 0 & | & 0 \\ 0 & 0 & 1 & | & 8 \end{bmatrix}$

In Exercises 23–36, solve each system of linear equations using augmented matrices. For dependent systems of equations, give the general solution and two particular solutions.

23. $x + 2y = 11$
    $2x + 5y = -11$

24. $x + 2y = -6$
    $3x - y = 17$

25. $2x + 3y = 17$
    $4x + y = -1$

26. $3x - 4y = 83$
    $6x + 3y = 78$

27. $2x + y = 7$
    $ y - z = 2$
    $x + z = 2$

28. $x + y = -2$
    $ -y + z = 2$
    $x - z = -1$

29. $x + 2y + z = 11$
    $-x - y + 2z = 1$
    $2x - y + z = 4$

30. $3x - y + 2z = 4$
    $2x + 2y - z = 10$
    $x - y + 3z = -4$

31 $x_1 + 2x_2 - 2x_3 = -7$
    $-2x_1 + 3x_2 - 17x_3 = -14$
    $4x_1 + 2x_2 + 10x_3 = -4$

32 $x_1 - 2x_2 + 7x_3 = 3$
    $2x_1 + 2x_2 - 3x_3 = -5$
    $x_1 - 11x_2 + 24x_3 = 11$

33 $5a + b - 2c = -3$
    $2a + 4b + c = -3$
    $-3a + 5b - 6c = -21$

34 $6a - 3b + 3c = 3$
    $3a + 3b - c = 5$
    $5a + 2b - 2c = 4$

35 $r + 2s - 5t = 4$
    $3r - s + 2t = 3$
    $r + 9s - 22t = 10$

36 $r - 3s + 2t = 1$
    $4r - 2s + t = 2$
    $2r + 4s - 3t = 0$

B

In Exercises 37 and 38, find a general solution and two particular solutions for each system.

37 $2a - b - 3c = -5$
    $3a + b - 2c = -10$

38 $3a - b + 5c = 4$
    $2a + 2b - 2c = 0$

39 The sum of three numbers is 108. The largest number is 16 less than the sum of the other two numbers. The sum of the largest and the smallest is twice the other number. Find the three numbers.

40 The largest of three numbers is seven times the second number. The second number is seven times the smallest number. The sum of the numbers is 285. Find the three numbers.

41 Use a single matrix to solve both of these systems.

$$x + 3y = 3 \qquad x + 3y = 7$$
$$2x + y = -4 \qquad 2x + y = 9$$

42 Solve the system $2x + 3y - 2z = 0$
    $4x + 5y - 4z = 0$
    $3x - 2y + 5z = 0$

C

43 Solve the system $x_1 + x_2 + 2x_3 - x_4 = 2$
    $2x_1 + 4x_2 - 6x_3 + x_4 = -4$
    $x_1 - 4x_2 + 5x_3 - 3x_4 = 11$

44 Solve the system $A + B = 3$
    $ - B + C = 0$
    $2A + B - C + D = 5$
    $ - B + C - D + E = -3$
    $A - C - E = 3$

In Exercises 45 and 46, solve each system of equations with the aid of a calculator or computer. Your answers should be accurate to the nearest hundredth.

45 $4 \ln x - \ln y = 2.45$
    $3 \ln x + 5 \ln y = 13.03$
    (*Hint*: Let $v = \ln x$ and $w = \ln y$.)

46 $11e^x + 6e^y = 35.90$
    $8e^x - 3e^y = 18.75$
    (*Hint*: Let $v = e^x$ and $w = e^y$.)

SECTION 6-3

Applications of Linear Systems

Section Objective

5 Solve word problems using systems of equations.

Systems of equations are used to solve many problems that ask for the value of more than one quantity. We shall continue to use the strategy for word problems given in Section 2-2.

Strategy for Solving Word Problems ▼

Step 1	a. Determine what you are asked to find.
	b. Identify what you are to find with variables.
Step 2	a. Form the *word equations*.
	b. Translate the word equations into algebraic equations.
Step 3	Solve the system of equations, and answer the question asked.
Step 4	Check the reasonableness of your answer.

EXAMPLE 1 An experiment requires that an animal be fed 108 grams of protein and 400 grams of carbohydrates daily. Food A contains 25% protein and 60% carbohydrates, and food B contains 15% protein and 80% carbohydrates. How many grams of food A and food B should be mixed to satisfy the experimental requirement?

SOLUTION Let a = Number of grams of food A

b = Number of grams of food B

The data of many applications may be put into tabular form.

Word Equation:	Ingredient in Food A	+	Ingredient in Food B	=	Total Amount of This Ingredient
Protein	$0.25a$	+	$0.15b$	=	108
Carbohydrates	$0.60a$	+	$0.80b$	=	400

The *word equation* is based on the mixture principle. The ingredient in each food is determined using the rate principle.

$$\begin{bmatrix} 0.25 & 0.15 & | & 108 \\ 0.60 & 0.80 & | & 400 \end{bmatrix} \xrightarrow{r_1' = 4r_1} \begin{bmatrix} 1 & 0.60 & | & 432 \\ 0.60 & 0.80 & | & 400 \end{bmatrix} \xrightarrow{r_2' = r_2 - 0.6r_1} \begin{bmatrix} 1 & 0.60 & | & 432.0 \\ 0 & 0.44 & | & 140.8 \end{bmatrix}$$

$$\xrightarrow{r_2' = \frac{1}{0.44}r_2} \begin{bmatrix} 1 & 0.60 & | & 432 \\ 0 & 1 & | & 320 \end{bmatrix} \xrightarrow{r_1' = r_1 - 0.6r_2} \begin{bmatrix} 1 & 0 & | & 240 \\ 0 & 1 & | & 320 \end{bmatrix}$$

Thus $a = 240$ and $b = 320$.

Answer Mix 240 grams of food A with 320 grams of food B. Does this answer check? ☐

EXAMPLE 2 It took a ferry 20 minutes to go 2 kilometers downstream. The return trip upstream required 1 hour. Determine the rate of the ferry in still water and the rate of the current. (Assume the ferry's still water rate was the same in both directions.)

SOLUTION Let f = Ferry's speed in still water in kilometers per hour

c = Rate of the current in kilometers per hour

The current aids the ferry when it is moving downstream, so it travels at a rate of $f + c$ km/h. The current opposes the ferry when it is moving upstream, so it travels at $f - c$ km/h.

Word Equation:	Rate · Time = Distance
Downstream	$(f + c) \cdot \frac{1}{3} = 2$
Upstream	$(f - c) \cdot 1 = 2$

$$\begin{cases} \frac{1}{3}(f + c) = 2 \\ f - c = 2 \end{cases} \xrightarrow{r_1' = 3r_1} \begin{array}{l} f + c = 6 \\ f - c = 2 \\ \hline 2f = 8 \\ f = 4 \end{array}$$

Add to eliminate c.

$$f - c = 2$$
$$4 - c = 2$$
$$c = 2$$

Substitute for f in the second equation.

Answer Rate of ferry in still water is 4 kilometers per hour; rate of current is 2 kilometers per hour. ☐

Does this answer check?

EXAMPLE 3 Three earth-moving machines can dig a hole for a new building in 10 days. If the "twenty-one" scraper is moved to another job, it will take the bulldozer and a crane with a drag bucket 20 days to dig the hole. If, instead, the crane is moved to another job, it will take the bulldozer and the scraper 15 days to dig the hole. How many days would it take each machine alone to dig the hole?

SOLUTION Let S, B, and C represent the time in days it would take the scraper, the bulldozer, and the crane, respectively, to dig the hole alone. Since $W = R \cdot T$ (work = rate · time), the rates of work for the scraper, the bulldozer, and the crane are $1/S$, $1/B$, and $1/C$, respectively.

SECTION 6-3 APPLICATIONS OF LINEAR SYSTEMS

Word Equation:	Work the Scraper Does	+	Work the Bulldozer Does	+	Work the Crane Does	= 1 Job
10-day job	$10 \cdot \frac{1}{S}$	+	$10 \cdot \frac{1}{B}$	+	$10 \cdot \frac{1}{C}$	= 1
20-day job			$20 \cdot \frac{1}{B}$	+	$20 \cdot \frac{1}{C}$	= 1
15-day job	$15 \cdot \frac{1}{S}$	+	$15 \cdot \frac{1}{B}$			= 1

$$\begin{cases} 10 \cdot \frac{1}{S} + 10 \cdot \frac{1}{B} + 10 \cdot \frac{1}{C} = 1 \\ 20 \cdot \frac{1}{B} + 20 \cdot \frac{1}{C} = 1 \\ 15 \cdot \frac{1}{S} + 15 \cdot \frac{1}{B} = 1 \end{cases} \longrightarrow \begin{bmatrix} 10 & 10 & 10 & | & 1 \\ 0 & 20 & 20 & | & 1 \\ 15 & 15 & 0 & | & 1 \end{bmatrix}$$

Solve for $1/S$, $1/B$, and $1/C$.

$$\xrightarrow[r'_2 = \frac{1}{20}r_2]{r'_1 = \frac{1}{10}r_1} \begin{bmatrix} 1 & 1 & 1 & | & \frac{1}{10} \\ 0 & 1 & 1 & | & \frac{1}{20} \\ 15 & 15 & 0 & | & 1 \end{bmatrix}$$

$$\xrightarrow{r'_3 = r_3 - 15r_1} \begin{bmatrix} 1 & 1 & 1 & | & \frac{1}{10} \\ 0 & 1 & 1 & | & \frac{1}{20} \\ 0 & 0 & -15 & | & -\frac{1}{2} \end{bmatrix}$$

$$\xrightarrow[r'_3 = -\frac{1}{15}r_3]{r'_1 = r_1 - r_2} \begin{bmatrix} 1 & 0 & 0 & | & \frac{1}{20} \\ 0 & 1 & 1 & | & \frac{1}{20} \\ 0 & 0 & 1 & | & \frac{1}{30} \end{bmatrix}$$

$$\xrightarrow{r'_2 = r_2 - r_3} \begin{bmatrix} 1 & 0 & 0 & | & \frac{1}{20} \\ 0 & 1 & 0 & | & \frac{1}{60} \\ 0 & 0 & 1 & | & \frac{1}{30} \end{bmatrix}$$

If $\frac{1}{S} = \frac{1}{20}$, $\frac{1}{B} = \frac{1}{60}$, and $\frac{1}{C} = \frac{1}{30}$, then $S = 20$, $B = 60$, and $C = 30$.

Answer It would take the scraper 20 days, the bulldozer 60 days, and the crane 30 days to dig the hole alone. □

The general form of the equation of a parabola that opens up or down is $y = ax^2 + bx + c$. We can determine the equation of a parabola of this form given any three of its points.

EXAMPLE 4 Find the equation of the parabola that opens vertically and passes through the points (1, 1), (2, 2), and (3, 1).

SOLUTION Substitute the x and y values of each point into the equation $y = ax^2 + bx + c$, and then solve for a, b, and c.

Points	Equations	Augmented Matrix
(1, 1)	$1 = a + b + c$	
(2, 2)	$2 = 4a + 2b + c$	$\begin{bmatrix} 1 & 1 & 1 & \vert & 1 \\ 4 & 2 & 1 & \vert & 2 \\ 9 & 3 & 1 & \vert & 1 \end{bmatrix}$
(3, 1)	$1 = 9a + 3b + c$	

which reduces to

$$\begin{bmatrix} 1 & 0 & 0 & \vert & -1 \\ 0 & 1 & 0 & \vert & 4 \\ 0 & 0 & 1 & \vert & -2 \end{bmatrix}$$

You may wish to verify this result by completing the intermediate steps omitted here.

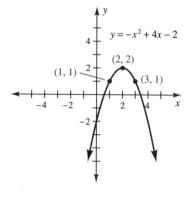

Thus $a = -1$, $b = 4$, and $c = -2$.

Answer $y = -x^2 + 4x - 2$

Break-Even Analysis

Two important functions related to the profitability of a product are the cost function and the revenue function. The cost function, C, represents the total manufacturing cost for a certain number of units, and the revenue function, R, indicates the total income from selling these units.

The graphs of these two functions on the same coordinate system, as in Figure 6-1, indicate how a business is doing.

1. The business is losing money when the graph of C is above the graph of R. The value of x is in the loss interval, since the cost is greater than the revenue.

2. The business is making a profit when the graph of C is below the graph of R. The value of x is in the profit interval, since the cost is less than the revenue.

3. The business is breaking even at the point of intersection of R and C. At this **break-even point**, cost equals revenue. The **break-even value** of x is the value separating the loss interval from the profit interval.

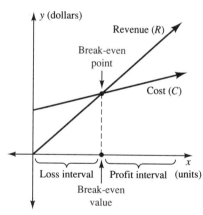

Figure 6-1 Cost and revenue functions

EXAMPLE 5 A bicycle manufacturer has fixed costs of $8000 per week and a cost per unit of $100. The wholesale price is $150 per bicycle.

(a) Find the cost function.
(b) Find the revenue function.
(c) Find the break-even value.
(d) Find the total cost and revenue at the break-even value.

Assume that the manufacturer sells all units produced.

SECTION 6-3 APPLICATIONS OF LINEAR SYSTEMS

SOLUTION Let b = Number of bicycles produced and sold.

(a) Total cost = Fixed cost + Variable cost
= (Fixed cost) + (Cost per unit)(Number of units produced)
$C = 8000 + (100)(b)$ The cost function

(b) Revenue = (Price per unit)(Number of units sold)
$R = 150b$ The revenue function

(c)   Cost = Revenue   At the break-even value, cost equals revenue.
$8000 + 100b = 150b$
$8000 = 50b$
$b = 160$ The break-even value

(d) $C = 8000 + 100b$
$= 8000 + 100(160)$ Substitute $b = 160$ into the cost function.
$C = \$24{,}000$ Total cost at break-even value
$R = 150b$
$= 150(160)$ Substitute $b = 160$ into the revenue function.
$R = \$24{,}000$ Total revenue at break-even value

▶ **Self-Check** ▼

Two different-sized pipes can fill a tank in 3 hours if they are both open. If both pipes are used for 1 hour and the larger pipe is shut off, it would then take the smaller pipe an additional 5 hours to finish the job. How many hours would it take each pipe to fill the tank alone? Give a system of equations describing this problem, using L and S to represent the filling times for the larger and smaller pipes, respectively.

Law of Supply and Demand

The market price, p, of an item helps determine the supply, S, and demand, D, of that item. A low market price usually causes the consumer demand to be higher than the supply, generating a shortage of the item. An increase in the price generally produces a decrease in demand and an increase in supply. The point at which supply equals demand is called the **equilibrium point**. The value of p at the equilibrium point is called the **equilibrium price**. This concept is illustrated in Figure 6-2.

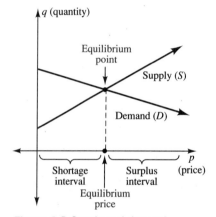

Figure 6-2 Supply and demand functions

▶ **Self-Check Answer** ▼

$$\begin{cases} \dfrac{3}{L} + \dfrac{3}{S} = 1 \\ \dfrac{1}{L} + \dfrac{6}{S} = 1 \end{cases}, \quad L = 5 \text{ hours}, \ S = 7.5 \text{ hours}$$

EXAMPLE 6 The supply and demand functions for a rubber ball are $S(p) = \frac{5}{4}p$ and $D(p) = -\frac{3}{4}p + 520$, respectively, where p is the price of the ball in cents.

(a) Determine the demand when the price of the ball is $1.60.
(b) Determine the equilibrium price.
(c) Determine the supply and demand at the equilibrium point.

SOLUTIONS

(a) $D(p) = -\frac{3}{4}p + 520$

$D(160) = -\frac{3}{4}(160) + 520 = 400$ Substitute 160 (cents) for p into the demand function.

Thus the demand is 400 when the price is $1.60.

(b) $S(p) = D(p)$ At the equilibrium price, supply equals demand.

$\frac{5}{4}p = -\frac{3}{4}p + 520$ Solve for p.

$2p = 520$

$p = 260$ 260 cents = $2.60

The equilibrium price is $2.60.

(c) $S(260) = \frac{5}{4}(260) = 325$ Substitute 260 (cents), the equilibrium price, into the supply and demand functions.

$D(260) = -\frac{3}{4}(260) + 520 = 325$

The supply and demand are both 325 at the equilibrium price of $2.60. □

EXERCISES 6-3

A

1. Tickets for a college play cost $1.50 for students and $3.00 for nonstudents. The income for one night from selling 1150 tickets was $2250. How many of each type of ticket were sold?

2. One of two complementary angles is 6° larger than twice the other angle. Determine the measure of each angle.

3. A furniture company makes a chair that wholesales at $200. If the fixed costs are $14,000 per week and the cost of making each chair is $120, how many chairs must be made each week to break even?

4. A manufacturer makes a baseball cap that wholesales at $3.20. If the fixed costs are $1520 and it takes $1.30 to make each cap, how many caps must be made to break even?

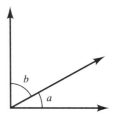

Figure for Exercise 2

SECTION 6-3 APPLICATIONS OF LINEAR SYSTEMS

5 The supply and demand functions for a new toy are $S(p) = 18p$ and $D(p) = 4400 - 2p$ where p is the price in dollars.
  a. Determine the demand when the price is $5.
  b. Determine the equilibrium price.
  c. Determine the supply and demand at the equilibrium price.

6 The supply and demand functions for a new pesticide are $S(p) = 12p$ and $D(p) = 5000 - 8p$ where p is the price in dollars.
  a. Determine the supply when the price is $15.
  b. Determine the equilibrium price.
  c. Determine the supply and demand at the equilibrium price.

7 A truck has an empty weight of 30,000 pounds. On the first trip the truck delivered two pallets of bricks and three pallets of concrete blocks. The gross weight ticket on this delivery was 48,000 pounds. On the second trip the truck delivered four pallets of bricks and one pallet of concrete blocks. The gross weight ticket on this delivery was 49,600 pounds. Use this information to determine the weight of a pallet of bricks and the weight of a pallet of concrete blocks.

8 A metallurgist wants to make an alloy containing 30 kilograms of metal A and 40 kilograms of metal B. He has purchased bars of two other alloys. A bar of the first alloy contains 1 kilogram of metal A and 1 kilogram of metal B, and a bar of the second alloy contains 1 kilogram of metal A and 2 kilogram of metal B. How many of each of these bars are required to make the desired alloy?

9 The sum of the measures of the interior angles of a triangle is 180°. In one triangle, the sum of angle A and twice angle B is 175°. Twice angle B plus angle C is 215°. Find the number of degrees in each angle.

10 A department store chain with three stores retails calculators of types A, B, and C. The table shows the number sold of each type of calculator and the total income from these sales at each store. Find the price of each type of calculator.

Figure for Exercise 9

	A	B	C	Total Sales
Store 1	1	2	5	$156
Store 2	2	4	8	$294
Store 3	3	3	7	$321

11 A tire store sells three types of tires: the belted model, the standard radial, and the deluxe radial. The table shows the number of each type of tire sold on Thursday, Friday, and Saturday, along with the total income these sales produced. How much does the store charge for each model of tire?

	Belted Model	Standard Radial	Deluxe Radial	Total Income
Thursday	10	6	8	$1342
Friday	12	8	4	$1158
Saturday	20	10	8	$1879

12. A boat took 30 minutes to deliver cargo 10 kilometers upstream and took 20 minutes to return. Find the rate of the boat and the current if both rates are constant throughout the trip.

13. An airplane takes two hours to fly 1200 kilometers with a tailwind. Another plane takes off at the same time and takes three hours to fly the identical route in the opposite direction. Find the speed of the wind and the speed of the planes in still air if they are flying at the same constant airspeed.

14. Find the equation of the parabola that opens vertically and passes through the points (1, 4), (−1, 6), and (2, 12).

15. Find the equation of the parabola that opens vertically and passes through the points (−1, 6), (0, 1), (1, −2).

16. An investor had her broker place $50,000 in certificates of deposit, stocks, and commodities. The broker invested twice as much in certificates of deposit as in commodities. After one year, the certificates of deposit earned 12%, the stock gained 10%, and the commodities lost 5%. If these investments yielded a total of $3900, how much was invested in each type?

17. A garden needs 9 pounds of nitrogen, 18 pounds of phosphoric acid, and 18 pounds of potash. The gardener has two fertilizers available. Fertilizer A is 3-12-12 (that is, it is 3% nitrogen, 12% phosphoric acid, and 12% potash). Fertilizer B is 10-10-10. How many pounds of each fertilizer must be spread to satisfy the requirements?

18. A supermarket is preparing an order of 400 pounds of hamburger that is 85% lean meat and 15% fat and filler. How many pounds of 78% lean and 86% lean should be mixed to fill this order?

B

19. Two pipes fill a tank, and one pipe empties the tank. If both inlets are open, they fill the tank in five hours. If the outlet pipe is also open, it takes 10 hours to fill the tank. The first inlet is open for three hours before the outlet is opened. After an additional two hours the second inlet is opened. Three hours later the tank is full. How many hours would it take the outlet pipe to empty the tank if the inlets were closed?

20. A machine shop has two models of the same machine. If each machine is used for eight hours, a total of 136 items can be produced. If machine A is used for five hours and machine B is used for six hours, a total of 97 items can be produced. How many items can each machine produce in one hour?

21. For the function $f(x) = x^3 + Ax^2 + Bx + C$, where A, B, and C are constants, find A, B, and C so that $f(1) = 1$, $f(2) = 4$, and $f(-2) = 4$.

22. The equation of a circle can be written in the form $x^2 + y^2 + Ax + By + C = 0$. Find the equation of the circle passing through the points (0, 2), (1, 4), and (2, 8).

23. A service station operator has computed the consumer demand in gallons for unleaded gas to be $d = 4175 - 12.5p$, where p is the retail price of gas in cents per gallon and $p \leq 250$. The company will supply the station with $q = 30p - 500$ gal of unleaded gas where $p \geq 50$.
    a. Find the demand when the price is 130 cents.
    b. Find the equilibrium price.
    c. Find the supply and demand at the equilibrium price.

24 Twenty miles from the state line, a state trooper clocks a speeding car at 80 miles per hour. One minute later, when his partner returns from assisting another motorist, they pursue the speeder at the rate of 90 miles per hour. Will the speeder be caught before he crosses the state line?

25 A student received a grade of 70 on a multiple-choice test having 25 questions. Each question answered correctly was worth four points, each question answered incorrectly cost one penalty point, and each question not attempted caused no change in the point total. The student omitted five questions. How many questions did the student answer correctly?

26 A company has a machine that makes a plastic part for an automobile at a cost of $0.50 each. They are considering buying a new machine at a cost of $5000 that can make these parts for $0.45 each. How many parts must the company make before the new machine will produce a savings? (Ignore other factors such as tax credits.)

SECTION 6-4

Nonlinear Systems of Equations

Section Objectives

6 Use the substitution method to solve a nonlinear system of equations.

7 Use the elimination method to solve a nonlinear system of equations.

Nonlinear systems of equations are used in business, engineering, and many other areas. For example, supply and demand functions are often nonlinear; thus the equilibrium price can sometimes be found by using a system of nonlinear equations.

A rough sketch of the graph of each relation in a nonlinear system of equations can sometimes reveal the number of real solutions, as shown in Figure 6-3. For example, a straight line and a circle have from 0 to 2 points of intersection in the real plane. Each point of intersection represents a real solution. Solutions with imaginary numbers may exist, but will not be indicated in the real plane; a complex solution may exist when there is no point of intersection.

Although graphs provide an excellent source of intuition and insight and can be used to approximate solutions, algebraic methods are often preferable since they produce exact solutions. The substitution method may be used when it is convenient to solve an equation for one variable in terms of the other variable. Since linear and parabolic functions contain a first-degree term, these equations are well suited for the substitution method.

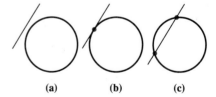

Figure 6-3 (a) No points of intersection; no real solution. **(b)** One point of intersection; one real solution. **(c)** Two points of intersection; two real solutions.

EXAMPLE 1 Solve the system $\begin{cases} x + y = -1 \\ x^2 + y^2 - 8x + 2y = 23 \end{cases}$.

SOLUTION

$$y = -1 - x$$ Solve the first equation for y.

$$x^2 + (-1-x)^2 - 8x + 2(-1-x) = 23$$ Substitute for y in the second quation.

$$x^2 + 1 + 2x + x^2 - 8x - 2 - 2x = 23$$ Simplify and solve for x.

$$2x^2 - 8x - 24 = 0$$

$$x^2 - 4x - 12 = 0$$

$$(x - 6)(x + 2) = 0$$

$x = 6$ or $x = -2$ Substitute for x in the first equation, $y = -1 - x$.

$y = -1 - 6$ $y = -1 - (-2)$

$y = -7$ $y = 1$

Answer $(6, -7)$ and $(-2, 1)$

> ▶ **Self-Check** ▼
>
> A circle and an ellipse can intersect at 0, 1, 2, 3, or 4 points. Illustrate each possibility with a sketch.

EXAMPLE 2 Solve the system $\begin{cases} y = x^2 + 5 \\ x^2 + y^2 = 1 \end{cases}$.

SOLUTION

$$x^2 + (x^2 + 5)^2 = 1$$ Substitute for y in the second equation and simplify.

$$x^2 + x^4 + 10x^2 + 25 = 1$$

$$x^4 + 11x^2 + 24 = 0$$ This results in an equation of quadratic form that could be solved by letting $z = x^2$. However, we factor.

$$(x^2 + 8)(x^2 + 3) = 0$$

$x^2 = -8$ or $x^2 = -3$ Solve for x^2.

$x = \pm i\sqrt{8} = \pm 2i\sqrt{2}$ $x = \pm i\sqrt{3}$ Solve for x.

$y = (\pm i\sqrt{8})^2 + 5$ $y = (\pm i\sqrt{3})^2 + 5$ Substitute for x in the first equation, $y = x^2 + 5$.

$ = -8 + 5$ $ = -3 + 5$

$y = -3$ $y = 2$

Answer $(2i\sqrt{2}, -3)$, $(-2i\sqrt{2}, -3)$, $(i\sqrt{3}, 2)$, and $(-i\sqrt{3}, 2)$

The elimination method is frequently used when it is difficult to solve an equation for one of the variables. In particular, we often use this method if there are second-degree terms in both variables.

▶ **Self-Check Answers** ▼

SECTION 6-4 NONLINEAR SYSTEMS OF EQUATIONS

EXAMPLE 3 Use the elimination method to solve $\begin{cases} x^2 + (y-4)^2 = 9 \\ x^2 + y^2 = 25 \end{cases}$.

SOLUTION

$\begin{aligned} x^2 + y^2 - 8y &= -7 \\ x^2 + y^2 &= 25 \end{aligned}$ Expand the first equation and simplify so that the constant term is in the right member.

$-8y = -32$ Subtract the second equation from the first equation.

$y = 4$ Solve for y.

$x^2 + (4)^2 = 25$ Substitute for y in the second equation.

$x^2 = 9$

$x = \pm 3$ Solve for x.

Answer $(-3, 4)$ and $(3, 4)$

▶ **Self-Check** ▼

Solve the system $\begin{cases} x - y = -2 \\ x^2 + y^2 = 2 \end{cases}$.

EXAMPLE 4 Approximate the solution of this system of equations to the nearest hundredth:

$$\begin{cases} \log(x-1) + y = 5 \\ \log(x+2) - y = -4 \end{cases}$$

SOLUTION

$\begin{aligned} \log(x-1) + y &= 5 \\ \log(x+2) - y &= -4 \end{aligned}$ Use the elimination method to eliminate the y variable.

$\log(x-1) + \log(x+2) = 1$ Add the two equations.

$\log(x^2 + x - 2) = 1$ Use the product rule for logarithms to rewrite the left member as a single logarithmic term.

$x^2 + x - 2 = 10^1$ Express in exponential form, base 10.

$x^2 + x - 12 = 0$ Simplify and then solve for x.

$(x+4)(x-3) = 0$

$x = -4$ or $x = 3$

$\log(-4-1) + y = 5$ $\log(3-1) + y = 5$ Substitute for x in the first equation.

-4 is an extraneous value $y = 5 - \log 2$

since $\log(-5)$ is undefined. $y \approx 5 - 0.3010$ Calculator approximation

$y \approx 4.70$ Round to nearest hundredth.

Answer The solution is approximately $(3.00, 4.70)$.

▶ **Self-Check Answer** ▼

$(-1, 1)$

EXAMPLE 5 Solve the system $\begin{cases} y = 2(5^x) + 3 \\ y = 5^{2x} - 4(5^x) + 8 \end{cases}$.

SOLUTION

$2(5^x) + 3 = 5^{2x} - 4(5^x) + 8$ Substitute for y in the second equation and combine like terms.

$0 = 5^{2x} - 6(5^x) + 5$ This equation is of quadratic form, since $5^{2x} = (5^x)^2$.

$0 = z^2 - 6z + 5$ Let $z = 5^x$ and solve for z.

$0 = (z - 5)(z - 1)$

$z = 5$ or $z = 1$

$5^x = 5$ $5^x = 1$ Replace z with 5^x.

$x = 1$ $x = 0$ Solve these exponential equations by inspection.

$y = 2(5^1) + 3$ $y = 2(5^0) + 3$ Substitute for x in the first equation.

$y = 13$ $y = 5$

Answer (1, 13) and (0, 5)

▶ **Self-Check** ▼

Solve the system of equations
$\begin{cases} 3x^2 + 2y^2 = 12 \\ x^2 + y^2 = 5 \end{cases}$.

The next example illustrates a method of solving systems of equations of the form

$$\begin{cases} A_1 x^2 + B_1 xy + C_1 y^2 = D_1 \\ A_2 x^2 + B_2 xy + C_2 y^2 = D_2 \end{cases}$$

The strategy is to first use the elimination method to form a new equation whose constant term is zero. We can then solve this quadratic equation for one of the variables.

EXAMPLE 6 Solve the system $\begin{cases} 2x^2 + 2xy + y^2 = 10 \\ 22x^2 + xy + 2y^2 = 50 \end{cases}$.

SOLUTION

$\begin{cases} 2x^2 + 2xy + y^2 = 10 \\ 22x^2 + xy + 2y^2 = 50 \end{cases} \xrightarrow{r_1' = -5r_1}$ $\begin{aligned} -10x^2 - 10xy - 5y^2 &= -50 \\ 22x^2 + xy + 2y^2 &= 50 \end{aligned}$ Multiply the first equation by -5.

$12x^2 - 9xy - 3y^2 = 0$ Add to produce a zero constant.

$3(4x^2 - 3xy - y^2) = 0$

$(4x + y)(x - y) = 0$ Factor. Solve this quadratic equation for y.

$y = -4x$ or $y = x$

$2x^2 + 2x(-4x) + (-4x)^2 = 10$ $2x^2 + 2x(x) + x^2 = 10$ Substitute for y in the first equation.

$2x^2 - 8x^2 + 16x^2 = 10$ $2x^2 + 2x^2 + x^2 = 10$ Solve each equation for x.

$10x^2 = 10$ $5x^2 = 10$

$x^2 = 1$ $x^2 = 2$

▶ **Self-Check Answer** ▼

$(\sqrt{2}, \sqrt{3}), (\sqrt{2}, -\sqrt{3}), (-\sqrt{2}, \sqrt{3}), (-\sqrt{2}, -\sqrt{3})$

SECTION 6-4 NONLINEAR SYSTEMS OF EQUATIONS

$x = \pm 1$ $x = \pm\sqrt{2}$

$x = 1$ or $x = -1$ $x = \sqrt{2}$ or $x = -\sqrt{2}$

$y = -4(1)$ $y = -4(-1)$ $y = \sqrt{2}$ $y = -\sqrt{2}$ Substitute $x = \pm 1$ into $y = -4x$ and
$y = -4$ $y = 4$ $x = \pm\sqrt{2}$ into $y = x$ and calculate y.

Answer $(1, -4), (-1, 4), (\sqrt{2}, \sqrt{2})$, and $(-\sqrt{2}, -\sqrt{2})$ □

Since some algebraic methods produce exact solutions, they are often preferred to graphical solutions. Not all systems can be solved exactly by algebraic methods, however. Thus it is very helpful to be able to approximate the real solutions to a system of two equations in two variables by estimating their points of intersection. The necessary graphs can be obtained by hand plotting of points or by using a graphics calculator or computer. In all cases it is wise to check your estimate by substituting it into both equations in the system.

EXAMPLE 7   Solve the system $\begin{cases} y = x^2 - 2x - 1 \\ 2x + y = 3 \end{cases}$ graphically.

SOLUTION   Graph the straight line and the parabola on the same coordinate system. Then determine the points of intersection.
The points of intersection are approximately $(-2, 7)$ and $(2, -1)$.

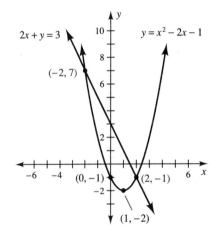

Check   Substitute these points into both equations.

$$\begin{array}{ll} & y = x^2 - 2x - 1 \\ (-2, 7) & 7 \stackrel{?}{=} (-2)^2 - 2(-2) - 1 \\ & 7 \stackrel{?}{=} 4 + 4 - 1 \\ & 7 = 7 \\ (2, -1) & -1 \stackrel{?}{=} 2^2 - 2(2) - 1 \\ & -1 \stackrel{?}{=} 4 - 4 - 1 \\ & -1 = -1 \end{array} \qquad \begin{array}{l} 2x + y = 3 \\ 2(-2) + 7 \stackrel{?}{=} 3 \\ -4 + 7 \stackrel{?}{=} 3 \\ 3 = 3 \\ 2(2) + (-1) \stackrel{?}{=} 3 \\ 4 - 1 \stackrel{?}{=} 3 \\ 3 = 3 \end{array}$$

Answer $(-2, 7)$ and $(2, -1)$ □

The next example is solved graphically utilizing a TI-81 graphics calculator. The strategy is to graph each equation and then to use the trace option to approximate the points of intersection. Thus the first window of values selected should be large enough to give an overall view that includes all points of intersection. Then the accuracy of each approximation can be refined by zooming in on a specific portion of the graph. A knowledge of the general shape of the graph of each function is important since this will help you select windows appropriate for each system of equations.

Approximating Points of Intersection with a Graphics Calculator ▼

Step 1	Enter the formulas for both functions.
Step 2	Select an appropriate viewing window for the first view and graph these functions. (The standard viewing rectangle is appropriate for many functions.)
Step 3	Zoom in on each point of intersection.
Step 4	Use the trace option to approximate the coordinates of each point of intersection.

EXAMPLE 8 Approximate to the nearest hundredth each coordinate of the point of intersection of $y = e^{x-2} + 1$ and $y = x^3 + 2x^2 - x - 2$.

SOLUTION

[1]

Enter the formulas for both functions into y_1 and y_2.

[2] ZOOM | Standard (option 6)

Select option 6, which is the standard viewing rectangle with $-10 \leq x \leq 10$ and $-10 \leq y \leq 10$.

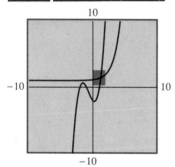

Zoom in on the shaded region.

[3] ZOOM | Box (option 1)

Select approximately $(0, 0)$ as one corner of the box by pressing ENTER. Then move the cursor to approximately $(2, 2.38)$ and press ENTER to fix the diagonal corner of the box.

Select option 1, the box option. Then fix the corners of the new viewing rectangle and zoom in on this rectangle.

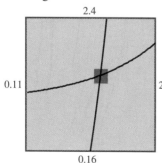

Zoom in on the shaded region.

SECTION 6-4 NONLINEAR SYSTEMS OF EQUATIONS

ZOOM **Box (option 1)**
Use the arrow keys and select approximately (1.10, 1.28) as one corner of the box by pressing **ENTER**. Then move the cursor to approximately (1.30, 1.53) and press **ENTER** to fix the diagonal corner of the box.

Select option 1, the box option. Then fix the corners of the new viewing rectangle and zoom in again.

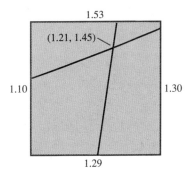

4 **TRACE**
Use the arrow keys to move the cursor over the point of intersection and determine the approximate solution for this system of equations.

The accuracy of the approximation can be observed by how much the coordinates change as the cursor moves from one side of the point of intersection to the other side.

Answer (1.21, 1.45)

A graphics calculator is restricted to graphing relations that are expressed as explicit functions in the form $y = f(x)$. Some equations that define relations like circles must first be decomposed into explicit functions before they can be graphed. This process is illustrated in the next example, where we consider only the function describing the upper semicircle instead of considering the circle defined by the given relation.

EXAMPLE 9 Approximate to the nearest hundredth each coordinate of the point of intersection of $x^2 + y^2 = 25$ and $y = x^2 - 3x + 3$ that lies in quadrant I.

SOLUTION Since $y > 0$ in quadrant I, we can solve $x^2 + y^2 = 25$ for y, obtaining $y = \sqrt{25 - x^2}$.

$x^2 + y^2 = 25$ is the equation of a circle.
$y = \sqrt{25 - x^2}$ is the equation of the upper semicircle.
Enter the formulas for both functions into y_1 and y_2.

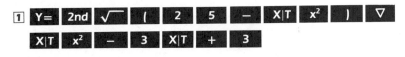

2 `ZOOM` `Standard (option 6)`

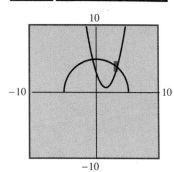

Select option 6, which is the standard viewing rectangle with $-10 \leq x \leq 10$ and $-10 \leq y \leq 10$.

Zoom in on the shaded region.

3 `ZOOM` `Box (option 1)`

Move the cursor to approximately (3.05, 3.02) and press `ENTER`. Then move the cursor to approximately (3.68, 4.60) and press `ENTER` to fix the diagonal corner of the box.

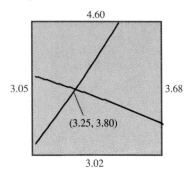

Select option 1, the box option. Then zoom in on the point of intersection in quadrant I.

4 `TRACE`

Use the arrow keys to move the cursor over the point of intersection and determine the approximate solution for this system of equations.

Answer (3.25, 3.80)

The accuracy of the approximation can be observed by how much the coordinates change as the cursor moves from one side of the point of intersection to the other side.

In summary, we have noted in this section that each real solution of a system of equations in two variables is represented geometrically by a point of intersection of the graphs of the equations. The algebraic and graphical methods of solving systems of equations each have advantages and disadvantages. Algebraic methods can sometimes be used to produce exact solutions; however, systems involving higher-degree polynomials or nonpolynomial relations are often beyond the scope of the substitution or the

▶ **Self-Check** ▼

Zoom in on the point of intersection in quadrant II of the graphs in Example 9 and approximate to the nearest hundredth each coordinate of this point of intersection.

▶ **Self-Check Answer** ▼

$(-0.55, 4.97)$

SECTION 6-4 NONLINEAR SYSTEMS OF EQUATIONS

elimination method. The graphical method can be used to produce approximate solutions to systems with relatively complicated functions. Relations that are not functions, however, cannot be graphed with a graphics calculator unless these relations are first split into separate explicit functions of the form $y = f(x)$. (See Example 9.)

EXERCISES 6-4

A

In Exercises 1–4, sketch an example of each possible number of points of intersection of the two graphs.

1. A line and a parabola
2. A line and an ellipse
3. A parabola and a circle
4. A parabola and an ellipse

In Exercises 5–24, solve each system of equations.

5. $x^2 + y^2 = 26$
   $x - y = 4$

6. $\dfrac{x^2}{4} + \dfrac{y^2}{1} = 1$
   $x + 2y = 2$

7. $x^2 + y^2 - 8x - 6y = -15$
   $x + 3y = 13$

8. $x^2 + y^2 + x + 2y = 5$
   $2x + y = -3$

9. $x^2 + y^2 = 24$
   $x^2 - y^2 = 8$

10. $x^2 + y^2 = 4$
    $x^2 + 2y^2 = 11$

11. $x^2 - y^2 = 6$
    $y = 5x$

12. $9x^2 - 4y^2 = 36$
    $3x - 4y = 6$

13. $x^2 + y^2 = 25$
    $x^2 + y^2 - 8y = 9$

14. $x^2 - 6x - y^2 = 16$
    $x^2 + y^2 = 4$

15. $x^2 + y^2 = 5$
    $4x + y^2 = 8$

16. $x^2 + y^2 = 9$
    $y = x^2 + 3$

17. $2x^2 + 3y^2 = 14$
    $x^2 - y^2 = 4$

18. $3x^2 + 4y^2 = 20$
    $x^2 + 9y^2 = 20$

19. $4x^2 - 3y^2 = 24$
    $3x^2 - 2y^2 = 19$

20. $4x^2 - 9y^2 = 36$
    $9x^2 - 4y^2 = 36$

21. $y = x^2 - 1$
    $x^2 + y^2 + 3y = -2$

22. $y = x^2 + 2$
    $x^2 + y^2 - 3y = 6$

23. $x^2 + y^2 = 18$ (*Hint:* Use substitution.)
    $xy = 9$

24. $x^2 + y^2 = 32$
    $xy = -16$

25. An inventor is marketing an energy-saving device. The supply and demand functions are $S(p) = 2p - 20$ and $D(p) = 1200/p$. Find the equilibrium price p (in dollars) and the corresponding number of units supplied.

26. A publishing company has found that the supply and demand functions of a certain book are $S(p) = p^2 + 79p + 200$ and $D(p) = 2200 - p$ for a suggested retail price p (in dollars). Determine the equilibrium price and the corresponding number of books that should be printed.

B

In Exercises 27–34, solve each system of equations for all real solutions.

27. $y = \log(x - 4) + 5$
    $y = -\log(x + 5) + 6$

28. $y = \log_4(2x + 8) + 6$
    $y = \log_4 x + 7$

29. $y = 2(3^x) + 4$
    $y = 3^{2x} + 1$

30. $y = 3^{2x} + 4 \cdot 3^x$
    $y = -3^{2x} + 3^x + 9$

31 $2x^2 - 5xy + 2y^2 = 20$
   $x^2 - 4xy + 8y^2 = 20$

32 $6x^2 + 6xy + 9y^2 = 27$
   $5x^2 + 7xy + 9y^2 = 27$

33 $y = 2^{2x+2} - 2^x + 3$
   $y = 2^{x+2} + 2$

34 $y = 5^{2x+2} - 5^x$
   $y = 5^{x+1} - 1$

35 A customer purchases carpets for two square rooms with a combined area of 74 yd² (square yards). The price of the first carpet was $12 per yd², and that of the second was $15 per yd². If the total cost was $1035, find the dimensions of each square room.

36 A rectangle has a perimeter of 20 meters and an area of 24 m². Find its dimensions.

37 A health spa has a circular room containing a circular swimming pool that touches the edge of the room. A total of 125π m² of the room is outside of the pool. Find the radii of the two circles if the distance between the two centers is five meters.

38 A farmer has a field that is a trapezoid with two right angles. The area of the field is 50,964 square meters, and the perimeter is 1044 meters. Find the values of x and y in the figure below.

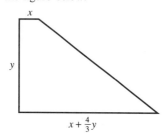

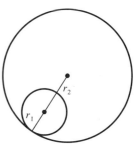

Figure for Exercise 37

C

In Exercises 39–41, eliminate the x^2 and y^2 terms. Then solve the resulting linear equation for x or y and back-substitute into one of the original equations.

39 $x^2 + y^2 + 5x + y = 26$
   $x^2 + y^2 + 2x - y = 15$

40 $x^2 + y^2 + 14x - 2y = -25$
   $x^2 + y^2 + 2x - 2y = 23$

41 $3x^2 - 10xy - 5y^2 = -3$
   $x^2 - 6xy - 3y^2 = -2$

42 A rectangular poster has a border five centimeters wide. Find the dimensions of the poster if the area of the entire poster is 2400 cm² and the area of the border is 1000 cm².

43 A purchasing agent was going to spend $144.00 on pencils, but he negotiated a price reduction of $0.04 for each pencil after the first 100 were purchased. If he spent $88.00 on pencils with this reduction included, how many pencils did he purchase and what was the price for each of the first 100 pencils?

In Exercises 44–46, use a graphics calculator to approximate to the nearest hundredth the real solutions of each system of equations.

44 $y = x(x + 2)(x + 1)(x - 3)$
   $y = x^2 + x - 8$

45 $y = e^{x-3} - 1$
   $y = \sqrt{25 + 2x - x^2}$

46 $y = \ln(x - 1)$
   $y = (x - 2)^2 - 1$

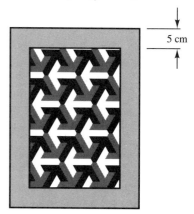

Figure for Exercise 42

SECTION 6-5

Systems of Inequalities

Section Objectives

8 Graph an inequality in two variables.

9 Graph a system of inequalities.

The line $Ax + By + C = 0$ in Figure 6-4 separates the plane into two regions called **half-planes**. The points in one half-plane will satisfy $Ax + By + C < 0$, whereas the points in the other half-plane will satisfy $Ax + By + C > 0$.

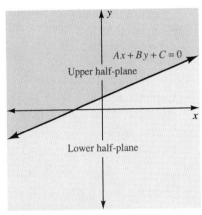

Figure 6-4 Half-planes

Graphing a Linear Inequality ▼

Step 1 Graph the equality $Ax + By + C = 0$.
  a. Use a solid line if the equality is included in the solution.
  b. Use a dashed line if the equality is not included in the solution.

Step 2 Choose an arbitrary test point not on the line [(0, 0) is often convenient]. Substitute this test point into the inequality.

Step 3 **a.** If this test point satisfies the inequality, shade the half-plane containing this test point.
  b. If the test point does not satisfy the inequality, shade the other half-plane.

EXAMPLE 1 Graph $2x + y \geq 4$.

SOLUTION

[1] Draw a solid line for $2x + y = 4$, since the original statement included the equality.

[2] Test the origin:
$$2x + y \geq 4$$
$$2(0) + 0 \geq 4 \qquad \text{Use (0, 0) as the test point, since the origin is not on this line.}$$
$$0 \geq 4 \text{ is false.}$$

[3] Shade the half-plane that does not include the origin. □

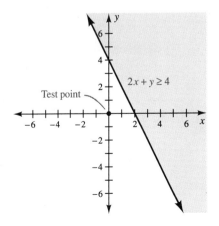

EXAMPLE 2 Graph $2x > y$.

SOLUTION

① Draw a dashed line for $2x = y$, since the equality is not part of the solution.

② Test the point $(1, 0)$: $2x > y$ Do not test $(0, 0)$, because the origin is on this line.
$$2(1) \overset{?}{>} 0$$
$$2 > 0 \text{ is true.}$$

③ Shade the half-plane that includes the point $(1, 0)$.   □

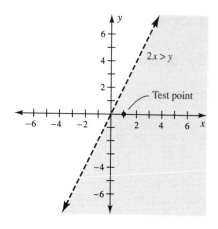

The procedure for graphing a linear inequality can be used for any inequality for which the graph of the corresponding equality separates the plane into two distinct regions.

EXAMPLE 3 Graph $y > x^2 - 4$.

SOLUTION

① Graph the equality $y = x^2 - 4$ with a dashed line. This graph is a parabola opening upward with a vertex shifted four units below the origin.

② Test the origin: $y > x^2 - 4$ Use $(0, 0)$ as the test point, since the origin is not on this parabola.
$$0 \overset{?}{>} 0^2 - 4$$
$$0 > -4 \text{ is true.}$$

③ Shade the region containing the origin.   □

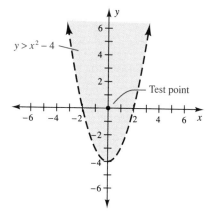

If the graph of the equality portion of an inequality separates the plane into more than two regions, test a point from each region. Then shade the regions containing the test points that satisfy the inequalities.

Arrows may be used instead of shading to indicate the solution of an inequality, as shown in Figure 6-5. This notation can be used to minimize the visual clutter produced by a system of inequalities. We will use arrows to denote the solution of the individual inequalities and then use shading for the solution of the system, which is the intersection of these individual solutions.

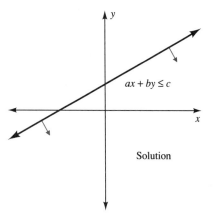

Figure 6-5 Arrow notation for inequalities

SECTION 6-5 SYSTEMS OF INEQUALITIES

EXAMPLE 4 Graph the solution of $\begin{cases} x + y \geq 2 \\ x - y < 3 \end{cases}$.

SOLUTION

[1] Graph each inequality on the same set of axes.

[2] Test point $(0, 0)$: $x + y \geq 2$ $x - y < 3$
$0 + 0 \geq 2$ is false. $0 - 0 < 3$ is true.

[3] Use arrows to denote these individual solutions. Shade the intersection of these regions to denote the solution of the system.

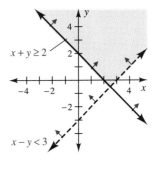

EXAMPLE 5 Graph the solution of $\begin{cases} x^2 + 9y^2 \leq 36 \\ x + y \geq 0 \end{cases}$.

SOLUTION Graph each inequality and shade the portion of the plane that satisfies both inequalities.

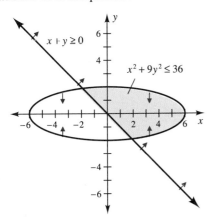

The inequality $x^2 + 9y^2 \leq 36$ yields the points inside the ellipse. Note that $(1, 1)$ can be used as a test point that both these inequalities satisfy.

▶ **Self-Check** ▼

Graph the inequalities in Exercises 1 and 2.

1 $x + y \leq 5$ **2** $x > 4$

3 Graph $x^2 + y^2 \geq 4$.

▶ **Self-Check Answer** ▼

1
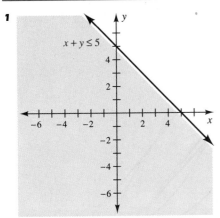

2
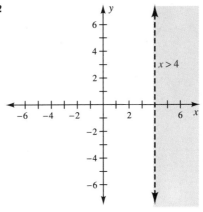

3
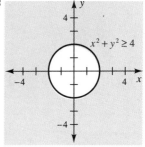

EXAMPLE 6 Graph the solution of $\begin{cases} x + y \leq 4 \\ x - y \leq 2 \\ x \geq 0 \\ y \geq 0 \end{cases}$.

SOLUTION Graph each of the four inequalities and then shade the portion of the plane that satisfies all these inequalities.

Note that (1, 1) can be used as a test point that all of these inequalities satisfy.

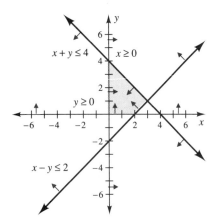

▶ **Self-Check** ▼

Graph the solution of each of these systems of inequalities.

1 $\begin{cases} 2x - y \leq 4 \\ x + 2y > 6 \end{cases}$ **2** $\begin{cases} y \geq -2 \\ y \leq 3 \end{cases}$

3 Solve graphically

$\begin{cases} x^2 + y^2 > 4 \\ x^2 + 25y^2 \leq 25 \end{cases}$.

▶ **Self-Check Answer** ▼

1

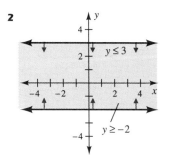

2x − y ≤ 4
x + 2y > 6

2

y ≤ 3
y ≥ −2

3

EXERCISES 6-5

A

In Exercises 1–4, match each inequality with its graph.

1 $y > 2x + 4$

2 $y \geq 2x + 4$

3 $y < 2x + 4$

4 $y \leq 2x + 4$

a.
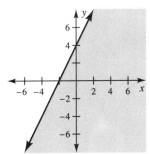

b.
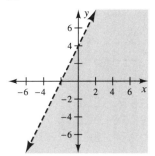

c.
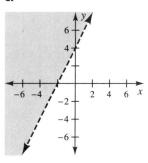

d.
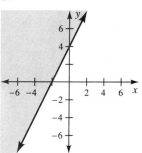

In Exercises 5–24, graph each inequality.

5 $x - y \geq 5$

6 $x + y < 4$

7 $2x - 3y > 9$

8 $3x - 2y \leq 4$

9 $x < 1$

10 $x \geq -2$

11 $y \leq 4$

12 $y > 5$

13 $x \geq 3y$

14 $3x < -y$

15 $\frac{1}{2}x + \frac{1}{3}y \leq 1$

16 $\frac{1}{4}x - \frac{1}{5}y \geq 1$

17 $x^2 + y^2 < 16$

18 $\frac{x^2}{4} + \frac{y^2}{9} > 1$

19 $y \geq x^2 - 1$

20 $y \leq -x^2 + 1$

21 $y < 2^x$

22 $y \leq \log_2 x$

23 $y > \ln x$

24 $y \geq 2^{-x}$

In Exercises 25–40, graph the solution of each system of inequalities.

25 $x - y \geq 4$
$2x - y < 6$

26 $x + 2y \leq 2$
$2x - y \geq 4$

27 $2x - 5y < 10$
$2x - y \geq 7$

28 $3x + 2y \geq 12$
$2x + 5y < 10$

29 $x \geq 1$
$x < 4$

30 $y < -2$
$y \geq 6$

31 $x^2 + y^2 \geq 16$
$x \geq y$

32 $x^2 + y^2 < 9$
$x - y > 1$

33 $x^2 + y^2 > 1$
$16x^2 + y^2 < 16$

34 $\frac{x^2}{16} + \frac{y^2}{25} \geq 1$
$x^2 + y^2 \leq 36$

35 $y < x^2 - 8x + 12$
$2x + 3y < 6$

36 $x^2 + y^2 \leq 64$
$x^2 + y^2 > 25$

37 $y > e^x$
$y < x + 2$

38 $y < \ln x$
$y > \frac{x - 5}{2}$

39 $-2 \leq x \leq 1$
$-4 \leq y \leq 7$

40 $|x| < 5$
$|y| < 1$

430 CHAPTER 6 SYSTEMS OF EQUATIONS AND INEQUALITIES

B

In Exercises 41–44, write a system of inequalities with the given solutions.

41
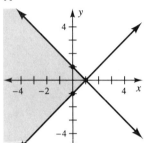

42
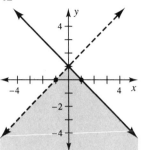

43
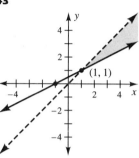

44
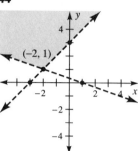

In Exercises 45–48, graph the solution of each system of inequalities.

45 $x + y \leq 5$
$x - y \leq 3$
$x \geq 0$
$y \geq 0$

46 $3x + 3y \leq 8$
$x + 2y \leq 3$
$x \geq 0$
$y \geq 0$

47 $2x - y \geq 2$
$3x + 4y \leq 12$
$2x - 3y \leq 12$

48 $y > x$
$y \leq x^3$

In Exercises 49 and 50, graph each inequality.

49 $x^2 + y^2 - 2x + 4y \leq -1$

50 $9x^2 + 4y^2 + 54x - 16y \leq 47$

C

51 Let $f(x) = 2x + 1$, and graph the system $\begin{cases} y \geq f(x) \\ y \leq f^{-1}(x) \end{cases}$.

In Exercises 52–55, graph each inequality.

52 $y > \dfrac{1}{x - 1}$

53 $y \leq \dfrac{1}{(x - 1)(x + 4)}$

54 $y > \dfrac{3}{(x + 4)^2}$

55 $x^2 - y^2 + 6x - 2y > -7$

56 Graph the simultaneous solution of the equation $x - y = 3$ and the inequality $x + y \leq 1$.

57 **a.** Graph $f(x) = x$ and $g(x) = \dfrac{2}{x - 2}$ on the same set of axes.

b. Determine the values of x for which $f(x) > g(x)$.

c. What relationship does this graph have to the rational inequality $\dfrac{2}{x - 2} < x$?

58 Graph $f(x) = \log_{0.5} x$ and $f^{-1}(x)$ on the same set of axes. Then determine the values of x for which $f(x) > f^{-1}(x)$.

59 **a.** Graph $f(x) = x(x - 1)(x - 2)$.

b. Graph the line $g(x)$ through $(0, f(0))$ and $(3, f(3))$.

c. Determine the values of x for which $f(x) > g(x)$.

SECTION 6-6

Linear Programming

Section Objective

10 Solve a linear programming problem using a graphing strategy.

Linear programming can be used to determine how to allocate limited resources in order to obtain an optimal outcome. New techniques for solving linear programming problems were developed as a means of allocating supplies during World War II. The basic terminology of linear programming is presented in the following box.

Linear Programming Terms ▼

Objective function:	The function to be maximized or minimized.
Constraints:	The linear inequalities in the statement of the problem.
Feasible region:	The graphic solution of the system of inequalities.
Corner points:	The vertices of the feasible region.

A typical linear programming problem is to maximize profit subject to certain conditions or constraints. These problems can typically be reduced to the following algebraic format and graphed as shown in Figure 6-6.

Maximize the profit: $P = 2x + y$ The objective function

subject to: $\begin{cases} x + y \leq 5 \\ 3x + y \leq 9 \\ x \geq 0 \\ y \geq 0 \end{cases}$ The constraints

The feasible region with corner points (0, 0), (3, 0), (2, 3), and (0, 5).

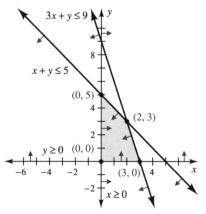

Figure 6-6 Graph of a linear programming problem

The profit function $P = 2x + y$ can be written as $y = -2x + P$, which is a straight line with slope -2 and y-intercept P. Our goal is to find the maximum profit satisfying the constraints. Thus we will examine four profit lines with slope -2 in search of the line that passes through the feasible region and has the greatest y-intercept (profit).

The profit lines diagrammed in Figure 6-7 show y-intercepts (profits) of 2, 4, 7, and 9. However, a profit of 9 is *not* possible, since this profit line does *not* contain any point in the feasible region and cannot therefore satisfy all the constraints. Thus the maximum profit is 7, and it is produced by the profit line passing through the corner point (2, 3). Any profit line with slope -2 and y-intercept greater than 7 will miss the feasible region.

This procedure suggests that the optimal values can always be found at a corner point. A strategy using this fact is outlined in the following box.

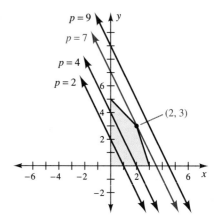

Figure 6-7 The optimal profit line

Solving Linear Programming Problems ▼

Step 1 Graph the feasible region.
Step 2 Find the corner points.
Step 3 Substitute the corner points into the objective function.
Step 4 The largest (smallest) value obtained in Step 3 is the maximum (minimum) value.

EXAMPLE 1 Given $P = 3x + 2y$, find the maximum and minimum value of P satisfying $x + 2y \leq 4$, $3x - y \leq 6$, $x \geq 0$, and $y \geq 0$.

SOLUTION

1 Graph the feasible region.

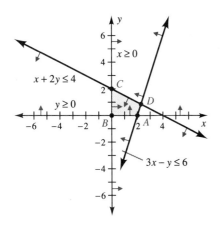

2 Find the corner points. The corner points $A(2, 0)$, $B(0, 0)$, and $C(0, 2)$ can be found by inspection. Corner point $D(\frac{16}{7}, \frac{6}{7})$ is found by determining the point of intersection of lines $x + 2y = 4$ and $3x - y = 6$.

SECTION 6-6 LINEAR PROGRAMMING

3 Substitute the corner points into the profit equation $3x + 2y = P$.
4 Determine the maximum value and the minimum value.

Corner Point	$3x + 2y = P$ Profit	
$A(2, 0)$	$3(2) + 2(0) =$	6
$B(0, 0)$	$3(0) + 2(0) =$	0 The minimum value
$C(0, 2)$	$3(0) + 2(2) =$	4
$D(\frac{16}{7}, \frac{6}{7})$	$3(\frac{16}{7}) + 2(\frac{6}{7}) =$	$\frac{60}{7}$ The maximum value

Answer The maximum value of $8\frac{4}{7}$ occurs when $x = \frac{16}{7}$ and $y = \frac{6}{7}$. The minimum value of 0 occurs when $x = 0$ and $y = 0$. □

EXAMPLE 2 Flowers Unlimited has two spring floral arrangements, the Easter Bouquet and the Spring Bouquet. The Easter Bouquet requires 10 jonquils and 20 daisies and produces a profit of $1.50. The Spring Bouquet requires 5 jonquils and 20 daisies and yields a profit of $1. How many of each type of arrangement should the florist make to maximize the profit if 120 jonquils and 300 daisies are available? (Assume that all bouquets will be sold.)

SOLUTION Let $x =$ Number of Easter Bouquets to be made
$y =$ Number of Spring Bouquets to be made

The following tables organize the data.

Word Equation:	Profit on Easter Bouquets	+	Profit on Spring Bouquets	=	Total Profit
Objective function	$1.50x$	+	$1.00y$	=	P

The *word equation* is based upon the mixture principle. The profit of each item is determined by using the rate principle.

Word Inequality:	Number in the Easter Bouquets	+	Number in the Spring Bouquets	$\leq$	Total Number Available
Constraint due to jonquils	$10x$	+	$5y$	$\leq$	120
Constraint due to daisies	$20x$	+	$20y$	$\leq$	300

We must

Maximize the profit: $P = 1.50x + 1.00y$ Objective function

subject to: $\begin{cases} 10x + 5y \leq 120 \\ 20x + 20y \leq 300 \\ x \geq 0 \\ y \geq 0 \end{cases}$

The constraints $x \geq 0$ and $y \geq 0$ are implied constraints, since a negative number of bouquets cannot be made.

Graph the feasible region. The graph indicates that $A(12, 0)$, $B(0, 0)$, and $C(0, 15)$ are corner points. The corner point $D(9, 6)$ is the intersection of the lines $10x + 5y = 120$ and $20x + 20y = 300$. These corner points are substituted into the objective function in the table below.

Corner Point	$1.50x + 1.00y = P$ Profit	
$A(12, 0)$	$1.50(12) + 1.00(0) =$	18.00
$B(0, 0)$	$1.50(0) + 1.00(0) =$	0
$C(0, 15)$	$1.50(0) + 1.00(15) =$	15.00
$D(9, 6)$	$1.50(9) + 1.00(6) =$	19.50 The maximum profit

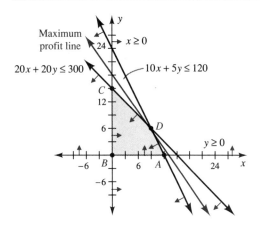

Answer Flowers Unlimited will make a maximum profit of $19.50 by making nine Easter Bouquets and six Spring Bouquets. □

EXAMPLE 3 A construction company needs to hire at least 100 employees for a project. They will need at least 30 more unskilled laborers than skilled laborers. At least 20 skilled laborers should be hired. The unskilled laborer earns $8 per hour, and the skilled laborer earns $15 per hour. How many employees should the company hire to minimize its hourly cost while satisfying all of the requirements?

▶ **Self-Check** ▼

Determine the maximum value of the function $P = 2x + 3y$ over the region determined by the corner points $(12, 0)$, $(0, 0)$, $(0, 15)$, and $(9, 6)$.

SOLUTION Let $x =$ Number of unskilled laborers to be hired
            $y =$ Number of skilled laborers to be hired

Objective function:

(Total cost) = (Cost for unskilled) + (Cost for skilled)   This equation is based on the mixture principle.
$C = 8x + 15y$ Each cost is determined using the rate principle.

▶ **Self-Check Answer** ▼

$P = 45$ at $(0, 15)$

Constraints:

$$\begin{cases} x + y \geq 100 \\ x \geq y + 30 \\ y \geq 20 \\ x \geq 0 \\ y \geq 0 \end{cases}$$

Hire at least 100 employees.
Hire at least 30 more unskilled employees.
Hire at least 20 skilled workers.
The implied constraints

Graph the feasible region. The corner point A is the intersection of $y = 20$ and $x + y = 100$. B is the intersection of $x + y = 100$ and $x = y + 30$. This feasible region is unbounded to the right. The value of $C = 8x + 15y$ will increase within this region as x or y increases. Thus the minimum cost will occur at one of the corner points A or B, both of which are substituted into the objective function in the table below.

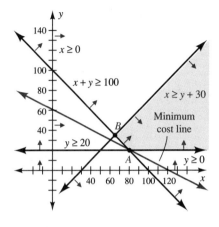

Corner Point	$8x$ + $15y$ = C Total Cost		
$A(80, 20)$	$8(80) + 15(20) =$	940	Minimum cost
$B(65, 35)$	$8(65) + 15(35) =$	1045	

Answer A minimum hourly cost of $940 is obtained by hiring 80 unskilled workers and 20 skilled workers. □

EXERCISES 6-6

A

In Exercises 1–4, determine the maximum value of each function over the feasible region determined by the corner points $(0, 0)$, $(8, 0)$, $(6, 3)$, $(2, 7)$, and $(0, 8)$.

1 $P = 17x + 24y$ **2** $P = 17x + 36y$ **3** $P = 2x + y$ **4** $P = x + y$

In Exercises 5–14, solve each linear programming problem using the strategy outlined in this section.

5 Maximize: $P = 3x + 2y$
subject to: $\begin{cases} x + y \leq 6 \\ x \geq y \\ x \geq 0 \\ y \geq 0 \end{cases}$

6 Maximize: $P = x + 6y$
subject to: $\begin{cases} x + 2y \leq 4 \\ x \leq 2y \\ x \geq 0 \\ y \geq 0 \end{cases}$

7 Minimize: $C = 2x - y$
subject to: $\begin{cases} 2x + 3y \geq 6 \\ 3x + 2y \leq 6 \\ x \geq 0 \\ y \geq 0 \end{cases}$

8 Minimize: $C = x - 3y$
subject to: $\begin{cases} 3x + 4y \leq 15 \\ -x + 2y \geq 5 \\ x \geq 0 \\ y \geq 0 \end{cases}$

9 Maximize: $P = 3x + y$
subject to: $\begin{cases} x + y \leq 4 \\ 3x + y \geq 2 \\ x \geq 0 \\ y \geq 0 \end{cases}$

10 Maximize: $P = 4x + 2y$
subject to: $\begin{cases} x + 2y \leq 6 \\ 2x + y \geq 1 \\ x \geq 0 \\ y \geq 0 \end{cases}$

11 Minimize: $C = 3x + 2y$

subject to: $\begin{cases} y + x \geq 2 \\ 3y - 4x \leq 7 \\ x \geq 0 \\ y \geq 0 \end{cases}$

12 Minimize: $C = x + 2y$

subject to: $\begin{cases} y - 2x \leq 4 \\ y - x \geq 1 \\ x \geq 0 \\ y \geq 0 \end{cases}$

13 Maximize: $P = 6x + 4y$

subject to: $\begin{cases} x - 2y \leq 3 \\ x + y \leq 6 \\ x \geq 0 \\ y \geq 0 \end{cases}$

14 Maximize: $P = 2x + 5y$

subject to: $\begin{cases} y \leq x + 5 \\ 7x + 2y \leq 28 \\ x \geq 0 \\ y \geq 0 \end{cases}$

B

In Exercises 15–18, solve each linear programming problem using the strategy outlined in this section.

15 Minimize: $C = 2x + y$

subject to: $\begin{cases} x + y \geq 6 \\ 2x + 3y \leq 24 \\ 2 \leq x \leq 6 \\ y \geq 0 \end{cases}$

16 Minimize: $C = 3x - 2y$

subject to: $\begin{cases} x + y \geq 6 \\ 2x + 3y \leq 24 \\ 2 \leq y \leq 6 \\ x \geq 0 \end{cases}$

17 Maximize: $P = 5x + 4y$

subject to: $\begin{cases} x + 2y \leq 15 \\ x - y \leq 6 \\ 2x + 3y \geq 12 \\ x \geq y \\ x \geq 0 \\ y \geq 0 \end{cases}$

18 Minimize: $P = 5x + 4y$

subject to: $\begin{cases} x + 2y \leq 15 \\ x - y \leq 6 \\ 2x + 3y \geq 12 \\ x \geq y \\ x \geq 0 \\ y \geq 0 \end{cases}$

19 The student scholarship fund has a total of $100,000 to distribute. The distribution of the awards is based on a combination of scholastic performance and need. The scholarship winners are determined by scholastic performance. Then the amount of money awarded is determined by need. The students in category I receive $1000 and those in category II receive $4000. The managers of the fund are bound by regulations to award at least 10 more scholarships to category II scholars than to category I scholars. How many scholarships of each type should be granted to maximize the number of students receiving scholarships?

20 A furniture manufacturer makes two types of coffee tables. A circular table requires 3 hours of skilled labor and 6 hours of unskilled labor and produces a profit of $50. A rectangular table requires 2 hours of each type of labor and yields a profit of $30. If 90 hours of skilled labor and 120 hours of unskilled labor are available for a week, determine how many of each type of table should be made in order to maximize the profit.

SECTION 6-6 LINEAR PROGRAMMING

C

21. A garment maker produces suits and blouses. Each suit requires 1 hour of cutting and 6 hours of sewing. Each blouse requires $\frac{1}{4}$ hour of cutting and 2 hours of sewing. Union contracts specify at least 50 hours of cutting and at least 360 hours of sewing per week. The cost of material and production is $70 for a suit and $20 for a blouse. How many suits and blouses should be made in order to minimize the total cost?

22. A company makes specialty calculators at two plants. The eastside plant can produce 80 nursing calculators and 40 business calculators in one day at a total cost of $500. The westside plant can produce 40 nursing calculators and 40 business calculators in one day at a total cost of $400. Find the number of days each plant should operate to minimize the cost of filling an order for 700 nursing calculators and 400 business calculators within 12 days.

23. A grocery store sells two different snack mixes, the Student Mix and the Hollywood Mix. The Student Mix requires 50 grams of peanuts for each 100 grams of raisins, and the ratio for the Hollywood Mix is 100 grams of peanuts to 100 grams of raisins. A 100-gram bag of Student Mix costs $0.70, and a 100-grams bag of Hollywood Mix costs $0.90. The store has 4 kilograms of peanuts and 5 kilograms of raisins. How many 100-gram bags of each mix should be made in order to maximize the store's income?

24. A company produces two containers from plastic and aluminum. Container A contains 400 grams of aluminum and 600 grams of plastic, and container B contains 300 grams of aluminum and 700 grams of plastic. The selling prices are $55 for container A and $60 for container B. How many of each container should be made to maximize the revenue if the company has 480 kilograms of aluminum and 840 kilograms of plastic on hand?

25. A computer company produces microcomputers, home computers, and printers at two sites. The number of each type produced per day at sites A and B is given in the table. The daily cost of operating is $4500 for site A and $7500 for site B. The company has orders on hand for 640 microcomputers, 480 home computers, and 360 printers, to be completed within 60 days. Find the number of days each site should operate in order to minimize the cost of filling the orders.

	Site A	Site B
Microcomputers	16	30
Home Computers	8	35
Printers	12	15

26. A company manufactures wall clocks, payroll clocks, and portable security clocks at their Princeton and Clinton plants. The Princeton plant can produce 50 wall clocks, 35 payroll clocks, and 15 security clocks per day at a total cost of $2000. The Clinton plant can produce 50 wall clocks, 15 payroll clocks, and 40 security clocks per day at a total cost of $2500. How many days should each plant operate to minimize the cost of filling an order for 250 wall clocks, 105 payroll clocks, and 120 security clocks within 10 days?

KEY CONCEPTS

1. Linear systems of equations:
   a. Independent and consistent systems have exactly one solution.
   b. Inconsistent systems have no solution.
   c. Dependent systems have an infinite number of solutions.

2. The linear system $\begin{cases} a_1x + b_1y = c_1 \\ a_2x + b_2y = c_2 \end{cases}$ is
   a. Independent and consistent if $\dfrac{a_1}{a_2} \neq \dfrac{b_1}{b_2}$.
   b. Inconsistent if $\dfrac{a_1}{a_2} = \dfrac{b_1}{b_2} \neq \dfrac{c_1}{c_2}$.
   c. Dependent if $\dfrac{a_1}{a_2} = \dfrac{b_1}{b_2} = \dfrac{c_1}{c_2}$.

3. Methods of solving linear systems of equations
   a. The substitution method,
   b. The elimination method,
   c. The augmented matrix method,
   d. The inverse matrix method (Section 7.2), and
   e. Cramer's Rule (Section 7.3).

4. Elementary row operations on augmented matrices
   1. Any two rows in the matrix may be interchanged.
   2. Any row in the matrix may be multiplied by a nonzero constant.
   3. Any row in the matrix may be replaced by the sum of this row and a constant multiple of another row.

5. Properties of the reduced form of a matrix
   1. The first nonzero entry in a row is a 1. All other entries in the column containing the leading 1 are zeros.
   2. All nonzero rows are above any rows containing only zeros.
   3. The first nonzero entry in a row is to the left of the first nonzero entry in the following row.

6. The line $Ax + By = C$ separates the plane into two half-planes. A test point can be used to determine which half-plane satisfies $ax + by < c$ and which satisfies $ax + by > c$.

REVIEW EXERCISES FOR CHAPTER 6

In Exercises 1–15, solve each system of equations. Use the substitution method for Exercises 1–3.

1. $3x - y = 7$
   $2x + 5y = -18$

2. $4x + y = -2$
   $y = x^2 + 3x + 10$

3. $xy = 12$
   $x^2 + y^2 = 25$

SECTION 6-6 LINEAR PROGRAMMING

Use the elimination method for Exercises 4–6.

4 $6x - 5y = 1$
$9x + 10y = 0$

5 $6x + 6y = 1$
$4x - 9y = 5$

6 $2x - 3y + 4z = 41$
$6x + 5y - 2z = -31$
$5x + 2y - 3z = -16$

7 $2x^2 + 3y^2 = 44$
$x^2 - y^2 = 12$

8 $y = \log(x^2 + 1) + 1$
$y = \log(x - 2) + 2$

Use the augmented matrix method for Exercises 9–12.

9 $3x - y + z = 6$
$x + 2y - 3z = -1$
$4x + y - 2z = 5$

10 $x - 5y + 7z = 58$
$4x + 3y - 8z = -17$
$2x - 2y + 5z = 47$

11 $2x - 3y + 4z = 9$
$3x + 2y - 4z = 9$
$5x - y = 9$

12 $5x - 3y + 7z = 1$
$2x - 4y + 5z = -5$

13 $\dfrac{1}{x} + \dfrac{1}{y} = -2$
$\dfrac{3}{x} - \dfrac{4}{y} = 43$

14 $x^2 + xy + y^2 = 1$
$x^2 - xy + y^2 = 3$

15 $x^2 + y^2 + 2x + y = 12$
$x^2 + y^2 + 3x - y = 6$

16 Find the equation of the parabola that opens vertically and passes through (2, 15), (1, 7), and (−1, 21).

17 An industrial spy observed that machines A and B together output 752 units in eight hours. On another day, machine A operated for only three hours, and machine B for only five hours. The production that day was 364 units. How many units should the spy report that each machine produces per hour?

18 A truck made three deliveries hauling loads of 1960, 1740, and 700 pounds, respectively. The first delivery contained 20 cement bags, 4 boxes of nails, and 2 buckets of joint compound. The second delivery contained 18 cement bags and 6 boxes of nails. The third delivery contained 10 boxes of nails and 5 buckets of joint compound. Determine the weight of a bag of cement.

19 A self-employed couple makes custom running shoes, which they sell for $150 a pair. If the fixed costs for one week are $450 and it takes $105 in materials to make each pair, what is the break-even value?

20 The demand for airline tickets on one route is approximated by $d = 426 - 2.5p$, where p is the price in dollars. The supply of seats that the airline will provide to a charter service is approximated by $q = 5p$. What is the equilibrium price? How many tickets can be sold at the equilibrium price?

In Exercises 21 and 22, use graphs to approximate to the nearest tenth the solution of each system of equations.

21 $y = \log_2 x$
$y = (x - 1)^2 - 1$

22 $\dfrac{x^2}{9} + \dfrac{y^2}{4} = 1$
$y = e^x$

In Exercises 23–26, graph the solution of each system of inequalities.

23 $2x + 3y < 6$
$-3x + 2y \geq 6$

24 $|x| \leq 2$
$|y| < 3$

25 $y > x^2 + 2x + 1$
$x^2 + y^2 \leq 25$

26 $7x + 3y \geq 21$
$2x + 6y \geq 18$
$x \geq 0$
$y \geq 0$

27 Give three particular solutions of $\begin{cases} 5x - 15y = 20 \\ 4x - 12y = 16 \end{cases}$.

28 Transform $\begin{bmatrix} 2 & 3 & -4 & 5 & | & 8 \\ 1 & 2 & -2 & 3 & | & 4 \\ -3 & 4 & -2 & 9 & | & 2 \end{bmatrix}$ into its reduced form.

29 Find k such that $\begin{cases} 2x + 5y = 6 \\ 3x + ky = 10 \end{cases}$ is an inconsistent system of equations.

30 Minimize: $C = 2x + 4y$

subject to: $\begin{cases} 5x + 3y \geq 10 \\ x + 2y \leq 8 \\ x \geq 0 \\ y \geq 0 \end{cases}$

31 Solve: $\begin{aligned} w + 5x + 2y + 7z &= -8 \\ w - 3x - 9y + 4z &= 22 \\ 3w + 7x - 4y + 14z &= 16 \\ 4w + 8x - y + z &= -2 \end{aligned}$

32 A manufacturer that has $400,000 available for upgrading its assembly line is considering two types of industrial robots. Robot A costs $5000 and can do 405 hours of work per week. Robot B costs $8000 and can do 540 hours of work per week. The total number of robots is limited to 60 by the space available on the assumbly line. How many robots of each type should be purchased to maximize factory productivity?

OPTIONAL EXERCISES FOR CALCULUS-BOUND STUDENTS

Solve each of the systems of equations in Exercises 1–5. In Exercise 5, give each coordinate to the nearest hundredth.

1
$\begin{aligned} x_1 - x_2 + 3x_3 - 4x_4 &= 1 \\ -2x_1 + 4x_2 - x_3 + x_4 &= -4 \\ 3x_1 + 2x_2 - 4x_3 + 5x_4 &= 7 \\ 2x_1 - x_2 - x_3 + 4x_4 &= 7 \end{aligned}$

2
$\begin{aligned} \frac{4}{x+3} + \frac{9}{y-1} + \frac{12}{5z+1} &= 9 \\ \frac{8}{x+3} + \frac{3}{y-1} - \frac{18}{5z+1} &= 1 \\ \frac{16}{x+3} + \frac{6}{y-1} - \frac{6}{5z+1} &= 7 \end{aligned}$

3
$\begin{aligned} x + y + z &= 4 \\ 2x - y + z &= 0 \\ 4x + y + 3z &= 8 \end{aligned}$

4
$\begin{aligned} 6x^2 - 5y^2 &= -8 \\ 2x^2 + 2y^2 &= 34 \end{aligned}$

5
$\begin{aligned} 9e^{2x} - 4e^{2y} &= 3 \\ 4e^{2x} - 3e^{2y} &= -6 \end{aligned}$

6 Find x and y so that $(2x + y) + 29i = 1 + (3x - 4y)i$

7 Use a single matrix to solve all three of these systems.

a. $\begin{aligned} 2x + y - 3z &= -1 \\ x + 3y - z &= 0 \\ -2x + 2y + 5z &= 3 \end{aligned}$
b. $\begin{aligned} 2x + y - 3z &= 7 \\ x + 3y - z &= 8 \\ -2x + 2y + 5z &= -3 \end{aligned}$
c. $\begin{aligned} 2x + y - 3z &= 7 \\ x + 3y - z &= 5 \\ -2x + 2y + 5z &= -7 \end{aligned}$

8 Find the equation of a parabola that opens vertically and passes through the points $(0, 2)$, $(-1, 7)$, and $(1, -1)$.

MASTERY TEST FOR CHAPTER 6

Solve each of the systems of equations in Exercises 9–15. In Exercises 14 and 15, use a graphics calculator to approximate to the nearest hundredth the real solutions of each system of equations.

9. $y = \log_2(x - 9) + 4$
   $y = \log_2(x + 3) + 6$

10. $\ln(x - 4) + \ln(y - 1) = 0$
    $\ln x \quad + \ln y \quad = 0$

11. $9x^2 + 15xy + y^2 = -5$
    $27x^2 + 9xy + 7y^2 = 25$

12. $x^2 + y^2 + x + 12y = -8$
    $2x^2 + 2y^2 - 4x + 9y = -4$

13. $x^2 - 4xy - 5y^2 = 11$
    $5x^2 - 8xy + 3y^2 = -33$

14. $y = (x + 3)(2x + 1)(x - 1)$
    $y = -x^3 + 6x^2 - 9x + 3$

15. $y = 2^x$
    $y = -(x + 1)(2x - 3)^2$

16. A rectangular garden is formed on a right-triangular lot by drawing perpendiculars to each leg from a point on the hypotenuse. Determine the dimensions of the garden if its area is 80 square meters and the legs of the triangle are 18 meters and 24 meters. (See the figure to the right.)

17. Graph the solution of $y > \dfrac{3}{(x + 4)^2}$.

In Exercises 18 and 19, graph the solution of each system of inequalities.

18. $x - y \geq -5$
    $2x - y \leq -2$
    $2x + y \geq -6$

19. $y^2 - x^2 \leq 1$
    $y < 3 - x^2$

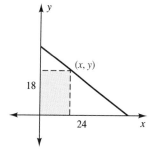

Figure for Exercise 16

20. Determine the values of x for which $f(x) > f^{-1}(x)$ given $f(x) = 2x - 5$.

MASTERY TEST FOR CHAPTER 6

Exercise numbers correspond to Section Objective numbers.

1. Solve the system $\begin{cases} x + y = 3 \\ 2x - 5y = 0 \end{cases}$ using the substitution method.

2. Solve the system $\begin{cases} 3x - 2y = 5 \\ 5x + 4y = 67 \end{cases}$ using the elimination method.

3. Determine the number of solutions for each system of linear equations.

   a. $\begin{cases} 3x - 4y = 5 \\ 6x + 4y = 10 \end{cases}$
   b. $\begin{cases} 6x - 8y = 10 \\ 9x - 12y = 15 \end{cases}$
   c. $\begin{cases} 9x - 12y = 15 \\ 12x - 16y = 16 \end{cases}$

4. Solve the system $\begin{cases} x - 3y + 5z = -17 \\ 2x + y - 7z = 2 \\ 2x - 5y + 3z = -6 \end{cases}$ using an augmented matrix.

5. An investor placed a total of $50,000 in silver, stocks, and money market funds for a period of one year. She invested one-fourth as much in silver as in the other two areas combined. The silver lost 7%, the stocks gained 5%, and the money market fund earned 10%. If the total earnings were $2300, how much was invested in each?

6 Solve the system of equations $\begin{cases} y = x + 3 \\ 2x^2 - y^2 + 2x + 2y = 5 \end{cases}$ using the substitution method.

7 Solve the system of equations $\begin{cases} 3x^2 + 2y^2 = 10 \\ x^2 - y^2 = 5 \end{cases}$ using the elimination method.

8 Graph the solution of $y < x^2 - 2x - 8$.

9 Graph the solution of $\begin{cases} x^2 + y^2 \le 36 \\ 3x - 2y > 12 \end{cases}$.

10 Maximize: $P = 2x + y$

subject to: $\begin{cases} 3x + 2y \le 12 \\ x + 2y \le 8 \\ x \ge 0 \\ y \ge 0 \end{cases}$

CHAPTER SEVEN

Matrices and Determinants

Chapter Seven Objectives

1 Apply the basic operations to matrices.

2 Find the inverse of a square matrix.

3 Solve a system of linear equations using the inverse of a matrix.

4 Compute the determinant of a square matrix.

5 Solve a 2×2 or 3×3 system of linear equations using Cramer's rule.

6 Use the properties of determinants.

Evariste Galois, 1811–1832

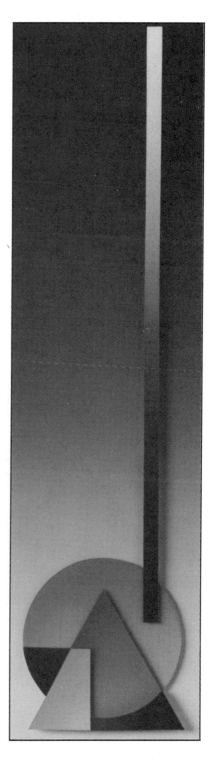

The Frenchman Galois is known for both his genius and his lack of good judgment. At the age of 21 he died in an ill-advised duel of honor. His collected works fill only 60 pages, yet they contained many new discoveries in analysis. This was typical of Galois, who was able to carry on even difficult mathematics entirely in his head and likewise wrote for conciseness rather than clarity.

Matrices provide a convenient and orderly means of organizing many data bases. The advent of computers has fostered the increased usage of the matrix form to represent information and of mathematics to manipulate and analyze this information. We can use one symbol to represent a matrix containing a large set of data and then use a computer to perform perhaps millions of computations with these data. The subject of matrices is so important that many texts are devoted exclusively to this topic. This chapter is an introduction to this important material.

SECTION 7-1

Matrix Operations

Section Objective

1 Apply the basic operations to matrices.

A matrix, with its horizontal rows and vertical columns, provides an excellent way to display and store information. Computers utilize matrices to store information for rapid access and to clearly display output. The matrix format is sometimes referred to as a table, chart, or array. Weather results are often displayed in matrix form, as in Table 7-1. One can note from the table, for example, that in Denver on March 28 the high temperature was 46° Fahrenheit and the low temperature was 28° Fahrenheit.

Table 7-1 Temperatures for March 28

City	High	Low
Atlanta	48	39
Boston	59	39
Chicago	37	22
Denver	46	28
Houston	68	48
Memphis	47	40
San Diego	62	57
Seattle	47	40

Capital letters are usually used to denote matrices. The **order of a matrix** is given by stating the number of rows and columns in the matrix. The matrix $A = \begin{bmatrix} 2 & -1 & 8 \\ 4 & 0 & 1 \end{bmatrix}$ has order 2×3, read "2 by 3," since it has 2 rows and 3 columns. The **entry** a_{ij} is located in the ith row and jth column. The entry in the first row and third column of A is denoted a_{13}. In this example, $a_{13} = 8$.

SECTION 7-1 MATRIX OPERATIONS

A matrix of order $m \times n$ can be written in the form

$$A = \begin{bmatrix} a_{11} & a_{12} & a_{13} & \cdots & a_{1n} \\ a_{21} & a_{22} & a_{23} & \cdots & a_{2n} \\ a_{31} & a_{32} & a_{33} & \cdots & a_{3n} \\ \vdots & \vdots & \vdots & & \vdots \\ a_{m1} & a_{m2} & a_{m3} & \cdots & a_{mn} \end{bmatrix}$$

The notation $A = [a_{ij}]_{m \times n}$ is a condensed method of representing this matrix.

Equality of Matrices ▼

Two matrices are equal if and only if they have the same order and their corresponding entries are equal.

Let $A = \begin{bmatrix} 1 & -1 & 2 \\ 3 & 4 & 0 \end{bmatrix}$, $B = \begin{bmatrix} 1 & -1 \\ 3 & 4 \end{bmatrix}$, $C = \begin{bmatrix} 1 & 2-3 & \frac{4}{2} \\ 3 & 1+3 & 0 \end{bmatrix}$,

and $D = \begin{bmatrix} 1 & -1 & -2 \\ 3 & 4 & 0 \end{bmatrix}$.

Then

- $A = C$, since A and C have order 2×3 and their corresponding entries are equal.
- $A \neq B$, since A has order 2×3 and B has order 2×2.
- $A \neq D$, since $a_{13} \neq d_{13}$; $a_{13} = 2$, but $d_{13} = -2$.

Operations on Matrices ▼

If $A = [a_{ij}]_{m \times n}$, $B = [b_{ij}]_{m \times n}$, and k is a real number, then

Addition: $A + B = [c_{ij}]_{m \times n}$, where $c_{ij} = a_{ij} + b_{ij}$
Scalar multiplication: $kA = [d_{ij}]_{m \times n}$, where $d_{ij} = ka_{ij}$
Subtraction: $A - B = A + (-1)B$

$A + B$ and $A - B$ are matrices with the same order as A and B. Each entry in $A + B$ is obtained by adding the corresponding entries in A and B. If A and B are of different orders, then $A + B$ and $A - B$ are undefined.

▶ Self-Check ▼

1. Write the expanded form for $A = [a_{ij}]_{3 \times 2}$.
2. State the order of $A = \begin{bmatrix} 1 & -1 \\ 2 & 0 \\ 3 & 5 \end{bmatrix}$ and $B = \begin{bmatrix} 1 & 2 & 1 \end{bmatrix}$.
3. Determine a_{21} and a_{12} in problem 2.
4. Determine x so that $\begin{bmatrix} 1 & 2 \\ 2 & x \end{bmatrix} = \begin{bmatrix} 1 & 2 \\ 1+1 & 1-2 \end{bmatrix}$.

▶ Self-Check Answers ▼

1. $A = \begin{bmatrix} a_{11} & a_{12} \\ a_{21} & a_{22} \\ a_{31} & a_{32} \end{bmatrix}$
2. A is a 3×2 matrix and B is a 1×3 matrix.
3. $a_{21} = 2$, $a_{12} = -1$
4. $x = -1$

$A + A = 2A$ for any matrix A. This illustrates that scalar multiplication is consistent with addition. For example,

$$\begin{bmatrix} 2 & -1 \\ 3 & 2 \end{bmatrix} + \begin{bmatrix} 2 & -1 \\ 3 & 2 \end{bmatrix} = \begin{bmatrix} 4 & -2 \\ 6 & 4 \end{bmatrix} = 2 \begin{bmatrix} 2 & -1 \\ 3 & 2 \end{bmatrix}$$

EXAMPLE 1 Let $A = \begin{bmatrix} 2 & 3 & -1 \\ 1 & 0 & 2 \end{bmatrix}$ and $B = \begin{bmatrix} 0 & -1 & 2 \\ 3 & -4 & 7 \end{bmatrix}$. Find

(a) $A + B$ (b) $-3A$ (c) $A - B$ (d) $2A - 3B$

SOLUTIONS

(a) $A + B = \begin{bmatrix} 2 & 3 & -1 \\ 1 & 0 & 2 \end{bmatrix} + \begin{bmatrix} 0 & -1 & 2 \\ 3 & -4 & 7 \end{bmatrix} = \begin{bmatrix} 2+0 & 3+(-1) & -1+2 \\ 1+3 & 0+(-4) & 2+7 \end{bmatrix}$

$= \begin{bmatrix} 2 & 2 & 1 \\ 4 & -4 & 9 \end{bmatrix}$

(b) $-3A = -3 \begin{bmatrix} 2 & 3 & -1 \\ 1 & 0 & 2 \end{bmatrix} = \begin{bmatrix} -3(2) & -3(3) & -3(-1) \\ -3(1) & -3(0) & -3(2) \end{bmatrix}$

$= \begin{bmatrix} -6 & -9 & 3 \\ -3 & 0 & -6 \end{bmatrix}$

(c) $A - B = \begin{bmatrix} 2 & 3 & -1 \\ 1 & 0 & 2 \end{bmatrix} - \begin{bmatrix} 0 & -1 & 2 \\ 3 & -4 & 7 \end{bmatrix}$

$= \begin{bmatrix} 2 & 3 & -1 \\ 1 & 0 & 2 \end{bmatrix} + \begin{bmatrix} 0 & 1 & -2 \\ -3 & 4 & -7 \end{bmatrix}$

$= \begin{bmatrix} 2 & 4 & -3 \\ -2 & 4 & -5 \end{bmatrix}$

(d) $2A - 3B = 2 \begin{bmatrix} 2 & 3 & -1 \\ 1 & 0 & 2 \end{bmatrix} - 3 \begin{bmatrix} 0 & -1 & 2 \\ 3 & -4 & 7 \end{bmatrix}$

$= \begin{bmatrix} 4 & 6 & -2 \\ 2 & 0 & 4 \end{bmatrix} + \begin{bmatrix} 0 & 3 & -6 \\ -9 & 12 & -21 \end{bmatrix}$

$= \begin{bmatrix} 4 & 9 & -8 \\ -7 & 12 & -17 \end{bmatrix}$ □

Any matrix that consists of all zero entries is called a **zero matrix**. The additive identity for real numbers is 0, since $a + 0 = a$ for all real numbers a. Similarly, $O_{m \times n}$, which is the way we denote the zero matrix of order $m \times n$, is the additive identity for all $m \times n$ matrices, since $A_{m \times n} + O_{m \times n} = A_{m \times n}$.

SECTION 7-1 MATRIX OPERATIONS

EXAMPLE 2 A local computer vendor has three outlets, stores A, B, and C, in one city. The unit sales for November and December are recorded for four items in the matrices N and D. Determine the total sales, T, for these two months, which is given by $T = N + D$.

$$N = \begin{bmatrix} 350 & 12 & 20 & 30 \\ 400 & 9 & 12 & 20 \\ 180 & 8 & 5 & 18 \end{bmatrix} \begin{matrix} \text{Store A} \\ \text{Store B} \\ \text{Store C} \end{matrix}$$

with columns: Diskettes, Printers, Disk drives, Computers

$$D = \begin{bmatrix} 325 & 10 & 20 & 25 \\ 385 & 10 & 15 & 21 \\ 200 & 10 & 7 & 20 \end{bmatrix} \begin{matrix} \text{Store A} \\ \text{Store B} \\ \text{Store C} \end{matrix}$$

SOLUTION

$$T = N + D = \begin{bmatrix} 675 & 22 & 40 & 55 \\ 785 & 19 & 27 & 41 \\ 380 & 18 & 12 & 38 \end{bmatrix} \begin{matrix} \text{Store A} \\ \text{Store B} \\ \text{Store C} \end{matrix}$$

Thus, for example, store B sold a total of 41 computers during these two months. □

EXAMPLE 3 An office supply store prices many of its items based on the quantity purchased. The pricing array, P_0, for three items is given below. The manager of the store has decided that all prices must be increased by 7% because of inflation. Determine the new pricing array, which is given by $P_1 = 1.07 P_0$.

$$P_0 = \begin{bmatrix} 1.19 & 17.80 & 45.95 \\ 1.08 & 17.00 & 45.60 \\ 1.03 & 16.85 & 44.90 \end{bmatrix} \begin{matrix} \text{0--5 items} \\ \text{6--49 items} \\ \text{50+ items} \end{matrix}$$

with columns: Pens, Desk pads, Chairs

SOLUTION

$$P_1 = 1.07 P_0 = \begin{bmatrix} 1.27 & 19.05 & 49.17 \\ 1.16 & 18.19 & 48.79 \\ 1.10 & 18.03 & 48.04 \end{bmatrix} \begin{matrix} \text{0--5 items} \\ \text{6--49 items} \\ \text{50+ items} \end{matrix}$$

Thus, for example, the new price for 63 desk pads is $18.03 each. □

▶ **Self-Check** ▼

Let
$$A = \begin{bmatrix} 1 & -2 \\ 3 & 1 \\ 4 & -2 \end{bmatrix}, B = \begin{bmatrix} 0 & -3 \\ -3 & 2 \\ -1 & 3 \end{bmatrix},$$

and $C = \begin{bmatrix} 5 & 7 & -3 \\ 4 & 2 & 8 \end{bmatrix}$. Find

1 $A + B$ **2** $4A$
3 $2B - 3A$ **4** $A - C$

▶ **Self-Check Answers** ▼

1 $A + B = \begin{bmatrix} 1 & -5 \\ 0 & 3 \\ 3 & 1 \end{bmatrix}$ **2** $4A = \begin{bmatrix} 4 & -8 \\ 12 & 4 \\ 16 & -8 \end{bmatrix}$ **3** $2B - 3A = \begin{bmatrix} -3 & 0 \\ -15 & 1 \\ -14 & 12 \end{bmatrix}$ **4** $A - C$ is undefined, since the orders of A and C are different.

448 CHAPTER 7 MATRICES AND DETERMINANTS

A $1 \times n$ matrix such as $[2 \quad -5 \quad 0 \quad 7]$ is called a **row vector**, and an $m \times 1$ matrix such as $\begin{bmatrix} 5 \\ -3 \\ 6 \end{bmatrix}$ is called a **column vector**. The product of two vectors, defined below, results in a scalar (a constant) and is therefore referred to as the **scalar product of two vectors**. This product is also known as the **dot product** or the **inner product**.

Scalar Product of Vectors ▼

Let $A = [a_1 \quad a_2 \quad \cdots \quad a_m]$ and $B = \begin{bmatrix} b_1 \\ b_2 \\ \vdots \\ b_m \end{bmatrix}$. The scalar product is

$$A \cdot B = a_1 b_1 + a_2 b_2 + \cdots + a_m b_m$$

Note that A is a row vector with m columns and B is a column vector with m rows. The scalar product is undefined if A and B do not have the same number of entries.

EXAMPLE 4 Let $A = [1 \quad 2 \quad 3]$ and $B = \begin{bmatrix} 4 \\ -5 \\ 6 \end{bmatrix}$. Find $A \cdot B$.

SOLUTION $[1 \quad 2 \quad 3] \cdot \begin{bmatrix} 4 \\ -5 \\ 6 \end{bmatrix} = 1(4) + 2(-5) + 3(6) = 12$

Row vector with three columns

Column vector with three rows

The scalar product

□

The product of two matrices A and B is defined in terms of the scalar product of the row vectors of A and the column vectors of B. Thus $A \cdot B$ is defined if and only if the number of columns of A equals the number of rows of B.

Multiplication of Matrices ▼

Let $A = [a_{ij}]_{m \times n}$ and $B = [b_{ij}]_{n \times p}$.

$$A \cdot B = [c_{ij}]_{m \times p}$$

where c_{ij} is the scalar product of the ith row of A and the jth column of B.

SECTION 7-1 MATRIX OPERATIONS

$$[a_{ij}]_{m \times n} \cdot [b_{ij}]_{n \times p} = [c_{ij}]_{m \times p}$$

- The n's must be equal.
- Order of the product is $m \times p$.
- Number of rows in A (from m on left of c_{ij}).
- Number of columns in B (from p on right of c_{ij}).

EXAMPLE 5 Let $A = \begin{bmatrix} -3 & -2 & -1 \\ 1 & 2 & 3 \end{bmatrix}$ and $B = \begin{bmatrix} 4 & 0 \\ -3 & -1 \\ 2 & 2 \end{bmatrix}$. Find $A \cdot B$.

SOLUTION $[a_{ij}]_{2 \times 3} \cdot [b_{ij}]_{3 \times 2} = [c_{ij}]_{2 \times 2}$

Equal

Product of order 2×2

Since the number of columns of A equals the number of rows of B, the product is defined. The answer will be a 2×2 matrix.

Thus

$$A \cdot B = \begin{bmatrix} [\text{row 1 of } A]\begin{bmatrix}\text{col. 1} \\ \text{of } B\end{bmatrix} & [\text{row 1 of } A]\begin{bmatrix}\text{col. 2} \\ \text{of } B\end{bmatrix} \\ [\text{row 2 of } A]\begin{bmatrix}\text{col. 1} \\ \text{of } B\end{bmatrix} & [\text{row 2 of } A]\begin{bmatrix}\text{col. 2} \\ \text{of } B\end{bmatrix} \end{bmatrix}$$

The c_{ij} entry of $A \cdot B$ is the scalar product of the ith row of A times the jth column of B.

$$= \begin{bmatrix} [-3 \; -2 \; -1]\begin{bmatrix}4 \\ -3 \\ 2\end{bmatrix} & [-3 \; -2 \; -1]\begin{bmatrix}0 \\ -1 \\ 2\end{bmatrix} \\ [1 \; 2 \; 3]\begin{bmatrix}4 \\ -3 \\ 2\end{bmatrix} & [1 \; 2 \; 3]\begin{bmatrix}0 \\ -1 \\ 2\end{bmatrix} \end{bmatrix}$$

$$= \begin{bmatrix} -3(4) + (-2)(-3) + (-1)(2) & -3(0) + (-2)(-1) + (-1)(2) \\ 1(4) + (2)(-3) + (3)(2) & 1(0) + (2)(-1) + 3(2) \end{bmatrix}$$

$$A \cdot B = \begin{bmatrix} -8 & 0 \\ 4 & 4 \end{bmatrix}_{2 \times 2}$$

Number of rows in A
Number of columns in B

The TI-81 graphics calculator is capable of performing the basic matrix operations of addition, subtraction, multiplication, and scalar multiplication. The multiplication of matrices A and B from Example 5 is illustrated below.

▶ **Self-Check** ▼

Let $A = [4 \; -3 \; -2 \; 6]$,

$B = [7 \; -5 \; 9]$, and $C = \begin{bmatrix} -2 \\ -7 \\ 8 \\ 5 \end{bmatrix}$.

Find

1 $A \cdot C$ 2 $B \cdot C$

▶ **Self-Check Answers** ▼

1 $A \cdot C = 27$ 2 $B \cdot C$ is undefined.

First enter matrices A and B using the **MATRX** key and the EDIT option. Then use these keystrokes: **2nd** **[A]** **×** **2nd** **[B]** **ENTER** to obtain the screen shown in Figure 7-1 to the right.

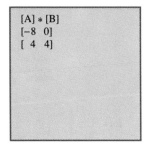

Figure 7-1

Thus $\begin{bmatrix} -3 & -2 & -1 \\ 1 & 2 & 3 \end{bmatrix} \cdot \begin{bmatrix} 4 & 0 \\ -3 & -1 \\ 2 & 2 \end{bmatrix} = \begin{bmatrix} -8 & 0 \\ 4 & 4 \end{bmatrix}.$

EXAMPLE 6 Let $A = \begin{bmatrix} -2 & -3 \\ 4 & 6 \\ -6 & -9 \\ 2 & 3 \end{bmatrix}$ and $B = \begin{bmatrix} 3 & -3 & 6 \\ -2 & 2 & -4 \end{bmatrix}$. Find $A \cdot B$.

SOLUTION $[a_{ij}]_{4 \times 2} \cdot [b_{ij}]_{2 \times 3} = [c_{ij}]_{4 \times 3}$

Equal

Product of order 4×3

The product is defined, and the answer will be a 4×3 matrix.

Thus

$A \cdot B = \begin{bmatrix} -2(3) + (-3)(-2) & -2(-3) + (-3)(2) & -2(6) + (-3)(-4) \\ 4(3) + 6(-2) & 4(-3) + 6(2) & 4(6) + 6(-4) \\ -6(3) + (-9)(-2) & -6(-3) + (-9)(2) & -6(6) + (-9)(-4) \\ 2(3) + 3(-2) & 2(-3) + 3(2) & 2(6) + 3(-4) \end{bmatrix}$

$= \begin{bmatrix} 0 & 0 & 0 \\ 0 & 0 & 0 \\ 0 & 0 & 0 \\ 0 & 0 & 0 \end{bmatrix}_{4 \times 3}$ — Number of rows in A
— Number of columns in B

$A \cdot B = O_{4 \times 3}$ This is the zero matrix of order 4×3.

▶ **Self-Check** ▼

Let $A = \begin{bmatrix} 1 & 2 \\ 3 & 4 \end{bmatrix}$ and $B = \begin{bmatrix} 5 & 6 \\ 7 & 8 \end{bmatrix}$.

1 Find $A \cdot B$.
2 Find $B \cdot A$.
3 Does $A \cdot B = B \cdot A$?

There are some major differences between the multiplication of matrices and the multiplication of real numbers. If a and b are real numbers and $ab = 0$, then $a = 0$ or $b = 0$. However, as Example 6 illustrated, it is possible for matrices A and B to have $A \cdot B = O_{m \times n}$ with neither A nor B equal to the zero matrix. It is also possible for $A \cdot B$ to be unequal to $B \cdot A$.

▶ **Self-Check Answers** ▼

1 $\begin{bmatrix} 19 & 22 \\ 43 & 50 \end{bmatrix}$ **2** $\begin{bmatrix} 23 & 34 \\ 31 & 46 \end{bmatrix}$ **3** No, $A \cdot B \neq B \cdot A$.

SECTION 7-1 MATRIX OPERATIONS

EXAMPLE 7 Use matrix multiplication to determine the gross income matrix G from the total sales matrix T and the pricing matrix P.

$$T = \begin{bmatrix} 675 & 22 & 40 & 55 \\ 785 & 19 & 27 & 41 \\ 380 & 18 & 12 & 38 \end{bmatrix} \begin{matrix} \text{Store A} \\ \text{Store B} \\ \text{Store C} \end{matrix}$$

with columns: Diskettes, Printers, Disk drives, Computers

and

$$P = \begin{bmatrix} 3 \\ 350 \\ 400 \\ 200 \end{bmatrix} \begin{matrix} \text{Diskette price} \\ \text{Printer price} \\ \text{Disk drive price} \\ \text{Computer price} \end{matrix}$$

SOLUTION (Number sold) · (Price per item) = Gross income   *Word equation*

$$T_{3 \times 4} \quad \cdot \quad P_{4 \times 1} \quad = \quad G_{3 \times 1} \qquad \text{Matrix equation}$$

Therefore

$$G = T \cdot P = \begin{bmatrix} 675 & 22 & 40 & 55 \\ 785 & 19 & 27 & 41 \\ 380 & 18 & 12 & 38 \end{bmatrix} \cdot \begin{bmatrix} 3 \\ 350 \\ 400 \\ 200 \end{bmatrix}$$

$$G = \begin{bmatrix} 36{,}725 \\ 28{,}005 \\ 19{,}840 \end{bmatrix} \begin{matrix} \text{Gross income store A} \\ \text{Gross income store B} \\ \text{Gross income store C} \end{matrix}$$

Thus, for example, gross income at store B is $28,005. □

The multiplicative identity for real numbers is 1, since $a \cdot 1 = 1 \cdot a = a$ for all real numbers a. The equalities

$$\begin{bmatrix} a_{11} & a_{12} \\ a_{21} & a_{22} \end{bmatrix} \cdot \begin{bmatrix} 1 & 0 \\ 0 & 1 \end{bmatrix} = \begin{bmatrix} a_{11} & a_{12} \\ a_{21} & a_{22} \end{bmatrix}$$

and

$$\begin{bmatrix} 1 & 0 \\ 0 & 1 \end{bmatrix} \cdot \begin{bmatrix} a_{11} & a_{12} \\ a_{21} & a_{22} \end{bmatrix} = \begin{bmatrix} a_{11} & a_{12} \\ a_{21} & a_{22} \end{bmatrix}$$

demonstrate that $I = \begin{bmatrix} 1 & 0 \\ 0 & 1 \end{bmatrix}$ is the multiplicative identity for 2×2 matrices.

A **square matrix** has the same number of rows as columns. Each of the 2 × 2 matrices above is a square matrix. In general there is an identity matrix I_n for each set of square matrices of order $n \times n$. Note in the following definition that I_n has ones on the **main diagonal** (entries a_{ij} with $i = j$) and zeros for all other entries.

Identity Matrix, I_n ▼

The identity matrix for square matrices $A_{n \times n}$ is the square matrix

$$I_n = \begin{bmatrix} 1 & 0 & 0 & \cdots & 0 \\ 0 & 1 & 0 & \cdots & 0 \\ 0 & 0 & 1 & \cdots & 0 \\ \vdots & \vdots & \vdots & \ddots & \vdots \\ 0 & 0 & 0 & \cdots & 1 \end{bmatrix}$$

$$A_{n \times n} \cdot I_n = I_n \cdot A_{n \times n} = A_{n \times n}$$

The **transpose** of a matrix A, denoted by A^t, is the matrix obtained by interchanging the rows and columns of A. That is, row 1 becomes column 1, row 2 becomes column 2, etc.

EXAMPLE 8 Find the transpose of each matrix.

SOLUTION

(a) $A = \begin{bmatrix} 1 & 2 & -1 \\ 0 & 1 & 3 \end{bmatrix}$ $A^t = \begin{bmatrix} 1 & 0 \\ 2 & 1 \\ -1 & 3 \end{bmatrix}$

(b) $B = \begin{bmatrix} 1 & -1 \\ 2 & -2 \end{bmatrix}$ $B^t = \begin{bmatrix} 1 & 2 \\ -1 & -2 \end{bmatrix}$

(c) $C = \begin{bmatrix} 1 & 4 \\ 2 & 5 \\ 3 & 6 \end{bmatrix}$ $C^t = \begin{bmatrix} 1 & 2 & 3 \\ 4 & 5 & 6 \end{bmatrix}$

(d) $D = \begin{bmatrix} 1 & 2 & 4 \end{bmatrix}$ $D^t = \begin{bmatrix} 1 \\ 2 \\ 4 \end{bmatrix}$

▶ **Self-Check** ▼

Let $A = \begin{bmatrix} 2 & -3 & 4 \\ 5 & 0 & 7 \\ 9 & 8 & -6 \end{bmatrix}$,

$O = \begin{bmatrix} 0 & 0 & 0 \\ 0 & 0 & 0 \\ 0 & 0 & 0 \end{bmatrix}$, and $I = \begin{bmatrix} 1 & 0 & 0 \\ 0 & 1 & 0 \\ 0 & 0 & 1 \end{bmatrix}$.

Find

1 $A + O$ **2** $A \cdot O$
3 $A \cdot I$ **4** $O \cdot I$

▶ **Self-Check Answers** ▼

1 $A + O = A$ **2** $A \cdot O = O$ **3** $A \cdot I = A$ **4** $O \cdot I = O$

SECTION 7-1 MATRIX OPERATIONS

EXERCISES 7-1

A

1. Give the order of $A = \begin{bmatrix} 5 & -3 \\ 4 & 7 \\ -9 & 8 \end{bmatrix}$, and list a_{12}, a_{21}, and a_{22}.

2. Give the order of $B = \begin{bmatrix} -6 & 9 & 4 & -7 \\ 8 & -3 & 0 & 11 \end{bmatrix}$, and list b_{13}, b_{22}, and b_{23}.

3. Write the expanded form of $A = [a_{ij}]_{2 \times 3}$.

4. Write the expanded form of $A = [a_{ij}]_{3 \times 3}$.

5. Write the 2×3 matrix with $a_{22} = 5$, $a_{11} = 7$, $a_{21} = 10$, $a_{12} = 4$, $a_{23} = 8$, and $a_{13} = 0$.

6. Write the 4×2 matrix with $a_{21} = -1$, $a_{22} = 3$, $a_{11} = 4$, $a_{32} = 5$, $a_{12} = 8$, $a_{41} = -9$, $a_{42} = 11$, and $a_{31} = 2$.

In Exercises 7–20, perform the operations indicated, using the matrices

$$A = \begin{bmatrix} 1 & -1 & 2 \\ 2 & -3 & 5 \end{bmatrix}, \quad B = \begin{bmatrix} -1 & 0 & 5 \\ 2 & 3 & 4 \end{bmatrix}, \quad C = \begin{bmatrix} 1 & -1 \\ 2 & 3 \\ 4 & 6 \end{bmatrix}.$$

7. $A + B$
8. $A - B$
9. $-2A$
10. $6B$
11. $3A - 2B$
12. $-4A + 3B$
13. $a_{23}B$
14. $b_{22}A$
15. A^t
16. $B + C$
17. $A - C^t$
18. $A^t + 2C$
19. $A^t - B$
20. $(2A - 5B)^t$

In Exercises 21–24, find x and y such that $A = B$.

21. $A = \begin{bmatrix} 2 & x \\ 3 & y \end{bmatrix}, B = \begin{bmatrix} 1+1 & \frac{8}{4} \\ 3+0 & 5+6 \end{bmatrix}$

22. $A = \begin{bmatrix} 2 & 8 \\ 4 & -5 \end{bmatrix}, B = \begin{bmatrix} 2x+1 & 8 \\ 4y-1 & -5 \end{bmatrix}$

23. $A = \begin{bmatrix} x & 1 \\ x+y & 2 \end{bmatrix}, B = \begin{bmatrix} y-2 & 1 \\ -6 & 2 \end{bmatrix}$

24. $A = \begin{bmatrix} 0 & 2x-3y \\ 3 & 6 \end{bmatrix}, B = \begin{bmatrix} 0 & 1 \\ x-3y & 6 \end{bmatrix}$

In Exercises 25–30, determine m and n such that the product $A \cdot B = C$ will be defined.

25. $A_{1 \times 4} \cdot B_{4 \times 3} = C_{m \times n}$
26. $A_{m \times n} \cdot B_{3 \times 7} = C_{6 \times 7}$
27. $A_{2 \times 3} \cdot B_{m \times n} = C_{2 \times 5}$
28. $A_{1 \times 6} \cdot B_{m \times n} = C_{1 \times 4}$
29. $A_{2 \times m} \cdot B_{n \times 3} = C_{2 \times 3}$
30. $A_{m \times 4} \cdot B_{4 \times n} = C_{3 \times 1}$

In Exercises 31–46, perform the operations indicated, using the matrices

$$A = \begin{bmatrix} 1 & -1 & 2 \\ 3 & -3 & 5 \end{bmatrix}, \quad B = \begin{bmatrix} -1 & 0 & 5 \\ 2 & 3 & 4 \end{bmatrix}, \quad C = \begin{bmatrix} 1 & -1 \\ 2 & 3 \\ 4 & 6 \end{bmatrix}, \quad D = [4 \; 0 \; 2].$$

31. $A \cdot C$
32. $C \cdot A$
33. $C \cdot (A + B)$
34. $(C \cdot A) + (C \cdot B)$
35. A^2
36. $(A \cdot C)^2$
37. $(B \cdot C)^2$
38. $D \cdot D^t$
39. $A \cdot A^t$
40. $C^t - 2B$
41. $(C^t)^t - 3A^t$
42. $4A + O$
43. $(A \cdot C) \cdot I$
44. $(C \cdot A) \cdot I$
45. $(A \cdot C) \cdot B$
46. $A \cdot (C \cdot B)$

B

47 Use addition of matrices to find the total at bats, runs scored, hits, and runs batted in for these eight New York Yankees in three games during July 1936.

$$
\begin{array}{c}
\begin{array}{c} \text{Powell} \\ \text{Rolfe} \\ \text{DiMaggio} \\ \text{Gehrig} \\ \text{Dickey} \\ \text{Selkirk} \\ \text{Crosetti} \\ \text{Lazzeri} \end{array}
\end{array}
\begin{array}{c} \text{Game 1} \\ \begin{array}{cccc} \text{AB} & \text{R} & \text{H} & \text{RBI} \end{array} \\ \begin{bmatrix} 3 & 0 & 1 & 0 \\ 4 & 1 & 1 & 1 \\ 4 & 0 & 1 & 1 \\ 3 & 1 & 2 & 0 \\ 0 & 0 & 0 & 0 \\ 2 & 0 & 0 & 0 \\ 4 & 0 & 0 & 0 \\ 0 & 0 & 0 & 0 \end{bmatrix} \end{array}
\begin{array}{c} \text{Game 2} \\ \begin{array}{cccc} \text{AB} & \text{R} & \text{H} & \text{RBI} \end{array} \\ \begin{bmatrix} 4 & 0 & 0 & 0 \\ 3 & 1 & 1 & 0 \\ 4 & 1 & 1 & 0 \\ 3 & 1 & 0 & 0 \\ 4 & 0 & 2 & 2 \\ 4 & 1 & 1 & 2 \\ 3 & 0 & 0 & 0 \\ 4 & 0 & 1 & 0 \end{bmatrix} \end{array}
\begin{array}{c} \text{Game 3} \\ \begin{array}{cccc} \text{AB} & \text{R} & \text{H} & \text{RBI} \end{array} \\ \begin{bmatrix} 4 & 0 & 0 & 0 \\ 5 & 1 & 3 & 0 \\ 4 & 2 & 2 & 2 \\ 2 & 0 & 0 & 0 \\ 4 & 1 & 0 & 0 \\ 4 & 1 & 4 & 3 \\ 5 & 0 & 0 & 0 \\ 3 & 0 & 0 & 0 \end{bmatrix} \end{array}
$$

48 Use addition of matrices to find the total number of points earned on the three tests by each student.

$$
\begin{array}{c} \\ \text{Alexander} \\ \text{Anderson} \\ \text{Bailey} \\ \text{Barrett} \end{array}
\begin{array}{c} \text{Test 1} \\ \begin{bmatrix} 92 \\ 76 \\ 57 \\ 83 \end{bmatrix} \end{array}
\begin{array}{c} \text{Test 2} \\ \begin{bmatrix} 89 \\ 82 \\ 51 \\ 73 \end{bmatrix} \end{array}
\begin{array}{c} \text{Test 3} \\ \begin{bmatrix} 95 \\ 72 \\ 68 \\ 85 \end{bmatrix} \end{array}
$$

49 A concessionaire sold hot dogs, candy bars, and soft drinks at three park district functions this month. The entries in matrix A denote the number of each item sold at each function, and the entries in matrix B represent the wholesale cost and the retail price of each item.

$$
A = \begin{array}{c} \begin{array}{ccc} \text{Hot dogs} & \text{Candy bars} & \text{Soft drinks} \end{array} \\ \begin{bmatrix} 211 & 325 & 473 \\ 121 & 252 & 360 \\ 261 & 352 & 521 \end{bmatrix} \end{array} \begin{array}{c} \text{Function 1} \\ \text{Function 2} \\ \text{Function 3} \end{array}
$$

$$
B = \begin{array}{c} \begin{array}{cc} \text{Wholesale} & \text{Retail} \\ \text{cost} & \text{price} \end{array} \\ \begin{bmatrix} 0.50 & 1.00 \\ 0.25 & 0.40 \\ 0.25 & 0.50 \end{bmatrix} \end{array} \begin{array}{c} \text{Hot dogs} \\ \text{Candy bars} \\ \text{Soft drinks} \end{array}
$$

Use matrix multiplication to determine the total cost and total revenue of each function.

SECTION 7-1 MATRIX OPERATIONS

50 Family A and family B are planning to buy the denoted amounts of the following staples:

	Family A	Family B
Gallons of Milk	1	4
Loaves of Bread	2	2
5-lb bags of Flour	1	4
5-lb bags of Sugar	2	2
Pounds of Hamburger	3	4

Although each family can go to only one store, each family has a choice of shopping at either store C or store D. The prices at each store are given below.

	Gallon of Milk	Loaf of Bread	5-lb Bag of Flour	5-lb Bag of Sugar	Pound of Hamburger
Store C	$1.95	$1.10	$1.15	$1.29	$1.48
Store D	$2.00	$1.07	$1.18	$1.25	$1.45

Denote the quantities purchased as a matrix and the prices as a matrix. Use matrix multiplication to determine the store at which each family should shop.

51 Matrix A gives the number of items in stock Monday morning, and matrix B gives the number of items sold on Monday. Calculate and interpret $A - B$.

$$A = \begin{bmatrix} 75 & 112 & 65 \\ 75 & 46 & 30 \\ 113 & 203 & 78 \end{bmatrix} \begin{matrix} \text{Item 1} \\ \text{Item 2} \\ \text{Item 3} \end{matrix}$$

with columns Small, Medium, Large.

$$B = \begin{bmatrix} 12 & 18 & 37 \\ 8 & 30 & 30 \\ 60 & 70 & 18 \end{bmatrix} \begin{matrix} \text{Item 1} \\ \text{Item 2} \\ \text{Item 3} \end{matrix}$$

52 Given matrices A and B representing the votes cast in precincts A and B, calculate and interpret $A + B$.

$$A = \begin{bmatrix} 12{,}483 & 18{,}376 & 3800 \\ 17{,}009 & 13{,}400 & 2051 \\ 14{,}507 & 16{,}688 & 1593 \end{bmatrix} \begin{matrix} \text{President} \\ \text{Governor} \\ \text{Secretary of state} \end{matrix}$$

with columns Democratic candidate, Republican candidate, Independent candidate.

$$B = \begin{bmatrix} 6317 & 9517 & 1957 \\ 8917 & 7218 & 1457 \\ 8436 & 7821 & 831 \end{bmatrix} \begin{matrix} \text{President} \\ \text{Governor} \\ \text{Secretary of state} \end{matrix}$$

53 Adjust the prices in this pricing matrix for a sale that discounts all items 20%. Show this process in a matrix equation.

$$P = \begin{bmatrix} 87.95 & 94.50 \\ 76.99 & 82.95 \\ 31.52 & 35.89 \\ 112.39 & 125.99 \end{bmatrix} \begin{matrix} \text{Item a} \\ \text{Item b} \\ \text{Item c} \\ \text{Item d} \end{matrix}$$

with columns labeled Regular and Deluxe.

54 Construct a 3×4 matrix A where each $a_{ij} = i + j$.

55 Construct a 4×3 matrix A where each $a_{ij} = 2i + 3j$.

56 Construct a 4×4 matrix A where $a_{ij} = 0$ if $i \neq j$ and $a_{ij} = 1$ if $i = j$.

57 If $A = \begin{bmatrix} 2 & 3 & -1 \\ 0 & 1 & 4 \end{bmatrix}$ and $B = \begin{bmatrix} -2 & 5 & 6 \\ 3 & -1 & -5 \end{bmatrix}$, find a matrix X such that $X - 2A = 3B$.

58 If $A = \begin{bmatrix} 1 \\ 1 \\ 4 \\ -3 \end{bmatrix}$ and $B = \begin{bmatrix} -2 \\ 5 \\ 6 \\ -4 \end{bmatrix}$, find a matrix X such that $X + 5A = -3B$.

C

In Exercises 59–62, use a calculator or a computer to perform the operations indicated, given that

$$A = \begin{bmatrix} 1.089 & -3.678 & 5.872 \\ 4.664 & 7.525 & 1.008 \\ 4.187 & -2.297 & -7.146 \end{bmatrix} \quad \text{and} \quad B = \begin{bmatrix} 8.463 & 5.498 & -0.289 \\ 7.025 & 3.061 & -1.095 \\ 6.962 & -1.474 & 1.113 \end{bmatrix}$$

59 $A + B$ **60** $A - B$ **61** $A \cdot B$ **62** $B \cdot A$

SECTION 7-2

The Inverse of a Square Matrix

Section Objectives

2 Find the inverse of a square matrix.

3 Solve a system of linear equations using the inverse of a matrix.

The multiplicative inverse of a nonzero real number a is a^{-1} if $a \cdot a^{-1} = a^{-1} \cdot a = 1$. Similarly, A^{-1} is the multiplicative inverse of the square matrix A if $A \cdot A^{-1} = A^{-1} \cdot A = I$. Although every nonzero real number has a multiplicative inverse, this is not true of matrices. Some nonzero square matrices have inverses, but others do not.

The inverse of a matrix is used in many mathematical processes. In this section we will utilize the inverse of a matrix to solve systems of equations.

SECTION 7-2 THE INVERSE OF A SQUARE MATRIX

Inverse of a Matrix ▼

A^{-1} is the inverse of the square matrix A if
$$A \cdot A^{-1} = A^{-1} \cdot A = I$$

EXAMPLE 1 Show that $\begin{bmatrix} 2 & -3 \\ -3 & 5 \end{bmatrix}$ is the inverse of $A = \begin{bmatrix} 5 & 3 \\ 3 & 2 \end{bmatrix}$.

SOLUTION
$$\begin{bmatrix} 5 & 3 \\ 3 & 2 \end{bmatrix} \cdot \begin{bmatrix} 2 & -3 \\ -3 & 5 \end{bmatrix} = \begin{bmatrix} 1 & 0 \\ 0 & 1 \end{bmatrix}$$

$$\begin{bmatrix} 2 & -3 \\ -3 & 5 \end{bmatrix} \cdot \begin{bmatrix} 5 & 3 \\ 3 & 2 \end{bmatrix} = \begin{bmatrix} 1 & 0 \\ 0 & 1 \end{bmatrix}$$

Thus $A^{-1} = \begin{bmatrix} 2 & -3 \\ -3 & 5 \end{bmatrix}$ is the inverse of A, since $A \cdot A^{-1} = I$ and $A^{-1} \cdot A = I$. □

EXAMPLE 2 Find the inverse of the matrix $A = \begin{bmatrix} 2 & 3 \\ 2 & 4 \end{bmatrix}$.

SOLUTION We can use the fact that $A \cdot A^{-1} = I$ to find A^{-1}.

Let
$$A^{-1} = \begin{bmatrix} x_1 & y_1 \\ x_2 & y_2 \end{bmatrix}$$
  Solve for $x_1, x_2, y_1,$ and y_2.

$$\begin{bmatrix} 2 & 3 \\ 2 & 4 \end{bmatrix} \cdot \begin{bmatrix} x_1 & y_1 \\ x_2 & y_2 \end{bmatrix} = \begin{bmatrix} 1 & 0 \\ 0 & 1 \end{bmatrix}$$
  $A \cdot A^{-1} = I$

$$\begin{bmatrix} 2x_1 + 3x_2 & 2y_1 + 3y_2 \\ 2x_1 + 4x_2 & 2y_1 + 4y_2 \end{bmatrix} = \begin{bmatrix} 1 & 0 \\ 0 & 1 \end{bmatrix}$$
  Multiply $A \cdot A^{-1}$.

$2x_1 + 3x_2 = 1 \quad 2y_1 + 3y_2 = 0$
$2x_1 + 4x_2 = 0 \quad 2y_1 + 4y_2 = 1$
  The corresponding entries of equivalent matrices are equal.

$$\begin{bmatrix} 2 & 3 & | & 1 & 0 \\ 2 & 4 & | & 0 & 1 \end{bmatrix} \xrightarrow{r'_1 = \frac{1}{2}r_1} \begin{bmatrix} 1 & \frac{3}{2} & | & \frac{1}{2} & 0 \\ 2 & 4 & | & 0 & 1 \end{bmatrix}$$

Form the augmented matrix $[A|I]$ for these two systems of equations with identical coefficients, following the procedure illustrated in Example 8 of Section 6-2. Note that the right side of the augmented matrix is the identity matrix. Then use elementary row operations to transform the augmented matrix into reduced form.

$$\xrightarrow{r'_2 = r_2 - 2r_1} \begin{bmatrix} 1 & \frac{3}{2} & | & \frac{1}{2} & 0 \\ 0 & 1 & | & -1 & 1 \end{bmatrix}$$

$$\xrightarrow{r'_1 = r_1 - \frac{3}{2}r_2} \begin{bmatrix} 1 & 0 & | & 2 & -\frac{3}{2} \\ 0 & 1 & | & -1 & 1 \end{bmatrix}$$

Thus $x_1 = 2, y_1 = -\frac{3}{2}; x_2 = -1, y_2 = 1$. Note that the matrix is now in the form $[I|A^{-1}]$.

Check $\begin{bmatrix} 2 & 3 \\ 2 & 4 \end{bmatrix} \cdot \begin{bmatrix} 2 & -\frac{3}{2} \\ -1 & 1 \end{bmatrix} = \begin{bmatrix} 1 & 0 \\ 0 & 1 \end{bmatrix}$

Answer $A^{-1} = \begin{bmatrix} 2 & -\frac{3}{2} \\ -1 & 1 \end{bmatrix}$ □

458 CHAPTER 7 MATRICES AND DETERMINANTS

The process in Example 2 can be applied to find the inverse, if it exists, of any square matrix.

Inverse of a Square Matrix A ▼

Step 1 Form the augmented matrix $[A \mid I]$.
Step 2 Convert $[A \mid I]$ to reduced form with elementary row reductions.
Step 3 **a.** The form $[I \mid A^{-1}]$ will be produced if A^{-1} exists.
b. A row of zeros will be produced on the left side if A^{-1} does not exist.

The elementary row operations that transform A into I will also transform I into A^{-1}. Thus to calculate A^{-1} from A, we start with A and I side by side and end with I and A^{-1} side by side.

$$[A \mid I] \quad \text{Starting format}$$
$$\downarrow \downarrow \quad \text{Several elementary row operations}$$
$$[I \mid A^{-1}] \quad \text{Ending format if } A^{-1} \text{ exists}$$

EXAMPLE 3 Find the inverse of $B = \begin{bmatrix} 1 & 2 & -1 \\ 2 & 5 & 0 \\ -4 & -8 & 8 \end{bmatrix}$.

SOLUTION

$\begin{bmatrix} 1 & 2 & -1 & | & 1 & 0 & 0 \\ 2 & 5 & 0 & | & 0 & 1 & 0 \\ -4 & -8 & 8 & | & 0 & 0 & 1 \end{bmatrix} \xrightarrow{\substack{r_2' = r_2 - 2r_1 \\ r_3' = r_3 + 4r_1}} \begin{bmatrix} 1 & 2 & -1 & | & 1 & 0 & 0 \\ 0 & 1 & 2 & | & -2 & 1 & 0 \\ 0 & 0 & 4 & | & 4 & 0 & 1 \end{bmatrix}$

Form the augmented matrix $[A|I]$, then use elementary row operations to transform the augmented matrix into reduced form.

$\xrightarrow{r_1' = r_1 - 2r_2} \begin{bmatrix} 1 & 0 & -5 & | & 5 & -2 & 0 \\ 0 & 1 & 2 & | & -2 & 1 & 0 \\ 0 & 0 & 4 & | & 4 & 0 & 1 \end{bmatrix}$

$\xrightarrow{r_3' = \frac{1}{4}r_3} \begin{bmatrix} 1 & 0 & -5 & | & 5 & -2 & 0 \\ 0 & 1 & 2 & | & -2 & 1 & 0 \\ 0 & 0 & 1 & | & 1 & 0 & \frac{1}{4} \end{bmatrix}$

$\xrightarrow{\substack{r_1' = r_1 + 5r_3 \\ r_2' = r_2 - 2r_3}} \begin{bmatrix} 1 & 0 & 0 & | & 10 & -2 & \frac{5}{4} \\ 0 & 1 & 0 & | & -4 & 1 & -\frac{1}{2} \\ 0 & 0 & 1 & | & 1 & 0 & \frac{1}{4} \end{bmatrix}$

Check $B \cdot B^{-1} = \begin{bmatrix} 1 & 2 & -1 \\ 2 & 5 & 0 \\ -4 & -8 & 8 \end{bmatrix} \cdot \begin{bmatrix} 10 & -2 & \frac{5}{4} \\ -4 & 1 & -\frac{1}{2} \\ 1 & 0 & \frac{1}{4} \end{bmatrix} = \begin{bmatrix} 1 & 0 & 0 \\ 0 & 1 & 0 \\ 0 & 0 & 1 \end{bmatrix}$

Answer $B^{-1} = \begin{bmatrix} 10 & -2 & \frac{5}{4} \\ -4 & 1 & -\frac{1}{2} \\ 1 & 0 & \frac{1}{4} \end{bmatrix}$ ☐

SECTION 7-2 THE INVERSE OF A SQUARE MATRIX

The TI-81 graphics calculator is capable of calculating the inverse of a square matrix. The keystrokes for calculating the inverse of matrix B in Example 3 are illustrated below.

First enter the matrix as matrix B using the **MATRX** key and the EDIT option. Then use these keystrokes: **2nd** **[B]** **x^{-1}** **ENTER** to obtain the screen shown in Figure 7-2.

Thus
$$\begin{bmatrix} 1 & 2 & -1 \\ 2 & 5 & 0 \\ -4 & -8 & 8 \end{bmatrix}^{-1} = \begin{bmatrix} 10 & -2 & \frac{5}{4} \\ -4 & 1 & -\frac{1}{2} \\ 1 & 0 & \frac{1}{4} \end{bmatrix}$$

```
[B] ⁻¹
[ 10  -2   1.25]
[ -4   1   -.5 ]
[  1   0    .25]
```

Figure 7-2

If we attempt to find an inverse of a square matrix that does not have an inverse, a row of zeros will appear on the left side of the augmented matrix, as shown in the diagram below.

$$[A \mid I] \qquad \text{First step}$$

$$\begin{bmatrix} * & * & \cdots & * & * & * & \cdots & * \\ * & * & \cdots & * & * & * & \cdots & * \\ \vdots & \vdots & & \vdots & \vdots & \vdots & & \vdots \\ 0 & 0 & \cdots & 0 & * & * & \cdots & * \end{bmatrix}$$

Several elementary row operations

Last step (a row of zeros)

EXAMPLE 4 Find the inverse of $A = \begin{bmatrix} 2 & 4 \\ 3 & 6 \end{bmatrix}$.

SOLUTION

$$\begin{bmatrix} 2 & 4 & | & 1 & 0 \\ 3 & 6 & | & 0 & 1 \end{bmatrix} \xrightarrow{r_1' = \frac{1}{2}r_1} \begin{bmatrix} 1 & 2 & | & \frac{1}{2} & 0 \\ 3 & 6 & | & 0 & 1 \end{bmatrix} \xrightarrow{r_2' = r_2 - 3r_1} \begin{bmatrix} 1 & 2 & | & \frac{1}{2} & 0 \\ 0 & 0 & | & -\frac{3}{2} & 1 \end{bmatrix}$$

Answer The two zeros on the left side of the second row indicate that A^{-1} does not exist. □

The inverse of a matrix may be used to solve a system of linear equations. Three matrices are associated with this system of equations:

$$\begin{cases} a_{11}x_1 + a_{12}x_2 + a_{13}x_3 = b_1 \\ a_{21}x_1 + a_{22}x_2 + a_{23}x_3 = b_2 \\ a_{31}x_1 + a_{32}x_2 + a_{33}x_3 = b_3 \end{cases}$$

These matrices are the **coefficient matrix** A, the **constant matrix** B, and the **variable matrix** X.

$$A = \begin{bmatrix} a_{11} & a_{12} & a_{13} \\ a_{21} & a_{22} & a_{23} \\ a_{31} & a_{32} & a_{33} \end{bmatrix}, \qquad B = \begin{bmatrix} b_1 \\ b_2 \\ b_3 \end{bmatrix}, \qquad X = \begin{bmatrix} x_1 \\ x_2 \\ x_3 \end{bmatrix}$$

▶ **Self-Check** ▼

Find the inverse of $A = \begin{bmatrix} 3 & 1 \\ 3 & 2 \end{bmatrix}$.

▶ **Self-Check Answer** ▼

$$\begin{bmatrix} \frac{2}{3} & -\frac{1}{3} \\ -1 & 1 \end{bmatrix}$$

The entries of the coefficient matrix are the coefficients of the variables on the left side of the equations. The entries of the constant matrix are the constants on the right side of the equations. The variable matrix is a column vector containing the variables in the equation. The system of equations can then be represented by the matrix equation $A \cdot X = B$.

$$A \cdot X = \begin{bmatrix} a_{11}x_1 + a_{12}x_2 + a_{13}x_3 \\ a_{21}x_1 + a_{22}x_2 + a_{23}x_3 \\ a_{31}x_1 + a_{32}x_2 + a_{33}x_3 \end{bmatrix} = \begin{bmatrix} b_1 \\ b_2 \\ b_3 \end{bmatrix} = B$$

The solution of a system of linear equations is found by solving the matrix equation $A \cdot X = B$ for the variable matrix X, if A^{-1} exists.

$A \cdot X = B$	The matrix form of a system of equations
$A^{-1} \cdot A \cdot X = A^{-1} \cdot B$	Multiply both sides by A^{-1}.
$I \cdot X = A^{-1} \cdot B$	Replace $A \cdot A^{-1}$ by I.
$X = A^{-1} \cdot B$	$I \cdot X = X$

Since $X = A^{-1} \cdot B$, we may solve a system of linear equations by multiplying the constant matrix B on the left by A^{-1}.

▶ **Self-Check** ▼

1 Write the matrix form of the system of linear equations.

$$3x + 2y - 3z = 5$$
$$x + y = 6$$
$$-2x - 2y + 3z = 1$$

2 Write the system of equations represented by the matrix equation $A \cdot X = B$, where

$$A = \begin{bmatrix} 1 & -1 \\ 2 & 5 \end{bmatrix}, X = \begin{bmatrix} x \\ y \end{bmatrix},$$

$$B = \begin{bmatrix} -1 \\ 2 \end{bmatrix}$$

EXAMPLE 5 Solve the system $\begin{cases} 2x + 3y = 6 \\ 2x + 4y = 5 \end{cases}$ using $X = A^{-1} \cdot B$.

SOLUTION $A \cdot X = B$ The matrix equation with $A = \begin{bmatrix} 2 & 3 \\ 2 & 4 \end{bmatrix}$,

$X = A^{-1} \cdot B$ $X = \begin{bmatrix} x \\ y \end{bmatrix}$, and $B = \begin{bmatrix} 6 \\ 5 \end{bmatrix}$

$\begin{bmatrix} x \\ y \end{bmatrix} = \begin{bmatrix} 2 & -\frac{3}{2} \\ -1 & 1 \end{bmatrix} \begin{bmatrix} 6 \\ 5 \end{bmatrix}$ $A^{-1} = \begin{bmatrix} 2 & -\frac{3}{2} \\ -1 & 1 \end{bmatrix}$ from Example 2 of this section

$\begin{bmatrix} x \\ y \end{bmatrix} = \begin{bmatrix} \frac{9}{2} \\ -1 \end{bmatrix}$ Thus $x = \frac{9}{2}$ and $y = -1$.

Answer $(\frac{9}{2}, -1)$ □

The inverse matrix method for solving systems of linear equations is particularly appropriate for calculators with matrix capabilities. The keystrokes for the solution of the system given in Example 5 are given on page 72.

▶ **Self-Check Answers** ▼

1 $\begin{bmatrix} 3 & 2 & -3 \\ 1 & 1 & 0 \\ -2 & -2 & 3 \end{bmatrix} \begin{bmatrix} x \\ y \\ z \end{bmatrix} = \begin{bmatrix} 5 \\ 6 \\ 1 \end{bmatrix}$ **2** $\begin{cases} x - y = -1 \\ 2x + 5y = 2 \end{cases}$

SECTION 7-2 THE INVERSE OF A SQUARE MATRIX

This method is also appropriate when several systems of linear equations all have the same coefficients, as in Example 6, which contains three systems of equations with the same coefficient matrix.

First enter matrices A and B using the **MATRX** key and the EDIT option. Then use these keystrokes: **2nd** **[A]** **x^{-1}** **×** **2nd** **[B]** **ENTER** to obtain the screen shown in Figure 7-3.

Thus $\quad\quad\quad x = \dfrac{9}{2}\quad$ and $\quad y = -1.$

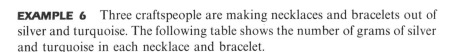

[A]$^{-1}$ * [B]
[4.5]
[−1]

Figure 7-3

EXAMPLE 6 Three craftspeople are making necklaces and bracelets out of silver and turquoise. The following table shows the number of grams of silver and turquoise in each necklace and bracelet.

	Necklace	Bracelet
Silver	63	38
Turquoise	21	19

The next table displays the quantity of silver and turquoise each craftsperson has available.

	First Craftsperson	Second Craftsperson	Third Craftsperson
Silver	631	618	518
Turquoise	242	225	217

Find the number of necklaces and bracelets each craftsperson should make to use all of the available material.

SOLUTION Let $n =$ Number of necklaces to be made
$\quad\quad\quad\quad\quad b =$ Number of bracelets to be made

The three systems of equations are

First Craftsperson   **Second Craftsperson**   **Third Craftsperson**
$63n + 38b = 631$ $\quad 63n + 38b = 618$ $\quad 63n + 38b = 518$

$21n + 19b = 242$ $\quad 21n + 19b = 225$ $\quad 21n + 19b = 217$

Word equations:

$\left(\begin{array}{c}\text{Silver in}\\\text{necklace}\end{array}\right) + \left(\begin{array}{c}\text{Silver in}\\\text{bracelet}\end{array}\right) = \text{Total silver}$

$\left(\begin{array}{c}\text{Turquoise in}\\\text{necklace}\end{array}\right) + \left(\begin{array}{c}\text{Turquoise in}\\\text{bracelet}\end{array}\right) = \left(\begin{array}{c}\text{Total}\\\text{turquoise}\end{array}\right)$

The solution of each system can be represented by the matrix equation

$$\begin{bmatrix} n \\ b \end{bmatrix} = \begin{bmatrix} 63 & 38 \\ 21 & 19 \end{bmatrix}^{-1} \cdot \begin{bmatrix} c_1 \\ c_2 \end{bmatrix}$$

where c_1 and c_2 are the constants.

Next the inverse of the coefficient matrix is computed.

$$\begin{bmatrix} 63 & 38 & | & 1 & 0 \\ 21 & 19 & | & 0 & 1 \end{bmatrix} \xrightarrow{r_1' = \frac{1}{63}r_1} \begin{bmatrix} 1 & \frac{38}{63} & | & \frac{1}{63} & 0 \\ 21 & 19 & | & 0 & 1 \end{bmatrix} \xrightarrow{r_2' = r_2 - 21r_1} \begin{bmatrix} 1 & \frac{38}{63} & | & \frac{1}{63} & 0 \\ 0 & \frac{19}{3} & | & -\frac{1}{3} & 1 \end{bmatrix}$$

$$\xrightarrow{r_2' = \frac{3}{19}r_2} \begin{bmatrix} 1 & \frac{38}{63} & | & \frac{1}{63} & 0 \\ 0 & 1 & | & -\frac{1}{19} & \frac{3}{19} \end{bmatrix}$$

$$\xrightarrow{r_1' = r_1 - \frac{38}{63}r_2} \begin{bmatrix} 1 & 0 & | & \frac{1}{21} & -\frac{2}{21} \\ 0 & 1 & | & -\frac{1}{19} & \frac{3}{19} \end{bmatrix}$$

Thus
$$\begin{bmatrix} 63 & 38 \\ 21 & 19 \end{bmatrix}^{-1} = \begin{bmatrix} \frac{1}{21} & -\frac{2}{21} \\ -\frac{1}{19} & \frac{3}{19} \end{bmatrix}$$

Therefore the solution of each system is

First Craftsperson

$$\begin{bmatrix} n \\ b \end{bmatrix} = \begin{bmatrix} \frac{1}{21} & -\frac{2}{21} \\ -\frac{1}{19} & \frac{3}{19} \end{bmatrix} \cdot \begin{bmatrix} 631 \\ 242 \end{bmatrix} = \begin{bmatrix} 7 \\ 5 \end{bmatrix}$$

Second Craftsperson

$$\begin{bmatrix} n \\ b \end{bmatrix} = \begin{bmatrix} \frac{1}{21} & -\frac{2}{21} \\ -\frac{1}{19} & \frac{3}{19} \end{bmatrix} \cdot \begin{bmatrix} 618 \\ 225 \end{bmatrix} = \begin{bmatrix} 8 \\ 3 \end{bmatrix}$$

Third Craftsperson

$$\begin{bmatrix} n \\ b \end{bmatrix} = \begin{bmatrix} \frac{1}{21} & -\frac{2}{21} \\ -\frac{1}{19} & \frac{3}{19} \end{bmatrix} \cdot \begin{bmatrix} 518 \\ 217 \end{bmatrix} = \begin{bmatrix} 4 \\ 7 \end{bmatrix}$$

Answer The first craftsperson should make seven necklaces and five bracelets, the second craftsperson should make eight necklaces and three bracelets, and the third craftsperson should make four necklaces and seven bracelets. Note: All three solutions can be obtained from the matrix equation

$$\begin{bmatrix} n_1 & n_2 & n_3 \\ b_1 & b_2 & b_3 \end{bmatrix} = \begin{bmatrix} \frac{1}{21} & -\frac{2}{21} \\ -\frac{1}{19} & \frac{3}{19} \end{bmatrix} \cdot \begin{bmatrix} 631 & 618 & 518 \\ 242 & 225 & 217 \end{bmatrix} = \begin{bmatrix} 7 & 8 & 4 \\ 5 & 3 & 7 \end{bmatrix} \quad \square$$

EXERCISES 7-2

A

In Exercises 1 and 2, write each system of linear equations in matrix form.

1 $2x + z = 5$
$ y - z = 1$
$x + y = 2$

2 $x + 2y + z = 1$
$2x - y = 2$
$w + y = 3$
$w + x = -7$

In Exercises 3 and 4, write the system of linear equations associated with $A \cdot X = B$ for the given matrices A, X, and B.

3 $A = \begin{bmatrix} 1 & 0 & -1 \\ 3 & 4 & -3 \\ 0 & 1 & 2 \end{bmatrix}$, $X = \begin{bmatrix} x \\ y \\ z \end{bmatrix}$, $B = \begin{bmatrix} -2 \\ 6 \\ -6 \end{bmatrix}$

4 $A = \begin{bmatrix} 1 & 0 & 0 & 1 \\ -1 & 1 & -1 & 0 \\ 2 & 0 & 5 & 0 \\ -2 & 1 & 0 & 0 \end{bmatrix}$, $X = \begin{bmatrix} x_1 \\ x_2 \\ x_3 \\ x_4 \end{bmatrix}$, $B = \begin{bmatrix} 1 \\ 8 \\ 0 \\ -6 \end{bmatrix}$

SECTION 7-2 THE INVERSE OF A SQUARE MATRIX

In Exercises 5–8, determine if $B = A^{-1}$ by calculating $A \cdot B$ and $B \cdot A$.

5 $A = \begin{bmatrix} 6 & 5 \\ 3 & 3 \end{bmatrix}$, $B = \begin{bmatrix} 1 & -\frac{5}{3} \\ -1 & 2 \end{bmatrix}$

6 $A = \begin{bmatrix} 8 & 10 \\ 2 & 3 \end{bmatrix}$, $B = \begin{bmatrix} \frac{3}{4} & -\frac{5}{3} \\ -\frac{1}{2} & 2 \end{bmatrix}$

7 $A = \begin{bmatrix} 2 & 1 & -1 \\ 0 & 1 & -1 \\ 4 & 2 & 0 \end{bmatrix}$, $B = \begin{bmatrix} \frac{1}{2} & -\frac{1}{2} & 0 \\ -1 & 1 & \frac{1}{2} \\ -2 & 0 & 1 \end{bmatrix}$

8 $A = \begin{bmatrix} -5 & 4 & -1 \\ 4 & -3 & 1 \\ -14 & 11 & -4 \end{bmatrix}$, $B = \begin{bmatrix} 1 & 5 & 1 \\ 2 & 6 & 1 \\ 2 & -1 & -1 \end{bmatrix}$

In Exercises 9–20, determine the inverse matrix, if it exists.

9 $\begin{bmatrix} 1 & 2 \\ 3 & 7 \end{bmatrix}$ **10** $\begin{bmatrix} 5 & 7 \\ 2 & 3 \end{bmatrix}$ **11** $\begin{bmatrix} 3 & 4 \\ 2 & 2 \end{bmatrix}$ **12** $\begin{bmatrix} 6 & 3 \\ 2 & 1 \end{bmatrix}$

13 $\begin{bmatrix} 4 & 6 \\ 2 & 3 \end{bmatrix}$ **14** $\begin{bmatrix} 4 & 4 \\ 4 & 3 \end{bmatrix}$ **15** $\begin{bmatrix} 1 & 1 & 1 \\ 1 & 2 & 3 \\ 1 & 3 & 6 \end{bmatrix}$ **16** $\begin{bmatrix} 1 & -2 & 2 \\ 0 & 1 & 1 \\ -1 & 2 & -1 \end{bmatrix}$

17 $\begin{bmatrix} 4 & 5 & 3 \\ 4 & 6 & 4 \\ 1 & 1 & 1 \end{bmatrix}$ **18** $\begin{bmatrix} 2 & -1 & 3 \\ 2 & 1 & 1 \\ 2 & 0 & 2 \end{bmatrix}$ **19** $\begin{bmatrix} 4 & -2 & 6 \\ 0 & 1 & -1 \\ 15 & 5 & 10 \end{bmatrix}$ **20** $\begin{bmatrix} 1 & -1 & -1 \\ 1 & 5 & 1 \\ 2 & 8 & 2 \end{bmatrix}$

In Exercises 21–28, use the inverse of the coefficient matrices calculated in Exercises 9–20 to solve each system of linear equations.

21 $x + 2y = 1$
$3x + 7y = -1$

22 $5x + 7y = 3$
$2x + 3y = -2$

23 $3x + 4y = -6$
$2x + 2y = 4$

24 $4x + 4y = -3$
$4x + 3y = 4$

25 $x + y + z = -2$
$x + 2y + 3z = 1$
$x + 3y + 6z = 3$

26 $a - 2b + 2c = 5$
$b + c = -6$
$-a + 2b - c = 0$

27 $4u + 5v + 3w = 2$
$4u + 6v + 4w = 1$
$u + v + w = -4$

28 $x - y - z = 0$
$x + 5y + z = 0$
$2x + 8y + 2z = 1$

B

In Exercises 29–32, use inverse matrices to solve the systems.

29 $r + 2t = 4$
$3r + 3s + 4t = -4$
$3r + 3s + 3t = 8$

30 $-3A + B - 2C = 3$
$8A + 3B + 2C = 0$
$4A + B + C = 6$

31 $3x_1 + 2x_2 + x_3 = 4$
$2x_1 + 2x_3 = -1$
$3x_1 + 8x_2 + 3x_3 = -4$

32 $5k + 2m + n = 16$
$2k + 3m + n = -8$
$4k + 2m + 2n = -1$

33 Use the inverse of the coefficient matrix and a single matrix multiplication to solve each system of equations.

a. $3x - 2y = -6$
$2x + 2y = 5$

b. $3x - 2y = 7$
$2x + 2y = 1$

c. $3x - 2y = 3$
$2x + 2y = -3$

34 Find A, if $A^{-1} = \begin{bmatrix} \frac{3}{2} & \frac{1}{2} \\ \frac{5}{2} & 1 \end{bmatrix}$.

35 If $A = \begin{bmatrix} a_{11} & a_{12} \\ a_{21} & a_{22} \end{bmatrix}$, show that $A^{-1} = \dfrac{1}{a_{11}a_{22} - a_{21}a_{12}} \begin{bmatrix} a_{22} & -a_{12} \\ -a_{21} & a_{11} \end{bmatrix}$ for $a_{11}a_{22} \neq a_{21}a_{12}$.

36 Find the inverse of $\begin{bmatrix} 8 & 7 \\ 6 & 9 \end{bmatrix}$ using the formula from Exercise 35.

C

In Exercises 37 and 38, find the inverse matrix, if it exists.

37. $\begin{bmatrix} 0.50 & -0.20 & -1.00 \\ 0 & 0.20 & -0.40 \\ -0.25 & -0.10 & 1.00 \end{bmatrix}$

38. $\begin{bmatrix} 3 & 3 & -1 & 2 \\ 0 & 2 & 3 & 0 \\ -2 & -4 & 1 & -2 \\ 1 & 2 & 0 & 1 \end{bmatrix}$

39. A tailor uses 6 yards of material and 11 buttons to make a man's suit and 5 yards of material and 8 buttons to make a woman's suit. She has three shipments of material and buttons on order. How many of each type of suit should be made after each shipment to use all of the material and buttons?

	Shipment		
	1	2	3
Yards of Material	122	170	92
Number of Buttons	212	293	157

40. The number of units of labor, machine time, and raw materials varies from day to day at a plastics factory because of labor contracts, scheduled maintenance of equipment, and the delivery schedule of raw materials. The number of units available each day is given in the following table.

	Monday	Tuesday	Wednesday	Thursday	Friday
Labor	340	330	310	385	335
Machine Time	220	200	200	230	230
Materials	340	280	300	320	380

The factory manufactures three products: chairs, tables, and stools. The number of units of labor, machine time, and raw materials needed for each of these products is shown in the following table.

	Chairs	Tables	Stools
Units of Labor	1	3	2
Units of Machine Time	1	2	1
Units of Materials	2	4	1

Determine the number of units that should be produced each day in order to fully utilize all the available resources.

SECTION 7-3

Determinants and Cramer's Rule

Section Objectives

4 Compute the determinant of a square matrix.

5 Solve a 2 × 2 or 3 × 3 system of linear equations using Cramer's rule.

Leibnitz and Cramer used determinants to solve systems of linear equations in the seventeenth and eighteenth centuries.

The determinant of a square matrix A is a number denoted by $|A|$. The determinant associated with a 2 × 2 matrix is defined in the following box.

Determinant of a 2 × 2 Matrix ▼

If $A = \begin{bmatrix} a_{11} & a_{12} \\ a_{21} & a_{22} \end{bmatrix}$, then the determinant of A is

$$|A| = a_{11}a_{22} - a_{12}a_{21} \qquad \begin{vmatrix} a_{11} & a_{12} \\ a_{21} & a_{22} \end{vmatrix} = a_{11}a_{22} - a_{12}a_{21}$$

EXAMPLE 1 Find $\begin{vmatrix} 1 & 2 \\ 4 & 3 \end{vmatrix}$.

SOLUTION $\begin{vmatrix} 1 & 2 \\ 4 & 3 \end{vmatrix} = \underbrace{1 \cdot 3}_{a_{11}a_{22}} - \underbrace{2 \cdot 4}_{a_{12}a_{21}} = 3 - 8 = -5$ Substitute the given values into the definition.

▶ **Self-Check ▼**

Find $\begin{vmatrix} 4 & -11 \\ 7 & \frac{1}{2} \end{vmatrix}$.

Determinants of 2 × 2 matrices can be used to evaluate the determinant of a 3 × 3 matrix. Each entry of a 3 × 3 matrix has an associated 2 × 2 determinant called its minor.

Minor M_{ij} ▼

If $A = \begin{bmatrix} a_{11} & a_{12} & a_{13} \\ a_{21} & a_{22} & a_{23} \\ a_{31} & a_{32} & a_{33} \end{bmatrix}$, then the minor of a_{ij}, denoted M_{ij}, is the determinant of the 2 × 2 matrix obtained by deleting the ith row and the jth column of A.

▶ **Self-Check Answer ▼**

EXAMPLE 2 For $A = \begin{bmatrix} 1 & 2 & 3 \\ 4 & 5 & 6 \\ 7 & 8 & 9 \end{bmatrix}$, determine the minors.

(a) M_{31}

(b) M_{23}

SOLUTIONS

(a) M_{31} is the minor of the entry $a_{31} = 7$. Delete row 3 and column 1 of A.

$$M_{31} = \begin{vmatrix} 2 & 3 \\ 5 & 6 \end{vmatrix} = 12 - 15 = -3$$

(b) M_{23} is the minor of the entry $a_{23} = 6$. Delete row 2 and column 3 of A.

$$M_{23} = \begin{vmatrix} 1 & 2 \\ 7 & 8 \end{vmatrix} = 8 - 14 = -6$$

Cofactor A_{ij} ▼

If $A = \begin{bmatrix} a_{11} & a_{12} & a_{13} \\ a_{21} & a_{22} & a_{23} \\ a_{31} & a_{32} & a_{33} \end{bmatrix}$, then the cofactor of a_{ij}, denoted A_{ij}, is given by $A_{ij} = (-1)^{i+j} M_{ij}$ where M_{ij} is the minor of a_{ij}.

EXAMPLE 3 For $A = \begin{bmatrix} 1 & 2 & 3 \\ 4 & 5 & 6 \\ 7 & 8 & 9 \end{bmatrix}$, determine the cofactors.

(a) A_{31}

(b) A_{23}

SOLUTIONS

(a) A_{31} is the cofactor of the entry $a_{31} = 7$.

$$A_{31} = (-1)^{3+1} \begin{vmatrix} 2 & 3 \\ 5 & 6 \end{vmatrix} = (-1)^4(-3) = (+1)(-3) = -3 \qquad \text{See Example 2 for the details on } M_{31} \text{ and } M_{23}.$$

(b) A_{23} is the cofactor of the entry $a_{23} = 6$.

$$A_{23} = (-1)^{2+3} \begin{vmatrix} 1 & 2 \\ 7 & 8 \end{vmatrix} = (-1)^5(-6) = (-1)(-6) = 6$$

SECTION 7-3 DETERMINANTS AND CRAMER'S RULE

Each cofactor A_{ij} is either $+M_{ij}$ or $-M_{ij}$. The "+" and "−" signs alternate from entry to entry, as indicated below. This pattern can be used to quickly determine the correct factor of M_{ij}.

$$\begin{bmatrix} A_{11} & A_{12} & A_{13} \\ A_{21} & A_{22} & A_{23} \\ A_{31} & A_{32} & A_{33} \end{bmatrix} = \begin{bmatrix} +M_{11} & -M_{12} & +M_{13} \\ -M_{21} & +M_{22} & -M_{23} \\ +M_{31} & -M_{32} & +M_{33} \end{bmatrix}$$

▶ **Self-Check** ▼

Find the minor and cofactor of each entry in the first row of

$$A = \begin{bmatrix} 1 & 2 & 3 \\ 4 & 5 & 6 \\ 7 & 8 & 9 \end{bmatrix}$$

The cofactors of any row or column can be used to evaluate a determinant. This method of finding a determinant is called expansion by cofactors.

Expansion by Cofactors ▼

The determinant of a 3 × 3 matrix is the sum, taken along any row or column, of the products of each entry and its cofactor.

EXAMPLE 4 Compute the determinant of the matrix $A = \begin{bmatrix} -2 & 1 & 0 \\ 3 & 2 & -2 \\ 0 & 1 & 4 \end{bmatrix}$ by

(a) expanding along the first row
(b) expanding along the second column

SOLUTIONS

(a) $|A| = a_{11}A_{11} + a_{12}A_{12} + a_{13}A_{13}$ Expand along the first row.

$$= -2 \cdot (-1)^{1+1} \begin{vmatrix} 2 & -2 \\ 1 & 4 \end{vmatrix} + 1 \cdot (-1)^{1+2} \begin{vmatrix} 3 & -2 \\ 0 & 4 \end{vmatrix}$$

$$+ 0 \cdot (-1)^{1+3} \begin{vmatrix} 3 & 2 \\ 0 & 1 \end{vmatrix}$$

$$= -2(1)(10) + 1(-1)(12) + 0$$

$$= -20 - 12 + 0$$

$|A| = -32$

(b) $|A| = a_{12}A_{12} + a_{22}A_{22} + a_{32}A_{32}$ Expand along the second column.

$$= 1 \cdot (-1)^{1+2} \begin{vmatrix} 3 & -2 \\ 0 & 4 \end{vmatrix} + 2 \cdot (-1)^{2+2} \begin{vmatrix} -2 & 0 \\ 0 & 4 \end{vmatrix}$$

$$+ 1 \cdot (-1)^{3+2} \begin{vmatrix} -2 & 0 \\ 3 & -2 \end{vmatrix}$$

$$= 1(-1)(12) + 2(1)(-8) + 1(-1)(4)$$

$$= -12 - 16 - 4$$

$|A| = -32$ □

▶ **Self-Check Answer** ▼

$M_{11} = -3, A_{11} = -3; M_{12} = -6, A_{12} = 6; M_{13} = -3, A_{13} = -3$

We can simplify the calculations required to compute a determinant if we expand along the row or column containing the most zeros.

EXAMPLE 5 Compute $|A|$ for $A = \begin{bmatrix} 11 & -16 & 20 \\ 8 & 17 & 0 \\ 6 & -25 & 0 \end{bmatrix}$.

SOLUTION $|A| = a_{13}A_{13} + a_{23}A_{23} + a_{33}A_{33}$ Expand along column 3, which has the most zeros. There is no need to evaluate A_{23} or A_{33}, since their coefficients are zeros.

$= 20(-1)^{1+3} \begin{vmatrix} 8 & 17 \\ 6 & -25 \end{vmatrix} + 0 + 0$

$= 20(+1)(-302)$

$|A| = -6040$ □

▶ **Self-Check** ▼

Compute the determinant of

$A = \begin{bmatrix} -2 & 1 & 0 \\ 3 & 2 & -2 \\ 0 & 1 & 4 \end{bmatrix}$ by

1 expanding along the third row
2 expanding along the first column

The TI-81 graphics calculator is capable of calculating the determinant of a square matrix. The keystrokes for calculating the determinant of matrix A in Example 5 are illustrated below.

First enter the matrix as matrix A using the **MATRX** key and the EDIT option. Then use these keystrokes: **MATRX** **5(det)** **2nd** **[A]** **ENTER** to obtain the screen shown in Figure 7-4.

Thus $\begin{vmatrix} 11 & -16 & 20 \\ 8 & 17 & 0 \\ 6 & -25 & 0 \end{vmatrix} = -6040$

Figure 7-4

Determinants of 4×4 or larger square matrices can also be calculated by expansion by cofactors. Each expansion produces cofactors of order one less than those in the preceding step. Thus any determinant can be reduced to a calculation involving determinants of 2×2 matrices.

EXAMPLE 6 Compute the determinant of $A = \begin{bmatrix} 1 & 2 & 0 & -1 \\ -1 & 0 & -2 & 0 \\ 0 & 1 & 3 & -2 \\ 2 & -1 & 1 & 0 \end{bmatrix}$.

SOLUTION $|A| = a_{14}A_{14} + a_{24}A_{24} + a_{34}A_{34} + a_{44}A_{44}$ Expand along column 4. The calculation of the two determinants in this step is left for you to verify.

$= (-1) \cdot (-1)^{1+4} \begin{vmatrix} -1 & 0 & -2 \\ 0 & 1 & 3 \\ 2 & -1 & 1 \end{vmatrix} + 0 + (-2) \cdot (-1)^{3+4} \begin{vmatrix} 1 & 2 & 0 \\ -1 & 0 & -2 \\ 2 & -1 & 1 \end{vmatrix} + 0$

$= (-1)(-1)(0) + (-2)(-1)(-8)$

$|A| = -16$ □

▶ **Self-Check Answer** ▼

-32 (the same answer as in Example 4)

SECTION 7-3 DETERMINANTS AND CRAMER'S RULE

Determinants may be used to solve a system of equations of the form
$\begin{cases} a_{11}x + a_{12}y = b_1 \\ a_{21}x + a_{22}y = b_2 \end{cases}$ if $\begin{vmatrix} a_{11} & a_{12} \\ a_{21} & a_{22} \end{vmatrix}$, the determinant of the coefficient matrix, is not zero. We will verify this by using the elimination method to solve this system.

$$a_{11}a_{22}x + a_{12}a_{22}y = b_1 a_{22}$$ Multiply the first equation by a_{22}.
$$-a_{12}a_{21}x - a_{12}a_{22}y = -b_2 a_{12}$$ Multiply the second equation by $-a_{12}$.
$$a_{11}a_{22}x - a_{12}a_{21}x = b_1 a_{22} - b_2 a_{12}$$ Add these equations.

$$x = \frac{b_1 a_{22} - b_2 a_{12}}{a_{11}a_{22} - a_{12}a_{21}}$$ Solve for x.

Thus
$$x = \frac{\begin{vmatrix} b_1 & a_{12} \\ b_2 & a_{22} \end{vmatrix}}{\begin{vmatrix} a_{11} & a_{12} \\ a_{21} & a_{22} \end{vmatrix}}$$

Replace $b_1 a_{22} - b_2 a_{12}$ by $\begin{vmatrix} b_1 & a_{12} \\ b_2 & a_{22} \end{vmatrix}$.

Replace $a_{11}a_{22} - a_{12}a_{21}$ by $\begin{vmatrix} a_{11} & a_{12} \\ a_{21} & a_{22} \end{vmatrix}$.

Similarly,
$$y = \frac{a_{11}b_2 - a_{21}b_1}{a_{11}a_{22} - a_{12}a_{21}}$$

Thus
$$y = \frac{\begin{vmatrix} a_{11} & b_1 \\ a_{21} & b_2 \end{vmatrix}}{\begin{vmatrix} a_{11} & a_{12} \\ a_{21} & a_{22} \end{vmatrix}}$$

Replace $a_{11}b_2 - a_{21}b_1$ by $\begin{vmatrix} a_{11} & b_1 \\ a_{21} & b_2 \end{vmatrix}$.

Gabriel Cramer is credited with publishing this result in 1750. By putting the solution into determinant notation, Cramer made it easier to remember and use the result. The conciseness of Cramer's rule, which uses only the coefficients and constants, is analogous to that of the quadratic formula for solving quadratic equations.

Cramer's Rule ▼

If $D \neq 0$, the solution $\begin{cases} a_{11}x + a_{12}y = b_1 \\ a_{21}x + a_{22}y = b_2 \end{cases}$ is $x = \dfrac{D_x}{D}$, $y = \dfrac{D_y}{D}$, where

$$D = \begin{vmatrix} a_{11} & a_{12} \\ a_{21} & a_{22} \end{vmatrix}$$ Determinant of the coefficient matrix

$$D_x = \begin{vmatrix} b_1 & a_{12} \\ b_2 & a_{22} \end{vmatrix}$$ Replace the x-coefficients with the constants b_1 and b_2.

$$D_y = \begin{vmatrix} a_{11} & b_1 \\ a_{21} & b_2 \end{vmatrix}$$ Replace the y-coefficients with the constants b_1 and b_2.

EXAMPLE 7 Solve $\begin{cases} 13a - 12b = 5 \\ 11a - 17b = 3 \end{cases}$ using Cramer's rule.

SOLUTION $D = \begin{vmatrix} 13 & -12 \\ 11 & -17 \end{vmatrix} = 13(-17) - 11(-12) = -89$

$D_a = \begin{vmatrix} 5 & -12 \\ 3 & -17 \end{vmatrix} = 5(-17) - 3(-12) = -49$

$D_b = \begin{vmatrix} 13 & 5 \\ 11 & 3 \end{vmatrix} = 13(3) - 11(5) = -16$

Thus $a = \dfrac{D_a}{D} = \dfrac{-49}{-89} = \dfrac{49}{89}$, $b = \dfrac{D_b}{D} = \dfrac{-16}{-89} = \dfrac{16}{89}$

Answer $\left(\dfrac{49}{89}, \dfrac{16}{89} \right)$ □

If $D = 0$, the equation will be either dependent or inconsistent. In this case, we suggest that you immediately use another method to analyze the system rather than calculate the other determinants.

Cramer's rule may also be applied to larger systems of linear equations. For example, if the determinant of the coefficient matrix D is not equal to 0 the solution of the system

$$\begin{cases} a_{11}x + a_{12}y + a_{13}z = b_1 \\ a_{21}x + a_{22}y + a_{23}z = b_2 \\ a_{31}x + a_{32}y + a_{33}z = b_3 \end{cases}$$

is $x = \dfrac{D_x}{D}$, $y = \dfrac{D_y}{D}$, and $z = \dfrac{D_z}{D}$, where D_x, D_y, and D_z are determinants obtained by replacing the appropriate column of the coefficient matrix with the constants b_1, b_2, and b_3.

▶ **Self-Check** ▼

Solve these systems using Cramer's rule.

1. $\begin{cases} 9r + 10t = 1 \\ 7r - 5t = 7 \end{cases}$

2. $\begin{cases} 21x - 39y = 12 \\ -35x + 65y = -20 \end{cases}$

EXAMPLE 8 Solve $\begin{cases} x + y + z = 5 \\ 2x - y + z = 6 \\ 2x + 2y - z = -1 \end{cases}$ using Cramer's rule.

SOLUTION $D = \begin{vmatrix} 1 & 1 & 1 \\ 2 & -1 & 1 \\ 2 & 2 & -1 \end{vmatrix} = 9$ Calculate D, D_x, D_y, and D_z. The steps are left for you to verify.

$D_x = \begin{vmatrix} 5 & 1 & 1 \\ 6 & -1 & 1 \\ -1 & 2 & -1 \end{vmatrix} = 11$

▶ **Self-Check Answers** ▼

1. $\left(\dfrac{15}{23}, -\dfrac{56}{115} \right)$ 2. $D = 0$. Since $-\dfrac{21}{35} = -\dfrac{39}{65} = -\dfrac{12}{20}$, the system is dependent.

SECTION 7-3 DETERMINANTS AND CRAMER'S RULE

$$D_y = \begin{vmatrix} 1 & 5 & 1 \\ 2 & 6 & 1 \\ 2 & -1 & -1 \end{vmatrix} = 1$$

$$D_z = \begin{vmatrix} 1 & 1 & 5 \\ 2 & -1 & 6 \\ 2 & 2 & -1 \end{vmatrix} = 33$$

Thus $x = \dfrac{D_x}{D} = \dfrac{11}{9}$, $y = \dfrac{D_y}{D} = \dfrac{1}{9}$, and $z = \dfrac{D_z}{D} = \dfrac{33}{9} = \dfrac{11}{3}$.

Answer $\left(\dfrac{11}{9}, \dfrac{1}{9}, \dfrac{11}{3}\right)$

Cramer's rule is frequently the method of choice for 2 × 2 or 3 × 3 systems of linear equations. This is especially true if the coefficients are not small integers and a programmable calculator or computer can be used in the computation process. The TI-81 graphics calculator is capable of working with up to a 6 × 6 matrix. Thus systems of up to six linear equations can be solved quite efficiently by using the inverse matrix method. For larger systems we recommend the augmented matrix method, which will run faster on a computer than Cramer's rule.

EXERCISES 7-3

A

In Exercises 1–9, evaluate the determinants.

1. a. $\begin{vmatrix} 5 & 2 \\ 3 & 6 \end{vmatrix}$ b. $\begin{vmatrix} 4 & -5 \\ 6 & 3 \end{vmatrix}$ c. $\begin{vmatrix} 6 & 1 \\ 5 & -3 \end{vmatrix}$ d. $\begin{vmatrix} 10 & -21 \\ -5 & 15 \end{vmatrix}$

2. a. $\begin{vmatrix} 4 & 1 \\ 2 & 8 \end{vmatrix}$ b. $\begin{vmatrix} 3 & 2 \\ -1 & 8 \end{vmatrix}$ c. $\begin{vmatrix} -8 & 5 \\ 6 & 2 \end{vmatrix}$ d. $\begin{vmatrix} -6 & 35 \\ 32 & -18 \end{vmatrix}$

In Exercises 3–6, use the matrix $A = \begin{bmatrix} 1 & -1 & 2 \\ 3 & 1 & -2 \\ 2 & 1 & 1 \end{bmatrix}$.

3. a. Find the minor and the cofactor for each entry in row 2.
   b. Compute $|A|$ by expanding along row 2.
4. a. Find the minor and the cofactor for each entry in row 3.
   b. Compute $|A|$ by expanding along row 3.
5. a. Find the minor and the cofactor for each entry in column 1.
   b. Compute $|A|$ by expanding along column 1.
6. a. Find the minor and the cofactor for each entry in column 3.
   b. Compute $|A|$ by expanding along column 3.

In Exercises 7–14, compute the value of each determinant.

7. $\begin{vmatrix} 2 & -1 & 5 \\ 6 & -2 & 0 \\ 8 & 7 & 0 \end{vmatrix}$
8. $\begin{vmatrix} 3 & 6 & -5 \\ 0 & 2 & 0 \\ 4 & 8 & 3 \end{vmatrix}$
9. $\begin{vmatrix} 3 & 6 & -4 \\ 10 & 18 & 15 \\ 0 & 3 & 0 \end{vmatrix}$
10. $\begin{vmatrix} 1 & -2 & 4 \\ 2 & -1 & 3 \\ 1 & -5 & 9 \end{vmatrix}$

11. $\begin{vmatrix} 2 & 1 & -3 \\ 1 & 3 & -2 \\ 2 & 1 & 1 \end{vmatrix}$
12. $\begin{vmatrix} 1 & -1 & 1 \\ 1 & 1 & 1 \\ -1 & -1 & 1 \end{vmatrix}$
13. $\begin{vmatrix} 2 & 5 & 1 \\ -1 & 3 & 1 \\ 8 & 9 & 1 \end{vmatrix}$
14. $\begin{vmatrix} 6 & -1 & 5 \\ 4 & 1 & 2 \\ -3 & 0 & 1 \end{vmatrix}$

In Exercises 15–24, use Cramer's rule to solve each system of linear equations.

15. $11x + 12y = 13$
    $7x - 15y = 6$
16. $17x_1 + 19x_2 = 1$
    $5x_1 + 8x_2 = 2$
17. $7a - 6b = 5$
    $8a - 11b = 6$
18. $8u - 13v = 6$
    $6u + 11v = 5$
19. $0.3m + 0.2n = 6$
    $0.7m - 0.3n = 1$
20. $0.4p - 0.2q = 2.1$
    $0.6p + 0.1q = 1.5$
21. $x + y = 6$
    $x - z = -3$
    $ y + 2z = 4$
22. $x + 2z = 8$
    $ 2y - z = 5$
    $2x + 3z = -6$
23. $x + y - 2z = 3$
    $2x - 3y + z = 4$
    $2x + 2y - 3z = 6$
24. $x + 2y + 3z = 7$
    $x - y + 3z = 3$
    $3x - 2y + 4z = 10$

B

25. **a.** Evaluate $\begin{vmatrix} 0 & 0 \\ 0 & 0 \end{vmatrix}$.
    b. Evaluate $\begin{vmatrix} 0 & 0 & 0 \\ 0 & 0 & 0 \\ 0 & 0 & 0 \end{vmatrix}$.
    c. Make a conjecture about the determinant of any zero matrix.

26. **a.** Evaluate $\begin{vmatrix} 1 & 0 \\ 0 & 1 \end{vmatrix}$.
    b. Evaluate $\begin{vmatrix} 1 & 0 & 0 \\ 0 & 1 & 0 \\ 0 & 0 & 1 \end{vmatrix}$.
    c. Make a conjecture about the determinant of any identity matrix.

27. **a.** Evaluate $\begin{vmatrix} a_{11} & 0 \\ a_{21} & a_{22} \end{vmatrix}$.
    b. Evaluate $\begin{vmatrix} a_{11} & a_{12} \\ 0 & a_{22} \end{vmatrix}$.
    c. Evaluate $\begin{vmatrix} a_{11} & a_{12} & a_{13} \\ 0 & a_{22} & a_{23} \\ 0 & 0 & a_{33} \end{vmatrix}$.
    d. Evaluate $\begin{vmatrix} a_{11} & 0 & 0 \\ a_{21} & a_{22} & 0 \\ a_{31} & a_{32} & a_{33} \end{vmatrix}$.
    e. A triangular matrix is a matrix with all zeros above or below the main diagonal. Make a conjecture about the determinant of a triangular matrix.

28. Find the determinant of $\begin{bmatrix} 1 & 0 & 0 \\ 2 & 3 & 4 \\ 5 & 6 & 7 \end{bmatrix}$ and of its transpose. Make a conjecture based on this problem.

In Exercises 29–32, solve for x.

29. $\begin{vmatrix} x & 2 \\ 3 & 6 \end{vmatrix} = 0$
30. $\begin{vmatrix} 4 & x \\ 2 & 9 \end{vmatrix} = 2$
31. $\begin{vmatrix} x & 0 & 2 \\ 1 & x & 3 \\ 1 & 1 & 2 \end{vmatrix} = 0$
32. $\begin{vmatrix} 1 & x & 3 \\ x & 1 & 3 \\ -1 & -1 & 6 \end{vmatrix} = 0$

SECTION 7-3 DETERMINANTS AND CRAMER'S RULE

In Exercises 33 and 34, compute the determinant of each matrix.

33 $\begin{bmatrix} 2 & -1 & 3 & 4 \\ 1 & -1 & 2 & 3 \\ 5 & 7 & 1 & 1 \\ 0 & 0 & 0 & 2 \end{bmatrix}$
34 $\begin{bmatrix} 1 & 2 & 3 & 1 \\ -1 & 0 & 1 & 1 \\ 1 & 0 & -1 & -1 \\ 2 & 3 & 4 & 1 \end{bmatrix}$

C

The area of a triangle formed by the vertices (x_1, y_1), (x_2, y_2), and (x_3, y_3) is the absolute value of

$$\frac{1}{2} \begin{vmatrix} x_1 & y_1 & 1 \\ x_2 & y_2 & 1 \\ x_3 & y_3 & 1 \end{vmatrix}$$

Use this result to determine the area formed by the vertices given in Exercises 35 and 36.

35 (2, 5), (2, 10), (14, 10)   **36** (3, 1), (6, 11), (5, 7)

In Exercises 37 and 38, use Cramer's rule and a calculator or computer to solve each system of equations.

37 $1.28x - 2.04y = -6.120$
$7.06x + 3.15y = 55.536$

38 $2.41x + 4.12y + 1.42z = -36.45$
$3.17x + 1.37y + 7.31z = 89.73$
$4.56x + 6.54y + 5.46z = -36.28$

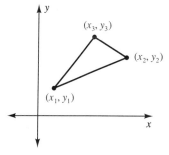

Figure for Exercises 35–36

39 Let $A = \begin{bmatrix} 1 & 2 \\ 3 & 4 \end{bmatrix}$.
  a. Evaluate $|A|$.
  b. Multiply row 1 of A by 2 and then evaluate $|A|$.
  c. Multiply row 2 of A by -3 and then evaluate $|A|$.
  d. Multiply column 1 of A by 4 and then evaluate $|A|$.
  e. Multiply column 2 of A by -5 and then evaluate $|A|$.
  f. Make a conjecture about the value of the determinant of a matrix after a row or column is multiplied by the constant k.

40 Let $A = \begin{bmatrix} 1 & 0 & 0 \\ 2 & 3 & 4 \\ 5 & 6 & 7 \end{bmatrix}$.
  a. Evaluate $|A|$.
  b. Interchange row 1 and row 2 of A and then evaluate the determinant of this matrix.
  c. Interchange row 1 and row 3 of A and then evaluate the determinant of this matrix.
  d. Interchange column 1 and column 2 of A and then evaluate the determinant of this matrix.
  e. Interchange column 2 and column 3 of A and then evaluate the determinant of this matrix.
  f. Make a conjecture about the value of the determinant of a matrix after two rows or columns are interchanged.

41 Let $A = \begin{bmatrix} 1 & 2 \\ 3 & 4 \end{bmatrix}$.

  a. Evaluate $|A|$.
  b. Find the determinant of the matrix resulting from the elementary row operation $r'_2 = r_2 - 3r_1$.
  c. Find the determinant of the matrix resulting from the elementary row operation $r'_2 = r_2 + 2r_1$.
  d. Find the determinant of the matrix resulting from the elementary row operation $r'_1 = r_1 + 2r_2$.
  e. Find the determinant of the matrix resulting from the elementary row operation $r'_1 = r_1 + 3r_2$.
  f. Make a conjecture about the value of the determinant of a matrix after a row is replaced with the sum of that row and a constant multiple of another row.

SECTION 7-4

Properties of Determinants

Section Objectives

6 Use the properties of determinants.

The properties of determinants introduced in this section can be used to simplify the calculation of determinants of square matrices. *The determinant of a matrix equals the determinant of its transpose.* Since transposing a matrix interchanges its rows and columns, all the properties of determinants involving rows are also true for columns.

EXAMPLE 1 Show that $|A| = |A^t|$ if $A = \begin{bmatrix} a_{11} & a_{12} \\ a_{21} & a_{22} \end{bmatrix}$.

SOLUTION $|A| = a_{11}a_{22} - a_{12}a_{21}$ Definition of a 2 × 2 determinant

$|A^t| = a_{11}a_{22} - a_{21}a_{12}$ $A^t = \begin{bmatrix} a_{11} & a_{21} \\ a_{12} & a_{22} \end{bmatrix}$

Thus $|A| = |A^t|$. $a_{11}a_{22} - a_{12}a_{21} = a_{11}a_{22} - a_{21}a_{12}$ □

The three properties on page 475 show the effects of elementary row operations on the determinant of a matrix.

Properties of Elementary Row (Column) Operations on Determinants ▼

1. $|B| = -|A|$ if matrix B is obtained by interchanging two rows (columns) of matrix A.
2. $|B| = k|A|$ if matrix B is obtained by multiplying a row (column) of matrix A by a constant k.
3. $|B| = |A|$ if matrix B is obtained by replacing a row (column) of matrix A by the sum of that row (column) and a constant multiple of another row (column).

EXAMPLE 2   Given $A = \begin{bmatrix} 1 & 2 & 0 \\ -2 & 3 & 2 \\ 2 & -3 & 0 \end{bmatrix}$, evaluate each of the following determinants.

(a) $|A|$

(b) $|B|$, if B is obtained by interchanging the first and second rows of A.

(c) $|C|$, if C is obtained by multiplying the second column of A by -2.

(d) $|E|$, if E is obtained by replacing row 3 of A with the sum of row 3 and row 2.

SOLUTIONS

(a) $|A| = \begin{vmatrix} 1 & 2 & 0 \\ -2 & 3 & 2 \\ 2 & -3 & 0 \end{vmatrix} = 0 \begin{vmatrix} -2 & 3 \\ 2 & -3 \end{vmatrix} - 2 \begin{vmatrix} 1 & 2 \\ 2 & -3 \end{vmatrix} + 0 \begin{vmatrix} 1 & 2 \\ -2 & 3 \end{vmatrix}$ Expand along column 3.

$= 0 - 2(-3 - 4) + 0$

$= -2(-7)$

$|A| = 14$

(b) $|B| = -|A|$ Interchanging two rows changes the sign of the determinant.

$|B| = -14$

(c) $|C| = -2|A|$ Multiplying the second column of A by -2 multiplies the determinant of A by -2.

$= -2(14)$

$|C| = -28$

(d) $|E| = |A|$ $r'_3 = r_3 + r_2$. Thus the value of the determinant is unchanged.

$|E| = 14$

□

The determinant $|A|$ was evaluated in Example 2 by expanding along column 3, which had two zeros and one nonzero entry. This simplified our computations, since we only had to evaluate one 2×2 determinant. One major use of the elementary row and column operations is to introduce zeros into a row or column in order to simplify the computation of a determinant.

EXAMPLE 3 Compute each of the determinants in parts (a) and (b) by first introducing two zeros into some row or column. In part (c), first introduce three zeros into some row or column.

(a) $\begin{vmatrix} 1 & -4 & 3 \\ -2 & 7 & 1 \\ 4 & -8 & 22 \end{vmatrix}$

(b) $\begin{vmatrix} 5 & -3 & 3 \\ 8 & -6 & 2 \\ 27 & -10 & 4 \end{vmatrix}$

(c) $\begin{vmatrix} 4 & -2 & 9 & -5 \\ 4 & -6 & 1 & -12 \\ -1 & 2 & -2 & 5 \\ -5 & 3 & -3 & 15 \end{vmatrix}$

SOLUTION

(a) $\begin{vmatrix} 1 & -4 & 3 \\ -2 & 7 & 1 \\ 4 & -8 & 22 \end{vmatrix} = \begin{vmatrix} 1 & -4 & 3 \\ 0 & -1 & 7 \\ 0 & 8 & 10 \end{vmatrix}$ Use the one in row 1 to produce two zeros in column 1 by applying these row operations: $r'_2 = r_2 + 2r_1;\ r'_3 = r_3 - 4r_1$

$= 1 \cdot (-1)^{1+1} \begin{vmatrix} -1 & 7 \\ 8 & 10 \end{vmatrix} + 0 + 0$ Expand along column 1.

$= 1(1)(-10 - 56)$

$= -66$

(b) $\begin{vmatrix} 5 & -3 & 3 \\ 8 & -6 & 2 \\ 27 & -10 & 4 \end{vmatrix} = \begin{vmatrix} -7 & 6 & 3 \\ 0 & 0 & 2 \\ 11 & 2 & 4 \end{vmatrix}$ Use the two in column 3 to produce two zeros in row 2 by applying these column operations: $c'_1 = c_1 - 4c_3;\ c'_2 = c_2 + 3c_3$

$= 0 + 0 + 2 \cdot (-1)^{2+3} \begin{vmatrix} -7 & 6 \\ 11 & 2 \end{vmatrix}$ Expand along row 2.

$= 2(-1)(-14 - 66)$

$= 160$

(c) $\begin{vmatrix} 4 & -2 & 9 & -5 \\ 4 & -6 & 1 & -12 \\ -1 & 2 & -2 & 5 \\ -5 & 3 & -3 & 15 \end{vmatrix} = \begin{vmatrix} 4 & 6 & 1 & 15 \\ 4 & 2 & -7 & 8 \\ -1 & 0 & 0 & 0 \\ -5 & -7 & 7 & -10 \end{vmatrix}$ Use the negative one in column 1 to produce three zeros in row 3: $c'_2 = c_2 + 2c_1;\ c'_3 = c_3 - 2c_1;\ c'_4 = c_4 + 5c_1$

$= (-1) \cdot (-1)^{3+1} \begin{vmatrix} 6 & 1 & 15 \\ 2 & -7 & 8 \\ -7 & 7 & -10 \end{vmatrix}$ Expand along row 3.

$= (-1) \begin{vmatrix} 6 & 1 & 15 \\ 44 & 0 & 113 \\ -49 & 0 & -115 \end{vmatrix}$ Use the one in row 1 to produce two zeros in column 2: $r'_2 = r_2 + 7r_1;\ r'_3 = r_3 - 7r_1$

SECTION 7-4 PROPERTIES OF DETERMINANTS

$$= (-1) \cdot (1) \cdot (-1)^{1+2} \begin{vmatrix} 44 & 113 \\ -49 & -115 \end{vmatrix} + 0 + 0 \qquad \text{Expand along column 2.}$$

$$= (-1)[(-1)(-5060 + 5537)]$$

$$= 477 \qquad \square$$

Elementary row and column operations can be used to factor a number out of a row or column of a determinant. This strategy can be used to simplify the values and thus simplify pencil-and-paper calculations.

EXAMPLE 4 Compute $\begin{vmatrix} 36 & 48 \\ 2 & 28 \end{vmatrix}$ by factoring 12 out of the first row.

SOLUTION
$$\begin{vmatrix} 36 & 48 \\ 2 & 28 \end{vmatrix} = 12 \begin{vmatrix} 3 & 4 \\ 2 & 28 \end{vmatrix} \qquad \text{Factor 12 out of row 1:} \\ 36 = 12 \cdot 3;\ 48 = 12 \cdot 4$$

$$= 12(84 - 8)$$

$$= 912 \qquad \square$$

Inspection may be used to evaluate some determinants whose values are zero.

Theorem ▼

The determinant of a square matrix is zero
1. if the matrix has a row or column of zeros
2. if the matrix has two identical rows or columns
3. if one row (column) is a multiple of another row (column)

Proof

1. Suppose A has a row (column) of zeros. The zero can be factored out of this row, and thus $|A| = 0|A| = 0$.
2. Suppose A has two identical rows (columns). Let A' be the matrix obtained by replacing one of the identical rows with the sum of that row and (-1) times the equivalent row. Then A' has a row of zeros, and thus $|A| = |A'| = 0$.
3. The proof of part 3 is similar to the proof of part 2 and is left to the reader.

▶ Self-Check ▼

Given $D = \begin{vmatrix} 2 & -4 & 3 \\ -3 & 2 & -2 \\ 3 & 5 & 1 \end{vmatrix}$,

1 introduce two zeros into row 3 of D.

2 introduce two zeros into column 3 of D.

Use inspection to explain why each of these determinants is zero.

3 $\begin{vmatrix} 2 & 0 & -3 \\ 4 & 0 & 11 \\ 7 & 0 & 18 \end{vmatrix}$

4 $\begin{vmatrix} 3 & -4 & 8 & 9 \\ 5 & 11 & 2 & 0 \\ 0 & 0 & 7 & 6 \\ 3 & -4 & 8 & 9 \end{vmatrix}$

5 $\begin{vmatrix} -7 & 4 & -6 \\ 12 & -10 & 15 \\ 16 & -2 & 3 \end{vmatrix}$

▶ Self-Check Answers ▼

Some possible answers are

1 $\begin{vmatrix} -7 & -19 & 3 \\ 3 & 12 & -2 \\ 0 & 0 & 1 \end{vmatrix}$
2 $\begin{vmatrix} -7 & -19 & 0 \\ 3 & 12 & 0 \\ 3 & 5 & 1 \end{vmatrix}$
3 Column 2 has all zeros.

4 Rows 1 and 4 are identical.
5 Column 2 is $(-\frac{2}{3})$ of column 3.

EXERCISES 7-4

A

In Exercises 1-4, perform the indicated row and column operations to complete the missing entries in each exercise.

1. $\begin{vmatrix} 1 & 2 & 5 \\ 3 & 5 & 6 \\ 5 & 8 & 2 \end{vmatrix} = \begin{vmatrix} 1 & 2 & 5 \\ & & \\ & & \end{vmatrix}$ $r'_2 = r_2 - 3r_1$
   $r'_3 = r_3 - 5r_1$

2. $\begin{vmatrix} 7 & 2 & 5 \\ 3 & 1 & 6 \\ 5 & 8 & 2 \end{vmatrix} = \begin{vmatrix} & & \\ 3 & 1 & 6 \\ & & \end{vmatrix}$ $r'_1 = r_1 - 2r_2$
   $r'_3 = r_3 - 8r_2$

3. $\begin{vmatrix} 1 & 2 & 5 \\ 3 & 5 & 6 \\ 5 & 8 & 2 \end{vmatrix} = \begin{vmatrix} 1 & & \\ 3 & & \\ 5 & & \end{vmatrix}$ $c'_2 = c_2 - 2c_1$
   $c'_3 = c_3 - 5c_1$

4. $\begin{vmatrix} 7 & 2 & 5 \\ 3 & 1 & 6 \\ 5 & 8 & 2 \end{vmatrix} = \begin{vmatrix} & 2 & \\ & 1 & \\ & 8 & \end{vmatrix}$ $c'_1 = c_1 - 3c_2$
   $c'_3 = c_3 - 6c_2$

5. Use column operations to introduce two zeros into row 2 of $\begin{vmatrix} 8 & 5 & 9 \\ 3 & -1 & -2 \\ 7 & 4 & -6 \end{vmatrix}$.

6. Use row operations to introduce two zeros into column 2 of $\begin{vmatrix} 8 & 5 & 9 \\ 3 & -1 & -2 \\ 7 & 4 & -6 \end{vmatrix}$.

In Exercises 7-13, each determinant is obtained from $\begin{vmatrix} 3 & -2 & 5 \\ -6 & 8 & 4 \\ -3 & 5 & 7 \end{vmatrix} = 18$

by using either the elementary row (column) operations or the theorem given in this section. Use this given information to evaluate each determinant by inspection.

7. $\begin{vmatrix} 3 & -6 & -3 \\ -2 & 8 & 5 \\ 5 & 4 & 7 \end{vmatrix}$

8. $\begin{vmatrix} 3 & -2 & 5 \\ -3 & 5 & 7 \\ -6 & 8 & 4 \end{vmatrix}$

9. $\begin{vmatrix} 5 & -2 & 3 \\ 4 & 8 & -6 \\ 7 & 5 & -3 \end{vmatrix}$

10. $\begin{vmatrix} 3 & -2 & 5 \\ -6 & 8 & 4 \\ -6 & 10 & 14 \end{vmatrix}$

11. $\begin{vmatrix} 3 & 6 & 5 \\ -6 & -24 & 4 \\ -3 & -15 & 7 \end{vmatrix}$

12. $\begin{vmatrix} 3 & -2 & 5 \\ 6 & 0 & 24 \\ -3 & 5 & 7 \end{vmatrix}$

13. $\begin{vmatrix} 3 & -12 & 5 \\ -6 & 0 & 4 \\ -3 & -9 & 7 \end{vmatrix}$

In Exercises 14-20, each determinant is obtained from $\begin{vmatrix} 2 & 1 & 4 \\ -5 & 4 & 3 \\ 2 & 1 & -6 \end{vmatrix} = -130$

by using either the elementary row (column) operations or the theorem given in this section. Use this given information to evaluate each determinant by inspection.

14. $\begin{vmatrix} 2 & 4 & 1 \\ -5 & 3 & 4 \\ 2 & -6 & 1 \end{vmatrix}$

15. $\begin{vmatrix} 2 & 1 & -6 \\ -5 & 4 & 3 \\ 2 & 1 & 4 \end{vmatrix}$

16. $\begin{vmatrix} 2 & -5 & 2 \\ 1 & 4 & 1 \\ 4 & 3 & -6 \end{vmatrix}$

17. $\begin{vmatrix} 2 & 2 & 4 \\ -5 & 8 & 3 \\ 2 & 2 & -6 \end{vmatrix}$

18. $\begin{vmatrix} 2 & -3 & 4 \\ -5 & -12 & 3 \\ 2 & -3 & -6 \end{vmatrix}$

19. $\begin{vmatrix} 2 & 1 & 10 \\ -5 & 4 & -12 \\ 2 & 1 & 0 \end{vmatrix}$

20. $\begin{vmatrix} 2 & 1 & 4 \\ -5 & 4 & 3 \\ 0 & 0 & -10 \end{vmatrix}$

SECTION 7-4 PROPERTIES OF DETERMINANTS

In Exercises 21–26, use the theorem on determinants given in this section to evaluate each determinant by inspection.

21. $\begin{vmatrix} 2 & 5 & 0 \\ -1 & 8 & 0 \\ 2 & 11 & 0 \end{vmatrix}$

22. $\begin{vmatrix} 6 & 6 & 6 \\ -5 & -5 & 3 \\ 4 & 4 & 2 \end{vmatrix}$

23. $\begin{vmatrix} 2 & 1 & -4 \\ -3 & 1 & 6 \\ 4 & 1 & -8 \end{vmatrix}$

24. $\begin{vmatrix} 3 & -7 & 1 \\ 10 & -12 & 15 \\ 0 & 0 & 0 \end{vmatrix}$

25. $\begin{vmatrix} 2 & -1 & 2 \\ 1 & 6 & 2 \\ 2 & -1 & 2 \end{vmatrix}$

26. $\begin{vmatrix} 2 & 4 & -6 \\ -1 & 8 & 3 \\ 3 & -9 & -9 \end{vmatrix}$

In Exercises 27–34, evaluate each determinant by first introducing two zeros into some row or column.

27. $\begin{vmatrix} 1 & 5 & 1 \\ -3 & -2 & -3 \\ 6 & -8 & -4 \end{vmatrix}$

28. $\begin{vmatrix} 3 & -1 & 8 \\ 2 & 4 & -6 \\ -1 & -2 & 3 \end{vmatrix}$

29. $\begin{vmatrix} 1 & 6 & 3 \\ 0 & 8 & -2 \\ 2 & 7 & 1 \end{vmatrix}$

30. $\begin{vmatrix} 1 & 3 & -5 \\ 2 & -6 & 7 \\ 1 & -9 & 12 \end{vmatrix}$

31. $\begin{vmatrix} 3 & 1 & 2 \\ 5 & 3 & 2 \\ -8 & -4 & -4 \end{vmatrix}$

32. $\begin{vmatrix} -3 & 3 & 4 \\ 6 & -5 & -1 \\ -2 & 8 & 2 \end{vmatrix}$

33. $\begin{vmatrix} -3 & 0 & 9 \\ 4 & 3 & -12 \\ 2 & 5 & 4 \end{vmatrix}$

34. $\begin{vmatrix} 6 & -3 & 5 \\ 4 & 4 & 12 \\ 2 & 3 & -8 \end{vmatrix}$

In Exercises 35–38, use Cramer's rule to solve each system of linear equations.

35. $\begin{aligned} 3x - 2y + z &= 10 \\ 2x + 2y - 3z &= 0 \\ -2x + 2y - z &= -8 \end{aligned}$

36. $\begin{aligned} x + y - 2z &= 5 \\ 3x - y + z &= -6 \\ 2x + 4y - 5z &= 1 \end{aligned}$

37. $\begin{aligned} 2x - 3y + 4z &= 5 \\ 3x + 4y - z &= 1 \\ 5x + 2y + 2z &= 2 \end{aligned}$

38. $\begin{aligned} 3x + 4y + 6z &= -7 \\ 2x + y - 3z &= 1 \\ 7x - 3y + 5z &= 8 \end{aligned}$

B

In Exercises 39–42, solve for x.

39. $\begin{vmatrix} 2 & 3 & -1 \\ 3 & 2 & 1 \\ 4 & x & 2 \end{vmatrix} = -5$

40. $\begin{vmatrix} 3 & 5 & 2 \\ 4 & x & 1 \\ -3 & 4 & -4 \end{vmatrix} = 103$

41. $\begin{vmatrix} x & x & 6 \\ 3 & x & 6 \\ 1 & 2 & 3 \end{vmatrix} = 18$

42. $\begin{vmatrix} x & -2 & -4 \\ -2 & x & 2 \\ 1 & 1 & 2 \end{vmatrix} = 20$

In Exercises 43–48, evaluate each determinant.

43. $\begin{vmatrix} 77 & 105 & 411 \\ 78 & 103 & 412 \\ 76 & 104 & 413 \end{vmatrix}$

44. $\begin{vmatrix} 13 & 16 & 19 \\ 28 & 34 & 40 \\ 27 & 33 & 39 \end{vmatrix}$

45. $\begin{vmatrix} -2 & 1 & 3 & 0 \\ 2 & 2 & -2 & 3 \\ 2 & -1 & 2 & -3 \\ 4 & 0 & -3 & 2 \end{vmatrix}$

46. $\begin{vmatrix} 4 & -3 & 2 & -2 \\ 0 & 2 & 1 & 0 \\ 2 & 4 & -3 & -2 \\ 3 & -2 & 4 & 5 \end{vmatrix}$

47. $\begin{vmatrix} 2 & 4 & 1 & 2 & 7 \\ 1 & -5 & -1 & -1 & 9 \\ 3 & 6 & 1 & 3 & 8 \\ 4 & 7 & -1 & -4 & -11 \\ 0 & 0 & 0 & 0 & 3 \end{vmatrix}$

48. $\begin{vmatrix} 8 & -5 & 4 & 3 & 2 \\ 2 & -3 & 4 & 2 & 0 \\ 2 & -5 & 4 & 3 & 2 \\ -1 & 2 & 4 & -6 & 0 \\ -2 & 1 & 3 & 4 & 0 \end{vmatrix}$

C

49 Show that interchanging the rows of $A = \begin{bmatrix} a_{11} & a_{12} \\ a_{21} & a_{22} \end{bmatrix}$ changes the sign of the determinant of A.

50 Show that multiplying row 1 of $A = \begin{bmatrix} a_{11} & a_{12} & a_{13} \\ a_{21} & a_{22} & a_{23} \\ a_{31} & a_{32} & a_{33} \end{bmatrix}$ by k multiplies the determinant of A by k.

KEY CONCEPTS

1 Matrix multiplication is not commutative.
It is possible to have matrices A and B such that $A \cdot B = 0$ but $A \neq 0$ and $B \neq 0$.

2 A^{-1} is the inverse of the square matrix A if $A \cdot A^{-1} = A^{-1} \cdot A = I$ where I is the identity matrix.

3 If $A \cdot X = B$ is the matrix equation representing a system of linear equations, then the solution of the system is $X = A^{-1}B$ if A^{-1} exists.

4 The determinant of the 2 × 2 matrix $A = \begin{bmatrix} a_{11} & a_{12} \\ a_{21} & a_{22} \end{bmatrix}$ is

$$\begin{vmatrix} a_{11} & a_{12} \\ a_{21} & a_{22} \end{vmatrix} = a_{11}a_{22} - a_{12}a_{21}$$

5 Determinants of 3 × 3 or larger square matrices can be determined by expansion by cofactors of any row or column.

6 Properties of determinants:
  1. $|B| = -|A|$ if matrix B is obtained by interchanging two rows (columns) of matrix A.
  2. $|B| = k|A|$ if matrix B is obtained by multiplying a row (column) of matrix A by a constant k.
  3. $|B| = |A|$ if matrix B is obtained by replacing a row (column) of matrix A by the sum of that row (column) and a constant multiple of another row (column).

7 The determinant of a square matrix is zero
  1. if the matrix has a row or column of zeros.
  2. if the matrix has two identical rows or columns.
  3. if one row (column) is a multiple of another row (column).

8 Cramer's rule for a 3 × 3 linear system of equations: If $D \neq 0$, the solution of the system is $x = \dfrac{D_x}{D}$, $y = \dfrac{D_y}{D}$, and $z = \dfrac{D_z}{D}$.

REVIEW EXERCISES FOR CHAPTER 7

Use $A = \begin{bmatrix} 2 & -1 & 4 & 3 \\ 0 & 5 & 4 & -3 \end{bmatrix}$, $B = \begin{bmatrix} 3 & -1 & -4 & 0 \\ 2 & 1 & -5 & -1 \end{bmatrix}$, and $C = \begin{bmatrix} 1 \\ 0 \\ -1 \\ 4 \end{bmatrix}$ to evaluate each item in Exercises 1–9.

1. The order of each matrix.
2. $a_{12}, b_{23},$ and c_{31}
3. $A + B$
4. $A - B$
5. $3A - 2B$
6. The scalar product $C^t \cdot C$
7. $C \cdot A$
8. $B \cdot A^t$
9. $(B^t)^t$
10. Find x and y so that $A = B$.

$$A = \begin{bmatrix} x^2 & 2y-1 \\ 3 & 11 \end{bmatrix}, B = \begin{bmatrix} 7x-12 & 9 \\ 3 & 11 \end{bmatrix}$$

11. Solve for a and b so that $A_{a \times 2} \cdot B_{2 \times b} = C_{3 \times 4}$.
12. Determine if $B = A^{-1}$.

$$A = \begin{bmatrix} 3 & 4 \\ -2 & 1 \end{bmatrix}, B = \begin{bmatrix} -3 & -4 \\ 2 & -1 \end{bmatrix}$$

Use the following information for Exercises 13–17. A small plant manufactures four types of utility trailers: compact (C), flat-bed (F), verstatile (V), and heavy-duty (H). The two matrices indicate the number of each type produced on a day shift and on a night shift.

$$\text{Day shift} \qquad \text{Night shift}$$

$$D = \begin{bmatrix} 25 \\ 19 \\ 15 \\ 12 \end{bmatrix} \begin{matrix} C \\ F \\ V \\ H \end{matrix} \qquad N = \begin{bmatrix} 19 \\ 14 \\ 11 \\ 9 \end{bmatrix} \begin{matrix} C \\ F \\ V \\ H \end{matrix}$$

13. Use matrix addition to compute the number of each type of trailer produced in 24 hours.
14. The day shift produced more trailers of each type than did the night shift. Use matrix subtraction to find how many more were produced.
15. Use scalar multiplication to determine the number of each kind of trailer produced on each shift in a five-day week.
16. Multiply the matrices D and N on the left by $[1 \ 1 \ 1 \ 1]$ to compute the total number of trailers produced on each shift.
17. The wholesale and suggested retail prices for the trailers are given in the matrix P.

$$\begin{matrix} & C & F & V & H & \end{matrix}$$
$$P = \begin{bmatrix} 200 & 450 & 475 & 650 \\ 250 & 525 & 575 & 675 \end{bmatrix} \begin{matrix} \text{Wholesale price} \\ \text{Retail price} \end{matrix}$$

Use matrix multiplication to compute the total wholesale and total retail revenue generated by selling all of the trailers made by the day shift.

18 Write in matrix form the system of linear equations

$$3x - 2y + z = -2$$
$$6x + 9y - 5z = 4$$
$$2x - 5y + 3z = 3$$

19 Let $A = \begin{bmatrix} 2 & 5 & -4 \\ 6 & -1 & 2 \\ 1 & 3 & 5 \end{bmatrix}$, $X = \begin{bmatrix} x \\ y \\ z \end{bmatrix}$, and $B = \begin{bmatrix} 12 \\ -10 \\ 5 \end{bmatrix}$. Write the system of linear equations associated with $A \cdot X = B$.

In Exercises 20 and 21, determine the inverse of each matrix.

20 $\begin{bmatrix} 3 & 6 \\ -2 & 1 \end{bmatrix}$

21 $\begin{bmatrix} 1 & 3 & -1 \\ 0 & 2 & 1 \\ 0 & 0 & 5 \end{bmatrix}$

In Exercises 22 and 23, solve each system of linear equations by using the inverse of the coefficient matrix.

22
$3a + 2b = 6$
$2a - b = 5$

23
$3r + 2s - t = 5$
$-2r + s + t = 6$
$2r + 2s - 3t = 7$

24 A small furniture factory makes tables with three types of surfaces: regular, laminated, and inlaid. The chart below shows the number of hours required for the three operations of shaping, assembling, and finishing.

	Table		
	Regular	Laminated	Inlaid
Hours of Shaping	2.0	2.0	3.0
Hours of Assembling	0.5	0.5	1.5
Hours of Finishing	3.0	2.5	5.0

Because of vacations, doctor's appointments, etc., the number of hours of labor for each operation varies each week during a month, as shown in the chart below.

	Labor Available			
	Week 1	Week 2	Week 3	Week 4
Hours for Shaping	44	54	52	54
Hours for Assembling	17	18	16	18
Hours for Finishing	70	78	70	79

Use the inverse of a matrix to determine the number of each type of table that can be made each week using all of the available labor.

REVIEW EXERCISES FOR CHAPTER 7

25 Evaluate $\begin{vmatrix} 5 & 6 \\ -2 & 3 \end{vmatrix}$.

26 Use Cramer's rule to solve $\begin{matrix} 3a + 2b = 6 \\ 5a - b = -5 \end{matrix}$.

27 Let $A = \begin{bmatrix} 2 & -1 & 3 \\ 4 & 0 & 5 \\ -3 & 2 & 1 \end{bmatrix}$

    a. Find the minor and the cofactor for each entry in row 1.
    b. Find the minor and the cofactor for each entry in column 2.
    c. Compute the determinant of A by expanding along row 1.
    d. Compute the determinant of A by expanding along column 2.

In Exercises 28–34, use $\begin{vmatrix} 3 & -5 & 8 \\ 2 & 4 & 7 \\ 1 & -3 & 2 \end{vmatrix} = -8$

and the properties of determinants to evaluate each determinant by inspection.

28 $\begin{vmatrix} 3 & 3 & 1 \\ -5 & 4 & -3 \\ 8 & 7 & 2 \end{vmatrix}$ **29** $\begin{vmatrix} 3 & -10 & 8 \\ 2 & 8 & 7 \\ 1 & -6 & 2 \end{vmatrix}$ **30** $\begin{vmatrix} 3 & -5 & 8 \\ 0 & 10 & 3 \\ 1 & -3 & 2 \end{vmatrix}$ **31** $\begin{vmatrix} 3 & -5 & 8 \\ 1 & -3 & 2 \\ 2 & 4 & 7 \end{vmatrix}$

32 $\begin{vmatrix} 3 & -5 & 2 \\ 2 & 4 & 3 \\ 1 & -3 & 0 \end{vmatrix}$ **33** $\begin{vmatrix} -5 & 3 & 8 \\ 4 & 2 & 7 \\ -3 & 1 & 2 \end{vmatrix}$ **34** $\begin{vmatrix} 3 & -5 & 8 \\ 2 & 4 & 7 \\ -3 & 9 & -6 \end{vmatrix}$

In Exercises 35–42, evaluate each determinant by inspection or by using the properties of determinants to obtain a row or column with at most one nonzero entry.

35 $\begin{vmatrix} 2 & 0 & 3 \\ -6 & 0 & -8 \\ 5 & 0 & 7 \end{vmatrix}$ **36** $\begin{vmatrix} 1 & 2 & -5 \\ 0 & 1 & 6 \\ 2 & -3 & 5 \end{vmatrix}$ **37** $\begin{vmatrix} 3 & 4 & -2 \\ 5 & -6 & 1 \\ -2 & 4 & 3 \end{vmatrix}$

38 $\begin{vmatrix} 6 & 5 & -7 \\ -8 & 4 & 3 \\ 6 & 5 & -7 \end{vmatrix}$ **39** $\begin{vmatrix} a_{11} & 0 & 0 \\ a_{21} & a_{22} & 0 \\ a_{31} & a_{32} & a_{33} \end{vmatrix}$ **40** $\begin{vmatrix} 3 & 3 & 2 & -6 \\ 2 & 4 & 1 & -8 \\ -1 & -1 & -1 & 2 \\ 1 & 5 & 1 & -10 \end{vmatrix}$

41 $\begin{vmatrix} 2 & 1 & 7 & 4 \\ 0 & 3 & 8 & 1 \\ 0 & -2 & 9 & 1 \\ 0 & 4 & -3 & 5 \end{vmatrix}$ **42** $\begin{vmatrix} 3 & 5 & -2 & 6 \\ 2 & 0 & 1 & 0 \\ 3 & 4 & -5 & -1 \\ 2 & -2 & 3 & -4 \end{vmatrix}$

43 Use Cramer's rule to solve $\begin{matrix} x_1 + x_2 + x_3 = 5 \\ x_2 - x_3 = -2. \\ x_1 - x_2 + x_3 = -3 \end{matrix}$

44 Write the equation of the circle, in the form $x^2 + y^2 + Ax + By + C = 0$, that passes through the points (1, 1), (2, 4), and (4, 3). Use Cramer's rule to solve the system of equations for A, B, and C.

OPTIONAL EXERCISES FOR CALCULUS-BOUND STUDENTS

1. Find two 3×3 matrices A and B such that $A \cdot B \neq B \cdot A$.
2. Find two nonzero 2×2 matrices A and B such that $A \cdot B = 0$.
3. If $A = \begin{bmatrix} 4 & -2 \\ 3 & -2 \end{bmatrix}$ and $B = \begin{bmatrix} 3 & 6 \\ 1 & 3 \end{bmatrix}$:
   a. Find A^{-1}. b. Find B^{-1}. c. Find $A \cdot B$. d. Find $(A \cdot B)^{-1}$.
   e. Find $B^{-1} \cdot A^{-1}$. f. Find $A^{-1} \cdot B^{-1}$. g. Compare the answers for parts d, e, and f.
4. Find the inverse of $\begin{bmatrix} 1 & -1 & 3 \\ 2 & -3 & 1 \\ 3 & -4 & 2 \end{bmatrix}$.
5. Use the inverse matrix method to solve $\begin{array}{r} x + y - 2z = 4 \\ x - 2y + 3z = -4 \\ 2x + 2y - z = 5 \end{array}$.
6. Find A, if $(A^t)^{-1} = \begin{bmatrix} 2 & 3 \\ 2 & 4 \end{bmatrix}$.
7. Prove the commutative property of addition for all 2×2 matrices.
8. Prove the associative property of addition for all 2×2 matrices.
9. a. Show that the area of triangle ABC in the figure is given by
$$A = \frac{1}{2} \begin{vmatrix} x_1 & y_1 & 1 \\ x_2 & y_2 & 1 \\ x_3 & y_3 & 1 \end{vmatrix}$$
(*Hint*: $A = $ Area $ADEC$ + Area $CEFB$ − Area $ADFB$.)

   b. In general, why must the area be given as the absolute value of the expression in part (a)?

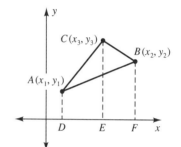

Figure for Exercise 9

10. Show that $\begin{vmatrix} x & y & 1 \\ x_1 & y_1 & 1 \\ x_2 & y_2 & 1 \end{vmatrix} = 0$ is the equation of a line containing the points (x_1, y_1) and (x_2, y_2).

11. Show that the row operation $r'_3 = r_3 + kr_2$ applied to
$$A = \begin{bmatrix} a_{11} & a_{12} & a_{13} \\ a_{21} & a_{22} & a_{23} \\ a_{31} & a_{32} & a_{33} \end{bmatrix}$$ does not change the value of the determinant.

12. Use Cramer's rule to solve $\begin{array}{r} x - y + 3z = 1 \\ 2x - 3y + z = -3 \\ 3x - 4y + 2z = 5 \end{array}$.

13. Solve $\begin{vmatrix} x & 2 & 3 \\ 3 & x & 4 \\ 1 & 0 & 2 \end{vmatrix} = 1$ for x.

MASTERY TEST FOR CHAPTER 7

Exercise numbers correspond to Section Objective numbers.

1. Given $A = \begin{bmatrix} 1 & 5 & -6 \\ 0 & 7 & -4 \end{bmatrix}$, $B = \begin{bmatrix} 2 & -1 & 3 \\ 4 & -5 & 7 \end{bmatrix}$, and $C = \begin{bmatrix} 3 & 2 \\ 0 & 2 \\ -1 & -3 \end{bmatrix}$,

   evaluate each of the following expressions.

   a. $A + B$ **b.** $A - B$ **c.** $-2A$ **d.** $3A - 4B$ **e.** $A \cdot C$ **f.** B^t

2. Find the inverse of the matrix $A = \begin{bmatrix} 1 & 2 & -1 \\ -1 & -1 & 2 \\ 4 & 6 & -4 \end{bmatrix}$.

3. Use the inverse of the coefficient matrix to solve $\begin{cases} 4x - 3y = 9 \\ 2x + 3y = 5 \end{cases}$.

4. Evaluate $\begin{vmatrix} 1 & 0 & -1 \\ 2 & -2 & 3 \\ -1 & 5 & 0 \end{vmatrix}$.

5. Solve the system $\begin{cases} x - 3y + 2z = -20 \\ 5x + 9y - 8z = -34 \\ 4x - 7y + z = -39 \end{cases}$ using Cramer's rule.

6. Evaluate these determinants.

   a. $\begin{vmatrix} 17 & 18 & 93 \\ 43 & 73 & 49 \\ 17 & 18 & 93 \end{vmatrix}$ **b.** $\begin{vmatrix} 17 & 18 & 93 \\ 53 & 53 & 278 \\ 17 & 19 & 91 \end{vmatrix}$ **c.** $\begin{vmatrix} 2 & -1 & 0 & 1 \\ 1 & -1 & 1 & 5 \\ -1 & 3 & 0 & -2 \\ 3 & 0 & -1 & 1 \end{vmatrix}$

CHAPTER EIGHT

Sequences and Counting Problems

Chapter Eight Objectives

1. Write the terms of a sequence given the general term.
2. Use summation notation and evaluate the series associated with a finite sequence.
3. Calculate the nth term of an arithmetic sequence.
4. Find the sum of a finite arithmetic sequence.
5. Calculate the nth term of a geometric sequence.
6. Find the sum of a finite geometric sequence.
7. Evaluate an infinite geometric series.
8. Use mathematical induction to prove statements true for all natural numbers n.
9. Evaluate factorial expressions.
10. Determine a given term of a binomial expansion.
11. Use the fundamental counting principle.
12. Determine the number of permutations or combinations of n objects taken r at a time.
13. Give the sample space for an experiment.
14. Give the set of occurrences for an event and the probability of an event.
15. Use the probability formulas for $P(\tilde{E})$, $P(A \cap B)$, and $P(A \cup B)$.

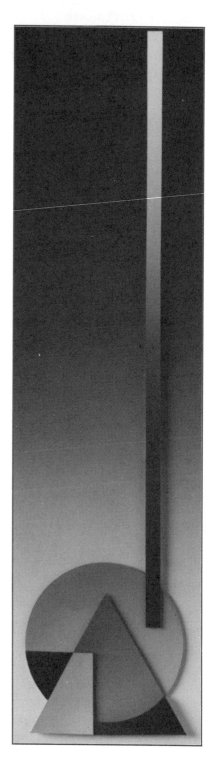

Johann Friederich Carl Gauss, 1777-1855

The German Gauss is known as the Prince of Mathematicians. Later mathematicians noted that before 1800 Gauss had foreseen much of nineteenth-century mathematics. The talents of this child prodigy were recognized early. At the age of 3, Gauss found an error in the weekly payroll prepared by his father. At the age of 10, he mentally computed the sum of an arithmetic progression—a problem that none of his fellow students answered correctly by the end of the hour. What he had recognized almost instantly was the general strategy applicable to all arithmetic progressions.

Events in nature frequently form regular and interesting patterns. Learning more about a particular subject often involves the recognition and analysis of these patterns. The ability to recognize patterns is such an important skill that many IQ and job placement tests contain questions designed to test this ability. For example, test-takers might be asked to write the next number in each of the following numerical patterns:

$$2, 5, 8, 11, 14,$$
$$2, 6, 18, 54, 162,$$
$$1, 1, 2, 3, 5, 8,$$
$$0, 3, 8, 15, 24, 35,$$

We will examine these sequences in this chapter.

SECTION 8-1

Sequences and Series

Section Objectives

1. Write the terms of a sequence given the general term.
2. Use summation notation and evaluate the series associated with a finite sequence.

Sequences are used in many fields, such as

- business—to enumerate the payments necessary to repay a loan
- biology—to describe the growth pattern of a living organism

- calculator design—to specify the sequence of terms used to calculate functions such as e^x and $\cos x$
- calculus—to give the areas of a sequence of rectangles used to approximate the area of a region

Formally, a **sequence** is a function whose domain is a set of consecutive natural numbers. For example, the sequence 2, 5, 8, 11, 14, 17 can be viewed as the function $\{(1, 2), (2, 5), (3, 8), (4, 11), (5, 14), (6, 17)\}$ whose domain is the set of natural numbers $\{1, 2, 3, 4, 5, 6\}$ (see Table 8-1). We usually do not write the domain and instead indicate it by listing the range elements in a specific order. The range elements are called the **terms** of the sequence. Throughout the rest of this chapter, the elements in the range will be restricted to the real numbers. A **finite sequence** has a last term; a sequence that continues without end is called an **infinite sequence**. A finite sequence with n terms can be denoted by $a_1, a_2, \ldots, a_n$, and an infinite sequence can be denoted by $a_1, a_2, \ldots, a_n, \ldots$. Since a_n can represent any term of a sequence, it is useful to have a formula that gives a_n. The formula for a_n is called the **general term** of the sequence. For example, if $a_n = 3n - 1$, then $a_1 = 3(1) - 1 = 2$, $a_2 = 3(2) - 1 = 5, \ldots, a_6 = 3(6) - 1 = 17$. Some sample sequences are

Table 8-1 Alternative notations for the sequence 2, 5, 8, 11, 14, 17

$1 \to 2$	$f(1) = 2$	$a_1 = 2$
$2 \to 5$	$f(2) = 5$	$a_2 = 5$
$3 \to 8$	$f(3) = 8$	$a_3 = 8$
$4 \to 11$	$f(4) = 11$	$a_4 = 11$
$5 \to 14$	$f(5) = 14$	$a_5 = 14$
$6 \to 17$	$f(6) = 17$	$a_6 = 17$

- 2, 6, 18, 54, 162, 486, which is a finite sequence with six terms. $a_1 = 2, a_2 = 6, \ldots, a_6 = 486$. The general term is $a_n = 2(3^{n-1})$.
- 0, 3, 8, 15, 24, 35, 48, $\ldots$, $n^2 - 1, \ldots$, which is an infinite sequence. The general term is $a_n = n^2 - 1$.

EXAMPLE 1 The general term of a finite sequence with five terms is $a_n = 7n + 3$. Write the sequence.

SOLUTION $a_n = 7n + 3$

$a_1 = 7(1) + 3 = 10$
$a_2 = 7(2) + 3 = 17$
$a_3 = 7(3) + 3 = 24$
$a_4 = 7(4) + 3 = 31$
$a_5 = 7(5) + 3 = 38$

Answer 10, 17, 24, 31, 38 □

EXAMPLE 2 Determine a_{50} if $a_n = n^2 + 4n - 3$.
SOLUTION $a_{50} = (50)^2 + 4(50) - 3 = 2697$ □

SECTION 8-1 SEQUENCES AND SERIES

The general term of a sequence is sometimes determined by one or more of the preceding terms. A sequence determined in this manner is said to be **defined recursively**. An example of a recursion formula is $a_n = 2a_{n-1}$.

EXAMPLE 3 Write the first five terms of the sequence defined by each recursion formula.

(a) $a_1 = 3$ and $a_n = 2a_{n-1}$ for $n > 1$

(b) The Fibonacci sequence having $a_1 = 1$, $a_2 = 1$, and $a_n = a_{n-2} + a_{n-1}$ for $n > 2$

SOLUTIONS

(a) $a_1 = 3$
$a_2 = 2a_1 = 2(3) = 6$
$a_3 = 2a_2 = 2(6) = 12$
$a_4 = 2a_3 = 2(12) = 24$
$a_5 = 2a_4 = 2(24) = 48$

Answer 3, 6, 12, 24, 48

(b) $a_1 = 1$ The first two terms are given.
$a_2 = 1$
$a_3 = a_1 + a_2 = 1 + 1 = 2$ For $n = 3$, a_n is a_3, a_{n-2} is a_1, and a_{n-1} is a_2.
$a_4 = a_2 + a_3 = 1 + 2 = 3$
$a_5 = a_3 + a_4 = 2 + 3 = 5$

Answer 1, 1, 2, 3, 5

EXAMPLE 4 Graph the first six terms of the infinite sequence whose general term is $a_n = 6n - n^2$.

SOLUTION

n	a_n
1	5
2	8
3	9
4	8
5	5
6	0

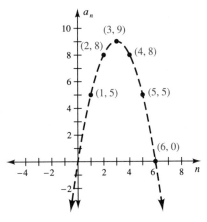

Since the domain consists only of the natural numbers, the graph consists only of discrete points and is not a solid curve. The dashed curve is shown solely to emphasize the overall pattern exhibited by this function.

▶ **Self-Check** ▼

Write the first five terms of each sequence.

1 $a_n = \dfrac{1}{2n}$

2 $a_1 = 1$, $a_n = na_{n-1}$ for $n > 1$

▶ **Self-Check Answers** ▼

1 $\dfrac{1}{2}, \dfrac{1}{4}, \dfrac{1}{6}, \dfrac{1}{8}, \dfrac{1}{10}$ **2** 1, 2, 6, 24, 120

Some sequences have no formula for their general term. For other sequences there can be many formulas for the general term. Since the variety of sequences is infinite, there is no set procedure for determining such a formula even if one exists. One way to approach this problem systematically is to graph several terms of the sequence in search of a familiar visual pattern. The graph may then suggest a linear, parabolic, exponential, or logarithmic relationship.

EXAMPLE 5 Determine the general term of the infinite sequence whose first five terms are $-3, 2, 7, 12,$ and 17.

SOLUTION

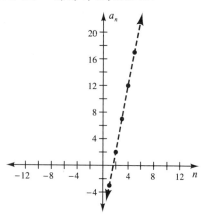

Graph these five ordered pairs and note that this relation appears linear. We can use any two points to determine the linear equation.

n	a_n
1	-3
2	2
3	7
4	12
5	17

$$m = \frac{2-(-3)}{2-1} = 5$$

Calculate the slope of the line using the points $(1, -3)$ and $(2, 2)$.

$$a_n - (-3) = m(n-1)$$

This is the point-slope form of the line with variables n and a_n and containing the point $(1, -3)$.

$$a_n + 3 = 5(n-1)$$

Substitute for m, the slope.

Answer The general term is $a_n = 5n - 8$.

The sum of the terms of a sequence is called a **series**. If $a_1, a_2, \ldots, a_n$ is a finite sequence, then the indicated sum $a_1 + a_2 + \cdots + a_n$ is the series associated with this sequence.

EXAMPLE 6 Find the value of the six-term series associated with the sequence whose general term is $a_n = n^2$.

SOLUTION The series is the sum

$$a_1 + a_2 + a_3 + a_4 + a_5 + a_6 = 1^2 + 2^2 + 3^2 + 4^2 + 5^2 + 6^2$$
$$= 1 + 4 + 9 + 16 + 25 + 36$$
$$= 91$$

Substitute the first six natural numbers into the formula $a_n = n^2$ to determine the first six terms and then add these terms.

SECTION 8-1 SEQUENCES AND SERIES

A convenient notation for denoting a series is called **summation notation**. This notation uses the greek letter Σ (called sigma, corresponding to S for sum) to denote the summation.

Summation Notation ▼

$$\sum_{i=1}^{n} a_i = a_1 + a_2 + \cdots + a_{n-1} + a_n$$

Index variable — Initial value of index — Formula for general term — Terminal value of index

This is read "the sum from i equals 1 to n, of a sub i."

Generally the index variable is denoted by $i, j, k, m,$ or n. The index variable is always replaced with successive integers from the initial value through the terminal value. For example, in $\sum_{k=1}^{5} a_k$, k is replaced with 1, 2, 3, 4, and then 5, to yield

$$\sum_{k=1}^{5} a_k = a_1 + a_2 + a_3 + a_4 + a_5.$$

EXAMPLE 7 Evaluate these series.

SOLUTIONS

(a) $\sum_{i=1}^{6} 2i$

$\sum_{i=1}^{6} 2i = 2(1) + 2(2) + 2(3) + 2(4) + 2(5) + 2(6)$

$= 2 + 4 + 6 + 8 + 10 + 12$

$= 42$

Replace the index variable i with the successive integers 1 through 6 and then sum these terms.

(b) $\sum_{j=4}^{7} \frac{1}{3j}$

$\sum_{j=4}^{7} \frac{1}{3j} = \frac{1}{3(4)} + \frac{1}{3(5)} + \frac{1}{3(6)} + \frac{1}{3(7)}$

$= \frac{105 + 84 + 70 + 60}{1260}$

$= \frac{319}{1260}$

Replace the index variable j with the successive integers 4 through 7 and then sum these terms.

(c) $\sum_{k=1}^{50} 8k^0$

$\sum_{k=1}^{50} 8k^0 = \sum_{k=1}^{50} 8 = \underbrace{8 + 8 + \cdots + 8}_{50 \text{ terms}}$

$= 50(8)$

$= 400$

This series is the sum of 50 terms with each term equal to 8. This example illustrates that the formula for a_k does not have to involve the index variable k.

□

In calculus the same series can often be represented in more than one manner. Example 8 illustrates one way to compare two series in order to determine whether they are the same.

EXAMPLE 8 Verify that $\sum_{i=1}^{5} 2^{i-1}$ and $\sum_{j=0}^{4} 2^{j}$ represent the same series.

SOLUTION
$$\sum_{i=1}^{5} 2^{i-1} = 2^0 + 2^1 + 2^2 + 2^3 + 2^4$$
$$= 31$$

and
$$\sum_{j=0}^{4} 2^{j} = 2^0 + 2^1 + 2^2 + 2^3 + 2^4$$
$$= 31$$

Thus $\sum_{i=1}^{5} 2^{i-1} = \sum_{j=0}^{4} 2^{j}$ ☐

▶ **Self-Check** ▼

Evaluate these series.

1 $\sum_{k=1}^{4} (k^2 + 2k)$ **2** $\sum_{j=3}^{6} (j + 10)$

Although one cannot actually add an infinite number of nonzero terms, the infinite series, as denoted below, is considered in calculus where this "sum" is defined in terms of partial sums. It is these partial sums that we will consider here.

$$\sum_{k=1}^{\infty} a_k = a_1 + a_2 + a_3 + a_4 + a_5 + a_6 + \cdots + a_k + \cdots$$

The partial sums of $\sum_{k=1}^{\infty} a_k$ are:

$$s_1 = a_1$$
$$s_2 = a_1 + a_2$$
$$s_3 = a_1 + a_2 + a_3$$
$$s_4 = a_1 + a_2 + a_3 + a_4$$
$$\vdots$$
$$s_n = a_1 + a_2 + a_3 + a_4 + \cdots + a_n$$

The **nth partial sum**, denoted by s_n, is the sum of the first n terms of the infinite series. The sequence $\{s_n\}$ is called the **sequence of partial sums**.

▶ **Self-Check Answers** ▼

1 50   **2** 58

SECTION 8-1 SEQUENCES AND SERIES

EXAMPLE 9 Given the infinite series $\sum_{k=1}^{\infty} \frac{1}{2^k}$,

(a) find $s_1, s_2, s_3,$ and s_4.
(b) find a formula for s_n, the nth partial sum.

SOLUTIONS

(a) s_1 $\quad s_1 = a_1 = \dfrac{1}{2^1} = \dfrac{1}{2}$

s_2 $\quad s_2 = a_1 + a_2 = \dfrac{1}{2^1} + \dfrac{1}{2^2}$

$\qquad \qquad = \dfrac{1}{2} + \dfrac{1}{4}$

$\qquad \qquad = \dfrac{3}{4}$

s_3 $\quad s_3 = a_1 + a_2 + a_3 = \dfrac{1}{2^1} + \dfrac{1}{2^2} + \dfrac{1}{2^3}$

$\qquad \qquad = \dfrac{1}{2} + \dfrac{1}{4} + \dfrac{1}{8}$

$\qquad \qquad = \dfrac{7}{8}$

s_4 $\quad s_4 = a_1 + a_2 + a_3 + a_4 = \dfrac{1}{2^1} + \dfrac{1}{2^2} + \dfrac{1}{2^3} + \dfrac{1}{2^4}$

$\qquad \qquad = \dfrac{1}{2} + \dfrac{1}{4} + \dfrac{1}{8} + \dfrac{1}{16}$

$\qquad \qquad = \dfrac{15}{16}$

Substitute 1, 2, 3, and then 4 into $\dfrac{1}{2^k}$, the formula for the kth term of the series.

(b) s_n $\quad s_n = \dfrac{2^n - 1}{2^n}$

The formula for s_n is determined by observing the pattern exhibited by the first four partial sums:

$s_1 = \dfrac{1}{2} = \dfrac{2^1 - 1}{2^1}, \quad s_2 = \dfrac{3}{4} = \dfrac{2^2 - 1}{2^2},$

$s_3 = \dfrac{7}{8} = \dfrac{2^3 - 1}{2^3},$ and

$s_4 = \dfrac{15}{16} = \dfrac{2^4 - 1}{2^4}.$

□

EXERCISES 8-1

A

In Exercises 1–14, write the first five terms of the infinite sequence whose general term is given.

1. $a_n = 9 - 2n$
2. $a_n = 5n + 3$
3. $a_i = i^3 - 2i^2 + 1$
4. $a_i = 3i^2 + 4i - 2$
5. $a_j = (-2)^j$
6. $a_j = (-3)^{-j}$
7. $a_k = \dfrac{1}{5k+1}$
8. $a_k = \dfrac{1}{2k-1}$
9. $a_i = \dfrac{i}{i+1}(-1)^{i+1}$
10. $a_i = \dfrac{i^2}{i+1}(-1)^{i-1}$
11. $a_1 = 2, a_n = a_{n-1} + 3$ for $n > 1$
12. $a_1 = -3, a_n = (-1)^n a_1$ for $n > 1$
13. $a_1 = 1, a_2 = 2, a_n = a_{n-2}(a_{n-1})$ for $n > 2$
14. $a_1 = 36, a_2 = 12, a_n = \dfrac{a_{n-2}}{a_{n-1}}$ for $n > 2$

In Exercises 15–18, evaluate the term requested from the given general term.

15. $a_n = \dfrac{n^3 - 1}{n^3 + 1}, a_{15} = ?$
16. $a_n = \dfrac{n^2}{n^2 + 1}, a_{20} = ?$
17. $a_i = (-1)^{i-1} i^3, a_{100} = ?$
18. $a_i = i^i, a_4 = ?$

In Exercises 19–32, evaluate the series given.

19. $3 + 5 + 7 + 9 + \cdots + 17$
20. $2 + 4 + 6 + 8 + \cdots + 20$
21. $a_1 + a_2 + \cdots + a_7$, given $a_n = 3n - 1$
22. $a_1 + a_2 + \cdots + a_5$, given $a_n = n^3$
23. $\sum_{i=1}^{5} (i^2 - 4)$
24. $\sum_{j=5}^{9} (2j + 3)$
25. $\sum_{k=3}^{6} 2^k$
26. $\sum_{k=4}^{6} \dfrac{1}{k}$
27. $\sum_{i=1}^{5} 10^{-i}$
28. $\sum_{i=1}^{100} 7$
29. $\sum_{j=1}^{1000} 3$
30. $\sum_{j=1}^{4} (\tfrac{2}{3})^{j-2}$
31. $\sum_{k=1}^{4} (-1)^k (3k + 1)$
32. $\sum_{k=1}^{5} (-1)^{k-1} (5k - 2)$

B

In Exercises 33 and 34, graph the first five terms of each sequence whose general term is given.

33. $a_n = 2n - 3$
34. $a_n = (-1)^n (n^2 - 2n)$

In Exercises 35 and 36, write out the series given and determine whether they are equal.

35. $\sum_{i=0}^{6} (2i + 1), \sum_{j=1}^{7} 2j$
36. $\sum_{i=1}^{4} \dfrac{1}{2^i}, \sum_{k=0}^{3} \dfrac{1}{2^{k+1}}$

In Exercises 37–40, find the partial sums s_1, s_2, s_3, and s_4 for each infinite series.

37. $\sum_{k=1}^{\infty} \dfrac{1}{k^2}$
38. $\sum_{k=1}^{\infty} \dfrac{1}{4^k}$
39. $\sum_{k=1}^{\infty} \left(\dfrac{2}{5}\right)^k$
40. $\sum_{k=1}^{\infty} \dfrac{1}{n(n+1)}$

In Exercises 41–45, write each series in summation notation.

41. $5 + 10 + 15 + 20 + 25$
42. $\tfrac{1}{2} + \tfrac{2}{3} + \tfrac{3}{4} + \tfrac{4}{5}$
43. $49 + 64 + 81 + 100 + 121$
44. $\tfrac{2}{7} - \tfrac{2}{8} + \tfrac{2}{9} - \tfrac{2}{10} + \tfrac{2}{11} - \tfrac{2}{12}$
45. $-3 + 5 - 7 + 9 - 11 + 13 - 15$

SECTION 8-2 ARITHMETIC SEQUENCES AND SERIES

C

In Exercises 46–48, graph the terms, determine the nature of the relationship, and then find the general term a_n.

46 $-6, 1, 8, 15, 22$

47 $4, 0, -2, -2, 0, 4$

48 $-10, -12, -12, -10, -6$

49 Find the partial sums $s_1, s_2, s_3,$ and s_4 for $\sum_{k=1}^{\infty} \frac{1}{3^n}$.

50 A small business purchased $720 worth of reference books and depreciated them over an eight-year period. The deductions for depreciation were $160 for the first year, $140 for the second year, and $120 for the third year. Give the depreciation for the last five years.

51 Write the first five terms of the sequence defined recursively by $P_1 = 20{,}000$ and $P_n = 1.0075 P_{n-1} - 200$. This sequence represents the remaining principal on a loan of $20,000 borrowed at 9% interest after a monthly payment of $200. Record each term to the nearest penny.

52 A biologist studying the reproductive capabilities of a culture puts 1.2 kilograms into a test vat. The test vat contains 2.4 kilograms after 8 hours, 4.8 kilograms after 16 hours, and 9.6 kilograms after 24 hours. Predict the amount that will be present 48 hours after the start of the experiment.

53 A compact body is carefully observed during free fall from the top of a skyscraper, as shown in the figure to the right. In the first second the body travels 4.89 meters. At the end of 2 seconds it has traveled a total of 19.56 meters, and at the end of 3 seconds it has traveled a total of 44.01 meters. Predict how far it will fall by the end of 4 seconds. (*Hint:* Factor out 4.89 from each term.)

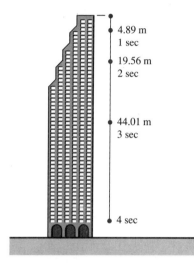

Figure for Exercise 53

In Exercises 54 and 55, use a calculator to approximate the series.

54 $\cos 1 \approx 1 - \dfrac{1}{1 \cdot 2} + \dfrac{1}{1 \cdot 2 \cdot 3 \cdot 4} - \dfrac{1}{1 \cdot 2 \cdot 3 \cdot 4 \cdot 5 \cdot 6} + \dfrac{1}{1 \cdot 2 \cdot 3 \cdot 4 \cdot 5 \cdot 6 \cdot 7 \cdot 8}$

55 $e \approx 1 + \dfrac{1}{1} + \dfrac{1}{1 \cdot 2} + \dfrac{1}{1 \cdot 2 \cdot 3} + \dfrac{1}{1 \cdot 2 \cdot 3 \cdot 4} + \dfrac{1}{1 \cdot 2 \cdot 3 \cdot 4 \cdot 5} + \dfrac{1}{1 \cdot 2 \cdot 3 \cdot 4 \cdot 5 \cdot 6}$

SECTION 8-2

Arithmetic Sequences and Series

Section Objectives

3 Calculate the nth term of an arithmetic sequence.

4 Find the sum of a finite arithmetic sequence.

An **arithmetic sequence** or **arithmetic progression** is a sequence in which consecutive terms differ by a constant amount. This constant amount is called the **common difference** and is denoted by d.

EXAMPLE 1 Determine which of these sequences are arithmetic. For those that are arithmetic, give the common difference, d.

SOLUTIONS

(a) 7, 11, 15, 19, 23, ... This sequence is arithmetic, since the difference between any two consecutive terms is 4.

$11 - 7 = 4$, $15 - 11 = 4$,
$19 - 15 = 4$, $23 - 19 = 4$

(b) 19, 12, 5, -2, -9, ... This sequence is arithmetic, with $d = 12 - 19 = -7$.

$12 - 19 = -7$, $5 - 12 = -7$,
$-2 - 5 = -7$, $-9 - (-2) = -7$

(c) 8, 13, 20, 25, 30, ... This sequence is not arithmetic, since $13 - 8 = 5$ but $20 - 13 = 7$.

The difference between consecutive terms in this sequence is not constant.

Since the common difference is given by $a_n - a_{n-1} = d$, the general term can be found by the recursive formula $a_n = a_{n-1} + d$ for $n > 1$. This allows us to rewrite the terms $a_1, a_2, a_3, a_4, \ldots, a_n$ as

$a_1 \qquad\qquad = a_1$
$a_2 = a_1 + d = a_1 + \qquad d$
$a_3 = a_2 + d = a_1 + \qquad 2d$
$a_4 = a_3 + d = a_1 + \qquad 3d$
$\vdots$
$a_n = a_{n-1} + d = a_1 + (n-1)d$

This pattern continues, with the common difference added once for each term except the first term.

▶ **Self-Check** ▼

Determine which of these sequences are arithmetic. For those that are arithmetic, give the common difference, d.

1 6, 1, -4, -9, ...
2 4, 12, 36, 108, 324, ...
3 7.2, 7.7, 8.2, 8.7, 9.2, ...

Arithmetic Sequence ▼

If $a_n - a_{n-1} = d$ for $n > 1$, then the sequence is arithmetic and the constant d is called the common difference. The formula for the nth term is $a_n = a_1 + (n-1)d$.

EXAMPLE 2 Write the first six terms of the arithmetic sequence with $a_1 = 9$ and $d = -6$.

SOLUTION
$a_1 = 9$
$a_2 = 9 + (-6) = 3$
$a_3 = 3 + (-6) = -3$
$a_4 = -3 + (-6) = -9$
$a_5 = -9 + (-6) = -15$
$a_6 = -15 + (-6) = -21$

Add the common difference, -6, to each term to obtain the next term.

Answer 9, 3, -3, -9, -15, -21, ...

▶ **Self-Check Answers** ▼

1 Arithmetic; $d = -5$ **2** Not arithmetic **3** Arithmetic; $d = 0.5$

SECTION 8-2 ARITHMETIC SEQUENCES AND SERIES

EXAMPLE 3 Find a_{23} in an arithmetic sequence if $a_1 = 9$ and $d = 11$.

SOLUTION $a_n = a_1 + (n-1)d$ Formula for the nth term
$a_{23} = 9 + (23-1)11$ Substitute in the given values.
$a_{23} = 251$

EXAMPLE 4 Find the first eight terms of an arithmetic sequence with $a_3 = 7$ and $a_6 = 13$.

SOLUTION $a_1,\quad a_2,\quad a_3,\quad a_4,\quad a_5,\quad a_6,\quad a_7,\quad a_8$
$a_3 - 2d,\; a_3 - d,\; 7,\; a_3 + d,\; a_3 + 2d,\; 13,\; a_3 + 4d,\; a_3 + 5d$ Express the terms as a_3 plus or minus multiples of d.
$a_6 = a_3 + 3d$
$13 = 7 + 3d$ Substitute in the given values.
$6 = 3d$ Solve for d.
$d = 2$

Answer 3, 5, 7, 9, 11, 13, 15, 17, ...   Use a_3 and d to generate the other terms. For example, $a_1 = a_3 - 2d = 7 - 2(2) = 3$.

EXAMPLE 5 Find d in the arithmetic sequence if $a_1 = 12$ and $a_{79} = 207$.

SOLUTION $a_n = a_1 + (n-1)d$ Formula for the nth term
$207 = 12 + (79 - 1)d$ Substitute in the given values.
$195 = 78d$ Solve for d.
$d = 2.5$

EXAMPLE 6 How many terms are in the arithmetic sequence $-8, 9, 26, \ldots, 1437$?

SOLUTION $d = 9 - (-8)$ First find d; $d = a_2 - a_1$.
$= 17$
$a_n = a_1 + (n-1)d$ Formula for the nth term
$1437 = -8 + (n-1)17$ Substitute $a_1 = -8$, $a_n = 1437$, and $d = 17$ into the formula for a_n.
$1445 = 17(n-1)$
$n - 1 = 85$ Solve for n.
$n = 86$ terms

▶ **Self-Check** ▼

Find a_1 in the arithmetic sequence with $d = \frac{1}{3}$ and $a_{37} = 81$.

▶ **Self-Check Answer** ▼

$a_1 = 69$

The terms in a finite arithmetic sequence between the first and last terms are called **arithmetic means**. A sequence with only three terms a, m, and l has only one arithmetic mean, m, which is called **the arithmetic mean**. In this case, $m - a = l - m$, so $2m = a + l$ and thus $m = \dfrac{a + l}{2}$. Therefore the arithmetic mean is an average of the first and last terms of the sequence.

EXAMPLE 7

(a) Find the arithmetic mean between 71 and 93.

(b) Insert four arithmetic means between 18 and 43.

SOLUTIONS

(a) $m = \dfrac{a + l}{2}$ Formula for arithmetic mean

$= \dfrac{71 + 93}{2}$ Substitute in the given values.

$= 82$

Answer   The mean is 82.

(b) $a_1 = 18$ and $a_6 = 43$ There are six terms.

$a_n = a_1 + (n - 1)d$ Formula for the nth term

$43 = 18 + (6 - 1)d$ Substitute in the known values.

$43 = 18 + 5d$ Solve for d.

$25 = 5d$

$d = 5$

The sequence is computed as follows: 18, $18 + 5$, $18 + 10$, $18 + 15$, $18 + 20$, $18 + 25$.

Answer   18, 23, 28, 33, 38, 43

▶ **Self-Check** ▼

Insert five arithmetic means between -8 and 7.

The arithmetic series denoted by $S_n = \sum\limits_{i=1}^{n} a_i$ is the sum of the n terms $a_1 + a_2 + \cdots + a_{n-1} + a_n$ of an arithmetic sequence. For a sequence with only a few terms, it is easy to obtain the sum by merely adding the terms. For example,

$$S_5 = \sum_{i=1}^{5} 3i = 3 + 6 + 9 + 12 + 15 = 45$$

▶ **Self-Check Answer** ▼

$-8, -5.5, -3, -0.5, 2, 4.5, 7$

SECTION 8-2 ARITHMETIC SEQUENCES AND SERIES

For a sequence with many terms, such as $\sum_{i=1}^{100} i$, however, it is useful to have a formula that allows us to calculate the sum without actually doing all the adding. To develop the logic needed to derive such a formula, let us first examine a simplified way of adding the terms of a long sequence such as $\sum_{i=1}^{100} i$. First we list the terms twice, once in increasing order and once in decreasing order, and then we add the two lines.

$\sum_{i=1}^{100} i = 1 + 2 + 3 + \cdots + 98 + 99 + 100$ Listed in increasing order.

$\sum_{i=1}^{100} i = 100 + 99 + 98 + \cdots + 3 + 2 + 1$ Listed in decreasing order.

$2\sum_{i=1}^{100} i = 101 + 101 + 101 + \cdots + 101 + 101 + 101$ Add corresponding terms.

$2\sum_{i=1}^{100} i = 100(101)$ Each of the 100 terms is 101.

$\sum_{i=1}^{100} i = \dfrac{100(101)}{2}$ Simplify and solve for the sum.

$\sum_{i=1}^{100} i = 5050$

To develop a formula for $S_n = \sum_{i=1}^{n} a_i$, we will use the same logic illustrated in the preceding paragraph. We will first add the terms from a_1 to a_n and will then add them in the reverse order from a_n to a_1.

(1) $S_n = a_1 + (a_1 + d) + (a_1 + 2d) + \cdots + [a_1 + (n-3)d] + [a_1 + (n-2)d] + [a_1 + (n-1)d]$
(2) $S_n = [a_1 + (n-1)d] + [a_1 + (n-2)d] + [a_1 + (n-3)d] + \cdots + (a_1 + 2d) + (a_1 + d) + a_1$

$2S_n = (a_1 + a_1 + (n-1)d) + (a_1 + a_1 + (n-1)d) + (a_1 + a_1 + (n-1)d) + \cdots + (a_1 + a_1 + (n-1)d)$ Add corresponding terms.

$2S_n = (a_1 + a_n) + (a_1 + a_n) + (a_1 + a_n) + \cdots + (a_1 + a_n)$ Substitute a_n for $a_1 + (n+1)d$.

$2S_n = n(a_1 + a_n)$ Note that $(a_1 + a_n)$ is added n times.

$S_n = \dfrac{n(a_1 + a_n)}{2}$ Solve for S_n.

$S_n = \dfrac{n[2a_1 + (n-1)d]}{2}$ Substitute $a_1 + (n-1)d$ for a_n for an alternative form of this formula.

Arithmetic Series ▼

The arithmetic series $S_n = \sum_{i=1}^{n} a_i$ is given by

$$S_n = \frac{n}{2}(a_1 + a_n) \quad \text{or} \quad S_n = \frac{n}{2}[2a_1 + (n-1)d]$$

An alternate form for S_n is $S_n = n\left(\frac{a_1 + a_n}{2}\right)$. Thus one can consider this sum as the average of the first and the last terms used n times.

EXAMPLE 8 Find the sum of the first 100 natural numbers using the formula for S_n.

SOLUTION $a_1 = 1$, $a_{100} = 100$, $n = 100$

$S_n = \frac{n}{2}(a_1 + a_n)$ Formula for S_n

$S_{100} = \frac{100}{2}(1 + 100)$ Substitute in the given values.

$= 50(101)$

$S_{100} = 5050$

EXAMPLE 9 Find the number of terms in an arithmetic sequence with $a_1 = 13$, $d = 4$, and $S_n = 12{,}075$.

SOLUTION $S_n = \frac{n}{2}[2a_1 + (n-1)d]$ Formula for S_n

$12{,}075 = \frac{n}{2}[2(13) + (n-1)4]$ Substitute in the given values.

$12{,}075 = n[13 + 2(n-1)]$

$12{,}075 = 2n^2 + 11n$

$2n^2 + 11n - 12{,}075 = 0$

$(2n + 161)(n - 75) = 0$

$n = -\frac{161}{2}$ or $n = 75$ The value $-\frac{161}{2}$ is rejected, since n must be a natural number.

Answer $n = 75$

SECTION 8-2 ARITHMETIC SEQUENCES AND SERIES

EXAMPLE 10 Approximate the area of the figure to the right by adding the areas of the 100 indicated rectangles.

SOLUTION

Area of first rectangle $a_1 = w \cdot l = 1 \cdot 1 = 1$
Area of second rectangle $a_2 = w \cdot l = 1 \cdot 2 = 2$
Area of third rectangle $a_3 = w \cdot l = 1 \cdot 3 = 3$
$\vdots$
Area of 99th rectangle $a_{99} = w \cdot l = 1 \cdot 99 = 99$
Area of 100th rectangle $a_{100} = w \cdot l = 1 \cdot 100 = 100$

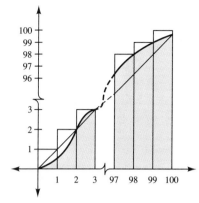

The total area is approximated by adding the areas of 100 rectangles.

$$A \approx 1 + 2 + 3 + \cdots + 99 + 100$$
$$\approx \tfrac{100}{2}(1 + 100)$$
$$A \approx 5050 \text{ square units}$$

$$S_n = \frac{n}{2}(a_1 + a_n)$$

Answer The approximate area of the figure is 5050 square units.

▶ **Self-Check** ▼

Find $S_{79} = \sum\limits_{i=1}^{79}(9i + 37)$.

EXERCISES 8-2

A

In Exercises 1 and 2, determine which of the sequences are arithmetic. For those that are arithmetic, give the common difference, d.

1 a. 5, 11, 17, 23, 29, ...   **b.** 35, 38, 41, 45, 48, ...   **c.** 23, 18, 13, 8, ...   **d.** $a_k = 2k + 1$

2 a. 3, 14, 25, 36, 47, ...   **b.** 17, 15, 13, 11, 8, ...   **c.** $\tfrac{1}{3}, \tfrac{5}{6}, 1\tfrac{1}{3}, 1\tfrac{5}{6}, \ldots$ **d.** $a_k = 2^k + 1$

In Exercises 3–12, write the first five terms of the arithmetic sequence satisfying the conditions.

3 $a_1 = 5, d = 9$
4 $a_1 = 9, d = 5$
5 $a_1 = 6, a_n - a_{n-1} = -\tfrac{2}{3}$
6 $a_1 = 12, a_n - a_{n-1} = -1\tfrac{1}{4}$
7 $a_2 = 7, a_3 = 5$
8 $a_4 = 11, a_5 = 12.7$
9 $a_2 = 11, a_4 = 17$
10 $a_2 = -8, a_6 = 16$
11 $a_i = \tfrac{1}{3}(2i - 1)$
12 $a_i = 19 - 3i$

In Exercises 13–22, find the value of the given arithmetic series.

13 $a_1 = 4, a_{81} = 204, S_{81} = ?$
14 $a_1 = 82, d = -3, S_{172} = ?$
15 $a_1 = 114, d = 5, S_{208} = ?$
16 $a_1 = 91, a_{119} = 858, S_{119} = ?$
17 $\sum\limits_{i=1}^{168}(3i + 7)$
18 $\sum\limits_{i=1}^{168}(7i - 3)$
19 $a_1 = 6, a_2 = 13, S_{73} = ?$
20 $a_{112} = -219, a_{113} = -221, S_{115} = ?$
21 $5 + 8 + 11 + \cdots + 26$
22 $14 + 9 + 4 + \cdots + (-26)$

▶ **Self-Check Answer** ▼

$S_{79} = 31{,}363$

In Exercises 23–36, find the indicated quantities from the information given about the arithmetic sequences.

23. $a_1 = 50, d = -17, a_{51} = ?$
24. $a_{63} = 19, d = 3, a_1 = ?$
25. $a_7 = 7, a_9 = 23, a_{11} = ?$
26. $a_{41} = -93, a_{45} = -49, a_{43} = ?$
27. $29, 25, 21, \ldots, -175; n = ?$
28. $a_{47} = 23, a_{52} = 3, d = ?$
29. $a_1 = 2, a_{40} = 80, S_{40} = ?$
30. $S_{77} = 45, d = -11, a_1 = ?$
31. $S_{50} = 3275, d = 3, a_1 = ?$
32. $S_6 = 18, a_6 = -2, d = ?$
33. $S_n = 136, a_n = 31, a_1 = 3, n = ?$
34. $S_7 = 59.5, a_7 = 16, a_1 = ?$
35. $S_{70} = 24,745, a_{70} = 664, d = ?$
36. $36.7, 37.6, 38.5, \ldots, 136.6; n = ?$
37. Find the arithmetic mean between 17 and 71.
38. Find the arithmetic mean between 29 and 97.
39. Insert five arithmetic means between -1 and 41.
40. Insert six arithmetic means between 80 and 59.

B

41. Rolls of carpet are stacked in a warehouse with 25 rolls on level one, 24 on level two, and so forth, until the top level has only 4 rolls. How many are in the stack?

42. Logs are stacked so that each layer after the first has one less log than the previous layer. If 153 logs are to be stacked in this manner with only 1 log on the top layer, how many logs should be placed at the bottom?

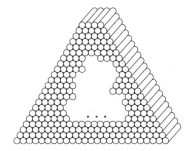

Figure for Exercise 41

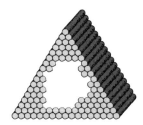

Figure for Exercise 42

46 cm

80 cm

Figure for Exercise 45

43. The productivity gain from installing a new robot on an assembly line has been estimated to be $5000 the first year, $7500 the second year, and $10,000 the third year. If this trend continues, what will be the total gain for the first 10 years?

44. If each student in a class of 25 students shakes hands with everyone else exactly once, how many handshakes take place?

45. The lengths of the rungs of a wooden ladder form an arithmetic sequence. If there are 17 rungs ranging in length from 80 centimeters to 46 centimeters, find the total length of all these rungs.

46. How many integers between 43 and 117 are divisible by 7?

47. Find the sum of the odd natural numbers less than 102.

C

48. Approximate the area of the figure by adding the areas of the 1000 indicated rectangles.

In Exercises 49 and 50, use a calculator to determine the quantities indicated.

49. $a_1 = 17.053, d = 2.741, a_7 = ?$
50. $a_1 = 8.69, a_{11} = 57.99, S_{11} = ?$

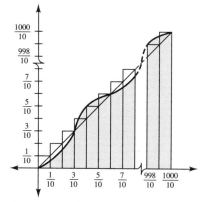

Figure for Exercise 48

51 Is the sequence log 3, log 9, log 27, log 81, ... an arithmetic sequence? Explain your answer.

52 If $a_1, a_2, \ldots, a_n$ and $b_1, b_2, \ldots, b_n$ are arithmetic sequences, show that $a_1 + b_1, a_2 + b_2, \ldots, a_n + b_n$ is also an arithmetic sequence.

SECTION 8-3

Geometric Sequences and Series

Section Objectives

5 Calculate the nth term of a geometric sequence.

6 Find the sum of a finite geometric sequence.

7 Evaluate an infinite geometric series.

A **geometric sequence** or **geometric progression** is a sequence in which consecutive terms form a constant ratio. This constant is called the **common ratio** and is denoted by r.

EXAMPLE 1 Determine which of these sequences are geometric. For those that are geometric, give the common ratio, r.

SOLUTIONS

(a) 10, 20, 40, 80, ... This sequence is geometric, since the ratio of any two consecutive terms is 2.

$$\frac{20}{10} = \frac{40}{20} = \frac{80}{40} = 2$$

(b) 108, −72, 48, −32, ... This sequence is geometric with $r = -\frac{2}{3}$.

$$\frac{-72}{108} = \frac{48}{-72} = \frac{-32}{48} = -\frac{2}{3}$$

(c) 7, 21, 63, 150, 300, ... This sequence is not geometric, since $\frac{21}{7} = 3$ but $\frac{300}{150} = 2$.

The ratio of consecutive terms is not constant in this sequence.

▶ **Self-Check** ▼

Since the common ratio is given by $\frac{a_n}{a_{n-1}} = r$, the general term can be found by the recursive formula $a_n = r a_{n-1}$ for $n > 1$. This allows us to rewrite the terms $a_1, a_2, a_3, a_4, \ldots, a_n$ as

Determine which of these sequences are geometric. For those that are geometric, give the common ratio, r.

1 160, 80, 40, 20, 10, ...

2 4, 8, 16, 24, 48, ...

3 64, −80, 100, −125, ...

$$\begin{aligned}
a_1 &= a_1 \\
a_2 &= a_1 r = a_1 r \\
a_3 &= a_2 r = a_1 r^2 \\
a_4 &= a_3 r = a_1 r^3 \\
&\vdots \\
a_n &= a_{n-1} r = a_1 r^{n-1}
\end{aligned}$$

This pattern continues, with the power of the common ratio increased by one for each term except the first term.

▶ **Self-Check Answers** ▼

1 Geometric; $r = \frac{1}{2}$ **2** Not geometric   **3** Geometric; $r = -\frac{5}{4}$

Geometric Sequence ▼

If $\dfrac{a_n}{a_{n-1}} = r$ for $n > 1$, then the sequence is geometric and the constant r is called the common ratio.
The formula for the nth term is $a_n = a_1 r^{n-1}$.

EXAMPLE 2 Write the first six terms of the geometric sequence with $a_1 = 3$ and $r = -2$.

SOLUTION
$a_1 = 3$
$a_2 = 3(-2) = -6$ Multiply each term by the common ratio, -2, to obtain the next term.
$a_3 = -6(-2) = 12$
$a_4 = 12(-2) = -24$
$a_5 = -24(-2) = 48$
$a_6 = 48(-2) = -96$

Answer $3, -6, 12, -24, 48, -96, \ldots$ ☐

EXAMPLE 3 Find a_{123} in a geometric sequence with $a_1 = -5$ and $r = -1$.

SOLUTION
$a_n = a_1 r^{n-1}$ Formula for the nth term
$a_{123} = (-5)(-1)^{123-1}$ Substitute in the given values.
$= -5$

Answer a_{123} is -5. ☐

EXAMPLE 4 Find the first eight terms of a geometric sequence with $a_3 = 6$ and $a_5 = 24$.

SOLUTION
$a_5 = a_3 r^2$ Express a_5 using a_3 and r.
$24 = 6r^2$ Substitute in the given values.
$r^2 = 4$ Solve for r.
$r = 2$ or $r = -2$ Write the sequences using both values of r.

Answer Both $1.5, 3, 6, 12, 24, 48, 96, 192, \ldots$ and $1.5, -3, 6, -12, 24, -48, 96, -192, \ldots$ satisfy the stated conditions.

Expressed with respect to a_3 the first eight terms are: $\dfrac{a_3}{r^2}, \dfrac{a_3}{r}, a_3, a_3 r, a_3 r^2, a_3 r^3, a_3 r^4,$ and $a_3 r^5$. ☐

SECTION 8-3 GEOMETRIC SEQUENCES AND SERIES

EXAMPLE 5 How many terms are in the geometric sequence $\frac{1}{16}, \frac{1}{8}, \frac{1}{4}, \ldots, 4096$?

SOLUTION $r = \dfrac{\frac{1}{8}}{\frac{1}{16}} = 2$ First find r.

$a_n = a_1 r^{n-1}$ Formula for the nth term

$4096 = \dfrac{1}{16}(2)^{n-1}$ Substitute in the given values.

$2^{n-1} = 65{,}536$ Multiply both members by 16.

$\ln 2^{n-1} = \ln 65536$ Take the natural log of both members.

$(n-1)\ln 2 = \ln 65536$ Simplify using the power rule for logarithms.

$n - 1 = \dfrac{\ln 65536}{\ln 2}$ Evaluate using a calculator.

$n - 1 = 16$

$n = 17$

Answer There are 17 terms.

▶ **Self-Check** ▼

Find a_1 in the geometric sequence with $r = \frac{1}{3}$ and $a_{10} = \frac{1}{81}$.

For a geometric sequence with terms a, m, and l, $\dfrac{m}{a} = \dfrac{l}{m}$. Thus $m^2 = al$ and so either $m = \sqrt{al}$ or $m = -\sqrt{al}$. Each of these values of m can be called the **geometric mean** of a and l.

EXAMPLE 6

(a) Find the geometric means between 3 and 147.

(b) Insert four geometric means between 48 and -1.5.

SOLUTIONS

(a) $m = \pm\sqrt{al}$ Formula for the geometric means

$= \pm\sqrt{3(147)}$ Substitute in the given values.

$= \pm\sqrt{441}$

$m = \pm 21$

Answer 3, 21, 147 or 3, -21, 147

▶ **Self-Check Answer** ▼

$a_1 = 243$

(b) $a_1 = 48$ and $a_6 = -1.5$ There are six terms.

$$a_n = a_1 r^{n-1}$$ Formula for the nth term

$$-1.5 = (48)r^{6-1}$$ Substitute in the known values.

$$-1.5 = 48r^5$$

$$r^5 = -0.03125$$ Divide both members by -48.

$$r = \sqrt[5]{-0.03125}$$ Then use the power key on your calculator to evaluate $+0.03125^{0.2}$. Change the sign of this result to obtain $\sqrt[5]{-0.03125}$.

$$r = -0.5$$ Use this ratio to write the terms of this geometric progression.

Answer $48, -24, 12, -6, 3, -1.5$

The **geometric series** denoted by $S_n = \sum_{i=1}^{n} a_i$ is the sum of the n terms $a_1 + a_2 + \cdots + a_{n-1} + a_n$ of a geometric sequence. For example,

$$S_5 = \sum_{i=1}^{5} 3^i = 3 + 9 + 27 + 81 + 243 = 363$$

For sequences with more terms, it is useful to have a condensed formula for S_n. To develop this formula for S_n, consider the expanded forms for S_n and rS_n as illustrated below.

(1) $S_n = a_1 + a_1 r + a_1 r^2 \quad + \cdots + a_1 r^{n-2} + a_1 r^{n-1}$ Multiply both sides by r and shift terms to the right to align similar terms.

(2) $rS_n = \quad\quad a_1 r + a_1 r^2 + a_1 r^3 + \cdots + a_1 r^{n-2} + a_1 r^{n-1} + a_1 r^n$

$S_n - rS_n = a_1 + 0 \quad + 0 \quad\quad + \cdots + 0 \quad\quad + 0 \quad\quad - a_1 r^n$ Subtract equation 2 from equation 1.

$$S_n(1 - r) = a_1(1 - r^n)$$ Factor both sides.

$$S_n = \frac{a_1(1 - r^n)}{1 - r} \quad \text{for } r \neq 1$$ Divide both members by $1 - r$. If $r \neq 1$, then $1 - r \neq 0$.

Geometric Series ▼

> For $r \neq 1$, the geometric series $S_n = \sum_{i=1}^{n} a_i$ is given by
>
> $$S_n = \frac{a_1(1 - r^n)}{1 - r} \quad \text{or} \quad S_n = \frac{a_1 - ra_n}{1 - r}$$

The simplest way to evaluate a geometric series with only a few terms may be to simply add all the terms. For series with more terms, however,

SECTION 8-3 GEOMETRIC SEQUENCES AND SERIES

this formula is quite useful. If $r = 1$, the formula cannot be used; however, if $r = 1$, the sequence of terms in the series is a constant sequence whose sum is na_1.

EXAMPLE 7 Find S_8 if the geometric sequence has $a_1 = 1$ and $r = \dfrac{1}{3}$.

SOLUTION $S_n = \dfrac{a_1(1 - r^n)}{1 - r}$ Formula for S_n

$S_8 = \dfrac{1[1 - (\frac{1}{3})^8]}{1 - \frac{1}{3}}$ Substitute in the given values.

$S_8 = \dfrac{\frac{3^8 - 1}{3^8}}{\frac{2}{3}}$ Simplify, using a calculator if necessary.

$S_8 = \dfrac{3280}{2187}$ ☐

EXAMPLE 8 Find the value of r in a geometric sequence if $a_1 = 3125$, $a_n = 12.8$, and $S_n = 5199.8$.

SOLUTION $S_n = \dfrac{a_1 - ra_n}{1 - r}$ Formula for S_n

$5199.8 = \dfrac{3125 - r(12.8)}{1 - r}$ Substitute in the given values.

$5199.8 - 5199.8r = 3125 - 12.8r$ Solve for r.

$2074.8 = 5187r$

$r = 0.4$ ☐

▶ **Self-Check** ▼

Find a_1 in the geometric sequence with $r = \frac{1}{3}$ and $a_{10} = \frac{1}{81}$.

EXAMPLE 9 One gimmick advertised in a newspaper was a "sure-fire" secret formula for becoming a millionaire. The secret discovered by those unwise enough to pay for it was as follows: "On the first day of the month save 1¢, on the second day save 2¢, on the third 4¢, etc." How many days would be required to save a total of at least $1,000,000? At this rate, what amount would be saved on the last day?

▶ **Self-Check Answer** ▼

$S_{11} = 9090.909091$

SOLUTION

$$S_n = \frac{a_1(1-r^n)}{1-r}$$ Use the formula for a geometric series.

$$1{,}000{,}000 \leq \frac{0.01(1-2^n)}{1-2}$$ Substitute into this formula the value $r = \dfrac{a_2}{a_1} = \dfrac{0.02}{0.01} = 2$. The sum must be at least \$1,000,000.

$$1{,}000{,}000 \leq \frac{0.01(1-2^n)}{-1}$$ Simplify and then solve for n.

$$1{,}000{,}000 \leq 0.01(2^n - 1)$$

$$100{,}000{,}000 \leq 2^n - 1$$

$$2^n \geq 100{,}000{,}001$$

$$\log 2^n \geq \log 100{,}000{,}001$$ Take the common log of both sides.

$$n \log 2 \geq \log 100{,}000{,}001$$ Power rule

$$n \geq \frac{\log 100{,}000{,}001}{\log 2}$$

$$n \geq 26.6 \text{ days}$$ 26 days will not be enough; the 27th day is needed to reach \$1,000,000.

$$n = 27 \text{ days}$$

$$a_{27} = a_1 r^{26} = 0.01(2^{26})$$ a_{27} is the amount to save on day 27.

$$a_{27} = \$671{,}088.64$$

Answer The system will produce a millionaire in 27 days, provided you can keep saving the large amounts in the latter days, such as \$671,088.64 on the 27th day. □

If $|r| \leq 1$, then the absolute values of the terms of the geometric sequence are decreasing. For example, the geometric sequence $\dfrac{1}{2}, \dfrac{1}{4}, \dfrac{1}{8}, \dfrac{1}{16}, \ldots, \dfrac{1}{2^n}, \ldots$ is a decreasing, infinite geometric sequence. We already have a formula for finding S_n, the sum of a finite geometric sequence. Can the infinite sum $\sum\limits_{i=1}^{\infty} \left(\dfrac{1}{2}\right)^i$ be meaningful? Although we could never actually add an infinite number of terms, we can adopt a new meaning for this sum. If the finite partial sum S_n approaches some limiting value S as n becomes large, then we will call this limit the **infinite sum**. Symbolically, $\sum\limits_{i=1}^{\infty} a_i = S$ if S_n approaches S as n increases. In Table 8-2, observe the pattern of the finite partial sums as n increases.

SECTION 8-3 GEOMETRIC SEQUENCES AND SERIES

Table 8-2 The infinite sum

n	1	2	3	4	5	6	7	8	9	10	...
a_n	$\frac{1}{2}$	$\frac{1}{4}$	$\frac{1}{8}$	$\frac{1}{16}$	$\frac{1}{32}$	$\frac{1}{64}$	$\frac{1}{128}$	$\frac{1}{256}$	$\frac{1}{512}$	$\frac{1}{1024}$	...
S_n	$\frac{1}{2}$	$\frac{3}{4}$	$\frac{7}{8}$	$\frac{15}{16}$	$\frac{31}{32}$	$\frac{63}{64}$	$\frac{127}{128}$	$\frac{255}{256}$	$\frac{511}{512}$	$\frac{1023}{1024}$	...

Note that as n increases, the sequence decreases, and in fact, a_n tends to 0. Also, the finite partial sums become closer and closer to 1; that is, S_n approaches 1 as n becomes large. Thus we say that the infinite sum $\sum_{i=1}^{\infty} (\frac{1}{2})^i$ equals 1.

Since an infinite sum S is defined to be the limit that S_n approaches, this sum is not meaningful if there is no limiting value. In particular, if $|r| \geq 1$ the terms will not approach 0, and thus S is not meaningful since S_n will not approach any limit.

A general formula for the infinite sum can be obtained by examining the formula for S_n.

$$S_n = \frac{a_1(1 - r^n)}{1 - r}$$

If $|r| < 1$, then $|r|^n$ approaches 0. Thus $S_n = \frac{a_1(1 - r^n)}{1 - r}$ approaches $\frac{a_1(1 - 0)}{1 - r}$; that is, S_n approaches $\frac{a_1}{1 - r}$. The infinite sum S is the limiting value, so $S = \frac{a_1}{1 - r}$.

Infinite Geometric Series ▼

If $|r| < 1$, then the sum of an infinite geometric sequence is

$$S = \frac{a_1}{1 - r}$$

If $|r| \geq 1$, this sum does not exist.

EXAMPLE 10 Find $S = \sum_{i=1}^{\infty} \left(\frac{1}{2}\right)^i$.

SOLUTION $S = \frac{a_1}{1 - r}$

$= \frac{\frac{1}{2}}{1 - \frac{1}{2}}$ Substitute $a_1 = \frac{1}{2}$ and $r = \frac{1}{2}$ into the formula for the sum of an infinite geometric progression. Then simplify and solve for S.

$S = 1$ Compare this result to Table 8-2. □

EXAMPLE 11 Write $0.454545\ldots$ as a fraction.

SOLUTION $0.454545\ldots = 0.45 + 0.0045 + 0.000045 + \cdots$

$$0.454545\ldots = \sum_{i=1}^{\infty} 45(0.01)^i \quad \text{This is an infinite geometric series with } a_1 = 0.45 \text{ and } r = 0.01.$$

$$S = \frac{a_1}{1-r} \quad \text{Formula for } S$$

$$S = \frac{0.45}{1 - 0.01} \quad \text{Substitute in the known values.}$$

$$S = \frac{0.45}{0.99}$$

$$S = \frac{5}{11}$$

Answer $0.454545\ldots = \dfrac{5}{11}$ Check this by dividing 5 by 11.

▶ **Self-Check** ▼

Write $0.\overline{693}$ as a fraction.

EXAMPLE 12 A hardball is dropped onto concrete from a height of 9 meters. Each time it hits the concrete, it rebounds to two-thirds of the height from which it fell. Find the total distance this bouncing ball travels.

SOLUTION Let d = Total distance traveled. Then

(Total distance traveled) = (Distance the ball falls) + (Distance the ball rises)   Word equation

$d = (9 + 6 + 4 + \cdots) + (6 + 4 + \cdots)$ The distances the ball rebounds (and then falls) form a geometric sequence.

$d = (9 + 6 + 4 + \cdots) + (9 + 6 + 4 + \cdots) - 9$ Add and then subtract 9.

$d = 2(9 + 6 + 4 + \cdots) - 9$

$d = 2\left(\dfrac{9}{1 - \frac{2}{3}}\right) - 9$ Evaluate the sum within the parentheses using the formula $S = \dfrac{a_1}{1-r}$, with $a_1 = 9$ and $r = \dfrac{2}{3}$.

$d = 2(27) - 9$

$d = 45$

Answer The ball travels 45 meters.

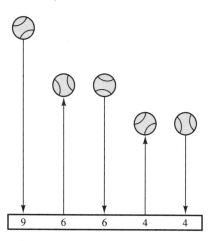

▶ **Self-Check Answer** ▼

$\dfrac{77}{111}$

SECTION 8-3 GEOMETRIC SEQUENCES AND SERIES

EXERCISES 8-3

A

In Exercises 1–4, determine which of the sequences are geometric. For those that are geometric, give the common ratio r.

1. **a.** 1, 10, 100, 1000, ... **b.** 1, −5, 25, −125, 625, ... **c.** 6, 3, 1, $\frac{1}{2}$, $\frac{1}{4}$, ... **d.** 0.15, 0.0015, 0.000015, ...
2. **a.** 3, −21, 147, −1029, ... **b.** 2, 10, 40, 200, 1000, ... **c.** 1, 1.3, 1.69, 2.197, 2.8561, ... **d.** $\frac{2}{7}, \frac{3}{7}, \frac{9}{14}, \frac{27}{28}, \frac{81}{56}, \ldots$
3. **a.** $a_i = 3^i$ **b.** $a_i = i^3$ **c.** 2, $2\sqrt{2}$, 4, $4\sqrt{2}$, ... **d.** 13, −13, 13, −13, 13, −13, ...
4. **a.** 7, 7, 7, 7, 7, ... **b.** 0.7, 0.07, 0.007, 0.0007, ... **c.** 4, $4\sqrt{3}$, 12, $12\sqrt{3}$, 36, ... **d.** $a_i = (\frac{2}{3})^i$

In Exercises 5–18, write the first five terms of the geometric progression satisfying the given conditions.

5. $a_1 = 5, r = 4$
6. $a_1 = 4, r = 5$
7. $a_1 = 7, a_n = -2a_{n-1}$
8. $a_1 = 8, a_n = -3a_{n-1}$
9. $a_2 = 24, r = -\frac{2}{3}$
10. $a_4 = 24, r = -\frac{3}{2}$
11. $a_1 = 0.15, r = 0.01$
12. $a_3 = 0.007, r = 0.1$
13. $a_i = (-\frac{2}{3})^i$
14. $a_i = (-\frac{2}{5})^i$
15. $a_1 = 3, a_3 = 75$
16. $a_2 = -1, a_4 = -4$
17. $a_3 = 3\sqrt{3}, r = \sqrt{3}$
18. $a_2 = \sqrt[3]{4}, r = \sqrt[3]{2}$

In Exercises 19–24, find the value of the given geometric series.

19. $a_1 = 3, r = 2, S_{10} = ?$
20. $a_1 = 7, r = 3, S_{11} = ?$
21. $a_1 = 12, a_2 = -9, S_9 = ?$
22. $a_8 = 9, a_9 = 9, S_{99} = ?$
23. $\sum_{i=1}^{11} 2^i$
24. $\sum_{i=1}^{8} (\frac{5}{4})^i$

In Exercises 25–28, (a) write the partial sums s_1, s_2, and s_3 and (b) evaluate the infinite geometric series.

25. $\sum_{i=1}^{\infty} (\frac{3}{8})^i$
26. $\sum_{i=1}^{\infty} (\frac{5}{16})^i$
27. $\sum_{i=1}^{\infty} (-\frac{5}{8})^i$
28. $\sum_{i=1}^{\infty} (-\frac{7}{16})^i$

In Exercises 29–40, find the indicated quantity from the information given about the geometric sequence.

29. $a_1 = \frac{1}{81}, r = -3, a_{10} = ?$
30. $a_1 = 64, r = -\frac{1}{2}, a_{10} = ?$
31. $a_1 = 1, a_n = 2048, r = 2, n = ?$
32. $a_1 = \frac{1}{9}, a_n = 243, r = 3, n = ?$
33. $a_{39} = 425, a_{41} = 17, r = ?$
34. $a_{81} = 288, a_{83} = 32, r = ?$
35. $S_5 = 0.22222, r = 0.1, a_1 = ?$
36. $S_7 = 381, r = 2, a_1 = ?$
37. $a_1 = 8, r = -0.5, S_7 = ?$
38. $a_1 = 81, r = -\frac{2}{3}, S_8 = ?$
39. $S_3 = 183, a_1 = 48, r = ?$
40. $S_3 = 74, a_1 = 32, r = ?$

In Exercises 41–48, write each repeating decimal as a fraction.

41. $0.818181\ldots$
42. $0.\overline{72}$
43. $0.777\ldots$
44. $0.\overline{654}$
45. $0.\overline{9}$
46. $0.\overline{621}$
47. $12.5\overline{6}$
48. $1.03\overline{5}$
49. Find the geometric means between 16 and 25.
50. Find the geometric means between 98 and 8.
51. Insert two geometric means between 27 and −8.
52. Insert three geometric means between $\frac{1}{4}$ and 4.

B

53. A machine originally worth $40,000 depreciates in value each year such that its value at the end of the year is only four-fifths of its value at the beginning of the year. Find the value of the machine after six years.

54. A sum of $100 is invested at the beginning of a year, and after one year the investment is worth $115. If the investment continues to grow at the same rate, how much will the investment be worth after 10 years?

55. A blacksmith attaches each horseshoe with eight nails. The blacksmith offers to charge by the nail for shoeing all four hooves. The first nail would be 1 cent, the second 2 cents, the third 4 cents, etc. At this rate, how much would the job cost?

56. A ball dropped from 2 meters rebounds to seven-tenths of its previous height on each bounce. How far does it travel?

57. A platform that is partially supported on springs vibrates up and down so that each upward vibration lifts the platform to three-eighths of its former height. If the platform is released and falls from a height of six centimeters, what is the total distance it travels? Assume that the platform falls to the same level each time.

58. A child's swing moves through an arc of four meters. On each subsequent swing it travels only seven-eighths of the distance of the precious arc. How far does the swing travel?

59. Approximate the area of the figure to the right by adding the areas of the five indicated rectangles.

60. A chain letter requires that a recipient send out five copies of the letter. If you start a chain letter by mailing out five such letters, how many letters will be in circulation after the sixth mailing?

61. An art design is formed by drawing a square with sides of 20 centimeters and then connecting the midpoints of the sides of the square to form a second square. (See the figure to the right.) If this process is continued indefinitely, determine the perimeter of the third square and the total perimeter of all the squares.

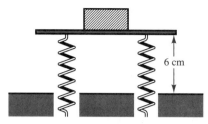

Figure for Exercise 57

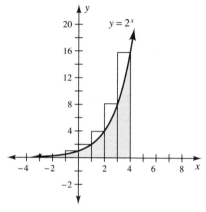

Figure for Exercise 59

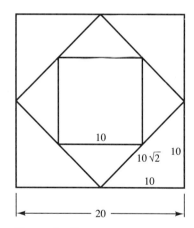

Figure for Exercise 61

C

62. A sum of $100 is invested at the end of each month at an interest rate of 12% per year. Write a sequence representing the value of each contribution by the end of the year. Let $A_1 = 100$ be the value of the last contribution; A_2 be the value of the next to last contribution, which draws interest for 1 month; and A_{12} be the value of the first contribution, which draws interest for 11 months.

63. Determine the common ratio of the geometric sequence in Exercise 62 and find the value of the investment by the end of the year by evaluating the associated geometric series.

64. A payment P is invested at the end of each period at an interest rate of r per period. Write a sequence representing the value of each contribution by the end of n periods. (*Hint*: Let A_1 be the value of the nth payment; A_2 be the value of the $(n-1)$st payment, which draws interest for 1 period; and A_n be the value of the first payment, which draws interest for $(n-1)$ periods.)

65. Determine the common ratio of the geometric sequence in Exercise 64 and find the value of the investment by the end of the n periods by evaluating the associated geometric series.

In Exercises 66 and 67, use a calculator to determine the quantity indicated. Your answers should be accurate to eight significant digits.

66 $a_1 = 5.8, r = 3.1, a_5 = ?$

67 $S_{10} = 3.1662674, r = 0.55, a_1 = ?$

68 If $a_1, a_2, \ldots, a_n$ and $b_1, b_2, \ldots, b_n$ are geometric progressions, show that $a_1 b_1, a_2 b_2, \ldots, a_n b_n$ is also a geometric progression.

69 If $a_1, a_2, \ldots, a_n$ is a geometric progression, show that $ka_1, ka_2, \ldots, ka_n$ is also a geometric progression for any real constant k.

70 For $\sum_{n=1}^{\infty} \left(\frac{3}{5}\right)^i$ write the partial sums $s_1, s_2, s_3,$ and s_n.

SECTION 8-4

Mathematical Induction

Section Objective

8 Use mathematical induction to prove statements true for all natural numbers n.

Many important laws or formulas in science and mathematics are based on astute observation of patterns exhibited. The general law then claims that this pattern will continue and that the formula applies not only to the cases observed but also to all cases. Although these particular observations can lead to the discovery of a formula and can make it seem reasonable, such observations do *not* prove that the formula always works. In fact, even well-known mathematicians have made false conjectures based on their observations. Many superstitions, such as those involving howling dogs and rings around the moon, are based on specific observations that are not true for the general case.

It is very important that a scientist or mathematician know that a formula will always work and therefore can be used with confidence. Thus proving formulas is a basic part of mathematics. **Mathematical induction** is a method of proving statements $S(n)$ true for *all* natural numbers n. Obviously this could never be accomplished by proving statements one at a time. Instead, we use the principle of mathematical induction given in the following box.

Principle of Mathematical Induction ▼

If $S(n)$ is a statement involving the natural number n and

1. $S(1)$ is true and
2. the truth of $S(k)$ implies the truth of $S(k + 1)$

then $S(n)$ is true for each natural number n.

Using the logic of mathematical induction is analogous to climbing a ladder with an infinite number of rungs. To climb each rung of the ladder requires two things.

1 **Initialization:** You must be able to get on the first rung. (Ladders on water towers are often placed high so that children cannot get on the first rung and thus are prevented from climbing the ladder.)
2 **Perpetuation:** You must be able to perpetuate the climb by advancing from any rung to the next rung. (Rungs missing from an old wooden ladder might prevent you from advancing even though you could get on the first rung.)

Every mathematical induction proof must contain both the initialization step and the perpetuation step. To show perpetuation, we do not claim that a given statement is true; rather, we check to see whether, if we are at some rung on the ladder, we can climb to the next rung. Mathematically we describe this by saying that **if** the statement $S(k)$ is true, **then** the next statement $S(k+1)$ must also be true. In Example 3, we will prove that the statement

$$S(n): \sum_{i=1}^{n} \frac{1}{i(i+1)} = \frac{n}{n+1}$$

is true for every natural number n. But first we examine the meaning of this statement.

EXAMPLE 1 $S(n)$ denotes the statement $\sum_{i=1}^{n} \frac{1}{i(i+1)} = \frac{n}{n+1}$. Write the meaning of $S(1)$, $S(2)$, $S(3)$, $S(k)$, and $S(k+1)$.

SOLUTION $S(1)$: $\quad \dfrac{1}{1(1+1)} = \dfrac{1}{1+1}$

$\dfrac{1}{2} = \dfrac{1}{2}$

$S(2)$: $\quad \dfrac{1}{1(1+1)} + \dfrac{1}{2(2+1)} = \dfrac{2}{2+1}$

$\dfrac{1}{2} + \dfrac{1}{6} = \dfrac{2}{3}$

$\dfrac{4}{6} = \dfrac{2}{3}$

$\dfrac{2}{3} = \dfrac{2}{3}$

SECTION 8-4 MATHEMATICAL INDUCTION

$$S(3): \quad \frac{1}{1(1+1)} + \frac{1}{2(2+1)} + \frac{1}{3(3+1)} = \frac{3}{3+1}$$

$$\frac{1}{2} + \frac{1}{6} + \frac{1}{12} = \frac{3}{4}$$

$$\frac{9}{12} = \frac{3}{4}$$

$$\frac{3}{4} = \frac{3}{4}$$

$$S(k): \quad \frac{1}{1(1+1)} + \frac{1}{2(2+1)} + \frac{1}{3(3+1)} + \cdots + \frac{1}{k(k+1)} = \frac{k}{k+1}$$

$$S(k+1): \quad \frac{1}{1(1+1)} + \frac{1}{2(2+1)} + \frac{1}{3(3+1)} + \cdots + \frac{1}{k(k+1)} + \frac{1}{(k+1)[(k+1)+1]} = \frac{k+1}{(k+1)+1}$$

$\square$

EXAMPLE 2 Write the meaning of $S(4)$ and $S(k+1)$ if $S(n)$ denotes the statement $\sum_{i=1}^{n} (2i-1) = n^2$.

SOLUTION
$S(4)$: $[2(1)-1] + [2(2)-1] + [2(3)-1] + [2(4)-1] = 4^2$
$\quad\ \ 1 \quad + \quad 3 \quad + \quad 5 \quad + \quad 7 \quad\ = 4^2$
$\qquad\qquad\qquad\qquad\quad 16 \qquad\qquad = 16$

$S(k+1)$: $[2(1)-1] + [2(2)-1] + \cdots + (2k-1) + [2(k+1)-1] = (k+1)^2$
$\quad\ \ 1 \quad + \quad 3 \quad + \cdots + (2k-1) + \quad (2k+1) \quad = (k+1)^2$

$\square$

To prove that a statement $S(n)$ is true for all natural numbers, we must (step 1) prove that $S(1)$ is true and (step 2) show that the truth of $S(k)$ implies the truth of $S(k+1)$. To initialize and perpetuate this proof, we suggest using the following outline.

Outline of a Mathematical Induction Proof ▼

To prove that $S(n)$ is a true statement for each natural number n:

Step 1 Initialize the proof by showing that $S(1)$ is true.

Step 2 Perpetuate the proof by:
  a. Writing out the statement for $S(k)$ and then leaving a few blank lines.
  b. Writing out the statement for $S(k+1)$.
  c. Showing in the blank lines that if $S(k)$ is true then $S(k+1)$ must also be true.

EXAMPLE 3 Prove that the statement $S(n): \sum_{i=1}^{n} \frac{1}{i(i+1)} = \frac{n}{n+1}$ is true for each natural number n.

Proof

1 $S(1)$: $\frac{1}{1(1+1)} \stackrel{?}{=} \frac{1}{1+1}$ *Show that $S(1)$ is true to initialize the process.*

$\frac{1}{2} = \frac{1}{2}$ is true.

Thus $S(1)$ is true.

2 If $S(k)$ is true, that is, *Write out $S(k)$ and try to perpetuate the process.*

if $\frac{1}{1(1+1)} + \frac{1}{2(2+1)} + \cdots + \frac{1}{k(k+1)} = \frac{k}{k+1}$

show that $S(k+1)$ must also be true. *Skip to the last line and write out $S(k+1)$.*

$\left(\frac{1}{2} + \frac{1}{6} + \cdots + \frac{1}{k(k+1)}\right) + \frac{1}{(k+1)[(k+1)+1]}$ *Add the $(k+1)$st term to both sides of $S(k)$ to obtain the left side of $S(k+1)$.*

$= \frac{k}{k+1} + \frac{1}{(k+1)[(k+1)+1]}$

$= \frac{k(k+2)+1}{(k+1)(k+2)}$ *Add using a common denominator to obtain a single term as in $S(k+1)$.*

$= \frac{(k+1)^2}{(k+1)(k+2)}$ *Factor the numerator.*

$= \frac{k+1}{k+2}$ *Reduce the fraction.*

then $\frac{1}{1(1+1)} + \frac{1}{2(2+1)} + \cdots + \frac{1}{k(k+1)} + \frac{1}{(k+1)[(k+1)+1]}$ *This statement is $S(k+1)$, which is what we wanted to show.*

$= \frac{k+1}{(k+1)+1}$

This completes the proof, since we have both initialized and perpetuated the process. Thus $\sum_{i=1}^{n} \frac{1}{i(i+1)} = \frac{n}{n+1}$ is true for each natural number n. ☐

Some statements can be proven true by more than one method. The formula for the arithmetic series in Example 4 can be proven by mathematical induction as well as by the methods covered in Section 8-2.

▶ **Self-Check** ▼

If $S(n)$ denotes the statement
$\sum_{i=1}^{n} (2i-1)^2 = \frac{1}{3}n(2n-1)(2n+1)$,
write the meaning of these statements.
1 $S(4)$ **2** $S(k)$ **3** $S(k+1)$

▶ **Self-Check Answers** ▼

1 $1^2 + 3^2 + 5^2 + 7^2 = \frac{4}{3}(7)(9)$; $84 = 84$ **2** $1^2 + 3^2 + \cdots + (2k-1)^2 = \frac{1}{3}k(2k-1)(2k+1)$
3 $1^2 + 3^2 + \cdots + (2k-1)^2 + (2k+1)^2 = \frac{1}{3}(k+1)(2k+1)(2k+3)$

SECTION 8-4 MATHEMATICAL INDUCTION

EXAMPLE 4 Use mathematical induction to prove that the sum of the first n odd integers is n^2.

Proof Let $S(n)$: $\sum_{i=1}^{n} (2i - 1) = n^2$, that is, $1 + 3 + 5 + \cdots + (2n - 1) = n^2$.

1 $S(1)$: $2(1) - 1 \stackrel{?}{=} 1^2$ *Show that $S(1)$ is true to initialize the process.*

$1 = 1$ is true.

Thus $S(1)$ is true.

2 If $S(k)$ is true, that is,

if $1 + 3 + 5 + \cdots + (2k - 1) = k^2$ *Write out $S(k)$ and try to perpetuate the process.*

show that $S(k + 1)$ must also be true.   *Skip to the last line and write out $S(k + 1)$.*

$1 + 3 + 5 + \cdots + (2k - 1) + [2(k + 1) - 1] = k^2 + [2(k + 1) - 1]$ *Add the $(k + 1)$st term to both sides of $S(k)$ to obtain the left side of $S(k + 1)$.*

$= k^2 + 2k + 1$ *Simplify the right side.*

then $1 + 3 + 5 + \cdots + (2k - 1) + [2(k + 1) - 1] = (k + 1)^2$ *This statement is $S(k + 1)$, which is what we wanted to show.*

By parts 1 and 2, $S(n)$ is true for all natural numbers. ☐

EXAMPLE 5 Prove that $4^n > n^2$ for each natural number n.

Proof

1 $S(1)$: $4^1 \stackrel{?}{>} 1^2$ *Initialize the proof by showing that $S(1)$ is true.*

$4 > 1$ is true.

Thus $S(1)$ is true.

2 If $S(k)$ is true, that is,   *Write out $S(k)$ here and $S(k + 1)$ near the end of the proof to perpetuate the proof.*

if $4^k > k^2$

show that $S(k + 1)$ must also be true.

$4 \cdot 4^k > 4k^2$ *Multiply both sides by 4 to produce the left side of $S(k + 1)$.*

$4^{k+1} > k^2 + 2k^2 + k^2$

$> k^2 + 2k + 1$ *Since $k \geq 1$, $2k^2 + k^2 \geq 2k + 1$.*

then $4^{k+1} > (k + 1)^2$ *Now factor to obtain $S(k + 1)$, the statement we wanted to imply from $S(k)$.*

This completes the proof, since we have both initialized and perpetuated the process. Thus $4^n > n^2$ is a true statement for each natural number n. ☐

▶ **Self-Check** ▼

Complete this mathematical induction proof by filling in part 2 of the proof. Prove that the sum of the first n even integers is $n(n + 1)$.

Proof   Let

$S(n)$: $2 + 4 + 6 + \cdots + 2n = n(n + 1)$.

1 $S(1)$: $2(1) \stackrel{?}{=} 1(1 + 1)$

$2 = 2$ is true.

Thus $S(1)$ is true.

2 If $S(k)$ is true, that is,
if $2 + 4 + 6 + \cdots + 2k = k(k + 1)$
show that $S(k + 1)$ must also be true.

$2 + 4 + 6 + \cdots + 2k + 2(k + 1)$
$= (k + 1)[(k + 1) + 1]$

▶ **Self-Check Answer** ▼

$2 + 4 + 6 + \cdots + 2k + 2(k + 1) = k(k + 1) + 2(k + 1)$
$= (k + 1)(k + 2)$
$= (k + 1)[(k + 1) + 1]$

EXERCISES 8-4

A

In Exercises 1–6, write out the statements $S(1)$, $S(2)$, $S(3)$, and $S(4)$.

1. $\sum_{i=1}^{n} (3i - 2) = \dfrac{3n^2 - n}{2}$

2. $\sum_{i=1}^{n} 2^i = 2^{n+1} - 2$

3. $\sum_{i=1}^{n} i(2i + 1) = \dfrac{4n^3 + 9n^2 + 5n}{6}$

4. $\sum_{i=1}^{n} (i^2 - i + 3) = \dfrac{n^3 + 8n}{3}$

5. $n < 2^n$

6. $(1 + x)^n \geq 1 + nx$ for $x > 0$

In Exercises 7–10, write out $S(k)$ and $S(k + 1)$ for each statement.

7. $\sum_{i=1}^{n} 2^i = 2^{n+1} - 2$

8. $\sum_{i=1}^{n} (3i - 2) = \dfrac{3n^2 - n}{2}$

9. $\sum_{i=1}^{n} (i^2 - i + 3) = \dfrac{n^3 + 8n}{3}$

10. $\sum_{i=1}^{n} i(2i + 1) = \dfrac{4n^3 + 9n^2 + 5n}{6}$

In Exercises 11–20, use mathematical induction to prove that each statement is true for all natural numbers n.

11. $\sum_{i=1}^{n} i = \dfrac{n(n + 1)}{2}$

12. $\sum_{i=1}^{n} (3i - 2) = \dfrac{3n^2 - n}{2}$

13. $\sum_{i=1}^{n} (4i + 1) = 2n^2 + 3n$

14. $\sum_{i=1}^{n} (3i - 1) = \dfrac{3n^2 + n}{2}$

15. $\sum_{i=1}^{n} 2^i = 2^{n+1} - 2$

16. $\sum_{i=1}^{n} 3^i = \dfrac{3^{n+1} - 3}{2}$

17. $\sum_{i=1}^{n} i^2 = \dfrac{2n^3 + 3n^2 + n}{6}$

18. $\sum_{i=1}^{n} i(2i + 1) = \dfrac{4n^3 + 9n^2 + 5n}{6}$

19. $\sum_{i=1}^{n} \dfrac{1}{2^i} = 1 - \dfrac{1}{2^n}$

20. $\sum_{i=1}^{n} \dfrac{1}{(3i - 2)(3i + 1)} = \dfrac{n}{3n + 1}$

Exercises 21–24 refer to the Fibonacci sequence defined by $a_1 = 1$, $a_2 = 1$, and $a_n = a_{n-2} + a_{n-1}$ for $n \geq 3$. Use mathematical induction to prove that each statement is true for all natural numbers n.

21. $a_1 + a_2 + \cdots + a_n = a_{n+2} - 1$

22. $a_1 + a_3 + \cdots + a_{2n-1} = a_{2n}$

23. $a_2 + a_4 + \cdots + a_{2n} = a_{2n+1} - 1$

24. $(a_{n+1})^2 - a_n a_{n+2} = (-1)^n$

B

25. Use an infinite row of dominoes to explain the logic of mathematical induction.

In Exercises 26–28, label each statement true (T) or false (F).

26. A mathematical induction proof must contain both the initialization step and the perpetuation step.

27. Mathematical induction can be used to prove statements true for all real numbers.

28. If applied properly, mathematical induction can sometimes prove false statements true.

In Exercises 29–33, determine which of the statements can be proven by mathematical induction. If the statement cannot be proven this way, state why not.

29. $n^2 - 3n + 2 = 0$ for $n = 1, 2, 3, \ldots$

30. $n = n + 1$ for $n = 1, 2, 3, \ldots$

31. $n^2 - n + 41$ is prime for $n = 1, 2, 3, \ldots$

32. $n^4 - 10n^3 + 35n^2 - 50n + 24 = 1$ for $n = 1, 2, 3, \ldots$

33. $x^2 - 1 = (x - 1)(x + 1)$ for all real numbers

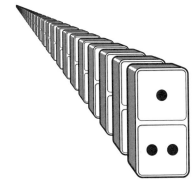

Figure for Exercise 25

SECTION 8-4 MATHEMATICAL INDUCTION

C

In Exercises 34–44, use mathematical induction to prove that each statement is true for all natural numbers n.

34 $\sum_{i=1}^{n} \frac{1}{3^i} = \frac{1}{2}\left(1 - \frac{1}{3^n}\right)$

35 $1^3 + 2^3 + 3^3 + \cdots + n^3 = \left(\frac{n(n+1)}{2}\right)^2$

36 $1(4) + 2(9) + 3(16) + \cdots + n(n+1)^2 = \frac{n(n+1)(n+2)(3n+5)}{12}$

37 $\left(1 + \frac{1}{1}\right)\left(1 + \frac{1}{2}\right)\left(1 + \frac{1}{3}\right) \cdots \left(1 + \frac{1}{n}\right) = n + 1$

38 $(1^2 + 1) + (2^2 + 1) + (3^2 + 1) + \cdots + (n^2 + 1) = \frac{2n^3 + 3n^2 + 7n}{6}$

39 $n < 2^n$

40 $\log n \leq n$

41 $1 + x + x^2 + \cdots + x^{n-1} = \frac{1 - x^n}{1 - x}$ for $x \neq 1$

42 A set with n elements has 2^n subsets.

43 $n^3 + 2n$ is divisible by 3.

44 $\sum_{i=1}^{n} \frac{1}{(2i-1)(2i+1)} = \frac{n}{2n+1}$

45 It is difficult to determine by visual inspection whether the point P is inside or outside the intricately drawn closed figure. However, if we pick a reference point R that is clearly outside of the figure and draw a line segment from P to R, we can easily tell whether P is inside or outside. If the line segment crosses the figure an odd number of times, P is inside; if it crosses an even number of times, P is outside. Prove this claim by mathematical induction.

Figure for Exercise 45

46 Examine the following "proof" and determine the error that exists.
$S(n)$: Any group of n horsemen will ride in a straight line (for $n > 2$).

Proof

1 $S(2)$ is trivially true since any two points determine a straight line.

2 Assume that $S(k)$ is true and show that $S(k + 1)$ must then be true.
If $k + 1$ are riding, consider the first k of them. By assumption, they are in a straight line.
Now consider the last k of them. By assumption, they must also be in a straight line.
However, these lines coincide in perhaps several points (see the arrows in the figure illustrating this); thus the lines must be the same line. Therefore all $(k + 1)$ horsemen are riding in a straight line.

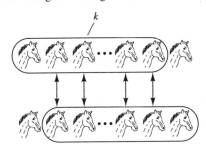

SECTION 8-5

Binomial Theorem

Section Objectives

9 Evaluate factorial expressions.

10 Determine a given term of a binomial expansion.

Binomial expansions of the form $(a + b)^n$ were first examined in this text in Section 1-4. We will now reexamine these expansions using factorial notation.

Factorial Notation ▼

$$n! = n \cdot (n - 1) \cdot (n - 2) \cdots 3 \cdot 2 \cdot 1$$
$$0! = 1$$

EXAMPLE 1 Evaluate each of these factorial expressions.

SOLUTIONS

(a) $4!$ $4! = 4 \cdot 3 \cdot 2 \cdot 1 = 24$

Evaluate each factorial by taking the product of the first n natural numbers.

(b) $6!$ $6! = 6 \cdot 5 \cdot 4 \cdot 3 \cdot 2 \cdot 1 = 720$

The TI-81 keystrokes are:

| 6 | MATH | ! (choice 5) | ENTER |

(c) $\dfrac{8!}{5!}$ $\dfrac{8!}{5!} = \dfrac{8 \cdot 7 \cdot 6 \cdot \cancel{5} \cdot \cancel{4} \cdot \cancel{3} \cdot \cancel{2} \cdot \cancel{1}}{\cancel{5} \cdot \cancel{4} \cdot \cancel{3} \cdot \cancel{2} \cdot \cancel{1}}$

Write the numerator and the denominator in expanded form and then divide both by all common factors.

$= 8 \cdot 7 \cdot 6$

$= 336$

An alternative format can be used to simplify the evaluation of expressions such as the one in Example 1(c). Replacing $5 \cdot 4 \cdot 3 \cdot 2 \cdot 1$ by $5!$, we can write $8!$ as $8 \cdot 7 \cdot 6 \cdot 5!$. Use of this condensed format is illustrated in the next example.

SECTION 8-5 BINOMIAL THEOREM

EXAMPLE 2 Evaluate each of these factorial expressions.

SOLUTIONS

(a) $\dfrac{8!}{5!}$ $\quad \dfrac{8!}{5!} = \dfrac{8 \cdot 7 \cdot 6 \cdot \cancel{5!}}{\cancel{5!}}$ Divide both the numerator and the denominator by 5!.

$\phantom{(a) \dfrac{8!}{5!} \quad} = 8 \cdot 7 \cdot 6$

$\phantom{(a) \dfrac{8!}{5!} \quad} = 336$

(b) $\dfrac{10!}{2!8!}$ $\quad \dfrac{10!}{2!8!} = \dfrac{10 \cdot 9 \cdot \cancel{8!}}{2 \cdot 1 \cdot \cancel{8!}}$ Divide both the numerator and the denominator by 8!

$\phantom{(b) \dfrac{10!}{2!8!} \quad} = \dfrac{\overset{5}{\cancel{10}} \cdot 9}{\cancel{2}}$ Then divide both by 2.

$\phantom{(b) \dfrac{10!}{2!8!} \quad} = 45$

(c) $\dfrac{7!}{0!4!}$ $\quad \dfrac{7!}{0!4!} = \dfrac{7 \cdot 6 \cdot 5 \cdot \cancel{4!}}{0!\cancel{4!}}$ Divide both the numerator and the denominator by 4!.

$\phantom{(c) \dfrac{7!}{0!4!} \quad} = \dfrac{7 \cdot 6 \cdot 5}{0!}$

$\phantom{(c) \dfrac{7!}{0!4!} \quad} = \dfrac{210}{1}$ Then replace 0! by 1.

$\phantom{(c) \dfrac{7!}{0!4!} \quad} = 210$

> **▶ Self-Check ▼**
>
> Evaluate these factorial expressions.
>
> **1** $5!$ **2** $\dfrac{9!}{8!}$ **3** $\dfrac{8!}{3!5!}$ **4** $\dfrac{7!}{0!7!}$

We will now reexamine the expansion of $(a + b)^n$, summarized in Chapter 1 as follows.

Expanding $(a + b)^n = a^n + na^{n-1}b + \cdots + nab^{n-1} + b^n$ ▼

Step 1 Write the exponents on all $(n + 1)$ terms: Start with a^n, decreasing the exponents on a by 1 and increasing the exponents on b by 1 until the last term is b^n.

Step 2 Write the coefficients of each term:
  a. Make the first coefficient 1 and the second n.
  b. Calculate each new coefficient by multiplying the current coefficient by the exponent on a and dividing by one more than the exponent on b.
  c. Use the symmetric pattern of coefficients, which starts with the first term and ends with the last term.

▶ Self-Check Answers ▼

1 120   **2** 9   **3** 56   **4** 1

EXAMPLE 3 Write the expansion of $(a + b)^5$ using factorial notation.
SOLUTION

$(a + b)^5 = \_a^5 + \_a^4b + \underline{}a^3b^2 + \underline{}a^2b^3 + \underline{}ab^4 + \underline{}b^5$ First write the exponents on all six terms.

$= 1a^5 + 5a^4b + \dfrac{5 \cdot 4}{2} a^3b^2 + \dfrac{5 \cdot 4 \cdot 3}{3 \cdot 2} a^2b^3 + \dfrac{5 \cdot 4 \cdot 3 \cdot 2}{4 \cdot 3 \cdot 2} ab^4 + \dfrac{5 \cdot 4 \cdot 3 \cdot 2 \cdot 1}{5 \cdot 4 \cdot 3 \cdot 2} b^5$ Then write the coefficients on each term.

$= \dfrac{5!}{0!5!} a^5 + \dfrac{5!}{1!4!} a^4b^1 + \dfrac{5!}{2!3!} a^3b^2 + \dfrac{5!}{3!2!} a^2b^3 + \dfrac{5!}{4!1!} ab^4 + \dfrac{5!}{5!0!} b^5$ Express each of these coefficients in factorial notation.

$\square$

Using the same technique exhibited in the previous example, we can write the first few terms of the general expansion of $(a + b)^n$ as

$$(a + b)^n = a^n + \dfrac{n}{1} a^{n-1}b + \dfrac{n(n - 1)}{1 \cdot 2} a^{n-2}b^2 + \dfrac{n(n - 1)(n - 2)}{1 \cdot 2 \cdot 3} a^{n-3}b^3$$

$$+ \dfrac{n(n - 1)(n - 2)(n - 3)}{1 \cdot 2 \cdot 3 \cdot 4} a^{n-4}b^4 + \cdots + nab^{n-1} + b^n$$

These coefficients can now be rewritten in factorial notation as

$$(a + b)^n = \dfrac{n!}{0!(n - 0)!} a^n + \dfrac{n!}{1!(n - 1)!} a^{n-1}b + \dfrac{n!}{2!(n - 2)!} a^{n-2}b^2$$

$$+ \cdots + \dfrac{n!}{r!(n - r)!} a^{n-r}b^r + \cdots + \dfrac{n!}{n!(n - n)!} b^n$$

Using summation notation, we can condense this to

$$(a + b)^n = \sum_{r=0}^{n} \dfrac{n!}{r!(n - r)!} a^{n-r}b^r$$

Binomial Theorem ▼

For any natural number n and complex numbers a and b,

$$(a + b)^n = \sum_{r=0}^{n} \dfrac{n!}{r!(n - r)!} a^{n-r}b^r$$

Note the relationship between the exponents and the factorials in the denominator of the coefficient. Although this notation is concise, it may be easier to continue to form a binomial expansion using the method outlined earlier in the box. If only one term of an expansion is needed, then the formula is quite useful. Note from the summation notation in the box that

SECTION 8-5 BINOMIAL THEOREM

the first term corresponds to $r = 0$, the second term corresponds to $r = 1$, etc. In general, r, the exponent on b, is always one less than the number of the term.

EXAMPLE 4 Find the seventh term of $(a + b)^{11}$.

SOLUTION $(r + 1)\text{st term} = \dfrac{n!}{r!(n-r)!} a^{n-r} b^r$ This formula is for the $(r + 1)$st term of a binomial expansion.

$$7\text{th term} = \dfrac{11!}{6!5!} a^5 b^6$$

For the seventh term, the exponent on b is $r = 7 - 1 = 6$. Thus the exponent on a is $n - r = 11 - 6 = 5$.

$$= 462 a^5 b^6$$

EXAMPLE 5 Find the ninth term of $(2x - y)^{12}$.

SOLUTION $(2x - y)^{12} = [2x + (-y)]^{12}$

$$(r + 1)\text{st term} = \dfrac{n!}{r!(n-r)!} a^{n-r} b^r$$

Substitute $2x$ for a and $-y$ for b in the formula for the $(r + 1)$st term of $(a + b)^n$.

$$9\text{th term} = \dfrac{12!}{8!4!} (2x)^4 (-y)^8$$

For the ninth term, $r = 9 - 1 = 8$ and $n - r = 12 - 8 = 4$.

$$= 495(16x^4)(y^8)$$

$$= 7920 x^4 y^8$$

▶ **Self-Check** ▼

Find the fourth term of $(3x - 5y)^{10}$.

Since the expansion of $(a + b)^n$ is appropriate for all natural numbers n and all complex numbers a and b, we will use this expansion to calculate the sixth term of the complex expression in the next example.

EXAMPLE 6 Find the sixth term of $(3 + 5i)^{10}$.

SOLUTION $(r + 1)\text{st term} = \dfrac{n!}{r!(n-r)!} a^{n-r} b^r$

Substitute 3 for a and $5i$ for b in the formula for the $(r + 1)$st term of $(a + b)^n$.

$$6\text{th term} = \dfrac{10!}{5!5!} (3)^5 (5i)^5$$

For the sixth term, $r = 6 - 1 = 5$ and $n - r = 10 - 5 = 5$.

$$= \dfrac{10 \cdot 9 \cdot 8 \cdot 7 \cdot 6 \cdot 5!}{5 \cdot 4 \cdot 3 \cdot 2 \cdot 1 \cdot 5!} (243)(3125 i^5)$$

Simplify, replacing i^5 by i.

$$= 2 \cdot 9 \cdot 2 \cdot 7 \cdot (243)(3125 i^5)$$

$$= 191{,}362{,}500 i$$

▶ **Self-Check Answer** ▼

$$\dfrac{10!}{3!7!} (3x)^7 (-5y)^3 = -32{,}805{,}000 x^7 y^3$$

The **binomial coefficient** $\dfrac{n!}{r!(n-r)!}$ is often denoted $\binom{n}{r}$. We will examine this notation further in Section 8-6 when we look at combinations.

EXAMPLE 7 Evaluate each of these binomial coefficients.

SOLUTIONS

(a) $\binom{8}{3}$ $\quad \binom{8}{3} = \dfrac{8!}{3!5!}$ $\qquad$ Substitute 8 for n and 3 for r in the binomial coefficient $\binom{n}{r} = \dfrac{n!}{r!(n-r)!}$.

$\qquad\qquad\qquad = \dfrac{8 \cdot 7 \cdot \cancel{6} \cdot \cancel{5!}}{\cancel{3} \cdot \cancel{2} \cdot \cancel{1} \cdot \cancel{5!}}$

$\qquad\qquad\qquad = 8 \cdot 7$

$\qquad\qquad\qquad = 56$

(b) $\binom{11}{7}$ $\quad \binom{11}{7} = \dfrac{11!}{7!4!}$ $\qquad$ Substitute 11 for n and 7 for r in the binomial coefficient $\binom{n}{r} = \dfrac{n!}{r!(n-r)!}$.

$\qquad\qquad\qquad = \dfrac{11 \cdot 10 \cdot \overset{3}{\cancel{9}} \cdot \cancel{8} \cdot \cancel{7!}}{\cancel{7!} \cdot \cancel{4} \cdot \cancel{3} \cdot \cancel{2} \cdot 1}$

$\qquad\qquad\qquad = 11 \cdot 10 \cdot 3$

$\qquad\qquad\qquad = 330$

▶ **Self-Check** ▼

Evaluate $\binom{9}{5}$.

EXERCISES 8-5

A

In Exercises 1–20, evaluate each expression.

1. $3!$
2. $7!$
3. $8!$
4. $9!$
5. $0!4!$
6. $3!5!$
7. $6!2!$
8. $0!6!$
9. $5!(6 \cdot 7)$
10. $6!(7 \cdot 8)$
11. $\dfrac{9!}{7!}$
12. $\dfrac{8!}{5!}$
13. $\dfrac{10!}{3!7!}$
14. $\dfrac{13!}{2!11!}$
15. $\dfrac{20!}{0!20!}$
16. $\dfrac{17!}{0!17!}$
17. $\binom{9}{4}$
18. $\binom{8}{4}$
19. $\binom{12}{12}$
20. $\binom{13}{0}$

In Exercises 21–28, write the binomial expansion of each expression.

21. $(x + 3)^4$
22. $(w - 2)^5$
23. $(2x - y)^5$
24. $(3v + 2w)^4$
25. $(v^2 - 5w)^4$
26. $(2a - 3b^2)^6$
27. $(3 + 2i)^4$
28. $(2 - 3i)^5$

▶ **Self-Check Answer** ▼

126

SECTION 8-6 COUNTING: PERMUTATIONS AND COMBINATIONS

In Exercises 29–40, write the indicated term for the binomial expansion of each expression.

29 The eighth term of $(v + w)^{17}$
30 The tenth term of $(v + w)^{17}$
31 The tenth term of $(x - y)^{20}$
32 The twelfth term of $(x - y)^{20}$
33 The eleventh term of $(2m^3 - n)^{14}$
34 The ninth term of $(2m^3 - n)^{14}$
35 The twelfth term of $(5a^4 - b)^{16}$
36 The thirteenth term of $(5a^4 - b)^{16}$
37 The middle term of $(x + 2y)^{10}$
38 The middle term of $\left(\dfrac{2a^2}{3} - \dfrac{3b^3}{2}\right)^6$
39 The seventh term of $(2 - i)^{11}$
40 The ninth term of $(5 + 2i)^{13}$

B

41 Using the binomial expansion of $(1 + 0.1)^6$, evaluate $(1.1)^6$.
42 Using the binomial expansion of $(1 - 0.1)^7$, evaluate $(0.99)^7$.
43 Solve $(n + 2)! = 30n!$
44 Solve $(n + 3)! = 60n!$

In Exercises 45–52, write the first four terms of the binomial expansion of each expression.

45 $(r + s)^{16}$
46 $(x - 3)^{12}$
47 $(3m - n)^{14}$
48 $(m - n)^{17}$
49 $(3 + 5i)^8$
50 $(3 - 2i)^{10}$
51 $(\sqrt{x} - y)^9$
52 $(\sqrt[3]{a} + b)^7$

C

53 Using mathematical induction, prove that $(a + b)^n = \sum\limits_{r=0}^{n} \dfrac{n!}{r!(n - r)!} a^{n-r} b^r$.

SECTION 8-6

Counting: Permutations and Combinations

Section Objectives

11 Use the fundamental counting principle.

12 Determine the number of permutations or combinations of *n* objects taken *r* at a time.

Some problems have so many possible outcomes that it is informative to list or to count these outcomes. This section is an introduction to some of the basic counting procedures. These topics are covered in more detail in courses in finite mathematics, probability and statistics, and combinatorial analysis.

One method for displaying all the possibilities for a sequence of choices is a **tree diagram**. Consider, for example, a sporting goods store that sells four models of baseball gloves. Each model comes in black or brown and in a left-hand or a right-hand glove. The tree diagram in Figure 8-1 on page 526 illustrates the 16 distinct gloves that this store stocks.

526 CHAPTER 8 SEQUENCES AND COUNTING PROBLEMS

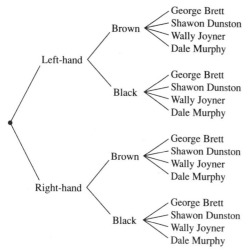

Figure 8-1 Tree diagram

Note that the number of distinct gloves can be obtained by multiplying as follows:

$$\begin{pmatrix}\text{Number}\\ \text{of hands}\end{pmatrix} \cdot \begin{pmatrix}\text{Number}\\ \text{of colors}\end{pmatrix} \cdot \begin{pmatrix}\text{Number}\\ \text{of models}\end{pmatrix} = \begin{pmatrix}\text{Total number of}\\ \text{possibilities}\end{pmatrix}$$
$$2 \quad \cdot \quad 2 \quad \cdot \quad 4 \quad = \quad 16$$

This observation can be formalized into a principle that allows us to count the number of possibilities without actually listing all of them in a tree diagram.

Fundamental Counting Principle ▼

> If the first of two successive events can occur m ways and the second can occur n ways, the joint event can occur $m \cdot n$ ways.

EXAMPLE 1 Gregor Mendel, 1822–1884, was a pioneer with his research in plant genetics. He crossed peas with various characteristics in order to observe the offspring. Each pea was either yellow or green and either smooth or wrinkled. How many distinct types of peas could there be? If two of these plants were crossed, how many experiments would be necessary to observe all cases?

SOLUTION
$$\begin{pmatrix}\text{Number of}\\ \text{textures}\end{pmatrix} \cdot \begin{pmatrix}\text{Number of}\\ \text{colors}\end{pmatrix} = \begin{pmatrix}\text{Number of}\\ \text{types of peas}\end{pmatrix}$$
$$2 \quad \cdot \quad 2 \quad = \quad 4$$

$$\begin{pmatrix}\text{Number of types}\\ \text{of the first pea}\end{pmatrix} \cdot \begin{pmatrix}\text{Number of types}\\ \text{of the second pea}\end{pmatrix} = \begin{pmatrix}\text{Number of experiments}\\ \text{necessary}\end{pmatrix}$$
$$4 \quad \cdot \quad 4 \quad = \quad 16$$

Each *word equation* is based on the fundamental counting principle.

SECTION 8-6 COUNTING: PERMUTATIONS AND COMBINATIONS

Answer There are four types of peas. If the different types of pea plants were crossed, 16 experiments would be necessary to observe all the possibilities. □

The fundamental counting principle can be generalized to cover any finite number of events. In the next example we use this principle for three successive events.

EXAMPLE 2 How many ways can 3 cards be turned up in succession in a blackjack game using a deck of 52 cards?

SOLUTION

$$\begin{pmatrix}\text{Number of}\\ \text{cards for the}\\ \text{first draw}\end{pmatrix} \cdot \begin{pmatrix}\text{Number of}\\ \text{cards for the}\\ \text{second draw}\end{pmatrix} \cdot \begin{pmatrix}\text{Number of}\\ \text{cards for the}\\ \text{third draw}\end{pmatrix} = \begin{pmatrix}\text{Number of ways 3}\\ \text{successive cards}\\ \text{can be turned up}\end{pmatrix}$$

$$52 \cdot 51 \cdot 50 = 132{,}600$$

This *word equation* is based on the fundamental counting principle.

Substitute the given values into the word equation.

Answer 132,600 ways □

EXAMPLE 3 Wendy's will fix its "Old Fashioned Hamburger" with any or all of the following items on it: catsup, onion, mustard, pickle, lettuce, tomato, mayonnaise, relish. How many ways can its ham-burgers be served?

SOLUTION We can consider the selection of a hamburger as eight successive events, each of which has two outcomes—including or not including each item on the hamburger. Therefore, by the fundamental counting principle, there are $2 \cdot 2 \cdot 2 \cdot 2 \cdot 2 \cdot 2 \cdot 2 \cdot 2 = 2^8 = 256$ ways to serve Wendy's hamburgers. See Figure 8-2 on the next page. □

▶ **Self-Check** ▼

How many distinct guesses could be made as to the correct key on a 10-question true-false test?

EXAMPLE 4 A business selling tools to mechanics wants to determine the shortest route that will allow a delivery van to visit five rural towns in one day. If it lists all possible routes that connect these five towns, how many routes will it list?

SOLUTION

$$\begin{pmatrix}\text{Number of}\\ \text{towns for}\\ \text{1st visit}\end{pmatrix} \cdot \begin{pmatrix}\text{Number of}\\ \text{towns left}\\ \text{for}\\ \text{2nd visit}\end{pmatrix} \cdot \begin{pmatrix}\text{Number of}\\ \text{towns left}\\ \text{for}\\ \text{3rd visit}\end{pmatrix} \cdot \begin{pmatrix}\text{Number of}\\ \text{towns left}\\ \text{for}\\ \text{4th visit}\end{pmatrix} \cdot \begin{pmatrix}\text{Number of}\\ \text{towns left}\\ \text{for}\\ \text{5th visit}\end{pmatrix} = \begin{pmatrix}\text{Number of}\\ \text{distinct}\\ \text{routes}\end{pmatrix}$$

$$5 \cdot 4 \cdot 3 \cdot 2 \cdot 1 = 120$$

The *word equation* is based on the fundamental counting principle.

Substitute the given values into the word equation.

Answer There are 120 possible routes. □

▶ **Self-Check Answer** ▼

$2^{10} = 1024$ guesses

Figure 8-2 We Fix Hamburgers 256 Ways at Wendy's

C—Catsup
L—Lettuce
M—Mustard
Ma—Mayonnaise
O—Onion
P—Pickle
Re—Relish
T—Tomato

1. Plain
2. C
3. O
4. M
5. P
6. L
7. T
8. Ma
9. Re
10. CO
11. CM
12. CP
13. CL
14. CT
15. CMa
16. CRe
17. OM
18. OP
19. OL
20. OT
21. OMa
22. ORe
23. MP
24. ML
25. MT
26. MMa
27. MRe
28. PL
29. PT
30. PMa
31. PRe
32. LT
33. LMa
34. LRe
35. TMa
36. TRe
37. MaRe
38. COM
39. COP
40. COL
41. COT
42. COMa
43. CORe
44. CMP
45. CML
46. CMT
47. CMMa
48. CMRe
49. CPL
50. CPT
51. CPMa
52. CPRe
53. CLT
54. CLMa
55. CLRe
56. CTMa
57. CTRe
58. CMaRe
59. OMP
60. OML
61. OMT
62. OMMa
63. OMRe
64. OPL
65. OPT
66. OPMa
67. OPRe
68. OLT
69. OLMa
70. OLRe
71. OTMa
72. OTRe
73. OMaRe
74. MPL
75. MPT
76. MPMa
77. MPRe
78. MLT
79. MLMa
80. MLRe
81. MTMa
82. MTRe
83. MMaRe
84. PLT
85. PLMa
86. PLRe
87. PTMa
88. PTRe
89. PMaRe
90. LTMa
91. LTRe
92. LMaRe
93. TMaRe
94. COMP
95. COML
96. COMT
97. COMMa
98. COMRe
99. COPL
100. COPT
101. COPMa
102. COPRe
103. COLT
104. COLMa
105. COLRe
106. COTMa
107. COTRe
108. COMaRe
109. CMPL
110. CMPT
111. CMPMa
112. CMPRe
113. CMLT
114. CMLMa
115. CMLRe
116. CMTMa
117. CMTRe
118. CMMaRe
119. CPLT
120. CPLMa
121. CPLRe
122. CPTMa
123. CPTRe
124. CPMaRe
125. CLTMa
126. CLTRe
127. CLMaRe
128. CTMaRe
129. OMPL
130. OMPT
131. OMPMa
132. OMPRe
133. OMLT
134. OMLMa
135. OMLRe
136. OMTMa
137. OMTRe
138. OMMaRe
139. OPLT
140. OPLMa
141. OPLRe
142. OPTMa
143. OPTRe
144. OPMaRe
145. OLTMa
146. OLTRe
147. OLMaRe
148. OTMaRe
149. MPLT
150. MPLMa
151. MPLRe
152. MPTMa
153. MPTRe
154. MPMaRe
155. MLTMa
156. MLTRe
157. MLMaRe
158. MTMaRe
159. PLTMa
160. PLTRe
161. PLMaRe
162. PTMaRe
163. LTMaRe
164. COMPL
165. COMPT
166. COMPMa
167. COMPRe
168. COMLT
169. COMLMa
170. COMLRe
171. COMTMa
172. COMTRe
173. COMMaRe
174. COPLT
175. COPLMa
176. COPLRe
177. COPTMa
178. COPTRe
179. COPMaRe
180. COLTMa
181. COLTRe
182. COLMaRe
183. COTMaRe
184. CMPLT
185. CMPLMa
186. CMPLRe
187. CMPTMa
188. CMPTRe
189. CMPMaRe
190. CMLTMa
191. CMLTRe
192. CMTMaRe
193. CMLMaRe
194. CPLTMa
195. CPLTRe
196. CPLMaRe
197. CPTMaRe
198. CLTMaRe
199. OMPLT
200. OMPLMa
201. OMPLRe
202. OMPTMa
203. OMPTRe
204. OMPMaRe
205. OMLTMa
206. OMLTRe
207. OMLMaRe
208. OMTMaRe
209. OPLTMa
210. OPLTRe
211. OPLMaRe
212. OPTMaRe
213. OLTMaRe
214. MPLTMa
215. MPLTRe
216. MPLMaRe
217. MPTMaRe
218. MLTMaRe
219. PLTMaRe
220. COMPLT
221. COMPLMa
222. COMPLRe
223. COMPTMa
224. COMPTRe
225. COMPMaRe
226. COMLTMa
227. COMLTRe
228. COMLMaRe
229. COMTMaRe
230. COPLTMa
231. COPLTRe
232. COPLMaRe
233. COPTMaRe
234. COLTMaRe
235. CMPLTMa
236. CMPLTRe
237. CMPLMaRe
238. CMPTMaRe
239. CMLTMaRe
240. CPLTMaRe
241. OMPLTRe
242. OMPLTRe
243. OMPLMaRe
244. OMPTMaRe
245. OMLTMaRe
246. OPLTMaRe
247. MPLTMaRe
248. COMPLTMa
249. COMPLTRe
250. COMPLT
251. COMPLMaRe
252. COMPTMaRe
253. COPLTMaRe
254. CMPLTMaRe
255. OMPLTMaRe
256. COMPLTMaRe

SECTION 8-6 COUNTING: PERMUTATIONS AND COMBINATIONS

A **permutation** of distinct objects is an arrangement of these objects in which the order of the arrangement does make a difference. Example 4 illustrated the number of permutations of five objects taken five at a time. The answer to Example 4 could also be expressed as 5!. (Factorial notation was introduced in Section 8-5.) In general, the number of permutations of n objects taken n at a time, denoted $P(n, n)$, is

$$P(n, n) = n(n - 1)(n - 2) \cdots (2)(1) = n!$$

where n choices, $n - 1$ choices left, $n - 2$ choices left, ..., Only 2 choices left, Only 1 choice left.

EXAMPLE 5 How many different ways can four horses finish a race?   This assumes no ties or photo finishes.

SOLUTION $P(4, 4) = 4! = 4 \cdot 3 \cdot 2 \cdot 1 = 24$ Each order of finish is a permutation.

□

The number of ways 3 successive cards can be turned up from a deck of 52 cards (see Example 2) is the number of permutations of 52 objects taken 3 at a time. If we denote this number by $P(52, 3)$, we have

$$P(52, 3) = 52 \cdot 51 \cdot 50 = \frac{52 \cdot 51 \cdot 50 \cdot 49 \cdot 48 \cdots 2 \cdot 1}{49 \cdot 48 \cdots 2 \cdot 1}.$$

Thus $P(52, 3) = \dfrac{52!}{49!}$. This problem generalizes to $P(n, r)$, which is the number of permutations of n objects taken r at a time.

$$P(n, r) = n(n - 1)(n - 2) \cdots (n - r + 1) \qquad \text{The product of } r \text{ factors}$$

$$= \frac{n(n - 1)(n - 2) \cdots (n - r + 1)(n - r) \cdots (2)(1)}{(n - r) \cdots (2)(1)}$$

$$= \frac{n!}{(n - r)!}$$

Number of Permutations ▼

> The number of permutations of n distinct objects taken r at a time, denoted by $P(n, r)$, where $r \leq n$, is
>
> $$P(n, r) = \frac{n!}{(n - r)!}$$

EXAMPLE 6 A ship can signal other ships using coded flags. How many distinct messages can be sent by displaying a sequence of two out of seven possible flags?

SOLUTION
$$P(7, 2) = \frac{7!}{(7-2)!}$$ Each distinct message is a permutation of seven objects taken two at a time.

$$= \frac{7!}{5!}$$

$$= \frac{7 \cdot 6 \cdot \cancel{5!}}{\cancel{5!}}$$ Divide both the numerator and the denominator by 5!.

$$= 7 \cdot 6$$

$$= 42$$

Answer 42 messages

The TI-81 graphics calculator is capable calculating the number of permutations of n objects taken r at a time. The keystrokes for calculating the $P(7, 2)$ from Example 6 are illustrated below.

First press **7** to enter 7, then use the MATH menu to complete the calculation of $P(7, 2)$ with the keystrokes **MATH** **PRB** **2 (nPr)** **2** **ENTER** to obtain the screen shown in Figure 8-3.

Thus $P(7, 2) = 42$

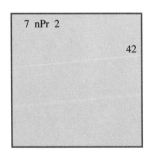

Figure 8-3

A selection of objects for which order is not considered is called a **combination**. The number of combinations of n objects selected r at a time is denoted by $C(n, r)$.

EXAMPLE 7 List all permutations and all combinations of the three letters $\{a, b, c\}$ taken two at a time.

SOLUTION The permutations are ab, ac, ba, bc, ca, and cb. The combinations are ab, ac, and bc.

The permutations ab and ba represent the same combination.

▶ **Self-Check** ▼

Twenty-six ping-pong balls are lettered A to Z. Three of these are randomly chosen to form part of a secret code for a bank card. How many such codes are possible?

Any combination of r objects can be ordered into $r!$ permutations. Therefore we can determine the number of permutations of r objects by multiplying the number of combinations by $r!$. Thus $P(n, r) = r![C(n, r)]$, and so $C(n, r) = \frac{P(n, r)}{r!}$. Substituting for $P(n, r)$, we have $C(n, r) = \frac{n!}{r!(n-r)!}$.

▶ **Self-Check Answer** ▼

$P(26, 3) = 15,600$

SECTION 8-6 COUNTING: PERMUTATIONS AND COMBINATIONS

Number of Combinations ▼

The number of combinations of n objects taken r at a time, denoted by $C(n, r)$, where $r \leq n$, is

$$C(n, r) = \frac{n!}{r!(n-r)!}$$

EXAMPLE 8 Determine the number of possible 5-card poker hands that can be dealt from a deck of 52 cards.

SOLUTION $\quad C(n, r) = \dfrac{n!}{r!(n-r)!}$ Each hand is a combination, since order is not considered.

$$C(52, 5) = \frac{52!}{5!(52-5)!}$$

There are 52 cards, taken 5 at a time.

$$= \frac{52 \cdot 51 \cdot \overset{10}{\cancel{50}} \cdot 49 \cdot \overset{2}{\cancel{48}} \cdot \cancel{47!}}{\cancel{5} \cdot \cancel{4} \cdot \cancel{3} \cdot \cancel{2} \cdot 1 \cdot \cancel{47!}}$$

$$= 52 \cdot 51 \cdot 10 \cdot 49 \cdot 2$$

$$= 2{,}598{,}960$$

Answer   2,598,960 possible poker hands

▶ **Self-Check** ▼

List all combinations of $\{a, b, c, d, e\}$ taken two at a time.

The TI-81 graphics calculator is capable of calculating the number of combinations of n objects taken r at a time. The keystrokes for calculating the $C(52, 5)$ from Example 8 are illustrated below.

First press [5] [2] to enter 52, then use the MATH menu to complete the calculation of $C(52, 5)$ with the keystrokes:

[MATH] [PRB] [3 (nCr)] [5] [ENTER]

to obtain the screen shown in Figure 8-4.

Thus $\quad C(52, 5) = 2{,}598{,}960$

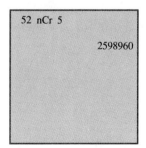

Figure 8-4

▶ **Self-Check Answer** ▼

ab, ac, ad, ae, bc, bd, be, cd, ce, de

The following examples illustrate the relationship between combinations and binomial coefficients:

- The number of ways we can select 4 chairs out of 10 to be occupied by 4 people if the order of the 4 is not important is

$$C(10, 4) = \frac{10!}{4!6!} = 210$$

- The number of ways we can select 4 blanks out of 10 to fill in with y's (the other 6 blanks will be filled in with x's) is $C(10, 4) = 210$.
- The number of ways we can select 4 factors of y and 6 factors of x for a product with 10 factors is $C(10, 4) = 210$.
- The binomial coefficient of $x^6 y^4$ in the expansion of $(x + y)^{10}$ is $C(10, 4) = 210$.

As mentioned in Section 8-5, the number of combinations of n objects taken r at a time is sometimes denoted by $\binom{n}{r}$. This notation is frequently used to represent the binomial coefficients in the binomial theorem. Using this notation,

$$(x + y)^n = \binom{n}{0} x^n + \binom{n}{1} x^{n-1} y + \binom{n}{2} x^{n-2} y^2$$
$$+ \cdots + \binom{n}{r} x^{n-r} y^r + \cdots + \binom{n}{n} y^n$$

EXAMPLE 9 Compute the coefficient of $x^7 y^5$ in the expansion of $(x + y)^{12}$.

SOLUTION Since $n = 12$ and $r = 5$, $\binom{n}{r} = \binom{12}{5}$.

$$\binom{12}{5} = \frac{12!}{5!7!}$$
$$= \frac{12 \cdot 11 \cdot 10 \cdot 9 \cdot 8 \cdot 7!}{5 \cdot 4 \cdot 3 \cdot 2 \cdot 1 \cdot 7!}$$
$$= 11 \cdot 9 \cdot 8$$
$$= 792 \qquad \square$$

Before applying the formula for either $P(n, r)$ or $C(n, r)$, be sure to determine whether the problem involves permutations or combinations. Remember that order makes a difference in permutations but does not make a difference in combinations.

Does order make a difference?	
Yes	No
Permutation	Combination

SECTION 8-6 COUNTING: PERMUTATIONS AND COMBINATIONS

EXAMPLE 10   Determine whether each of the following items is a permutation or a combination.

SOLUTIONS

(a) The answer key to a multiple-choice test that has 5 possible choices for each of the 25 questions

Permutation — Order makes a difference. The answers must be given in the same order as the questions.

(b) A group of 10 objects randomly selected for testing from among the 500 objects produced on an assembly line

Combination — Order does not make a difference. All objects selected will be given the same test regardless of the order in which they are selected.

(c) The access code to a digital-control combination lock

Permutation — This problem is misleading since the term "combination lock" is a misnomer. The order of the digits entered makes a difference. Perhaps a better name for this type of lock would be a "permutation lock."

▶ **Self-Check** ▼

How many committees of 4 people can be selected from a club with 20 members?

EXERCISES 8-6

A

In Exercises 1–6, evaluate each expression.

1. a. $P(6, 6)$ b. $P(6, 4)$ c. $C(6, 6)$ d. $C(6, 4)$
2. a. $P(7, 5)$ b. $C(7, 5)$ c. $P(5, 0)$ d. $C(5, 0)$
3. a. $P(7, 1)$ b. $C(7, 1)$ c. $P(7, 0)$ d. $C(7, 0)$
4. a. $P(10, 3)$ b. $C(10, 3)$ c. $P(7, 6)$ d. $C(7, 6)$
5. a. $\binom{8}{2}$ b. $\binom{8}{6}$ c. $\binom{8}{4}$ d. $\binom{8}{0}$
6. a. $\binom{9}{3}$ b. $\binom{9}{6}$ c. $\binom{9}{4}$ d. $\binom{9}{9}$

7. List all possible permutations and combinations of the four letters {a, b, c, d} taken two at a time.

8. List all possible permutations and combinations of the five digits {1, 2, 3, 4, 5} taken two at a time.

In Exercises 9–12, draw a tree diagram to represent the sequence of choices.

9. A traveler has a choice of two airlines from Champaign to Chicago and four airlines from Chicago to Washington. Show all choices available on this route.

▶ **Self-Check Answer** ▼

$C(20, 4) = 4845$

10 A family has three children. Show all possible sequences of girls and/or boys.

11 A steak house offers five main dinners and three types of dessert. Show all possible choices consisting of a main dinner and a dessert.

12 Illustrate all sequences in which exactly 25¢ can be inserted into a vending machine using any mixture of quarters, dimes, and nickels.

13 How many ways can a disc jockey play 3 different records from a collection of 500 singles?

14 A shoe store sells five styles of running shoes. Each style is stocked in nine sizes and two colors. How many distinct possibilities are stocked?

15 A manufacturer makes three styles of boots. Each style comes in seven sizes and four colors. How many distinct boots can be produced?

16 From a class of 140 males and 150 females, one couple is getting married. This couple is one of how many possible matches?

17 Each of five questions on a multiple-choice test has four possible answers. How many distinct ways could the correct answers be arranged?

18 How many distinct arrangements of 3 letters can be made from the 26 letters in the alphabet if (a) repetition of letters is allowed, (b) repetition is not allowed?

19 A manufacturer sells digital control combination locks. Each code consists of a sequence of four distinct digits chosen from the digits 0, 1, ..., 9. How many codes are possible?

20 Two dice, one red and one green, are tossed simultaneously. How many distinct results are possible? (The faces of each die are numbered from 1 to 6.)

21 A computer generates a quiz by randomly selecting 5 questions from a bank of 100 questions. How many possible quizzes can the computer generate?

22 A state lottery game randomly selects six of the natural numbers 1 through 44. How many different selections are possible?

23 Participants in a daily state lottery game pick four digits from 0 through 9 (repeated digits are allowed.) How many different selections are possible?

24 The last four digits of a telephone number are distinct. If you remember the digits but forget the correct order, how many ways could you arrange these digits in search of the correct number?

25 Three contestants on a game show select one door each from three doors that hide prices. How many ways can they select the doors?

26 Six distinct coins (dollar, half-dollar, quarter, dime, nickel, and penny) are flipped. How many ways can these coins land heads and tails?

27 The order in which chemicals are added to a mix sometimes affects the results. In how many different ways can five chemicals be added one at a time?

28 Two light bulbs out of 1000 are randomly selected for testing. How many different selections are possible?

B

29 A line is determined by two points. How many distinct lines can be determined by ten points if no three of the points lie on the same line?

30. How many distinct lines are determined by the vertices of a pentagon? How many of these lines are diagonals?

31. How many distinct lines are determined by the vertices of a hexagon? How many of these lines are diagonals?

32. How many games are necessary for all eight softball teams in a round-robin tournament to play each other exactly once?

33. How many ways can all the letters of the word MATH be arranged?

34. How many ways can all the letters of the word COMPUTER be arranged?

35. How many ways can a 5-card heart flush be formed from a 52-card deck? (A deck has 13 hearts, and a heart flush contains only hearts.)

36. How many ways can 2 of 4 aces be selected from a 52-card deck?

37. How many ways can 3 of 4 kings be selected from a 52-card deck?

38. How many ways can a 5-card hand contain a full house consisting of 2 aces and 3 kings?

39. How many possible 13-card bridge hands can be dealt from a 52-card deck?

40. There are eight possible toppings that can be ordered for a pizza. How many possible pizzas contain exactly three of these toppings?

C

In Exercises 41–47, verify each statement.

41. $P(n, 1) = n$
42. $P(n, 0) = 1$
43. $P(n, n) = P(n, n - 1)$
44. $C(n, n) = 1$
45. $C(n, r) = C(n, n - r)$
46. $C(n, 0) = 1$
47. $C(n, r) = C(n - 1, r - 1) + C(n - 1, r)$

Each bit of a computer can represent two different values, since it can be "on" or "off." Bits are generally grouped together to form convenient units of storage called words. If one bit of each word is reserved for special purposes and all other bits are used to represent a number, how many distinct numbers can be represented by the following words?

48. An 8-bit word
49. A 16-bit word
50. A 32-bit word

SECTION 8-7

Probability

Section Objectives

13	Give the sample space for an experiment.
14	Give the set of occurrences for an event and the probability of an event.
15	Use the probability formulas for $P(\bar{E})$, $P(A \cap B)$, and $P(A \cup B)$.

If a coin is flipped 50 times on our planet Earth, we can be sure it will land 50 times but we cannot be sure how many heads or tails will occur. Many phenomena of nature, including the law of gravity, behave according to completely determined principles for which we have constructed algebraic models. Many other events, including the weather, the winners in lotteries, the behavior of atoms, and the claims on insurance policies, behave according to the laws of probability. An insurance company cannot determine which vehicle will be involved in an accident, but it can predict the probability of an accident in order to establish the premium for each policy.

This section will introduce some of the basic concepts of probability theory. This theory was first studied as early as 1526 in relation to games of chance. Some of the fundamental concepts are still easiest to explain using examples from such games.

Intuitively, we expect the probability of an event to be a measure of the fractional portion the occurrence of this event represents in relationship to all possible outcomes. For example, the probability of drawing an ace from a deck of 52 cards is $\frac{4}{52}$, or $\frac{1}{13}$, since 4 of the 52 cards are aces. The set of all possible outcomes of an experiment is called the **sample space**. Any subset of this sample space is called an **event**. The probability of an event E is denoted by $P(E)$. If the sample space can be separated into distinct mutually exclusive outcomes with each outcome equally likely, then $P(E)$ is the ratio of the number of outcomes favorable to E to the total number of outcomes.

Probability ▼

The probability $P(E)$ of an event E in a finite sample space S of distinct equally likely outcomes is the ratio of the number of outcomes favorable to E to the total number of outcomes:

$$P(E) = \frac{n(E)}{n(S)}$$

$n(E)$ represents the number of outcomes favorable to event E. $n(S)$ represents the number of outcomes in the sample space.

Two events deserve special attention. If an event E is certain, then $E = S$ and thus $P(E) = \frac{n(E)}{n(S)} = \frac{n(S)}{n(S)} = 1$. If an event E is impossible, then $n(E) = 0$ and thus $P(E) = \frac{n(E)}{n(S)} = 0$. Therefore the probability of an event is always a number between 0 and 1 inclusive; that is, $0 \leq P(E) \leq 1$.

The listing of a sample space can be used to compute probabilities as illustrated below.

- The sample space for flipping three coins and recording heads (H) or tails (T) is {HHH, HHT, HTH, THH, HTT, THT, TTH, TTT}.
- The event of getting exactly two heads on three flips is {HHT, HTH, THH}.
- The probability of getting exactly two heads on three flips is
$$P(2H) = \frac{n(2H)}{n(S)} = \frac{3}{8}.$$

Three of the eight outcomes in the sample space have exactly two heads.

SECTION 8-7 PROBABILITY

EXAMPLE 1 Find the probability that the roll of a die will result in a multiple of 3.

SOLUTION $S = \{1, 2, 3, 4, 5, 6\}$ The sample space (six outcomes)

$E = \{3, 6\}$ The set of favorable outcomes (two outcomes)

$$P(E) = \frac{n(E)}{n(S)} = \frac{2}{6} = \frac{1}{3}$$

▶ **Self-Check** ▼

Determine the probability of drawing a face card (jack, queen, or king) from a deck of 52 cards.

Before we can apply the formula $P(E) = \dfrac{n(E)}{n(S)}$, we must be sure that the outcomes in S are equally likely. Consider the experiment of tossing a pair of dice and recording the sum of the two dice. The sums that can result are $\{2, 3, 4, 5, 6, 7, 8, 9, 10, 11, 12\}$; however, these totals are not equally likely. In particular, $P(2) \neq P(7)$. The 36 equally likely outcomes in the sample space for rolling a white die and a colored die are given in Figure 8-5.

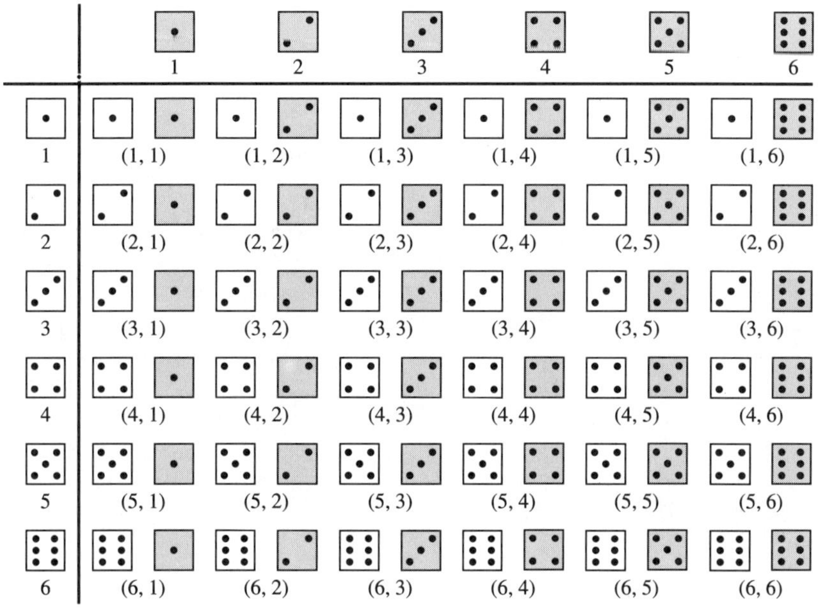

Figure 8-5

▶ **Self-Check Answer** ▼

$\dfrac{3}{13}$

EXAMPLE 2 A pair of dice are rolled. Determine the probability that the sum of the numbers on the upturned faces will be

SOLUTIONS

(a) 2   A total of 2 is produced only by the one pair (1, 1). Thus $P(2) = \frac{1}{36}$.

(b) 7   A total of 7 is produced by the six pairs $\{(6, 1), (5, 2), (4, 3), (3, 4), (2, 5), (1, 6)\}$. Thus $P(7) = \frac{6}{36} = \frac{1}{6}$.

(c) 13   The set of outcomes favorable to a total of 13 is the null set. Thus $P(13) = \frac{0}{36} = 0$.   □

Since the probability of an event is determined by $P(E) = \dfrac{n(E)}{n(S)}$, it is usually more important to count the number of outcomes than to list them. Thus the formulas for the number of permutations or combinations are important tools in probability theory.

EXAMPLE 3 Determine the probability of being dealt a heart flush (all hearts) in a five-card hand.

SOLUTION

$P(\text{heart flush}) = \dfrac{\text{Number of possible heart flushes}}{\text{Number of possible 5-card hands}}$ *Word equation*

$P(\text{heart flush}) = \dfrac{C(13, 5)}{C(52, 5)}$ The number of possible heart flushes is the number of combinations of 13 hearts taken 5 at a time. The number of possible hands is the number of combinations of 52 cards taken 5 at a time. Substitute these values into the formula $C(n, r) = \dfrac{n!}{r!(n-r)!}$

$= \dfrac{13!}{5!8!} \div \dfrac{52!}{5!47!}$

$= \dfrac{13!}{5!8!} \cdot \dfrac{5!47!}{52!}$ *Word equation*

$= \dfrac{13 \cdot 12 \cdot 11 \cdot 10 \cdot 9 \cdot 8! \cdot 47!}{8! \cdot 52 \cdot 51 \cdot 50 \cdot 49 \cdot 48 \cdot 47!}$ *Simplify by dividing out the common factors.*

$= \dfrac{11 \cdot 3}{4 \cdot 17 \cdot 5 \cdot 49 \cdot 4} = \dfrac{33}{66{,}640}$

$P(\text{heart flush}) \approx 0.0004952$

▶ **Self-Check** ▼

Determine the probability of getting exactly three heads when five coins are tossed.

▶ **Self-Check Answer** ▼

$P(3 \text{ heads}) = \dfrac{C(5, 3)}{2^5} = \dfrac{5}{16} = 0.3125$

SECTION 8-7 PROBABILITY

EXAMPLE 4 Determine the probability of picking win, place, and show (first, second, and third) in a seven-horse race.

SOLUTION $P(\text{win, place, show}) = \dfrac{\text{Number of ways one can pick correctly}}{\text{Number of ways horses can finish}}$ *Word equation*

$P(\text{win, place, show}) = \dfrac{1}{P(7, 3)}$

There is only one favorable outcome. The number of possibilities is the number of permutations of seven objects taken three at a time.

$= \dfrac{1}{\dfrac{7!}{4!}}$

$= \dfrac{4!}{7!} = \dfrac{1 \cdot \cancel{4!}}{7 \cdot 6 \cdot 5 \cdot \cancel{4!}}$ Simplify by dividing out the common factors.

$= \dfrac{1}{7 \cdot 6 \cdot 5} = \dfrac{1}{210}$

$P(\text{win, place, show}) \approx 0.0047619$

There are several formulas that allow us to calculate the probability of desired events from the known probabilities of related events. The first of these formulas involves complementary events. The complement of event E, denoted $\tilde{E}$, in the sample space S is the set of all outcomes in S that are not in set E. Since $P(S) = 1$, $P(E) + P(\tilde{E}) = 1$, or $P(\tilde{E}) = 1 - P(E)$.

EXAMPLE 5 Determine the probability that in a family with three children at least one child will be a boy.

SOLUTION $P(\text{at least 1 boy}) = 1 - P(\text{no boys})$ *Word equation*

$= 1 - \dfrac{1}{8}$

By the fundamental counting principle, there are $2 \cdot 2 \cdot 2 = 8$ outcomes. Only one outcome has three girls and no boys.

$= \dfrac{7}{8}$

If the occurrence or nonoccurrence of event A does not affect the occurrence or nonoccurrence of event B, then we say that A and B are **independent events**. For independent events A and B, the **probability of A and B**, denoted $P(A \cap B)$, is given by $P(A \cap B) = P(A) \cdot P(B)$. The **probability of A or B**, denoted $P(A \cup B)$, is given by $P(A \cup B) = P(A) + P(B) - (A \cap B)$.

▶ **Self-Check** ▼

Determine the probability that a student who guesses each answer on a five-question multiple-choice test will guess at least one answer correctly. Each answer has four choices.

▶ **Self-Check Answer** ▼

$P(\text{at least 1 correct answer}) = 1 - (\tfrac{3}{4})^5 \approx 0.7626953$

Probability Formulas ▼

If A, B, and E are events in a sample space S, then

$P(\tilde{E}) = 1 - P(E)$

$P(A \cap B) = P(A) \cdot P(B)$ where A and B are independent

$P(A \cup B) = P(A) + P(B) - P(A \cap B)$

EXAMPLE 6 A single card is drawn from a deck of 52 cards. If K represents the event of a king and H represents the event of a heart, determine

SOLUTIONS

(a) $P(K)$ $P(K) = \dfrac{4}{52} = \dfrac{1}{13}$ There are 4 kings in a deck of 52 cards.

(b) $P(H)$ $P(H) = \dfrac{13}{52} = \dfrac{1}{4}$ There are 13 hearts in a deck of 52 cards.

(c) $P(K \cap H)$ $P(K \cap H) = P(K) \cdot P(H)$

$= \dfrac{1}{13} \cdot \dfrac{1}{4} = \dfrac{1}{52}$ Substitute the values calculated in parts (a) and (b). Only one card is both a king and a heart.

(d) $P(\tilde{K})$ $P(\tilde{K}) = 1 - P(K)$

$= 1 - \dfrac{1}{13} = \dfrac{12}{13}$ Twelve of the 13 ranks are not kings.

(e) $P(K \cup H)$ $P(K \cup H) = P(K) + P(H) - P(K \cap H)$ We must subtract $\dfrac{1}{52}$ to avoid counting the king of hearts twice.

$= \dfrac{4}{52} + \dfrac{13}{52} - \dfrac{1}{52}$

$= \dfrac{16}{52} = \dfrac{4}{13}$

▶ **Self-Check** ▼

A coin is tossed and a die is rolled. Determine

1 P(head and a 3)

2 P(head or a 3)

EXERCISES 8-7

A

In Exercises 1–8, list the sample space of each experiment.

1 The order of boys and girls in a family with two children
2 The order of boys and girls in a family with three children
3 The outcome of a four-question true-false (T-F) test
4 The outcome when a coin is tossed and a die is rolled

▶ **Self-Check Answers** ▼

1 $\dfrac{1}{12}$ **2** $\dfrac{7}{12}$

SECTION 8-7 PROBABILITY

5. The selection of a ball glove that comes in five models and can be left-handed or right-handed
6. The order of finish of horses A, B, C, and D in a four-horse race
7. The signals that can be sent by displaying two flags one above the other, the flags available being white (W), red (R), blue (B), green (G), and yellow (Y) (Both flags can be the same color.)
8. The selection of one of four sites (A, B, C, D) for a branch bank and the selection of one of three styles (E, F, G) for the architecture of the building

A multiple-choice quiz has five questions with four choices each. The key is BCADC. In Exercises 9–18, determine the probability of each event for a student who randomly guesses each answer.

9. The first answer is correct.
10. The second answer is wrong.
11. The first answer is correct and the second answer is wrong.
12. Both the first and second answers are correct.
13. The first or the second answer is correct.
14. The first or the second answer is wrong.
15. All five answers are correct.
16. All five answers are wrong.
17. At least one answer is correct.
18. At least one answer is wrong.

In Exercises 19–30, determine the probability of each of the events described, given that the family in question has five children.

19. The first child is a girl.
20. The second child is a boy.
21. The first two children are girls.
22. The last two children are boys.
23. All five children are girls.
24. All five children are boys.
25. At least one child is a girl.
26. At least one child is a boy.
27. The first three children are girls and the last two are boys.
28. The first two children are girls and the last three are boys.
29. The first child is a boy or the second child is a boy.
30. The first child is a girl or the second child is a girl.

B

A red die and a white die are tossed simultaneously. In Exercises 31–42, determine the probability of each of the events described.

31. A total of 5
32. A total of 6
33. A total of 7
34. A total of 4
35. A total of 12
36. A total of 2
37. A total of 1
38. A total that is even
39. A total of 3 or less
40. A total of 11 or more
41. A total that is a multiple of 5
42. A total of either 2 or 11

A single card is drawn from a deck of 52 cards. In Exercises 43–58, determine the probability that the card is as described.

43. A queen
44. Black
45. Black and a queen
46. Not a queen
47. Black or a queen
48. Black and not a queen
49. A club
50. A club or a queen
51. Not black
52. Not a club
53. Not black and not a queen
54. Not a club and not a queen
55. A jack or an ace
56. A jack and an ace
57. A queen and an ace
58. A one-eyed jack (two of the jacks are one-eyed)

C

A bin of 25 electrical switches contains 5 defective switches. A random sample of 4 switches is tested from this bin. In Exercises 59–66, determine the probability of each of the events described.

59 The first one selected is defective.
60 The first one selected is not defective.
61 The first two selected are defective.
62 The first two selected are not defective.
63 All four are defective.
64 All four are not defective.
65 At least one is defective.
66 At least one is not defective.

67 For a state lottery, six of the natural numbers 1 through 44 are randomly selected each week. Express as a fraction the probability that the numbers selected next week will be 4, 11, 17, 27, 31, 42.

68 A state lottery has a daily pick-four game in which participants pick four digits from 0 through 9 (repeated digits are allowed). What is the probability that tomorrow's winning number will be 2161?

KEY CONCEPTS

1 Summation notation:

$$\sum_{i=1}^{n} a_i = a_1 + a_2 + \cdots + a_{n-1} + a_n$$

and

$$\sum_{i=1}^{\infty} a_i = a_1 + a_2 + a_3 + \cdots + a_n + \cdots$$

Partial sum, s_n:

For $\sum_{i=1}^{\infty} a_i$, the nth partial sum is $s_n = a_1 + a_2 + a_3 + \cdots + a_n$.

2 Arithmetic sequences:
  a. Common difference: $d = a_n - a_{n-1},\ n > 1$
  b. nth term: $a_n = a_1 + (n-1)d$
  c. Series of n terms: $S_n = \dfrac{n}{2}(a_1 + a_n) = \dfrac{n}{2}[2a_1 + (n-1)d]$
  d. Arithmetic mean: $m = \dfrac{a+l}{2}$

3 Geometric sequences:
  a. Common ratio: $r = \dfrac{a_n}{a_{n-1}},\ n > 1$
  b. nth term: $a_n = a_1 r^{n-1}$
  c. Geometric mean: $m = \pm\sqrt{al}$
  d. Series of n terms: $S_n = \dfrac{a_1(1-r^n)}{1-r} = \dfrac{a_1 - ra_n}{1-r}$ for $r \neq 1$.

KEY CONCEPTS

4 Infinite geometric series:

   a. If $|r| < 1$, $S = \dfrac{a_1}{1 - r}$.

   b. If $|r| \geq 1$, the sum does not exist.

5 Principle of mathematical induction:

   If $S(n)$ is a statement involving the natural number n and

   a. $S(1)$ is true and

   b. the truth of $S(k)$ implies the truth of $S(k + 1)$ then $S(n)$ is true for each natural number n.

6 Factorial notation:

$$n! = n \cdot (n - 1) \cdot (n - 2) \cdots 3 \cdot 2 \cdot 1, \; n \geq 1$$

$$0! = 1$$

7 Binomial theorem:

$$(a + b)^n = \sum_{r=0}^{n} \binom{n}{r} a^{n-r} b^r$$

$$= \sum_{r=0}^{n} \frac{n!}{r!(n - r)!} a^{n-r} b^r$$

8 Fundamental counting principle:

   If the first of two successive events can occur m ways and the second can occur n ways, the joint event can occur $m \cdot n$ ways.

9 Counting formulas:

   a. Number of permutations of n things r at a time: $P(n, r) = \dfrac{n!}{(n - r)!}$

   b. Number of combinations of n things r at a time:

$$C(n, r) = \frac{n!}{r!(n - r)!} \quad \text{or} \quad \binom{n}{r} = \frac{n!}{r!(n - r)!}$$

10 Probability:

   a. Probability of event E in an equally likely sample space S: $P(E) = \dfrac{n(E)}{n(S)}$

   b. Probability of the complement of E: $P(\tilde{E}) = 1 - P(E)$

   c. Probability of A and B (if A and B are independent events): $P(A \cap B) = P(A) \cdot P(B)$

   d. Probability of A or B: $P(A \cup B) = P(A) + P(B) - P(A \cap B)$

REVIEW EXERCISES FOR CHAPTER 8

In Exercises 1–8, write the first five terms of each sequence described.

1. $a_n = \dfrac{1}{3^{n-1}}$
2. $a_n = \dfrac{1}{3^n - 1}$
3. $a_n = (-1)^n(3 + n)$
4. $a_{n+2} = a_{n+1} - a_n$, $a_1 = 3$, $a_2 = 5$
5. Arithmetic sequence with $a_2 = 5$ and $a_4 = 9$
6. A geometric sequence with $r = 1.1$ and $a_1 = 1000$
7. The sequence of digits of the number e
8. $a_1 = 1$, $a_n = na_{n-1}$ (Also write an alternative notation for a_n.)

In Exercises 9–13, determine whether each of the sequences is arithmetic, geometric, both, or neither.

9. $5, 3, 1, -1, -3, -5$
10. $5, -3, 1.8, -1.08, 0.648$
11. $1, -1, 1, -1, 1, -1$
12. $7, 7, 7, 7, 7, 7$
13. $7, -7, -7, 7, 7, -7, -7$

In Exercises 14–22, evaluate each of the series described.

14. $\sum_{i=1}^{7} (2i - 7)$
15. $\sum_{j=3}^{8} (j^2 - 2j)$
16. $\sum_{k=4}^{9} (-1)^k(k^2 + 3)$
17. $\sum_{i=1}^{37} 4$
18. $n = 6$, $a_1 = 3$, $a_n = 2a_{n-1} + 5$, $S_6 = ?$
19. $a_1 = 3$, $d = 3$, $S_{50} = ?$ (arithmetic)
20. $1 + 0.25 + 0.0625 + 0.015625 + 0.00390625 + \cdots$
21. $a_n = 2n$, $S_{5000} = ?$
22. s_5 the fifth partial sum of $\sum_{i=1}^{\infty} 4(0.1)^i$

In Exercises 23–26, determine the number of terms in each of the sequences.

23. $a_1 = 7$, $d = -3$, $a_n = -23$, $n = ?$ (arithmetic)
24. $a_1 = 5$, $d = \frac{3}{4}$, $S_n = 262.5$, $n = ?$ (arithmetic)
25. $a_1 = 1000$, $r = 0.3$, $a_n = 2.43$, $n = ?$ (geometric)
26. $a_1 = 1250$, $r = \frac{3}{5}$, $S_n = 2979.2$, $n = ?$ (geometric)

In Exercises 27–30, find the value requested in each sequence.

27. $a_1 = 86$, $a_{86} = 1$, $d = ?$ (arithmetic)
28. $a_1 = 160$, $a_5 = 10$, $r = ?$ (geometric)
29. $a_6 = 8$, $r = \frac{1}{3}$, $a_1 = ?$ (geometric)
30. $a_7 = 8$, $d = \frac{1}{3}$, $a_1 = ?$ (arithmetic)

In Exercises 31 and 32, expand each binomial.

31. $(x - 3y)^4$
32. $(2x^2 + 5y)^6$

In Exercises 33–37, evaluate each expression.

33. $P(6, 6)$
34. $P(25, 2)$
35. $C(8, 6)$
36. $C(11, 0)$
37. $\binom{12}{9}$

A roulette wheel contains 38 slots numbered 00, 0, 1, 2, ..., 35, 36. Half the slots from 1 through 36 are red, and half are black; 00 and 0 are neither black nor red. In Exercises 38–44, determine the probability that a ball will randomly land in a slot as described.

38. Black slot
39. Slot 27
40. Odd-numbered slot
41. Slot 0 or slot 00
42. Slot 17 twice consecutively
43. Red slot twice consecutively
44. Red one time and black the next time

A red die and a white die are tossed. In Exercises 45–49, determine the probability of the event described.

45 The red die is a 6. **46** The white die is a 6. **47** Both are 6's

48 At least one 6 occurs. **49** The total is a multiple of 5.

50 How many lines can be formed by the vertices of an octagon? How many of these line segments are diagonals?

51 Draw a tree diagram to illustrate all possible routes from A to C through B, given that there are three distinct routes from A to B and four routes from B to C.

52 A piece of paper is 0.15 millimeter thick. If the paper could be folded 20 times so that it doubled in thickness with each folding, how thick would the folded piece be?

53 List the sample space for a two-digit security code used by a pharmacist to confirm orders received over the phone. The tens' digit must be a prime (2, 3, 5, or 7), and the units' digit must be odd.

54 A sum of $100 invested at the beginning of one year compounds annually at 8% for five years. Write the sequence describing the value of this investment at the end of each of the five years.

55 Because of its shape, a football stadium cannot have the same number of seats in each row. There are 45 seats in the twentieth row, 43 in the nineteenth row, 41 in the eighteenth row, etc. How many seats are in this section of 20 rows?

56 Insert five arithmetic means between 83 and 125.

57 Insert three geometric means between $\frac{3}{5}$ and $\frac{243}{3125}$.

58 A ball is dropped from a height of 20 meters. Each bounce results in a rebound three-fifths of the previous height. Determine the total distance the ball travels.

In Exercises 59 and 60, use mathematical induction to prove that each of the statements is true for every natural number n.

59 $\sum_{i=1}^{n} 3i = \frac{3}{2}n(n+1)$

60 $1^2 + 3^2 + 5^2 + \cdots + (2n-1)^2 = \frac{4n^3 - n}{3}$

MASTERY TEST FOR CHAPTER 8

Exercise numbers correspond to Section Objective numbers.

1 Write the first five terms of the infinite sequence whose general term is given.

  a. $a_i = 3i + 5$
  b. $a_i = (-2)^{i+1}$
  c. $a_i = \dfrac{i-1}{i+1}$
  d. $a_i = (i^2 - 5)(-1)^i$
  e. $a_{n+1} = 2a_n - 3$, $a_1 = 5$
  f. $a_{n+1} = a_n(a_n - 3)$, $a_1 = -2$

2 Evaluate the series given below.

  a. $\sum_{i=1}^{6} (3i - 5)$
  b. $\sum_{j=4}^{8} j(6 - j)$
  c. $\sum_{k=1}^{80} 11$
  d. $\sum_{n=1}^{5} \left(\dfrac{n}{n+1}\right)(-1)^n$

3 Find the indicated quantities from the information given about these arithmetic sequences.
   a. $a_1 = 20$, $d = 3$, $a_{60} = ?$
   b. $a_8 = 90$, $a_{10} = 84$, $d = ?$
   c. $a_{70} = 40$, $d = 0.5$, $a_1 = ?$
   d. $11, 15, 19, \ldots, 163$; $n = ?$

4 Find the values of these arithmetic series.
   a. $9 + 14 + 19 + \cdots + 114$
   b. $a_1 = 43$, $a_{47} = 227$, $S_{47} = ?$
   c. $\sum_{i=1}^{83} 6i$
   d. $a_{11} = -3$, $a_{13} = -7$, $S_{20} = ?$

5 Find the indicated quantities from the information given about these geometric sequences.
   a. $a_1 = 36$, $r = -\frac{2}{3}$, $a_5 = ?$
   b. $a_1 = 12{,}345$, $a_5 = 1.2345$, $r = ?$
   c. $a_4 = 8$, $a_6 = 32$, $r = ?$
   d. $a_7 = 192$, $a_8 = 384$, $a_1 = ?$

6 Find the values of these geometric series.
   a. $a_1 = 3$, $r = 0.1$, $S_6 = ?$
   b. $a_1 = 45$, $r = -1$, $S_{100} = ?$
   c. $\sum_{i=1}^{5} (-\frac{3}{2})^i$
   d. $a_2 = 15$, $a_3 = 45$, $S_5 = ?$

7 Write these repeating decimals as fractions.
   a. $0.\overline{54}$
   b. $2.\overline{8}$

8 Use mathematical induction to prove that
$$\sum_{i=1}^{n} (2i-1)^2 = \frac{n(2n+1)(2n-1)}{3}$$

9 Evaluate $\dfrac{12!}{4!\,8!}$.

10 Find the fifth term of $(x - 3y)^{12}$.

11 A hardware store stocks cans of paint in 3 sizes. The paint is available in 11 colors, and each color has 2 possible finishes. How many distinct cans must be stocked to allow for all possibilities?

12 a. A soft-drink company randomly selects 2 bottles of every 5000 for quality control testing. How many such selections are possible?
   b. Six of 12 planes left on an aircraft carrier are aligned for takeoff. How many alignments are possible?

13 A baseball team has four games left in its season. Give the sample space for the results, using W for a win and L for a loss.

14 List the occurrences for the event of exactly three wins in Exercise 13. Then determine the probability of exactly three wins, assuming that the team is equally likely to win or lose each game.

15 Determine these probabilities associated with the rolling of a single die.
   a. The result is less than 7.
   b. The result is 7.
   c. The result is 5.
   d. The result is not 5.
   e. The result is a multiple of 2.
   f. The result is a multiple of 3.
   g. The result is a multiple of 2 and of 3.
   h. The result is a multiple of 2 or of 3.

APPENDIX A

Calculators

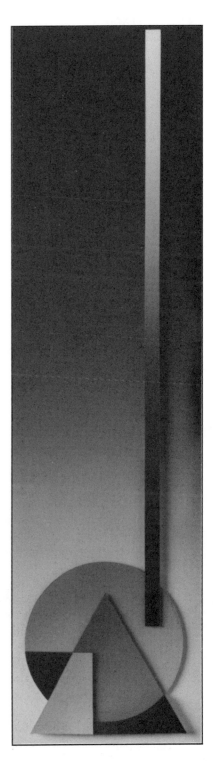

The Role of Calculators

The author fully supports the position of the Illinois Council of Teachers of Mathematics* that the use of calculators and computers should be an integral part of the education process, rather than an artificial adjunct to it. Students who integrate calculators into their coursework will be better prepared for their careers, and instructors who use calculators to enhance their presentations of subjects, such as inverse functions, will give students a fuller understanding of these topics. In addition, use of the calculator in the classroom provides an opportunity for students to get a balanced perspective on the calculator from a professional mathematics instructor. Students need to learn that calculators are not a panacea and that they do not always produce acceptable answers.

Types of Calculators

Most calculators use either AOS (Algebraic Operating System) or RPN (Reverse Polish Notation). The chief difference between AOS and RPN is in the order in which data and operations are entered. This text assumes use of AOS, in which the order of operations is the same as in algebra. Thus, many expressions can be entered into the calculator symbol by symbol, exactly as they appear. Typically RPN calculators have an **ENT** key, rather than an **=** key. Although many professionals believe RPN to be superior to AOS, we recommend that the average student use AOS when first learning the material in this text.

---

* *Report of the Task Force on Microcomputers in the Schools*, a special publication of the Illinois Council of Teachers of Mathematics, 1981.

There are many special-purpose calculators that perform operations with fractions, perform base conversions for programmers, or do statistical computations. We advise that the average student wait until a specific need arises before using these calculators.

Selecting a Calculator

Since calculators now occur on everything from watches to telephones, choosing the proper tool is not always a simple matter. We suggest that you consider the following factors in selecting a calculator.

1. **Size.** Although you want something you can carry, don't get a calculator so small your fingers have trouble depressing the keys one at a time.
2. **Display visibility.** Models with a liquid crystal display and a tilt display are generally easiest to read. Check the visibility of the display under a variety of lighting conditions.
3. **Keys.** You may want to get "click" keys, which have a definitive feel when an entry has been made, rather than "soft" keys. With soft keys, you have to look at the display to determine whether an entry has been made. You also want to make sure that the keys are well placed for your fingers.
4. **Model.**
   a. *Scientific model.* A scientific model will have the main features needed for this course, as it will usually have keys for squares, square roots, reciprocals, powers, logarithms, and the trigonometric functions. (When keystrokes are illustrated in this book, those for scientific models are denoted by S.) A four-function calculator without a square or square root key is too limited to perform all the computations needed in this text. A more specialized calculator that is programmable or has several functions for each key has more raw power, but it is also harder for a novice to use. We recommend that you choose a model that allows square roots to be taken with a single keystroke, unless you are preparing for more advnaced courses, in which case you many want a model with more features. Whatever the model, check the operating system to be sure the calculator follows AOS. For example, $2 + 3 \cdot 4$ should be evaluated as $2 + 12 = 14$, not $5 \cdot 4 = 20$.
   b. *Graphics model.* A graphics calculator will generally have all the main features of a scientific calculator plus more. (When keystrokes are illustrated in this book, those for the Texas

APPENDIX A

Instruments TI-81, a graphics calculator, will be denoted by G.) Specifically, a graphics calculator will have the capability to graph most of the elementary functions found in college algebra, trigonometry, and calculus. In addition, these calculators are designed to be able to obtain information about the graphs shown on the display. In particular, these calculators can be used to determine intercepts of a function or points of intersection of two functions. These calculators are certainly more powerful than scientific calculators, but they are also slightly more complex to operate and more expensive to purchase. In the opinion of the author, the added expense of a graphics calculator is justified for a student taking college algebra, and especially for students who will be taking subsequent mathematics courses. The TI-81 is illustrated in this text because, at the time of publication, it seemed superior to other calculators with similar capabilities with respect to a balance among cost, ease of operation, and size of display.

Advantages of Graphics Calculators

1. The ability to easily graph a function is such a powerful capability that it is hard to overstate this advantage. Many algebraic questions can be answered directly by examining the graph of a function.
2. Graphics calculators display the numbers and operations entered, as well as the result of these keystrokes. This means you can visually proof your keystrokes—something you cannot do with an ordinary scientific calculator.
3. Keystroke errors, even those that prevent execution on the calculator, can often be corrected by using the editing feature, which includes an insert key **INS** and a delete key **DEL**.

General Suggestions on Calculator Usage

Some general suggestions on calculator usage follow. For specific instructions, consult the manual with your calculator or a "how-to" guide such as *The Graphics Calculator Supplement* by Lawrence R. Huff and David R. Peterson (PWS-KENT Publishing Company, Boston, 1992). If you run into problems, ask your instructor for help.

1. Go through a problem with simple values whose result is known before you undertake the calculation of similar problems whose results are unknown.

2 For calculations use as many digits of accuracy as your calculator will allow. Since most calculators store more digits than they display, you can increase accuracy by observing the following guidelines:

a. Leave intermediate values in the calculator rather than copying down the display digits and then reentering these values. Learn to use the memory and parentheses keys.

b. To enter fractional values that result in repeating decimals, use the `1/x` key or divide the numerator by the denominator

(with calculators that work directly with fractions this is not a problem).

c. Enter `π` and `e` using the special keys if they are provided on your calculator. Such entries will typically be accurate to two or three more digits than are shown on the display.

Calculator Memories

Using the memory feature available on your calculator can increase the accuracy of your results by reducing round-off error. For many computations the use of the memory feature can also significantly speed up input and lessen the risk of keystroke error. We will illustrate the use of the memory feature below with an example of synthetic division. First we should mention that different models differ significantly as to the labeling of the keys and the keystrokes needed to access the memory feature. Three typical types are illustrated below. You should consult your owner's manual for specifics on your calculator.

1 *Calculators with a single memory.* A value can be stored in memory by pressing `STO` or a similar key such as `M in` (for into memory). A value that has been stored in memory can be recalled by pressing `RCL` or a similar key such as `M out` (for out of memory).

2 *Calculators with multiple memories.* To use a memory on a calculator with multiple memories, one must first specify which memory is to be used. These memories are generally specified either by number or by letter.

a. The TI-55 II has eight user data memories, which are numbered 0–7. A value can be stored in memory 5 on a TI-55 II by entering the value and then pressing `STO` `5`. To recall this value, press `RCL` `5`.

APPENDIX A

b. The TI-81 has 27 standard memories, which are labeled A–Z and θ. A value can be stored in memory A on a TI-81 by entering the value and then pressing [STO→] [ALPHA] [A]. To recall this value, press [ALPHA] [A] [ENTER].

In the following example, [STO] and [RCL] are used to store and retrieve data from a calculator memory. As indicated above, you may need to make modifications to do this problem on your calculator.

EXAMPLE Use synthetic division and a calculator to divide

$$P(x) = 2x^4 + 6.48x^3 - 28.8384x^2 - 0.6016x + 32.7168$$

by $x + 5.68$. Express the answer in the form $P(x) = Q(x)(x - a) + R(x)$.

SOLUTION

```
       -5.68 |  2    6.48   -28.8384   -0.6016    32.7168
                                                  -32.7168
              ─────────────────────────────────────────────
              2    -4.88    -1.12       5.76 ←       0
```

Step 1 [5] [.] [6] [8] [+/−] [STO] [×] [2] [+] [6] [.] [4] [8] [=]

Step 2 [DISPLAY] [×] [RCL] [+] [2] [8] [.] [8] [3] [8] [4] [+/−] [=]

Step 3 [DISPLAY] [×] [RCL] [+] [0] [.] [6] [0] [1] [6] [+/−] [=]

Step 4 [DISPLAY] [×] [RCL] [+] [3] [2] [.] [7] [1] [6] [8] [=] [FINAL DISPLAY]

Answer $P(x) = (2x^3 - 4.88x^2 - 1.12x + 5.76)(x + 5.68)$

The quotient is $2x^3 - 4.88x^2 - 1.12x + 5.76$ with remainder 0. Thus -5.68 is a zero of $P(x)$.

□

Synthetic Division by Calculator ▼

To find $P(x) \div (x - a)$, enter

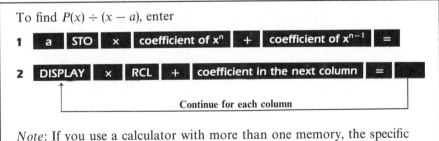

Note: If you use a calculator with more than one memory, the specific memory used must also be entered.

Answers to Odd-Numbered Section Exercises and all Review Exercises, Optional Exercises, and Mastery Tests

CHAPTER ONE

Exercises 1-1

1 a. 5.23 b. 5.23 c. 0 d. 12 **3** a. = b. <
5 a. < b. > **7** −7 **9** 0 **11** $\sqrt{16}$
13 11, 13, 17, 19, 23 **15** 0 **17** 1 **19** $\mathbb{Q}, \mathbb{R}$
21 $\mathbb{N}, \mathbb{W}, \mathbb{I}, \mathbb{Q}, \mathbb{R}$ **23** $\mathbb{W}, \mathbb{I}, \mathbb{Q}, \mathbb{R}$ **25** $\tilde{\mathbb{Q}}, \mathbb{R}$
27 $\mathbb{Q}, \mathbb{R}$ **29** a. $\frac{17}{5}$ b. $\frac{16}{5}$ c. $\frac{2}{3}$ d. $\frac{5}{1}$

31 a. number line from −4 to 4, interval [2, 4)
  b. number line from −4 to 4, interval (2, 4)
  c. number line from −4 to 4, interval (−∞, 2)
  d. number line from −4 to 4, interval [2, ∞)

33 a. number line, (−∞, −3) ∪ (2, ∞)
  b. number line, (−1, 1)
  c. number line, (−3, −2] ∪ (1, 3]
  d. number line from −1 to 7, [2, 6)

35 a. [−5, 3) b. (−∞, 5) c. (−∞, −4] ∪ [4, +∞)
37 a. [2, 7) b. (−3, 5] c. [−2, ∞) **39** Yes, 2
41 No; no matter what number is selected as the smallest real number greater than 1, there is a real number halfway between 1 and that number.
43 No; see the answer to Exercise 41.
45 No; no matter what number is selected as the largest real number less than 3, there is a real number halfway between that number and 3.
47 F
49 a. No; the number of nonzero decimal places for many rational numbers exceeds the accuracy of a calculator.
  b. 0.000 000 003 33 (answers may vary); calculators may store more digits of $\frac{1}{3}$ than the display can show; no.

Exercises 1-2

1 $(r + s) + t$ **3** r **5** (vw) **7** 0 **9** $a > z$
11 $a = b$ **13** $(a + b)y$ **15** w
17 Multiplicative inverse
19 Commutative property of addition
21 Associative property of addition
23 Associative property of multiplication
25 Commutative property of multiplication
27 Distributive property **29** $40w$ **31** $c + 15$
33 $-y$ **35** 1
37 Yes. For every real number n, there is a value $-n$ that may be added to n, resulting in a sum of 0.
39 $a = 1, b = 2$ (or any example with $b > |a|$)
41 $a = 5, b = -5$ (or any example with $b = -a$)
43 a. T b. T c. F d. F **45** F **47** F
49 a. If a is any real number, b. then $-(-a) = a$.
51 Step 1: Reflexive property
    Step 2: Substitution property
53 See *Student Solutions Manual*.

Exercises 1-3

1 −16 **3** 2 **5** 28 **7** 1 **9** $\frac{10}{7}$
11 117 **13** $\frac{1}{81}$ **15** 34 **17** 15 **19** −108
21 −13 **23** $-\frac{1}{2}$ **25** $-\frac{13}{24}$ **27** 64 **29** $\frac{1}{y^9}$

31 $\dfrac{1}{x-y}$ **33** $v^{12}w^8$ **35** $\dfrac{25c^{26}}{16a^2b^{18}}$ **37** $\dfrac{1}{16x^2y^{12}}$

39 $\dfrac{2}{v}+\dfrac{3}{w}=\dfrac{3v+2w}{vw}$ **41** 1   **43** x^{5m} **45** x^{m+2}

47 $\dfrac{9a^{2m+8}}{4}$ **49** $x-2y$ **51** 3,000,000,000

53 0.000 000 000 000 000 000 001 673
55 2.998×10^{10} **57** 6.02×10^{23}
59 0.000 000 000 000 4917   **61** $-1{,}209{,}870{,}000$
63 **Keystrokes**
S: 8 . 7 6 5 EE 1 2
G: 8 . 7 6 5 EE 1 2 ENTER
Display
8.765   12
8.765 E   12

65 b   **67** 3.2×10^{-3} g, 2, ten-thousandths of a gram
69 1.234×10^0 g, 4, thousandths of a gram
71 4.2×10^{27} m   **73** 1.842×10^6 m   **75** 21.04
77 6.155   **79** $9,321.91

Exercises 1-4

1 Trinomial   **3** Binomial   **5** $5v^2-8v+13$
7 $9x^2y+2xy^2$ **9** -27 **11** $\tfrac{3}{7}$ **13** Fourth
15 $-x^2+2xy+12y^2-6$ **17** $-3x-20y-15z$
19 $7w^4-58w^3-19w^2-7w+76$ **21** $3mn-4n^2$
23 $54x^9y^{12}$ **25** $-15m^4n^3+10m^3n^3-15m^2n^3$
27 $15v^2-vw-28w^2$ **29** $6a^3+a^2-a-21$
31 $5y^3-31y^2+11y$ **33** $2xy$ **35** $24ab$
37 $x^3-6x^2+12x-8$ **39** $28x^{-2}-x^{-1}y^{-1}-15y^{-2}$
41 y^2-2y+3 **43** $3x^2-6xy-5yz$ **45** $b-2$
47 $5n-3+\dfrac{5}{2n-1}$ **49** $7x^2+9$
51 $a^4+a^3+a^2+a+1$ **53** $-3x^{2m+9}+2x^{m+2}$
55 $x^{4n}-1$ **57** $9x^{2n}+3x^n+1$ **59** $3x+2z$
61 x^2
63 $3x^2+4x-5$ and $-3x^2+2x+7$ (second-degree terms must be opposites)
65 Fifth   **67** Second
69 a. $-$50   b. $-$24   c. $42   d. 0
71 a. $P=8t^2-18t-8$ b. $P=t^3-3t^2+t-25$

Optional Exercises

1 9   **2** 8   **3** $x^4+8x^3y+24x^2y^2+32xy^3+16y^4$
4 $81x^4-108x^3y+54x^2y^2-12xy^3+y^4$
5 $x^5-10x^4+40x^3-80x^2+80x-32$
6 $x^6+18x^5+135x^4+540x^3+1215x^2+1458x+729$
7 $15{,}625v^6+37{,}500v^5+37{,}500v^4+20{,}000v^3+6000v^2+960v+64$

8 $243v^5-2025v^4+6750v^3-11{,}250v^2+9375v-3125$
9 $x^8-4x^6+6x^4-4x^2+1$
10 $x^{10}+5x^8y^3+10x^6y^6+10x^4y^9+5x^2y^{12}+y^{15}$
11 $v^{15}+15v^{14}w+105v^{13}w^2+455v^{12}w^3+\cdots+w^{15}$
12 $v^{11}+33v^{10}+495v^9+4455v^8+\cdots+177{,}147$

Exercises 1-5

1 $(5a+4b)(5a-4b)$ **3** $(5a-1)^2$
5 $(x+1)(x+5)$ **7** $3(4m-5)(m-1)$
9 $4ay(y-1)$ **11** $(z+1)(5c-7)$
13 $4x(x-5)(x^2+5x+25)$
15 $23a(x^2+1)(x+1)(x-1)$
17 $(2x-y)(4x^2+2xy+y^2)$ **19** $(5y-3z)^2$
21 $3a(x+8)(x+3)$ **23** Prime
25 $4b(x-2)(x^2+2x+4)$ **27** $(2x^5+3y^3)^2$
29 $(x+y)(a-b)$ **31** $7st(s-t)(s+t)(s^2+t^2)$
33 $(c+d)(x+y)$ **35** $3xy(3x+4y)^2$
37 $4a(a^2+2b)(a^4-2a^2b+4b^2)$
39 $(c+d)^2(3c+2d)$ **41** $4(2t+3)^3(2t+1)(t+1)$
43 $(3x+2y)(6x-5y)$ **45** $7ab(3a-5)(3a+5)$
47 $-8a(x-1)(x+1)$ **49** $6(k+j)(x-1)$
51 $(x-y)(x^2+xy+y^2+1)$
53 $(3x+5y-1)(3x-5y-1)$
55 $3a(x-y+1)(x+y-1)$ **57** $(v+1)^3$
59 $3b(a+b)^3$ **61** $(x+6)^3$ **63** $9x^2$
65 $(v^2-v+3)(v^2+v+3)$
67 $(x^2+2x+2)(x^2-2x+2)$
69 $(y^2-3y+5)(y^2+3y+5)$
71 $x^{-2}-9x^{-4}=x^{-4}(x^2-9)=\dfrac{(x+3)(x-3)}{x^4}$
73 $2u^{-2}-7u^{-3}+3u^{-4}=u^{-4}(2u^2-7u+3)=\dfrac{(2u-1)(u-3)}{u^4}$
75 $(a-b)(a^4+a^3b+a^2b^2+ab^3+b^4)$
77 $(x^m-y^n)(x^m+y^n)$ **79** $(2x^m+5y^n)^2$
81 a. Odd   b. Odd   c. Odd   d. Odd
e. Even   f. Even   g. Even   h. Odd

Exercises 1-6

1 8   **3** $2,-5$ **5** $\dfrac{3a}{4b}$ **7** $6xy-19y^2$
9 $\dfrac{-(5x+2)}{7}$ **11** $\dfrac{x+y}{13}$ **13** $\dfrac{5(a-1)}{a+6}$
15 $\dfrac{x+y}{s+t}$ **17** $\dfrac{a+1}{9}$ **19** $\dfrac{2(1-x)}{3}$ **21** $\dfrac{-18x^5}{5y^5}$
23 $\dfrac{10y^2(2x+3)}{7x(5x-2)}$ **25** $\dfrac{y^3}{xz^3}$ **27** $\dfrac{3(4x^2-6x+9)}{5(x-4)}$

ANSWERS

29 $-\dfrac{3y(x-y)}{14}$ **31** $\dfrac{4y^2(x-2y)}{7x^2(x-y)}$ **33** $\dfrac{7xy}{15}$
35 $3x - 3y$ **37** $3s^2 - 3st + 3t^2$ **39** -2
41 $\dfrac{2(x^2+1)}{(x+1)(x-1)}$ **43** $\dfrac{x^2+14x-2}{(x-3)(x+4)}$ **45** $-\dfrac{5}{p+2}$
47 $-\dfrac{b}{a^2-ab+b^2}$ **49** $\dfrac{9x(x-7)}{10y^2(x+2)}$ **51** $\dfrac{x}{2x+1}$
53 $\dfrac{2a+1}{2a^2+a+2}$ **55** $\dfrac{x-1}{x}$ **57** $\dfrac{1}{2}$ **59** $-\dfrac{1}{mn}$
61 $\dfrac{5x-3y}{4xy}$ **63** $\dfrac{1}{(x+2)(x-5)}$ **65** T **67** F
69 $\dfrac{(a+b)^2}{a^2-ab+b^2}$ **71** $\dfrac{v^2-3v+4}{7}$ **73** $\dfrac{v-1}{2v-1}$

Exercises 1-7

1 a. $7^{1/2}$ **b.** $5^{1/3}$ **c.** $x^{1/2}$ **d.** $-y^{3/5}$
3 a. $\sqrt{11}$ **b.** $\sqrt[4]{13}$ **c.** $\sqrt[3]{x^2}$ **d.** $\sqrt[4]{y^3}$
5 a. 7 **b.** 5 **c.** 2 **d.** -2
7 a. 6 **b.** 3 **c.** $\dfrac{5}{12}$ **d.** $\dfrac{9}{4}$ **9** $z^{4/5}$ **11** $2^{3/20}$
13 $z^{1/5}$ **15** $\dfrac{vw^3}{3}$ **17** $\dfrac{y}{z}$ **19** $4x, 4|x|$
21 $2x^3y^2\sqrt{3}, 2|x|y^2\sqrt{3}$ **23** $3\sqrt{2}$ **25** $11\sqrt[3]{x}$
27 $\sqrt{3}$ **29** $-9\sqrt{3x}$ **31** 4   **33** $4x - 12\sqrt{x} + 9$
35 $a - 1$ **37** $3\sqrt{5}$ **39** $2(2 - \sqrt{3})$
41 $-4(\sqrt{7} + \sqrt{2})$ **43** $\dfrac{\sqrt{ab}+b}{a-b}$ **45** $5x\sqrt{2x}$
47 $3x^2y^4\sqrt[3]{2x^2y}$ **49** $\dfrac{\sqrt{3x}}{2y}$ **51** $\dfrac{\sqrt[3]{50vw}}{5w}$
53 $\sqrt{5vw}$ **55** $\sqrt[4]{ab^3}$ **57** If $x = -8, \sqrt{x^2} \neq -8$.
59 c   **61** a   **63** b   **65** 900   **67** $\dfrac{2\sqrt{3}}{3}$
69 $-38\sqrt{5} - 2\sqrt[3]{2}$ **71** $70x\sqrt[3]{2y} - 20y\sqrt[3]{5x}$
73 $\dfrac{\sqrt[3]{9}-5\sqrt[3]{3}+25}{128}$ **75** 7.263

Exercises 1-8

1 $a = 18, b = -5$ **3** $a = 7, b = 13$
5 $a = 0, b = -1$ **7** $a = 2, b = -4$
9 $a = -\sqrt{17}, b = 0$ **11** $9 - i$ **13** $3 + 5i$
15 $13i\sqrt{3}$ **17** $-1 - 2i$ **19** $2 + 2i\sqrt{2}$
21 $-1 - i$ **23** -7 **25** $-23 + 3i$ **27** 53

29 $34 + 27i$ **31** $24 + 10i$ **33** $2 - 2i$
35 $\dfrac{15}{17} - \dfrac{8}{17}i$ **37** $7 + 6i$ **39** -2 **41** $2 + 3i$
43 $-2i$

45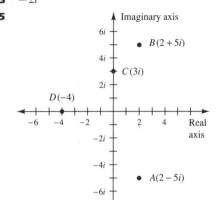

47 $-11 + 2i$ **49** 1   **51** $2i\sqrt{2}$ **53** $\dfrac{1}{6} - \dfrac{\sqrt{11}}{6}i$

55 $(2-i)^2 - 4(2-i) + 5 \stackrel{?}{=} 0$
$4 - 4i + i^2 - 8 + 4i + 5 \stackrel{?}{=} 0$
$4 - 1 - 8 + 5 \stackrel{?}{=} 0$
$0 = 0$
Therefore $2 - i$ is a solution.

57 $\pm 5i$
59 $(-1 + i\sqrt{3})^3 = (-1)^3 + 3(-1)^2(i\sqrt{3})^1 + 3(-1)(i\sqrt{3})^2$
$+ (i\sqrt{3})^3$
$= -1 + 3i\sqrt{3} - 9i^2 + 3i^3\sqrt{3}$
$= -1 + 3i\sqrt{3} + 9 - 3i\sqrt{3}$
$= 8$
Therefore $-1 + i\sqrt{3}$ is a cube root of 8.

61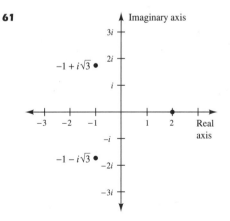

Review Exercises for Chapter 1

1. -3
2. -6
3. $4\frac{1}{4}$
4. 346
5. $-8\sqrt{13} + 46\sqrt{5}$
6. $-4\sqrt{3} + 18\sqrt{5}$
7. 13
8. $3\sqrt{19} + 3\sqrt{14}$
9. $16 - 30i$
10. $6 - 4i$
11. $-2i$
12. 2
13. $24xy$
14. $3v^2w + 3vw^2$
15. $-32a^3b + 24a^2b^2 - 8ab^3$
16. $8x^2y^2$
17. $\dfrac{-1}{2v^3w^9}$
18. $2a - 3b + 5c^3$
19. $15a^2b^2(2a - 3b)$
20. $7v(v + 3)(v - 3)$
21. $(5y + 2)(25y^2 - 10y + 4)$
22. $(5a + 7b)(2a - 3b)$
23. $(7m + 10n)^2$
24. $4ax(x - 6)(x - 1)$
25. $(a - b)(2x + 3y)$
26. $(x - 5)(x - y)$
27. $(x + y)^3$
28. $(x + y - 3)(x - y + 3)$
29. $(a^2 - 2a + 3)(a^2 + 2a + 3)$
30. $(a + 3b + 5)(a + 3b + 1)$
31. $(x^m + 5)(x^m - 5)$
32. $\dfrac{11m}{3n}$
33. w
34. $\dfrac{1}{a-1}$
35. $-\dfrac{m}{n(2m - 3n)}$
36. $\dfrac{w^2+1}{w}$
37. $x - 1$
38. 2
39. $-\dfrac{c(a+b)}{(b-c)(a-c)}$
40. $x + y$
41. $5x^2y^3z^4\sqrt{3xyz}$
42. $3x^3y^4\sqrt[3]{2}$
43. $x\sqrt{x} + x\sqrt{y} + y\sqrt{x} + y\sqrt{y}$
44. $33\sqrt{5x}$
45. 6
46. $\dfrac{x^{21}y^4}{z^2}$
47. $[-3, 5)$
48. $(-\infty, 6)$
49. $[-4, +\infty)$
50. $(-3, 3)$
51. $(-\infty, -\pi] \cup [\pi, +\infty)$
52. $[1, 7]$
53. a. 2 b. $0, 2$ c. $-93, 0, 2$ d. $-93, -9.3, 0, 2, 5.\overline{7}, \frac{117}{19}$ e. $-\pi, -\sqrt{2}, 1 + \sqrt{2}, e$ f. $-93, -9.3, -\pi, -\sqrt{2}, 0, 2, 1 + \sqrt{2}, e, 5.\overline{7}, \frac{117}{19}$ g. $\sqrt{-3}, 5i$ h. $2 + 5i$
54. $67, 71$
55. -3
56. $x = 3, y = -1$ (any pair with opposite signs)
57. $x = 3$ and $y = 4$
58. $x = \frac{1}{2}$ (any x such that $0 < x < 1$ or $x < -1$)
59. $x = \frac{1}{2}$ (any x such that $0 < x < 1$)
60. Commutative property of addition
61. Commutative property of addition
62. Distributive property of multiplication over addition
63. Commutative property of multiplication
64. Zero factor theorem
65. No
66. Yes
67. 35.39
68. -9.252×10^{25}
69. $-3x^{-2} + 6x^{-3} = -3x^{-3}(x - 2) = -\dfrac{3(x-2)}{x^3}$
70. $2 - 8x^{-2} = 2x^{-2}(x^2 - 4) = \dfrac{2(x+2)(x-2)}{x^2}$

Mastery Test for Chapter 1

1. a. 53 b. $\frac{22}{7}$ c. $1, 2$ d. An integer
2. a. $(-1, 3]$ b. $[2, 7]$ c. $(-\infty, 5)$ d. $(3, 4) \cup [6, +\infty)$
3. a. Commutative property of multiplication
   b. Distributive property of multiplication over addition
   c. Commutative property of addition
   d. Transitive property of less than
4. a. 28 b. 8 c. $\frac{61}{84}$ d. $-\frac{49}{50}$
5. a. $32x^{10}y^{15}$ b. $\dfrac{x^4}{9y^8}$ c. $128x$ d. $\dfrac{9}{16x^2y^4}$
6. a. 39 b. 5 c. -271
7. a. 4.513×10^6 b. 1.76×10^{-4}
8. a. 4 b. 5 c. Trinomial d. -18
9. a. $4v^2 - 52v + 169$ b. $25v^2 - 49$ c. $-36a + 4b + 32c$ d. $2x^3 + 11x^2 - 8x - 80$ e. $2x - 5$ f. $a^3 - 6a^2b + 12ab^2 - 8b^3$
10. a. $5a(2b - 1)(b + 3)$ b. $10b(7x - 2)(7x + 2)$ c. $-7(y + 2)(y - 2)$ d. $(2w - 1)(4w^2 + 2w + 1)$ e. $(3x + 5y - 1)(3x - 5y - 1)$ f. $a(v^2 - v + 2)(v^2 + v + 2)$
11. a. $\dfrac{a+5b}{a-b}$ b. $\dfrac{3(x+y)}{4(x-y)}$
12. a. $\dfrac{3(x+2)}{x-1}$ b. $4(5w + z)$ c. $\dfrac{m^2 - n^2}{mn}$
13. a. 15 b. $2\sqrt[3]{3}$ c. -5
14. a. $2\sqrt{6}$ b. $\sqrt{2}$ c. $\frac{2}{3}$ d. $\dfrac{5\sqrt{3}}{3}$
15. a. $42\sqrt{5}$ b. $10 + 2\sqrt{6}$ c. -43 d. $60 + 15\sqrt{11}$
16. a. $-7 + 33i$ b. 89 c. $\frac{48}{25} + \frac{36}{25}i$ d. $-7 - 24i$
17.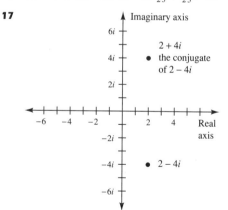

CHAPTER 2

Exercises 2-1

1. 4
3. 20
5. Conditional; 0
7. Contradiction; no solution.
9. Identity; every real number is a solution.
11. Conditional; 25
13. Conditional; -7.2

ANSWERS

15 Conditional; 3
17 Identity; every real number is a solution.
19 $\frac{5}{4}$ **21** No solution **23** $\frac{60}{11}$ **25** 2
27 $p - a - c$ **29** $\frac{2A}{h}$ **31** $\frac{n}{1-2n}$ **33** $\frac{2A-bh}{h}$
35 $\frac{y^2 - 2xy}{x^2 - 3xy^2}$ (for $x^2 - 3xy^2 \neq 0$) **37** 39
39 No solution **41** 2 **43** No solution
45 $2n + 6 = 2 + n$; $n = -4$ **47** $\frac{n+7}{9} = 3$; $n = 20$
49 $y = kx$; $y = 30$ **51** $z = kxy$; $z = 11$
53 $w = \frac{k}{d^2}$; $w \approx 144.6$ pounds
55 a. 3 **b.** $\frac{x-3}{(x+1)(x-1)}$ **57 a.** 1 **b.** $\frac{t-1}{t+2}$
59 Let $a = 1, b = 2$, and $c = 0$; then ac and bc equal 0, but $a \neq b$
61 No **63** $150,000
65 The product of seven and a number is fourteen.
67 The sum of five and twice a number is six times the number.
69 One-half of a number is nine.
71 The product of two and the sum of two numbers is the same as the sum of twice the first number plus twice the second number. **73** 357.13

Exercises 2-2

1 6.51 cm² **3** 53.91 cm² **5** 10.00 meters
7 20.6 centimeters **9** $9,259.26 **11** 45°, 135°
13 $-2, -1, 0$ **15** 31, 33, 35 **17** 800 tabletops
19 12,400 tickets **21** $90,000
23

a.	Pounds of Sand	3	6	15	1	s
	Pounds of Clay	5	10	25	$\frac{5}{3}$	$\frac{5}{3}s$
	Pounds of Mixture	8	16	40	$\frac{8}{3}$	$\frac{8}{3}s$

b. (Pounds of sand) + (Pounds of clay) = Pounds of mixture
c. 1350 pounds of sand, 2250 pounds of clay

25 a.

Larger Number	27	20	19	$14 + s$
Smaller Number	13	6	5	s
Their Difference	14	14	14	14
Their Sum	40	26	24	$14 + 2s$

b. 14, 28
27 4 inches **29** $6,250 at 6%, $1,250 at 6.5%
31 240 liters of A, 640 liters of B **33** $240,000
35 5.6 meters **37** $(n+2)^2 - n^2 = 52$
39 $s^2 + (s+2)^2 = 290$ **41** $\pi(2r-3)^2 - \pi r^2 = 57.75$
43 $s^2 + (s+7)^2 = (s+9)^2$
45 $200\pi r^2 - 200\pi(r-1)^2 = 15{,}700$

Exercises 2-3

1 $4n^2 - 11n - 3 = 0$
  $a = 4, b = -11, c = -3$
3 $2m^2 - 4 = 0$
  $a = 2, b = 0, c = -4$
5 ± 8 **7** $\pm 4i$ **9** $-2, \frac{14}{3}$ **11** $-\frac{3}{5}, 2$
13 $-5, -1$ **15** $0, 7$ **17** $-\frac{3}{2}, 5$
19 $b^2 - 4ac = -32$; two imaginary solutions
21 $b^2 - 4ac = 0$; a double real root
23 $b^2 - 4ac = -96$; two imaginary roots **25** $-\frac{4}{3}, \frac{5}{3}$
27 $2 + \sqrt{5}, 2 - \sqrt{5}$ **29** $-1 + i, -1 - i$
31 $5\sqrt{2}, -5\sqrt{2}$ **33** $0, \frac{2}{5}$ **35** $\frac{3}{2} - \frac{i}{2}, \frac{3}{2} + \frac{i}{2}$
37 $\frac{1}{2} - \frac{7}{2}i, \frac{1}{2} + \frac{7}{2}i$ **39** $\frac{-7 + \sqrt{21}}{2}, \frac{-7 - \sqrt{21}}{2}$
41 $-3, 6$ **43** $x^2 - 9 = 0$ **45** $7x^2 + 11x - 6 = 0$
47 $x^2 - 2 = 0$ **49** $-1 + \sqrt{5}, -1 - \sqrt{5}$
51 $\frac{1}{2}, \frac{3}{2}$ **53** $-y, 2y$ **55** $x^2 + 4 = 0$
57 $x^2 - 2x + 2 = 0$ **59** $-18, 18$ **61** $\frac{1}{9}, 1$
63 $-\frac{3}{2}, 0, \frac{2}{3}$ **65** $-5, -2, 2$ **67** $-\frac{2x}{3}, 5x$
69 $-13, -11$, and $11, 13$ **71** 6.8 meters by 13.2 meters
73 9.3 centimeters **75** $r = \frac{\sqrt{\pi V h}}{\pi h}$
77 $-2y + y\sqrt{3}, -2y - y\sqrt{3}$ **79** $-8.375, 25.63$

Exercises 2-4

1 $z = x^2, z^2 + 6z + 5 = 0$ **3** $z = \sqrt{y}, z^2 - 5z - 6 = 0$
5 $z = \frac{v-2}{v}, z^2 = 2z + 15$ **7** $z = \frac{1}{w}, z^2 + z - 2 = 0$
9 $z = r^{1/3}, z^2 - 2z - 35 = 0$ **11** $\pm 3, \pm 1$
13 $\pm\sqrt{2}, \pm\frac{\sqrt{2}}{2}$ **15** $\frac{1}{4}, 16$ **17** $\pm\frac{1}{2}$
19 $-1, -\frac{1}{2}, 2, \frac{3}{2}$ **21** $\pm 2, \pm 2i$ **23** $16, 1$
25 $-\frac{1}{32}, -32$ **27** $\pm\frac{\sqrt{13}}{3}$ **29** $\frac{13}{12}, \frac{1}{2}$
31 $-\frac{1}{4}, -2$ **33** $-\frac{5}{3}$ **35** $\frac{1}{2}$ **37** $0, \frac{2}{3}$ **39** 4
41 $\frac{1}{3}$ **43** 4 **45** 0 **47** No solution
49 a. $(x+4)(x+1)(x-1)(x-4)$ **b.** $-4, -1, 1, 4$
c. $-4, -1, 1, 4$ **51** $5, -3$ **53** -4
55 $\frac{gP^2}{4\pi^2}$ **57** $\pm\sqrt{c^2 - a^2}$ **59** 30 miles per hour
61 -6.703 **63** 6 **65** 3 **67** 2, 1 **69** 1

Exercises 2-5

1

	Unit Value of Coins	·	Number of Coins	=	Value of Coins
Dimes	0.10	·	d	=	.10d
Nickels	0.05	·	$6d$	=	0.3d
Pennies	0.01	·	$3(6d)$	=	.18d

There are 65 dimes.

3

	Fee Rate	·	Base Payment	=	Amount of Fee
New Policies	0.12	·	p	=	0.12p
Old Policies	0.03	·	$26{,}250 - p$	=	$03(26{,}250 - p)$

Payments for new policies were $6,250.

5 $125,000 was invested at 6.75% and $225,000 was invested at 7%.

7

	Rate	· Time	= Distance
Slower Train	r km/h	· 4	= $4r$
Faster Train	$(r + 30)$ km/h	· 4	= $4(r + 30)$

The slower train traveled 85 kilometers per hour and the faster train traveled 115 kilometers per hour.

9 141 kilometers per hour, 147 kilometers per hour

11

	Percent of Disinfectant	·	mL Solution	=	mL Disinfectant
60% Solution	0.60	·	$82.5 - s$	=	$0.6(82.5 - s)$
15% Solution	0.15	·	s	=	$0.15s$
Total Mixture	0.20	·	82.5	=	$0.2(82.5)$

Approximately 9.17 liters of 60% solution and 73.3 liters of 15% solution should be used.

13 90 hours **15** 18 hours, 9 hours

17

	Amount of Gross Sales	÷	Price Per Item	=	Number Sold
First Week	10,000	÷	p	=	$\dfrac{10{,}000}{p}$
Second Week	12,000	÷	$p - 5$	=	$\dfrac{12{,}000}{p - 5}$

The original price of the product was $25.

19 a. $A = (w + 3)(w + 5)$ **b.** $A = 6w + 15$ **c.** 8 inches
21 9 meters, 6 meters **23** 315 adults
25 19 miles per hour
27 15 kilograms 20% mix, 25 kilograms 36% mix
29 280 liters **31** 5, 11
33 44 centimeters, 56 centimeters
35 a. 500 miles **b.** 0.5 hours

Exercises 2-6

1 $(-1, +\infty)$
3 $[3, +\infty)$
5 $(-15, +\infty)$
7 $(-2, +\infty)$
9 $(-\infty, +\infty)$
11 $(-\infty, -2)$
13 $(4, 7)$
15 $(-\infty, 4] \cup (5, +\infty)$
17 $(-\infty, -1.5] \cup [1.5, +\infty)$
19 $[-7, +\infty)$
21 $[-3, 0)$ **23** $(-1, 2]$
25 $(-\infty, -3] \cup (-1, +\infty)$ **27** No solution
29 $[-\tfrac{5}{4}, 9]$ **31** 2, 8 **33** $(-6, 2)$
35 $(-\infty, 1] \cup [2, +\infty)$ **37** $\tfrac{2}{3}$, 3
39 $(-\infty, -5) \cup (\tfrac{25}{3}, +\infty)$ **41** $(-\infty, +\infty)$
43 No solution **45** $|x| \leq 3$ **47** $|x - 7| > 3$
49 $|x - 3| < 5$
51 a. Any $m > 0$ **b.** Any $m < -\tfrac{1}{2}$
53 a. Any $x > 0$ **b.** Any $x < 0$ **c.** $x = 0$
55 $|x| \leq 4$ **57** $|x| > 4$ **59** $|x + 1| \leq 4$
61 $|x - 2| > 7$ **63** $[6, 10)$ **65** 81 or more; yes
67 $|l - 160| < 0.5$ **69** $|b - 850| \leq 50$
71 $20° < C < 25°$
73 a. x and y have opposite signs
 b. x and y have the same signs **75** $[-1.161, 1.914]$

Exercises 2-7

1 $(-2, 1)$ **3** $[-7, 4]$ **5** $(-2, 1) \cup (4, +\infty)$
7 $[-5, +\infty)$ **9** $(-\infty, -\tfrac{1}{2}) \cup [\tfrac{3}{2}, +\infty)$
11 $(-\infty, -4] \cup [6, +\infty)$ **13** $(-\infty, -5] \cup [7, +\infty)$
15 $(-\infty, 0) \cup (\tfrac{1}{2}, +\infty)$ **17** $(-4, 1] \cup [3, +\infty)$
19 $(-\infty, 2) \cup (3, 5)$ **21** $[-1, 0) \cup [6, +\infty)$

ANSWERS

23 $(-\infty, -\frac{3}{2}] \cup (-\frac{1}{2}, \frac{1}{3}]$ **25** $(\frac{6}{5}, 2)$
27 $[-\frac{2}{3}, 1) \cup (1, \frac{8}{3}]$ **29** $(-2, -1) \cup (0, 2) \cup (3, +\infty)$
31 $(-\infty, -\frac{1}{2}] \cup [\frac{5}{3}, +\infty)$ **33** $(-\infty, -4] \cup [4, +\infty)$
35 a. (10, 50) **b.** 10, 50 **c.** $[0, 10) \cup (50, +\infty)$
37 (0.05 seconds, 20.36 seconds)
39 $(-5, -3) \cup (3, +\infty)$ **41** $[-4, -3] \cup [-2, -1]$
43 $(-\infty, \frac{7}{4}] \cup [\frac{5}{2}, +\infty)$
45 $(-\infty, -3) \cup (-3, -2) \cup (-2, 2) \cup (3, +\infty)$
47 $(-\infty, 9.24) \cup (9.71, +\infty)$

Review Exercises for Chapter 2

1 $-\frac{1}{2}$ **3** 12 **5** $-1, 5$ **7** 11 **9** $-\frac{1}{5}, 5$
11 $-11, 3$ **13** $-1 \pm 2i$ **15** $\frac{11}{3}$ **17** -1
19 ± 11 **21** 9 **23** $\pm \frac{\sqrt{2}}{2}, \pm 5$ **25** $-7, 0, 5$
27 $\frac{3V}{\pi r^2}$ **29** $\frac{S - 2\pi r^2}{2\pi r}$ **31** $x = y + 7$
$x = y - 7$
33 $15x^2 - 16x - 15 = 0$ **35** $x^2 - 4x + 13 = 0$
37 $x^3 - 5x^2 - 9x + 45 = 0$ **39** A double root
41 $w \le \frac{20}{3}$ **43** $[-3, 8]$ **45** $(-\infty, -\frac{7}{2}) \cup (\frac{2}{3}, +\infty)$
47 $[-2, 0] \cup [4, +\infty)$
49 $(-\infty, -2) \cup (-2, 5) \cup (7, +\infty)$ **51** 8 meters
53 $40,000
55 4.5 milliliters of 5% solution, 13.5 milliliters of 25% solution
57 30 minutes **59** 214 chairs

Optional Exercises for Calculus-Bound Students for Chapters 1 and 2

1 $-15x^{-4} + 7x^{-3} + 2x^{-2} = x^{-4}(-15 + 7x + 2x^2)$
$= \frac{2x^2 + 7x - 15}{x^4}$
$= \frac{(2x - 3)(x + 5)}{x^4}$

2 $x^{3/2} - 25x^{-1/2} = x^{-1/2}(x^2 - 25) = \frac{(x + 5)(x - 5)}{\sqrt{x}}$

3 $y' = -\frac{7}{2x - 3y}$ **4** $y' = 4y$ or $y' = -y$
5 $b = 2a$ ($b \ne -a$ since $b > a > 0$)
6 i. For $a \ne 0$, there is no x for which $\frac{a}{0} = x$, because $a \ne 0 \cdot x$.
ii. For $a = 0$, $a = 0 \cdot x$ for any value of x. Thus there is not a unique value of x to assign to $\frac{a}{0} = x$.
7 $7(y - 3)(y + 3)(y^2 + 9)$ **8** $(2w^2 - 1)(4w^4 + 2w^2 + 1)$

9 0, 1, 6 **10** $-3, -2, 3$
11 $(x^2 - xy + y^2)(x^2 + xy + y^2)$
12 $(2x^2 - 3xy - y^2)(2x^2 + 3xy - y^2)$
13 $-\frac{1}{2} \pm \frac{\sqrt{3}}{2}i$ **14** $\frac{2\sqrt[3]{5x^2}}{x}$ **15** $\sqrt{x + h} + \sqrt{x}$
16 $\frac{\sqrt[3]{x^2} + \sqrt[3]{xy} + \sqrt[3]{y^2}}{x - y}$ **17** $(-\infty, -3] \cup [4, +\infty)$
18 a. $(-\infty, -8) \cup (8, +\infty)$ **b.** $-8, 8$ **c.** $(-8, 8)$
19 $|x - 11| < 6$ **20** $0, \frac{1}{2}, 1$ **21** $-1, \frac{1}{3}$
22 $(-\infty, -4) \cup (-2, 3) \cup (5, +\infty)$ **23** $[-3, 0] \cup [1, 2]$
24 (0.05 second, 8.93 seconds) **25** 324.3 m²
26 0.196 inch
27 a. 123 cm² **b.** 123,000 cm³ **c.** 1.15 cm
28 a. 5 meters **b.** $1\frac{2}{3}$ **c.** 9 seconds

Mastery Test for Chapter 2

1 a. Conditional equation **b.** Identity
c. Conditional equation **d.** Contradiction
2 a. -3 **b.** 14 **c.** No solution
3 a. $T = \frac{I}{PR}$ **b.** $c = 2s - a - b$ **c.** $y' = \frac{y}{x + y}$
4 240,000 cm³
5 a. $-13, -12$ or 12, 13
b. $2,500 at 6% and $1,500 at 8%
6 a. $-2, \frac{5}{3}$ **b.** $2 + \sqrt{7}, 2 - \sqrt{7}$ **c.** $2 + i, 2 - i$
7 a. Two distinct real roots **b.** Two imaginary roots
c. A double real root
8 a. $35x^2 + 11x - 6 = 0$ **b.** $x^2 - 10x + 25 = 0$
c. $x^2 - 3 = 0$ **d.** $x^2 - 6x + 10 = 0$
9 a. $\pm 1, \pm 10$ **b.** $-\frac{26}{9}, -2$ **c.** $\frac{1}{32}, 1$
10 a. $-\frac{9}{5}$ **b.** $-\frac{1}{2}$
11 a. The messenger would drive $2\frac{3}{4}$ hours to overtake the first driver.
b. It would take the slower painter 60 hours working alone.
12 a. $[\frac{10}{13}, +\infty)$ **b.** $(-\infty, 3) \cup [15, +\infty)$ **c.** $(-\frac{5}{4}, 3)$
13 a. $(-\infty, -\frac{1}{2}) \cup (0, 2)$
b. $(-\infty, -2] \cup (-1, 1) \cup [\frac{4}{3}, +\infty)$

CHAPTER THREE

Exercises 3-1

1 a. Not a function **b.** Function **c.** Not a function
d. Function
3 a. Function **b.** Not a function
5 a. Function **b.** Not a function **c.** Not a function
d. Function **e.** Function **f.** Not a function
g. Function

7 $D = \{3, -9, 5\}$, $R = \{7, 11, -2\}$
9 $D = \{6, 11, 17\}$, $R = \{-1, 0, 4\}$
11 a. $R = \{\frac{1}{2}, 2, 3\}$
b.
D	R
$\frac{1}{4}$	$\frac{1}{2}$
4	2
9	3

c. $\{(\frac{1}{4}, \frac{1}{2}), (4, 2), (9, 3)\}$

d.

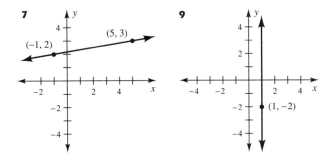

13 a. $\{(-4, 1), (-3, 2), (-1, 3), (2, -1)\}$
b. $-4 \to 1, -3 \to 2, -1 \to 3, 2 \to -1$
c. $\{-4, -3, -1, 2\}$ **d.** $\{1, 2, 3, -1\}$
15 a. $\{(-2, -7), (-1, -4), (0, -1), (1, 2), (2, 5)\}$
b. $-2 \to -7, -1 \to -4, 0 \to -1, 1 \to 2, 2 \to 5$
c. $\{-2, -1, 0, 1, 2\}$ **d.** $\{-7, -4, -1, 2, 5\}$
17 a. -21 **b.** $-4x - 5$ **c.** $4h + 7$ **d.** $4h - 2$
e. $8h - 1$ **f.** $8h - 9$ **g.** $4h$
19 a. $-\frac{4}{5}$ **b.** -9 **c.** $-\frac{17}{52}$ **d.** undefined
21 $D = [-2, 4)$, $R = [-1, 4]$
23 $D = [-\pi, \pi)$, $r = [-1, 1]$
25 $D = \mathbb{R}$, $R = \{2\}$, constant on $\mathbb{R}$
27 $D = \mathbb{R}$, $R = [0, +\infty)$ increasing on $[0, +\infty)$, decreasing on $(-\infty, 0]$
29 $D = (-4, 4]$, $R = [-1, 4]$ increasing on $[-3, -2]$, $[0, 3]$ decreasing on $(-4, -3]$, $[-2, 0]$, $[3, 4]$
31 $\mathbb{R}$ **33** $\mathbb{R}$ **35** $\mathbb{R} \sim \{-\frac{1}{3}\}$ **37** $\mathbb{R} \sim \{-3, 3\}$
39 $[-1, +\infty)$
41 Mapping (arrow) notation, ordered-pair notation, table of values, functional notation, graphs
43 Temperature may be the same at several different times; for example,

45 5 **47** $2x + h - 1$ **49** 25
51 No real solution; even root of negative number
53 $3x^2 + 3xh + h^2$ **55** $D = (-\infty, -1] \cup [6, +\infty)$
57 $D = (-\infty, -3) \cup [1, +\infty)$
59 $(-10, 381), (-9, 317), (-8, 259), (-7, 207), (-6, 161),$
$(-5, 121), (-4, 87), (-3, 59), (-2, 37), (-1, 21), (0, 11),$
$(1, 7), (2, 9), (3, 17), (4, 31), (5, 51), (6, 77), (7, 109),$
$(8, 147), (9, 191), (10, 241)$

Exercises 3-2

1 a. 2 **b.** 0 **c.** undefined
3 a. 3, increasing **b.** $-\frac{2}{3}$, decreasing **c.** 0, constant
5 a. undefined **b.** $-\frac{2}{3}$, decreasing **c.** -3, decreasing

7 **9**

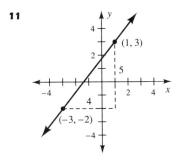

11

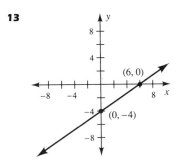

13

15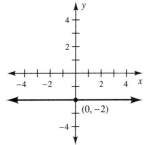

17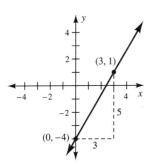

19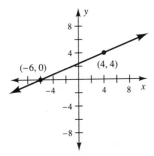

21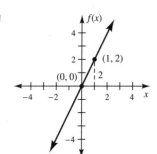

23 Perpendicular **25** Parallel **27** Neither
29 $2x - y - 5 = 0$ **31** $2x - 4y - 3 = 0$
33 $y + 6 = 0$ **35** $x + 4 = 0$ **37** $y + 7 = 0$
39 $x + y - 1 = 0$ **41** $x - y + 5 = 0$

43 $2x - y - 1 = 0$ **45** $x + 5y - 19 = 0$
47 $5x + 7y - 25 = 0$

49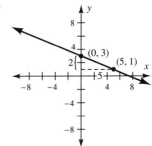

51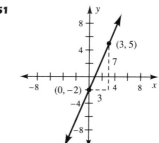

53 **a.** x-intercept $= 3$ **b.** no x-intercept
   y-intercept $= 2$ y-intercept $= -3$
   slope $= -\frac{2}{3}$ slope $= 0$
c. x-intercept $= -5$
   no y-intercept
   slope undefined
55 Yes, vertices form right triangle
57 $V(n) = -320n + 4000$
59 **a.** 300 **b.** 10 **c.** increasing **d.** $10/unit
61 $r(x) = \dfrac{x}{2\pi}$

63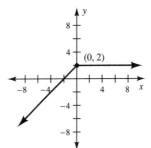

Increasing on $(-\infty, 0]$
Constant on $[0, +\infty)$

65

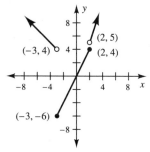

Decreasing on $(-\infty, -3]$
Increasing on $[-3, +\infty)$

67 a. $l = 24 - 2w$ **b.** $A = w(24 - 2w)$
69 a. $w(h) = 48 - 2h$ **b.** $A(h) = h(48 - 2h)$

Exercises 3-3

1 a.

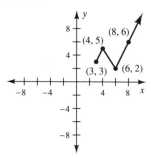

b.

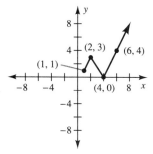

c.

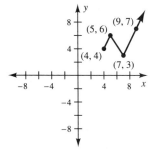

d.

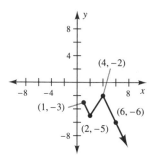

3 a. $(2, 6)$ **b.** $(-3, 3)$ **c.** $(8, -6)$ **d.** $(2, -2)$

5

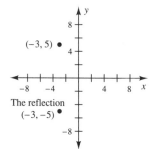

The reflection

7 a.

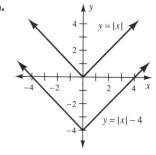

b.

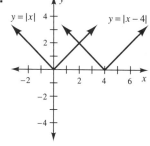

c.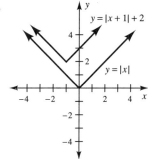

d.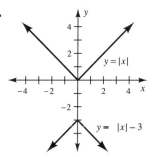

9 a.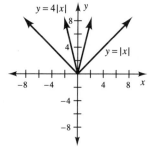

b.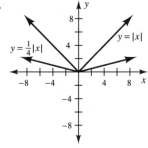

c.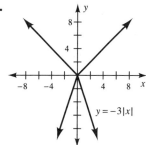

11 a.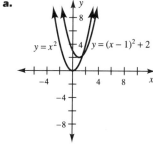

b.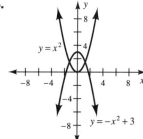

c.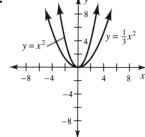

13 a. Upward **b.** y-intercept $(0, -10)$
c. x-intercept $(-5, 0)$ and $(2, 0)$

15 a. Upward **b.** y-intercept $(0, 4)$
c. x-intercept $(3 \pm \sqrt{5}, 0)$

17 **a.** Upward **b.** y-intercept = (0, 6) **c.** No x-intercept
19 **a.** (0, 2) **b.** (3, 5) **c.** (2, −7)

21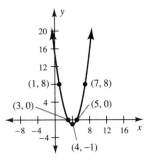

23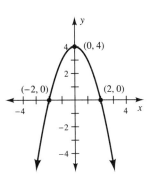

25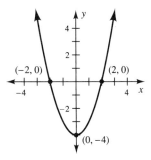

27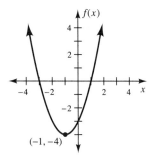

29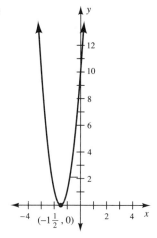

31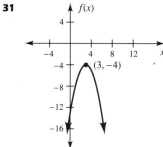

33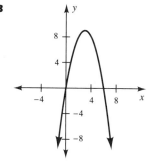

35 $y = x^2 - 5$ **37** $y = (x + 2)^2 - 2$
39 **a.** 35 windmills **b.** $625 **c.** increasing [1, 35]
  d. decreasing [35, +∞)
41 **a.** 78.4 meters **b.** (0, 3) seconds **c.** (3, 7) seconds
43 24.5 m²

45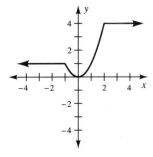

Increasing on [0, 2]
Decreasing on [−1, 0]
Constant on (−∞, −1] and [2, +∞]

47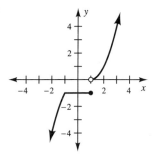

Increasing on (−∞, −1] and (1, +∞)
Constant on [−1, 1]

49 a. left **b.** right   **51** 8

53 $f\left(-\dfrac{b}{2a}\right) = a\left(-\dfrac{b}{2a}\right)^2 + b\left(-\dfrac{b}{2a}\right) + c$ **55** 8, 8

$= a \cdot \dfrac{b^2}{4a^2} - \dfrac{b^2}{2a} + c$

$= \dfrac{b^2 - 2b^2 + 4ac}{4a}$

$= \dfrac{4ac - b^2}{4a}$

Exercises 3-4

1 a. x-axis **b.** x-axis, y-axis, origin **c.** none
3 a. y-axis **b.** y-axis **c.** y-axis

5 a.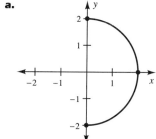

b. 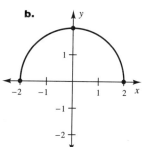 **c.**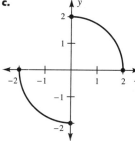

7 a. 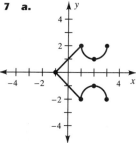 **b.**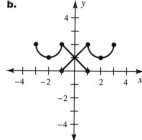

c.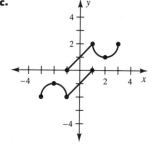

A-14 ANSWERS

9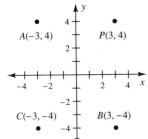

11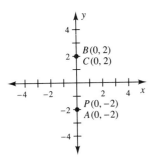

13 a. Neither **b.** Even **c.** Odd

15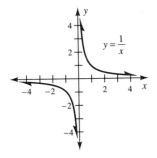

17 Symmetric about the y-axis

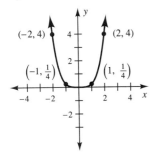

19 distance = 5, midpoint = $(-1, \frac{1}{2})$

21 distance = 5, midpoint = $\left(a - 1, \dfrac{2b + 3}{2}\right)$

23 30 **25** Yes, these are the vertices of a right triangle

27 The points are collinear **29** $2\sqrt{2}$

31 a. 9 **b.** -8 **c.** 6 **d.** -3

33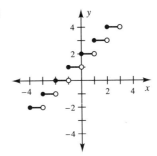

35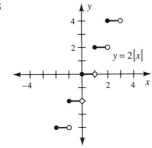

37 a.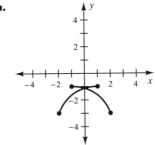

b.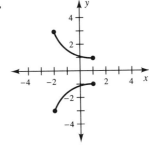

c.

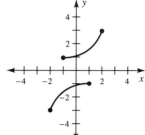

d.

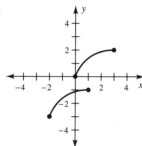

39 -7.389 **41** All four sides are 17 units in length.
43 $-10, 4$ **45** 5   **47** 6   **49** 2.44288

51

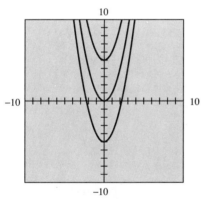

53

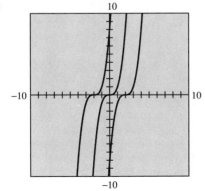

Exercises 3-5

1 a. 12 **b.** -6 **c.** 27 **d.** $\frac{1}{3}$
3 a. 80 **b.** 11 **c.** 8 **d.** 23
5 $f \circ g = \{(0, 12), (4, 4)\}$
$g \circ f = \{(-1, 1), (0, -3), (1, 1)\}$
7 $f \circ g = \{(-3, 4), (-1, -3), (4, 1)\}$
$g \circ f = \{(-4, 1), (1, 2), (2, 3), (3, -2)\}$
9 $f + g = 2x^2 + x - 6$, $\mathbb{R}$
$f - g = 2x^2 - 3x$, $\mathbb{R}$
$f \cdot g = 4x^3 - 8x^2 - 3x + 9$, $\mathbb{R}$
$\dfrac{f}{g} = x + 1$, $\mathbb{R} \sim \{\frac{3}{2}\}$
$f \circ g = 8x^2 - 26x + 18$; $\mathbb{R}$

11 $f + g = \dfrac{x^2 + x + 1}{x(x + 1)}$, $\mathbb{R} \sim \{-1, 0\}$

$f - g = \dfrac{x^2 - x - 1}{x(x + 1)}$, $\mathbb{R} \sim \{-1, 0\}$

$f \cdot g = \dfrac{1}{x + 1}$, $\mathbb{R} \sim \{-1, 0\}$

$\dfrac{f}{g} = \dfrac{x^2}{x + 1}$, $\mathbb{R} \sim \{-1, 0\}$

$f \circ g = \dfrac{1}{x + 1}$, $R \sim \{-1, 0\}$

13 $f \neq g$, domain $f \neq$ domain g
15 $f \neq g$, domain $f \neq$ domain g **17** $f = g$

19

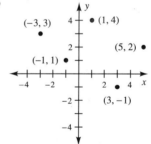

21

23 a. $5,000 fixed cost, regardless of number of units produced
b. $25,025,000 **c.** $25,030,000 **d.** $5,006
e. $C(u) = u^2 + 5u + 5000$ **f.** $A(u) = \dfrac{u^2 + 5u + 5000}{u}$

25 $(f \circ g)(x) = 12x + 5$, $\mathbb{R}$
$(g \circ f)(x) = 12x - 13$, $\mathbb{R}$

27 $(f \circ g)(x) = |x - 8|$, $\mathbb{R}$
$(g \circ f)(x) = |x| - 8$, $\mathbb{R}$

29 $(f \circ g)(x) = \dfrac{1}{x^3 + 1}$, $\mathbb{R} \sim \{-1\}$

$(g \circ f)(x) = \dfrac{x^3 + 1}{x^3}$, $\mathbb{R} \sim \{0\}$

31 a. $V(b) = 10b$ **b.** $F(b) = 3000$ **c.** $C(b) = 10b + 3000$
d. $A(b) = \dfrac{10b + 3000}{b}$ **e.** $R(b) = 12b$
f. $P(b) = 2b - 3000$ **g.** $P(1000) = -1000$
h. $P(1500) = 0$ **i.** $P(2000) = 1000$ **j.** 1500 units

33 a. $1,900, cost to manufacture 5000 doses of vaccine
b. $2,850, cost to wholesaler for 5000 doses of vaccine
c. $0.45d + 600$, cost to wholesaler for d doses of vaccine
d. $2,850, cost to wholesaler for 5000 doses of vaccine

35 $(f \circ g)(x) = x$

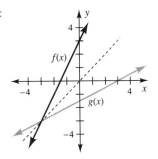

37 $h(x) = (f \circ g)(x)$

39 $\dfrac{g}{f}(x) = \dfrac{\sqrt{x}}{x^2 + 1}$

41 a. $A(w) = 576 - w^2$ **b.** $w(x) = 24 - 2x$
c. $A(x) = 96x - 4x^2$; the area of the border as a function of x.

43. $(f \circ g)(x) = \dfrac{1}{4x^2 - 10x - 6}$, $\mathbb{R} \sim \left\{-\dfrac{1}{2}, 3\right\}$

$(g \circ f)(x) = \dfrac{2}{x^2 - 5x - 6}$, $\mathbb{R} \sim \{6, -1\}$

45 $(f \circ g)(x) = \sqrt[3]{x^3 + 8}$, $\mathbb{R}$
$(g \circ f)(x) = x + 8$, $\mathbb{R}$

47 $f(x) = \sqrt[3]{x}$, $g(x) = x^2 + 4$
49 $f(x) = x^2 + 3x + 5$, $g(x) = x + 2$ **51** 11.3876

53 0.0878146
55

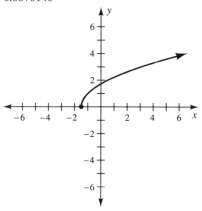

Exercises 3-6

1 a. One-to-one function **b.** Not a one-to-one function
c. Not a one-to-one function
3 a. One-to-one function **b.** Not a one-to-one function
c. Not a one-to-one function **d.** One-to-one function
5 5
7 Inverse: $\{(1, 0), (4, 1), (7, 2)\}$
Domain of the function: $\{0, 1, 2\}$
Domain of the inverse: $\{1, 4, 7\}$
9 Inverse: $\{(0, 1), (1, 2), (2, 4), (3, 8), (4, 16), (5, 32)\}$
Domain of function: $\{1, 2, 4, 8, 16, 32\}$
Domain of inverse: $\{0, 1, 2, 3, 4, 5\}$

11 $f^{-1}(x) = \dfrac{x - 3}{5}$; $\mathbb{R}$; $\mathbb{R}$ **13** $h^{-1}(x) = 3(x + 8)$; $\mathbb{R}$; $\mathbb{R}$

15 $y = \dfrac{20 - 5x}{4}$; $\mathbb{R}$; $\mathbb{R}$ **17** $y = \dfrac{x + 3}{x}$; $\mathbb{R} \sim \{1\}$; $\mathbb{R} \sim \{0\}$

19 $y = \dfrac{5x}{1 - x}$; $\mathbb{R} \sim \{-5\}$; $\mathbb{R} \sim \{1\}$

21 $f^{-1}(x) = \dfrac{5x + 3}{2 - 4x}$; $\mathbb{R} \sim \left\{-\dfrac{5}{4}\right\}$, $\mathbb{R} \sim \left\{\dfrac{1}{2}\right\}$

23 $y = \sqrt{x}$; $[0, +\infty)$; $[0, +\infty)$ **25** $y = x^3$; $\mathbb{R}$; $\mathbb{R}$

27

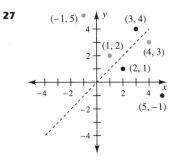

ANSWERS

29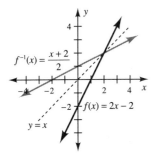

31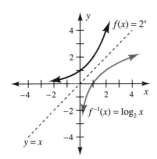

33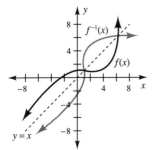

35 **a.** Inverses **b.** Not inverses **c.** Inverses

37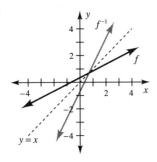

39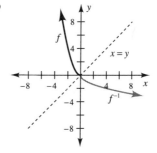

41 One-to-one function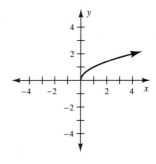

43 Not a one-to-one function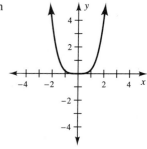

45 One-to-one function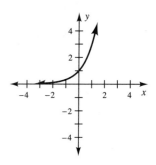

47 $(g \circ f)(x) = x$, $(f \circ g)(x) = x$
49 $(g \circ f)(x) = x$, $(f \circ g)(x) = x$

51 Inverses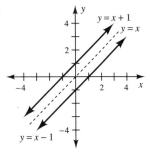

53 Inverses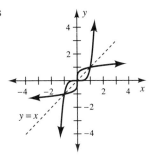

55 $y = \sqrt[n]{x}$ **57.** $y = x^n$

Exercises 3-7

1 $x^2 + y^2 = 100$
$x^2 + y^2 - 100 = 0$
3 $(x - 5)^2 + (y + 4)^2 = 16$
$x^2 + y^2 - 10x + 8y + 25 = 0$
5 $(x - 2)^2 + (y - 6)^2 = 2$
$x^2 + y^2 - 4x - 12y + 38 = 0$
7 $(x + 3)^2 + (y - 8)^2 = \frac{1}{25}$
$25x^2 + 25y^2 + 150x - 400y + 1824 = 0$
9 $x^2 + (y - \frac{1}{2})^2 = \frac{1}{4}$
$x^2 + y^2 - y = 0$
11 $(0, 0)$, 12; $x^2 + y^2 = 144$

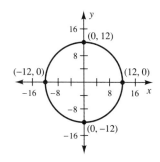

13 $(-5, 4)$, 8; $(x + 5)^2 + (y - 4)^2 = 64$

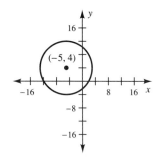

15 $(3, 0)$, 3; $x^2 + y^2 - 6x = 0$

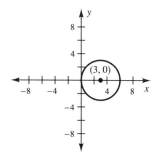

17 $(1, -5)$, 2; $x^2 + y^2 - 2x + 10y + 22 = 0$

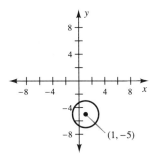

19 $(1, 5)$, $\frac{1}{2}$; $4x^2 + 4y^2 - 8x - 40y + 103 = 0$

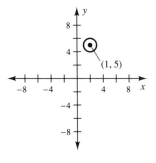

ANSWERS

21 $(0, 0)$, 12, 8; $\dfrac{x^2}{36} + \dfrac{y^2}{16} = 1$

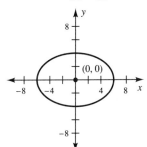

23 $(1, -2)$, 14, 6; $\dfrac{(x-1)^2}{9} + \dfrac{(y+2)^2}{49} = 1$

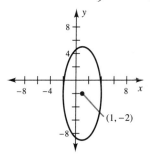

25 $(-5, -3)$; 6, 1; $\dfrac{(y+3)^2}{9} + \dfrac{4(x+5)^2}{1} = 1$

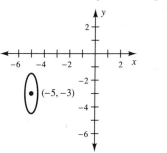

27 $(0, 0)$, 14, 12; $36x^2 + 49y^2 = 1764$

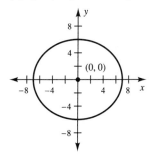

29 $(-1, -3)$, 8, 6; $16x^2 + 9y^2 + 32x + 54y - 47 = 0$

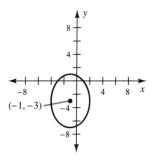

31 $\dfrac{x^2}{81} + \dfrac{y^2}{25} = 1$ **33** $\dfrac{(x-3)^2}{4} + \dfrac{(y-4)^2}{1} = 1$

35 $\dfrac{(x-6)^2}{9} + \dfrac{(y+2)^2}{25} = 1$ **37** $\dfrac{(x-5)^2}{16} + \dfrac{y^2}{4} = 1$

39 $\dfrac{(x+3)^2}{4} + \dfrac{(y+4)^2}{36} = 1$

41
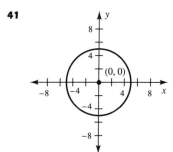

43
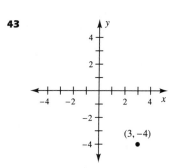

45 No points to graph **47** $x^2 + y^2 - 4y = 0$
49 $x^2 + y^2 - 8x + 8y + 16 = 0$
51 $(x-2)^2 + (y+3)^2 = 25$
53 $(x+4)^2 + (y+4)^2 = 16$ **55** 20π **57** 80π

59

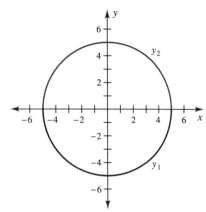

9 Center $(\frac{1}{2}, -\frac{2}{3})$
$a = 1, b = 1$
Hyperbola opens
horizontally

11 $\dfrac{y^2}{25} - \dfrac{x^2}{16} = 1$

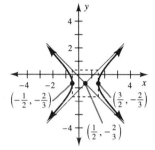

Exercises 3-8

1 Center $(0, 0)$
$a = 5, b = 9$
Hyperbola opens
horizontally

3 Center $(0, 0)$
$a = 2, b = 1$
Hyperbola opens
horizontally

13 $\dfrac{x^2}{9} - \dfrac{y^2}{49} = 1$ **15** $\dfrac{(y-5)^2}{81} - \dfrac{(x-2)^2}{225} = 1$

17 c   **18** f   **19** a   **20** g   **21** h   **22** e
23 b   **24** d   **25** Ellipse   **27** Parabola
29 Parabola   **31** No points   **33** Hyperbola

35 Parabola   **37** $\dfrac{x^2}{7} + \dfrac{(y-3)^2}{16} = 1$

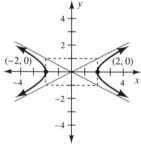

39

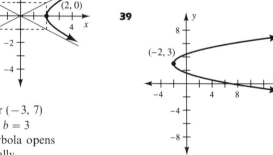

5 Center $(4, 6)$
$a = 7, b = 3$
Hyperbola opens
horizontally

7 Center $(-3, 7)$
$a = 4, b = 3$
Hyperbola opens
vertically

41

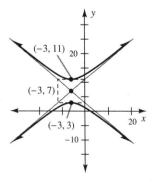
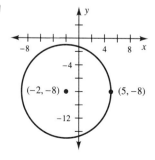

ANSWERS

43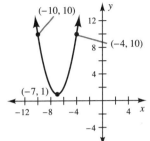

38 a. 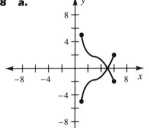   **b.**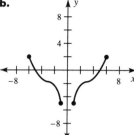

c.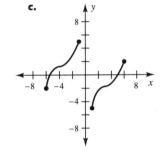

Review Exercises for Chapter 3

1 Midpoint: $(\frac{3}{2}, 8)$; distance $= 25$; slope $= \frac{24}{7}$
2 52   **3** 1   **4** 0   **5** -200 **6** -40
7 $6x^2 - 11x - 35$
8 $6x^2 + 12hx + 6h^2 + 11x + 11h - 35$ **9** 2
10 Function   **11** Function   **12** Not a function
13 Not a function   **14** Function
15 $2x - 3y + 23 = 0$ **16** $3x + 4y - 1 = 0$
17 $3x + 7y - 42 = 0$ **18** $3x + 7y - 18 = 0$
19 $x - 2 = 0$ **20** $y + 5 = 0$ **21** $2x - y - 5 = 0$
22 $x + 2y = 0$ **23** $y + 1 = 0$ **24** $y + 1 = 0$
25 $2x - 3y - 1 = 0$ **26** $x - y = 0$
27 $x^2 + y^2 = \frac{1}{4}$ **28** $x^2 + y^2 = 16$ **29** $\frac{x^2}{16} + \frac{y^2}{9} = 1$
30 $\frac{(x-3)^2}{\frac{1}{4}} + \frac{(y-1)^2}{4} = 1$
31 $(x-2)^2 + (y+2)^2 = 25$
32 $\frac{(x-1)^2}{25} + \frac{(y+3)^2}{4} = 1$
33 $D = \{3, 5, 7\}$; $R = \{-2, 9, 11\}$ **34** $D = \mathbb{R} \sim \{2, 7\}$
35 $D = [-2, +\infty)$
36 a. $D = [-4, 3]$; $R = [-1, 3]$ **b.** $D = [-4, 4]$;
  Increasing on $[-2, 1]$ $R = [-1, 10]$
  Decreasing on $[-4, -2]$ Increasing on
  Constant on $[1, 3]$ $[-4, 0]$
  Decreasing on $[0, 4]$
  c. $D = [-4, 4]$; $R = [1, 6]$
  Increasing on $[-4, -1]$, $[1, 3]$
  Decreasing on $[-1, 1]$
  Constant on $[3, 4]$
37 a. Even; symmetric about the y-axis.
  b. Even; symmetric about the y-axis.
  c. Neither; not symmetric to either axis or to the origin.
  d Odd; symmetric about the origin.

39 a. 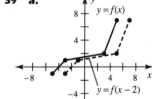   **b.**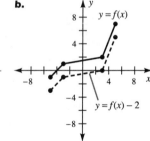

c. 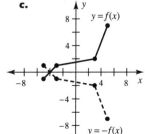   **d.**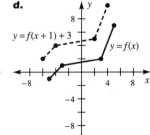

40 **a.** $(f + g)(x) = 2x^2 - 4x - 16$, $\mathbb{R}$
b. $\left(\dfrac{g}{f}\right)(x) = \dfrac{1}{2x + 3}$, $\mathbb{R} \sim \left\{-\dfrac{3}{2}, 4\right\}$
c. $(f \circ g)(x) = 2x^2 - 21x + 40$, $\mathbb{R}$
d. $f(-x) = 2x^2 + 5x - 12$, $\mathbb{R}$

41 f **42** n **43** b **44** o **45** i
46 a **47** l **48** k **49** e **50** j
51 d **52** m **53** c **54** h **55** g

56 Not equal. The domain of f does not equal the domain of g.

57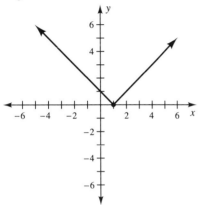

58 One-to-one function
Inverse: $\{(6, 1), (3, -2), (1, -4), (-2, 4)\}$
Domain of inverse: $\{-2, 1, 3, 6\}$

59 Not a one-to-one function

60 One-to-one function
Inverse: $\{(8, 7), (9, 8), (7, 9)\}$
Domain of inverse: $\{7, 8, 9\}$

61 One-to-one function
$f^{-1}(x) = 4x + 4$
Domain of $f^{-1} = \mathbb{R}$

62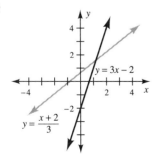

63 $22\frac{1}{2}$ seconds

64 **a.** a, d; $k > 1$ in a and $k < -1$ in d **b.** c
c. b, e; $k = 1$ in b; $k = -1$ in e

65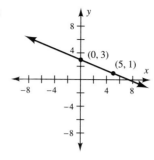

66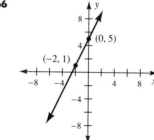

67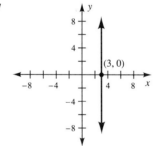

68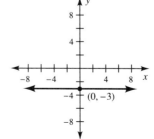

ANSWERS

69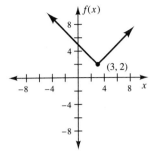

70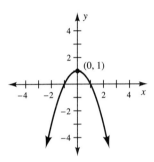

71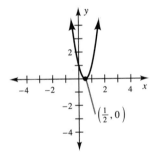

72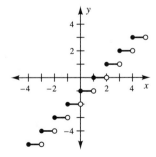

73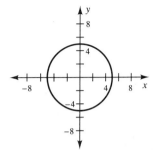

74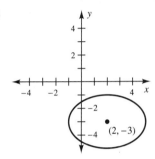

75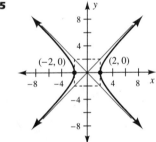

Optional Exercises for Calculus-Bound Students for Chapter 3

1 $4x + 2h - 5$

2 **a.** $-\dfrac{b}{a}$ **b.** $y - b = \dfrac{-b}{a}(x - 0)$

$$y - b = \dfrac{-bx}{a}$$

$$\dfrac{bx}{a} + y = b$$

$$\dfrac{x}{a} + \dfrac{y}{b} = 1$$

3 $x - y = 0$, yes.   **4** $h = f + g$ **5** $h = \dfrac{f}{g}$

6 $h = g \circ f$ **7** $h = f \circ g$
8 $f(x) = \sqrt{x}$, $g(x) = x^2 - 3x - 1$
9 $f(x) = \dfrac{1}{x}$, $g(x) = |3x + 5|$

26 $P(x) = (50 - x)(x - 10)$; $30

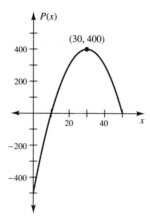

10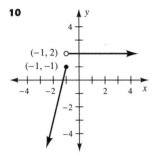
Increasing on $(-\infty, -1]$
Constant on $(-1, +\infty)$

11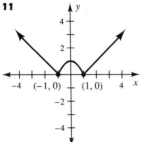
Increasing on $[-1, 0]$ and $[1, +\infty)$
Decreasing on $(-\infty, -1]$ and $[0, 1]$

27 $P(x) = [40 + 2(80 - x)](x - 50)$; $75

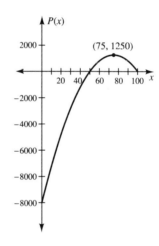

12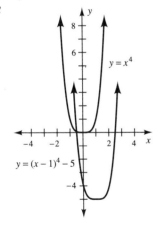

13 $f^{-1}(x) = \sqrt{x} + 1$ for $x \geq 0$ **14** $x - y = 0$
15 $y = k$ **16** $x = 3$
17 Points do not form a right triangle. **18** -2
19 3 **20** $(x - 3)^2 + (y + 4)^2 = 25$ **21** $-3, 5$
22 $8\sqrt{2}$, yes. The sides are equal, and adjacent sides are perpendicular.
23 -2 **24** $(\tfrac{17}{3}, 12)$ **25** $s = d$

Mastery Test for Chapter 3

1 **a.** Function **b.** Not a function **c.** Not a function
  d. Function **e.** Function **f.** Not a function
2 **a.** 6 **b.** 31 **c.** 0 **d.** 5
  e. $x^3 + 3x^2h + 3xh^2 + h^3 - 2$ **f.** $-x^3 - 2$
3 **a.** $D = \{2, 7, -3\}$; $R = \{5, -11, 8\}$
  b. $D = \mathbb{R}$; $R = [0, +\infty]$
  c. $D = \{4, 3, 5, 0\}$; $R = \{6, 7, -2, 9\}$
  d. $D = \{-5, -2, 2, 5\}$; $R = \{-4, -2, 1, 4\}$
  e. $D = [-2, 5]$; $R = [-1, 4]$ **f.** $D = \mathbb{R} \sim \{-2, 2\}$
  g. $D = [-5, +\infty)$

4 a. 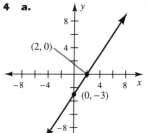 **b.**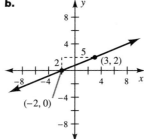

c. 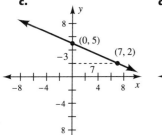 **d.**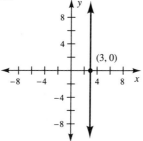

5 a. $m = 7$ **b.** $m = \frac{2}{3}$ **c.** Slope undefined **d.** $m = 0$
   e. $m = 1$
6 Perpendicular
7 a. $y = 7x - 11$ **b.** $x = 2$ **c.** $y = 2x + 2$ **d.** $y = 2$

8 a. 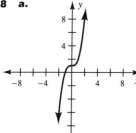 **b.**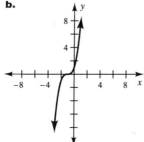

c.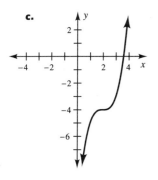

9 a. 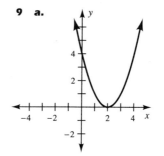 **b.**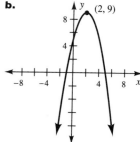

10 10 couplings
11 $f(x) = x^2 + 2$, symmetry with respect to y-axis.
   $g(x) = x^3 - x$, symmetry with respect to origin.
   None of these are symmetric to the x-axis.
12 a. distance $= 13$ **b.** midpoint $= (\frac{1}{2}, 2)$
13 a. $(f + g)(x) = x^2 - 9$; $D = \mathbb{R}$
   b. $(f - g)(x) = x^2 - 2x - 3$; $D = \mathbb{R}$
   c. $(f \cdot g)(x) = x^3 - 4x^2 - 3x + 18$; $D = \mathbb{R}$
   d. $\left(\dfrac{f}{g}\right)(x) = x + 2$; $D = \mathbb{R} \sim \{3\}$
14 a. $(f \circ g)(x) = f(g(x)) = \sqrt{3x^2 + 2}$; $D = \mathbb{R}$
   b. $(g \circ f)(x) = g(f(x)) = 3x$; $D = [\frac{1}{3}, +\infty)$
15 a. One-to-one **b.** Not one-to-one
   c. Not one-to-one **d.** One-to-one
   e. Not one-to-one (horizontal line test)
   f. One-to-one
16 a. $\{(1, 10), (2, 100), (3, 1000)\}$ **b.** $f^{-1}(x) = \dfrac{x + 3}{4}$
   c. $g^{-1}(x) = 10^x$ **d.** $y = \ln x$

17 a. 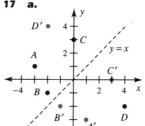 **b.**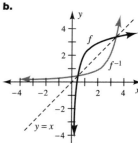

18 a. $(x + 3)^2 + (y - 6)^2 = 16$
   b. $(x - 5)^2 + (y - 2)^2 = 16$
   c. $\dfrac{x^2}{16} + \dfrac{y^2}{9} = 1$ **d.** $\dfrac{(x + 3)^2}{\frac{1}{16}} + \dfrac{(y + 4)^2}{64} = 1$

19 a. 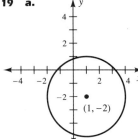   **b.**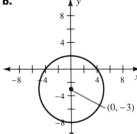

c. 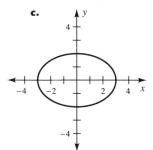   **d.**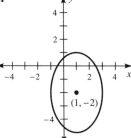

20 a. $\dfrac{x^2}{16} - \dfrac{(y-2)^2}{36} = 1$ **b.** $\dfrac{(x-2)^2}{4} - \dfrac{(y-5)^2}{100} = 1$

21 a. 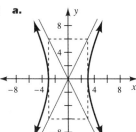   **b.**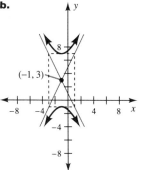

22 a. Circle   **b.** Parabola   **c.** Ellipse   **d.** Hyperbola

b. 0 (multiplicity 2), tangent to x-axis
$-\frac{5}{2}$ (multiplicity 1), crosses x-axis
$\frac{2}{5}$ (multiplicity 1), crosses x-axis

c. 3 (multiplicity 1), crosses x-axis
-2 (multiplicity 2), tangent to x-axis

3 a. -3, even multiplicity   **b.** -2, even multiplicity
-1, odd multiplicity   0, odd multiplicity
                             4, odd multiplicity

c. -3, even multiplicity
-1, even multiplicity
2, odd multiplicity

5 a. Negative   **b.** Positive   **c.** Negative
d. Positive

7 a. Above   **b.** Above   **c.** Below

9 a. Above   **b.** Above   **c.** Below   **d.** Above

11 a. Points down far left   **b.** Points down far left
Points up far right   Points down far right

c. Points up far left
Points down far right

13 a. Odd degree
Leading coefficient positive

b. Even degree
Leading coefficient negative

c. Odd degree
Leading coefficient negative

15 c   **17** c

19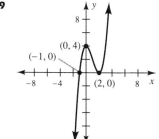

21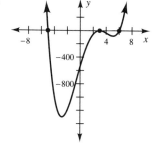

CHAPTER FOUR

Exercises 4-1

1 a. -2 (multiplicity 1), crosses x-axis
3 (multiplicity 2), tangent to x-axis
5 (multiplicity 1), crosses x-axis

A-27

23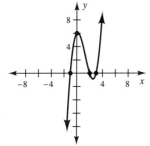

25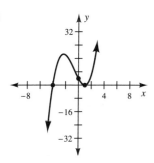

27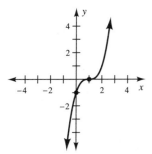

29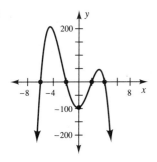

31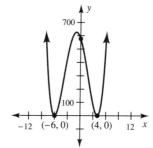

33 a. $(-3, 0) \cup (3, +\infty)$ **b.** $(-\infty, -3] \cup [0, 3]$
35 $P(x) = (x - 3)^2(x + 1)$ **37** $P(x) = -(x - 2)(x + 3)^2$
39 $P(x) = (x - 4)^2(x + 1)(x - 2)$ **41** c   **43** a

45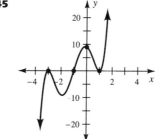

47 $y = (x + 2)(x + 1)(x - 1)$

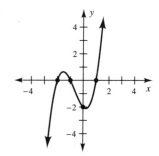

49 $y = 4(x + \frac{1}{2})(x - \frac{1}{2})(x + 1)(x - 1)$ or
$y = (2x + 1)(2x - 1)(x + 1)(x - 1)$

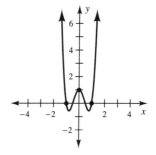

Exercises 4-2

1 $(x^2 - x)(x - 5) - 3$
3 $(3a^4 - 3a^3 + 2a^2 - 2a + 5)(a + 1) - 13$
5 $(6p^2 + 2p - \frac{4}{3})(p - \frac{1}{3}) + \frac{41}{9}$
7 $(x^2 - 2x + 4)(x + 2) + 0$
9 $[x^2 + (2 + i)x + 2 + 3i][x - (1 + i)] + (-6 + 5i)$
11 $0, P(x) = (x^2 - 7x + 11)(x - 1)$ **13** 678
15 -10 **17** $0, P(x) = (x^2 - 2x + 4)(x + 2)$
19
$$\begin{array}{r|rrrr} -4 & 1 & 5 & -16 & -80 \\ & & -4 & -4 & 80 \\ \hline & 1 & 1 & -20 & 0 \end{array}$$
Depressed factor $v^2 + v - 20$

21
$$\begin{array}{r|rrrrr} 5 & 1 & 0 & -5 & -6 & -7 \\ & & 5 & 25 & 100 & 470 \\ \hline & 1 & 5 & 20 & 94 & 463 \end{array}$$
Remainder $= 463 \ne 0$

23 $(t + 6)(t - 5)(t - 1)$ **25** $4, \frac{3}{2}, -4$
27 $-2, 3 \pm \sqrt{2}$ **29** $-6, 5, -2 \pm 4i$
31 $P(-1) = 1 - 32 + 16 + 5 + 10 = 0$; thus $x + 1$ is a factor of $P(x)$
33 $v^2 + \sqrt{2}v - 1 + \sqrt{2}; 0$ **35** -6 **37** $-6 - 5i$
39 b^3 **41** $P(-a) = 0$
43 Let $P(x) = -x^5 + x^4 - 7x^3 + 11x^2 + 19$ and let $k > 0$. Then
$$P(-k) = -(-k)^5 + (-k)^4 - 7(-k)^3 + 11(-k)^2 + 19$$
$$= k^5 + k^4 + 7k^3 + 11k^2 + 19 > 0$$
Thus $-k$ cannot be a zero for any $k > 0$
45 Let $P(x) = -x^8 - 9x^4 - 17$
Then $P(k) = -k^8 - 9k^4 - 17 < 0$ since $k^8 \ge 0$ and $k^4 \ge 0$. Thus k is not a zero of $P(x)$ for any real number k.
47 Let $P(x) = x^n + a^n$
Then $P(-a) = (-a)^n + a^n = 0$
Thus $x + a$ is a factor of $x^n + a^n$
49 $P(v) = 0; 50v^2 + 117.3v + 258.3$ **51** 43,500
53 $P(x) = (x + 7)(5x - 2)(2x - 81)$

Exercises 4-3

1 a. $2 - 7i, -2i$; degree 4
 b. $2 - i\sqrt{3}$, (multiplicity 2); degree 6
 c. $5 - 6i, -i$ (multiplicity 3); degree 12
3 a. $-3(2), 2(5)$ **b.** $-2(2), 0(3), \frac{1}{2}(2)$ **c.** $-2(2)$
5 a. $5(3), -1(1), -3(1)$
 b. $1(1), \dfrac{-1}{2} + i\dfrac{\sqrt{3}}{2}(1), -\dfrac{1}{2} - i\dfrac{\sqrt{3}}{2}(1)$
 c. $-2(1), 1 + i\sqrt{3}(1), 1 - i\sqrt{3}(1)$

7 $x^2 + 4$ **9** $x^2 - 2x + 2$
11 $x^3 - 6x^2 + 13x - 10$
13 $x^4 + 5x^3 - 5x^2 - 85x - 156$
15 $x^5 - 2x^4 + 21x^3 - 32x^2 + 80x$
17 $3x^3 - 33x^2 + 123x - 153$
19 $10x^4 - 40x^3 + 40x^2 - 40$ **21.** $i, -i$
23 $-2i, -\frac{4}{3}, \frac{3}{2}$ **25** $1 - i, -3, 2$
27 $\pm i\sqrt{5}$ **29** $(x + 1)(2x - 3)(x - 6); -1, \frac{3}{2}, 6$
31 $-5, 4, \pm i$ **33** $(u - 1)^3$; 1 (multiplicity 3)
35 $\pm 5, \pm 1$
37 All complex zeros of a real polynomial occur in conjugate pairs. A polynomial of odd degree has an odd number of zeros, counting multiplicity; thus it must have at least one real zero.
39 $P(x) = x^3 - (4 + 2i)x^2 + (5 + 6i)x - (2 + 4i)$
41 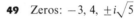 $-1 - i\sqrt{3}, -1 + i\sqrt{3}$
43 Yes; $P(x) = \left[x - \left(-\dfrac{1}{2} + i\dfrac{\sqrt{3}}{2}\right)\right]\left[x - \left(-\dfrac{1}{2} - i\dfrac{\sqrt{3}}{2}\right)\right]$
45 $P(x) = 4x^8 + 32x^6 + 64x^4$ **47** $\pm i$
49 Zeros: $-3, 4, \pm i\sqrt{5}$

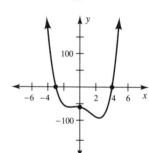

Exercises 4-4

1 $-6, -3, -2, -1, 1, 2, 3, 6$
3 $-10, -5, -\frac{5}{2}, -2, -1, -\frac{1}{2}, \frac{1}{2}, 1, 2, \frac{5}{2}, 5, 10$
5 $-20, -10, -\frac{20}{3}, -5, -4, -\frac{10}{3}, -\frac{5}{2}, -2, -\frac{5}{3}, -\frac{4}{3},$
$-1, -\frac{5}{6}, -\frac{2}{3}, -\frac{1}{2}, -\frac{1}{3}, -\frac{1}{6}, \frac{1}{6}, \frac{1}{3}, \frac{1}{2}, \frac{2}{3}, \frac{5}{6}, 1, \frac{4}{3}, \frac{5}{3}, 2, \frac{5}{2},$
$\frac{10}{3}, 4, 5, \frac{20}{3}, 10, 20$
7 b **9** d
11 a. $-12, -6, -4, -3, -2, -\frac{3}{2}, -1, -\frac{3}{4}, -\frac{1}{2}, -\frac{1}{4}, \frac{1}{4},$
$\frac{1}{2}, \frac{3}{4}, 1, \frac{3}{2}, 2, 3, 4, 6, 12$
 b.
$$\begin{array}{r|rrrr} 12 & 4 & -48 & -1 & 12 \\ & & 48 & 0 & -12 \\ \hline & 4 & 0 & -1 & 0 \end{array}$$
$(x - 12)(4x^2 - 1) = 0$
$(x - 12)(2x + 1)(2x - 1) = 0$
 c. Zeros: $-\frac{1}{2}, \frac{1}{2}, 12$

ANSWERS

A-29

13 a. $-15, -\frac{15}{2}, -5, -3, -\frac{5}{2}, -\frac{3}{2}, -1, -\frac{1}{2}, \frac{1}{2}, 1, \frac{3}{2}, \frac{5}{2}, 3,$
  $5, \frac{15}{2}, 15$
  b. $\underline{15|}\ \ 2\ \ -31\ \ \ \ 14\ \ \ \ 15$
  $\ \ \ \ \ \ \ \ \ 30\ \ -15\ \ -15$
  $\ \ \overline{\ \ 2\ \ -1\ \ \ -1\ \ \ \ \ 0}$
  $(x - 15)(2x^2 - x - 1) = 0$
  $(x - 15)(2x + 1)(x - 1) = 0$
  c. Zeros: $-\frac{1}{2}, 1, 15$

15 $-2, 2$ (multiplicity 2)   **17** $-3, -2, 3$

19 $-\frac{1}{2}, \frac{1}{3}, \frac{1}{2}$ **21** $-\frac{1}{2} - \frac{\sqrt{5}}{2}, -\frac{1}{2} + \frac{\sqrt{5}}{2}, -2$

23 $-1, \frac{2}{3}, 2, 3$
25 a. Positive **b.** Negative **c.** Negative
27 a. Positive **b.** Positive **c.** Negative
  d. Negative
29 $(-1, 2) \cup (4, +\infty)$ **31** $(1, +\infty)$
33 $(-\infty, -3) \cup (-2, 3)$ **35** $[-\frac{1}{2}, \frac{1}{3}] \cup [\frac{1}{2}, +\infty)$
37 Zeros: $-\frac{5}{4}, \frac{2}{3}, \frac{3}{2}$,
39 x-intercepts: $-1.5, -\sqrt{2}, \sqrt{2}$

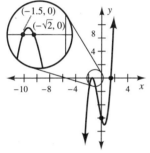

41 $P(\sqrt{2}) = 0$; thus $\sqrt{2}$ is a zero of $P(x)$. ± 1 and ± 2 are the only possible rational zeros of $P(x)$. $P(\pm 1) = -1 \neq 0$, and $P(\pm 2) = 2 \neq 0$. Since there are no rational zeros of $P(x)$, $\sqrt{2}$ is irrational.

43 Zeros: $-1, -\frac{2}{3}, 2$ (multiplicity 2)
45 Zeros: $-3, -2$ (multiplicity 2), $\frac{1}{2}, 2$
47 Possible rational roots: $-2, -1, 1, 2$
  $P(-2) = 18, P(-1) = 3, P(1) = 9, P(2) = 42$
49 4 centimeters
51 $(-2, 2), (3.5, +\infty)$

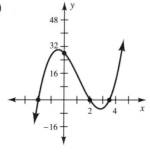

Exercises 4-5

1 a. $P(0) = -10$ and $P(1) = 30$
  b. $P(-2) = 3$ and $P(-3) = -8$
3 a. 1 positive zero, 2 or no negative zeros
  b. 2 or no positive zeros, 1 negative zero
  c. no positive zeros, 3 or 1 negative zeros
  d. 3 or 1 positive zeros, no negative zeros

5 $\underline{7|}\ \ 1\ \ -2\ \ -19\ \ \ \ 14\ \ \ \ 84$
  $\ \ \ \ \ \ \ \ 7\ \ \ \ \ 35\ \ 112\ \ 882$
  $\ \overline{\ \ 1\ \ \ \ 5\ \ \ \ 16\ \ 126\ \ 966}$

Since $7 > 0$ and the numbers in the last row of this synthetic division are nonnegative, 7 is an upper bound on the real zeros of $P(x)$.

7 $\underline{-6|}\ \ 1\ \ \ \ 1\ \ -25\ \ -19\ \ 144$
  $\ \ -6\ \ \ \ \ 30\ \ -30\ \ 294$
  $\ \overline{\ \ 1\ \ -5\ \ \ \ \ 5\ \ -49\ \ 438}$

Since $-6 < 0$ and the numbers in the last row of this synthetic division alternate in sign, -6 is a lower bound on the real zeros of $P(x)$.

9 a. Possible rational zeros: $-84, -42, -28, -21,$
  $-14, -12, -\frac{21}{2}, -7, -6, -4, -\frac{7}{2}, -3, -2, -\frac{3}{2},$
  $-1, -\frac{1}{2}, \frac{1}{2}, 1, \frac{3}{2}, 2, 3, \frac{7}{2}, 4, 6, 7, \frac{21}{2}, 12, 14, 21, 28,$
  $42, 84$ (32 possibilities)
  b. $P(-x)$ has no sign changes; thus all negative possibilities can be eliminated.
  c. $P(0) = -84, P(1) = 66$ [sign of $P(x)$ changes between 0 and 1]
  d. 0.5, 7, 12

11 a. Possible rational zeros: $-210, -105, -70, -42,$
  $-35, -30, -21, -15, -14, -10, -7, -6, -5,$
  $-3, -2, -1, 1, 2, 3, 5, 6, 7, 10, 14, 15, 21, 30, 35,$
  $42, 70, 105,$ and 210 (32 possibilities)
  b. 6 is an upper bound
  c. All rational possibilities greater than 6 can be eliminated, since 6 is an upper bound. The others have all been tested. Thus there are no positive zeros of $P(x)$.
  d. $-6, -\sqrt{35}, \sqrt{35}$

13 Possible rational roots: $-20, -10, -5, -4, -2, -1,$
  $1, 2, 4, 5, 10, 20$
  By synthetic division, $P(1) = -20 < 0$ and $P(4) = 220 > 0$. Thus 1 and 4 are not roots. $P(2) = -8$ and $P(3) = 52$. Thus there is, by the location theorem, a zero in this interval.

15 $P(x)$ has no sign change, which indicates that there are no positive real zeros. $P(-x)$ has one sign change; thus there is one negative real zero. $P(x)$ has five zeros counting multiplicity; thus four of them are not real numbers.

17 Zeros are between $-1, 0$; $0, 1$; and $2, 3$.
19 Zeros are between $0, 1$; $1, 2$; and $3, 4$.

21
$$\underline{5\,|}\begin{array}{rrrrr} 1 & -1 & -18 & 52 & -40 \\ & 5 & 20 & 10 & 310 \\ \hline 1 & 4 & 2 & 62 & 270 \end{array}$$

The last row of this synthetic division is all nonnegative. Thus 5 is an upper bound.

23 a. $P(1) = 2$, $P(2) = 9$ **b.** $P(\frac{3}{2}) = 0$, $P(\frac{5}{4}) = 0$
c. Two zeros cause two sign changes, thus the sign change is not apparent.

25 Zeros and solutions: $2, 3, \frac{9}{2}$
$P(t) = (t - 2)(2t - 9)(t - 3)$

27 Zeros and solutions: $-2, -\frac{2}{3}, \frac{2}{3}$
$P(x) = (3x - 2)(3x + 2)(x + 2)$

29 Zeros and solutions: $-\sqrt{5}, \frac{1}{3}, \sqrt{5}$
$P(t) = (3t - 1)(t - \sqrt{5})(t + \sqrt{5})$

31 $-\frac{7}{2}, -2, -\frac{1}{4} \pm i\frac{\sqrt{3}}{4}$

33 $-\frac{3}{2}, \frac{3}{4}, 1$ (multiplicity 3)

35 $\pm i$ (each of multiplicity 2), $\pm \frac{\sqrt{6}}{2}, \pm \frac{\sqrt{6}}{3}i, \pm i\sqrt{2\sqrt{3}}$

37 $y = 2x^3 + 5x^2 - 6x - 15$

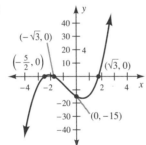

Exercises 4-6
1 3.7   **3** 1.7   **5** 3.2   **7** -0.4 **9** 5.45
11 2.27   **13** -0.3 **15** 2.4
17 3

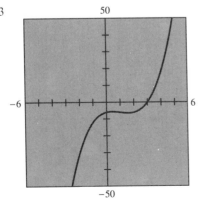

19 $-6, -1, 1$

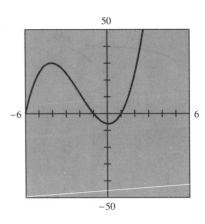

21 -1.33 **23** 1.20   **25** 0.6, 1.2
27 $-3.0, -2.0, 2.0, 3.0$ **29** $-0.21, 4.55$

Exercises 4-7
1 a. $\mathbb{R} \sim \{-\frac{5}{3}\}$ **b.** $\mathbb{R} \sim \{\frac{4}{3}, -\frac{1}{2}\}$
3 a. -3 **b.** 4
5 a. $y = 0$; $x = -\frac{7}{2}$ (odd) **b.** $y = \frac{3}{4}$; $x = -\frac{3}{4}$ (odd)
7 a. $y = 2$; $x = -4$ (odd), $x = -3$ (odd)
b. $y = 0$; $x = 0$ (odd), $x = 5$ (even)

9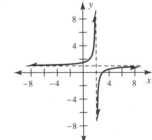

11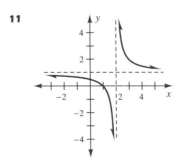

ANSWERS

13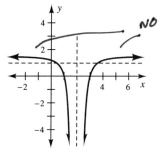

15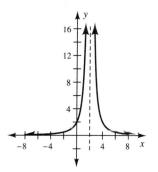

17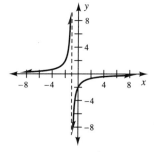

19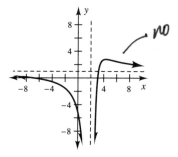

21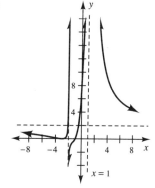

23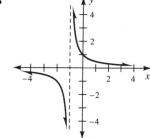

25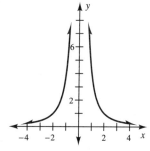

27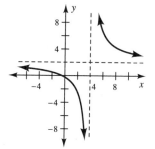

29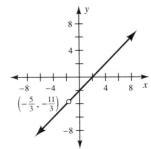

31 $y = \dfrac{x+2}{x-3}$ **33** $y = \dfrac{x-4}{(x+2)^2}$

35 a   **36** e   **37** d   **38** c
39 b. ii   **40** d   **41** e   **42** a
43 b   **44** c   **45** Yes. $y = \dfrac{2}{x-5}$

47 No. There is only one horizontal line the graph can approach as $|x|$ gets larger. This will be either $y = 0$ or $y = \dfrac{a_n}{b_n}$, depending on the degrees of the numerator and the denominator.

49 No. The function is undefined at the x-value of the vertical asymptote.

51 Yes. $y = \dfrac{1}{x}$

53 Every polynomial function $y = P(x)$ can be written as $y = \dfrac{P(x)}{1}$ with the constant polynomial 1 as a denominator.

55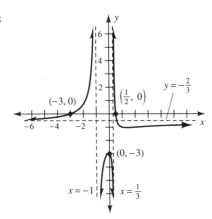

57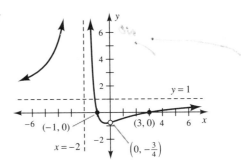

59 $A(t) = \dfrac{200(5t+6)}{t(48-t)},\ 0 < t \le 24$

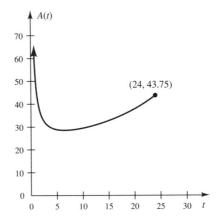

61 $y = \dfrac{2x^2 - 5x - 3}{x^2 - 5x + 6}$

63 Oblique asymptote: $y = x - \tfrac{1}{2}$

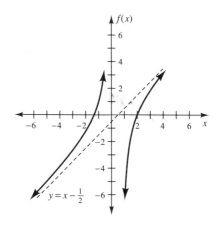

65 Oblique asymptote: $y = 2x - 5$

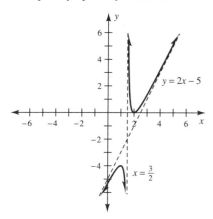

67 No. If the degree of the numerator is one more than the degree of the denominator, the rational function will have an oblique linear asymptote. If the degree of the numerator is less than or equal to the degree of the denominator, then the rational function has a horizontal asymptote.

69 $y = \dfrac{x^2 + x - 1}{x - 1}$

71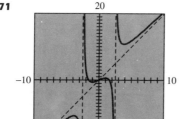

73 $-0.2, 1.2$

Review Exercises for Chapter 4

1 $3y^3 - 6y^2 + 10y - 20 + \dfrac{41}{y + 2}$

2 $t^3 - (1 + 3i)t^2 - (1 + 3i)t - 2 - 3i$

3 $P(5) = 0$, $P(x) = (x^4 - 2x^2 + x - 5)(x - 5)$

4 $P(-2b) = -3b^3$ **5** -15

6 $-2x^3 + 6x^2 + 20x - 48$

7 $-x^4 + 4x^3 + 8x^2 - 12x - 15$

8 $-x^3 + 8x^2 - 22x + 20$

9 -5 (multiplicity 3), 0 (multiplicity 2), 1 (multiplicity 4)

10 0 (multiplicity 2), $\frac{1}{2}$ (multiplicity 2)

11 -5 (multiplicity 1), $-\frac{3}{2}$ (multiplicity 2)

12 $1 \pm \sqrt{2}, 2 \pm i$

13 $-30, -15, -10, -\frac{15}{2}, -6, -5, -\frac{10}{3}, -3, -\frac{5}{2}, -2, -\frac{5}{3}, -\frac{3}{2}, -1, -\frac{5}{6}, -\frac{2}{3}, -\frac{1}{2}, -\frac{1}{3}, -\frac{1}{6}, \frac{1}{6}, \frac{1}{3}, \frac{1}{2}, \frac{2}{3}, \frac{5}{6}, 1, \frac{3}{2}, \frac{5}{3}, 2, \frac{5}{2}, 3, \frac{10}{3}, 5, 6, \frac{15}{2}, 10, 15, 30$

14
$$\begin{array}{r|rrrrr} 5 & 1 & -2 & -10 & 14 & -3 \\ & & 5 & 15 & 25 & 195 \\ \hline & 1 & 3 & 5 & 39 & 192 \end{array}$$

The last row of this synthetic division is nonnegative. Thus 5 is an upper bound.

$$\begin{array}{r|rrrrr} -3 & 1 & -2 & -10 & 14 & -3 \\ & & -3 & 15 & -15 & 3 \\ \hline & 1 & -5 & 5 & -1 & 0 \end{array}$$

The last row of this synthetic division alternates in sign. Thus -3 is a lower bound on the zeros of $P(x)$.

15 Possible rational roots: $-10, -5, -\frac{5}{2}, -2, -1, -\frac{1}{2}, \frac{1}{2}, 1, 2, \frac{5}{2}, 5, 10$

$$\begin{array}{r|rrrrr} 1 & 2 & -1 & 9 & 8 & 10 \\ & & 2 & 1 & 10 & 18 \\ \hline & 2 & 1 & 10 & 18 & 28 \end{array}$$

The last row of this synthetic division is nonnegative. Thus 1 is an upper bound on the zeros of $P(x)$.

$$\begin{array}{r|rrrrr} -1 & 2 & -1 & 9 & 8 & 10 \\ & & -2 & 3 & -12 & 4 \\ \hline & 2 & -3 & 12 & -4 & 14 \end{array}$$

The last row of this synthetic division alternates in sign. Thus -1 is a lower bound on the zeros of $P(x)$.

$$\begin{array}{r|rrrrr} \frac{1}{2} & 2 & -1 & 9 & 8 & 10 \\ & & 1 & 0 & \frac{9}{2} & \frac{25}{4} \\ \hline & 2 & 0 & 9 & \frac{25}{2} & \frac{65}{4} \end{array}$$ $\frac{1}{2}$ is not a zero

$$\begin{array}{r|rrrrr} -\frac{1}{2} & 2 & -1 & 9 & 8 & 10 \\ & & -1 & 1 & -5 & -\frac{3}{2} \\ \hline & 2 & -2 & 10 & 3 & \frac{17}{2} \end{array}$$ $-\frac{1}{2}$ is not a zero

Thus, no rational roots exist.

16 Possible rational roots: $-7, -1, 1, 7$
$P(\pm 7) = 42$, $P(\pm 1) = -6$. Thus no rational roots.
$P(\sqrt{7}) = 0$, so $\sqrt{7}$ must be irrational.

17 One positive real zero; two or no negative real zeros.

18 Four, two, or no positive real zeros; no negative real zeros.

19 $P(0) = -8 < 0$ and $P(1) = 15 > 0$

20 $P(2) = -1 < 0$ and $P(3) = 4 > 0$

21 $-5, -2, 1$ **22** $-5, -2, \frac{1}{2}, 4$

23 $-\dfrac{5}{2}, \dfrac{2}{3}, -\dfrac{1}{2} \pm i\dfrac{\sqrt{3}}{2}$

24 -2 (multiplicity 3), $\frac{4}{3}$, 2

25 $-4, -\frac{2}{3}, \frac{5}{4}$ **26** $(-4, -\frac{3}{2}) \cup (5, +\infty)$

27 $(-\infty, -\frac{5}{2}]$ **28** 0.2 **29** $\mathbb{R}$
30 $\mathbb{R} \sim \{\frac{2}{3}\}$ **31** $\mathbb{R} \sim \{\frac{3}{4}, -\frac{7}{2}\}$
32 Horizontal asymptote: $h(s) = 0$
Vertical asymptote: $s = 1$
33 Horizontal asymptote: $y = -1$
Vertical asymptote: $x = 3$
34 Horizontal asymptote: $S(t) = \frac{1}{4}$
Vertical asymptote: $t = -\frac{1}{2}, t = \frac{3}{2}$
35 Horizontal asymptote: $R(t) = 2$
No vertical asymptotes

36 d. $y = (x+1)^2(3-2x)$ **37 f.** $y = \dfrac{2x-3}{x+1}$

38 k. $y = \dfrac{2x^2 - x - 3}{2x - 3}$ **39 g.** $y = \dfrac{x^2 + 2x + 1}{2x^2 - x - 3}$

40 e. $y = \dfrac{x+1}{2x-3}$ **41 h.** $y = \dfrac{x+1}{4x^2 - 6x + 9}$

42 b. $y = (x+1)(2x-3)^2$ **43 j.** $y = \dfrac{x^2 + 2x + 1}{2x^2 + x - 6}$

44 c. $y = (x+1)^2(2x-3)$ **45 a.** $y = (x+1)(2x-3)$

46 i. $y = \dfrac{x-1}{2x+3}$

47 Increasing $(-\infty, 1)$
Decreasing $(1, +\infty)$

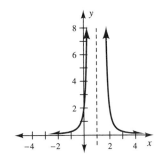

48 $\dfrac{x+2}{x-2} > 0$ in the intervals $(-\infty, -2) \cup (2, +\infty)$

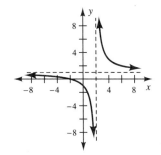

49

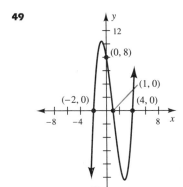

50

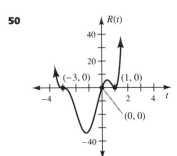

51

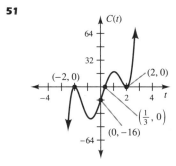

52

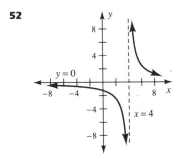

53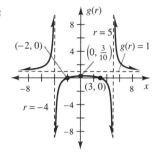

54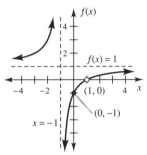

55 $y = \dfrac{2x-1}{x^2-x-2}$

3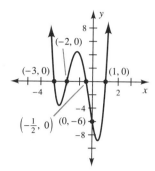

4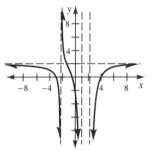

Detail of curve at different scale

5 $(-0.5, 0.5) \cup (1, +\infty)$

6 $P(x) = x^{17} + a^{17}$
$P(-a) = 0$, thus $x + a$ is a factor of $P(x)$.

7 $-6, 3 - \sqrt{7}, 2, 3 + \sqrt{7}$ **8** $(-\infty, -2) \cup (-2, +\infty)$

Optional Exercises for Calculus-Bound Students for Chapter 4

1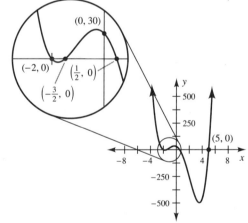

2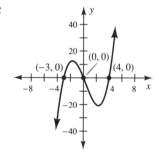

9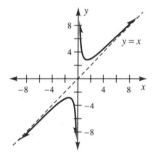

10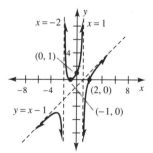

11 0.7 **12** $-1.2, 3.2$

Mastery Test for Chapter 4

1 a.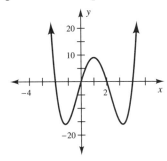

b. $P(x) = (x - 5)(x + 1)^2$

2 $2x^3 - 4x^2 + 3x - 5 + \dfrac{5}{x + 2}$ **3** 115

4 $x^4 - 2x^3 - 2x^2 + 6x + 5$

5 Possible numerators: $\pm 1, \pm 3, \pm 5, \pm 15$
Possible denominators: $\pm 1, \pm 2, \pm 3, \pm 6$
Possible rational zeros: $-15, -\frac{15}{2}, -5, -3, -\frac{5}{2}, -\frac{5}{3},$
$-\frac{3}{2}, -1, -\frac{5}{6}, -\frac{1}{2}, -\frac{1}{3}, -\frac{1}{6}, \frac{1}{6}, \frac{1}{3}, \frac{1}{2}, \frac{5}{6}, 1, \frac{3}{2}, \frac{5}{3}, \frac{5}{2}, 3, 5,$
$\frac{15}{2}, 15$

6
```
3 |  1   -2   -10    14   -3
          3     3   -21  -21
      1   1    -7    -7  -24      P(3) = -24

4 |  1   -2   -10    14   -3
          4     8    -8   24
      1   2    -2     6   21      P(4) = 21
```
There is a zero between 3 and 4, since $P(3)$ and $P(4)$ have different signs.

7 No, two, or four positive zeros; one negative zero

8 5 is an upper bound; -3 is a lower bound

9 $3, -\frac{1}{3}, 1 \pm \sqrt{5}$ **10** 3.7

11 Horizontal asymptote: $y = \frac{4}{2} = 2$
Vertical asymptote: $x = \frac{3}{2}$ and $x = -3$

12 a. $y = \dfrac{2x - 1}{x + 2}$

Horizontal asymptote: $y = \frac{2}{1} = 2$
Vertical asymptote: $x = -2$

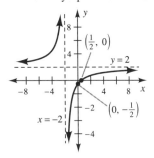

b. $y = \dfrac{x - 1}{x^2 - x - 2} = \dfrac{x - 1}{(x - 2)(x + 1)}$
Horizontal asymptote: $y = 0$
Vertical asymptote: $x = 2, x = -1$

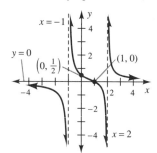

CHAPTER FIVE

Exercises 5-1

1 a. 6 **b.** 2 **c.** $\frac{3}{2}$ **d.** -2
3 a. $\frac{3}{2}$ **b.** -3 **c.** -1 **d.** $\frac{1}{3}$
5 a. $\frac{1}{3}$ **b.** ± 2 **c.** -1
7 a. 2 **b.** $\frac{4}{5}$ **c.** $-8, 2$
9 a. 20.0855 **b.** 0.0497871 **c.** 23.1407 **d.** 2980.96

11

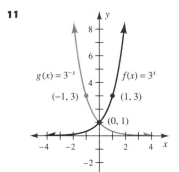

13

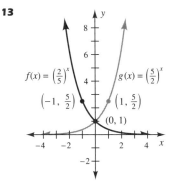

15 $851.41 **17** $1,613.71 **19** $1,185.30
21 12,200 bacteria **23** 78.0 milligrams **25** b
27 c **29** 4.5 **31** d **33** a **35** c
37 b **39 a.** 3 **b.** 5 **c.** -2 **d.** $\frac{1}{3}$
41 a. Straight line **b.** Parabola
c. Exponential function
43 $0.22 **45** $a = 2, b = 3, x = 0$
47

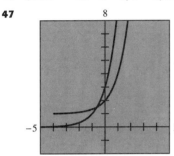

49

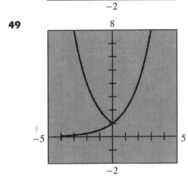

51 1.6

Exercises 5-2

1 a. $8^2 = 64$ **b.** $23^0 = 1$ **c.** $(\frac{7}{5})^{-2} = \frac{25}{49}$ **d.** $7^{2.3} = x$
3 a. $\log_{27} 3 = \frac{1}{3}$ **b.** $\log_{\frac{3}{7}}(\frac{49}{7}) = -2$ **c.** $\log_6 1 = 0$
d. $\log_b x = 11$ **5** 2 **7** 5 **9** -2
11 $\frac{1}{2}$ **13** Does not exist **15** Does not exist
17 $-\frac{1}{3}$ **19** 0 **21** 14.5 **23** $\frac{1}{2}$ **25** 2
27 2 **29** 0 **31** $\frac{3}{2}$ **33** $\frac{5}{3}$ **35** 5 **37** -2
39 1 **41** 51 **43** 11 **45** 47
47

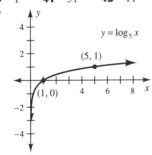

49

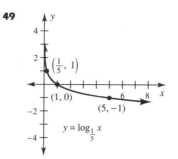

51 1.6 **53** e **55** b **57** d **59** a
61 b **63** 7.4 **65** 5 **67** 3

69

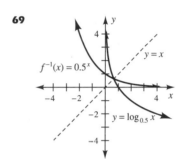

Exercises 5-3

	Logarithmic Form	Verbal Form	Exponential Form
1 a.	$\log w = z$	The common log of w is z.	$10^z = w$
b.	$\ln v = w$	The natural log of v is w.	$e^w = v$
c.	$\ln n = m$	The natural log of n is m.	$e^m = n$

3 a. 5 **b.** 7 **c.** -3 **d** -7
5 a. 0.918555 **b.** 2.11505 **c.** 2.96755 **d.** 0.105254
7 a. -7.83863 **b.** -18.0491 **c.** 5.65599 **d.** 7.92837
9 a. 505.138 **b.** 1.32879 **c.** 13.1971 **d.** 0.367519
11 $3 \log x + \log y$ **13** $\log(x + 3) - \log(y - 3)$
15 $3 \ln(x - 9) - \frac{1}{2} \ln(y + 1)$
17 $\frac{1}{5}(2 \ln x + 3 \ln y) - 4 \ln z$
19 $3[\ln x + 2 \ln(y + 5) - 4 \ln(z - 3)]$
21 $\log x^3 y^2$ **23** $\ln \frac{(x + 5)^3}{y^4}$ **25** $\log x^2 \sqrt[7]{y^3}$
27 $\ln \sqrt{\frac{(5x + 2)}{(x - 7)}}$ **29** 2.327 **31** -4.402 **33** c
35 a **37** d **39** $y = e^{x \ln 3}$ **41** $y = e^{-x \ln 5}$
43 1.1133 **45** -0.2519 **47** 1
49 $e^{-\ln x} = (e^{\ln x})^{-1} = x^{-1} = \dfrac{1}{x}$

51 $\log_7 x \stackrel{?}{=} \dfrac{\ln x}{\ln 7} \approx \dfrac{\ln x}{1.945910149}$
$\approx 0.5138983 \ln x$

53 $\log 24^x - \log 60^x - \log 0.2^{2x}$
$\stackrel{?}{=} x \log 24 - x \log 60 - 2x \log 0.2$
$= x(\log 24 - \log 60 - \log(0.2)^2)$
$= x(\log 24 - \log 60 - \log 0.04)$
$= x \cdot \log\left(\dfrac{24}{(60)(0.04)}\right)$
$= x \cdot \log 10$
$= x \cdot 1$
$= x$

55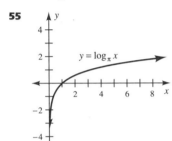

Exercises 5-4

1	-2	**3**	-4	**5**	3	**7**	2	**9**	-2
11	0.860952	**13**	1.06791	**15**	-1.38129				
17	0.987541	**19**	1.63456	**21**	± 3.48633				
23	6.69416	**25**	6.96031	**27**	15.6945	**29**	1, 3		
31	8	**33**	6	**35**	No solution	**37**	1, 3		
39	4	**41**	3	**43**	-1	**45**	10^{100}	**47**	1, 100
49	$10^{10,000,000,000}$	**51**	± 148.413	**53**	13.4 years				
55	$0.29	**57**	139 years	**59**	138 days				
61	17.3 years	**63**	6.6 years	**65**	$10^{1.4} \approx 25$ times				
67	90 decibels	**69**	6.4	**71**	$x = 2 - y$				
73	$x = 100 - y$	**75**	1.1	**77**	-0.4				

Review Exercises for Chapter 5

1 $x^y = 10$ **2** $10^z = y$ **3** $\log_7 y = x + 3$
4 $\ln w = z$ **5** -3 **6** 0 **7** 1 **8** $\tfrac{1}{2}$
9 $\tfrac{2}{3}$ **10** 16 **11** ± 3 **12** $1, e^2$ **13** 32
14 $\tfrac{64}{125}$ **15** -2 **16** $\tfrac{1}{2}$ **17** 1 **18** 1
19 4 **20** $\tfrac{1}{25}$ **21** 5 **22** 113 **23** 19
24 ± 6 **25** 7 **26** 11 **27** 2 **28** ± 4
29 2.45254 **30** -5.54678 **31** -11.7343
32 156.269 **33** 0.500000 **34** 1.30000
35 9.00000 **36** 1.41598
37 $\log(x - 3) + 2 \log(x + 4) - \log(x - 7)$
38 $\tfrac{1}{2} \ln(x + y) - \ln z$ **39** $\ln \dfrac{x^3}{(x + 5)(x - 7)}$
40 $\log \dfrac{x^3}{\sqrt{y}}$ **41** 1.08262 **42** 0.358774
43 0.478525 **44** $-1.00000, 8.00000$ **45** 8.18352
46 4.00000 **47** 7.00000, 11.0000 **48** No solution
49 20.0000
50 A one-to-one function, $f^{-1}(x) = \ln x - 2$ domain $f^{-1} = (0, +\infty)$
51 A one-to-one, $f^{-1}(x) = \dfrac{3^x + 1}{2}$, domain $f^{-1} = \mathbb{R}$

52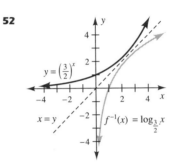

53 (0, 1) **54** (1, 0) **55** a, c, d, g
56 $e^{-x \ln 3} = (e^{\ln 3})^{-x} = 3^{-x} = (\tfrac{1}{3})^x$
57 $x \log_3 e = x\left(\dfrac{\ln e}{\ln 3}\right) = x\left(\dfrac{1}{\ln 3}\right) = \dfrac{x}{\ln 3}$ **58** 7.325 years
59 19,970 days **60** e **61** k **62** p **63** g
64 c **65** n **66** b **67** i **68** f **69** o
70 m **71** l **72** h **73** a **74** j **75** d

Optional Exercises for Calculus-Bound Students for Chapter 5

1 $(g \circ f)(x) = g[3x - 2] = \dfrac{(3x - 2) + 2}{3} = \dfrac{3x}{3} = x$

$(f \circ g)(x) = f\left[\dfrac{x + 2}{3}\right] = 3\left(\dfrac{x + 2}{3}\right) - 2$
$= (x + 2) - 2 = x$

2 $f^{-1}(x) = -\dfrac{7x + 5}{2x - 3}$ **3** $1 + \sqrt{34}$

4 $x = 3y + 100$ **5** $x = \dfrac{\log 5}{\log 2} + 3y$ **6** 0.01, 10

7 2, 3
8 Equal for $x = 1$ and $x = 100$; not equal for all other positive real numbers.
9 4 **10** 6 **11** $x = -3$ **12** $y = 2$

ANSWERS

13 $e^{-x \ln 2} = (e^{\ln 2})^{-x} = 2^{-x} = (\frac{1}{2})^x = (0.5)^x$

14 $\log 45^x - \log 9^x + \log 0.5^{-x}$
$= x \log 45 - x \log 9 - x \log 0.5$
$= x(\log 45 - \log 9 - \log 0.5)$
$= x\left(\log \dfrac{45}{9(0.5)}\right)$
$= x(\log 10)$
$= x$

15 Let $x = b^m$ and $y = b^n$, and thus $\log_b x = m$ and $\log_b y = n$. Then $\dfrac{x}{y} = \dfrac{b^m}{b^n} = b^{m-n}$

$\log_b\left(\dfrac{x}{y}\right) = \log_b b^{m-n}$

$\log_b\left(\dfrac{x}{y}\right) = m - n$

$\log_b\left(\dfrac{x}{y}\right) = \log_b x - \log_b y$

16 Let $\log_b a = y$, then $b^y = a$.
$b^{(x \log_b a)} = (b^{\log_b a})^x = (b^y)^x = a^x$
Thus $a^x = b^{(x \log_b a)}$.

17 $\sinh x + \cosh x = \dfrac{e^x - e^{-x}}{2} + \dfrac{e^x + e^{-x}}{2} = \dfrac{2e^x}{2} = e^x$

18 $x = \ln(y + \sqrt{y^2 + 1})$

19

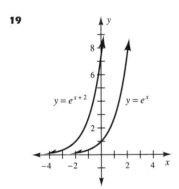

20

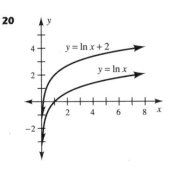

21

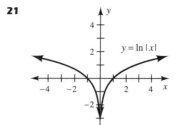

22 $-0.833, 0.833$ **23** $-1.511, 1.511$
24 $g(x) = x^2$; $g(-1) = g(1)$; $-1 \neq 1$ **25** 2.9 hours

Mastery Test for Chapter 5

1 a.
b.

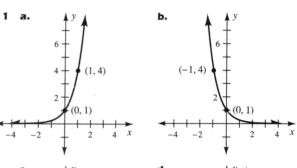

c.
d.

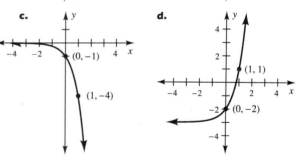

2 a. $462.61 **b.** $467.06 **c.** $133.82
3 a. 2 **b.** $-\frac{1}{2}$ **c.** -4 **d.** Not defined
  e. $\frac{2}{3}$ **f.** $-\frac{3}{2}$
4 a. 1 **b.** 7 **c.** 9 **d.** 0 **e.** -1 **f.** 33
5 a.
b.

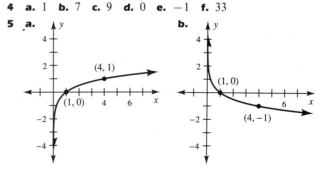

c.

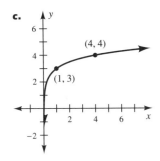

6 a. 2.9748 **b.** −7.4585 **c.** 4.2971 **d.** 13.117
7 a. 1.2091 **b.** 6.3219 **c.** 1.6055
8 a. $3 \log x + 2 \log y - \log z$
   b. $\frac{1}{2} \log x + \frac{1}{2} \log y - \log z$ **c.** $\log \frac{x^3}{\sqrt{y}}$
   d. $\log(x + y)^2 z^3$
9 a. 0.57405 **b.** 3 **c.** 11.6 years **d.** 4.9 years

CHAPTER SIX

Exercises 6-1

1 $4(-6) - 3(5) = -39$
   $2(-6) + 5(5) = 13$

3

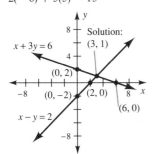

5 Every point on the line is a solution.

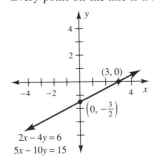

7 One **9** None **11** Infinitely many solutions
13 One **15** (5, 4)

17 General solution: $(1 - 2a, a)$; three particular solutions: $(1, 0), (-1, 1), (-3, 2)$
19 No solution **21** $(-1, -1)$ **23** $(10, 4)$
25 $(-5, 6)$
27 General solution: $\left(a, \dfrac{4a - 2}{5}\right)$; three particular solutions: $\left(0, -\dfrac{2}{5}\right), (-2, -2), (3, 2)$
29 $(-\frac{1}{2}, \frac{1}{3})$
31 General solution: $(5a + 3, a)$; three particular solutions: $(3, 0), (8, 1), (13, 2)$
33 $(0, 0)$ **35** No solution **37** $(\frac{1}{2}, -\frac{1}{2})$
39 $(1.5, 0.5, 2.5)$ **41** $(3.5, -1.5, 1)$ **43** $(59, 18)$
45 \$75, \$37.50 **47** 16 liters of 15%; 24 liters of 8%
49 $(1, 7)$ **51** $(\frac{1}{2}, \frac{1}{2}, -\frac{1}{2})$
53 $(-3, -\sqrt{7}), (-3, \sqrt{7}), (3, -\sqrt{7}), (3, \sqrt{7})$
55 $(31, 43)$
57 $(5 - 2a, -3a + 2, a), (5, 2, 0), (3, -1, 1), (1, -4, 2)$

Exercises 6-2

1 $\begin{bmatrix} 2 & 3 & | & 5 \\ 6 & -4 & | & 2 \end{bmatrix}$ **3** $\begin{bmatrix} 1 & 1 & 1 & 1 & | & 1 \\ 2 & 0 & -1 & 0 & | & 0 \\ 0 & 1 & 0 & 0 & | & 0 \end{bmatrix}$

5 $x + 3z = 7$
   $4y - 2z = 9$

7 $\begin{bmatrix} 1 & 2 & 4 & | & 3 \\ 3 & 5 & 7 & | & 1 \\ 4 & 9 & 2 & | & 8 \end{bmatrix}$ **9** $\begin{bmatrix} 1 & 3 & 5 & | & 11 \\ 0 & 1 & -1 & | & -9 \\ 4 & 8 & 3 & | & 7 \end{bmatrix}$

11 Reduced form
13 Not reduced form; reduced form is
$\begin{bmatrix} 1 & 0 & 8 & -2 & 0 & | & 1 \\ 0 & 1 & 0 & 0 & 0 & | & 0 \\ 0 & 0 & 0 & 0 & 0 & | & 0 \end{bmatrix}$

15 Not reduced form; reduced form is
$\begin{bmatrix} 1 & 0 & -2 & 0 & | & 1 \\ 0 & 1 & 1 & 0 & | & 1 \\ 0 & 0 & 0 & 1 & | & 0 \end{bmatrix}$

17 Not reduced form; reduced form is
$\begin{bmatrix} 1 & 2 & 0 & 0 & | & 0 \\ 0 & 0 & 1 & 0 & | & 0 \\ 0 & 0 & 0 & 1 & | & 1 \end{bmatrix}$

19 Reduced form
21 a. $(2, 5, -3)$ **b.** No solution
   c. General solution: $(2 - 4a, 5 - 2a, a)$; two particular solutions: $(2, 5, 0), (-2, 3, 1)$

ANSWERS

23 (77, −33) **25** (−2, 7) **27** (3, 1, −1)
29 (2, 3, 3)
31 General solution: (1 − 4a, −4 + 3a, a); two particular solutions: (1, −4, 0), (−3, −1, 1)
33 ($\frac{1}{2}$, −$\frac{3}{2}$, 2) **35** No solution
37 General solution: (c − 3, −c − 1, c); two particular solutions: (−3, −1, 0), (−2, −2, 1)
39 26, 36, 46 **41** (−3, 2), (4, 1) **43** (10, −4, 3, 10)
45 (3.00, 7.00)

Exercises 6-3

1 800 student tickets; 350 nonstudent tickets
3 175 chairs
5 **a.** 4390 toys **b.** $220 **c.** 3960 toys
7 4080 pounds of bricks; 3280 pounds of blocks
9 35°, 70°, 75°
11 Belted, $33.50; standard, $50.50; deluxe, $88.00
13 100 kilometers per hour, 500 kilometers per hour
15 $y = x^2 - 4x + 1$
17 100 pounds of fertilizer A; 60 pounds of fertilizer B
19 10 hours **21** $A = 0, B = -4, C = 4$
23 **a.** 2550 gallons **b.** 110 cents **c.** 2800 gallons
25 18 questions

Exercises 6-4

1
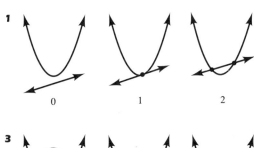

3
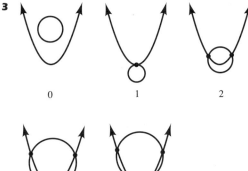

5 (−1, −5), (5, 1) **7** (1, 4), (7, 2)
9 (4, 2$\sqrt{2}$), (4, −2$\sqrt{2}$), (−4, 2$\sqrt{2}$), (−4, −2$\sqrt{2}$)
11 ($\frac{1}{2}i$, $\frac{5}{2}i$), (−$\frac{1}{2}i$, −$\frac{5}{2}i$) **13** ($\sqrt{21}$, 2), (−$\sqrt{21}$, 2)
15 (3, 2i), (3, −2i), (1, 2), (1, −2)
17 $\left(\frac{\sqrt{130}}{5}, \frac{\sqrt{30}}{5}\right), \left(\frac{-\sqrt{130}}{5}, \frac{-\sqrt{30}}{5}\right), \left(\frac{\sqrt{130}}{5}, \frac{-\sqrt{30}}{5}\right),$
$\left(-\frac{\sqrt{130}}{5}, \frac{\sqrt{30}}{5}\right)$
19 (3, 2), (3, −2), (−3, 2), (−3, −2)
21 (0, −1), (i$\sqrt{2}$, −3), (−i$\sqrt{2}$, −3)
23 (3, 3), (−3, −3) **25** $30, 40 units **27** (5, 5)
29 (1, 10) **31** (6, 2), (−6, −2), (2, −1), (−2, 1)
33 (−2, 3), (0, 6)
35 5 yards × 5 yards, 7 yards × 7 yards
37 10 meters, 15 meters **39** (1, 4), (3, 1)
41 ($\frac{1}{2}$, $\frac{1}{2}$), (−$\frac{1}{2}$, −$\frac{1}{2}$), ($\frac{1}{2}$, −$\frac{3}{2}$), (−$\frac{1}{2}$, $\frac{3}{2}$)
43 1500 pencils, $0.096 **45** (4.54, 3.67)

Exercises 6-5

1 c **3** b

5
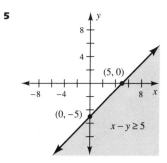

7
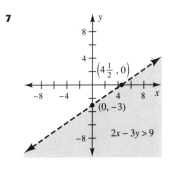

9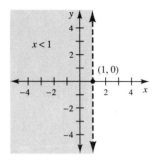

11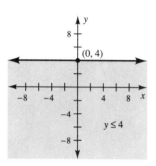

13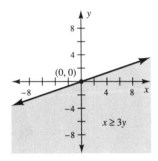

15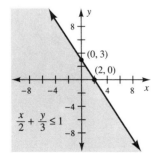

17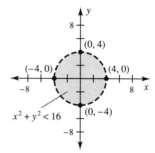

19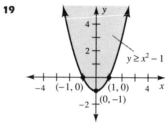

21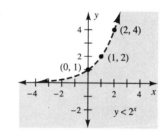

23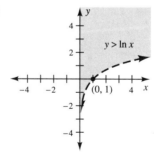

25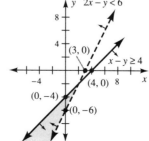

ANSWERS

27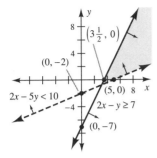

29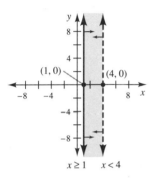

31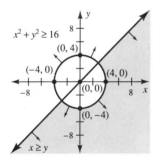

33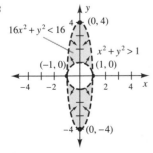

35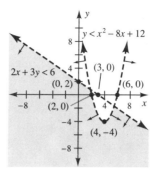

37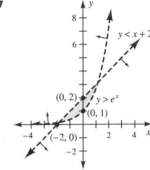

39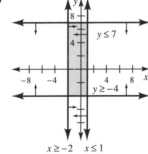

41 $x + y \leq 1$
$x - y \leq 1$

43 $x - 2y \leq -1$
$x > y$

45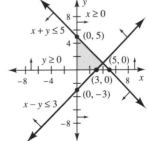

47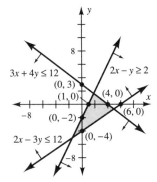

49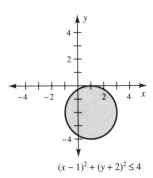

51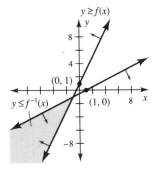

53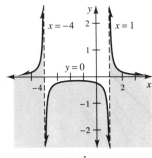

55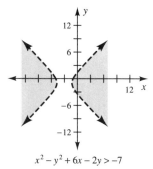

$x^2 - y^2 + 6x - 2y > -7$

57 a.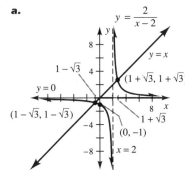

b. $(1 - \sqrt{3}, 2) \cup (1 + \sqrt{3}, +\infty)$
c. The x-coordinates of the solution of $f(x) > g(x)$

59 a. and b.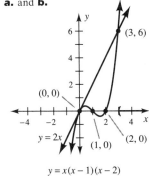

c. $(3, +\infty)$

Exercises 6-6

1 202 **3** 16 **5** 18 **7** -3 **9** 12
11 4 **13** 34 **15** 8 **17** 57
19 Category I, 12; category II, 22
21 20 suits, 120 blouses
23 20 bags of Student Mix; 30 bags of Hollywood Mix
25 Site A, 10 days; site B, 16 days

ANSWERS

Review Exercises for Chapter 6

1. $(1, -4)$
2. $(-4, 14), (-3, 10)$
3. $(4, 3), (-4, -3), (3, 4), (-3, -4)$
4. $(\frac{2}{21}, -\frac{3}{35})$
5. $(\frac{1}{2}, -\frac{1}{3})$
6. $(2, -7, 4)$
7. $(4, 2), (4, -2), (-4, 2), (-4, -2)$
8. $(7, \log 500), (3, 2)$
9. $(\frac{11}{7} + \frac{1}{7}a, -\frac{9}{7} + \frac{10}{7}a, a)$
10. $(8, -3, 5)$
11. No solution
12. $(-\frac{13}{14}a + \frac{19}{14}, \frac{11}{14}a + \frac{27}{14}, a)$
13. $(\frac{1}{5}, -\frac{1}{7})$
14. $(1, -1), (-1, 1)$
15. $(-\frac{22}{5}, \frac{4}{5}), (0, 3)$
16. $y = 5x^2 - 7x + 9$
17. 53 units per hour for machine A; 41 units per hour for machine B
18. 84 pounds
19. 10 pairs
20. $56.80, 284 tickets
21. $(0.6, -0.8), (2.5, 1.3)$
22. $(-3.0, 0.0), (0.7, 1.9)$

23.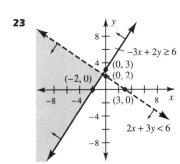

24.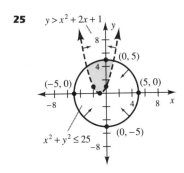

25.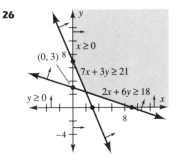

26.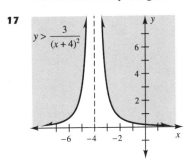

27. $(4, 0), (7, 1), (10, 2)$

28. $\begin{bmatrix} 1 & 0 & 0 & -1 & | & \frac{1}{2} \\ 0 & 1 & 0 & 1 & | & 0 \\ 0 & 0 & 1 & -1 & | & -\frac{7}{4} \end{bmatrix}$

29. 7.5
30. $C = 4$
31. $(-113, 53, -38, -12)$
32. 27 of robot A; 33 of robot B

Chapter 6 Optional Exercises for Calculus-Bound Students

1. $(2, 0, 1, 1)$
2. $(1, \frac{5}{2}, 1)$
3. $(-\frac{2}{3}a + \frac{4}{3}, -\frac{1}{3}a + \frac{8}{3}, a)$
4. $(\sqrt{7}, \sqrt{10}), (\sqrt{7}, -\sqrt{10}), (-\sqrt{7}, \sqrt{10}), (-\sqrt{7}, -\sqrt{10})$
5. Exact values $(\frac{1}{2}\ln 3, \frac{1}{2}\ln 6)$; approximate values $(0.55, 0.90)$
6. $x = 3, y = -5$
7. $(1, 0, 1), (1, 2, -1),$ and $(\frac{8}{7}, \frac{6}{7}, -\frac{9}{7})$
8. $y = x^2 - 4x + 2$
9. No solution
10. No solution
11. $(\frac{1}{3}, -2), (-\frac{1}{3}, 2), (1, -1),$ and $(-1, 1)$
12. $(3, -2)$ and $(-\frac{8}{29}, -\frac{20}{29})$
13. $(i, -2i), (-i, 2i), \left(\dfrac{3\sqrt{22}}{4}i, \dfrac{\sqrt{22}}{4}i\right)$, and $\left(-\dfrac{3\sqrt{22}}{4}i, -\dfrac{\sqrt{22}}{4}i\right)$
14. $(0.91, -0.98)$
15. $(-1.02, 0.49)$
16. 18.11 meters $\times$ 4.42 meters and 5.89 meters $\times$ 13.58 meters both satisfy the given conditions.

17.

18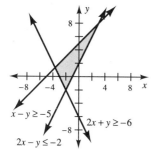

19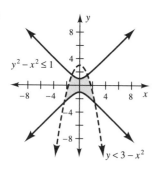

20 $x > 5$

Mastery Test for Chapter 6

1 $(\frac{15}{7}, \frac{6}{7})$ **2** $(7, 8)$
3 a. One solution   **b.** Infinitely many solutions
   c. No solution
4 $(-13, -7, -5)$
5 The amounts invested in silver, stocks, and money market funds were $10,000, $20,000, and $20,000, respectively.
6 $(4, 7), (-2, 1)$
7 $(2, i), (2, -i), (-2, i), (-2, -i)$

8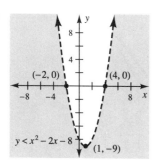

9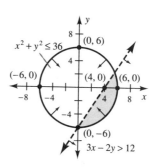

10 The maximum value of 8 occurs when $x = 4$ and $y = 0$.

CHAPTER SEVEN

Exercises 7-1

1 3×2; $a_{12} = -3$, $a_{21} = 4$, $a_{22} = 7$

3 $A = \begin{bmatrix} a_{11} & a_{12} & a_{13} \\ a_{21} & a_{22} & a_{23} \end{bmatrix}$ **5** $\begin{bmatrix} 7 & 4 & 0 \\ 10 & 5 & 8 \end{bmatrix}$

7 $\begin{bmatrix} 0 & -1 & 7 \\ 4 & 0 & 9 \end{bmatrix}$ **9** $\begin{bmatrix} -2 & 2 & -4 \\ -4 & 6 & -10 \end{bmatrix}$

11 $\begin{bmatrix} 5 & -3 & -4 \\ 2 & -15 & 7 \end{bmatrix}$ **13** $\begin{bmatrix} -5 & 0 & 25 \\ 10 & 15 & 20 \end{bmatrix}$

15 $\begin{bmatrix} 1 & 2 \\ -1 & -3 \\ 2 & 5 \end{bmatrix}$ **17** $\begin{bmatrix} 0 & -3 & -2 \\ 3 & -6 & -1 \end{bmatrix}$

19 Not defined   **21** $x = 2, y = 11$
23 $x = -4, y = -2$ **25** $m = 1, n = 3$
27 $m = 3, n = 5$ **29** $m = n =$ any natural number

31 $\begin{bmatrix} 7 & 8 \\ 17 & 18 \end{bmatrix}$ **33** $\begin{bmatrix} -5 & -1 & -2 \\ 15 & -2 & 41 \\ 30 & -4 & 82 \end{bmatrix}$

35 A^2 is not defined   **37** $\begin{bmatrix} 1105 & 1550 \\ 1200 & 1705 \end{bmatrix}$

39 $\begin{bmatrix} 6 & 16 \\ 16 & 43 \end{bmatrix}$ **41** $\begin{bmatrix} -2 & -10 \\ 5 & 12 \\ -2 & -9 \end{bmatrix}$

43 $\begin{bmatrix} 7 & 8 \\ 17 & 18 \end{bmatrix}$ **45** $\begin{bmatrix} 9 & 24 & 67 \\ 19 & 54 & 157 \end{bmatrix}$

ANSWERS

47
$$\begin{bmatrix} 11 & 0 & 1 & 0 \\ 12 & 3 & 5 & 1 \\ 12 & 3 & 4 & 3 \\ 8 & 2 & 2 & 0 \\ 8 & 1 & 2 & 2 \\ 10 & 2 & 5 & 5 \\ 12 & 0 & 0 & 0 \\ 7 & 0 & 1 & 0 \end{bmatrix}$$

49
$$\begin{array}{cc} \text{Total} & \text{Total} \\ \text{cost} & \text{revenue} \end{array}$$
$$\begin{bmatrix} 305.00 & 577.50 \\ 213.50 & 401.80 \\ 348.75 & 662.30 \end{bmatrix} \begin{array}{l} \text{Function 1} \\ \text{Function 2} \\ \text{Function 3} \end{array}$$

51
$$\begin{array}{cccc} & \text{Small} & \text{Medium} & \text{Large} \\ & \begin{bmatrix} 63 & 94 & 28 \\ 67 & 16 & 0 \\ 53 & 133 & 60 \end{bmatrix} & \begin{array}{l} \text{Item 1} \\ \text{Item 2} \\ \text{Item 3} \end{array} \end{array}$$

Each entry represents the number of items in stock on Tuesday morning.

53
$$0.80P = \begin{array}{cc} \text{Regular} & \text{Deluxe} \\ \begin{bmatrix} 70.36 & 75.60 \\ 61.59 & 66.36 \\ 25.22 & 28.71 \\ 89.91 & 100.79 \end{bmatrix} & \begin{array}{l} \text{Item a} \\ \text{Item b} \\ \text{Item c} \\ \text{Item d} \end{array} \end{array}$$

55 $A = \begin{bmatrix} 5 & 8 & 11 \\ 7 & 10 & 13 \\ 9 & 12 & 15 \\ 11 & 14 & 17 \end{bmatrix}$ **57** $X = \begin{bmatrix} -2 & 21 & 16 \\ 9 & -1 & -7 \end{bmatrix}$

59 $A + B = \begin{bmatrix} 9.552 & 1.820 & 5.583 \\ 11.689 & 10.586 & -0.087 \\ 11.149 & -3.771 & -6.033 \end{bmatrix}$

61 $A \cdot B = \begin{bmatrix} 24.26 & -13.93 & 10.25 \\ 99.35 & 47.19 & -8.47 \\ -30.45 & 26.52 & -6.65 \end{bmatrix}$

Exercises 7-2

1 $\begin{bmatrix} 2 & 0 & 1 \\ 0 & 1 & -1 \\ 1 & 1 & 0 \end{bmatrix} \begin{bmatrix} x \\ y \\ z \end{bmatrix} = \begin{bmatrix} 5 \\ 1 \\ 2 \end{bmatrix}$

3 $\begin{cases} x - z = -2 \\ 3x + 4y - 3z = 6 \\ y + 2z = -6 \end{cases}$

5 $AB = BA = I$; thus $B = A^{-1}$

7 $AB \neq I$; thus $B \neq A^{-1}$ **9** $\begin{bmatrix} 7 & -2 \\ -3 & 1 \end{bmatrix}$

11 $\begin{bmatrix} -1 & 2 \\ 1 & -\frac{3}{2} \end{bmatrix}$ **13** No inverse

15 $\begin{bmatrix} 3 & -3 & 1 \\ -3 & 5 & -2 \\ 1 & -2 & 1 \end{bmatrix}$

17 $\begin{bmatrix} 1 & -1 & 1 \\ 0 & \frac{1}{2} & -2 \\ -1 & \frac{1}{2} & 2 \end{bmatrix}$ **19** No inverse

21 $(9, -4)$ **23** $(14, -12)$ **25** $(-6, 5, -1)$
27 $(-3, \frac{17}{2}, -\frac{19}{2})$ **29** $(28, -\frac{40}{3}, -12)$
31 $(\frac{41}{16}, -\frac{5}{16}, -\frac{49}{16})$
33 a. $(-0.2, 2.7)$ **b.** $(1.6, -1.1)$ **c.** $(0, -1.5)$
35 See *Student Solutions Manual*

37 $\begin{bmatrix} 16 & 30 & 28 \\ 10 & 25 & 20 \\ 5 & 10 & 10 \end{bmatrix}$

39
$$\begin{array}{cc} & \text{Shipment} \\ & \begin{array}{ccc} 1 & 2 & 3 \end{array} \\ \begin{array}{l} \text{Number of men's suits} \\ \text{Number of women's suits} \end{array} & \begin{bmatrix} 12 & 15 & 7 \\ 10 & 16 & 10 \end{bmatrix} \end{array}$$

Exercises 7-3

1 a. 24   **b.** 42   **c.** -23 **d.** 45
3 a. Minors: $-3, -3, 3$
      Cofactors: $3, -3, -3$
   b. 12
5 a. Minors: $3, -3, 0$
      Cofactors: $3, 3, 0$
   b. 12
7 290   **9** -255 **11** 20   **13** 0
15 $(\frac{89}{83}, \frac{25}{249})$ **17** $(\frac{19}{29}, -\frac{2}{29})$ **19** $(\frac{200}{23}, \frac{390}{23})$
21 $(-8, 14, -5)$ **23** $(\frac{13}{5}, \frac{2}{5}, 0)$
25 a. 0   **b.** 0
   c. The determinant of a zero matrix is zero.
27 a. $a_{11}a_{22}$ **b.** $a_{11}a_{22}$ **c.** $a_{11}a_{22}a_{33}$ **d.** $a_{11}a_{22}a_{33}$
   e. The determinant of a triangular matrix is the product of the entries on the main diagonal.
29 $x = 1$ **31** $x = \frac{1}{2}, x = 2$ **33** -6 **35** 30
37 $(5.1, 6.2)$
39 a. $|A| = -2$ **b.** -4 **c.** 6   **d.** -8 **e.** 10
   f. The determinant is k times the original determinant.

41 a. $|A| = -2$ b. -2 c. -2 d. -2 e. -2
   f. The value of this determinant does not change.

Exercises 7-4

1 $\begin{bmatrix} 1 & 2 & 5 \\ 0 & -1 & -9 \\ 0 & -2 & -23 \end{bmatrix}$ **3** $\begin{bmatrix} 1 & 0 & 0 \\ 3 & -1 & -9 \\ 5 & -2 & -23 \end{bmatrix}$

5 $\begin{bmatrix} 23 & 5 & -1 \\ 0 & -1 & 0 \\ 19 & 4 & -14 \end{bmatrix}$ **7** 18 **9** -18 **11** -54

13 18 **15** 130 **17** -260 **19** -130
21 0 **23** 0 **25** 0 **27** -130 **29** -50
31 0 **33** -90 **35** $(2, -2, 0)$
37 $\left(-\frac{38}{3}, \frac{43}{3}, \frac{55}{3}\right)$ **39** $x = 3$ **41** $x = 6, x = 1$
43 -1779 **45** -42 **47** -162
49 $\begin{vmatrix} a_{21} & a_{22} \\ a_{11} & a_{12} \end{vmatrix} = a_{21}a_{12} - a_{11}a_{22} = -|A|$

19 $2x + 5y - 4z = 12$
        $6x - y + 2z = -10$
        $x + 3y + 5z = 5$

20 $\frac{1}{15}\begin{bmatrix} 1 & -6 \\ 2 & 3 \end{bmatrix}$ **21** $\begin{bmatrix} 1 & -1.5 & 0.5 \\ 0 & 0.5 & -0.1 \\ 0 & 0 & 0.2 \end{bmatrix}$

22 $\left(\frac{16}{7}, -\frac{3}{7}\right)$ **23** $\left(-\frac{20}{17}, \frac{69}{17}, -\frac{7}{17}\right)$

24

	Week			
	1	2	3	4
Regular	10	6	0	8
Laminated	0	12	20	10
Inlaid	8	6	4	6

25 27 **26** $\left(-\frac{4}{13}, \frac{45}{13}\right)$
27 a. $-10, -10; 19, -19; 8, 8$ b. $19, -19; 11, 11; -2, 2$
   c. 23 d. 23
28 -8 **29** -16 **30** -8 **31** 8 **32** -8
33 8 **34** 24 **35** 0 **36** 57 **37** -150
38 0 **39** $a_{11}a_{22}a_{33}$ **40** 0 **41** 452
42 -94 **43** $(-5, 4, 6)$
44 $7x^2 + 7y^2 - 33x - 31y + 50 = 0$

Review Exercises for Chapter 7

1 A is 2×4, B is 2×4, C is 4×1
2 $a_{12} = -1, b_{23} = -5, c_{31} = -1$
3 $\begin{bmatrix} 5 & -2 & 0 & 3 \\ 2 & 6 & -1 & -4 \end{bmatrix}$ **4** $\begin{bmatrix} -1 & 0 & 8 & 3 \\ -2 & 4 & 9 & -2 \end{bmatrix}$
5 $\begin{bmatrix} 0 & -1 & 20 & 9 \\ -4 & 13 & 22 & -7 \end{bmatrix}$ **6** 18
7 Undefined **8** $\begin{bmatrix} -9 & -21 \\ -20 & -12 \end{bmatrix}$ **9** B
10 $(3, 5)(4, 5)$ **11** $a = 3, b = 4$ **12** $B \neq A^{-1}$

13 $\begin{bmatrix} 44 \\ 33 \\ 26 \\ 21 \end{bmatrix} \begin{matrix} C \\ F \\ V \\ H \end{matrix}$ **14** $\begin{bmatrix} 6 \\ 5 \\ 4 \\ 3 \end{bmatrix} \begin{matrix} C \\ F \\ V \\ H \end{matrix}$

15 $5D = \begin{bmatrix} 125 \\ 95 \\ 75 \\ 60 \end{bmatrix} \begin{matrix} C \\ F \\ V \\ H \end{matrix}$ $5N = \begin{bmatrix} 95 \\ 70 \\ 55 \\ 45 \end{bmatrix} \begin{matrix} C \\ F \\ V \\ H \end{matrix}$

16 71 days, 53 nights
17 \$28,475 wholesale, \$32,950 retail
18 $\begin{bmatrix} 3 & -2 & 1 \\ 6 & 9 & -5 \\ 2 & -5 & 3 \end{bmatrix} \begin{bmatrix} x \\ y \\ z \end{bmatrix} = \begin{bmatrix} -2 \\ 4 \\ 3 \end{bmatrix}$

Optional Exercises for Calculus-Bound Students for Chapter 7

1 $A = \begin{bmatrix} 0 & 0 & 1 \\ 0 & 0 & 0 \\ 0 & 0 & 0 \end{bmatrix}$, $B = \begin{bmatrix} 0 & 0 & 0 \\ 0 & 0 & 0 \\ 0 & 1 & 0 \end{bmatrix}$

2 $\begin{bmatrix} 1 & 2 \\ 3 & 6 \end{bmatrix} \cdot \begin{bmatrix} -2 & 2 \\ 1 & -1 \end{bmatrix} = \begin{bmatrix} 0 & 0 \\ 0 & 0 \end{bmatrix}$ (Answers may vary.)

3 a. $\begin{bmatrix} 1 & -1 \\ 1.5 & -2 \end{bmatrix}$ b. $\begin{bmatrix} 1 & -2 \\ -\frac{1}{3} & 1 \end{bmatrix}$
   c. $\begin{bmatrix} 10 & 18 \\ 7 & 12 \end{bmatrix}$ d. $\begin{bmatrix} -2 & 3 \\ \frac{7}{6} & -\frac{5}{3} \end{bmatrix}$ e. $\begin{bmatrix} -2 & 3 \\ \frac{7}{6} & -\frac{5}{3} \end{bmatrix}$
   f. $\begin{bmatrix} \frac{4}{3} & -3 \\ \frac{13}{6} & -5 \end{bmatrix}$ g. $(AB)^{-1} = B^{-1}A^{-1}$
   $(AB)^{-1} \neq A^{-1}B^{-1}$

4 $\begin{bmatrix} -1 & -5 & 4 \\ -0.5 & -3.5 & 2.5 \\ 0.5 & 0.5 & -0.5 \end{bmatrix}$

5 $(1, 1, -1)$ **6** $\begin{bmatrix} 2 & -1 \\ -\frac{3}{2} & 1 \end{bmatrix}$

ANSWERS

7 $\begin{bmatrix} a_1 & b_1 \\ c_1 & d_1 \end{bmatrix} + \begin{bmatrix} a_2 & b_2 \\ c_2 & d_2 \end{bmatrix} = \begin{bmatrix} a_1 + a_2 & b_1 + b_2 \\ c_1 + c_2 & d_1 + d_2 \end{bmatrix}$

$= \begin{bmatrix} a_2 + a_1 & b_2 + b_1 \\ c_2 + c_1 & d_2 + d_1 \end{bmatrix}$

$= \begin{bmatrix} a_2 & b_2 \\ c_2 & d_2 \end{bmatrix} + \begin{bmatrix} a_1 & b_1 \\ c_1 & d_1 \end{bmatrix}$

8 $\left(\begin{bmatrix} a_1 & b_1 \\ c_1 & d_1 \end{bmatrix} + \begin{bmatrix} a_2 & b_2 \\ c_2 & d_2 \end{bmatrix}\right) + \begin{bmatrix} a_3 & b_3 \\ c_3 & d_3 \end{bmatrix}$

$= \begin{bmatrix} a_1 + a_2 & b_1 + b_2 \\ c_1 + c_2 & d_1 + d_2 \end{bmatrix} + \begin{bmatrix} a_3 & b_3 \\ c_3 & d_3 \end{bmatrix}$

$= \begin{bmatrix} (a_1 + a_2) + a_3 & (b_1 + b_2) + b_3 \\ (c_1 + c_2) + c_3 & (d_1 + d_2) + d_3 \end{bmatrix}$

$= \begin{bmatrix} a_1 + (a_2 + a_3) & b_1 + (b_2 + b_3) \\ c_1 + (c_2 + c_3) & d_1 + (d_2 + d_3) \end{bmatrix}$

$= \begin{bmatrix} a_1 & b_1 \\ c_1 & d_1 \end{bmatrix} + \begin{bmatrix} a_2 + a_3 & b_2 + b_3 \\ c_2 + c_3 & d_2 + d_3 \end{bmatrix}$

$= \begin{bmatrix} a_1 & b_1 \\ c_1 & d_1 \end{bmatrix} + \left(\begin{bmatrix} a_2 & b_2 \\ c_2 & d_2 \end{bmatrix} + \begin{bmatrix} a_3 & b_3 \\ c_3 & d_3 \end{bmatrix}\right)$

9 a. Area of a trapezoid $= \frac{1}{2}h(b_1 + b_2)$
Area $ADEC = \frac{1}{2}(x_3 - x_1)(y_1 + y_3)$
$= \frac{1}{2}(x_3 y_1 + x_3 y_3 - x_1 y_1 - x_1 y_3)$
Area $CEFB = \frac{1}{2}(x_2 - x_3)(y_2 + y_3)$
$= \frac{1}{2}(x_2 y_2 + x_2 y_3 - x_3 y_2 - x_3 y_3)$
Area $ADFB = \frac{1}{2}(x_2 - x_1)(y_1 + y_2)$
$= \frac{1}{2}(x_2 y_1 + x_2 y_2 - x_1 y_1 - x_1 y_2)$
Area A = Area $ADEC$ + Area $CEFB$ − Area $ADFB$
$A = \frac{1}{2}(x_3 y_1 - x_1 y_3 + x_2 y_3 - x_3 y_2 - x_2 y_1 + x_1 y_2)$
$A = \frac{1}{2}[(x_2 y_3 - x_3 y_2) - (x_1 y_3 - x_3 y_1) + (x_1 y_2 - x_2 y_1)]$

$A = \frac{1}{2}\begin{vmatrix} x_1 & y_1 & 1 \\ x_2 & y_2 & 1 \\ x_3 & y_3 & 1 \end{vmatrix}$ (expanded about column three)

b. The determinant could be negative.

10 Let (x, y) be a point on the line containing (x_1, y_1) and (x_2, y_2). Then the three points form a degenerate triangle with area equal to zero. Thus from Exercise 9,

$\begin{vmatrix} x & y & 1 \\ x_1 & y_1 & 1 \\ x_2 & y_2 & 1 \end{vmatrix} = 0$

11 $\begin{vmatrix} a_{11} & a_{12} & a_{13} \\ a_{21} & a_{22} & a_{23} \\ a_{31} + ka_{21} & a_{32} + ka_{22} & a_{33} + ka_{23} \end{vmatrix}$

$= (a_{31} + ka_{21})A_{31} + (a_{32} + ka_{22})A_{32}$
$\quad + (a_{33} + ka_{23})A_{33}$
$= a_{31}A_{31} + a_{32}A_{32} + a_{33}A_{33} + k(a_{21}A_{31}$
$\quad + a_{22}A_{32} + a_{23}A_{33})$

$= |A| + k\left(a_{21}\begin{vmatrix} a_{12} & a_{13} \\ a_{22} & a_{23} \end{vmatrix} - a_{22}\begin{vmatrix} a_{11} & a_{13} \\ a_{21} & a_{23} \end{vmatrix}\right.$

$\left. + a_{23}\begin{vmatrix} a_{11} & a_{12} \\ a_{21} & a_{22} \end{vmatrix}\right)$

$= |A| + k(a_{12}a_{21}a_{23} - a_{13}a_{21}a_{22} - a_{11}a_{22}a_{23}$
$\quad + a_{13}a_{21}a_{22} + a_{11}a_{22}a_{23} - a_{12}a_{21}a_{23})$
$= |A| + k(0)$
$= |A|$

12 $(34, \frac{45}{2}, -\frac{7}{2})$ **13** $x = -1$ or $x = \frac{5}{2}$

Mastery Test for Chapter 7

1 a. $A + B = \begin{bmatrix} 3 & 4 & -3 \\ 4 & 2 & 3 \end{bmatrix}$

b. $A - B = \begin{bmatrix} -1 & 6 & -9 \\ -4 & 12 & -11 \end{bmatrix}$

c. $-2A = \begin{bmatrix} -2 & -10 & 12 \\ 0 & -14 & 8 \end{bmatrix}$

d $3A - 4B = \begin{bmatrix} -5 & 19 & -30 \\ -16 & 41 & -40 \end{bmatrix}$

e. $A \cdot C = \begin{bmatrix} 9 & 30 \\ 4 & 26 \end{bmatrix}$ **f.** $B^t = \begin{bmatrix} 2 & 4 \\ -1 & -5 \\ 3 & 7 \end{bmatrix}$

2 $A^{-1} = \begin{bmatrix} -4 & 1 & \frac{3}{2} \\ 2 & 0 & -\frac{1}{2} \\ -1 & 1 & \frac{1}{2} \end{bmatrix}$ **3** $(\frac{7}{3}, \frac{1}{9})$ **4** -23

5 $(-\frac{180}{13}, -\frac{46}{13}, -\frac{109}{13})$ **6 a.** 0 **b.** 309 **c.** -23

CHAPTER EIGHT

Exercises 8-1

1 7, 5, 3, 1, -1 **3** 0, 1, 10, 33, 76
5 $-2, 4, -8, 16, -32$ **7** $\frac{1}{6}, \frac{1}{11}, \frac{1}{16}, \frac{1}{21}, \frac{1}{26}$
9 $\frac{1}{2}, -\frac{2}{3}, \frac{3}{4}, -\frac{4}{5}, \frac{5}{6}$ **11** 2, 5, 8, 11, 14
13 1, 2, 2, 4, 8 **15** $\frac{1687}{1688}$ **17** $-1,000,000$
19 80 **21** 77 **23** 35 **25** 120 **27** 0.11111
29 3000 **31** 6

33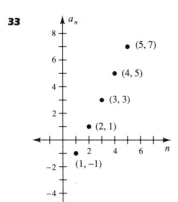

35 $\sum_{i=0}^{6} 2(i+1) = 2(1) + 2(2) + 2(3) + 2(4) + 2(5) + 2(6) + 2(7)$
$= \sum_{j=1}^{7} 2j$

37 $s_1 = 1, s_2 = \frac{5}{4}, s_3 = \frac{49}{36}, s_4 = \frac{205}{144}$

39 $s_1 = \frac{2}{5}, s_2 = \frac{14}{25}, s_3 = \frac{78}{125}, s_4 = \frac{406}{625}$

41 $\sum_{i=1}^{5} 5i$ **43** $\sum_{i=7}^{11} i^2$ **45** $\sum_{i=1}^{7} (-1)^i (2i+1)$

47 A parabola; $a_n = n^2 - 7n + 10$

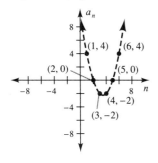

49 $s_1 = \frac{1}{3}, s_2 = \frac{4}{9}, s_3 = \frac{13}{27}, s_4 = \frac{40}{81}$
51 $20,000; $19,950; $19,899.63; $19,848.88; $19,797.75
53 78.24 meters   **55** 2.7180556

Exercises 8-2

1 a. Arithmetic, $d = 6$ **b.** Not arithmetic
  c. Arithmetic, $d = -5$ **d.** Arithmetic, $d = 2$
3 5, 14, 23, 32, 41   **5** 6, $5\frac{1}{3}$, $4\frac{2}{3}$, 4, $3\frac{1}{3}$
7 9, 7, 5, 3, 1   **9** 8, 11, 14, 17, 20
11 $\frac{1}{3}$, 1, $1\frac{2}{3}$, $2\frac{1}{3}$, 3   **13** 8424   **15** 131,352
17 43,764   **19** 18,834   **21** 124   **23** -800
25 39   **27** 52   **29** 1640   **31** -8 **33** 8

35 9   **37** 44   **39** 6, 13, 20, 27, 34   **41** 319
43 $162,500   **45** 1071 centimeters   **47** 2601
49 33.499   **51** Yes; $d = \log 3$

Exercises 8-3

1 a. Geometric, $r = 10$ **b.** Geometric, $r = -5$
  c. Not geometric   **d.** Geometric, $r = 0.01$
3 a. Geometric, $r = 3$ **b.** Not geometric
  c. Geometric, $r = \sqrt{2}$ **d.** Geometric, $r = -1$
5 5, 20, 80, 320, 1280   **7** 7, -14, 28, -56, 112
9 $-36, 24, -16, \frac{32}{3}, -\frac{64}{9}$
11 0.15, 0.0015, 0.000015, 0.000 000 15, 0.000 000 0015
13 $-\frac{2}{3}, \frac{4}{9}, -\frac{8}{27}, \frac{16}{81}, -\frac{32}{243}$
15 3, 15, 75, 375, 1875, or 3, -15, 75, -375, 1875
17 $\sqrt{3}, 3, 3\sqrt{3}, 9, 9\sqrt{3}$ **19** 3069
21 $S_9 \approx 7.372$ **23** 4094
25 $s_1 = \frac{3}{8} = 0.375; s_2 = \frac{33}{64} \approx 0.516; s_3 = \frac{291}{512} \approx 0.568;$
  $S = \frac{3}{5} = 0.6$
27 $s_1 = -\frac{5}{8} = -0.625; s_2 = -\frac{15}{64} \approx -0.234;$
  $s_3 = -\frac{245}{512} \approx -0.479; S = -\frac{7}{23} \approx -0.304$
29 -243 **31** 12   **33** $-\frac{1}{5}$ or $\frac{1}{5}$ **35** 0.2
37 5.375   **39** $-\frac{9}{4}$ or $\frac{5}{4}$ **41** $\frac{9}{11}$ **43** $\frac{7}{9}$
45 $\frac{1}{1}$ or 1   **47** $\frac{377}{30}$ **49** ± 20 **51** $-18, 12$
53 $10,485.76   **55** $42,949,672.95
57 13.2 centimeters   **59** 31
61 40 centimeters $(160 + 80\sqrt{2})$ centimeters
63 1.01, $1268.25
65 $1 + \frac{r}{100}, -\frac{100P}{r}\left[1 - \left(1 + \frac{r}{100}\right)^n\right]$ **67** 1.4284385
69 For $k \neq 0, r' = \frac{ka_i}{ka_{i-1}} = r$; thus the sequences have the same ratio. For $k = 0$, the new geometric sequence is $0, 0, 0, \cdots, 0$.

Exercises 8-4

1 $S(1): 3 \cdot 1 - 2 = \frac{3 \cdot 1^2 - 1}{2}$
  $S(2): (3 \cdot 1 - 2) + (3 \cdot 2 - 2) = \frac{3 \cdot 2^2 - 2}{2}$
  $S(3): (3 \cdot 1 - 2) + (3 \cdot 2 - 2) + (3 \cdot 3 - 2) = \frac{3 \cdot 3^2 - 3}{2}$
  $S(4): (3 \cdot 1 - 2) + (3 \cdot 2 - 2) + (3 \cdot 3 - 2) + (3 \cdot 4 - 2)$
  $= \frac{3 \cdot 4^2 - 4}{2}$

ANSWERS

3 $S(1)$: $1(2 \cdot 1 + 1) = \dfrac{4 \cdot 1^3 + 9 \cdot 1^2 + 5 \cdot 1}{6}$

$S(2)$: $1(2 \cdot 1 + 1) + 2(2 \cdot 2 + 1) = \dfrac{4 \cdot 2^3 + 9 \cdot 2^2 + 5 \cdot 2}{6}$

$S(3)$: $1(2 \cdot 1 + 1) + 2(2 \cdot 2 + 1) + 3(2 \cdot 3 + 1)$
$= \dfrac{4 \cdot 3^3 + 9 \cdot 3^2 + 5 \cdot 3}{6}$

$S(4)$: $1(2 \cdot 1 + 1) + 2(2 \cdot 2 + 1) + 3(2 \cdot 3 + 1) + 4(2 \cdot 4 + 1)$
$= \dfrac{4 \cdot 4^3 + 9 \cdot 4^2 + 5 \cdot 4}{6}$

5 $S(1)$: $1 < 2^1$
$S(2)$: $2 < 2^2$
$S(3)$: $3 < 2^3$
$S(4)$: $4 < 2^4$

7 $S(k)$: $2^1 + 2^2 + \cdots + 2^k = 2^{k+1} - 2$
$S(k+1)$: $2^1 + 2^2 + \cdots + 2^{k+1} = 2^{k+2} - 2$

9 $S(k)$: $(1^2 - 1 + 3) + (2^2 - 2 + 3) + \cdots + (k^2 - k + 3)$
$= \dfrac{k^3 + 8k}{3}$

$S(k+1)$: $(1^2 - 1 + 3) + (2^2 - 2 + 3) + \cdots$
$+ [(k+1)^2 - (k+1) + 3] = \dfrac{(k+1)^3 + 8(k+1)}{3}$

11–23 See the *Student Solutions Manual*.
25 To knock over each domino requires:
a. Initialization: You must be able to knock over the first domino.
b. Perpetuation: You must be able to perpetuate the fall by having each domino knock over its successor.
27 F **29** No, only true for $n = 1$ and $n = 2$.
31 No, this statement is false for $n = 41$.
33 No; $x^2 - 1 = (x - 1)(x + 1)$ is not a statement about natural numbers.
35–45 See the *Student Solutions Manual*.

Exercises 8-5

1 6 **3** 40,320 **5** 24 **7** 1440 **9** 5040
11 72 **13** 120 **15** 1 **17** 126 **19** 1
21 $x^4 + 12x^3 + 54x^2 + 108x + 81$
23 $32x^5 - 80x^4y + 80x^3y^2 - 40x^2y^3 + 10xy^4 - y^5$
25 $v^8 - 20v^6w + 150v^4w^2 - 500v^2w^3 + 625w^4$
27 $-119 + 120i$ **29** $19{,}448v^{10}w^7$
31 $-167{,}960x^{11}y^9$ **33** $16{,}016m^{12}n^{10}$
35 $-13{,}650{,}000a^{20}b^{11}$ **37** $8064x^5y^5$ **39** $-14{,}784$
41 1.771561 **43** 4
45 $r^{16} + 16r^{15}s + 120r^{14}s^2 + 560r^{13}s^3 + \cdots$
47 $4{,}782{,}969m^{14} - 22{,}320{,}522m^{13}n + 48{,}361{,}131m^{12}n^2$
$- 64{,}481{,}508m^{11}n^3 + \cdots$
49 $6561 + 87{,}480i - 510{,}300 - 1{,}701{,}000i + \cdots$

51 $x^4\sqrt{x} - 9x^4y + 36x^3\sqrt{xy^2} - 84x^3y^3 + \cdots$
53 See the *Student Solutions Manual*.

Exercises 8-6

1 a. 720 **b.** 360 **c.** 1 **d.** 15
3 a. 7 **b.** 7 **c.** 1 **d.** 1
5 a. 28 **b.** 28 **c.** 70 **d.** 1
7 Permutations: ab, ac, ad, ba, bc, bd, ca, cb, cd, da, db, dc; combinations: ab, ac, ad, bc, bd, cd
9 Chicago is C, Washington is W; 8 possible choices

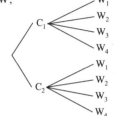

11 Main dinner is m, dessert is d; 15 possible choices

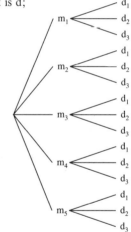

13 124,251,000 ways **15** 84 boots **17** 1024 ways
19 5040 codes **21** 75,287,520 quizzes
23 10,000 selections **25** 6 ways **27** 120 selections
29 45 lines **31** 15 lines, 9 of which are diagonal
33 24 ways **35** 1287 ways **37** 4 ways
39 635,013,559,600 bridge hands
41 $P(n, 1) = \dfrac{n!}{(n-1)!} = \dfrac{n(n-1)!}{(n-1)!} = n$

43 $P(n, n-1) = \dfrac{n!}{[n - (n-1)]!} = n! = P(n, n)$

45 $C(n, r) = \dfrac{n!}{r!(n-r)!} = \dfrac{n!}{(n-r)!r!} = C(n, n-r)$

47 $C(n-1, r-1) + C(n-1, r)$

$$= \frac{(n-1)!}{(r-1)![(n-1)-(r-1)]!} + \frac{(n+1)!}{r!(n-1-r)!}$$

$$= \frac{(n-1)!}{(r-1)!(n-r)!} + \frac{(n-1)!}{r!(n-1-r)!}$$

$$= \frac{r(n-1)! + (n-r)(n-1)!}{r!(n-r)!}$$

$$= \frac{n(n-1)!}{r!(n-r)!} = \frac{n!}{r!(n-r)!} = C(n, r)$$

49 2^{15} numbers = 32,768 numbers

Exercises 8-7

1 Boy, boy; boy, girl; girl, boy; girl, girl

3 {TTTT, TTTF, TTFT, TFTT, FTTT, TTFF, TFTF, TFFT, FTTF, FTFT, FFTT, FFFT, FFTF, FTFF, TFFF, FFFF}

5 Model is M, left is L, right is R:
{$(M_1, L), (M_2, L), (M_3, L), (M_4, L), (M_5, L), (M_1, R), (M_2, R), (M_3, R), (M_4, R), (M_5, R)$}

7 $\begin{cases} \text{W W W W W} & \text{R R R R R} & \text{B B B B B} \\ \text{W R B G Y} & \text{W R B G Y} & \text{W R B G Y} \\ \text{G G G G G} & \text{Y Y Y Y Y} & \\ \text{W R B G Y} & \text{W R B G Y} & \end{cases}$

9 $\frac{1}{4}$ **11** $\frac{3}{16}$ **13** $\frac{7}{16}$ **15** $\frac{1}{1024}$ **17** $\frac{781}{1024}$
19 $\frac{1}{2}$ **21** $\frac{1}{4}$ **23** $\frac{1}{32}$ **25** $\frac{31}{32}$ **27** $\frac{1}{32}$
29 $\frac{3}{4}$ **31** $\frac{1}{9}$ **33** $\frac{1}{6}$ **35** $\frac{1}{36}$ **37** 0
39 $\frac{1}{12}$ **41** $\frac{7}{36}$ **43** $\frac{1}{13}$ **45** $\frac{1}{26}$ **47** $\frac{7}{13}$
49 $\frac{1}{4}$ **51** $\frac{1}{2}$ **53** $\frac{6}{13}$ **55** $\frac{2}{13}$ **57** 0
59 $\frac{1}{5}$ **61** $\frac{1}{30}$ **63** $\frac{1}{2530}$ **65** $\frac{1561}{2530}$
67 $\frac{1}{7,059,052}$

Review Exercises for Chapter 8

1 $1, \frac{1}{3}, \frac{1}{9}, \frac{1}{27}, \frac{1}{81}$ **2** $\frac{2}{2}, \frac{1}{8}, \frac{1}{26}, \frac{1}{80}, \frac{1}{242}$
3 $-4, 5, -6, 7, -8$ **4** $3, 5, 2, -3, -5$
5 $3, 5, 7, 9, 11$ **6** $1000, 1100, 1210, 1331, 1464.1$
7 $2, 7, 1, 8, 2$ **8** $1, 2, 6, 24, 120; a_n = n$
9 Arithmetic **10** Geometric **11** Geometric
12 Both **13** Neither **14** 7 **15** 133
16 -39 **17** 148 **18** 474 **19** 3525 **20** $\frac{4}{3}$
21 25,005,000 **22** 0.44444 **23** 11 **24** 21
25 6 **26** 6 **27** -1 **28** $\pm\frac{1}{2}$ **29** 1944
30 6 **31** $x^4 - 12x^3y + 54x^2y^2 - 108xy^3 + 81y^4$

32 $64x^{12} + 960x^{10}y + 6000x^8y^2 + 20{,}000x^6y^3 + 37{,}500x^4y^4 + 37{,}500x^2y^5 + 15{,}625y^6$
33 720 **34** 600 **35** 28 **36** 1 **37** 220
38 $\frac{9}{19}$ **39** $\frac{1}{38}$ **40** $\frac{9}{19}$ **41** $\frac{1}{19}$ **42** $\frac{1}{1444}$
43 $\frac{81}{361}$ **44** $\frac{81}{361}$ **45** $\frac{1}{6}$ **46** $\frac{1}{6}$ **47** $\frac{1}{36}$
48 $\frac{11}{36}$ **49** $\frac{7}{36}$
50 28 lines, 20 of which are diagonals
51 12 possible routes

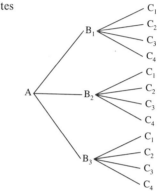

52 157,286.4 millimeters
53 {21, 23, 25, 27, 29, 31, 33, 35, 37, 39, 51, 53, 55, 57, 59, 71, 73, 75, 77, 79}
54 $100(1.08), 100(1.08)^2, 100(1.08)^3, 100(1.08)^4, 100(1.08)^5$
55 520 seats **56** 90, 97, 104, 111, 118
57 $\frac{9}{25}, \frac{27}{125}, \frac{81}{625}$ or $-\frac{9}{25}, \frac{27}{125}, -\frac{81}{625}$ **58** 80 meters
59–60 See the *Student Solutions Manual*.

Mastery Test for Chapter 8

1 a. 8, 11, 14, 17, 20 **b.** 4, -8, 16, -32, 64
  c. 0, $\frac{1}{3}, \frac{1}{2}, \frac{3}{5}, \frac{2}{3}$ **d.** 4, -1, -4, 11, -20
  e. 5, 7, 11, 19, 35 **f.** -2, 10, 70, 4690, 21,982,030
2 a. 33 **b.** -10 **c.** 880 **d.** $-\frac{37}{60}$
3 a. 197 **b.** -3 **c.** 5.5 **d.** 39
4 a. 1353 **b.** 6345 **c.** 20,916 **d.** -40
5 a. $\frac{64}{9}$ **b.** ± 0.1 **c.** ± 2 **d.** 3
6 a. 3.33333 **b.** 0 **c.** $-\frac{165}{32}$ **d.** 605
7 a. $\frac{6}{11}$ **b.** $\frac{26}{9}$
8 Show that $S_n: \sum_{i=1}^{n} (2i - 1)^2 = \frac{n(2n + 1)(2n - 1)}{3}$ is true for all natural numbers n.

I. $S(1)$: $[2(1) - 1]^2 \stackrel{?}{=} \frac{[2(1) + 1][2(1) - 1]}{3}$

$$1 = \frac{3(1)}{3}$$

Thus $S(1)$ is true.

II. If $S(k)$ is true, that is,

$$[2(1) - 1]^2 + [2(2) - 1]^2 + [2(3) - 1]^2 + \cdots + (2k - 1)^2$$
$$= \frac{k(2k + 1)(2k - 1)}{3}$$

is true, then show $S(k + 1)$ is true.

$$[2(1) - 1]^2 + [2(2) - 1]^2 + [2(3) - 1]^2 + \cdots + (2k - 1)^2 + [2(k + 1) - 1]^2$$
$$= \frac{k(2k + 1)(2k - 1)}{3} + [2(k + 1) - 1]^2$$

$$[2(1) - 1]^2 + [2(2) - 1]^2 + [2(3) - 1]^2 + \cdots + (2k - 1)^2 + (2k + 1)^2$$
$$= \frac{k(2k + 1)(2k - 1)}{3} + (2k + 1)^2$$

$$[2(1) - 1]^2 + [2(2) - 1]^2 + [2(3) - 1]^2 + \cdots + (2k - 1)^2 + (2k + 1)^2$$
$$= (2k + 1)\frac{k(2k - 1) + 3(2k + 1)}{3}$$

$$[2(1) - 1]^2 + [2(2) - 1]^2 + [2(3) - 1]^2 + \cdots + (2k - 1)^2 + (2k + 1)^2$$
$$= \frac{(2k + 1)(2k^2 + 5k + 3)}{3}$$

$$[2(1) - 1]^2 + [2(2) - 1]^2 + [2(3) - 1]^2 + \cdots + (2k - 1)^2 + (2k + 1)^2$$
$$= \frac{(k + 1)(2k + 3)(2k + 1)}{3}$$

$$[2(1) - 1]^2 + [2(2) - 1]^2 + [2(3) - 1]^2 + \cdots + (2k - 1)^2 + [2(k + 1) - 1]$$
$$= \frac{[k + 1][2(k + 1) + 1)][2(k + 1) - 1]}{3}$$

By parts I and II, $S(n)$ is true for all natural numbers n.
9 495 **10** $40,095x^8y^4$ **11** 66 cans
12 a. 12,497,500 selections **b.** 665,280 alignments
13 {WWWW, WWWL, WWLW, WLWW, LWWW, WWLL, WLLW, LLWW, WLWL, LWLW, LWWL, LLLW, LLWL, LWLL, WLLL, LLLL}
14 E = {WWWL, WWLW, WLWW, LWWW};
$P(E) = \frac{4}{16} = \frac{1}{4}$
15 a. 1 **b.** 0 **c.** $\frac{1}{6}$ **d.** $\frac{5}{6}$ **e.** $\frac{1}{2}$ **f.** $\frac{1}{3}$ **g.** $\frac{1}{6}$ **h.** $\frac{2}{3}$

Index

a_{ij}, entry, 444
Abel, Niels Henrik, 263
Abscissa, 155
Absolute value
    equations, 130
    functions, 176
    inequalities, 130
    notation, 7
Accuracy, 23
Addends, 28
Addition method, 391
Additive identity, 13
Additive inverse, 13
Algebra, 2
Algebraic fraction, 44
Algorithm, division, 276
Amortization, 27
Analytic geometry, 154
Antilogarithm, 362
AOS, 547
Approximation of zeros, 313
    algebraic, 314
    geometric, 315
Arbitrary value, 390
Area
    of an ellipse, 238
    of triangle, 473, 484
Argument of a function, 151
Arithmetic
    mean, 498
    progression, 495
    sequence, 495
    series, 498, 500
Associative properties, 12
Asymptotes
    even, 321, 326
    of a graph, 320
    horizontal, 321, 324
    of a hyperbola, 238
    oblique, 334
    odd, 321, 326
    vertical, 321

Augmented matrix, 396
Average cost, 187, 214
Avogadro's number, 26
Axiom, 11, 389
Axis, 69
    of an ellipse, 231
    of a parabola, 243

Back-substitution, 391
Base
    change of, 365
    of an exponential function, 342
    of a logarithm, 353
Binomial, 28
    coefficient, 524, 532
    expansion, 33, 521
    square of, 30
    theorem, 522
Bounded intervals, 6
Bounds on real zeros, 306
Break-even analysis, 410
Break-even point, 410
Break-even value, 410

Carbon-14 dating, 349
Cardan formula, 52
Cardano, Girolamo, 263
Cartesian coordinate system, 155
Catherine the Great, 2
Center
    of a circle, 229
    of an ellipse, 233
    of a hyperbola, 240
Change of base
    for exponents, 365
    for logarithms, 365
Circle, 229, 243
    center, 229
    radius, 229
    standard form, 229
Closure properties, 12
Coefficient, 28

    binomial, 524
    leading, 295
    matrix, 459
    numerical, 28
Cofactor, 466
Collinear, 203
Column matrix, 448
Combination, 530
Combined translations, 177
Common denominator, 46
Common difference, 495
Common factor, 36
Common logarithms, 359
Common ratio, 503
Commutative properties, 12
Complementary events, 539
Complete factorization theorem, 290
Completing the square, 41, 103
Complex fraction, 48
Complex number, 64
    conjugates, 67
    equality of, 65
    imaginary term, 64
    operations with, 65
    real term, 64
    standard form of, 64
Complex plane, 69
Complex rational expression, 48
Composite function, 208
Composite number, 5
Compound inequalities, 5, 127
Compound interest, 347
Concavity, 181
Conclusion, 14
Conditional equation, 77
Conics, 242, 246
    circle, 229, 243
    ellipse, 231, 244
    general form of, 247
    hyperbola, 238, 247
    parabola, 181, 242
    summary of, 242

INDEX

Conjugates
   complex numbers, 67
   radicals, 58
   zeros theorem, 287
Consistent system, 386
Constant function, 156
Constant matrix, 459
Constant of variation, 84
Constraints, 431
Continuous compounding formula, 349
Continuous decay, 373
Continuous growth, 373
Continuous interest, 349
Contradiction, 77
Coordinate, 2
Coordinate system, 155
Corner point, 431
Correspondence, 153
Counting formulas, 526
Covertices, 233
Cramer, Gabriel, 469
Cramer's rule, 469
CRAY computer, 26
Critical values, 134
Cube of a difference, 30
Cube of a sum, 30
Cube root, 52

Decay, 344, 373
Decibel, 377
Decompose, 212, 216
Decreasing function, 156
Defined recursively, 489
Degree
   of a monomial, 28
   of a polynomial, 28
del Ferro, Scipione, 263
Demand, 411
Dependent system, 386, 403
Dependent variable, 151
Depreciation, straight-line, 170
Depressed factor, 281
Descartes, Rene, 148, 154
Descartes' rule of signs, 305
Determinant, 465
   properties of, 475, 477
Dice, 537
Difference
   of two cubes, 36

   of two functions, 205
   of two matrices, 445
   of two squares, 36
Directly proportional, 84
Directrix, 243
Direct variation, 84
Discriminant, 106
Distinct solutions, 100
Distance
   between two points, 197
   as rate times time, 117
Distributive property, 12
Dividend, 277
Division
   algorithm, 276
   of polynomials, 31
   synthetic, 278
   by zero, 13, 81
Divisor, 277
Domain, 149, 154
Dot product, 448
Double root, 100
Dummy variable, 220

e, 2, 346
Earthquakes, 374
Echelon form, 398
Elementary row operations, 396
Elimination method, 391
Ellipse, 231
   center, 233
   covertices, 231
   foci, 231
   major axis, 231
   minor axis, 231
   standard form, 233
   vertices, 231
Ellipsis notation, 3
Empty set, 78
ENIAC computer, 26
Equal functions, 207
Equality, properties of, 11
Equal matrices, 445
Equations, 77
   absolute value, 130
   conditional, 77
   containing fractions, 80
   contradiction, 77
   equivalent, 78

   exponential, 369
   first-degree, 78
   identity, 77
   linear, 78
   logarithmic, 370
   quadratic, 99
   quadratic in form, 109
   radical, 81, 111
   systems of, 385
Equilibrium
   point, 411
   value, 411
Equivalent equations, 78
Equivalent inequalities, 126
Equivalent systems, 391
Éuler, Leonard, 2
Even asymptote, 321, 326
Even function, 194
Even intercept, 266
Even multiplicity, 326
Events, 536
   complementary, 539
   independent, 539
Excluded value, 44, 136
Expansion by cofactors, 467
Exponential decay, 344
Exponential equation, 369
Exponential function, 342
Exponential growth, 344
Exponential notation, 18
Exponents
   integer, 18
   irrational, 343
   natural number, 18
   negative, 18
   properties of, 20
   rational, 54
   real, 343
   zero, 18
Extraction of roots, 100
Extraneous value, 81, 111

Factor
   depressed, 281
   greatest common, 36
   principle, 15, 134
   theorem, 281
   trivial, 36
Factorial notation, 520

Factoring, 36
  by completing the square, 37, 41
  by grouping, 37, 40
  by trial and error, 37, 38
  using special products, 36
Feasible region, 431
Ferrari, Ludovico, 263
Fibonacci sequence, 489
Finite sequence, 488
First-degree equation, 78
FOIL, 29
Focus
  of an ellipse, 233
  of a hyperbola, 238
  of a parabola, 243
Formula, 89, 153
  quadratic, 104
Function, 149, 150
  absolute value, 176
  argument of, 151
  composite, 208
  constant, 156
  decay, 344, 373
  decomposing, 212, 216
  decreasing, 156
  dependent variable, 151
  difference of, 205
  domain of, 149, 154
  equal, 207
  even, 194
  exponential, 342
  greatest integer, 198
  growth, 344, 373
  increasing, 156
  independent variable, 151
  INT, 199
  inverse, 217, 224
  linear, 163
  logarithmic, 226, 352
  machine, 153, 208
  mapping notation, 150
  odd, 194
  one-to-one, 217
  ordered-pair notation, 150
  parabola, 181
  piecewise, 171
  polynomial, 264
  product of, 206
  profit, 206
  quotient of, 206
  range of, 149
  rational, 318
  sequence, 488
  sum of, 205
  value of, 151
  vertical line test, 157
Functional notation, 151
Fundamental counting principle, 526
Fundamental rectangle, 239
Fundamental theorem of algebra, 290

Galois, Evariste, 445
Gauss, Carl Friedrich, 290, 487
General form
  of a conic section, 247
  of a linear equation, 163
General solution, 390
General term of a sequence, 488
Geometric approximations of zeros, 315
Geometric
  infinite series, 509
  mean, 505
  progression, 503
  sequence, 503
  series, 506
Graph
  of a linear function, 163
  of a polynomial function, 267
  of a rational function, 326
  of a sequence, 489, 490
  sign, 134, 265
Graphical method of solution, 385, 420
Graphics calculator, 200, 316, 420, 449, 459, 469
Greater than, 5
Greatest common factor, 36
Greatest integer function, 198
Grouping, factoring by, 40
Growth
  continuous, 373
  exponential, 344
  periodic, 373

Half-life, 376
Half-plane, 425
Heron of Alexandria, 263
Heron's formula, 144

Hole, 319
Hooke's Law, 85
Horizontal asymptote, 321, 324
Horizontal line, 165
Horizontal line test, 218
Horizontal translation, 177
Hyperbola, 238, 247
  asymptotes, 238
  center, 240
  foci, 238
  fundamental rectangle, 239
  standard form, 240
Hyperbolic cosine, 382
Hyperbolic sine, 382
Hypothesis, 14

i, 63
  powers of, 66
Identity
  additive, 13
  equation, 77
  logarithmic, 355
  matrix, 452
  multiplicative, 13
Imaginary axis, 69
Imaginary numbers, 63
Imaginary term, 64
Inconsistent system, 386, 403
Increasing function, 156
Independent events, 539
Independent system, 386
Independent variable, 151
Index
  of a radical, 52
  of a summation, 491
Inequalities, 5
  absolute value, 130
  compound 5, 127
  equivalent, 126
  linear, 126
  nonlinear, 134
  polynomial, 299
  properties of, 126
  systems of, 425
Infinite
  geometric series, 509
  interval, 6
  number of solutions, 387
  sequence, 488
  series, 492

INDEX

Infinite, (*continued*)
   set, 3
   sum, 492, 508
Infinity symbol, 6
Initialization, 514
Inner product, 448
INT, 199
Integers, 3
Intercepts
   even, 266
   odd, 266
Interest
   compound, 346
   continuous, 349
   simple, 93
Intersection, 3
Interval
   bounded, 6
   infinite, 6
   loss, 410
   notation, 6
   profit, 410
   unbounded, 6
Inverse
   additive, 13
   function, 217, 224
   of a matrix, 457
   multiplicative, 13
   relation, 217
   variation, 84
Irrational numbers, 3
Irreducible polynomial, 39

Joint variation, 84

Law of supply and demand, 411
LCD, 46
Leading coefficient, 295
Leading term, 268
Least common denominator, 46
Leibnitz, 465
Less than, 5
Like radicals, 56
Like terms, 28
Linear
   equations, 78
   function, 163
   inequalities, 126, 425
   programming, 431
   system of equations, 385, 396
   term, 99

Lines
   equation of, 484
   horizontal, 165
   oblique, 334
   parallel, 167
   perpendicular, 167
   vertical, 165
Location theorem, 304
Logarithm
   base of, 353
   common, 359
   natural, 359
   properties of, 363
Logarithmic equations, 370
Logarithmic function, 226, 352
Logarithmic identities, 355
Long division of polynomials, 31
Loss interval, 410
Lower bounds, 306
Lower half-plane, 425
Lowest terms, 44

Machine, function, 153, 208
Main diagonal, 452
Major axis, 233
Mapping notation, 150
Mathematical induction, 513
   initialization, 514
   perpetuation, 514
Mathematical model, 77, 120
Matrices, 396, 444
   a_{ij}, entry, 444
   addition of, 445
   augmented, 396
   coefficient, 459
   cofactor, 466
   column, 448
   constant, 459
   determinant of, 465
   dot product, 448
   echelon form, 398
   equal, 445
   identity, 452
   inner product, 448
   inverse, 457
   main diagonal, 452
   minor, 465
   multiplication of, 448
   order of, 444
   product of, 448

   reduced form, 398
   row, 448
   row operations on, 396
   scalar multiplication of, 445
   scalar product of, 448
   square, 452
   subtraction of, 445
   transpose of, 452
   variable, 459
   zero, 446
Maximum value, 186, 432
Mean
   arithmetic, 498
   geometric, 505
Mendel, Gregor, 526
Midpoint formula, 198
Mill, John Stuart, 154
Minimum value, 186, 432
Minor, 465
Minor axis, 233
Mixture principle, 115
Monomial, 28
Multiplication of matrices, 448
Multiplicative identity, 13
Multiplicative inverse, 13
Multiplicity
   of an asymptote, 321
   even, 266
   odd, 266
   of a solution, 100
   of a zero, 266, 289
Mutually exclusive events, 536

Napier, John, 341
Natural base e, 346
Natural logarithm, 359
Natural numbers, 3
Negative coordinate, 3
Negative exponent, 18
Newton's cooling law, 382
n factorial, 520
Noether, Emmy, 77
Nonlinear inequality, 134
Nonlinear system of equations, 415
Norman window, 175
nth partial sum, 492
nth root, 54
nth term, 496, 504
Null set, 78

Number
　of combinations, 531
　of permutations, 529
Number line, 2
Numbers
　complex, 64
　composite, 5
　imaginary, 64
　integers, 3
　irrational, 3
　natural, 3
　prime, 5
　rational, 3
　real, 3
　whole, 3
Numerical coefficient, 28

Objective function, 431
Oblique asymptote, 334
Odd asymptote, 326
Odd function, 194
Odd intercept, 266
Odd multiplicity, 326
One-to-one function, 217
Orbits, 248
Order
　of a matrix, 444
　of operations, 19
Order relations, 5
Ordered pair, 150
Ordinate, 155
Origin, 2, 155

Parabola, 181, 242
　axis of symmetry, 243
　concavity, 183
　directrix, 243
　focus, 243
　intercepts, 183
　maximum value, 186
　minimum value, 186
　vertex, 181, 186
Parallel lines, 167
Partial sum, 492
Particular solution, 390
Pascal's triangle, 32
Periodic compounding formula, 347
Periodic growth, 373
Permutation, 529
Perpendicular lines, 167

Perpetuation, 514
pH, 375
Piecewise function, 171
Pioneer, 10
Plane, 155
Point-slope form, 167
Points of equality, 136
Polynomial, 28
　binomial, 28
　degree of, 28
　division, 31
　factoring, 36
　function, 265
　inequalities, 299
　irreducible, 39
　monomial, 28
　prime, 36
　real, 287
　remainder, 277
　sign changes of, 299
　standard form, 28
　trinomial, 28
　zero, 28
Population, 374
Positive coordinate, 3
Power rule, 82
　for exponents, 20
　for logarithms, 363
Powers of i, 66
Precision, 23
Prime number, 5
Prime over the integers, 36, 288
Prime notation, 84
Prime polynomial, 36
Principal square root, 52, 100
Principal root, 52
Principle
　of mathematical induction, 513
　mixture, 115
　rate, 115
Probability, 536
　formulas, 540
Product
　of binomials, 29
　of matrices, 448
　rule for exponents, 20, 363
　rule for logarithms, 363
　special, 30
　of two functions, 206
Profit interval, 410

Programming, linear, 431
Progression
　arithmetic, 495
　geometric, 503
Proof, 14
Proof by mathematical induction, 515
Properties, 10
　of determinants, 475
　of equality, 11
　of exponents, 20
　of exponential functions, 345
　of inequality, 11
　of logarithms, 363
　of radicals, 56
　of real numbers, 12
Pure imaginary, 64
Pythagorean theorem, 11, 166

Quadrant, 155
Quadratic
　equation, 99
　form, 109
　formula, 104
　inequalities, 134
　term, 99
Quotient, 277
　polynomial, 277
　rule for exponents, 20
　rule for logarithms, 363
　of two functions, 206

Radical
　equations, 81, 111
　index of, 52
　like, 56
　notation, 52
　properties of, 56
　rationalizing, 58
　reducing the order of, 60
　sign, 52
　simplifying, 55
Radicand, 52
Radioactive decay, 376
Radius of a circle, 229
Ramanujan, Srinivas, 384
Range, 149
Rate principle, 115
Rate of work, 119
Rational exponent, 54
Rational expression, 43

Rational function, 318
Rationalizing the denominator, 58
Rational numbers, 3
Rational zeros theorem, 295
Real
    axis, 69
    exponents, 343
    numbers, 3
    polynomial, 287
    term, 64
Rectangular coordinate system, 155
Recursive formula, 489, 496
Reduced form of a matrix, 398
Reducing
    the order of a radical, 60
    a rational expression, 44
Reflection, 178
Reflexive property, 11
Relation, 150
    inverse, 217
Remainder, 277
    theorem, 280
Repeating decimal, 510, 511
Richter scale, 374
Right triangle, 11, 198
Root
    of an equation, 77
    of multiplicity two, 100
    principle nth, 52
Rounding, 24
Row matrix, 448
Row operations, 396
RPN, 547

Sample space, 536
Scalar multiplication, 445
Scalar product, 448
Scientific notation, 22
Seismologists, 374
Semicircle, 231
Sequence, 488
    arithmetic, 495
    defined recursively, 489
    Fibonacci, 489
    finite, 488
    geometric, 503
    infinite, 488
    of partial sums, 492
    terms of, 488
Series, 490
    arithmetic, 498, 500
    geometric, 506
    infinite, 492, 509
Sets
    empty, 78
    infinite, 3
    intersection of, 3
    notation, 3
    null, 78
    solution, 77
    union of, 3
Shortage interval, 411
Shrinking, 179
Sigma notation, 491
Sign changes, 299
Sign graph, 134, 265
Significant digits, 23
Slope, 164
Slope-intercept form, 168
Solution
    distinct, 100
    elimination method, 391
    of an equation, 77
    general, 390
    graphical, 385
    multiplicity of, 100
    particular, 390
    set, 77
    substitution method, 389
    of a system of equations, 385
Special products, 30, 36
Specified variable, 83
Speed of light, 26
Square matrix, 452
Square of a binomial, 30
Square root, 52
Standard form
    of a circle, 229
    of a complex number, 64
    of an ellipse, 233
    of a hyperbola, 240
    of a polynomial, 28
    of a quadratic equation, 99
Steinmetz, Charles, 63
Straight-line depreciation, 170
Strategy for word problems, 90
Stretching, 178
Substitution
    method, 389
    property, 11
Sum
    of two cubes, 36
    of two functions, 205
    of two matrices, 445
Summation notation, 491
Supply and demand, 411
Surplus interval, 411
Symmetric property, 11
Symmetry
    with respect to the line $x = y$, 223
    with respect to the origin, 192
    with respect to the x-axis, 192
    with respect to the y-axis, 192
    tests for, 193
Synthetic division, 278
System of equations, 385
    consistent, 386
    dependent, 386
    equivalent, 391
    inconsistent, 386
    independent, 386
    linear, 385
    nonlinear, 415
System of inequalities, 425

Tartaglia, Niccolo, 263
Terms, 28
    general, 488
    leading, 268
    like, 28
    of a sequence, 488
Tests for symmetry, 193
Theorem, 11, 14
    addition theorem of equality, 15
    binomial, 522
    complete factorization, 290
    conclusion, 14
    conjugate zeros, 287
    Descartes's rule of signs, 305
    double-negative, 15
    factor, 281
    fundamental theorem of
        algebra, 290
    hypothesis, 14
    location, 304
    multiplication theorem of
        equality, 15
    power rule, 82
    proof, 14

Theorem, (*continued*)
    Pythagorean, 11, 166
    rational zeros, 295
    remainder, 280
    upper and lower bounds, 307
    zero-factor, 14
Tolerance, 131
Train axle, 144
Transformations, 391
Transitive property, 11
Translation, 176
    combined, 177
    horizontal, 177
    vertical, 177
Transpose, 452
Transposing an equation, 83
Tree diagram, 525
Tree rings, 145
Trial and error factoring, 38
Trichotomy property, 11
Trinomial, 28
Trivial factors, 36
Turning point, 267

Unbounded interval, 6
Undefined, 13, 18, 81, 209, 353, 379
Union, 3

Upper and lower bounds theorem, 307
Upper bounds, 306
Upper half-plane, 425

Value of a function, 151
Variable
    dependent, 151
    dummy, 220
    independent, 151
    matrix, 459
    specified, 83
Variation
    direct, 84
    inverse, 84
    joint, 84
    in sign, 305
Vector
    column, 448
    row, 448
Vertex
    of an ellipse, 233
    of a parabola, 181, 186
Vertical asymptote, 321
Vertical line, 165
Vertical line test, 157
Vertical translation, 177
Vieta, 263

Wendy's, 528
Whole number, 3
Word equation, 77, 89
Word problem strategy, 90, 115
    mixture principle, 115
    rate principle, 115
Work formula, 119
    rate of work, 119

x-axis, 155
x-intercept, 163, 181

y-axis, 155
y-intercept, 163, 181

Zero, 3
    approximation of, 313
    bounds on, 306
    complex, 287
    division by, 13, 81
    exponent, 18
    factor principle, 15, 101
    factor theorem, 14
    matrix, 446
    of multiplicity n, 266, 289
    polynomial, 28
    of a polynomial, 265
    rational, 295

Binomial Theorem

$$(a + b)^n = \sum_{r=0}^{n} \binom{n}{r} a^{n-r} b^r$$

Formulas from Geometry

	Area		Volume
Square	$A = s^2$	**Cube**	$V = s^3$
Rectangle	$A = lw$	**Parallelepiped**	$V = lwh$
Parallelogram	$A = bh$	**Cylinder**	$V = \pi r^2 h$
Triangle	$A = \frac{1}{2} bh$	**Cone**	$V = \frac{1}{3} \pi r^2 h$
Trapezoid	$A = \frac{h}{2}(b_1 + b_2)$	**Sphere**	$V = \frac{4}{3} \pi r^3$
Circle	$A = \pi r^2$		

Methods of Solving Linear Systems

Substitution method
Elimination method
Augmented matrix method
Cramer's rule

Determinant of a 2 × 2 Matrix

$$\begin{vmatrix} a_{11} & a_{12} \\ a_{21} & a_{22} \end{vmatrix} = a_{11} a_{22} - a_{12} a_{21}$$

Determinants of higher order can be determined by expansion by cofactors.

Properties of Determinants

$|B| = -|A|$ if matrix B is obtained by interchanging two rows (columns) of matrix A.

$|B| = k|A|$ if matrix B is obtained by multiplying a row (column) of matrix A by a constant k.

$|B| = |A|$ if matrix B is obtained by replacing a row (column) of matrix A by the sum of that row (column) and a constant multiple of another row (column).

The determinant of a square matrix is zero if the matrix has a row or column of zeros.

The determinant of a square matrix is zero if one row (column) is a multiple of another row (column).

Constants

$\pi \approx 3.14159\ 26535\ 89793$
$e \approx 2.71828\ 18284\ 59045$

Some Pythagorean Triples ($a^2 + b^2 = c^2$)

a	b	c
3	4	5
5	12	13
7	24	25
8	15	17
9	40	41